ISBN 978-1-334-48938-9
PIBN 10652515

Forgotten Books is a registered trademark of FB &c Ltd.
Copyright © 2018 FB &c Ltd.
FB &c Ltd, Dalton House, 60 Windsor Avenue, London, SW19 2RR.
Company number 08720141. Registered in England and Wales.

For support please visit www.forgottenbooks.com

HENRY COPSON PEAKE,

PRESIDENT OF THE INSTITUTION OF MINING ENGINEERS, 1899-1900.

TRANSACTIONS

OF

THE INSTITUTION

OF

MINING ENGINEERS.

VOL. XXVII.—1903-1904.

EDITED BY M. WALTON BROWN, SECRETARY.

NEWCASTLE-UPON-TYNE: PUBLISHED BY THE INSTITUTION.

PRINTED BY ANDREW REID & CO., LIMITED, NEWCASTLE-UPON-TYNE.
1905.

ADVERTIZEMENT.

———

The Institution, as a body, is not responsible for the statements and opinic advanced in the papers which may be read or in the discussions which may place at the meetings of the Institution or of the Federated Institutes.

— —

CONTENTS OF VOL. XXVII.

GENERAL MEETINGS.

THE INSTITUTION OF MINING ENGINEERS.

89809.

CONTENTS.

APPENDICES.

APPENDICES.—*Continued.*

APPENDICES.—*Continued.*

CONTENTS.

APPENDICES.—*Continued.*

APPENDICES.—*Continued.*

APPENDICES. – *Continued.*

APPENDICES.—*Continued.*

APPENDICES.—*Continued.*

LIST OF PLATES:—

ANY PUBLICATION OF A FEDERATED INSTITUTE MAY BE PLACED AT THE END
OF THE VOLUME, *i.e.,* "ANNUAL REPORT." "LIST OF MEMBERS," ETC.

THE INSTITUTION OF MINING ENGINEERS.

OFFICERS, 1903-1904.

Past-Presidents *(ex-officio)*.

Mr. WILLIAM NICHOLAS ATKINSON, H.M. Inspector of Mines, Barlaston, Stoke-upon-Trent.
Mr. JAMES STEDMAN DIXON, Fairleigh, Bothwell, N.B.
Mr. GEORGE LEWIS, Albert Street, Derby.
Sir WILLIAM THOMAS LEWIS, Bart., Mardy, Aberdare.
Mr. JOHN ALFRED LONGDEN, Stanton-by-Dale, near Nottingham.
Mr. GEORGE ARTHUR MITCHELL, 5, West Regent Street, Glasgow.
Mr. HENRY COPSON PEAKE, Walsall Wood Colliery, Walsall.
Mr. ARTHUR SOPWITH, Cannock Chase Collieries, near Walsall.
Sir LINDSAY WOOD, Bart., The Hermitage, Chester-le-Street.

President.

MR. JAMES COPE CADMAN, Madeley, Newcastle, Staffordshire.

Vice-Presidents.
* Deceased.

*Mr. HENRY AITKEN, Darroch, near Falkirk, N.B.
Mr. WILLIAM ARMSTRONG, Wingate, Co. Durham.
Mr. GEORGE ELMSLEY COKE, 65, Station Street, Nottingham.
Mr. JOHN DAGLISH, Rothley Crag, Cambo, R.S.O., Northumberland.
Mr. JAMES TENNANT FORGIE, Mosspark, Bothwell, N.B.
Mr. JOHN GERRARD, H.M. Inspector of Mines, Worsley, Manchester.
Mr. WILLIAM BIRKENHEAD MATHER JACKSON, Ringwood, Chesterfield.
Mr. GEORGE ARTHUR MITCHESON, Market Place, Longton, Staffordshire.
Mr. ROBERT THOMAS MOORE, 142, St Vincent Street, Glasgow.
Mr. HORACE BROUGHTON NASH, 23, Victoria Road, Barnsley.
Mr. JOHN BELL SIMPSON, Bradley Hall, Wylam-upon-Tyne.
Mr. ADDISON LANGHORNE STEAVENSON, Durham.
Mr. JOHN GEORGE WEEKS, Bedlington, R.S.O., Northumberland.
Mr. ROBERT SUMMERSIDE WILLIAMSON, Cannock Wood House, Hednesford, Staffordshire.

Councillors.
* Deceased

Mr. FREDERICK ROBERT ATKINSON, Duffield, near Derby.
Mr. JAMES BARROWMAN, Staneacre, Hamilton, N.B.
Mr. JOHN BATEY, St. Edmunds, Coleford, Bath.
Mr. THOMAS WALTER BENSON, 24, Grey Street, Newcastle-upon-Tyne.
Mr. GEORGE JONATHAN BINNS, Duffield House, Duffield, Derby.
Mr. BENNETT HOOPER BROUGH, 28, Victoria Street, London S.W.
Mr. MARTIN WALTON BROWN, 10, Lambton Road, Newcastle-upon-Tyne.
Mr. W. FORSTER BROWN, Cefn Coed, Malpas, near Newport, Monmouthshire.
Mr. JOHN SEACOME BURROWS, Green Hall, Atherton, near Manchester.
Mr. WILLIAM HENRY CHAMBERS, Conisborough, Rotherham.
Mr. WILLIAM FREDERICK CLARK, The Poplars, Aldridge, near Walsall.
Mr. MAURICE DEACON, Whittington House, near Chesterfield.
Mr. ROBERT WILSON DRON, 55, West Regent Street, Glasgow.
Mr. THOMAS EMERSON FORSTER, 3, Eldon Square, Newcastle-upon-Tyne.
Mr. GEORGE CARRINGTON FOWLER, The Cliff, Cinder Hill, Nottingham.
Mr. WILLIAM EDWARD GARFORTH, Snydale Hall, near Pontefract.
Mr. REGINALD GUTHRIE, Neville Hall, Newcastle-upon-Tyne.
Mr. JAMES HAMILTON, 208, St. Vincent Street, Glasgow.

Mr. ARTHUR HASSAM, King Street, Newcastle, Staffordshire.
*Mr. JAMES HASTIE, 4, Athole Gardens, Uddingston, N.B.
Mr. HENRY RICHARDSON HEWITT, H.M. Inspector of Mines, Breedon Hill Road, Derby.
Mr. JAMES A. HOOD, Rosewell, Mid Lothian, N.B.
Mr. GEORGE P. HYSLOP, The Shelton Iron, Steel and Coal Company, Limited, Stoke-upon-Trent.
Mr. THOMAS EDGAR JOBLING, Bebside, Northumberland.
Mr. PHILIP KIRKUP, Leafield House, Birtley, R.S.O., Co. Durham.
Mr. CHARLES CATTERALL LEACH, Seghill Colliery, Northumberland.
Mr. GEORGE ALFRED LEWIS, Albert Street, Derby.
Mr. WILLIAM LOGAN, Langley Park, Durham.
Mr. HENRY LOUIS, 11, Windsor Terrace, Newcastle-upon-Tyne.
Mr. ROBERT McLAREN, H.M. Inspector of Mines, 19, Morningside Park, Edinburgh.
Mr. JOHN HERMAN MERIVALE, Togston Hall, Acklington, Northumberland.
Mr. THOMAS WILFRED HOWE MITCHELL, Mining Offices, Regent Street, Barnsley.
Mr. CHARLES ALGERNON MOREING, 20, Copthall Avenue, London, E.C.
Mr. JOHN MORISON, Cramlington House, Northumberland.
Mr. DAVID M. MOWAT, Summerlee Iron Works, Coatbridge, N.B.
Mr. ROBERT DOUGLAS MUNRO, 111, Union Street, Glasgow.
Mr. JOHN NEVIN, Littlemoor House, Mirfield.
Mr. HENRY PALMER, Medomsley, R.S.O., Co. Durham.
Mr. WILLIAM HENRY PICKERING, 6, Dacres Lane, Calcutta, India.
Mr. RICHARD AUGUSTINE STUDDERT REDMAYNE, The University, Birmingham.
Mr. HENRY RICHARDSON, 89, Ashley Gardens, Westminster, London, S.W.
Mr. ALEXANDER SMITH, 3, Newhall Street, Birmingham.
*Mr. CHARLES SEBASTIAN SMITH, Shipley Collieries, Derby.
Mr. THOMAS TURNER, Caledonia Works, Kilmarnock, N.B.
Mr. GARDNER FREDERICK WILLIAMS, De Beers Consolidated Mines, Limited, Kimberley, South Africa.
Mr. JOHN ROBERT ROBINSON WILSON, H.M. Inspector of Mines, West Hill, Chapeltown Road, Leeds.

Auditors.

Messrs. JOHN G. BENSON AND SONS, Newcastle-upon-Tyne.

Treasurers.

Messrs. LAMBTON AND COMPANY, The Bank, Newcastle-upon-Tyne.

Secretary.

Mr. MARTIN WALTON BROWN, Neville Hall, Newcastle-upon-Tyne.

THE INSTITUTION OF MINING ENGINEERS.

FOUNDED JULY 1ST, 1889.

BYE-LAWS

As revised at Council Meeting held on May 29th, 1902.

I.—CONSTITUTION.

1.—The Institution of Mining Engineers shall consist of all or any of the societies interested in the advancement of mining, metallurgy, engineering and their allied industries, who shall from time to time join together and adhere to the Bye-Laws.

2.—The Institution shall have for its objects—

(a) The advancement and encouragement of the sciences of mining, metallurgy, engineering, and their allied industries.

(b) The interchange of opinions, by the reading of communications from members and others, and by discussions at general meetings, upon improvements in mining, metallurgy, engineering, and their allied industries.

(c) The publication of original communications, discussions, and other papers connected with the objects of the Institution.

(d) The purchase and disposal of real and personal property for such objects.

(e) The performance of all things connected with or leading to the purpose of such objects.

3.—The offices of the Institution shall be in Newcastle-upon-Tyne, or such other place as shall be from time to time determined by resolution of the Council.

4.—The year of the Institution shall end on July 31st in every year.

5.—The affairs and business of the Institution shall be managed and controlled by the Council.

II.—MEMBERSHIP.

6.—The original adherents or founders are as follows:—

(a) Chesterfield and Midland Counties Institution of Engineers, Chesterfield.

(b) Midland Institute of Mining, Civil and Mechanical Engineers, Barnsley.

(c) North of England Institute of Mining and Mechanical Engineers, Newcastle-upon-Tyne.

(d) South Staffordshire and East Worcestershire Institute of Mining Engineers, Birmingham.

7.—Written applications from societies to enter the Institution shall be made to the Council, by the President of the applying society, who shall furnish any information that may be desired by the Council.

8.—A.—If desired by the Council, any of the Federated Institutes shall revise their Bye-Laws, in order that their members shall consist of Ordinary Members, Associate Members, and Honorary Members, with Associates and Students, and section B following shall be a model Bye-Law to be adopted by any society when so desired by the Council.

B.—"The members shall consist of Ordinary Members, Associate Members and Honorary Members, with Associates and Students:—

(a) Each Ordinary Member shall be more than twenty-three years of age, have been regularly educated as a mining, metallurgical, or mechanical engineer, or in some other branch of engineering, according to the usual routine of pupilage, and have had subsequent employment for at least two years in some responsible situation as an engineer; or if he has not undergone the usual routine of pupilage, he must have been employed or have practised as an engineer for at least five years.

(b) Each Associate Member shall be a person connected with or interested in mining, metallurgy, or engineering, and not practising as a mining, metallurgical, or mechanical engineer, or some other branch of engineering.

(c) Each Honorary Member shall be a person who has distinguished himself by his literary or scientific attainments, or who may have made important communications to any of the Federated Institutes.

(d) Associates shall be persons acting as under-viewers, under-managers, or in other subordinate positions in mines or metallurgical works, or employed in analogous positions in other branches of engineering.

(e) Students shall be persons who are qualifying themselves for the profession of mining, metallurgical, or mechanical engineering, or other branch of engineering, and such persons may continue Students until they attain the age of twenty-five years."

9.—The Ordinary Members, Associate Members and Honorary Members, Associates and Students shall have notice of, and the privilege of attending, the ordinary and annual general meetings, and shall receive all publications of the Institution. They may also have access to, and take part in, the general meetings of any of the Federated Institutes.

10.—The members of any Federated Institute, whose payments to the Institution are in arrear, shall not receive the publications and other privileges of the Institution.

11.—After explanations have been asked by the President from any Federated Institute, whose payments are in arrear, and have not been paid within one month after written application by the Secretary, the Council may decide upon its suspension or expulsion from the Institution; but such suspension or expulsion shall only be decided at a meeting attended by at least two-thirds of the members of the Council by a majority of three-fourths of the members present.

III.—SUBSCRIPTIONS.

12.—Each of the Federated Institutes shall pay fifteen shillings per annum for each Ordinary Member, Associate Member, Honorary Member, Associate and Student, or such other sum, and in such instalment or instalments as may be determined from time to time by resolution or resolutions of the Council. Persons joining any of the Federated Institutes during the financial year of the Institution shall be entitled to all publications issued for that year, after his election is notified to the Secretary, and the instalment or instalments due on his behalf have been paid.

IV.—ELECTION OF OFFICERS AND COUNCIL.

13.—The officers of the Institution, other than the Secretary and Treasurer, shall consist of Councillors elected annually prior to August in each year, by and out of the Ordinary Members and Associate Members of each Federated Institute, in the proportion of one Councillor per forty Ordinary Members or Associate Members thereof; of Vice-Presidents elected by and from the Council at their first meeting in each year on behalf of each Institute, in the proportion of one Vice-President per two hundred Ordinary Members or Associate Members thereof; and of a President elected by and from the Council at their first meeting in each year; who, with the Local Secretaries of each Federated Institute and the Secretary and Treasurer, shall form the Council. All Presidents on retiring from that office shall be ex-officio Vice-Presidents so long as they continue Ordinary Members or Associate Members of any of the Federated Institutes.

14.—In case of the decease, expulsion, or resignation of any officer or officers, the Council may, if they deem it requisite, fill up the vacant office or offices at their next meeting.

V.—DUTIES OF OFFICERS AND COUNCIL.

15.—The Council shall represent the Institution and shall act in its name, and shall make such calls upon the Federated Institutes as they may deem necessary, and shall transact all business and examine accounts, authorise payments and may invest or use the funds in such manner as they may from time to time think fit, in accordance with the objects and Bye-Laws of the Institution.

16.—The Council shall decide the question of the admission of any society, and may decree the suspension or expulsion of any Federated Institute for non-payment of subscriptions.

17.—The Council shall decide upon the publication of any communications.

18.—There shall be three ordinary meetings of the Council in each year, on the same day as, but prior to, the ordinary or annual general meetings of the members.

19.—A special meeting of the Council shall be called whenever the President may think fit, or upon a requisition to the Secretary signed by ten or more of its members, or by the President of any of the Federated Institutes. The business transacted at a special meeting of the Council shall be confined to that specified in the notice convening it.

20.—The meetings of the Council shall be called by circular letter, issued to all the members at least seven days previously, accompanied by an agenda-paper, stating the nature of the business to be transacted.

21.—The order in which business shall be taken at the ordinary and annual general meetings may be, from time to time, decided by the Council.

22.—The Council may communicate with the Government in cases of contemplated or existing legislation, of a character affecting the interests of mining, metallurgy, engineering, or their allied industries.

23.—The Council may appoint Committees, consisting of members of the Institution, for the purpose of transacting any particular business, or of investigating any specific subject connected with the objects of the Institution.

24.—A Committee shall not have power or control over the funds of the Institution, beyond the amount voted for its use by the Council.

25.—Committees shall report to the Council, who shall act thereon and make use thereof as they may elect.

26.—The President shall take the chair at all meetings of the Institution, the Council, and Committees at which he may be present.

27.—In the absence of the President, it shall be the duty of the senior Vice-President present to preside at the meetings of the Institution. In case of the absence of the President and of all the Vice-Presidents, the meeting may elect any member of Council, or in case of their absence any Ordinary Member or Associate Member to take the chair at the meeting.

28.—At meetings of the Council six shall be a quorum.

29.—Every question shall be decided at the meetings of the Council by the votes of the majority of the members present. In case of equal voting, the President, or other member presiding in his absence, shall have a casting vote. Upon the request of two members, the vote upon any question shall be by ballot.

30.—The Secretary shall be appointed by and shall act under the direction and control of the Council. The duties and salary of the Secretary shall be fixed and varied from time to time at the will of the Council.

31.—The Secretary shall summon and attend all meetings of the Council, and the ordinary and annual general meetings of the Institution, and shall record the proceedings in the minute book. He shall direct the administrative and scientific publications of the Institution. He shall have charge of and conduct all correspondence relative to the business and proceedings of the Institution, and of all committees where necessary, and shall prepare and issue all circulars to the members.

32.—One and the same person may hold the office of Secretary and Treasurer.

33.—The Treasurer shall be appointed annually by the Council at their first meeting in each year. The income of the Institution shall be received by him, and shall be paid into Messrs. Lambton & Co.'s bank at Newcastle-upon-Tyne, or such other bank as may be determined from time to time by the Council.

34.—The Treasurer shall make all payments on behalf of the Institution, by cheques signed by two members of Council, the Treasurer and the Secretary after payments have been sanctioned by Council.

35.—The surplus funds may, after resolution of the Council, be invested in Government securities, in railway and other debenture shares such as are allowed for investment by trustees, in the purchase of land, or in the purchase, erection, alteration, or furnishing of buildings for the use of the Institution. All investments shall be made in the names of Trustees appointed by the Council.

36.—The accounts of the Treasurer and the financial statement of the Council shall be audited and examined by a chartered accountant, appointed by the Council at their first meeting in each year. The accountants' charges shall be paid out of the funds of the Institution.

37.—The minutes of the Council's proceedings shall at all times be open to the inspection of the Ordinary Members and Associate Members.

VI.—GENERAL MEETINGS.

38.—An ordinary general meeting shall be held in February, May and September, unless otherwise determined by the Council; and the ordinary general meeting in the month of September shall be the annual general meeting at which a report of the proceedings, and an abstract of the accounts of the previous year ending July 31st, shall be presented by the Council. The ordinary general meeting in the month of May shall be held in London, at which the President may deliver an address.

39.—Invitations may be sent by the Secretary to any person whose presence at discussions shall be thought desirable by the Council, and persons so invited shall be permitted to read papers and take part in the proceedings and discussions.

40.—Discussion may be invited on any paper published by the Institution, at meetings of any of the Federated Institutes, at which the writer of the paper may be invited to attend. Such discussion, however, shall in all cases be submitted to the writer of the paper before publication, and he may append a reply at the end of the discussion.

VII.—PUBLICATIONS.

41.—The publications may comprise :—

(a) Papers upon the working of mines, metallurgy, engineering, railways and the various allied industries.

(b) Papers on the management of industrial operations.

(c) Abstracts of foreign papers upon similar subjects.

(d) An abstract of the patents relating to mining and metallurgy, etc.

(e) Notes of questions of law concerning mines, manufactures, railways, etc.

42.—Each paper (with complete drawings, if any, to scale), to be read at any meeting of the Institution or of any of the Federated Institutes, shall be placed in the hands of the Secretary at least fourteen days before the date of the meeting at which the paper is to be read, and shall, subject to the approval of the Council, be printed, together with any discussion or remarks thereon.

43.—The Council may accept communications from persons who are not members of the Institution and allow them to be read at the ordinary or annual general meetings.

44.—No paper which has already been published (except as provided for in Bye-Law 41) shall appear in the publications of the Institution.

45.—A paper in course of publication cannot be withdrawn by the writer.

46.—Proofs of all papers and reports of discussions forwarded to any person for revision must be returned to the Secretary within seven days from the date of their receipt, otherwise they will be considered correct and be printed off.

47.—The copyright of all papers accepted for printing by the Council shall become vested in the Institution, and such communications shall not be published for sale or otherwise without the written permission of the Council.

48.—Twenty copies of each paper and the accompanying discussion shall be presented to the writer free of cost He may also obtain additional copies upon payment of the cost to the Secretary, by an application attached to his paper. These copies must be unaltered copies of the paper as appearing in the publication of the Institution, and the cover shall state that it is an "Excerpt from the Transactions of The Institution of Mining Engineers."

49.—The Federated Institutes may receive copies of their own portion of the publications in respect of such of their members as do not become members of the Institution, and shall pay 10s. per annum in respect of every copy so supplied ; and similar copies for exchanges shall be paid for at cost price.

50.—The Local Secretary of each Federated Institute shall prepare and edit all papers and discussions of such Institute, and promptly forward them to the Secretary, who shall submit proofs to the Local Secretary before publication.

51.—A list of the members, with their last known addresses, shall be printed in the publications of the Institution.

52.—The publications of the Institution shall only be supplied to members, and no duplicate copies of any portion of the publications shall be issued to any member or Federated Institute unless by order of the Council.

53.—The annual volume or volumes of the publications may be sold, in the complete form only, at such prices as may be determined from time to time by the Council :—to non-members for not less than £3 ; and to members who are desirous of completing their sets of the publications, for not less than 15s.

54.—The Institution, as a body, is not responsible for the statements and opinions advanced in the papers which may be read or in the discussions which may take place at the meetings of the Institution or of the Federated Institutes.

VIII.—MEDALS AND OTHER REWARDS.

55.—The Council, if they think fit in any year, may award a sum not exceeding sixty pounds, in the form of medals or other rewards, to the author or authors of papers published in the *Transactions*.

IX.—PROPERTY.

56.—The capital fund shall consist of such amounts as shall from time to time be determined by resolution of the Council.

57.—The Institution may make use of the following receipts for its expenses :—
(*a*) The interest of its accumulated capital fund ;
(*b*) The annual subscriptions ; and
(*c*) Receipts of all other descriptions.
58.—The Institution may form a collection of papers, books and models.
59.—Societies or members who may have ceased their connexion with the Institution shall have no claim to participate in any of its properties.

60.—All donations to the Institution shall be acknowledged in the annual report of the Council.

X.—ALTERATION OF BYE-LAWS.

61.—No alteration shall be made in the Bye-Laws of the Institution, except at a special meeting of the Council called for that purpose, and the particulars of every such alteration shall be announced at their previous meeting and inserted in the minutes, and shall be sent to all members of Council at least fourteen days previous to such special meeting, and such special meeting shall have power to adopt any modification of such proposed alteration of the Bye-Laws, subject to confirmation by the next ensuing Council meeting.

THE INSTITUTION OF MINING ENGINEERS.

SUBJECTS FOR PAPERS.

The Council of The Institution of Mining Engineers invite original communications on the subjects in the following list, together with other questions of interest to mining and metallurgical engineers.

Assaying.
Boiler explosions.
Bore-holes and prospecting.
Boring against water and gases.
Brickmaking by machinery.
Brine-pumping.
Canals, inland navigation, and the canalization of rivers.
Coal-getting by machinery.
Coal-washing machinery.
Coke manufacture and recovery of bye-products.
Colliery leases, and limited liability companies.
Compound winding-engines.
Compressed-air as a motive-power.
Corrosive action of mine-water on pumps, etc.
Descriptions of coal-fields.
Diamond-mining.
Distillation of oil-shales.
Drift and placer-mining.
Duration of coal-fields of the world.
Electric mining lamps.
Electricity and its applications in mines.
Electro-metallurgy of copper, etc.
Engine-counters and speed-recorders.
Explosions in mines.
Explosives used in mines.
Faults and veins.
Fuels and fluxes.
Gas-producers, and gaseous fuel and illuminants.
Gas, oil and petroleum engines.
Geology and mineralogy.
Gold-recovery plant and processes.
Graphite: its mining and treatment.
Haulage in mines.
Industrial assurance.
Inspection of mines.
Laws of mining and other concessions.
Lead-smelting.
Light railways.
Lubricating value of grease and oils.
Lubrication of trams and tubs.
Maintenance of canals in mining districts.
Manufacture of fuel-briquettes.
Mechanical preparation of ores and minerals.

Mechanical ventilation of mines, and efficiency of the various classes of ventilators.
Metallurgy of gold, silver, iron, copper, lead, etc.
Mining and uses of arsenic, asbestos, bauxite, mercury, etc.
Natural gas, conveyance and uses.
Occurrence of mineral ores, etc.
Ore-sampling machines.
Petroleum-deposits.
Preservation of timber.
Prevention of over-winding.
Pumping machinery.
Pyrometers and their application.
Quarries and methods of quarrying.
Rock-drills.
Safety-lamps.
Salt-mining, etc.
Screening, sorting and cleaning of coal.
Shipping and discharge of coal-cargoes.
Sinking, coffering and tubbing of shafts.
Sleepers of cast-iron, steel and wood.
Spontaneous ignition of coal and coal-seams.
Stamp-milling.
Steam-condensation arrangements.
Steam-power plants.
Submarine coal-mining.
Subsidences caused by mining-operations.
Surface-arrangements at mines.
Surveying.
Tin-mining.
Transport on roads.
Tunnelling, methods and appliances.
Utilization of dust and refuse coal.
Utilization of sulphureous gases resulting from metallurgical processes.
Ventilation of coal-cargoes.
Water as a motive-power in mines.
Water-tube boilers.
Watering coal-dust.
Water-incrustations in boilers, pumps, etc.
Winding arrangements at mines.
Winning and working of mines at great depths.

For selected papers, the Council may award prizes. In making awards, no distinction is made between communications received from members of the Institution or others.

4 0 0

TRANSACTIONS

OF

THE INSTITUTION

OF

MINING ENGINEERS.

THE NORTH OF ENGLAND INSTITUTE OF MINING
AND MECHANICAL ENGINEERS.

GENERAL MEETING,
HELD IN THE WOOD MEMORIAL HALL, NEWCASTLE-UPON-TYNE,
FEBRUARY 13TH, 1904.

MR. W. O. WOOD, PRESIDENT, IN THE CHAIR.

The SECRETARY read the minutes of the last General Meeting, and reported the Proceedings of the Council at their meetings on January 30th and that day.

The following gentlemen were elected, having been previously nominated:—

MEMBERS—

MR. HENRY ARNOLD ABBOTT, H.M. Inspector of Mines, 1, Highbury, West Jesmond, Newcastle-upon-Tyne.

MR. HAROLD VIVIAN ATHERON, Colliery Manager, Hall Lane, Hindley, near Wigan, Lancashire.

MR. FRANCIS BOWMAN, Mechanical Engineer, Ouston Colliery, Chester-le-Street, County Durham.

MR. VINCENT CORBETT, Mining Engineer, Blackett Colliery, Haltwhistle, Northumberland.

MR. JOHN DEAN, Mining Engineer, The Wigan Coal and Iron Company, Limited, Wigan, Lancashire.

MR. ARTHUR STANLEY DOUGLAS, Mining Engineer, Tudhoe House, Tudhoe, near Spennymoor, County Durham.

MR. HAROLD FLETCHER, Mining Engineer, 23, Old Jewry, London, E.C.

VOL. XXVII.—1903-1904. 1

MR. JOSEPH WILLIAM FORSTER, Constructional Engineer, New Kleinfontein Company, P.O., Benoni, Transvaal.

MR. JAMES GIBSON, Surveyor, Assayer, etc., P.O. Box 1026, Johannesburg, Transvaal.

MR. HOWE HEWLETT, Mining Engineer, Clock Face Colliery, Sutton Oak, St. Helen's, Lancashire.

MR. MARTIN STANGER HIGGS, Mine-manager and Consulting Engineer, Southern Main Reef Estates, Limited, Venterskroon, Transvaal.

MR. NORMAN STANLEY HOLLIDAY, Colliery Manager, Acklington Colliery, Amble, Northumberland.

MR. JAMES HOOPER, Mining Engineer, The Poplars, Mount Charles, St. Austell, Cornwall.

MR. DAVID HOWELLS, Mining Engineer, 12, David Price Street, Aberdare, Glamorganshire.

MR. JOHN JAMES, Mining Engineer, c/o The Bank of Egypt, Assuan, Upper Egypt.

MR. HENRY HOWARD JOHNSON, Mechanical Engineer, The Village Main Reef Gold-mining Company, Limited, P.O. Box 1091, Johannesburg, Transvaal.

MR. JOHN EVAN JORDAN, Engineer, 3, Bussey Buildings, Johannesburg, Transvaal.

MR. FREDERICK NEWBERY, Mining Engineer, P.O. Box 1305, Johannesburg, Transvaal; and 230, Camden Road, London, N.W.

MR. JOHN RAMSAY, Colliery Manager, Tursdale Colliery, Ferryhill, County Durham.

MR. WALKER OSWALD TATE, Colliery Manager, Shotton Colliery, Castle Eden, R.S.O., County Durham.

MR. HENRY LEIGH TROTMAN, Colliery Manager, Moorland House, Aspull, near Wigan, Lancashire.

MR. GEORGE HERBERT TWEDDELL, Electrical Engineer, 48, Percy Park, Tynemouth, Northumberland.

ASSOCIATE MEMBERS—

MR. JOHN ECCLES ASPINALL, Dundee, Natal, South Africa.

MR. JAMES RUSSELL, Westgate Road, Newcastle-upon-Tyne.

ASSOCIATES—

MR. SILAS SCRAFTON COWLEY, Deputy-overman, 10, Vane Terrace, New Seaham, via Sunderland, County Durham.

MR. WILLIAM SIMM ROCHESTER, Back-overman, Wellington Terrace, Edmondsley, Chester-le-Street, County Durham.

MR. RICHARD CHARLTON SIMPSON, Overman, Wellington Terrace, Edmondsley, Chester-le-Street, County Durham.

MR. NICHOLAS STOKER, Enginewright, North Walbottle, Newburn, R.S.O., Northumberland.

MR. HUGH WILSON, Under-manager, 18, Grange Villa, Chester-le-Street, County Durham.

STUDENTS—

MR. RICHARD FORSTER LAWSON, Mining Student, Daisy Hill, Edmondsley, Chester-le-Street, County Durham.

MR. HENRY MASON PARRINGTON, Mining Student, Hill House, Monkwearmouth, County Durham.

DISCUSSION OF MR. J. B. SIMPSON'S PAPER ON "THE PROBABILITY OF FINDING WORKABLE SEAMS .... BELOW THE BROCKWELL SEAM."*

Mr. J. H. MERIVALE said that a bore-hole had been put down by the owners of Broomhill collieries on Coquet Island, to a depth of about 600 feet below the approximate horizon of the Brock-well seam. So far, two seams, each about 2 feet thick, had been found, and the Mountain Limestone had been reached, but the hole had not reached any of the coal-seams of that series. There was one point of interest to which attention was drawn by the result of the Chopwell boring, and that was the deterioration of the seams to the east. He thought it was generally accepted that the coal-seams in what Mr. Simpson called the "regular Coal-measures" deteriorated to the east, along the coast-line of Northumberland and Durham; and from his own observations they did not deteriorate in a vertical plane, but in one hading to the west, that is, the lower the seams in the series the farther the deterioration extended westward. For 2 or 3 miles along the coast, for example, south of Amble, the upper seams were of good quality up to the shore-line, and then began to deteriorate as they passed eastward under the sea. Whereas, in the lower seams, the deterioration began some distance west-ward of the shore-line. If this deterioration of the seams, as they approached the sea, continued in the same plane, hading to the west, by the time that the Bernician series was reached the coal-seams would have lost their good quality far inland, and this appeared to be the case at Chopwell. A desire had been expressed that the Coal Commission should give valuable information about the Mountain Limestone seams, but he was inclined to think that any information which Mr. Simpson did not possess, would also be unknown to the Coal Commission. It was an interesting and important problem, and he would like to see some wealthy coal-owners combining together and putting down a bore-hole to a depth of 2,000 or 3,000 feet below the Brockwell seam, and thus solving the question so far as one particular locality was concerned.

—————

* *Trans. Inst. M.E.*, 1902, vol xxiv., page 549.

DISCUSSION OF MR. WILLIAM CHARLTON'S PAPER ON THE "USE OF RATCHET AND OTHER HAND-MACHINE DRILLS IN THE CLEVELAND MINES."[*]

Mr. W. CHARLTON said that the use of ratchet and other hand-machine drills in the Cleveland mines commenced in 1886, and from 0·50 per cent. of the ironstone won by this class of machine in that year, it had increased in 1901 to 39·55 per cent.; in 1902 to 46·92 per cent., and in 1903 to 64 per cent.[†] It would thus be recognized that in the Cleveland mines, the use of the old jumper-drill had been discarded in favour of power or ratchet-drills. There was a misprint in the last line of a paragraph which should read " power-machines, but by the use of hand-machines."[‡]

Prof. HENRY LOUIS asked for information as to the date of the introduction of the rotary or ratchet-drill into the North of England generally, as Mr. Charlton gave 1880 as the date of its introduction into the Cleveland district. The earliest dates that he had been able to obtain, so far, were 1869 at Usworth colliery, and about 1872 at Hedley Hill colliery; but he would be glad of more definite information. So far as he could learn, such drills were first devised for use in the Scotch oilshale-fields about 1865; the use seemed to have extended southward about 4 or 5 years later, and they were used for boring in stone before they were tried in coal. It was very curious that in this comparatively brief period they should have so completely supplanted jumper-drills, that it was scarcely possible to find a coal-miner able to use the latter. He had seen a crude rotary drilling-machine, known locally as a " garibaldi," used in the soft-coal region of Pennsylvania, while the stone in the same mine was drilled with a hammer and drill; he thought that this crude machine must have been of independent local invention, and had not been imported.

Mr. T. E. FORSTER said that hand-machine drills, for boring in coal, were introduced into Northumberland about 1879. They were in use considerably before that date for driving stone-drifts.

---

[*] *Trans. Inst. M.E.*, 1902, vol. xxiv., page 526.
[†] *Ibid.*, page 534.   [‡] *Ibid.*, page 537, line 14.

Mr. J. H. MERIVALE said that at Broomhill collieries, so far as he could remember, hand-drills were used for boring in coal about 1884 and for boring in stone, about three years later.

Mr. T. E. FORSTER asked whether the rotary drill had ever been used in Cleveland for boring in any stone harder than ironstone.

Mr. W. CHARLTON stated that Cleveland ironstone was much harder than coal, and a ratchet-machine could drill through the "dogger-band" in the ironstone, although it was much harder.

Mr. C. H. STEAVENSON said that he had had experience with ratchet drilling-machines in Cleveland mines; so when he came to Tyneside he brought a ratchet-drill with him, and had it tried in both stone and coal, but he was sorry to say that it was not successful. The fixing of the machine was difficult where the post-stone was too hard to drive in a wedge for the purpose of attaching the machine. In the case of ironstone, the wedge was driven into the rock at the side, and the end of the machine was fixed against this wedge. A hole was made for the entrance of the drill, and when the man started to work the ratchet the pressure kept the drill in position. In the post-stone of North of England collieries he found that the wedge could not be driven in far enough to make a sufficiently strong support for the machine, and in the case of blue metal or shale, the stone was too soft. Altogether he found that the ratchet drilling-machine was not so successful as the ordinary stone-man's or hewer's hand drilling-machine.

Mr. W. CHARLTON said that his paper showed the great advantage of the Cleveland method of fixing hand drilling-machines, and how that advantage had promoted their use in the ironstone-mines. Where that method could not be adopted, they could adopt the old-fashioned method of making a prop serve as a support for the machine.

Mr. WILLIAM SEVERS said that, at one of his collieries, a drift, 12 feet wide by 6 feet high, was being driven with ratchet-machines, and a prop was used as a support for the machine. Hand-drilling cost £10 per yard, but with ratchet-machines the cost was reduced to £7 per yard.

Mr. C. H. STEAVENSON said that his remarks referred to the type of drilling-machine used in Cleveland. Ratchet-machines were in use on Tyneside, but they were of a different type from the drills used in Cleveland.

The PRESIDENT (Mr. W. O. Wood) said that drilling-machines were introduced into the collieries of the North of England about 1872, and they were now universally used. It was rare to see a man using a hammer and drill, or even a jumper. At his own collieries, no difficulty was found in fixing the machine as Mr. Steavenson seemed to indicate: sometime props were used, and at other times, a proper stand.

Mr. WILLIAM SEVERS remarked that the men used the machine with a stand at first; but now it was discarded altogether, and a prop was used.

Mr. F. I. LESLIE DITMAS wrote that the displacement of the old method of jumping in the holes, naturally showed the advance that the Cleveland district had made in the reduction of the working costs of the mines, and the gradual evolution of man-power to that of man-and-machine-power, which was in its turn giving place to that of power-machines, driven by compressed air or electricity.

At one of the Rosedale mines, practically all the drilling was done by an electric drill, which differed in many respects from the other electric drills in use in Cleveland. It was found, when using electrically-driven drills, that a certain percentage of holes must of necessity be bored which were quite useless (this was a matter of slight moment, as the holes were drilled so rapidly) and which hand-machines would avoid, as, in this latter case, firing took place immediately after each hole was put in, and the next hole could be placed in the most advantageous position.

With hand-drills, the ratchet and Elliott machine-stand were used, but there were still a number of the older miners, who liked to be supplied with jumper-drills in addition. They argued and in most cases rightly, that for certain positions (especially in removing the pillars), it was an easier and a faster method to jump the holes, than to fix the Elliott telescopic machine-frame in a place 9 or 10 feet high, if the timbering were not quite suitable, and only a " pop-hole " was required. The physical labour involved in jumping a hole was greater, and more time was required than when it was drilled by a machine ; and it was simply a matter of time when jumpers would cease to exist, except for occasional and exceptional conditions. The younger miners, now, hardly knew how to obtain the best results from using the jumper-drill, nor did they care to learn how to use it. Of course, the power-drill (electric-driven for preference) would in time displace the best hand-machine drills used in most iron-stone-mines.

DISCUSSION OF MR. C. C. LEACH'S PAPER ON " SUPER-HEATED STEAM AT SEGHILL COLLIERY."*

Mr. C. C. LEACH said that he had carried out some further experiments to ascertain the loss of steam in steam-pipes. The experiments were made on a Sunday, when all the engines were stopped, except the donkey-pump, which fed the boilers. The test was continued from 7 a.m. to 5 p.m., the boiler-pressure was maintained to 80 pounds per square inch, and no steam was blown off. The temperature of the air was $57.8°$ Fahr. He found that 2,450 pounds of water per hour went into the boilers. Of this weight, the donkey-pump used for pumping the water into the boilers consumed 49 pounds, and 608 pounds were caught at leaks and steam-traps, leaving 1,793 pounds per hour to be accounted for. The coal burned per hour was 240 pounds, and this showed a loss of 9·6 per cent. on the ordinary day's coal-work. The area of the surface of the pipes, beneath the covering, was 3,036 square feet.

Mr. W. C. BLACKETT asked whether much water was found standing in the cylinders.

Mr. C. C. LEACH said that he had no idea, but steam was passing through the cylinder of the big winding-engine, as it felt warm to the touch of the hand.

Mr. J. H. MERIVALE said that these experiments were exceedingly interesting, but before the members could form a definite idea they would require considerably more information. The loss of water seemed enormous, if all the engines were

* *Trans. Inst. M.E.*, 1902, vol. xxiv., page 538.

placed near the boilers; but if some of them were placed under-ground and some distance inbye from the shaft-bottom, a very different construction would be put upon it. A loss of 10 per cent. of the quantity of water used was, after all, not so very high a figure; and with every engine going at full work the loss would probably not exceed 2 or 3 per cent.

Mr. C. C. Leach said that the loss of 9·6 per cent. referred to full work. The distance from the engines to the boilers varied from 46 feet to 618 feet, and there were no engines in the mine. The most important matter was the number of square feet exposed to the air, and this depended on the length of pipes, and their outside diameter. No doubt a great deal of steam was wasting through the two old-fashioned engines, but wherever a drop of water could be detected, a bucket was placed to catch it.

Mr. W. C. Blackett said that it might console Mr. Leach for the very great trouble that he had taken over his paper to know that one of the fruits of his missionary-work had been to induce him (Mr. Blackett) to go more closely into the question, and to make experiments in the same direction. At one of his collieries, where the boilers were fired by the waste-gases from the coke-ovens, and were capable of producing pressures much greater than some of the other boilers, it occurred to him that if he could confine all the steam in the boilers capable of work-ing at a high pressure, and not allow it to be used until it attained that pressure, he would get a superheated action on the low-pressure pipes when it was allowed to go through them. He accordingly put in a valve, which would not allow the steam to pass through it, until the correct pressure was attained. The use of this valve had an excellent effect: the superheat in the high-pressure steam became available; and, when it passed into the low-pressure pipes, the steam was dried, and there was a notable economy of fuel.

Mr. C. C. Leach said that one of the most remarkable matters in connection with the superheat in steam was the manner in which it disappeared before reaching the engine. In the ex-periments referred to, the steam entering the pipes was super-heated by 150° or 200° Fahr.

Mr. HENRY LAWRENCE said that many years ago he had charge of works using superheated steam, but he could not arrive at any saving, because he was firing with blast-furnace gases. There were 6 boilers in a row, one of which was held in reserve. He used the spare boiler as a superheater, and passed all the ordinary steam out of the various boilers into the superheater before it went away to the blast-engines. A very high amount of superheat, from 700° to 750° Fahr. was attained: it was sometimes stopped, when it rose too high, and the steam began to arrive at the engine with a temperature sufficient to injure the packing. There was a safety-valve on the superheater, and when steam was blowing off no vapour could be seen coming out of the safety-valve—it was almost colourless. The steam was very dry, and a hand could be passed through the vapour coming from the safety-valve on the superheater without being scalded, on account of there being no moisture in the steam. Judging from the temperature at the engines, he gathered that the superheat was much less, and that saturated steam and not dry steam arrived at many of the engines. He asked whether Mr. Leach had ascertained the temperature of the gases at the bottom of the chimney from the boilers when producing ordinary saturated steam, and the lower temperature of the gases after they had been passed through the superheater.

Mr. C. C. LEACH said that he did not notice any particular appearance when steam was blowing off at the safety-valve upon the superheater. The steam, however, was absolutely super-heated, as it had not time to lose it. The temperature of the flue-gases must be greater at the superheater than at the chimney; and they would probably lose about 500° Fahr. when they reached the chimney-bottom. The temperature ranged from 1,000° to 1,100° Fahr. at the superheater, and the diagram (Fig. 3, Plate XVI.)* shewed that there were considerable variations.

The PRESIDENT said that it appeared that Mr. Leach put 2,450 pounds of water into the boilers, and a very large propor-tion was lost. The solution of the mystery would no doubt provide occupation for some little time to come.

---

* *Trans. Inst. M.E.*, 1902, vol. xxiv., page 542.

DISCUSSION OF MR. W. H. BORLASE'S " DESCRIPTION OF THE LEAD-ORE WASHING-PLANT AT THE GREENSIDE MINES, PATTERDALE."*

Mr. ARTHUR E. NORTHEY (Talybont, R.S.O., Cardiganshire) wrote asking for the following information:—(1) What are the dimensions of the rough ore-stuff when it leaves the stone-breaker (Fig. 2, Plate XIII.†)?; (2) how many tons of ore-stuff are treated by the first three sets of crusher-rolls per hour, or, per day of 10 hours?; (3) what is the speed of the crusher-rolls?; and (4) what is the daily approximate finished output for the total labour-cost of 10s. 6d. per ton: that is, what is the daily finished output of the dressing-plant, when treating crude stuff containing 7 per cent. of lead-ore?

Mr. W. H. BORLASE (Glenridding, near Penrith), wrote that:—(1) The stuff was broken by the stone-breaker to about 1½ inches cube. (2) The tonnage varied according to the stuff supplied, and whether or not it had passed through the stone-breaker, from 40 to 50 tons per day of 10 hours. (3) The speed of the fluted rolls was 8 revolutions per minute, and of the plain rolls, 16 revolutions per minute. And (4) about 6 tons of dressed ore were produced daily. It must be understood that the crude stuff contained 7 per cent. of lead-ore, and this was relieved at the picking grates of much gangue and waste, making the stuff, supplied to the crushers, probably contain about 15 per cent. of lead-ore.

DISCUSSION OF MR. T. ADAMSON'S PAPER ON " WORKING A THICK COAL-SEAM IN BENGAL, INDIA."‡

Prof. HENRY LOUIS said that, on comparing the method described in the paper with other methods in use in India, although the former was not particularly safe, it would be difficult to suggest anything better. In all the coal-fields of India, coal was being wrought at an extraordinarily cheap rate. Hewing only cost 7d. per ton in some places, and the selling-price of coal at the pit was only 1s. 6d. to 2s. 6d. per ton. People who were mining coal systematically had to face competition from the native coal-owners who worked the outcrops of coal-seams,

---

* *Trans. Inst. M.E.*, 1903, vol. xxv., page 331.   † *Ibid.*, page 338.
‡ *Ibid.*, 1903, vol. xxv., pages 10, 192 and 396 ; and vol. xxvi., page 19.

in some cases 25 to 30 feet thick, taking out what they could and leaving the rest to be crushed in the small pillars that were left, the method of working being at the same time very unsafe. Even other methods of working the mines were not always what one would desire, and in one instance he knew of a manager, who made his pillars of pyramidal form, tapering upwards, because he said that was the safest shape to make a pillar. Were it not for a special providence in the shape of a splendid roof, accidents in the Indian coal-fields would be numerous. He (Prof. Louis) condemned the practice in the author's collieries of blasting with dynamite in the large chambers; the coal made a lot of dust, and he thought that they might easily adopt a safer explosive. He could not suggest any better method of working, under the circumstances, and he thought that much might be urged in mitigation of the criticism which had been passed on the method of working described in Mr. Adamson's paper.

Mr. J. B. ATKINSON (H.M. Inspector of Mines) said that at a previous meeting he had agreed with Mr. Steavenson that the method described in the paper seemed to be a somewhat dangerous one. The conditions prevailing, particularly the want of timber, perhaps made it the only possible way of working; but as regards the danger, that might possibly be decided by statistics.

DISCUSSION OF DR. J. S. HALDANE'S PAPER ON "MINERS' ANÆMIA, OR ANKYLOSTOMIASIS."*

Mr. W. C. BLACKETT said that it occurred to him that ankylostomiasis was being rather lightly treated in this part of the country at the present time. From a recent remark that Dr. Haldane had added to one of his publications, it seemed that the mines were not in so secure a position as many thought they might be. Many managers had taken comfort from the fact that the worm did not develop, except at a temperature of about 68° Fahr.; but Dr. Haldane had pointed out that further observations tended to show that the larvæ had developed at considerably lower temperatures. If that was the case, it seemed to him that they wanted to pay a little more regard to the subject

* *Trans. Inst. M.E.*, 1903, vol. xxv., page 643.

than they had been doing.   Although the temperature of many
of the mines did not reach 68° Fahr., there were some where it
did; and it should be borne in mind that many mines which
naturally reached that temperature were " dry and dusty mines,"
and were being, at the present time, carefully watered in many
parts for the purpose of laying the dust, and the water would
produce conditions favourable to the development of the larvæ.

He (Mr. Blackett) was of opinion that the best way to
deal with the threatened invasion of the disease was to show
the miners, by lectures illustrated with lantern-slides, the nature
of, and the remedy for, the parasite, and leave, very largely to
their good sense, the methods by which the fouling of the mine
could be avoided.   Doubtless there would be among them
selfish and irresponsible persons who, careless of the general
welfare, would take little trouble in the matter: but these men
could be dealt with in the course of time; and at any rate the
effort would serve to draw attention and constitute a beginning
for what might be a better state of things in future.

He (Mr. Blackett) further suggested that such earth-closets
as were constructed at the pit-bottoms, etc., could be made to
use up some of the dust of the mine, and this would "kill two
birds with one stone."

Another interesting point that occurred to him in connection with the medical part of the report was this: Dr. Haldane
had stated that the *Strongylus* or *Ankylostoma* only invaded the
small intestine.   Had he (Mr. Blackett) been in a position to
question Dr. Haldane, he would have said that there was considerable doubt about this, because it seemed to him that
the *Strongylus* which attacked the human being was almost
identical with that which invaded the horse.   They each produced exactly the same condition: there was the same anæmia,
the same paleness, as could be seen in the eyelids and nostrils in
the horse, which became incapable of work and listless, lost its
healthy appearance of coat, and, in fact, became exactly as had
been described in the report referring to man.   He also found
that practically similar remedies were used for the horse as for
the human subject.   The new remedy which had been tried
in the Cockerill Hospital, "terpine," was the essential part of
turpentine; the remedy given to the horse was a repeated dose
of turpentine and oil, with sometimes a little opium to kill the

pain. The same remedies were, therefore, given in both cases to get rid of the worm from the mucous membrane, after which other medicines were given, such as iron, etc., for a tonic. In the case of the horse, one judged from its appearance whether the worms had been got rid of, but one could never be quite sure. The symptoms and treatment were, therefore, almost identical. With respect to the small intestine, he (Mr. Blackett) had seen horses cut open, and had been present at *post-mortem* examinations, and in one very bad case—which he thought possible to be paralleled in man—he had seen the worm distributed throughout the intestines, and even within the mesenteric artery itself. It therefore seemed to him a pity to jump to the conclusion that the worm, in man, would only infest the small intestine.

Mr. C. C. LEACH said that in one of the return-airways at Seghill colliery the temperature was steady at only 57° Fahr., and therefore he thought that such a mine had not the slightest chance of propagating the disease. The only place in which any person was liable to contamination was in the intake-airways, these were very much cooler than the return airways, and fatal to the worm if ever it got there.

Mr. J. B. ATKINSON (H.M. Inspector of Mines) said that there were a number of deep collieries in the county of Durham where the temperature of the working-face was higher than 70° Fahr.; but they were almost always in such a condition of dryness that probably the worm would not develop. He had recently received a communication from the Home Office stating that at 61° Fahr. the worm might be developed. He should think that temperature was frequently exceeded in Durham mines; but perhaps not so often in Northumberland.

Mr. J. H. MERIVALE said that at the Broomhill collieries the temperature was lower than that which would propagate the worm. At the same time, he thought that they were the only collieries in this district which provided conveniences for the men belowground, and he had done so before this disease was heard of.

Mr. T. E. FORSTER asked how many conveniences were provided and how many men were employed.

Mr. C. H. MERIVALE stated that only four conveniences were provided, and that shifters had the task of cleaning them.

Mr. J. H. NICHOLSON said that it had been shewn that the principal danger arose from the employment of foreign workmen. He understood that Italian miners had been introduced into the lead-mines of the Alston district. The conditions in metalliferous mines were more likely to develop the worm than those of coal-mines, and Italian workmen, perhaps drafted from some of the large tunnels on the Continent, were very probable "worm-carriers." He suggested that it would be wise to keep such men under observation, until it was proved that they were free from the disease.

The PRESIDENT (Mr. W. O. Wood) said that not a single case of the disease had been discovered in the counties of Durham and Northumberland. It had been observed in Cornwall, where the conditions were exceedingly favourable to its propagation, but the interest in this district arose rather from an academical or physiological point of view. Although the roads were watered, the working-places were exceedingly dry, and it was impossible for any worm to flourish there.

Mr. M. WALTON BROWN suggested that the people, who especially required watching are the Poles, not only on account of ankylostomiasis, but also on account of the especially virulent form of *tænia* to which the Polish race is subject. It was also important that earth-closets should be provided for the use of the workmen, in preference to any other form of closet. He also suggested that the eggs would be destroyed by mine-waters, containing, say, chlorides of sodium and calcium, salts of iron, free sulphuric acid, etc.

Mr. J. C. B. HENDY believed that it was a medical axiom that, if a person who was at all sensitive thought very much about a certain disease, he was more liable to contract it than he would otherwise be. He thought that if they debated this question for any length of time rumours of it might reach the coal-hewers, and they might begin to think too much about it and so induce the disease in that way.

Mr. J. B. ATKINSON (H.M. Inspector of Mines) understood that the miners were considering the disease, and as they found the name difficult to pronounce it had acquired a new appellation, namely, " Hanky-panky."

Dr. J. S. HALDANE wrote that, with regard to the important question of temperature raised by Mr. Blackett and other speakers, he had relied on Westphalian colliery-statistics and German laboratory-experiments for the statement in his paper that the development of the infective stage of the larva probably required a temperature of over 70° Fahr. Since then, however, Dr. Boycott and he had re-investigated the question, and found that if sufficient time were given (3 or 4 weeks) a temperature of 61° Fahr. was quite sufficient.* It appeared also from a table given by Dr. Barbier, at the end of a paper read by him before Section IV. of the International Congress of Hygiene, Brussels, 1903, that a number of collieries in the Liége district had been infected very seriously, although their temperatures at the working-face were well below 70° Fahr., and even as low as 60° Fahr. in one or two cases. Probably, however, the infection spread much more rapidly in warm and damp collieries, and this might be the reason why the cooler and shallower pits had hitherto escaped in Westphalia.

He (Dr. Haldane) agreed that particular care should be taken when Italian miners were employed, as in some parts of Italy about half of the population seemed to be slightly infected; but Cornish miners, and English miners returning from tropical countries, might also introduce infection. He did not think that any danger from tape-worm infection could arise in mines, as the tape-worm required an intermediate host, such as an ox, sheep or pig for its development to the infective stage, and the infection was carried by uncooked meat from one of these animals.

---

* " Ankylostomiasis," No. II., by Drs. A. E. Boycott and J. S. Haldane, *The Journal of Hygiene*, 1904, vol. iv., page 82.

THE SOUTH STAFFORDSHIRE AND EAST WORCESTERSHIRE INSTITUTE OF MINING ENGINEERS.

GENERAL MEETING,
HELD AT THE UNIVERSITY, BIRMINGHAM, DECEMBER 7TH, 1903.

Mr. ISAAC MEACHEM, JUN., PRESIDENT, IN THE CHAIR.

The minutes of the last General Meeting and of Council Meetings were read and confirmed.

The following gentlemen were elected :—

STUDENTS—
Mr. GEORGE CYRIL BLAKE, Westfield, Llanelly.
Mr. S. H. CASHMORE, Barr, Birmingham.
Mr. CLAUDE TRYON, Dudley, Worcestershire.

Mr. THOMAS WARTH read a paper upon "Gold-mining in Southern Rhodesia, 1896-1902."

THE MINING INSTITUTE OF SCOTLAND.

GENERAL MEETING,
HELD IN THE ROOMS OF THE CHRISTIAN INSTITUTE, GLASGOW, FEBRUARY 10TH, 1904.

MR. JAMES S. DIXON, PAST-PRESIDENT, IN THE CHAIR.

The minutes of the last General Meeting were read and confirmed.

THE NEW PRESIDENT.

Mr. ROBERT THOMAS MOORE was elected President in the room of Mr. Henry Aitken, deceased.

The CHAIRMAN (Mr. J. S. Dixon), in introducing Mr. Moore, referred to his being the son of a former much esteemed President, and a rising and wellknown mining-engineer and coal-master, and expressed, on behalf of the Institute, the hope that he would have a long and successful presidentship.

Mr. R. T. MOORE, in taking the chair, thanked the members for the honour conferred on him, and stated that he would do his best to work towards the success of the Institute.

The following gentlemen were elected:—

MEMBERS—
MR. ANDREW JAMIESON, Young's Oil Company, Addiewell, West Calder.
MR. RONALD JOHNSTONE, JUN., 190, West George Street, Glasgow.

ASSOCIATE MEMBER—
MR. JOHN POLLOCK, 40, St. Enoch Square, Glasgow.

Office-bearers for the session 1904-1905 were nominated, and auditors were appointed to examine the Treasurer's accounts for the past year.

Notice of motion was given for an alteration of the Rules at the next General Meeting.

DISCUSSION OF MR. J. BALFOUR SNEDDON'S "DESCRIPTION OF THE DUDDINGSTON SHALE-MINES AND THE NIDDRIE CASTLE CRUDE-OIL WORKS."*

Mr. C. A. CARLOW (Leven) said that, in Fife, they had always used the Lancashire boiler. He understood that an ordinary-sized Stirling boiler was fitted with 245 little iron tubes; and he should like to know whether any difficulty had been experienced with the joints.

Mr. J. B. SNEDDON replied that the Stirling boilers had given no trouble, and had required nothing more than ordinary attention; and no repairs had been made to the tubes. There were three boilers, one of which was held in reserve, and each boiler was cleaned once in three weeks, that was, after working for two weeks.

The discussion was closed, and a cordial vote of thanks was accorded to Mr. Sneddon for his paper.

———

DISCUSSION OF MR. ALEXANDER FAULDS' PAPER ON "MINE-DAMS."†

Mr. ROBERT CRAWFORD (Loanhead), in further reference to the dam described by Mr. Hamilton,‡ wrote that the bricks were wetted before being built, and the water used for that purpose, as well as for the mortar, was free from grease and mud. In the building of the dam, the lower part of the brick-work was kept in advance, so as to allow time for it to subside before closing the space at the roof. Special care was taken to place the bricks and cement tightly and close to the roof. The builder used a piece of hard wood in the form of a wedge, for driving cement and small pieces of brick into all spaces between the roof and the brickwork. In the outer length of 9 feet of the dam, small pipes, 1 inch in diameter, with perforated sides, were built into every square foot of the area, so that in the event of the dam leaking through the inner length of 10 feet, the water would pass through these small pipes in

* *Trans. Inst. M.E.*, 1903, vol. xxvi., pages 122 and 176.
† *Ibid.*, 1903, vol. xxvi., pages 134 and 179.   ‡ *Ibid.*, page 180.

the outer length of 9 feet, and thereby not deteriorate in any way the brickwork in that portion, which would consequently maintain its maximum strength. A temporary wood-and-clay dam was erected on the inside of the site of the permanent dam, the water being conducted therefrom, until the brickwork hardened, through a cast-iron pipe, built in tightly about 1 foot from the bottom of the dam. One end of this pipe was made of a conical shape, into which a metal plug was fitted. This pipe was plugged 5 weeks after the completion of the dam. Two bags of sawdust were emptied into the inbyeside of the dam, before it was plugged, the idea of doing so being that in case of water leaking through any part of the dam, the sawdust would be carried by the water into the interstices and close them up.

Mr. ROBERT McLAREN (Edinburgh) said that the most important consideration in the matter of mine-dams was the situation. Not many years ago, a very handsome and strong dam was constructed in the shale-measures, but under a moderate pressure the water poured through the strata, and the dam, although substantial, was of no use, owing to the situation, in that instance, being defective. In Scotland, however, dams were as a rule well built, and easily withstood the pressure. A serious accident occurred at Devon colliery, some years ago, by which 6 lives were lost. In that case, the dam was built to force water, by its own static pressure, through a pipe to a pump, situated some distance up a dook. It appeared that the men, after closing the door, closed the pipe by means of a valve, with the result that the pressure of water broke the door in the centre. He (Mr. McLaren) found on subsequent examination that the dam was about twice as strong as necessary, but the door was too light and was unable to withstand the full pressure of the water. Dams, as a rule, seldom gave way, as they were substantially built, and usually much stronger than necessary. The dam, at Loanhead colliery, described by Mr. James Hamilton and Mr. Robert Crawford, was placed in a good position—a hard, bastard limestone. In Linlithgowshire, a dam was erected to withstand a high pressure, and it was subsequently taken out when the water was drained. He (Mr. McLaren) was very much surprised to find on examination that the dam was not more than half the thickness that he would have employed. The dam, however, was well built of brick and cement, and it was solidly set in good, hard rock.

Mr. J. M. RONALDSON (Glasgow) said that, before contemplating the construction of a dam, colliery-officials should seriously consider whether there was no way of doing without it. He thought that the fewer dams there were about a colliery the better it was for all concerned; and when any waste-water was shut off, there was always an uneasy feeling among the workmen.

The PRESIDENT (Mr. R. T. Moore) said that he remembered the accident at Alloa to which Mr. McLaren had referred. Although the dam was of sufficient strength, the door was too weak to withstand the pressure exerted against it. The accident demonstrated the importance of calculating the strength of a dam before it was erected, and it was essential that every part should have a considerable factor of safety.

The discussion was adjourned.

The SECRETARY (Mr. James Barrowman) then read the following paper on "Miners' Phthisis":—

MINERS' PHTHISIS.

By JAMES BARROWMAN, Mining Engineer.

The conditions of the miners' occupation, as affecting health, must always have an interest for those connected with mines, for there can be no doubt that, regarded merely as a work-producer, the healthy man in a healthy environment is the most economical; and we would be neglecting one of the main objects of our existence as an Institute if we did not keep always before us the attainment of the most favourable health-conditions possible in mines.

The subject of miners' phthisis is one upon which there has been considerable uncertainty of views, largely owing to the absence of systematic and authoritative investigation. Individual medical men and others have, in the face of great difficulties, caused by imperfect data and otherwise, studied the subject and written to a limited extent upon it, but their observations are apt to become lost in the pages of their own technical journals, and the lessons to be learned from them are not brought before mining engineers with sufficient emphasis to command attention in the sphere of the miners' work.

Phthisis, known under the common name of consumption, has, as one common, outstanding feature, a tuberculous condition of the lung, whereby the substance of that organ breaks down and wastes away. What is called " miners' phthisis " has come to have a different significance, because, in miners, a tuberculous state of the lungs is the exception rather than the rule; and the term now applies to that condition of the lungs which is characterized by altered colour and the formation of a fibrous tissue, encroaching on their porous structure, causing difficulty of respiration and shortness of breath.

There are varieties of the disease depending on the nature of the dust that invades the lungs, and to these varieties particular names are given. In cases where the dust is from stone, such as is got in quarrying and metalliferous mining, the term is " silicosis," and from coal-mining, " anthracosis."

The prevalence of silicosis among the miners in the gold-mines of the Witwatersrand led H.M. High Commissioner, Lord Milner, to appoint a commission of nine members, four of whom were medical men, to inquire into the extent and cause of this disease, and to recommend preventive and curative measures. An indication of the state of matters is given by the Transvaal Government Mining Engineer in his annual report for the year ending May, 1902, where it is stated that, out of 1,377 rockdrill-miners in the Witwatersrand mines prior to the war, 225, or 16·34 per cent. died during the 2½ years immediately preceding the war.

The Commission was appointed in December, 1902. They reported in May, 1903,* and their report forms, perhaps, the most valuable contribution to the subject that has yet appeared.

The number of miners working underground in the gold-mines of the Witwatersrand was found to be 4,403; and 1,210 of them were examined by the Commission. Of the 1,210 men examined, 187, or 15·4 per cent., were found to be affected by phthisis and an additional 7 per cent. were suspected cases. While it may not be safe to say that these percentages would have been obtained had the whole of the 4,403 miners been examined, still the number actually ascertained to be affected is strikingly large, especially when it is stated that their average age was only 35·5 years. Of the 187 men affected, 91·98 per cent. had been employed on rock-drills, the average time of their employment in that capacity having been 6·49 years. It was found that some of these men had worked rock-drills in other countries than the Transvaal, but the conclusions of the Commission are not weakened thereby. There was no reason to believe that any appreciable number of the miners examined had suffered from phthisis before going to the Transvaal.

For the purpose of coming to a conclusion as to the cause of the exceptional prevalence of miners' phthisis, the Commission directed their attention to the conditions of the miners' work under these four heads:—(1) The character and properties of the dust in the mine-atmosphere; (2) mine-ventilation; (3) mine-sanitation, and (4) life-conditions of the miners.

* *Report of the Miners' Phthisis Commission, with Minutes of Proceedings and Minutes of Evidence*, 1902-1903, Pretoria.

(1) *The Dust in the Mine-atmosphere.*—The evidence makes it clear that the principal sufferers are the rock-drillers, especially those who are engaged in the boring of upward holes in the process termed " raising," and in some of the holes in advance-mines or drives, these holes being usually bored dry, causing much dust. The condition is aggravated by the exhaust-air from the drilling-machines, which, agitating the atmosphere in an otherwise poorly-ventilated place, keeps the particles of dust in suspension.

It was found by experiment that the fine dust in each cubic foot of air, close to a drilling-machine boring a dry hole, amounted on an average to 0·1136 grain. Estimating the quantity of air breathed by a miner at 21 cubic feet per hour it follows that a miner engaged in dry drilling may inhale 2·38 grains of dust per hour. Nature has provided a very complete set of apparatus in the mouth, nose and throat for the interception of foreign matters in the process of breathing; but despite every obstacle dust does find its way into the substance of the lungs, and works havoc there, so much so that, in the words of one of the medical experts examined by the Commission, " he believes that every miner who has worked for a year or two on rock-drills is affected." The particles of dust though very small, are angular, sharp-edged and irritating, and cause hardening and enlargement of the lung-substance, seriously interfering with the breathing, causing a harassing cough and emaciation, and hastening the miner into his grave.

Respirators as a protection for the lungs have not found favour with the miners. It is difficult to get a perfect fit, without which they are useless, and the men are averse to the discomfort and restriction in speaking and smoking that their use imposes.

As to curative measures, it appears that this form of phthisis is more amenable to treatment in the earlier stages than tubercular phthisis, and improvement of health usually follows if the workman takes to another occupation.

(2) *Mine-ventilation.*—The gold-mines on the Rand are ventilated chiefly by natural means. The large use of compressed-air drilling-machines has encouraged a certain amount of carelessness in the carrying out of systematic ventilation,

seeing that the exhaust-air has been useful in keeping the atmosphere of the faces cool; but the presence of a high percentage of both carbon monoxide and carbon dioxide in the air proves that the ventilation throughout the mines is bad.

During the six months prior to June, 1902, the amount of nitroglycerine-explosives used in the mines was equal to 0·75 pounds per shift for each underground workman. The extensive use of these explosives, some of them of inferior quality, is an important element in the contamination of the air of the mines, particularly if a charge is imperfectly detonated, or too great for its work. It is not uncommon for workmen to be " gassed " in badly-ventilated places. The Government inspector of mines in his report for 1903 stated that 2 Europeans and 12 natives were suffocated during the year by poisonous gases in the mines. These facts point to the desirability of using stronger detonators and only good explosive materials. It has also been found that the use of low-flash-point lubricants in the air-cylinders of the compressors has had the effect of contaminating the air of the mines with carbon monoxide to a dangerous extent.

(3) *Mine-sanitation.*—The sanitary condition of the mines leaves much to be desired, and no doubt tends to lower the vitality of the underground workmen. Regulations against fouling of the workings, and for the providing of proper conveniences, are in force in a few cases, and seem capable of general adoption.

(4) *Life-conditions of the Miners.*—With regard to the facilities for changing wet clothing, it is pointed out that at a few of the mines conveniences exist for changing at the pithead; but for the most part the miners go home to their quarters in their working-clothes, and have no proper means of getting them dried, the consequence being that they frequently put them on as wet as they were taken off. A regular system of providing changing-houses at the mines, where the miners' working-clothes could be properly dried, would materially lessen the risks of pulmonary disease.

While the evidence goes to shew that the chief cause of miners' phthisis is the dry dust of the mines, and that immediate reform in that respect is called for, the Commission direct special attention to the other contributory causes as well. They recommend that it is urgently necessary:—(1) To prevent the discharge of minute, hard, angular particles of dust into the mine-atmosphere; (2) to provide adequate ventilation throughout the mines; (3) to provide a suitable sanitary system, whereby the workings may be kept clean; (4) to provide change-houses, where the miners may change and dry their clothes; and (5) to avoid the use of inferior lubricants in the air-cylinders of the compressors.

That the conclusions of the Commission with regard to the highly-dangerous nature of the dry dust of these mines are sound is corroborated to a striking degree in the experience of those engaged in the construction of the Simplon tunnel. Arrangements are there made whereby all drill-holes are bored wet, and systematic measures are carried out to lay the dust. Cases of phthisis are rare among the workmen there, whereas in the St. Gothard tunnel-works where the same precautions were not taken, the mortality from phthisis was very high. It is much to be desired that the results of the investigations of this Commission should be known as widely as possible, in order that the urgency of reform may be brought home to all concerned.

The report is, on the whole, a most interesting and somewhat startling contribution to an important subject; but members should not infer, therefrom, that death is being dealt out in similar fashion in the mines of this country. So far as Scotland is concerned, the lead-mines of Wanlockhead and Leadhills are the only metalliferous mines of considerable extent, and the working conditions there are quite different from those of the gold-mines of the Rand, the drilling being done entirely by hand and water being abundant. And, as for coal-mining, practical experience and expert research make it clearer every day that lung-disease as a result of breathing coal-dust is not a prevalent disease among colliers. Unlike the sharp and irritating dust produced in the mining and tunnelling of hard and gritty rocks, the dust of coal is softer, more rounded and comparatively harmless. The only conditions in coal-mining at all approaching those of the Rand gold-mines may be on the rare occasions where, in a dry colliery, long stone-mines or drifts are driven in gritty rock, without arrangements being carried out for watering the drill-holes. No doubt the dust of coal-workings always has a certain admixture of gritty particles derived from the strata connected with the coal, and this is an element worthy of some consideration from a sanitary point of view.

In former papers, the writer has pointed out from information provided in the successive census-returns that coal-mining occupies a favourable position as compared with many other occupations.* In regard to phthisis, the death-rate among coal-miners is considerably under the average of all males in the population, and even in dust-producing occupations the coal-miner is freer from diseases of the respiratory system taken as a whole than the mason, the hair-dresser, or the linen-manufacturer.

The observations of former inquirers have been lately supplemented by a symposium of medical men,† led off by Dr. Thomas Oliver of the Royal Infirmary, Newcastle-upon-Tyne, who has made a special study of this subject. The following references to Dr. Oliver's paper indicate the present position of medical opinion as to the real character of miners' phthisis, and the influence of environment on the coal-miner's health. As to the notion that miners' phthisis may be due to the gases given off by explosives, by the coal itself, or from the exhalations of men and horses underground, Dr. Oliver expresses an unhesitating opinion that with the exception of nitric peroxide, which may be formed after blasting with the higher explosives by the union of nitrous oxide gas with oxygen, and is a cause of pulmonary congestion, none of the other gases plays any part in the production of the disease.

With reference to the influence of coal-dust on the lungs, Dr. T. Oliver expressed his views as follows :—

In Britain, because phthisis is in miners a comparatively rare disease, there are medical men who are of the opinion that anthracosis and pulmonary tuberculosis are antagonistic to each other, that coal-dust and soot exercise, in other words, a protective influence upon the lung so far as tuberculosis is concerned. In the north of England, as elsewhere, we certainly find that miners' wounds, even though

* *Trans. Inst. M.E.*, 1896, vol. xi., page 240; and 1898, vol. xiv., page 484.
† *The British Medical Journal*, 1903, vol. ii., page 568.

begrimed with coal-dust, heal remarkably well. I am not disposed to argue from this fact as to any antiseptic influence exercised by the coal-dust, so much as simply to assert the remarkable freedom of coal from morbific germs, and to support the statement of the slightly irritating properties exhibited by coal-dust when imbedded in the tissues.*

And with regard to the occurrence of the tuberculous condition along with the fibrous enlargement of the lungs, Dr. Oliver said:—

As a matter of experience, we do find tubercle in the lungs in miners' phthisis; but instead of regarding the excessive fibrosis of the lungs as a consequence of tubercle, I am disposed to regard the fibrosis as the result of irritation caused by dust, and that tubercle is a secondary and therefore accidental infection; and I base that opinion upon clinical experience, pathological observation and experimental investigation.†

We arrive, therefore, at the conclusion founded on the latest investigations, that, while the dry dust of the gold-mines of the Witwatersrand has had widespread and deadly effect on the miners there, and that gritty dry dust is distinctly deleterious: the dust incident to coal-mining is comparatively harmless, and is not in itself an important factor in the conditions adversely affecting health.

———

Dr. J. Livingstone Loudon (Medical Officer of Health, Hamilton) wrote that miners' phthisis or fibroid phthisis was, he had no hesitation in saying, a rare disease in this neighbourhood (Hamilton). He had had many opportunities of coming into contact with miners of all ages, in his capacity of colliery-doctor, medical officer of health and medical officer to the poorhouse. The few genuine cases that he had seen were amongst old colliers, who worked under the old régime.

Briefly speaking, the disease consists of a hypertrophic inflammation of the connective tissue of the lungs, this inflammation being induced by the inhalation of irritants such as coal-dust, soot from the lamps, and gases produced from the explosives used in mines. In the old days, black-powder reek must have played a considerable part in this disease, as large quantities remained after a blast, on account of deficient ventilation. The initial condition of hypertrophy of the connective tissue is later on followed by an atrophy or shrinkage of the

* *The British Medical Journal*, 1903, vol. ii., page 571.

† *Ibid.*, page 572.

same tissue, and consequently the lung-substance proper is
compressed and the available air-capacity is encroached upon.
As a result of this change, the ensuing symptoms are those of
bronchitis, peribronchitis, emphysema and bronchiectasis, the
emphysema being to a large extent compensatory. He would
particularly draw attention to the fact that the lung-tissues
being, as abovementioned, weakened and ill-aerated are
rendered particularly vulnerable to the attacks of, and form
a particularly suitable breeding-ground for, the bacilli of tuberculosis. Consequently, a case starting as a fibroid condition
of the lungs, in many cases terminates as phthisis pulmonalis
or ordinary tuberculous consumption.

As Mr. Barrowman stated in his paper, the collier occupied
a high place in the list of occupational diseases, much higher
than would be expected. It must be borne in mind, however,
that apart from improved methods of working in mitigating
dust-particles, there had been a considerable diminution in the
number of hours worked per day, week and year, in a coal-pit,
and the collier now-a-days does not hold by six days per week.

The coal-dust deposited in the lung-tissues remains there
for ever, and on the supervention of an acute attack of
bronchitis, black spit is in evidence. He had seen such black
spit from an old collier, who had not been down a pit for
20 years.

The cases that terminate as tuberculous phthisis occur mostly
among the younger colliers, and other deteriorating influences
must be borne in mind in those juvenile cases : the most potent,
although perhaps the least suspected, being alcoholic excess,
which frequently fosters a condition of chronic pneumonia
peculiarly liable to tuberculous invasions.

He had seen a much heavier incidence of fibroid phthisis
among stone-masons than among coal-miners, and, moreover, this incidence is much heavier in masons who have worked
on a close-grained, hard stone, than among those hewing the
softer red stone.

The whole pathological process in fibroid phthisis is
extremely similar to that seen in certain diseased conditions
of other organs, such as the kidneys and liver. In the last-named
organ, the irritant is alcohol, but it plays precisely the same
part as coal-dust, etc., does in the lung; and the same hyper-

trophy of connective tissue occurs, followed later by a similar shrinkage and consequent compression of the essential working portion of the organ.

The PRESIDENT (Mr. R. T. Moore) said that it was some consolation to those who were associated with coal-pits to know that coal-dust possessed antiseptic properties. He was not surprised to learn that so large a proportion of the Europeans on the Rand were the victims of phthisis. The Kaffirs did the work in the mines, while Europeans superintended the Kaffirs and many of them looked after the rock-drills. The rock-drills were used in hard quartz, and the dust was very angular and injurious to the lungs. A large number of the men affected were those working rock-drills, and as these formed a large proportion of the Europeans it was natural that a large proportion of Europeans should be affected—much larger than in home-mines, where all the ordinary labour which was not subject to the deleterious influence of dust, was also done by Europeans. In Cornwall, when rock-drills were used, it was customary to play a jet of water on the hole, and this kept down the dust: this system was not in use ten years ago on the Rand, but it might be now. It seemed to him to be a most important precaution to adopt, and one that was more likely to be of service than any form of filter or respirator to be worn by the miner.

Mr. GEORGE A. MITCHELL (Glasgow) asked whether or not the Kaffirs were affected by miners' phthisis.

Mr. BARROWMAN said that the report had reference to whites only, but the minutes of evidence showed clearly that large numbers of the natives were afflicted with phthisis, probably partly on account of mine-dust and partly from the changed conditions of living at the mines.

Mr. ROBERT McLAREN (Edinburgh) said that at the Leadhills mines practically no dust arose from the boring of holes. The men used small hand-drills and hammers, and the rock gave off water. It should be remembered that coal-cutting machines of the disc-type produced large quantities of dust. He had sat behind one of these machines, and he had to confess that he was inconvenienced by the clouds of dust that arose. The only remedy was to keep the ventilation as perfect as possible, so that the air might carry away the dust as soon as it was made. The removal of the dust from coal-cutting machines required early consideration, or, sooner or later, the health of the workmen, who constantly breathed the dust-laden atmosphere, would be affected.

Mr. GEORGE A. MITCHELL said that he was prepared to acknowledge that there might be an element of danger, owing to the prevalence of dust, if the coal-cutting machine was cutting through hard and gritty material. He was not quite sure that he coincided with Mr. McLaren's view with regard to the provision of ample ventilation. If the ventilation was too strong it kept the dust suspended, and carried it towards the men who were working behind the machine. If it was found necessary to deal with the dust, it might be better to keep it down by the use of water. The dust from coal-cutting machinery had not yet caused any serious trouble, but with the more general introduction of coal-cutters into the collieries of Scotland, difficulty might arise. He (Mr. Mitchell) asked whether Mr. Barrowman could adduce any statistics to show how far phthisis prevailed in different parts of the United Kingdom. They all knew, for instance, that Welsh mines were much dustier than the mines of Scotland.

Mr. JAMES BARROWMAN replied that particulars of the comparative mortality of males from phthisis, contained in his paper on "Mining Mortality,"* were as follows:—

England and Wales—

Tin-miner	508
Lead-miner	380
All males, 25 to 65 years of age	192
Coal-miner, Yorkshire, West Riding	123
Do., Monmouthshire and South Wales ...	107
Do., Lancashire	102
Do., all districts	97
Do., Durham and Northumberland ...	94
Do., Staffordshire	83
Do., Derbyshire and Nottinghamshire ...	69
Ironstone-miner	90

Scotland—

Ironstone-miner	269
All males, 25 to 65 years of age	168
Coal- and shale-miners	100

* *Trans. Inst. M.E.*, 1898, vol. xiv., pages 488, 489 and 492.

Mr. G. A. MITCHELL said that the fact, that miners' phthisis was not so prevalent now as it used to be, seemed to indicate that it was attributable to bad ventilation in the past and not to dust. He thought that as the mines were getting deeper there was more dust than in bygone days, and yet there was less phthisis.

Dr. WILLIAM GRANT (Blantyre) wrote that miners' phthisis was now seldom met with by medical men practising in mining districts, unless as a secondary complication to some other form of lung-trouble. This confirmed the opinion expressed by medical authorities that the inhalation of coal-dust did not produce active lung-trouble, and was certainly not the cause of consumption among miners: indeed miners were less liable to tuberculous disease than any other class of workmen.

There could be no doubt that the continued inhalation of dust-particles tended to reduce the resisting power of the lung-tissue to invading organisms, more particularly those of an irritating form or substance; and, as a matter of fact, the stone-mason, (especially the hewer) was more liable to consumption than the metal-worker, and the coal-miner least of all. This was undoubtedly due to the fact that the coal-miner was inhaling dust of vegetable origin, and in very fine particles. There could be no doubt, however, that many coal-miners ultimately developed cough and dyspnœa, due to the gradual encroachment on the lung-tissue of coal-dust. Moreover, should he be seized with any acute lung-trouble, such as bronchitis, pneumonia or pleurisy, he was not able to throw it off so easily as one who had not been breathing dust-contaminated air, as already the resisting power of the lung-tissue was thereby diminished.

During the past 12 months, he (Dr. Grant) had had two such cases under his care: both made slow recoveries, and expectorated large quantities of mine-dust. Originally, they were lead-miners, but he did not think that their previous occupation had any bearing on the anthracosis. If, however, say, instead of being miners, the patients had been masons, he was bound to confess that he would have been more apprehensive for their recovery, and would have certainly been afraid of their developing tuberculous disease.

It is now clearly established that the inhalation of dust-particles produces a consolidation of the lung-tissue by excessive growth of fibrous tissue: the extent of the change, however, being dependent upon the physical character of the dust inhaled. In this respect, the environment of the coal-miner helps him, more especially the present-day miner, as it is an undoubted fact that the introduction of better systems of ventilation in mines has led to improvement in the general health of the miner, as dust in suspension and stationary (especially in a moist atmosphere) is more dangerous than dust in a dry condition and in suspension in an atmosphere moving with considerable velocity.

The discussion was adjourned.

DISCUSSION OF DR. J. S. HALDANE'S PAPER ON "MINERS' ANÆMIA OR ANKYLOSTOMIASIS."[*]

Mr. ROBERT McLAREN (Edinburgh) said that ankylostomiasis had been receiving attention of late, but perhaps not more than the subject deserved. The symptoms had been fully described by Dr. Haldane in his paper, and in his *Report* .... *on an Outbreak of Ankylostomiasis in a Cornish Mine.* The only redeeming feature was that the disease was not contagious in the same way as smallpox or fever, and a person could only contract the disease in the mine by direct contact. This, it appeared, might be due to at least three causes: inhaling the dry dust laden with larvæ; or allowing the hands to be soiled with the mud in which the fæces were mixed, and then taking food; or by drinking water after it had percolated through the waste, in which excreta had been deposited. It appeared that the worm could most readily propagate and thrive in an atmosphere with a temperature of about 70° Fahr.; and, in that case, there was little to be feared so far as the mines in Scotland were concerned, as very few of them had so high a temperature.

Those who had to travel old abandoned roads or airways, situated near a main haulage-way, knew the filthy condition which these roads acquired for some distance, owing to excreta being deposited; and it was on such roads, with a warm atmo-

* *Trans. Inst. M.E.*, 1903, vol. xxv., page 613; and 1904, vol. xxvii., page 11.

sphere, that the larvæ were to be found if any one of the men who visited such a road was a worm-carrier. He had often thought, when passing along such roads, that even from a purely sanitary point of view, the management should put down with a firm hand the habit of depositing excreta in old roads.

In deep mines in the east of Scotland, it was quite a common thing for the men to carry a can containing water with them into the mine, to slake their thirst while at work so that they might have no need to partake of pit-water. There was no harm, however, in drinking water that came direct from the metals or shales, before it reached the pavement; but harm could only come from drinking if the water in percolating through the waste on the pavement came into contact with fæces containing larvæ. It was important, therefore, that there should be in addition to good ventilation a system whereby any excreta could be deposited in places provided for that purpose and removed thence to the surface. A system of pails, placed in different parts of the mine, seemed best calculated to effect this end. The pails should be placed wherever a body of men had occasion to congregate during the shift, such as drawers at a lye for tubs, at the inbye-end of horse and haulage-roads, incline-planes, shaft-bottom, etc. There did not appear to be the same need for pails to be provided for facemen, as the excreta could be safely deposited on the débris in the cundie or goaf and covered. It might not generally be known that the pail-system was not a new thing in Scotland, as it was in use at Fence colliery, Lesmahagow, when that colliery was in operation prior to its abandonment in 1898.

No one would deny that the disease was a filthy one, but it was possible to make too much of it, and bring about a scare; and while it was well to guard against its introduction into the mines, still there was nothing meantime to cause alarm. One word of warning might be given to colliery medical men, for whom he had the highest respect, and on whom he had no desire to cast reflections: from the very nature of the disease it was possible for mistakes to be made, unless very great care was exercised, and a man might be diagnosed to be suffering from ankylostomiasis, when it was something else, but once having diagnosed the case and being absolutely certain, no time should be lost in communicating with the colliery-owner.

Mr. J. BARROWMAN said that this valuable paper by Dr. Haldane and the discussion thereon seemed to leave no doubt as to the cause of the presence of the worm, the cure of those who were affected, and the means of prevention. Difference of view as to the causes and prevention had been expressed by one medical expert in Austria,* but there seemed no good reason to question Dr. Haldane's conclusions.

It was worth while noticing that where indiscriminate fouling of the workings took place, the watering of dust which was a measure of safety in blasting and a measure of health in preventing the breathing of injurious particles, increased the risks of ankylostomiasis..

Mr. T. H. MOTTRAM (Glasgow) said that, in his experience, fæcal deposits were not so much made in the working-places, as at the termination of haulage-roads and at the airways between one district and another. If pails were provided at these places, and if medical certificates were required from men coming from suspected districts showing their freedom from the disease —as suggested by Dr. Haldane—there would be little chance of the disease spreading into this country. It appeared that moisture and a temperature exceeding 68° Fahr. were conditions favourable to the development of the worm, and therefore it was only to such mines where these conditions existed that special attention need be directed.

Mr. ROBERT McLAREN said he had been informed that there had been one case of ankylostomiasis in the Hamilton district.

Mr. JAMES BARROWMAN replied that the patient came from a Lanarkshire colliery, but he had been abroad immediately before his illness. The case was reported by Prof. Ralph Stockman, Glasgow University, who had the case under treatment in the Western Infirmary, Glasgow.†

Mr. GEORGE A. MITCHELL said that the symptoms were not always very apparent. Consequently, a colliery-manager would not invariably be able to distinguish the disease successfully.

* Dr. Iberer, *Mining Journal*, 1903, vol. lxxiv., page 235.

† "A Case of Ankylostomiasis in Scotland," *The British Medical Journal,* 1903, vol. ii., page 189.

Mr. J. M. Ronaldson (Glasgow) said that, so far as Scotland was concerned, there was no great fear of the disease spreading to any extent. When the members considered the fact that a temperature of 69° to 70° Fahr. was required before the worm could thrive, they knew that comparatively few collieries in Scotland had such a temperature, consequently, on the whole, there was no very great fear. With regard to the deposition of the excrement of the men in the mine, he had always thought that, simply by getting workmen into the habit of attending to such needs on the surface, a great deal might be done towards having the air of our mines kept in a more sanitary condition. He believed that this suggestion was easy of accomplishment, because it was simply a matter of habit with the men, whose working-day was now much shorter than formerly. There could be no doubt that the deposition of excrement on the surface instead of underground would produce a great improvement in the sanitary condition of mines.

The President (Mr. R. T. Moore) said that, probably, if workmen would cover up the deposits in the waste of longwall workings, the nuisance would be to a large extent overcome. He thought that the suggestion of placing pails at the pit-bottom and main-landings was well worthy of attention, not only in regard to this disease, but in regard to many other diseases. It seemed to him that the matter of sanitation in mines was one that should be very carefully looked into.

Dr. J. S. Haldane wrote that, in connection with some of the remarks made by Messrs. McLaren, Mottram and Ronaldson, he would like to take the opportunity of modifying the statements that he had made in his paper and in a recent blue-book, to the effect that a temperature of about 70° Fahr. was necessary for the development of the infective stage of the larval worm. He regretted to say that recent experiments by Dr. Boycott and himself showed that if sufficient time were given (about three weeks) a temperature of 61° Fahr. was sufficient. It also appeared from data furnished by Dr. Barbier, in connection with the Belgian Government enquiry, that a number of mines in the Liége district had been infected, although the temperature of the. workings was under 68° Fahr., or even under 62° Fahr. He, therefore, thought that precautionary measures ought to be taken even in cool and shallow mines.   He did not think that there was any risk of infection by dust, as drying killed both the larvæ and ova with great rapidity.

DALMUIR SEWAGE-PURIFICATION WORKS OF THE CORPORATION OF GLASGOW.

The members were received by Bailie Anderson, convener of the Sewage Committee of the Corporation; Mr. A. B. McDonald, City Engineer: and the local officials.   At the close of the inspection, on the motion of Mr. R. T. Moore, a vote of thanks was accorded to the Glasgow Corporation and their representatives for their kindness.

THE MIDLAND COUNTIES INSTITUTION OF ENGINEERS AND THE MIDLAND INSTITUTE OF MINING, CIVIL AND MECHANICAL ENGINEERS.

JOINT MEETING,
HELD AT THE ROYAL VICTORIA STATION HOTEL, SHEFFIELD, JANUARY 30TH, 1904.

MR. H. B. NASH, PRESIDENT OF THE MIDLAND INSTITUTE OF MINING, CIVIL AND MECHANICAL ENGINEERS, IN THE CHAIR.

The following gentlemen were elected to the Midland Counties Institution of Engineers, having been previously nominated:—

MEMBERS—

MR. JOSEPH CLARKE, Rosekenwyn, Pinxton.
MR. GEORGE HERBERT FOWLER, Mining Engineer, Hall End, Tamworth.
MR. S. H. McCONNEL, Colliery Owner, Stretton House, Alfreton.

ASSOCIATES—

MR. J. O. COOPER, Under-manager, Tinsley Park Colliery, near Sheffield.
MR. JAMES KNIGHTON, Under-manager, Tinsley Park Colliery, near Sheffield.
MR. JOHN PARKINS, Colliery Surveyor, Clay Cross, Chesterfield.
MR. JOSEPH SWAIN, Under-manager, Clay Cross, Chesterfield.

STUDENTS—

MR. ARTHUR SELBY, Mining Pupil, Tinsley Park Colliery, near Sheffield.
MR. PERCY WHITE, Mining Pupil, Tinsley Park Colliery, near Sheffield.

The following gentlemen were elected to the Midland Institute of Mining, Civil and Mechanical Engineers, having been previously nominated:—

MEMBERS—

MR. HUGH B. PLAYER, Electrical Engineer, 59 and 61, Blonk Street, Sheffield.
MR. THOMAS COOK, Colliery Manager, Mount Pleasant Colliery, Wollongong, New South Wales.

Mr. JOSIAH STEPHENSON WARD, Colliery Under-manager, Tankersley, near Barnsley.

Mr. SAMUEL H. GIBSON, Mechanical Engineer, Wheldon Road, Fryston, Castleford.

Mr. SAMUEL WANE, Mine Surveyor, 29, Bank Street, Lodge, Brymbo, near Wrexham.

Mr. THOMAS SEAMAN, Colliery Manager, The Gables, Lodge, Brymbo, near Wrexham.

Mr. THOMAS TAYLOR, Mechanical Engineer, New Moss Colliery, Audenshaw, near Manchester.

Mr. PHILIP BUCKLEY, Mechanical Engineer, Ashfield Road, Morley, near Leeds.

Mr. PERCY C. GREAVES read the following paper on " An Electrical Heading-Machine ":—

AN ELECTRICAL HEADING-MACHINE.

By PERCY C. GREAVES.

In venturing to read a paper on a heading-machine the writer claims no special success, but he simply makes a plain record of what it has done for the colliery which he represents. Some three years ago, he had occasion to lay out the pit, to win an area of coal which had been recently added to the coal-field; and, to do this, it was necessary to drive strait work for a considerable distance. The men were working on a price-list, which had been settled a few years previously with the miners union; and, when the men were ordered to do this class of work, they refused and demanded higher payment. The price-list specified that they were to be paid as follows:—For strait work, 5 feet wide, end on, 1s. 7d. per ton; cutting per yard, 2s. 2d.; widening to 9 feet, 6d. per yard; *plus* 47½ per cent.; and a bonus of 4s. if a man did 6 yards in a week. These prices work out to 9s. 7d. per yard of heading, 9 feet wide. The men demanded 1s. 10d. per ton, 2s. 6d. per yard, 2d. per tub for filling muck, and 6d. per yard for widening. These prices represent 11s. 3d. per yard of heading, 9 feet wide.

The writer is unable to give the average wages of men working in strait places, because very little of that kind of work had been done of recent years. Previous to the "good times" no trouble had been experienced in getting such work done, and the men were readily satisfied. Of course, the management could not agree to so large an increase in cost, the men, thereupon, refused to drive any more strait work, and went on strike.

As it was imperative that the work should be driven, and as electric power was used in the mine for hauling and cutting coal, the writer's thoughts naturally turned to that power, but he found that there was not a single British electric header on the market. There were several headers driven by air-power, but that means of driving could not be adopted owing to the great capital cost for the small amount of work to be done.   The writer heard of two American machines, the Jeffrey and the Morgan-Gardner, and it was arranged to make a trial with the Jeffrey electric heading-machine.

The section of the seam is as follows:—

			Ft. Ins.	Ft. Ins.
Hard black scale, with ironstone-balls	...		—	
COAL	...		2 0	
COAL and dirt,	... 12 inches to		1 2	
COAL, full of pyrites, "Dicks"	...	.	0 3	
			—	3 5
White stone	...		—	

The Jeffrey machine was started on April 1st, 1903, and has been at work ever since.   The machine is 18½ inches high, but it is 21 inches high on the runners, and weighs approximately 30 cwts.   It consists of a moving frame, round which the chain travels, with the motor and gearing fixed on it.   This frame travels through guides on a stationary frame, 9 feet long, which is fixed in the direction of the cut by means of two jacks; one at the front is placed against the coal-face and set at an angle, in order to take the thrust of the machine as the chain is entering the undercut; the other jack is set from the back of the stationary frame to the roof: and the two jacks hold the frame firmly in position.   The motor, of 15 horsepower, runs at a speed of about 900 revolutions per minute; it lies at right angles to the coal-face, and drives through bevel-wheels a vertical shaft, carrying a worm and a small pinion.   The latter gears into a large toothed wheel on a shaft, fixed below the motor, and here also is fixed the driving-sprocket for the chain.   The worm on the first vertical shaft drives two worm-wheels on shafts set at an angle, and at the other end of these shafts are two more worms, which gear into worm-wheels running loosely on a shaft fixed across the frame.   The worm-wheels are fitted with slots in the side with two or three steel driving-pins, and they are geared to run in opposite directions.   A pin-clutch, sliding on a key, is placed between the worm-wheels, and the machine-driver can therefore run this cross-shaft in either direction.   On the ends of the cross-shaft two small pinions are keyed: and they gear into racks fixed on the sides of the stationary frame, so that the machine may be fed in or out and may also be allowed to run idle. The rate of feed for cutting was 2 feet per minute, and it has been reduced to 1 foot per minute, but the withdrawal of the machine can be made in 30 seconds.

The chain is composed of links and tool-holders, the latter being set at different angles, so that straight tools can be used, but it was found advisable to bend the top cutters and make them pick-pointed, the section of the steel being $\frac{3}{4}$ inch by $\frac{1}{4}$ inch. There are 45 tool-holders, and the chain runs at a speed of about 300 feet per minute. The tools are fastened into the holders by means of set-screws.

The motor is of the four poled type, with two shunt-coils, for a current of 500 volts. The motor is of the open-running type, which is not desirable; but the writer has arranged to convert it into a closed motor, although a little trouble may be caused by its becoming heated. The connections for the trailing cable are plugs, fitted with a bayonet-socket arrangement so as to prevent them from being inadvertently pulled off. The trailing cable is attached to a diamond switch-box.

As the chain runs from right to left, it is advisable to work the machine from left to right, and in this way it works with the tools on a loose side, after the first cut. As soon as the machine has made the first cut, the jacks are released and it is pushed, by means of iron bars, into a position ready for the next cut, and the jacks are re-fastened. After the machine has made three cuts, it is lifted by means of an ingenious jack, the runners are moved back a little, and it is pulled back a few feet from the face by means of a Sylvester chain, fastened to a prop, so as to enable the shots to be fired in safety and the coal to be filled. The depth of the undercut averages 5 feet 7 inches, the width is 3 feet 8 inches, and the thickness of the cut is $4\frac{1}{4}$ inches. The cut is made in the dirt lying immediately above the "dicks."

It was found inadvisable to cut the holing as fast as the machine was originally speeded to do, as it took too much current. At the present rate of cutting, the machine takes 26 ampères at 480 volts, or say $16\frac{1}{2}$ horsepower. It will be seen that the motor is somewhat overloaded, and in the writer's opinion it would be advantageous to adopt a motor of larger size. Of course, as no run exceeds 6 minutes, the motor has plenty of time to cool.

The writer found that, in the three months ending December 31st, 1903, the machine made 226 cuts to an average depth of 5 feet 7 inches, 223 shifts had been worked, and 415 yards had been driven.   Two men are employed at 6s. 6d. per yard, without percentage, the driver earns about 8s. per day, the assistants are paid 5s. per day, and in addition, the men are paid for shifting the machine from one heading into another.

After the cut has been made, as much as possible of the loose dirt is removed, then the electric drill is fixed so as to make a hole at either side of the heading, and it takes 5 minutes to set the machine and drill a hole, 6 feet long.   This drill weighs 170 pounds, and is driven by a motor of 3 horsepower; and when running it takes 5 ampères at 480 volts.   The drill is easily moved by two men, and after the second hole is drilled, it is removed a few feet back so as to be placed out of danger.   The shots are then fired, the coal is filled, the "dicks" taken up, and the machine drawn forward for another cut.   The coal, worked by the machine, is very much larger than that formerly produced by hand-labour.

The machine is moved from one heading to another, every 150 or 200 feet, and during the time that the machine is working in the back heading, ripping is being done in the main heading. The slits are also driven by the machine.   The machine is only worked two shifts daily, because a greater distance is driven per shift than when three shifts are worked daily.

A test has been made to ascertain the length of time required to make the cut, and, though the writer places it on record, he prefers to take the results of three months' working to that of any single day.   The machine lay 8 feet from the face, and the men spent $8\frac{1}{2}$ minutes in placing it in position, 5 minutes were taken to prick the roof for the back jack, 2 minutes to fix the machine and run the motor light, 7 minutes to make the undercut and withdraw the chain, $6\frac{1}{4}$ minutes to move over and fix the machine for the second cut, $8\frac{1}{4}$ minutes to make the second cut, $5\frac{1}{2}$ minutes to move over and fix the machine for the third cut, 7 minutes to make the third cut, 6 minutes to remove the machine from the face, the total time being $55\frac{1}{2}$ minutes.

The machine saved the sum of 4s. 9d. per yard, taking the prices demanded by the men, towards power, interest on capital, depreciation, and repairs; or 3s. 1d. per yard, taking the price-list as a basis.   It is difficult to ascertain the cost of power, because the machine is actually running for less than 30 minutes per shift.   The machine has not been running long enough to determine the interest and depreciation.   The total repairs, exclusive of sharpening the cutter-teeth, have amounted to £9 13s. 9d. since April, 1903; but this amount does not include the wages of the colliery-mechanic employed in replacing broken wheels.   An average of 20 teeth required sharpening after each shift.

In conclusion, the writer would like to say that he has no financial interest in this or any other heading-machine, but he cannot close his paper without thanking Mr. R. Hood Haggie for his general assistance, and Mr. Waterhouse, the manager of the pit, for collecting the data upon which the writer has founded this paper.

Mr. J. GERRARD (H.M. Inspector of Mines) moved a vote of thanks to Mr. Greaves for bringing the subject before the members in so excellent a paper.

Mr. W. B. M. JACKSON observed that he had had some experience with the Morgan-Gardner—a machine of this type.   Their machine was one of the first of the kind introduced into this country, and they had obtained excellent results from it.   Mr. Greaves stated that the coal worked by the machine was much larger than that formerly produced by hand-labour; but he (Mr. Jackson) had not found that to be the case.   The coal had been undercut and then blown down, without cutting the side; and he asked the writer whether the coal had been nicked or sheared after it had been undercut by the machine.

Mr. PERCY C. GREAVES replied that the coal was undercut and then blown down, and the improvement of the sample was marked.   Some slits had been driven by hand, and the difference in quality was so great that it could be readily recognized. At another colliery, with which he was connected, the seam was thin, and the men cut everything to slack with the pick; but with the machine, very respectable coal was obtained, even after shots were fired—the coal was not so good as that produced from banks, but it was much better than slack.

Mr. W. B. M. JACKSON said that much depended on the condition of the seam. He had a seam 4 feet 2 inches thick, and when the coal was efficiently cut by hand, it produced a larger quantity of round coal than it had done since the electric cutter was used.

Mr. H. RHODES remarked that he had used a Stanley heading machine, and the methods of working which he had adopted were somewhat similar to those described by Mr. Greaves. The seam was nearly 6 feet thick; but, owing to a band of dirt in the centre of it, sometimes containing large lumps of iron-pyrites as big as a football, the working of the Stanley machine was abandoned. It was found that it could do two or three times as much work, with an increased cost of about 50 per cent., compared with hand-labour. If a heading machine were again introduced into that seam, it would be a machine of the Jeffrey class, even if it were driven with compressed air. The Stanley heading-machine had to face every portion of the seam at some period of its cut, and if it happened that a boulder appeared in the middle band at each side of the heading, there was no other resource than to stop the machine and to cut round the boulders by hand-labour.

Mr. R. HOLIDAY (Ackton Hall Colliery) said that Mr. Greaves' paper exactly confirmed his own experience. They put in a Jeffrey heading-machine, worked by a three-phase motor, although the makers stated that a three-phase motor could not run the machine, and that a small fortune had been spent in America in the attempt to make it do so successfully. It was found that the feed was exactly twice as fast as it should be; the motor was of 15 horsepower, and to run the machine with the feed as arranged would require 30 horsepower. In consequence, the machine was fed at half the rate, just the same as Mr. Greaves did. It was also found that the feed of the drill was twice as fast as it should be. The machine was worked in a seam, 2 feet thick, and it did extremely well. There was one question as to how far it mattered on which side the cut should be made; and Mr. Greaves suggested having a free side after the first cut. He thought that the cut should be made so that the chain would bring out the stuff. By this arrangement, the sliding frame was less likely to become jammed, and there was less liability of the machine sticking. The coal was taken down in the same way as described by Mr. Greaves, but the holing was made a little farther into one side than the other, so as to make the coal come down somewhat more easily.

Mr. W. HAY (Shirebrook Collieries) said that they had cut about 27,000 feet of heading with a Jeffrey heading-machine. The cost of minor repairs, over 9 years, including a new armature, was about £150. The machine saved an average of about 6s. per yard over hand-labour, and the round coal was increased by 20 per cent. The cut was made on one side and the shot placed in the other. The machine had given great satisfaction, and it was worked at the original speed of the gearing.

Mr. G. BLAKE WALKER (Wharncliffe Silkstone Collieries) said that, in driving wide places, they found that the Jeffrey heading-machine did more economical work than in narrow places, 6 feet wide, because there was less time spent in removals in proportion to the work done. The speed of the cut was about 1 foot per minute. There had been no trouble with the machine, which seemed to be well made, and would cut across a place, 36 feet wide, in about an hour, or a little over. The one drawback from their point of view was that this machine required a width of about 14 feet to work in; this was rather a large spread of roof to uphold unless the roof was exceptionally good; and some little delay took place in setting the timber behind the machine. Further, every time the machine was shifted, two or three props had to be taken out and replaced, if the roof would not stand without propping.

Mr. M. W. WATERHOUSE (Exhall) remarked that, in his experience, he had found an increase in the percentage of round coal. At a Warwickshire colliery, with which he was connected, a considerable reduction of cost was effected as compared with hand-labour—probably more than Mr. Greaves mentioned, because in their case the cost of hand-labour was very great.

Mr. R. HOOD HAGGIE (Derby), quoting from the price-list for 1903 in Indiana, U.S.A., said that the cost of hand-labour, including loading, was 3s. 9d. per ton. Cutting, with a machine of the Jeffrey type, cost 5d. per ton; and with a machine of the Sullivan or Ingersoll-Sargeant type, 7¼d. per ton. Loading, etc., using hand-drills, cost 2s. 1½d. per ton; and with machine-drills, 2s. per ton. Drilling by hand cost 1½d. per ton; and by machine, 1¼d. per ton. The cost with a punching-machine, with hand-drills was 2s. 10¼d., and with machine-drills, 2s. 8½d. per ton; and with a machine of the Jeffrey type, 2s. 8d. with hand-drills and 2s. 6¼d. per ton with machine-drills. Consequently, there was a saving of 1s. 2¾d. per ton by using a machine of the Jeffrey type, and machine-drills over hand-labour. In Indiana, the machine-men removed all sprags, took out the cutters from the machine and got it ready for removal to another place. If the room was wet the men were paid 1¼d. per ton extra; but that did not affect the saving, because the same allowance was made in respect of hand-labour. With regard to the rate of feed of the Jeffrey machines, every-one had learnt of late years that different holings required different rates of feed. In many seams, a machine would cut 6 feet under in 3 or 3½ minutes, while in others it did not cut more than 1 foot a minute or perhaps less than that. The proper feed could be best decided by trial, but it entailed no structural alteration in the Jeffrey machine.

Mr. W. H. PICKERING (H.M. Inspector of Mines) enquired whether Mr. Greaves had found any difference in the number of shots required in headings driven by machine and hand-holing. Although they were now well used to machines for coal-getting, he did not think that the advantages of machine-heading were sufficiently appreciated at the present time.

Mr. PERCY C. GREAVES, replying to the discussion, said that Jeffrey machines had been air-driven, and it was a remarkable circumstance that in another pit, with which he was connected, an air-driven machine had been taken out because it produced so much sparking at the chain. There was a little pyrites in the seam, where it was cutting, and it produced a stream of sparks along the length of the chain; and as it was a fiery mine, they durst not continue the working of the machine. It must be admitted that more shots were needed in a heading cut with the machine than with hand-holing. It meant two shots were fired in each length of 5 feet 7 inches, or practically one shot for every 3 feet; but with hand-holing he did not think that it would average a shot for every 10 or 15 feet; he could not, however, give the exact number, as a record had not been kept. The increase in the number of shots was a disappointing feature, but it could not be overcome, until they had a machine which would shear at one side and probably cut a little bit at the roof as well, because the coal adhered very strongly to the roof in some thin seams.

The CHAIRMAN (Mr. H. B. Nash) agreed with Mr. Pickering that sufficient attention had not been given in England in the past to the use of heading-machines. Coal-cutters had been chiefly confined to the longwall-face: but in mines which could not be worked on the longwall system or headings which it was difficult to work by hand-labour owing to the hardness of the coal, etc., the heading-machine would successfully and economically replace the men. He agreed with Mr. Greaves that it was disappointing to find that the use of the heading-machine had led to an increase in shot-firing, because that was one of the dangers which should be curtailed to the greatest extent possible. If a machine could be made to cut on one side of the road, as well as undercut the coal, it would enable the coal to be brought down in the ordinary way by wedges, and shot-firing could be abolished. He had pleasure in seconding the vote of thanks to Mr. Greaves for his paper.

The resolution was cordially adopted.

DISCUSSION OF DR. J. S. HALDANE'S PAPER ON "MINERS' ANÆMIA, OR ANKYLOSTOMIASIS."*

The CHAIRMAN (Mr. H. B. Nash) said that, fortunately, up to now British coal-mines had been free from this disease; but the fact that disastrous experiences had occurred in Germany and other countries, made it very desirable that the subject should be brought before the members. He believed that there would be an improvement in the sanitation of their mines, and this would render the possibility of the dissemination of the disease still more remote. In West Yorkshire, at several collieries, steps had already been taken for providing places at the pit-bottom and at the pass-byes for the use of the men. The

* *Trans. Inst. M.E.*, 1903, vol. xxv., page 643; and 1904, vol. xxvii., pages 11 and 32.

pail-system had been advocated as being an effective method, but it was thought better to make places similar to refuge-holes, well cemented on the sides and floor; they could be emptied and disinfected every day, and the contents would be removed with the stable-manure. If that were done, they would render the pit-bottom and the main roads more healthful and pleasant for the men, and at the same time they had taken the most effective steps possible for guarding against the development of the disease if it were unfortunately imported into the mine.

Mr. G. ELMSLEY COKE (Nottingham) thought that this paper could only be fully discussed by medical experts, but the precautions clearly explained by Dr. Court could be appreciated and adopted.* Some managers appeared to think that there was no danger of this disease invading our district—but he (Mr. Coke) hoped that no reasonable precautions would be omitted, more especially when the conditions in any part of the mine are favourable to the propagation of the worm. In his experience, few of the deep mines are absolutely dry, water being usually met with at faults, and a wet place with a temperature of over 68° Fahr. might become a centre of infection. By adopting proper sanitary arrangements, and by prohibiting anyone from an infected district from working underground, one might reasonably hope to secure immunity.

Mr. ISAAC HODGES (Whitwood Collieries), speaking of West and South Yorkshire collieries, said that they had little to fear, as in most of the shallow mines the temperature was below 68° Fahr., and in most of the deep mines there was not much moisture. At a recent well-attended meeting of colliery-managers of the district, it was felt that it was necessary to improve the sanitary condition of the mines, and many managers, at the present time, were providing underground closets. A difficulty arose from the fact that no one cared to provide a portable water-closet or even a portable pail, as the removal of the excreta was a labour that no one was particularly anxious to perform, and consequently there was some

---

* "The Worm-disease: What it is and How to Prevent it," *The World's Work*, 1903, vol. ii., page 611 ; and "Ankylostomiasis, or Miners' Worm-disease," *The Iron and Coal Trades' Review*, 1904, vol. lxviii., page 248.

difficulty in getting workmen to attend to this duty. At the pit-bottom, the stables were the best position, as a stall, with a concrete floor and sides covered with flue- or peat-moss dust, made an excellent closet, and the excreta could be sent out of the pit with the horse-manure. At congested places on the haulage-planes, a manhole, with a concrete floor, could be fitted as a closet; and, nowadays, the inbye-stables, at the end of the haulage-planes, could be utilized in the same manner. Such a system of closets would greatly improve the present methods at pit-bottoms and on haulage-roads. At the faces, there was not much difficulty, as the men usually selected a dry place in the goaf, and the offensive matter became buried under the falling roof in due course. In German mines, owing to the seams being steeply inclined and worked from various levels the infected fæces, deposited on a higher level, were carried down by the complete system of watering, common to German mines, to the lower levels, and the disease was thus disseminated and communicated to the workmen in a way which could not possibly occur in Yorkshire mines, where the seams lie at flatter gradients. Personally, he was glad that the fear of the outbreak of ankylostomiasis had focussed attention on the subject of underground sanitation, because it was one to which British colliery-managers must give great consideration. They had been trained under certain conditions, and they were apt to regard those conditions as those that would always apply; but he could not see why congested places underground should not be kept in a clean and sanitary condition just as on the surface. With the improvements now being introduced, he certainly believed that Yorkshire mines had no reason to fear the outbreak of such an epidemic as had occured in Germany.

Mr. H. RHODES said that he had assisted Mr. Pickering to take the temperatures in several mines, varying in depth from 2,000 to 2,400 feet, and he had only found one mine, in which there was any danger of the spreading of the disease, if it were introduced. The temperature, in the case of the mines tested, varied from 78° to 83° Fahr. in the main return-airways.

Mr. G. J. BINNS (Derby) said that he had the privilege of being present at the Home Office conference, when they were favoured with a number of details indicating the extreme incon-

venience, to the workmen especially, of carrying out the precautions which were observed in Germany; and Dr. J. S.
Haldane amplified his paper very materially. There had been
a discussion in the German Reichstag on the subject, and he
noticed that one of the labour members suggested the establishment of an 8 hours' day as a cure for the disease.*

Mr. C. CHETWYND ELLISON (New Monckton Collieries)
pointed out that the stables were generally the warmest place
that they could find in a mine, and a closet placed there would
be favourably disposed for the propagation of the disease. As
this disease had already made its appearance in this country,
in Scotland and Cornwall, whence men were likely to come in
search of work, the best plan that they could adopt would be
to insist upon receiving a sort of character or bill-of-health with
every man, so that they would know whence he came. If
they wanted to stop the disease coming into this part of the
country, the only recourse was to stop the employment of men
coming from infected parts. No one thought of engaging a
domestic servant without a character, and although it might be
difficult to carry out in the case of miners, still if the disease
were so serious, it would be well worth trying. Moreover, it
would be a great advantage, in more ways than one, to the
colliery-manager to know the man whom he was employing,
and whence he came.

Mr. G. J. BINNS remarked that the case reported to have
occurred in Scotland had not been identified—at least it had
not been so at the time of the Home Office Conference.

Dr. J. S. HALDANE wrote that he was glad to see, from the
remarks of the Chairman and other members, that steps had
already been taken in many collieries to limit the pollution of
the ground. He would like to take this opportunity of correcting the statement in his paper that the larvæ of *Ankylostoma*
probably did not develop to the infective stage at temperatures
under about 70° Fahr. This statement was based on data collected in Westphalian collieries and on laboratory-experiments

* *Glückauf*, 1904, vol. xl., No. 4, supplement; and abstract in *Colliery
Guardian*, 1904, vol. lxxxvii., pages 241 and 298.

made in connection with the outbreak of the disease there. Recent experiments by Dr. Boycott and himself had, however, shown that the infective larvæ will develop at as low a temperature as 61° Fahr.; and from the report giving the results of the Belgian Government enquiry it appeared that, out of 41 infected mines in the Liége district, the temperature at the working-face was under 68° (60° to 68°) Fahr. in 16 cases. The percentage of infected men in these sixteen mines varied from 4 to 52. It seemed, therefore, that the disease might break out even in cool and shallow mines.

DISCUSSION OF MR. W. H. HEPPLEWHITE'S PAPER ON "THE BEARD-MACKIE GAS-INDICATOR."*

The CHAIRMAN (Mr. H. B. Nash) asked Mr. Hepplewhite whether he considered that it would be of practical value to the miners themselves, if this gas-indicator was fitted into each safety-lamp, and whether with the ordinary knocking about which a lamp received, the indicator would not be liable to be broken, and thus become inoperative? Of course, when a deputy used this indicator, when testing for gas, he would take more care than an ordinary miner; but at the same time he was doubtful whether it would be safer to trust to that indicator with the possibility of the wires becoming broken, than to rely on the old-fashioned method of drawing down the wick and noting the length of cap upon the reduced flame, combined of course with the practical knowledge which came with long experience. With the liability to danger which was ever present, they were always on the look out for anything that would assist in detecting gas; and if this instrument was likely to prove of any benefit in that direction, then the British colliery-manager would quickly adopt it, but if not, he would preferably continue to use the old methods.

Mr. G. H. ASHWIN (Sheffield) said that he had had an opportunity of trying this lamp, but the results were different from those which Mr. Hepplewhite had described. He had tried the lamp in a return-airway where he could find no trace of gas with an ordinary lamp, and he found with this lamp an elongation of the flame before a glow appeared on the first

* *Trans. Inst. M.E.*, 1903, vol. xxvi., page 214.

of the wires; while Mr. Hepplewhite appeared to have seen a glow on the wires, but no elongation of the flame. He had tried the lamp in many places, but in each case there was an elongation of the flame before the glow appeared on the wires. At one place, he got an elongation of the flame extending to the height of the third wire, and a red glow, not incandescence, on the fourth wire of the ladder; but he could not perceive any elongation of the flame on an ordinary lamp. He thought that the lamp would find very small quantities of gas, and he was of opinion that it would prove a very useful lamp for managers and under-managers when ascertaining the condition of the air in return-airways.

Mr. G. J. BINNS observed that the use of platinum-wire for detecting small quantities of gas was very old. Many years ago, a description was published of the Liveing fire-damp indicator.[*]   The cost was £7 10s., and he had, at one time, 12 of them. He was not wilfully extravagant in getting so many, but he was at the time in the service of a Colonial Government, and being asked whether he knew of any device for detecting a small quantity of gas, he looked up the matter, and ordered two.   In the office of the Ministry of Mines, however, the "two" was altered into "twelve." He had used one of the indicators, which was very heavy, and contained two spirals of platinum-wire, one on each side of a photometer: one spiral being placed in free air, and the other in the atmosphere of the mine. The instrument was cumbersome, taking several hands to work it, and finally the spiral wires were burnt out and it was thrown aside.

Mr. W. H. HEPPLEWHITE said that the fact that the wires became blackened was a very great drawback to this gas-indicator, but this partly depended upon the kind of oil which was used, and the oil used at some pits was very unsuitable. Good results were obtained if sperm or best refined colza-oil was used, with a round wick. The lamp exhibited was fitted with a flat wick, ½ inch wide, in a round burner. He thought that Mr. Ashwin must have had over 3 per cent. of gas when the flame was elongated, and that all the wires would then be glowing.

* *Trans. N.E. Inst.*, 1878, vol. xxvii., page 287; and 1879, vol. xxviii., page 167.

Mr. G. H. Ashwin replied that, in the place referred to, with an ordinary lamp he could find no trace of gas; but with this lamp, burning sperm-oil, there was an elongation of the flame.

Mr. W. H. Hepplewhite said that the lamp would be most useful for making examinations in return-airways or in old workings, as it would enable anyone to detect a quantity of gas which was too small to show on the flame of an ordinary lamp. He would not advise that the gas-indicator should be put into workmen's lamps or even into those of every official, but it would be valuable for use by managers or chief deputies.

Mr. W. E. Garforth said, with regard to the principle that a platinum-wire burnt with greater brilliancy in an atmosphere of gas than in pure air, it seemed to him, that the American inventor had built his lamp on the knowledge obtained by Prof. E. H. Liveing, who was really entitled to the credit of the application of this discovery.

MIDLAND INSTITUTE OF MINING, CIVIL AND MECHANICAL ENGINEERS.

GENERAL MEETING,
HELD AT THE QUEEN'S HOTEL, LEEDS, APRIL 16TH, 1904.

MR. H. B. NASH, PRESIDENT, IN THE CHAIR.

The following gentlemen were elected, having been previously nominated:—

MEMBERS—

MR. MAURICE GEORGI, B.Sc., Electrical Engineer, Edinburgh.

MR. HENRY VERNON HAIGH, Mining Engineer, Bruntcliffe Collieries, near Leeds.

MR. CHARLES E. SMITH, B.Sc., Mining Engineer, 20, Baker Street, London, W.

STUDENT—

MR. PETER BOOTH, Mining Engineer's Pupil, 45, Richmond Avenue, Headingley, Leeds.

Mr. W. McD. MACKEY read the following paper on "Slack-washing: Preliminary Treatment for the Extraction of Fine Dust":—

SLACK-WASHING: PRELIMINARY TREATMENT FOR THE EXTRACTION OF FINE DUST.

By W. McD. MACKEY.

Introduction.—During the last few years there has been a decided increase in the amount of small coal, containing " duff " or fine dust, sent to the coal-washer; and, as riddles are generally being taken out of the pits, coal-owners are becoming more and more aware that in fine coal there is a valuable smudge suitable for coking, if it can be economically washed.

As a consequence, the washing-department has to contend with the disposal of the slurry in a more aggravated form than obtained, say, 10 years ago. The requirements of the altered conditions have not yet been met, and the problem is beset with many difficulties, as coal, probably the cheapest (or, at any rate, one of the cheapest) materials that is washed, must be dealt with not only in large quantity but with great simplicity in handling, if the washing is to be commercially successful. Efficient washing includes not only the production of a properly washed coal, and a dirt containing no coal or at most a negligible quantity, but also the recovery of what may be termed the " float-coal " in a condition suitable for coking or firing, and—what is included in the recovery of the fine coal—the clarifying of the water, whether it is run to waste or used over again in the washery.

The importance of obtaining a clear water for further washing—apart from the question of a clear effluent, if any is run to waste—is sometimes strangely overlooked, especially in small plants. The clarifying of the water that is to be re-used in the washer is vital to good washing, and should always be aimed at; but, when dealing with smudge containing much fine dust, this is often a matter of great difficulty.

Slurry, which is formed when smudge containing much fine dust is washed, is usually found, at least as far as the writer's experience goes, to contain as high a percentage of dirt as the original unwashed coal, and often more. The fine dust in coal

doubtless usually consists of coal mixed with fine shale, in about
the same proportion as in the bulk.   But, in washing, a large
proportion of the fine shale or dirt seems to be carried forward
with the water, and thus it passes on, along with the fine coal-
dust, and forms slurry or sludge.   This dirt is largely masked,
when, by any suitable system of filtering the wash-water through
a layer of the larger washed coal, the bulk of the fine dust in
suspension is caught.   The fact that much fine dirt, even of high
specific gravity, is carried forward with the water is easily
demonstrated, when there is much fine pyrites in the coal, by an
examination of the trough by which the water flows to the well
or settling-pond.

The points that have been urged may be summarized as
follows:—(a) The importance of clarified water for re-use in
the washer; (b) the difficulty of obtaining clarified water, when
the smudge contains much fine dust; (c) the imperfect washing
of fine dust; and (d) to these may be added the trouble of
dealing successfully with slurry, when once it is formed.
Further, the washing of smudge, containing much fine dust,
is under any circumstances necessarily tedious, and unduly
takes up what may be called the "washing capacity" of any
given plant.

The writer is aware that these considerations do not apply
equally to all coals, but his experience has led him to the opinion
that, in many cases, at least, it would be a decided improvement
if the very fine dust were extracted before washing, and either
added dry to the bulk of the washed smudge, or, possibly better,
burned as boiler-fuel at the colliery by any of the methods for
utilizing coal in the form of a dry powder.   The attempts
hitherto made to burn sludge seem to have been more of the
nature of getting rid of a nuisance, than of actually obtaining
benefit from it as a fuel.

Extraction of Fine Dust.—Riddling or sieving is not a
practicable method of removing the fine dust.   To take a concrete instance, a smudge recently came under the writer's notice,
of which considerably over 25 per cent. was capable of passing
through mesh of $\frac{1}{16}$ inch, and smudges of this class are not
at all uncommon.   It is impossible, in practice, to use a riddle
so fine as this, and in any case it would be undesirable to riddle
out such a large proportion.   A specimen of dust is on the table,

double riddled through mesh of $\frac{1}{24}$ inch and over mesh of $\frac{1}{30}$ inch. It will be recognized that it is granular, and capable of success-ful washing, being far removed from the impalpable dust that gives serious trouble in the form of slurry.

The next method that suggested itself was the employment of a fan, as used in cleaning grain. This method has often been proposed, and the theoretical advantages are obvious, as in such a method a certain amount of separation of the coal from the heavier dirt-particles might be effected. No doubt, the reason why it has not been put in practice—at least within the author's knowledge—lies in the difficulty of catching the dust when floating in a current of air, and the danger of explosion under such conditions.

Dust-separator.—Recently, while the writer was experiment-ing in connection with another matter, the following method suggested itself, but it is put forward with diffidence, as the author has not yet had an opportunity of testing its value on a large scale. The specimens of dust, here submitted, are the results of experiments made with a small apparatus roughly erected in a laboratory.

The suggested apparatus or plant has, at any rate, the merit of being very simple. It consists of a short endless band or belt, E, made of canvas or similar material (Figs. 1 and 2, Plate I.) This belt, of any convenient width, say, 5 feet, travels round pulleys, D, placed at a distance apart of about 4 feet, and set so that the upper surface of the belt travels towards the top pulley at an angle of 50 degrees: that being the angle at which the experiments were made. In practice, the angle would be varied till the best results were obtained. The smudge is dropped or fed from the bunker, A, in a thin stream, upon the belt, E, moving towards the top pulley, D_2; and while the granular particles run down the belt, are caught on a shoot, H, and conveyed to the washer, the impalpable dust adheres to the belt, is carried round the top pulley and either falls off, or is removed by a stationary brush, F, into a shoot, G, which may convey it to the boot of the washed-coal elevator, or elsewhere as desired.[*]

[*] British patent, No. 3,233, 1904, William McDonnell Mackey, 33, Chancery Lane, London : improvements in apparatus for sorting and separating minerals and other materials or substances.

The results of experiments indicate that the degree of fineness of the dust extracted can be controlled by varying the angle and surface of the belt; and, on the other hand, the degree to which the smudge can be freed from dust depends on the careful distribution of the smudge in a thin and regular stream on the surface of the belt. In order to lessen any tendency to the formation of a dust-cloud, the stationary brush for removing the dust adhering to the under surface of the travelling-belt would be enclosed, and the belt would travel at as slow a rate as might be found convenient.

The wear-and-tear of the surface of the belt would be considerable; but it may be pointed out that, at the angle suggested, the granular smudge after the first contact with the surface rolls off, pressing very slightly on the belt. And, from experiments, it appears that it is the point of first contact with the belt that is effective in causing a separation of the dust from the more granular coal; and therefore no object is served in using a long belt.

The samples of dust submitted are as follows:—

(1) 2·82 per cent. of dust extracted, all of which goes through mesh of $\frac{1}{36}$ inch; it contains 15·80 per cent. of ash. (2) 3·18 per cent. of dust extracted, nearly all of which goes through mesh of $\frac{1}{36}$ inch; it contains 16·78 per cent. of ash. In both samples (1 and 2), the ash in the extracted dust approximates to that of the unwashed smudge. (3) 6 per cent. of dust extracted, all of which goes through mesh of $\frac{1}{36}$ inch; it contains 20·50 per cent. of ash, and unwashed smudge contains 14 per cent. of ash. This is a smudge formerly washed by the writer on a large scale. The ash in the slurry taken from the well was found to rise as high as 26 per cent. in an air-dried sample.

———

The PRESIDENT (Mr. H. B. Nash) said that the removal of fine dust from slack was very difficult; and it was a very different matter experimenting with small quantities in a laboratory to dealing, in actual work, with several hundred tons per day. There was considerable novelty in Mr. Mackey's suggestions for effecting this separation, and it might prove an incentive to further experiments being made on a working scale to prove their practical value.

The Institution of Mining Engineers.
Transactions, 1903-1904

VOL. XXVII, PLATE I.

To illustrate Mr. W. McD. Mackey's Paper on "Slack-washing."

FIG. 1.—SIDE ELEVATION.

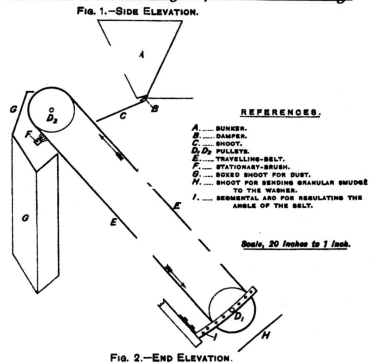

REFERENCES.

A. ..... BUNKER.
B. ..... DAMPER.
C. ..... SHOOT.
D₁ D₂ PULLEYS.
E. ..... TRAVELLING-BELT.
F. ..... STATIONARY-BRUSH.
G. ..... BOXED SHOOT FOR DUST.
H. ..... SHOOT FOR SENDING GRANULAR SMUDGE
TO THE WASHER.
I. ..... SEGMENTAL ARC FOR REGULATING THE
ANGLE OF THE BELT.

Scale, 20 inches to 1 inch.

FIG. 2.—END ELEVATION.

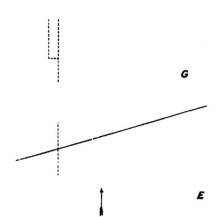

Mr. W. H. CHAMBERS said that he had given thought, for a considerable time, to the subject of Mr. Mackey's paper. He had discussed the dust difficulty with the makers of several coal-washing machines, and, at the beginning, they all stated that their machines would be able to wash the dust along with the other coal; but when they tried to do so in their machines, they found that it was impossible from some peculiarity of the substance. He had, recently, in Germany, practically demonstrated to the maker of a coal-washing machine that the washing of the dust, they produced, was impracticable, as, when mixed with the larger coal in the washing machine, the dust only created mud and rendered the washing of the other coal less efficient. He might say that, at the pits with which he was connected, although they had had coal-washing machines of large capacity—in fact, they washed fully half of the output—they had extracted the fine coal through a mesh of 3 millimetres, and that had just provided them with sufficient fuel for the boiler-furnaces. As their output was increasing, and as they were economizing the consumption of coal, the dust-problem might create a difficulty in the future. It had been running through his mind, therefore, to extract the finest dust before washing the coal, and he was glad to have the opportunity of considering Mr. Mackey's method along with others. He had considered the idea of inducing a current of air through the bars of the screens, and collecting the dust in that manner as the coal passed over the screens. This arrangement would, he thought, prevent the formation of the great clouds of dust, which usually existed about the pit-mouth, and, although it was not desirable, a great proportion of this dust was taken down the shaft by the air-current, which ventilated the mine.

Mr. ROSLYN HOLIDAY quoted an instance where the idea mentioned by Mr. Chambers was in operation. The current of air extracted the dust as it fell, removed it from the coal, and as the heavier dust fell first and the lighter afterwards, it was automatically separated. The dust, he said, was being sold for £1 1s. a ton at the foundries.

Mr. WALTER HARGREAVES thought that the question of the clarification of the water was the most important point connected with coal-washing; and, if they intended to wash the coal,

d over was very dirty, it
·k, because it would pro‑
ing the percentage of ash.

his experience, it was
ıss coke, if the fine dust
·ing washed, as it always
He was endeavouring to
a new screen, for collect‑
ipping the coal, and pre‑
ıd also for improving the
rendering it easier and
By this means he hoped
‑dust from going to the
he was going to extract
ı heavier specific gravity,
· air-current.

n his experience, if they
ashers, they would over‑
ᵥmall dust. He thought
ay, and that they should
ıt there was no difficulty
. and it had been proved
could be used for this

ience, by having plenty
could be passed through
with which he was con‑
ıff or fine coal should be
ıf 3 millimetres, in order
d; but, by providing an
ıe duff could be passed
ncreasing the percentage
ı·, admit that, since this
· was lost in the waste‑
ınt that 11 per cent. of
ımetres was carried away

respective of the dimen‑
. could not be prevented

efficiently clarified water must be used. The idea put forward
with regard to the slurry was also fairly new, and it would be
interesting to ascertain the amount of slurry that the dust was
likely to form; and knowing that they could make a calculation with the view of taking out the dust and adding it to the
washed coal. He should also like to know how the addition
of that dust would affect the amount of ash in the coke.

Mr. I. HODGES said that Mr. Mackey premised that " the
fine dust in coal doubtless usually consists of coal mixed with fine
shale, in about the same proportion as in the bulk,"* but the percentage of shale varied in different coals. In some coals, the proportion of dirt in the dust was three times greater than the dirt in
the average sample of coal; and it was obvious, if such unwashed
dust was added to the washed smudge, that the ash in the coke
would be appreciably increased and the coke rendered less
valuable. The calorific power of so dirty a dust made it useless
as boiler-fuel, and, as it could not be used for that purpose as
Mr. Mackey had suggested, there was no alternative but to wash
such dust. The necessity for clarifying the water was one of
the great difficulties experienced in coal-washing. At one of
his coal-washing plants, the water was used continuously, filtering through large washed-smudge draining-tanks, and there
was no overflow or leakage of water beyond the moisture carried
away in the washed smudge to the coke-ovens. At another coal-
washing plant, the conditions were so favourable that clean water
was always used, and test washings of the same smudge in these
plants shewed that the clean-water plant was the most efficient.

Mr. A. LUPTON said that the particular method of dealing
with any slack depended on the nature of the seam. Mr.
Mackey's system depended simply upon gravity and adhesion;
the smaller pieces, having less gravity in proportion to the
cube of their diameters, were more likely to adhere to the
cloth than the larger lumps, and, by diminishing the angle, the
force of gravity could be diminished. He understood, of course,
that the force of adhesion remained constant however the force
of gravity varied, so that Mr. Mackey's apparatus had a nice
adjustment, and that seemed to be one of the great features

* _Trans. Inst. M.E._, 1904, vol. xxvii., page 55.

appertaining to it. If the dust carried over was very dirty, it should not be added to the washed slack, because it would produce an unsatisfactory result by increasing the percentage of ash.

Mr. C. C. ELLISON said that, in his experience, it was absolutely impossible to make high-class coke, if the fine dust was put into the coke-ovens without being washed, as it always contained a high percentage of ash. He was endeavouring to make arrangements, by fitting a hood to a new screen, for collecting the fine dust, which was made on tipping the coal, and preventing it from going down the shaft. and also for improving the light and ventilation at the screens, rendering it easier and pleasanter for the men to clean the coal. By this means he hoped to prevent a certain percentage of coal-dust from going to the coal-washer; but he did not see how he was going to extract the fine shale-dust, which, being of much heavier specific gravity, would not rise with the coal-dust in the air-current.

Mr. T. W. H. MITCHELL said that, in his experience, if they provided a sufficient number of small washers, they would overcome the difficulty of the slurry and the small dust. He thought that they should give the washers fair play, and that they should not overload them. He also thought that there was no difficulty in burning the slurry in boiler-furnaces, and it had been proved at Cadeby colliery that the wet slurry could be used for this purpose.

Mr. J. NEAL said that, in his experience, by having plenty of room in the wash-boxes, the fine coal could be passed through the washer. In a coal-washing plant, with which he was connected, it was found desirable that the duff or fine coal should be taken out through a sieve, with a mesh of 3 millimetres, in order that the coal might be efficiently washed; but, by providing an additional wash-box and rewashing, the duff could be passed through the washer, without materially increasing the percentage of ash in the coke. He must, however, admit that, since this change had been made, more fine dust was lost in the waste-water. He had ascertained by experiment that 11 per cent. of the duff passed through a mesh of 3 millimetres was carried away by the waste-water.

Mr. W. H. CHAMBERS said that, irrespective of the dimensions of the wash-boxes, the fine coal could not be prevented

from forming slurry. They had been driven to try and extract
the fine coal without putting it into the water at all, and he was
peculiarly interested in Mr. Mackey's method of removing it.

Mr. W. H. PICKERING said that, as one of H.M. inspectors of
mines, he was interested in the removal of the floating dust; and
he hoped that Mr. Ellison and Mr. Chambers would enter into a
friendly rivalry in its removal by air-extraction.

Mr. W. McD. MACKEY, replying to the discussion, said that
he had examined several plants, where the water had not been
adequately clarified, and in each case he had found that the washing was not properly performed. The amount of coal lost by
not attending to the water was a matter which, he thought, was
sometimes not present to the minds of those washing the coal.
Coal and dirt were not separated as efficiently with muddy as
with clear water, and he would like to emphasize that important
point. He had frequently analysed slurry, and he invariably
found that it contained as much or more ash per cent. than the
original unwashed coal. Thus, with unwashed coal, containing
14 per cent. of ash, he had found 26 per cent. of ash in the
slurry. He spoke with diffidence as to the quantity of coal that
could be treated in the apparatus, but he thought, without going
into details, that a belt, 5 feet wide, should separate the dust
from 300 tons in a working day of 10 hours.

DISCUSSION ON MR. W. H. PICKERING'S "NOTES ON SYSTEMATIC TIMBERING."*

The PRESIDENT (Mr. H. B. Nash) said that the Coal-mines
Regulation Acts clearly defined that it was the duty of each
collier to protect himself and to keep his working-place safe by
setting all timber necessary for this purpose; and the suggested
rules were only intended to carry out this object in a systematic
instead of a haphazard fashion. It was impossible to draw any
hard-and-fast rule, as to the distance that props should be set
apart, either in one direction or the other, because the roof of
no two seams was alike; and the variations of the roof in different
working-places in the same seam were so great that the manager
must have discretionary power to vary the ordinary rules, when
circumstances required it.

* *Trans. Inst. M.E.*, 1902, vol. xxiv., page 95.

Mr. P. C. GREAVES said that they had had systematic timbering in operation for the past 15 years, and when the new rules were framed, they had only to post the notice. They experienced many advantages from systematic timbering: the only drawback being that the men had to be constantly warned about maintaining the exact distance.

Mr. W. H. PICKERING (H.M. Inspector of Mines) said that systematic timbering was a most important subject, both from the point of view of economy and that of safety. Were they going to have for ever these falls and consequent fatal accidents or were they going to make some effort to stop them? In Yorkshire, last year, there were 69 fatal accidents from falls. The most important thing that they had to do was to educate opinion amongst the miners on the question of proper pitmanship. They could all do that, and he thought that they were under a moral obligation to do it.

Mr. JOHN GERRARD (H.M. Inspector of Mines) sincerely hoped that the members would, at some early date, seriously consider in what way the frightful number of fatalities from falls of ground could be reduced. It was one of the principal objects of the Institution to consider how to prevent accidents in mines, and those from falls of ground, forming so large a proportion, claimed first and serious consideration. He could not believe that the members were either content with the present conditions, or that they were of opinion that nothing could be done. Mr. Pickering's excellent suggestions should form the basis of an earnest discussion.

About 130 of the members visited the engineering-works of Messrs. Graham, Morton and Company, Hunslet, and were subsequently entertained to lunch.

THE MINING INSTITUTE OF SCOTLAND.

ANNUAL GENERAL MEETING,
HELD IN THE HALL OF THE INSTITUTE, HAMILTON, APRIL 14TH, 1904.

MR. ROBERT THOMAS MOORE, PRESIDENT, IN THE CHAIR.

The minutes of the last General Meeting were read and confirmed.

The report of the Council was read as follows:—

ANNUAL REPORT OF THE COUNCIL, 1903-1904.

In this, the twenty-sixth annual report, the Council have pleasure in submitting a very favourable record of work for another year.

The number of members on the roll at this date is as follows:—

Honorary Members	4
Life Members	9
Life Associate Member	1
Members (subscription £2 2s.)	157
Members (subscription £1 5s.)	250
Associate Members	23
Associates	12
Students	16
Non-federated Life Member	1
Non-federated Members (subscription £1 1s.)	13
Non-federated Members (subscription 10s.6d.)	6
Total	492

This number compares with that of last year as follows:—

On the roll at April, 1903		482
Added during the year		47
Total		529
Died	16	
Retired	5	
Cut-off through non-payment of subscriptions	16	37
At present on the roll		492

This gratifying increase may well be exceeded, if individual members will interest themselves in commending the Institute to such as are not connected with it.

It is with great regret that the Council record the death of Mr. Henry Aitken, the President, in November last. Mr. Aitken always took the greatest interest in the proceedings of the Institute, and he was held in high esteem by the members. A memorial notice of Mr. Aitken prepared by the Council, has appeared in the *Transactions*.

Reference must also be made to the death of Mr. James Hastie, who was an active member, and an office-bearer of the Institute almost from its commencement, and who did much to advance its interests. Only a year ago, the Institute conferred upon Mr. Hastie the distinction of honorary membership. The Council have engrossed in the minutes their sense of the loss sustained by the Institute in the decease of Mr. Hastie.

The following papers have been read during the year and published along with the discussions thereon : —

"The Riedler Pump." By Mr. Harry D. D. Barman.
"Miners' Phthisis. By Mr. James Barrowman.
"Calcination of Blackband Ironstone at Dumbreck." By Mr. Mark Brand.
"Mine-dams." By Mr. Alexander Faulds.
"Stuffing-boxes dispensed with on Engines and Pumps." By Mr. Andrew Watson.
"Description of the Duddingston Shale-mines and Niddrie Castle Crude-oil Works." By Mr. J. Balfour Sneddon.

More members might well take part in the discussions and thereby enhance the value of the proceedings.

The excursion in October, 1903, to Winchburgh gave the members the opportunity of inspecting the Duddingston shale-mines and Niddrie Castle crude-oil works. The whole of the mechanical arrangements are driven by electricity: the first application of the kind in Scotland. The day's proceedings were most enjoyable.

On the occasion of the Glasgow meeting in February, 1904, a visit was paid to the sewage construction-works of the Glasgow Corporation at Dalmuir. A description of these works has not been printed in the *Transactions*, seeing that the works are not yet in operation.

The Library continues to receive valuable additions by

presentation, exchange and purchase. A set of geological lantern-slides are available to members for lecturing purposes. The donations to the Library during the past year, in addition to those received by exchanges are as follows:—

DONORS.	DONATIONS.
Mr. J. M. Ronaldson.	Mines' Reports and Statistics for 1902, 8 volumes.
Mr. Robert McLaren.	Mines' Reports and Statistics for 1902, 3 volumes. Report on Ankylostomiasis. By Dr. J. S. Haldane.
National Boiler and General Insurance Company, Ltd.	Notes on Material, Construction and Design of Land-boilers.
Mr. C. T. Clough.	10s. for books.
Patent Office.	Subject-list of Works on the Mineral Industry in the Library of the Patent Office.
Mr. R. D. Munro.	Presidential Address to the Technical College Scientific Society on the Education of the Engineer.
Mr. Bennett H. Brough.	Cantor Lectures on the Mining of Non-metallic Minerals.
Mr. Robt. Thos. Moore.	£1 6s. for books.
Manchester Steam-users' Association.	Memorandum by the Chief Engineer on Water-softeners.

The Treasurer's accounts for the year show that the Institute is in a good position financially.

There have been seven meetings of the Council during the year.

———

The report was unanimously adopted.

———

The SECRETARY read the annexed abstract of the Treasurer's accounts for the year, which was adopted.

THE TREASURER IN ACCOUNT WITH THE MINING INSTITUTE OF SCOTLAND.

FOR THE SESSION 1903-1904.

RECEIPTS.		£	s.	d.
To Balance brought forward ...		388	16	5½
" Subscriptions—				
5 Members, Session 1902-1903	at 42s. 0d.	10	10	0
4 Members, "	" 25s. 0d.	5	0	0
1 Member, "	" 4s. 0d.	0	4	0
1 Non-federated Member, "	" 21s. 0d.	1	1	0
1 Life Member, "	£25	25	0	0
148 Members, Session 1903-1904	" 42s. 0d.	310	16	0
1 Member, "	" 12s. 6d.	0	12	6
237 Members, "	" 25s. 0d.	296	5	0
1 Member, "	" 10s. 0d.	0	10	0
1 " , "	" 17s. 0d.	0	17	0
24 Associate Members, "	" 42s. 0d.	50	8	0
12 Associates, "	" 25s. 0d.	15	0	0
18 Students, "	" 25s. 0d.	22	10	0
14 Non-federated Members, "	" 21s. 0d.	14	14	0
6 " , "	" 10s. 6d.	3	3	0
3 Members, Session 1904-1905	" 42s. 0d.	6	6	0
1 Member, "	" 29s. 6d.	1	9	6
3 Members, "	" 25s. 0d.	8	15	0
1 Associate Member, "	" 42s. 0d.	2	2	0
1 Student, "	" 25s. 0d.	1	5	0
" Donations to Library ...		1	16	0
" Interest on Deposits ...		8	14	2
" *Transactions*, &c., sold ...		14	1	0
" Rents of Halls, &c. ...		5	10	6
		£1,190	**6**	**1¼**

PAYMENTS.	£	s.	d.	£	s.	d.
By The Institution of Mining Engineers ...				443	5	4
" Printing and Stationery ...				27	1	4
" Books and Book-binding ...				23	12	6
" Rents of Halls ...				29	13	0
" Gas and Assessments ...				5	14	10½
" Stamps, Telegrams and Carriages ...				11	8	2½
" Sundry Payments ...				3	19	0½
" Semi-jubilee, Expenses ...				15	0	7
" Cleaning Halls ...				8	9	0
" Salaries ...				96	8	0
" Expenses of Council ...				31	3	0
" Cash in Bank ...	487	7	11			
" in the Treasurer's Hands ...	7	3	4	494	11	3
				£1,190	**6**	**1¼**

March 28rd, 1904.—Examined, compared with vouchers, and found correct.

WILLIAM HOWAT, } AUDITORS.
THOS. JAMIESON,

Messrs. W. Howat and T. Jamieson were thanked for their services as auditors.

———

The following gentlemen were elected:—

MEMBERS—

Mr. JAMES AITKEN, Summerlee-Kirkwood Colliery, Coatbridge.
Mr. JAMES BRASH, 93, Hope Street, Glasgow.
Mr. THOMAS BROWN, 208, St. Vincent Street, Glasgow.
Mr. ARCHIBALD CLACHER, Braidhurst Colliery, Motherwell.
Mr. JOHN GRACIE, Anderson Street, Burnbank, Hamilton.
Mr. WILLIAM LANG, Wellsgreen Colliery, Windygates.
Mr. ROBERT ROBERTSON, Swinhill Colliery, Stonehouse.
Mr. WALTER SOMMERVILLE, Quarter Colliery, Hamilton.

ASSOCIATE MEMBER—

Mr. JAMES MITCHELL, Auchengray, Caldercruix.

———

ELECTION OF OFFICE-BEARERS, 1904-1905.

THE PRESIDENT (Mr. R. T. Moore) declared the following office-bearers elected for the session 1904-1905:—

PRESIDENT.
Mr. ROBERT THOMAS MOORE.

VICE-PRESIDENTS.

Mr. JAMES HAMILTON.   Mr. DAVID M. MOWAT.
Mr. JAMES A. HOOD.   Mr THOMAS THOMSON.

COUNCILLORS.

Mr. THOMAS ARNOTT.   Mr. JOHN MENZIES.
Mr. JAMES BAIN.   Mr. T. H. MOTTRAM.
Mr. ADAM BROWN.   Mr. WILLIAM SMITH.
Mr. ROBERT W. DRON.   Mr. J. BALFOUR SNEDDON.
Mr. DOUGLAS JACKSON.   Mr. THOMAS STEVENSON.
Mr. JAMES M'PHAIL.   Mr. WILLIAM WILLIAMSON.

———

The PRESIDENT (Mr. R. T. Moore) delivered the following address:—

PRESIDENTIAL ADDRESS.

By ROBERT T. MOORE.

My first duty in meeting you at the opening of a new Session is to thank you very cordially for the honour which you have done me in electing me President of the Institute. I can assure you that it is an honour which I value very highly, and I trust that, with the assistance of the able Vice-Presidents and Councillors whom you have elected, the business of the Institute will be carried on in accordance with the traditions of the past.

There have now been so many Presidential Addresses, that it is somewhat difficult to find a subject which has not been fully dealt with by some of one's predecessors in office; but it has occurred to me to take for a subject a short review of the development of the output of the various Scotch mining districts during the past 31 years. It is always interesting and instructive to take stock of one's position, to note the changes which have taken place, to consider the causes of these changes, and to speculate as to whither they lead. The period that we are about to consider is not a long one—only the length of a Scotch mining-lease—but during that time many changes have occurred.

Thirty years ago, the coal-trade had just passed through a period of unprecedented prosperity—prices had gone up by leaps and bounds, and large profits had been made. People said, "the coal-trade is now on a new basis, there will always be such a demand for coal that never again shall we have low prices," and they straightway began to look for fresh coal-fields in which to invest their profits. It was this feeling of confidence in the continued prosperity of the coal-trade that led to the opening-up of the pits in the Hamilton coal-field.

We have again emerged from a period of unprecedented prosperity—even better than that of the early "seventies" for the men worked steadily and they did not do so in the former "good times." People are again talking of the coal-trade being on a different basis from what it used to be. "Coal is scarce," they say, "and must be had, and we shall never again see the low prices which used to be," and they are setting about opening up new coal-fields, this time in Fife and the Lothians. It is a complete cycle—from good times to good times—and in considering a cycle, we are considering the whole trade.

The first thing to look at is the output:—In 1873, it was 16¾ million tons; in 1903, it was 35 million tons—rather more than double. But it is not enough to consider the total output of Scotland; much more useful information may be got by look-ing into the outputs from the different districts. The published annual statistics have been arranged so as to shew the output from each county. This is an arbitrary division, and does not give much information—either as to the geological divisions of the coal-fields, or as to the markets to which the coal goes.

For the sake of a more interesting comparison, I have grouped together the outputs from districts where the conditions of mining and the uses to which coal is put are somewhat similar; and, with this object in view, I have divided the Scotch output into the following five groups:—

(1) The Ayrshire coal-field, comprising the counties of Ayr, Argyll and Dumfries. The coals worked are principally the thick seams of the Upper Coal-measures. They are used for household coal, for steam coal, and for blast-furnaces, and are almost wholly disposed of for use in the district and for shipment to Ireland.

(2) The Lanarkshire coal-field, with which I have included Renfrew. The great bulk of the coals from this division have come, during the period under review, from the thick seams of the Upper Coal-measures. They have been used for house-hold and manufacturing purposes, for the supply of the city of Glasgow and the industries of Lanarkshire, while they have been exported from Ardrossan, Greenock and Glasgow, on the west, and Bo'ness and Grangemouth, on the east.

(3) The Central district, comprising Dumbarton, Stirling and Linlithgow. The coals worked in this division are mostly the thin seams of Scotland. In Dumbarton, they are found in

the Carboniferous Limestone Series, and in Stirling and Linlith-gow they occur principally in the lower series of the Upper Coal-measures. It is from this district that most of the Scotch steam or bunker coal has been got. The markets for this coal have been the shipping ports of Glasgow, on the west, and Bo'ness, Grangemouth and Leith, on the east. There has also been the local market of the oil-works, which have consumed a large quantity of coal.*

(4) The Fife division, comprising the counties of Fife, Clackmannan, Kinross and Perth. The coals worked in this dis-trict are almost all thick seams, and occur in the Coal-measures and the Carboniferous Limestone Series, the larger proportion being worked from the latter. They supply the local demand, sending coal to Dundee and the north of Scotland, and a large and increasing shipping trade for foreign export.

(5) The Edinburgh district, consisting of Edinburgh and Haddington. The coals worked in this district are thick seams occurring in both the Coal-measures and the Carboniferous Limestone Series, the larger proportion being worked from the latter. The market for the coals in this district has been largely a local one, the surplus having been shipped at Leith.

In all the districts except the Central district, during the period under review, most of the coal has come from thick seams, or coals of not less than about 4 feet in thickness.

I have shewn in Plate II. the outputs from the whole of Scotland and from the various districts, for the years from 1873 to 1903. At the beginning and the end of the period they were as follows:—

District.				1873. Tons	1903. Tons.
1. Ayr...	...	..	...	3,348,283	4,184,242
2. Lanark	...	...	...	9,377,686	17,359,123
3. Central	...	...	...	1,579,827	4,339,670
4. Fife	...	...	...	1,787,027	6,913,194
5. Edinburgh...	...	...	...	761,698	2,191,355
Scotland	...	...	...	16,857,772†	34,992,240†

* This district does not represent the whole of the thin seams worked in Scotland, a quantity about the same as the whole output of the Central district being worked from Lanarkshire, and included in the quantities given for that district.

† A small output, from an outlying coal-field in Sutherland, is included in in the total quantities.

When we trace the total output from year to year, it will
be seen that it rose suddenly from 1874 to 1875, then declined
to 1879, when there was another sudden rise to 1881, followed
by a comparatively steady period to 1886, after which there
has been a steady and rapid rise—-interrupted only by the miners'
strike in 1894.   Taking the districts individually, it is found
that in all of them, with the exception of Ayrshire, the output
has been steadily increasing.

The most striking feature in the outputs is the predominant
position of Lanarkshire, which has furnished more than half
of the entire output, and during the last 30 years this county has
practically governed the coal-trade of Scotland.

In 1873, the pits which had been started to open up the
Hamilton coal-field were still sinking.   They began to work
coal about 1875, and immediately there was a great increase
in the output.   There was no district in Scotland where the
coal was so favourably situated as in the new collieries.   The
coal-fields were flat and regular, and there was little water
to pump.   The Ell coal-seam, to which the pits were sunk,
was from 5 to 7 feet thick, and it was easily opened up and
cheaply worked.   When the pits in this district, therefore,
began to be productive, the market was at once flooded with
cheap coal, and for the next 4 years there was a bitter struggle
for markets.   The effect of this was that several of the older
pits, which were less favourably situated, were closed down
and the output decreased.   But the dullness of trade set people
seeking for fresh markets and efforts were made to dispose of
the surplus output abroad—ports were found at Bo'ness and
Grangemouth on the east, and at Glasgow on the west, and an
export-trade started.

It seems to me that it is to this development of the Hamilton
Ell coal-seam that we owe our Scotch export-trade.   It was a
coal of good quality; it was easily worked and could be cheaply
sold.   Very shortly, the foreigner, from the Baltic to the
Mediterranean, began to use it, and Hamilton Ell became the
Scotch standard shipping coal.   Its quality made it liked, its
cheap price enabled its sale to be pushed in competition with
English coals, and the position of Scotch coal became firmly
established in foreign markets.   It is true that it competed
with other Scotch coals, but the markets opened up by the

Ell coal, took other coals also and the ultimate effect was to benefit greatly all the other Scotch districts. This is well shewn by the way in which the increase of output from Fife, which is largely a shipping county, has steadily followed the increase of output from Lanark. The Scotch export-trade, which in 1873 was 1½ million tons, went on increasing until in 1903 it was over 7 million tons.

Though the cheap prices which were caused by the flood of coal were unfavourable to the coal-masters, yet they were good for the district. Cheap coal encourages industry. It was found that coal could be got in Lanarkshire 1s. a ton cheaper than in any other district in the kingdom. The district had for long been a centre of the iron-industry. There were railways and a large population at Glasgow, and there was a good harbour there.

People began to realize that it was more than usually well situated for works where coal was required, and steel-works sprang up—steel-ships were built on the Clyde—steel-bridge-building and many other industries started—all of them, because there was cheap coal.

The output soon sprang up to meet the demand, and as there were ample facilities for increasing it in the Hamilton collieries there was again over-production. The demand caused by the increased industries and export again overtook the output which went on increasing rapidly until 1892, when a new factor appeared. So long as the Hamilton collieries were working the three thick seams of the district, it was easy to keep up large outputs and increase them, if necessary. But in 1892, these seams had been developed to their full extent. Following the usual Scotch practice, the Hamilton coal-fields had been let in small leaseholds of 200 and 300 acres, and these began to get worked out. It was necessary to keep up the output from thinner seams, and, as these were more difficult to work, the increase of output, after 1892, became comparatively slow. Leaving out the years 1893 and 1894, which were affected by a strike, the output from 1895 to 1903 only increased at the rate of 173,000 tons a year (or 1 per cent. of the 1895 output) as against 646,000 tons a year (or 5½ per cent. of the 1886 output) for the period from 1886 to 1892 when the thick seams only were being worked. And this, notwithstanding the fact that, during the "good times," the thin seams have been largely opened up and the output from them pushed by the extensive use of coal-cutting machinery.

In 1892, in a paper* I read before the Philosophical Society of Glasgow, I estimated that the thick coal-seams in Lanarkshire would last another 24 years, and that while the output might be increased slightly until that time by the employment of machinery to increase the yield from the thin seams, after the thick seams were exhausted, the output would decrease. What I said then has been borne out during the 11 years which have ensued, and I have no doubt that, 13 years hence, the output of Lanarkshire will be a decreasing one.

It is worthy of note, that in 1902, for the first time in the recent history of Scotch mining, Lanarkshire supplied less than half of the total output of Scotland.

The outputs from the Central, Fife and Edinburgh districts have increased steadily, being affected by the same influences as the Lanarkshire district.

The most notable feature is the rapid increase of the Fife district from 1896 onwards. This increase is at the rate of 421,000 tons per annum (or 10½ per cent. of the 1896 output) which compares with the most rapid rate of increase of Lanarkshire—between 1886 and 1892—of 646,000 tons a year (or 5½ per cent. of the 1886 output). The beginning of this increase corresponds with the time when the Lanarkshire coal-field was ceasing to expand so rapidly, owing to the fact that the thick seams were fully developed. The output from the other districts also began to increase more rapidly about the same time. It is evident that the increased demand is being largely met from the Fifeshire collieries.

An interesting feature is the struggle for existence of the thin seams (some of them less than 2 feet thick) against the thick Lanarkshire seams, for it will be seen that the Central district, where these are principally worked, has been steadily increasing along with the other districts.

This has been due to a number of causes: Where they were worked, they were near the surface and the capital-outlay was

* *Transactions of the Philosophical Society of Glasgow*, 1892, vol. xxiv., page 51.

not great. The coal-fields were flat, and the conditions of the seams (as to roofs, and pavements and holing) were favourable The lordships were smaller than for the thick seams. Some of the seams were of good quality, and sold at high prices for household coals. But the principal condition in their favour was the dross.

While in the thick seams the proportion of dross was large, varying from 40 to 50 per cent. of the output, in many of the thin seams it was not more than 20 per cent. In some cases, such as the coking coal in Dumbartonshire and the Slamannan districts, the dross was suitable for coke-making, and in the bunker-coals, the screens were small and but a small proportion was taken out. With all these circumstances in their favour, the thin coals were able to live in competition with the thicker and more easily worked seams.

It is somewhat singular that Ayrshire, which contains a large quantity of coal in thick seams, has made so little progress. Probably the fact that, over a large part of the district, the seams are much faulted and interfered with by intrusive rocks, has prevented the rapid development of the coal-field. There was no great demand for the coal outside the district, for so long as there was plenty of cheap Lanarkshire coal to supply the centres of population and industry there, Ayrshire coal could not compete, paying as it did higher rates of railway-carriage. The harbours also were not suitable for the larger vessels in which the coal was exported, so that, although coal was exported from Ayrshire at an early date, the shipping was largely confined to a local trade with Ireland. Whatever be the cause, the output of the district has increased but little.

As I have observed, we are now nearing the end of our cheaply-worked Lanarkshire thick seams. It may be said that it was a pity to have worked them out so quickly, and to have thrown coal away at such small prices, but as we have seen the district benefited, for the cheap coal brought industries to it. It is often sound policy to take the best first. Twenty years ago, when the Hamilton collieries were working nothing but the Ell coal-seam, people thought that the Main and Splint seams would never pay, and one coal-master used to give up his leases when the Ell coal-seam was exhausted. Nowadays people are glad to get the despised Main and Splint seams. Circumstances are altered: there is little Ell coal to compete, and the Main and Splint seams and even the Pyotshaw, which would not have paid 20 years ago, are now profitable subjects. It may be that in another 20 years, Hamilton coal-masters will talk of the famous Main and Splint seams as we now do of the Ell seam.

But with the exhaustion of the Lanarkshire coal-field, we have to face the question of the continuance of the numerous industries which have been established in the district. Once an industry has been established it takes a good deal to move it. The works have been erected, houses have been built, men have been collected, and none of these are easily moved. I do not think that any difficulty need be experienced in keeping the industries going, even though no coal were being got in Lanarkshire.

A similar situation was faced by the Coatbridge ironmasters, whose iron-works were built to work the Airdrie blackband-ironstone seam found in the neighbourhood. When that seam was exhausted, it was found that the works could be kept going on iron-ore brought from Spain, and Coatbridge is still as important an ironmaking centre as it was when the blackband-ironstone was being worked.

The Lanarkshire works need not go so far for a supply of coal: there being plenty of it in Ayrshire, Fife and the Lothians. There are already (as I have shewn) extensive developments in the two last-named coal-fields, and the output from them is rapidly increasing. It is worthy of note that the combined output from Edinburgh and Fife was last year over 9 million tons, or almost exactly the output from Lanarkshire in 1873.

It is from these coal-fields that the future increase of the Scotch output is to be expected, and there is no reason why the output from them should not be as great as it was from Lanarkshire; and the coal from them will be brought to the Lanarkshire industries. All that is required is that cheap railway-rates should be given and these will come. There is no reason why coal should not be brought from the Edinburgh coal-field to Glasgow for 2s. or even 1s. 6d. per ton, and if this were the

rate, the price of coal at Glasgow need not be higher than it is
now. There is no doubt that the Scotch coal-trade is much
handicapped by the exorbitant rates charged by our railway-
companies. They are even higher now than they were in 1844,
for I see from a book by Mr. Matthias Dunn (on the *Winning
and Working of Collieries*) that the rate between Dalkeith and
Edinburgh was then 1d. per ton per mile, exclusive of hire
of waggons which cost $\frac{1}{10}$ penny to $\frac{1}{5}$ penny per ton per mile
more, while it is now $1\frac{3}{4}$d. per ton per mile. The rate on the
Hull and Selby railway at the same time was $\frac{1}{2}$d. per ton per
mile, inclusive of waggons. American railways carry coal for
300 miles for 3s. 4d. a ton, or less than 0˙14 penny per ton per
mile, and our Scotch railways should not charge more.

I do not know whether it lies " within my brief " to criticize
the management of the railways, but there are many obvious
points in which the working of the mineral-traffic is defective.
Our Scotch railway-companies have got into the habit of drawing small train-loads, and have made their sidings and other
arrangements for so doing. The consequence is that their lines
are blocked, and yet do not carry nearly the traffic of which
they are capable. If their trains averaged 500 tons, the average
of the Pennsylvania railway (instead of under 100 tons as they
do at present), they would have only one-fifth of the number of
trains on the line, and there would be plenty of time for five
times the amount of traffic. They tell us that Britain is a
small place, and people are in a greater hurry for their traffic
in this country than in America. I can assure them that there
is no saving of time by their methods; if, as one of their officials
stated in evidence, it takes a week, on an average, for a waggon
to make a return journey from Hamilton to Glasgow—a distance
of 13 miles—it would pay the shareholders to adopt American
methods. What profits they would make if they could work at
the same cost as the Pennsylvania railway, which pays a dividend
of 6 per cent. per annum from one-fifth of the rates per mile
charged by our Scotch railways! But they are monopolies, and
nothing can be done to convince them without some form of
competition. Perhaps in this connection it is interesting to
consider the feasibility of carrying our local supplies by road.

At present, the railway-companies charge $1\frac{3}{4}$d. to 2d. per ton
per mile, and even more when terminal charges are added,

for taking the coal into depôts, where it is loaded into carts
and carted to the works. Motor-waggons are now being constructed to carry at 2d. per ton per mile, which is as cheap as
the railway-rate; and, if they were used, the coal might be taken
direct from the colliery to the works, and the depôt-charges and
cost of carting would be saved. The electric tramways have
shewn us how roads may be used to compete successfully with
the railways.

We have been cutting down expenses at the collieries, and
this freight is an important item in the cost of fuel which has not
been touched. It is as important an item as the colliers' price.
The average rate paid to the North British Railway Company
per ton of mineral carried (inclusive of all special rates) is 1s. 5d.
and this is as much as in many cases is paid to the collier.
I speak thus strongly because I am convinced that the prosperity
of the district depends on cheap railway-rates being given,
and it is necessary to rouse both the public and the railway-
companies.

I shall only refer in a word to some of the changes which
have taken place in the working of our mines in the period that I
have reviewed.   In 1872, H.M. inspector of mines for the
eastern district of Scotland noted, with satisfaction, that there
were then in his district 6 ventilating fans either erected or in
course of erection. Now nobody thinks of using anything but a
fan for ventilation. The same gentleman noted in 1876, that
there were as much as 136,000 tons put out of one shaft—now
there have been 450,000 tons put out of a shaft in a year.
Thirty years ago, there was no handy method of conveying power
into the workings, and dip workings with water were a night-
mare. Now we can convey power to any part of the workings
with electricity, and pump water out of dooks without difficulty.
Thirty years ago, there was little coal-picking and less coal-
washing; now the coal is cleaned, washed and sized before being
sent to the markets. These and many other improvements will
occur to all of you.

We are a little apt to underrate what was done by our predecessors and we forget that much of what is now a common-
place was only got by their struggles.   The progress that has

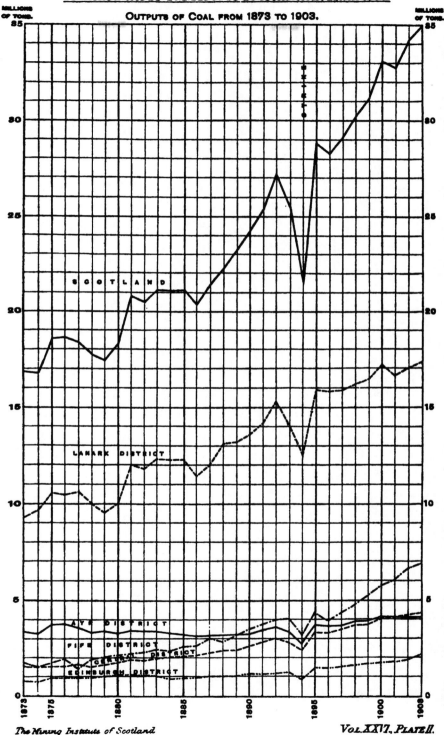

OUTPUTS OF COAL FROM 1873 TO 1903.

been made should be an incentive to us to try to improve matters so that we may leave the science of mining better than we found it.

Mr. JAMES S. DIXON (Glasgow), in moving a vote of thanks to Mr. Moore for his admirable address, said that the members must congratulate their President most cordially on having struck upon something new. The subject was of great interest to himself (Mr. Dixon) because within the last few weeks he had been dealing with this very matter in his report to the Royal Commission on Coal-supplies, which was in draft and no doubt would be laid before the Institute in the course of a few months. He might say, however, that his conclusions were very similar to those adduced by Mr. Moore in his address, and, consequently, two independent minds had arrived at practically identical conclusions on the same subject. The great coal-field of the future was to be found in Fife and the Lothians, and the comparison of the thick and thin seams which Mr. Moore had brought forward in his paper shewed clearly that during the past few years the Lanarkshire coal-field had been denuded, to a large extent, of its thicker and more valuable seams.

The resolution was cordially approved.

The discussion of Mr. Alexander Faulds's paper on " Mine Dams " was closed after a hearty vote of thanks had been awarded to the author.

DISCUSSION OF MR. JAMES BARROWMAN'S PAPER ON " MINERS' PHTHISIS."[*]

Mr. ROBERT McLAREN (H.M. Inspector of Mines, Edinburgh) wrote that miners' phthisis was not now so prevalent as it was 30 years ago : at that time, and prior to that, it was quite common to see men about 50 years of age suffering with a hollow cough and asthma. The cause was not far to seek, and was due to the badly ventilated mines then existing. The state of most mines was then anything but creditable, and men had perforce to work in an atmosphere fouled by noxious gases evolved from the strata, by blasting, from the smoke given off from their lights, and generated also from their breathing; indeed to keep the

[*] *Trans. Inst. M.E.*, 1904, vol. xxvii., page 21.

atmosphere as pure as possible, tallow lamps displaced those which burned oil. This condition of things has to a very large extent been rectified, and while much of it is due to Government legislation and inspection, it has largely been brought about by scientific methods adopted by the managers in providing good ventilation, and the common sense displayed by them and the men, the latter very properly refusing to work where the ventilation is defective. At the present time, the system which prevailed many years ago would not be tolerated.

Notwithstanding the favourable results obtained by a better system of ventilation, it is evident that men inhaling dust in a dust-laden atmosphere must suffer from anthracosis, and to lessen its bad effects some method should be adopted to prevent the atmosphere from being charged with dust: this might be done by a system of water-spraying such as would render the air humid, but from a sanitary point of view perhaps other ills might result.

For the proper working of a mine, both as to health and economy, the first and second recommendations laid down by the Commission* are absolutely necessary.

In connection with this subject, Dr. Thomas Oliver of New-castle-upon-Tyne delivered a lecture at the Mount Vernon Hospital for Consumption in London in October, 1903, and in describing the results to workers employed in several mineral industries he said :—

Since there are many dusty trades that give rise to pneumoconiosis, it is impossible to give illustrative examples of all . . . the pulmonary lesion is, practically speaking, the same in all, namely, an overgrowth of fibro-connective tissue accompanied by the deposition of a large amount of pigment in the lung, which varies in colour and in character with the peculiar nature of the dust which has been inhaled. There is, I think, no doubt that the pulmonary lesion is largely the result of local irritation caused by dust. The amount of coal-miners' phthisis in Britain to-day is only a very small fraction of what it used to be. That this favourable condition of things is largely a consequence of the improvement in the ventilation of coal-pits required by Act of Parliament is shown by the high mortality from phthisis that still exists among lead-miners. Lead-mines do not come so completely within the scope of the Mines Acts as do coal-pits, and as a consequence no steps are taken in some places to remove the foul air from the workings. While dampness of soil may play a part in the development of phthisis, there is not the least doubt that the breathing of an impure air for several hours per day reduces the resistance of an individual to tubercle. As regards lead-miners and their predisposition to lung disease, we cannot

* *Report of the Miners' Phthisis Commission, with Minutes of Proceedings, and Minutes of Evidence*, 1902-1903, Pretoria, page 25.

overlook the effect of the men working in wet clothes, and, when fatigued, of the long walk across a bleak moorland country after a hard day's work, also the influence of neglected catarrhs. Exposure to cold when fatigued and the wearing of damp clothes, inattention to mild attacks of bronchial catarrh, coupled with alcoholic intemperance, pave the way alike for the harmful operation of dust upon the lungs and the incidence of tubercle. The influences of exposure to cold upon men when fatigued, and of dust in causing lung disease, were well shown during the construction of the Mont Cenis and St. Gothard railway-tunnels. Many of the miners died from acute affections of the lung and from pulmonary phthisis.

Dr. W. M. Robertshaw, medical officer of health of Stocksbridge, Sheffield, drew attention at the meeting of the British Medical Association in Manchester, 1902 (also in the *Journal of the Sanitary Institute*, April, 1900), to the high death-rate from phthisis among ganister-miners and crushers. Ganister is a hard close-grained mineral, consisting mainly of silica, of which it contains 95 per cent. Dr. Robertshaw found an annual death-rate of 17·2 per 1,000 ganister-workers, the average age at death being 39 years. Dr. C. L. Birmingham, on the other hand, shows that the occupation is even more unhealthy than has just been stated, for he gives as the average annual death-rate from ganister-disease 22·29 per 1,000 workers. If proof were required of the part played by dust in causing lung disease it would be only necessary to compare the relative dangers run by the two different classes of ganister-workers, namely, miners and surface-men. Taking a particular mine in the Stocksbridge district, and summarizing the deaths from phthisis and respiratory diseases for the 10 years 1891-1900, it was found that while 11 surface-workers died from these diseases 61 miners had succumbed during the same period. In the 5 years 1891-1895, the average death-rate, at the age-period from 35 to 45 years, of males in England and Wales from phthisis and respiratory diseases, was, according to Dr. J. F. W. Tatham, 0·52 per cent.; while for the same period the average death-rate of ganister-miners from the same diseases was 3·9 per cent. Writing upon this subject, Dr. T. M. Legge, in drawing attention to' the difference in the mortality from phthisis between surface-workers and ganister-miners and the fact that the incidence of phthisis is ten times as great as in the general male population at corresponding ages, says that, although ganister-workers are short-lived, yet many of them had been employed for at least 10 or 15 years.

Coal-miners' phthisis, formerly of common occurrence, is now rarely met with, owing to the improved ventilation of the mines. So far as lung disease is concerned, coal-mining in this country is, comparatively speaking, a healthy occupation. When colliers become the subjects of bronchial and pulmonary catarrh they bring up large quantities of a black spit; it is thin and inky in colour, and on microscopical examination is found to contain large cells studded with granules of coal. An interesting point in regard to anthracosis is that the lungs may be deeply pigmented for years, without any signs of ill-health, and that there are some forms of coal-dust which do very little harm to the lungs compared with others. The larger the amount of stony particles in coal the greater is the tendency for the lung to suffer. Virgin coal contains no micro-organisms. How far the particles of carbon exercise, as some writers maintain, an antiseptic influence upon the lung, I am not prepared to say.*

v3쥬 * "Occupation as a Cause of Lung Disease," *The Lancet*, 1903, vol. clxv., pages 1,345-1,351.

Mr, ROBERT MARTIN (Niddrie) wrote that it would be interesting to know the cost and expense of a changing house at the mine fitted to deal with, say, 500 miners changing their working clothes in a very short space of time, and the number of attendants required to dry and mend (and mending is as important as drying) such for next day's work.   It would also be difficult to make the use of a changing house compulsory in view of the natural reluctance and aversion of a large number of men to leave their home and working clothes in the same room with others.   It seems that a much simpler, cheaper, and more efficacious remedy is wanted.   If pulmonary disease is to be avoided by a miner, it is too late to wait until he comes to the surface.   The best place probably for a miner to change his clothes, after working some hours in a highly-heated atmosphere, is just before he leaves the warm part of the mine and enters into the main intake air-ways, in which a strong current of cold air is blowing, or approaches the shaft down which there is a rush of air.   This may be done by changing his underclothing and carrying home the moist and wet articles under his arm.   It was quite a common thing in the steep stoop-and-room workings in the Pentland shale-mines for a miner to get a severe cold holing out the last cut, when he holed his place through in the level above, in the strong current of air which swept through the new opening, the place previously having been too warm and free from strong draughts.   The sudden change induced a chill.   The words "that they frequently put them on (that was, their pit-clothes) as wet as they were taken off" did not apply to Scotland.

Mr. WM. SMITH (Dalmellington Iron-works) said that, in reading over Mr. Barrowman's paper, he was struck with the shortness of life of the men employed at the rock-drills; and this to his mind was all the more unfortunate, when it could be so easily overcome, especially where the motive power of the drills was compressed air.

He had occasion recently to drive a mine through a volcanic neck, and the shot-holes were drilled by machine-drills.   The whin or conglomerate was very dry, and in boring the holes near the roof, especially those slanting upward, the men had to work in a perfect cloud of dust.   The workmen placed a thin cloth or handkerchief over their mouths.

However, he (Mr. Smith) had a cylindrical tank made, about 4 feet long and 2 feet in diameter, and mounted on a bogie-frame, so that it could be filled with water, where most convenient. Two cocks were fitted on the top of the cylindrical tank and at one end was a small mud-hole door. A connection was made from the air-main to one of the cocks, and from the second cock a hose-pipe was led nearly to the face of the mine. This second cock was fitted to a pipe leading from near the bottom of the tank, so that all the water put into it could be drained away. When the tank was filled with water and the compressed air turned on, the water could be squirted to the back of the hole. Of course, it was necessary that the mouth of the drill should be about 1 inch greater in diameter than the stem, so as to allow of the water passing freely to the inner end of the hole. This water-spray also had the advantage that, when boring through hard stone at the rate of 12 inches a minute, the drill kept its temper better, and the machine was more easily handled.

Mr. T. H. MOTTRAM (Glasgow) wrote that he had seen a statement to the effect that the life of a rock-driller in Great Britain was as long again as it was in South Africa, but whether this was true or not he was hardly prepared to say. It seemed, however, that rock-drillers in South Africa worked by contract, and earned from £50 to £70 per month; and that, in their anxiety to earn as much as possible, they returned to their working places before the dust had had time to settle and before the air had been cleared of the poisonous fumes produced by blasting-explosives.

The *Report of the Miners' Phthisis Commission* quoted by Mr. Barrowman made out a strong case for reform, and, if preventive and curative measures consisted chiefly in the simple methods of laying the dust or watering the drill-holes and in better ventilation, no time ought to be lost in giving effect to the recommendations of the Commissioners.

In a paper read by Dr. Thomas Oliver, at the Mount Vernon Hospital for Consumption in London, in October, 1903, he stated that in Great Britain the mortality had been reduced in most occupations: a diminution in keeping, to some extent, with the reduction of the death-rate from phthisis in the population generally. This reduction, however, was greatest among youths and females under the age of 25, a circumstance, somewhat suggesting that the improvement was more largely due to domestic than to industrial causes, to which latter males were more particularly exposed. In the same lecture, Dr. Oliver quoted Messrs. Lewis and Balfour to show that in colliers, wool-workers and cotton-spinners the reduction had been from 12 to 29 per cent. This favourable condition, he (Dr. Oliver) attributed to the improved ventilation of the mines brought about by Act of Parliament, which he said was borne out by the high mortality from phthisis still existing among lead-miners. The inference to be drawn from this was that if metalliferous mines were better ventilated, the high mortality would decrease, as it had done in coal-mines. Bearing in mind, however, that the dust in metalliferous mines was more dangerous, better ventilation alone could not be expected to produce the same satisfactory results as those achieved in coal-mines, unless accompanied by the systematic sprinkling of the drill-holes in the absence of water in the strata.

Of the 89,736 underground workers in Scotland under the Coal-mines Regulation Acts, 5,653 were employed in working minerals other than coal, less than 2,000 of these being employed in the ironstone-mines. According to the comparative mortality-returns, quoted by Mr. Barrowman, the ironstone-miner in England and Wales stood at 90 and in Scotland at 269, or three times as high. So marked a difference would suggest that there must be some error, as ironstone-mining was similar in both countries and governed by the same statute.

The close-grained grit, known as ganister, consisting mainly of silica, was only worked to a very limited extent in Scotland for the purpose of mixing with clays in the manufacture of fire-clay goods, and, consequently, had little bearing on the mortality-returns. Outside actual mining, however, there were about 14,000 persons employed as quarrymen and masons in and about quarries, under the Quarries Act; and as a considerable number of these were engaged in working and dressing rocks containing silica, and as the death-rate from phthisis among quarrymen and masons was considerably higher than among coal-miners there appeared to be more room here for the consideration of " preventive and curative measures " than even among ironstone workers. He (Mr. Mottram) asked whether Mr. Barrowman could offer any explanation as to the great difference existing in the comparative mortality of males (ironstone-miners) from phthisis in England and Wales and in Scotland, the comparative figures being: Scotland, 269; and England and Wales, 90.

Mr. BARROWMAN read from the *Glasgow Herald* of March 21st 1904, the following statement, made by Mr. William Mather, late general-secretary of the Transvaal Miners' Association, as the representative of the trades-unionists in South Africa, in connection with the proposed importation of Chinese labour:— " Mr. Mather contends that sufficient native labour is not likely to be found in the next five years, probably not in ten, even allowing for increase of population. It is a physical impossibility that the mines can be worked remuneratively by white labour, as the average life of a miner, owing to the inhalation of sharp particles of dust produced by drilling machines and blasting operations, is not more than seven years; whereas this does not operate to the same extent on the natives, whose lungs are much coarser, whose dietary is different, and whose average contracts being only for six to twelve months enable them to recuperate whilst living in the kraals." If this was a correct report, one felt that Mr. Mather was hardly doing justice to the miners whom he represented by accepting the situation in regard to dust as inevitable, and it was not a very worthy solution of the difficulty to propose the substitution of black for white labour.

As to the alleged coarser lungs of the blacks, he had been told by a medical man, who had lived among the Zulus for a quarter of a century, that they could understand wounds and bruises, and usually recovered from them, although much more severe than non-fatal cases of the kind in this country; but internal troubles, phthisis among others, which seemed to have no external cause were attributed to the influence of evil spirits, and they died in spite of the best medical treatment.

As to Mr. Mottram's enquiry, he had no explanation to give of the discrepancy, except that the number of ironstone-miners in Scotland was too small to yield satisfactory mortality-results.

The discussion was closed, and a vote of thanks was awarded to Mr. Barrowman.

———

Mr. J. R. WILSON read the following paper on a " Proposed Method of Sinking through Soft Surface ":—

PROPOSED METHOD OF SINKING THROUGH SOFT SURFACE.

By JAMES R. WILSON.

Within recent years, many sinking operations have been made in Scotland through soft surface. These operations have been confined chiefly to the coal-basins of the Forth valley and Midlothian, where the surface consists of approximately 100 feet of soft alluvial matter. Considerable difficulty has been experienced in sinking both circular and rectangular shafts through the soft strata and in reaching the rock-head.

In the case of rectangular shafts, the difficulty chiefly occurred in hanging the column of wooden lining from the surface, and in preventing the silt from rising inside the shaft. With circular shafts, on the other hand, where iron or steel cylinders were adopted, the abovementioned difficulties were accentuated by the greater difficulty experienced in keeping these cylinders in a truly vertical position while descending, and in overcoming the excessive skin-friction, when the depth exceeded 80 feet. This method of sinking with cylinders, while giving absolutely satisfactory results when successfully accomplished, has been, in the past, very uncertain in its results, owing to the great difficulty of keeping the cylinder truly vertical, and the presence of occasional boulders possibly nullifying all the efforts of the management. In addition to the care and worry, the cost of the cylinders have formed an exceedingly heavy item in the equipment of a colliery.

In has, therefore, occurred to the writer that a less expensive method might yield equally successful results, and though his ideas, so far as he is aware, have not been put to the test of practical experience, he ventures to hope that circumstances may yet occur to prove the value of his suggestion. This system would be found equally suitable for either rectangular or circular shafts. It is embodied in the idea of putting down a series of bore-holes—a little larger than, but in the shape of, the finished shaft. These bore-holes would be sunk into the rock-head, and would be tubed and fitted with malleable-iron rods. The spaces between the rods would be filled tightly with oaken wedges, to these the barring would afterwards be bolted, and, in this manner, each set of barring would be self-supporting.

If we assume, in detail, the sinking of a rectangular shaft, say, 29 feet long by 11 feet wide (Fig. 1, Plate III.), the first operation would consist in the laying down upon the surface of a wooden template—the shape of, and slightly larger than, the actual finished size of the shaft. These beams, E and F, should be of pitch-pine 12 inches broad and 6 inches deep, and through these, holes would be bored, 7 inches in diameter, to take the bore-rods. Into these holes would be inserted a cast-iron pipe, A, 6 inches in diameter and 9 feet long, perfectly straight, and put down to such a depth as to leave a length of 3 feet above the surface (Fig. 4, Plate III.). This pipe would be fixed in an absolutely vertical position and a tube, B, would be inserted therein with an internal diameter slightly larger than the size of the bore-rods, thus forming a starting-point for the boring operations: it being essential to the success of the proposed system that the bore-holes should be put down in as vertical a position as possible. The bore-holes would then be started, and might be put down at an average rate of 25 feet per day, until they are sunk to a depth of 4 feet into the rock-head. Greater speed could be attained by making a number of bore-holes simultaneously.

The bore-holes would, in the process of boring, be lined with tubing, C, and the tubing filled with malleable-iron rods, I, 18 feet long, fitted with male and female joints of a sufficient diameter to fit accurately into the tubing, C. These rods would extend 4 feet into the rock-head and 18 inches above the template, E, at the surface.

Immediately above the templates, E and F, and inside the area enclosed by the malleable-iron rods, I, a set of ground-beams, G and H, would be put in position (Figs. 2 and 3, Plate III.), these beams, made of pitch-pine, 18 inches deep and 12 inches broad, would extend 6 feet longitudinally past each end of the pit. The malleable-iron rods could now be firmly fixed, both at the surface and at the rock-head, and arrangements could then be made to start with the excavation of the soft material inside the shaft.

So soon as 9 inches of the surface-soil have been taken out, the driving-in of previously prepared oaken wedges between the iron uprights can be started. In the first place, two pieces of oak, d and e, 12 inches long, 2 inches broad and 9 inches deep, are placed in the intervening space, one against each iron rod, and into the vacant aperture between them, an oaken wedge, f, of the necessary dimensions is firmly driven in (Figs. 3 and 5, Plate III.). This process is continued round the perimeter of the shaft; and it is then in a position to have the first set of barring, g, bolted on, of pitch-pine, 5 inches broad and 9 inches high. This is attained by driving a number of rag-bolts, h, through the barring, g, into the oaken wedges, f. The barring is, therefore, now self-supporting. An additional depth of 9 inches of material is removed (or if the material is such as to admit of a depth of 18 inches, it would be so much the better); then a second set of wedges are put in position in like manner to the first, and another set of barring is attached. This process is repeated until the rock-head is reached, thus dispensing with the hanging of the wooden barring from the ground-beams at the surface, as usually adopted; and, in consequence, the disadvantage of the heavy weights bearing upon the surface of the ground in close proximity to the shaft is entirely obviated, as in this process each set of wooden barring is self-supporting, and the utmost rigidity is secured.

Further, there would not be the same tendency for the soft material or silt to well up inside the shaft. It should also be remembered that the rods would offer considerable resistance to the inflow of soft material, and consequently the tendency of the silt to rise inside the shaft would be greatly reduced.

This system of sinking would seem to be quite feasible for any depth not exceeding, say, 100 feet, and the primary difficulties might be summed up as follows:—(1) The putting-down of the bore-holes in a perfectly plumb position; (2) the presence of occasional boulders in the strata might cause some of the bore-holes to depart from the perpendicular; and (3) the

...nking through Soft Surface.

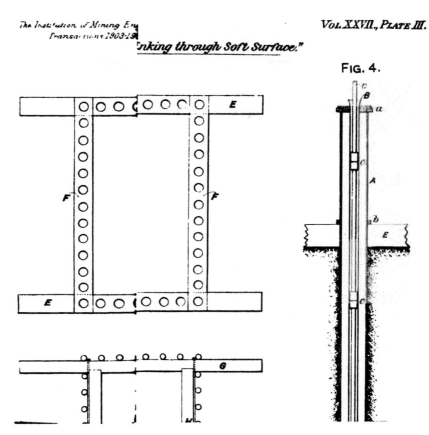

FIG. 4.

tendency of the soft material to rise inside the shaft. The writer is, however, of opinion that, with the exercise of reasonable professional skill, the first of these objections could be surmounted. With regard to the second objection, if the boulders were of sufficient size, the boring-chisel would make a hole through them; but, in any case, their presence would be much more serious in the case of a sinking-cylinder. This system would with every prospect of success be adopted for the sinking of circular shafts: the *modus operandi* being exactly the same.

With regard to the question of cost, it may be taken that, for a pit of the dimensions and shape indicated in the illustrations (Plate III.) the total cost would be, approximately, £25 per foot. The question of cost would, however, be affected by the quantity of water given off by the strata, and, should water be met with, it would be removed in the usual manner.

In conclusion, it may be claimed for the proposed method of sinking that it does away with the uncertainty of success or failure inseparably connected with the sinking-cylinder, that the question of surface-weight with its attendant difficulties is much reduced, and that it greatly diminishes the liability of the silt to rise in the shaft.

———

Mr. JAMES BARROWMAN asked whether Mr. Wilson had considered the probability of the wedges displacing the rods. How were the wedges to be made tight and the rods to be kept from bulging?

Mr. J. S. DIXON understood that Mr. Wilson proposed to put down the malleable-iron tube within the cast-iron pipe, and as the bore-holes had a greater diameter than the malleable-iron tube, he would like to know how the chisel was withdrawn.

Mr. J. R. WILSON stated that the pipe would be withdrawn every time that the rod was changed.

Mr. J. S. DIXON said that the proposed method seemed to him to be one of piling with iron tubes instead, as was usually done, with wood. He had never heard of wooden piling being driven down to such great depths, but the system suggested by

Mr. Wilson appeared to be very similar. The idea no doubt was a novel one, and without any actual trial, it was difficult to say what would be the result. Probably the freezing system was the most successful method of sinking through soft surface, but it was a very expensive process. If any system, less expensive and quite as effective, could be introduced, it would be worthy of serious consideration.

Mr. J. T. FORGIE (Bothwell) said that the proposed method was certainly novel, but he did not know that the difficulties of boring had been thoroughly considered by Mr. Wilson. A bore-hole, through soft surface, to a depth of 100 feet was not such an easy matter. It would take an immense number of bores under Mr. Wilson's proposal, to enclose a shaft of ordinary dimensions, whether rectangular or circular. There was, also, great difficulty in boring vertical holes; and with the most perfect con-trivances, it did not necessarily follow that the bore-hole would be vertical when it reached the rock-head. Personally, he thought that the freezing process was the most efficient to use. It might be expensive, and it was especially so when only a few feet had to be sunk. The initial expenses of the freezing system were not heavy, as the plant could be hired. The time taken in sinking through soft strata by the freezing system, and par-ticularly if the strata were heavily watered, was less than by circular shaft-cylinders or even square shafts. He visited a sink-ing in the North of France, a few years ago, where they had 300 feet of soft and heavily-watered strata. There were 100 feet of soft mud at the surface, covering 200 feet of chalk, which contained a large quantity of water. The shaft was about 16 feet in diameter, enclosed by 23 bore-holes, and the usual freezing apparatus was employed. The sinking was done more expeditiously and cheaply than it had been at some of the neigh-bouring collieries, where they had pumped out the water and sunk through the chalk in the ordinary way. He found that four of the bore-holes were actually 4 feet from the plumb-line at a depth of 200 feet, and 4½ feet at 280 feet.

Mr. J. R. WILSON said that he had never experienced difficulty in putting down vertical bore-holes. He had occasion recently to put down two bore-holes. A bore-hole was put down in deepening a pit, with the view of taking off the water. It was found, however, after sinking through 30 feet, that more water was produced than the bore-hole could carry away. A second bore-hole was then drilled, adopting the arrangement that he had described in his paper, of using a pipe to afford proper support to the rods. When the hole went down, it was only 1 inch from the plumb-line.

Mr. R. T. MOORE said that Mr. Wilson had been fortunate in getting the bore-holes to go down so accurately; and many members had experience of the difficulty of boring vertical holes.

The discussion was adjourned.

———

Mr. R. W. DRON read the following paper on " The Occurrence of Calcareous Coal in a Lanarkshire Coal-field ":—

THE OCCURRENCE OF CALCAREOUS COAL IN A LANARKSHIRE COAL-FIELD.

By ROBERT W. DRON, A.M.Inst.C.E.

The term "kingle" is used by Scottish miners to denote very hard ferruginous or calcareous sandstone. Beds of this hardened sandstone are common throughout most of the Scottish mining districts, and on account of their hardness are sometimes confused with igneous rock. Coal is frequently affected in the same way by the presence of lime and iron, and when this occurs it is almost universally described by the miners as "burnt coal."

The object of the present paper is to describe an area in which not only kingle sandstone, but calcareous or kingle coal is met with, in a Lanarkshire coal-field. The coal has always been described on the colliery-plans as "burnt coal," but the fact that no intrusive whinstone is found anywhere in the district led the writer to investigate more carefully the so-called "burning."

The seams worked extend from the Ell down to the Kiltongue coal-seam. Fig. 1 (Plate IV.) shows the district where this phenomenon occurs. The exact limits of the affected area are not yet definitely ascertained, but it measures at least 100 acres. Fig. 2 (Plate IV.) is a section through the area.

An analysis of the kingle coal has been made by Mr. William McConnachie, of the Coltness Ironworks, and his report is as follows:—

	Per cent.
Combustible carbonaceous matter	32·70
Carbonate of lime	32·03
Carbonate of magnesia	15·51
Ferric oxide	3·28
Alumina	8·39
Alkalies	1·20
Silica	1·04
Sulphur	0·49
Moisture	5·36

The results obtained on coking varied widely, from 25 up to 53 per cent. of volatile matter being recorded, doubtless owing to the carbonaceous matter and the carbonic acid acting on each other. Similarly the ash-results varied, owing to the greater or lesser amount of carbonic acid left on burning off, but the actual ash may be taken at 42 per cent. without serious error. The iron probably exists as carbonate of iron.

From the writer's observations, the alteration of the seams rarely extends beyond 300 feet from the surface. For instance,. at the fault, X (Fig. 2, Plate IV.), the Main and Splint coal-seams on the south or shallow side of the fault, have been altered, but the Virtuewell coal-seam, underneath, shows no. signs of alteration. On the north side of the fault, the Ell coal-seam, where it is found at the same level as the Splint coal-seam, is also altered, but the underlying Splint coal-seam at the same level as the Virtuewell coal-seam is unaltered.

North of the fault, Y, the alteration extends down to the Main coal-seam. The alteration of the seams does not appear to follow any rule that can be recognized.

In the Splint coal-seam, between the fault, X, and the river Clyde, there is an area of about 30 acres in which the alteration is confined to about 6 or 8 inches in the top part of the seam, while the remainder of the coal is perfectly clean and has been mostly worked out. To the south-west of this area, following the banks of the Clyde, the whole of the seam is altered and the coal is unworkable. At other points, the alteration occurs as irregular nodules throughout the seam, and occasionally areas. extending to a few hundred square feet are rendered totally unworkable. Where the greater part of the seam has been altered, it is noticeable that the thickness is considerably reduced.

In the Splint coal-seam, there is a band of cannel-coal, about 22 inches thick; and it is rather remarkable that this cannel-coal is frequently unaltered, even when the coal, both above and below it, is rendered useless.

In the Ell coal-seam, on the north side of the fault, X, the alteration affects only the lower part of the seam, so far at least as it has yet been proven. The clean coal in the upper part of the seam is 5 feet thick, and the altered coal beneath is from 2 to 2½ feet thick.

In the Main coal-seam, at a point north of the fault, Z (Fig. 1, Plate IV.), a place was driven for a distance of 50 feet, in order to prove an area of altered coal. At the commencement of this place, the seam was in its normal condition, and its thickness was 4½ feet. When the place was stopped, the seam was. altered from top to bottom, and was only 1½ feet thick.

Although there is no difficulty in arriving at the negative conclusion that this kingle coal has not been altered by the action of intrusive whinstone, it is very difficult to formulate a theory to account for the presence of the lime, magnesia and iron. No seams of limestone are found in the strata overlying the coal-seams, but there are several bands of ironstone. The local occurrence of the alteration, and the fact that it does not extend to the lower seams would lead to the conclusion that the lime has been deposited in the coal by infiltration from above, at a period subsequent to the deposition of the coal-seams. It is also noticeable that the alteration in question is, generally, greater in the vicinity of faults.

One possible theory, which might be put forward, is that during the deposition of the strata overlying the coal-seams colonies of mollusca inhabited certain limited areas, such as the one referred to in this paper, and their shells accumulated among the beds of sandstone and shale. By the circulation of underground waters, at a period subsequent to the faulting of the strata, these shells were dissolved, the carbonate of lime being carried down in solution and redeposited in the coal-seams and associated sandstones. The subject is of considerable economic and scientific interest, and the writer would suggest that members who can give details of similar alterations of coal-seams should record the facts in the *Transactions*.

Mr. T. LINDSAY GALLOWAY (Glasgow) said that, in the Campbeltown coal-field, a similar impurity occurred in the coal. It seemed difficult to account for a band of 6 or 8 inches of altered coal occurring in the middle of good coal. One could hardly believe that it was caused by contact with whin, or by any other igneous phenomenon. A few years ago, he (Mr. Galloway) sent a sample of the Campbeltown coal to Mr. B. N. Peach of the Geological Survey, and the latter at once propounded the same theory as that adduced by Mr. Dron. He submitted a piece of the altered coal from Campbeltown, and he desired the members to compare it with the specimens shown by Mr. Dron. It burned with difficulty, and left a white ash of almost the same shape as the original piece of coal. It effervesced on being tested with acid; and when a small quantity was ground to powder and treated with acid in a small tube it effervesced violently. It had not been his experience that the alteration had the effect of reducing the thickness of the coal-seam. The coal at Campbel-

The Institution of Mining Engineers.
Transactions 1903-1904.

VOL. XVII., PLATE IV.

To illustrate Mr. R. W. Dron's Paper on "The Occurrence of Calcareous Coal" etc.

FIG. 1.—PLAN.

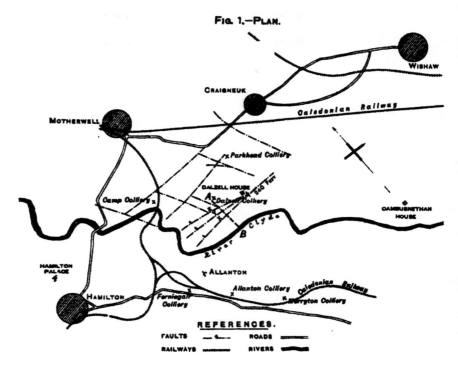

REFERENCES.

FAULTS		ROADS	
RAILWAYS		RIVERS	

Scale, 1 Mile to 1 Inch.

FIG. 2.—SECTION ALONG LINE A B OF FIG. 1.

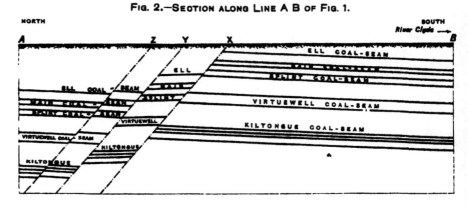

Scale, 528 Feet to 1 Inch.

And Reid & Comp Ltd Newcastle on Tyne

town was worked from a thick seam, but the impure beds were not generally more than 6 inches to 1 foot thick, and they did not extend horizontally more than 5 or 6 feet. He had not observed any alteration in the thickness or in the character of the adjoining coal.

Mr. JAMES HAMILTON (Glasgow) said that he had often seen what Mr. Dron called " calcareous coal " in the Splint seam. Wherever it was found, the blaes overlying the coal was strongly tinged of a red colour. Seeing that iron was one of the constituents of " calcareous coal," the colouring matter in the blaes was no doubt oxide of iron. Similar discolouration was seen in the specimen of sandstone produced by Mr. Dron.

In Home Farm colliery, this red coloration of the blaes was very marked where the Splint coal-seam was " burnt; " and in Rosehall colliery, the officials considered that its appearance was an infallible indication that the workings were approaching " burnt " coal. The first appearance was in the top of the brushing, the coal being clean; it then gradually descended until it reached the coal, which was " burnt," when red blaes rested on the top of it. The coal never, so far as he had seen, shewed any sign of red coloration.

Mr. J. S. DIXON said that mining-engineers in Lanarkshire were well acquainted with " burnt " coal. He knew of hundreds of acres of it, but he had not yet had the curiosity to submit specimens of that altered coal to chemical analysis. He could corroborate what Mr. Hamilton had said as to the red coloration of the strata above the coal-seam. He remembered that a squad of men driving a place through a tract of coal of this description came up the pit as red as if they had been working iron-ore. The occurrence of oxide of iron was recognized, but its relation to the burning of the coal was unknown. Messrs. Dron and Galloway had mentioned that certain seams were " burnt " at the bottom and in the centre, and not at the top. In his experience, the seams had been " burnt " from the top downwards. A large tract of the Main coal-seam in the Carmyle estate was affected in this way. The seam was about 6 feet thick, and probably a third of it was " burnt " in this fashion. It was difficult to understand how the infiltration, as suggested by Mr. Dron, could come from above, and destroy the coal at the bottom without interfering with the coal at the top of the seam. It seemed probable that this process must have been proceeding while the coal was being deposited, because they could scarcely imagine that lime could filter through a seam of coal and leave no indication that it had passed through. He (Mr. Dixon) had never heard of the altered coal being described as " kingle " : " burnt " coal was the common expression in Lanarkshire.

Mr. JAMES BARROWMAN said that it seemed to him that there was an analogous phenomenon in the case of sandstone. In almost every sandstone-quarry, more or less nodular pieces were encountered of a hard rock which was called " whin." These nodules had no connection with igneous rocks, and, being extremely hard, they were rejected by the quarry-master.

Mr. J. S. DIXON said that similar nodules occurred in coal-seams. In the Main coal-seam in the Hamilton district, there were " lunkers," which were agglomerates of carbonate of lime, etc., round centres, and they were probably formed during the deposition of the coal. These " lunkers " were never very large : being simply lumps of coal, impregnated with extraneous substances.

The PRESIDENT (Mr. R. T. Moore) suggested that the term " burnt " coal was somewhat loosely applied to any coal which was run together, no matter what was the cause; indeed, the term seemed to be applied to any coal that would not burn. Thus, there was burnt coal at faults, burnt coal in contact with kingle, and burnt coal in the neighbourhood of whin. As Mr. Dixon had remarked, it was very curious that, in the case mentioned by Mr. Dron, the bottom of the upper seam should be affected while the top of the lower one was untouched. It scarcely looked as if the alteration had been caused by infiltration. The analysis was interesting, and he thought that further information might be obtained by analysing samples of seams where similar phenomena occurred. There were many things in connection with the effect of whin on coal-seams which were difficult of explanation. There was for instance the curious result that a white-ash seam often became a brown-ash one, when it was slightly burnt by whin.

The discussion was adjourned.

SOUTH STAFFORDSHIRE AND EAST WORCESTER-SHIRE INSTITUTE OF MINING ENGINEERS.

GENERAL MEETING,
HELD AT THE UNIVERSITY, BIRMINGHAM, FEBRUARY 1ST, 1904.

MR. ISAAC MEACHEM, JUN., PRESIDENT, IN THE CHAIR.

The minutes of the last General Meeting and of Council Meetings were read and confirmed.

The following gentlemen were elected:—

MEMBERS—

MR. WILLIAM HILL, Mining Engineer, Wellington Road, Edgbaston, Birmingham.
MR. F. C. SWALLOW, Mining Engineer, Glenroy, Nuneaton.

Mr. J. H. JACKSON read the following "Notes on Early Mining in Staffordshire and Worcestershire":—

NOTES ON EARLY MINING IN STAFFORDSHIRE AND WORCESTERSHIRE.

By J. H. JACKSON.

Although the art of mining is an ancient one in England, the scarcity of records dealing with our greatest industry renders it a difficult subject to follow. The historian, when dealing with it, has, in some cases, obtained his information at second-hand, and in others he has viewed it in the light of one unfamiliar with the practical details of mining life. Brief reference has been made to the method of working mines at different periods; then occurs a lapse of years during which history is silent, although the industry no doubt advanced slowly but surely. It had become a source of extraordinary revenue to the capitalist long before it was deemed necessary to keep records of underground workings, and not until an Act of Parliament was passed to enforce it did the keeping of such records become general. Evidences of the antiquity of metal-mining in England have been preserved through the means of pigs of lead bearing inscriptions of the different periods in which they were made. The art of coal-mining, however, has been recorded for the most part by abandoned workings, which in some cases have brought calamity upon those who were unfortunate enough to discover them.

The coal-fields of North and South Staffordshire, especially the latter, when in their entirety were particularly adapted to the inadequate means that the old-time miner had at his disposal for the working of coal. Over many acres in the southern part of the county, the Thick coal-seam lies near the surface, and, to a great extent, the different strata lie in a plane identical with that of the horizon. There the old-time miner found his working-ground. Wherever the outcrops or wash-outs had favoured him, there he had left evidences of his labour. In addition to the abundance of fuel possessed by the ancient Britons in their vast forests, they appear to have been acquainted with the use of coal, quantities of which have been found in pre-Saxon deposits.

Both John Whitaker* and Thomas Pennant† have cited instances to prove that coal was known to the aborigines of Great Britain, the former depending in a great measure on the derivation of the word "coal," and the latter on the flint-axe found embedded in a seam of coal in Monmouthshire. Perhaps the most interesting record of ancient working is that left by Mr. Mammatt, who stated that "In Measham [near Ashby], where the bed was not more than forty or fifty feet from the surface, indications of ancient workings were found, in stone hammer heads, and large wedges of flint with hazel withes round them; also wheels of solid wood about eighteen inches in diameter."‡ In Lancashire, workings in the Arley mine have been ascribed to the Romans. It is not improbable that the Romans found the exposed outcrops of the richest measures of our Carboniferous system useful for their needs.

A deed, relating to the Heronville family, of the year 1315 describes a piece of land at Wednesbury as "lying near Bradeswalle against the coal pits," and John Leland, the antiquary, in 1538 writes:—"There are secoles at Weddesbyrie, a village near Walsall." W. Camden, who wrote his *Britannia* about the year 1586, says:—"The south part of Staffordshire hath coles digged out of the earth, and mines of iron, but whether more to their commodity or hindrance I leave to the inhabitants who do or shall better understand it."§ The register of burials of the Parish Church at Wednesbury for the year 1577 contains the following entry:—"Christopher Daly was buried; he was killed in the ryddinge, in the coal pit." The early mining industry of Wednesbury has been mentioned by Dr. Plot and also by — Wilkes, both of whom have recorded the fires that existed in their day in the old open works there.

By an old deed of the 46th year of the reign of Edward III., it appears that both coal and ironstone were raised at Amblecote, for T. Nash, referring to the Lyttelton manuscripts, states:—" By this deed it seems to me, that Betecote was then reckoned within the manor of Old Swinford, and that this land lay in that

---

* *The History of Manchester*, 1771, vol. i., pages 302-305.

† *Tours in Wales*, second edition, 1810, vol. i., page 25.

‡ *A Collection of Geological Facts and Practical Observations, intended to Elucidate the Formation of the Ashby Coal-field, etc.*, by Mr. Edward Mammatt, 1834, page 9.

§ *Britannia*, by Mr. William Camden, 1789, vol. ii., page 375.

part which was called Betecote, as it was bounded with Amble-
.cote, and some part was within the manor of Amblecote. Coal
and iron stone were then got in these parts, and what we now call
.pit-coal was at that time called sea-coal."*

In Dr. Plot's time, workings in the Thick coal-seam and
Stourbridge fire-clay were in existence, the latter mineral being
used for the manufacture of glasshouse pots. The historian in
1686 says : —" But the Clay that surpasses all others of this County,
.is that at Amblecot, on the bank of Stour, in the Parish of old
Swynford yet in Staffordshire, in the lands of that judicious and
.obliging Gent. the Worshipfull Harry Gray of Enfield Esq;
. . . . . I say the most preferrable clay of any, is that of
Amblecot, of a dark blewish colour, whereof they make the best
pots for the Glass-houses of any in England : Nay so very good
is it for this purpose, that it is sold on the place for sevenpence
the bushell, whereof Mr. Gray has sixpence, and the Workman
.one penny."†

Mention has already been made of the fires in the old open
works at Wednesbury, and as early as 1686 the spontaneous
.combustion of coal had attracted attention. Dr. Plot says : —
"All these things (I say) being put together, what can there
else be concluded but that some Coal-pits may and doe take
fire of themselves; as 'tis unanimously agree'd they doe at
Wednesbury (where the coal-works now on fire take up eleven
Acres of ground) Cosley, Etingsall, and Pensnet in this County,
.as Mr. Camden will have it, whereas indeed the place He
mentions then on fire, was Broadhurst on Pensnet in the Parish
.of Dudley and County of Worcester, where He says a Colepit
was fired by a Candle through the negligence of a Groover; and
so possibly it might; but as for the rest (which are in Staffordshire) 'tis agree'd they all fired natural of themselves, as they
.expect the shale and small-cole in the hollows and deads of all
the old works, will doe and have done, beyond all memory."‡

The following is an account of the method of working the
Thick coal in Dud Dudley's day. "In these Pits, after you have
made or hit the uppermost measures of Cole, and sink or digged

* *The History and Antiquities of Worcestershire*, by Mr. T. Nash, 1782,
vol. ii., page 208.

† *The Natural History of Stafford-shire*, by Dr. Robert Plot, 1686,
page 121.

‡ *Ibid.*, 141.

thorow them, the Colliers getting the nethermost part of the Coles first, about two yards in height or more, and when they have wrought the Crutes or Staules (as some Colliers call them) as broad and as far in under the ground, as they think fit, they throw the small Coles (fit to make Iron) out of their way on heaps to raise them up so high, to stand upon, that they may, with the working of their Picks or Maundrills over their heads, and at the one end of the Coles so far in as their Tool will permit, and so high as their working cometh unto a parting in the measure of Cole, the which Coles, to the parting by his self clogging and pondrous weight, fall often many Tuns of coles, many yards high down at once."* One traces by this, evidence of a system of undergoing, cutting and falling as practised later in the square-work system once so vigorously applied in getting the Thick coal, when the latter was more plentiful in the district.

In some of the old workings around Amblecote, however, another system seems to have been in practice, where the old-time miner has in some cases merely removed the Spires and White coal and the Slipper and Sawyer, leaving the intermediate measures behind; probably, the writer submits, on account of their unfitness for the purposes for which the fuel was then required. It is interesting to note that the divisions of the Thick coal still bear the same nomenclature as that which they did in Dud Dudley's day, their names holding good after the lapse of two and a half centuries.

But the record of mining left by the old maker of iron with pit-coal does not account for the whole of the rather extensive area which the old-time miner worked over along the Netherton anti-cline. For 6 miles, the Thick coal-seam lies in close proximity to the surface along the flanks of the ridge, and for three parts of that distance the miner working to-day finds traces of work of another period. A broken pick, a smoking pipe, an earthen drinking-mug or bottle, rudely glazed, are silent witnesses to the existence of those who delved underground in the dark days when England was plunged in internecine warfare and religious strife. In the latter end of the seventeenth century, these smoking pipes were made at a place called Salt-water Pool, at Bromley in the Chase of Pensnett.

* *Dud Dudley's Mettallum Martis*, 1665, page 36.

In the time of Dud Dudley, some of the shafts up which coal was drawn were constructed square and timbered, while others were circular in shape and bricked. At Wednesbury, a few were lined with special bricks made to fit the radius of the shafts, which were about 6 feet in diameter.

On the eastern flank of the Netherton anticline, near the Lodge Farm, the measures in places dip rapidly away from a bank of uptilted Silurian rocks, until they become intercepted by a downthrow fault which runs parallel to the strike of the strata. At the line of fault, the Thick coal-seam faces the White ironstone-measures. After leaving the dislocation, the strata at first incline towards the east at an angle of 22 degrees, and as far as the Grains and Gubbin ironstone-measures lie near the surface they have been worked by a number of old bell-shafts. These shafts are about 6 feet in diameter at the surface, and gradually increase to 8 feet where they reach the Gubbin iron-stone, then they " bell out " until the Upper Heathen coal is met with, at which point each is about 13 feet in diameter. The Heathen coal is never pierced by them. Six hundred feet away, could, at one time, be seen several heaps of cinders or slag, the refuse from smelting operations, now totally covered over by Messrs. Doulton's spoil-bank.

At various angles, bore-holes $2\frac{1}{2}$ inches in diameter are met with in the present open work, some of which have been traced for a number of feet in length. It appears as if the ancient mineral-prospector had been boring for ironstone. In the old bell-shafts, the marks of the pick are still to be seen as plainly visible in the Gubbin measures as the day when they were left there many generations ago. Both shafts and bore-holes have been put down through the burnt measures of the Thick coal-seam, which at this place has been consumed by fire for some distance along the outcrop.

Bell-shafts are not altogether a proof of the antiquity of mining, for the late Sir Warington W. Smyth, referring to the method of working the ironstone at Tankersley, Yorkshire, in 1853, says : —" Wherever the courses of ironstone-nodules come up to within a short distance of the surface they are vigorously attacked by the small shafts termed bell pits."[*] In the case

* "Iron-ores of the Northern and North-midland Counties of England : General Description," by Mr. W. W. Smyth, *Memoirs of the Geological Survey of Great Britain, and of the Museum of Practical Geology : The Iron-ores of Great Britain*, 1856, page 33.

of the Lodge Farm workings, however, beech-trees over 7 feet in girth have grown on the land about the shafts which have been filled up with rubbish. The writer submits that the pig-iron made here was afterwards conveyed to Cradley Forge to be worked into bars, etc., and believes that the work belongs to the seventeenth century. This hypothesis is not at all improbable, because two mugs found in one of the shafts resemble those often picked up in the workings around Amblecote, which are probably contemporaneous with Dud Dudley. Smoking pipes have also been found here, proving that the work at all events is not anterior to 1585, the year when tobacco was first introduced into England.

Mining records are wanting of all the " old men's work " in the South Staffordshire district, although in the section on Longimetra in the second edition of Digges's old black letter-work published in 1591, the thirty-sixth chapter is devoted to mine-surveying. Digges also mentions a system of plotting surveys on parchment or paper, proving that subterraneous surveying was practised in England over 300 years ago.[*]

It has been written elsewhere of the failure of Savary to drain the mines at Broadwaters near Wednesbury; but the first successful attempt made in the district to pump water by the application of steam appears to have been made as early as the year 1712, for J. T. Desaguliers says:—" Thomas Newcomen, iron-monger, and John Calley, glazier, of Dartmouth, made the several experiments in private, and having brought the engine to work with a piston, etc., they, in the latter end of the year 1711, made proposals to draw the water at Griff, in Warwickshire; but their invention meeting not with reception, in March following, through the acquaintance of Mr. Potter, of Bromsgrove, in Worcestershire, they bargained to draw water for Mr. Bach, of Wolverhampton, where, after a great many laborious attempts, they did make the engine work."[†]

The old-fashioned method of draining mines by chain-wheels actuated by horses, water-wheels, or manual labour was a clumsy contrivance. In Cornwall, this arrangement was known as the rag-and-chain pump, and, when worked by hand-power,

[*] *A Geometrical Practical Treatise named Pantometria*, by Mr. Thomas Digges, second edition, 1591, pages 51-55.

[†] *History of the Iron Trade*, by Mr. Harry Scrivenor, second edition, 1854, page 90.

required six men every 6 hours to pump water from a depth of 20 feet. Mr. Pryce states that:—" The men work at it naked, excepting their loose trowsers, and suffer much in their health and strength from the violence of the labour, which is so great that I have been witness to the loss of many lives by it."[*] The pumps in which the rag-and-chain worked were made of wood, some of which had bores 4 inches in diameter.

A system of railroads was in use in the Newcastle-upon-Tyne district, as early as the year 1676. Lord Keeper Guildford thus describes it:—" The manner of the carriage is by laying rails of timber from the colliery down to the river, exactly straight and parallel; and bulky carts are made with rowlets, fitting these rails, whereby the carriage is so easy, that one horse will draw down four or five chaldron of coals."[†] The earliest satisfactory mention of iron-rails being used in connection with mining refers to the year 1767, when 6 tons were cast at the iron-works of Colebrookdale, Shropshire.[‡] The late Mr. Keir was the first to notice the use of railroads in South Staffordshire, and this was about the year 1798. We find railways mentioned again in the year 1817, when they were used for conveying coal from the Wyrley Bank colliery to the turnpike-road for sale and use.

About 1680, coal was worked at Beaudesert, at a depth of 240 feet from the surface " which was the best cannel coal known except that of Lancashire." Workings were also in progress in the north of the county, in the Peacock coal, at Hanley Green near Newcastle-under-Lyme. At Biddulph, the historian Dr. Plot was shown some rearer-work, and there was an adit or footril belonging to Sir John Bowyer, at Apedale, which extended along the strike for 600 feet. The settlement of the pottery-trade near Burslem gave rise in a certain measure to the North Staffordshire coal-industry. " In the time of Plot," writes — Aiken, " coal was as low as 2d. the horse load," which at eight horse-loads to the ton (the usual estimation) amounts to only 16d. per ton. Coal had been worked at Beaudesert as early as the third year of the reign of Edward VI., for Stebbing Shaw states:— " According to a curious rental, etc. of Sir William Paget's

---

[*] *Mineralogia Cornubiensis*, by Mr. William Pryce, 1778, pages 150 and 151.

[†] *The History and Description of Fossil Fuel, the Collieries and Coal-trade of Great Britain*, [by Mr. John Holland], 1835, page 354.

[‡] *Ibid.*, page 353.

estate, the Manor of Longdon in 1550 was worth £40 per annum which then consisted of the Manor of Beaudesert with the appurtenances, valued at £13 10s. 0d. (particularly a mine of coal below the park then valued at £4 per annum and a smithe mill situate in the antient park)."[*]

In the year 1719, Lord Macclesfield, who owned an estate near Burslem, entered into an agreement with the owners of the low meadow-lands adjacent to cut a sough or gutter for the purpose of draining his mines. This freed a quantity of coal to the rise, which lasted 60 years. The proprietors then introduced horse-gins, and shortly afterwards steam-engines. The deepest engine-pit in the Burslem district in 1811, according to one writer, was a little over 300 feet from the surface.

Meanwhile, since 1680, coal had risen in value from 16d. per ton to 3s. 8d. in 1780 and afterwards to 6s. in 1795; in 1815, it was 8s. 4d. per ton at the pit-mouth. At Audley, about this time, the coal-mines employed many hands, and the price of coal was 9s. per heap of from 25 cwts. to 30 cwts. "At Kidcrew," writes W. Pitt, " in the north of the Parish of Wolstantan is to be seen some of the most improved kind of machinery for raising coal."[†] A fair mine-rent or royalty was considered to be one-sixth part of the price of the coal as sold at the pit-head.

The use of gunpowder for mining was first introduced into Staffordshire about the year 1680. This was at the copper-mine at Ecton, which was first proved and afterwards abandoned by Lord Devon, Sir Richard Fleetwood and some Dutchmen. Dr. Plot noticed the occurrence, for he says " they broke the rocks with Gunpowder."[‡] Up to this time, rocks had been broken by the application of fire, and the same writer mentions that " the first sort whereof, which [damp] arises from the smoak of coal itself, is said to happen only in such grooves where they make use of great fires to soften the rocks to make them yeild to the pick-Axe, ... And of this sort perhaps are the damps of the coalworks about Chedle."[§] Clearly it was to prevent accidents arising from the vitiation of the little air that did circulate that Article XL was framed for the lead-miners of the Peak district, at the Barmote Court, held in the Wapentake of Wirksworth in

---

[*] *History of Staffordshire*, by Mr. Stebbing Shaw, 1798, vol. i., page 212.

[†] *History of Staffordshire*, by Mr. W. Pitt, 1817, page 349.

[‡] *The Natural History of Stafford-shire*, by Dr. Robert Plot, 1686, page 165.

[§] *Ibid.*, page 134.

1665:—" We say, that any Miner in an open Rake, may kindle and light his fire after Four o'Clock in the Afternoon, giving his Neighbour lawful warning thereof."[*]

About the period that the Ecton mine proved a failure, Dr. Plot records the following method of working what he terms a "hanging mine." This was at a place called Hardingswood, in North Staffordshire. " I went down into one of these hanging mines at Hardingswood belonging to the aforemention'd Mr. Poole of this County, where He shewed me a level of 35 yards of roach as it lay in an oblique diping line above the water, which came to 35 foot perpendicular, diping one foot to every yard: in this Level He had five wallings or Stauls, out of which they dug the coal in great blocks; between the wallings there were ribbs left, and passages through them called thurlings, which give convenience of Air, and passage for the coal out of one walling into another, which in all coal mines stand thick or thin, partly according to the substance of the coal; . . . . . where there is a strong rock next the coal, and no bass, they will then venture their roof so far sometimes as to make their wallings 8 or 9 yards wide."[†]

From the above extract, there seems to have been in practice a certain system of ventilation, and, from what follows, we may infer that the necessity of sinking two shafts had already become apparent. " The second sort of damp occasion'd by smoak, they dispel either by water, where they have no Air pits, and in winter time; but cheifly by fire, which they let down in an Iron cradle, they call their Lamp, into the shaft or by pit next to that they intend to work, ... Which very way they use about Chedle, and 'tis a secure one too, but very chargeable."[‡] Dr. Plot further states that, in order to obviate the expense of sinking two shafts, an expedient devised and practised then at Liége might be adopted. This system of ventilation was communicated to the Royal Society by Sir Robert Moray or Murray. It consisted of a chimney 30 feet high, erected at the mouth of an adit whence an air-pipe was carried into the backs. In the chimney, a fire-grate was fixed and the returns passed through the air-pipes and into the chimney.[§]

    * *The Miner's Guide or Complete Miner*, 1810, page 48.

    † *The Natural History of Stafford-shire*, by Dr. Robert Plot, 1686, pages 147 and 148.

    ‡ *Ibid.*, page 138.

    § *Mineralogia Cornubiensis*, by Mr. William Pryce, 1778, page 148.

In Cornwall, 150 years ago, to save the cost of driving in hard ground, they carried a staging 1 foot above the floor of the level that was being driven, upon which boards were laid, their joints carefully stopped with clay, pitch and oakum, with turfs placed over all. This acted as an intake for the air-current, and was called a "saller." Traces of a kindred system, the writer of this paper has been told, have been found in some of the old workings in South Staffordshire.

The low price of Swedish ore had crippled the copper-mine of Ecton; but, in 1720, a Cornish company, after spending £13,000, discovered a rich lode of ore which eventually proved a bonanza. Such was the extent of it that in 1789, nearly 70 years later, it produced about £10,000 worth of copper, clear of all expenses, and found employment at the mines for 300 men.

Looking back to the middle of the seventeenth century, one finds the mining industry of Staffordshire more extensive than at first might be supposed. Tobacco pipe-clays were worked at Cannock, Lichfield, Darlaston, Monway-field and on Pensnett Chase in the south, and at Hanley Green in the north. At Betecote, now called Bedcote, near Stourbridge; at Amblecote and the Lye, the Old-mine fire-clay was worked, and already was in great demand in London. Coal had been traced around Amblecote, Netherton, Wednesbury, Gornal, Sedgley, Dudley, Coseley, Etingshall, on Pensnett Chase, and at Hasco Bridge now Askew Bridge, near Himley. In the north of the county, Apedale, Hanley, Burslem and Cheadle had their coal-mines. Limestone was being worked at Dudley, Sedgley and Walsall. Stone, for building purposes and grindstones, was quarried at Bilston, Coseley and Gornal. In fact, the old-time miner, 250 years ago, had already proved in part the hidden resources of our present coal-fields, which merely awaited the successful manufacture of iron with coal, the satisfactory application of steam, and the introduction of canals and railways, to become opened out to their present extent.

———

The PRESIDENT (Mr. I. Meachem, Jun.,) proposed a hearty vote of thanks to the author for his notes.

Mr. T. H. BAILEY seconded the resolution, which was cordially approved.

NORTH STAFFORDSHIRE INSTITUTE OF MINING AND MECHANICAL ENGINEERS.

GENERAL MEETING,
HELD AT STOKE-UPON-TRENT, MARCH 7TH, 1904.

MR. A. M. HENSHAW, PRESIDENT, IN THE CHAIR.

The minutes of the last General Meeting were read and confirmed.

The following gentlemen, having been previously nominated, were elected:—

MEMBERS—

MR. J. G. ALEXANDER, Sneyd Colliery, Burslem.
MR. H. H. B. DEANE, Hunton House, Gravelly Hill, Birmingham.
MR. WILLIAM LOCKETT, Norton Colliery, Smallthorne.

ASSOCIATE MEMBER--

MR. JOHN MAYER, Sneyd Colliery, Burslem.

STUDENT--

Mr. W. G. THOMPSON, Ball Green, Norton-in-the-Moors.

Mr. G. E. LAWTON read the following paper on "Fires in Mines, with Particular Reference to Seams in the North Staffordshire Coal-field":—

FIRES IN MINES, WITH PARTICULAR REFERENCE TO SEAMS IN THE NORTH STAFFORDSHIRE COAL-FIELD.

By G. E. LAWTON, F.G.S.

In bringing this subject before the members, the writer proposes to deal with it in a general way, and to place on record some of the experiences obtained in dealing with pit-fires, and the conclusions deduced therefrom. One of the most difficult problems with which a mining-engineer has to deal, and one which probably causes more anxiety than any other, is to carry out the operations necessary to extinguish a fire in a gassy mine.

Fires in mines may be divided into two classes, according as they are caused by spontaneous combustion, or not. The principal causes of these fires are spontaneous combustion of the coal, spontaneous combustion of the material in the goaf, accidental ignition of timber by open lights, explosions of fire-damp, underground boilers, furnaces, steam-pipes, machinery, electric appliances and shot-firing. As is well known, the commonest cause of pit-fires is spontaneous combustion, and there is no doubt that, directly and indirectly, more loss to those interested in collieries can be attributed to fires originating from this cause than to any other; but the most deplorable features in connection with these occurrences are the disasters which they sometimes cause, resulting in the loss of many valuable lives.

Papers have been read on this subject by Mr. Joel Settle[*] and Mr. Arthur Hassam,[†] in which they deal with fires occurring in particular seams at particular collieries.

Many coal-fields are liable to spontaneous combustion, but perhaps there is none that contains more seams of coal subject to take fire spontaneously than the North Staffordshire coal-

---

[*] *Trans. Inst. M.E.*, 1893, vol. v., page 10.

[†] *Ibid.*, 1894, vol. viii., page 332.

field.　The seams in which fires have occured in this coal-field and from this cause are:—The Great Row, Cannel Row, Yard, Rowhurst, Ragman, Rough Seven-feet, Hams, Ten-feet, Seven-feet Banbury, Eight-feet Banbury or Cockshead, and Bullhurst coal-seams.　Possibly there are others.　Fires have occurred in the Great Row, Cannel Row and Rowhurst coal-seams in the Hanley district; in the Yard coal-seam in the Longton and Fenton districts; in the Ten-feet coal-seam in the Burslem district; in the Ragman, Rough Seven-feet and Hams coal-seams in the Leycett district; in the Seven-feet Banbury coal-seam in the north-western part of the coal-field; and in the Eight-feet Banbury or Cockshead seam in the Longton and Talke districts. The Bullhurst seam, which is notorious for the frequency of fires, appears to be liable to take fire, more or less, over the greater portion of the coal-field.

TABLE I.—SECTIONS OF THE COAL-SEAMS AND THE ADJOINING STRATA.

GREAT ROW COAL-SEAM.			RAGMAN, ROUGH SEVEN-FEET AND HAMS COAL-SEAMS.			EIGHT-FEET BANBURY OR COCKSHEAD COAL-SEAM.		
(1) Hanley District.						**(7) Longton District.**		
	Ft.	In					Ft.	In.
Dark marl	13	6	**(5) Silverdale District.**			Bass	...	—
COAL ...	9	6		Ft.	In.	COAL, inferior Billies .	1	0
Dark marl	4	6	Light-grey bass ...			COAL, top	3	0
Rock		—	Cannel bass ..	0	9	COAL, bottom ...	4	10
CANNEL ROW COAL-SEAM.			Parting, charcoal or smuts .		—	**(8) Kidsgrove District.**		
(2) Hanley District.			COAL, Ragman				Ft.	In.
	Ft.	In.	seam ...	3	5	Shale		—
Black bass	6	3	Stone, full of			COAL, inferior	0	8
COAL ...	1	2	pyrites	0	0½	Hustle	0	3
Cannel ...	1	2	Loose shale, with			COA	8	0
COAL ...	4	0	thin streaks of			COAL, inferior	1	0
Shale ...	1	2	coal, Pous ...	4	0	BULLHURST COAL-SEAM.		
Cannel ...	0	6	COAL, Rough			**(9) Kidsgrove District.†**		
COAL ...	1	0	Seven-feet seam	7	0		Ft.	In.
Rock	7	0	Loose shale, with			Rock		—
YARD COAL-SEAM.			thin streaks of			COAL ...	8	0
(3) Fenton District.			coal and pyrites,			Rock		—
	Ft	In	Yardley ...	2	0	BULLHURST COAL-SEAM.		
Shale ...		-	COAL, Hams			**(10) General Section.**		
COAL ...	2	3	seam	4	0		Ft.	In.
Dirt ...	0	10	SEVEN-FEET BANBURY			Rock . ..		—
COAL ...	4	6	COAL-SEAM.			Hustle, or carbona-ceous shale,		
Shale		—	**(6) Kidsgrove District.**			nil to ..	2	0
ROWHURST COAL-SEAM.				Ft.	In.	COAL ...	2	0
(4) Hanley District			Rock*		—	Hustle, nil to	3	0
	Ft	In	COAL	7	0	COAL ...	3	0
Rider coal	3	2	Rock		—	Hustle, nil to	3	0
Black bass	10	0				COAL ..	4	0
Blue metal	2	0				Hustle, nil to	1	0
COAL ...	6	6				Rock ...		—
Marl ...	13	6						

* There is frequently several feet of shaly material between the overlying rock and the coal.
† This section was taken where the coal took fire, as described, at the Moss pits; and hustle was not noticed in the coal at this place.

The sections (Table I.), varying greatly in diverse localities, are taken from districts where the seams are liable to spontaneous ignition. On looking over the sections, it immediately strikes one that they contain most of the thick seams of the coal-field, and experience appears to teach that thick seams are more inclined to take fire spontaneously than thin seams. Perhaps some of the reasons for this ignition are that thick seams have a larger area of coal exposed to the oxidizing influence of the air. The goaves being of a larger cubic capacity, they cannot be so tightly packed as thin seams, therefore they take a longer time before they become solidified and impervious to air-currents. Thick seams, generally speaking, are more friable, and usually there is greater rock-pressure on the coal, causing it to become broken and comminuted, thus giving it a greater power for the absorption of oxygen.

It appears to be the fact that some of these seams are more prone to take fire in some localities than in others. There is no apparent reason for this, but it may be due to a difference in the physical and chemical constitution of the coal or matter associated with the coal, and in some instances it may probably be due to the method of working.

The origin of spontaneous combustion of coal or of goaf-material does not appear to be perfectly understood. Small coal constitutes one of the most important factors in the origin of spontaneous combustion, but at the same time there are many other important agencies which assist in bringing about combustion. One theory is that combustion is brought about by physico-chemical actions and reactions of gases, liquids and solids. A résumé of this theory is as follows:—(1) The physical and chemical constitution of the coal; (2) dissociation of hydro-carbons occluded in the coal, and the probable liberation of pure carbon; (3) the gases and vapours ignite first, and not the coal; (4) the exudation of the occluded vapours and gases into the open space increases the pressure, temperature and chemical action; (5) the friction of gases as well as solids in the goaf; and (6) the probable generation of electric stresses in gob-heaps.

Other agencies which appear to propagate combustion are:— The oxidation of coal or surrounding material and, according to Mr. Busse, the tendency of coal to weather, which in the first place is produced by the oxidation of hydrogen, then by the oxidation of carbon, and finally by the absorption of oxygen.

Rock-pressure plays an important part in generating fires, and when pillars of coal are subject to great rock-pressure, the grinding and slipping movements which take place in the ground may possibly, by this frictional action, in themselves generate sufficient heat to ignite coal. It is only reasonable to suppose that when a pillar of coal in a highly-inclined seam slips *en bloc* to the dip of the mine (as sometimes occurs in steep mines), intense heat is created, and no doubt some fires have been caused by such occurrences.

Many fires have taken place near to, and have been attributed to, faults. Perhaps this may be due to the fact that extra rock-pressure is often experienced in the region of faults, and the moisture which often exudes in such regions probably promotes chemical action.

It appears to have been established by experience that the tendency to the development, or at any rate to the extension, of combustion in coal, is increased by the presence of certain shales which yield inflammable liquids on exposure to heat.

Pyrites is often assumed to play an unimportant part in exciting fires, but it appears to be a fact that, when it is brought into contact with air and moisture, it is decomposed and the chemical reaction liberates heat. Of course, the old idea that the presence of pyrites is a sure indication of gob-fires occurring has been exploded, as experience shows that some coals poor in pyrites are subject to spontaneous combustion; and on the other hand, some seams rich in it are immune from fires.

Dr. J. S. Haldane, a short time since, experimented with a piece of Bullhurst coal. The piece of coal lay in his laboratory for three years, and at that time the lump had swollen and broken up into minute fragments, owing evidently to the formation of a white crystalline substance between the lamellæ of the coal. The white substance was strongly acid, possessed the astringent taste of iron salts, and gave the chemical reaction of a mixture of ferrous and ferric sulphate along with free sulphuric acid. The crystals were undoubtedly the products of oxidation of iron-pyrites. The writer has frequently seen in the workings of the Bullhurst seam, in large patches on the coal, the white crystalline substance referred to by Dr. Haldane, which appeared after the coal had been exposed to the air for some time. It would seem that this experiment proves pretty conclusively that pyrites has something to do with propagating fires in this seam.

Gob-fires of the worst type have been produced where slack has been left behind in the goaf, and these fires are greatly facilitated by allowing slight currents of air to pass through the goaf.

Some fires have been attributed to leaving timber in the goaf. Seeing that ordinary Norway-timber fires at a temperature of 500° Fahr., the surrounding material must in these cases have attained a high ignition-point, as coal can be fired at a temperature of 284° Fahr.

Other fires have been caused by leaving roof-coal, and some have been produced by rider-coals lying above and below and in close proximity to the working seam.

The spontaneous combustion of some coal is due to its columnar structure, and to the circumstance that the fissures in the coal are filled with fine dust. In such cases it often happens that this fine dust frequently fires in a very short time, without any evolution of smoke or heat having been noticed. Possibly explosions of untraced origin have been caused in this way, but perhaps this subject does not come within the purview of this paper.

Prof. P. P. Bedson succeeded in igniting coal-dust at a temperature of 291° Fahr.[*] If the increase of temperature be 1° Fahr. for every 70 feet of increase in depth below 50 feet and if the normal temperature is 50° Fahr. at that depth (the increase may be slightly higher or lower than these figures according to conditions), and if coal-dust will fire at a temperature of 291° Fahr., then it would appear that at a depth of 16,870 feet, or a little over 3 miles, the natural temperature of the strata will be such as to ignite coal-dust.

As regards the greater likelihood of spontaneous combustion in mines at great depths, it appears probable that in some seams fires will be more frequent, while in others they will be rarer.

The seams that will perhaps take fire more frequently are thick seams which cannot be adequately packed, and in which the principal factors operating to cause ignition are rock-pressure

---

[*] *Trans. N.E. Inst.*, 1888, vol. xxxvii., page 212.

and oxidation of the coal. As, naturally, at great depths the rock-pressure will be heavier and the temperature higher, consequently oxidation of the coal will be more rapid, for the activity with which oxygen is absorbed increases as the temperature rises.

The seams which may not be expected to take fire more frequently are those lying at moderate inclinations, which can be thoroughly packed with non-inflammable material, and in which the gob-material only takes fire after long exposure to air: because, in these cases, the extra rock-pressure would tend to solidify the gob more quickly, rendering it less susceptible to air-currents, thus probably preventing oxidation in the goaf.

Some thick seams of coal, possibly, may be unworkable at very great depths, in consequence of their greater liability to fires, unless some artificial means of reducing the temperature can be devised; but it is not unreasonable to expect that some practicable arrangement may be invented which can be applied to bring about this desirable state of affairs. After all, however, the factor which will determine the question of working coal at great depths will be one of cost of production and not of practicability.

The first gob-fire with which the writer was associated occurred some 20 years ago in the Rowhurst seam at Hanley. The coal lay at an inclination of about 1 in 4, and was worked on the longwall system. Without any indication that combustion was proceeding, smoke in rather large volumes came out of a pack-wall, A, on the high side of a main level, which was also the main-intake airway. This happened during the working-shift, and the workmen left the district in a panic. The first step taken was to ram dirt, which was lying near at hand, into the interstices in the pack-wall, out of which the smoke was exuding. This had the effect of temporarily checking the issue of smoke. A brattice-cloth, J, was placed longitudinally along the level to keep the wind off the fire, and a brick-wall, B, was built against the pack-wall and packed behind with sand. Afterwards, a brick arch, CD, was built along the roadway, and also packed round with sand. The arch about the region of the fire, A, after a short time, became so hot that a new level roadway, EFG, was driven round the affected length of roadway, the old roadway being packed solid with sand, CD, and terminated at each end with brick-and-mortar stoppings, H and I. These operations proved effectual, and no further trouble was experienced at this point (Fig. 1, Plate V.).

A description of another fire in this seam, first noticed on May 5th, 1891, may be of some interest. The fire was in a waste next to a jig, and about 45 feet from the working-face, in the neighbourhood of a pillar of coal which had been left against a fault, and probably the fault and the coal-pillar caused the mischief. Gob-stink was first perceived at the working-face at 10 p.m. on May 5th, and smoke was discovered in the waste soon afterwards. This waste, A, and the adjoining one were immediately built off with rubble-wall and sand stoppings, B and C, (Fig. 2, Plate V.). As this fire appeared to be within small bounds, and as the district had only been recently opened, it was decided to attempt to load out the fire, and with this object in view an opening, D, was made through a pack-wall into the waste, where the fire was burning. When this opening penetrated into the waste, the latter was found to be full of smoke, and the débris in it was highly heated, but no actual fire was observed. Damp sand was thrown over the heated material from time to time, as the loading-out operations proceeded. These loading-out operations were going on in a satisfactory manner, when a fall of roof in the waste probably liberated and brought down some gas, causing an explosion which burnt 7 men, of whom one died afterwards from his injuries. On an examination being made of the place after the explosion, gas was found burning in the roof in the waste. This gas-flame was extinguished by throwing wet sand into it. After this incident, the opening, D, through the pack-wall, was closed, strong stoppings were erected in the main levels further out-bye, and the district was abandoned for a time.

Many fires have occurred at the Harecastle collieries, principally in the Bullhurst seam, which is a very gassy one at these collieries. The dip of these seams ranged from nearly horizontal to an inclination of 60 degrees. The seams were worked on the pillar-and-stall system, and they were divided into panels or small districts. The seams were cut up by numerous faults, but although the faults, as is usually the case, increased the difficulties of working the coal, there was a redeeming feature attached to them, for the panels or districts were so arranged that they were bounded in most cases by faults. Since advantage could be taken of the faults in this manner, it was not necessary to leave as much coal against districts which had to be sealed off in consequence of fire, as otherwise would have been the case if the panels had not been bounded by faults. Moreover, the points in the roadways intersected by the faults were generally suitable sites for the erection of permanent or temporary stoppings.

Several fires in the Seven-feet Banbury seam at the Moss pits of these collieries did not get beyond the incipient stage. When coal, which had "rated off" the sides of pillars, was allowed to remain and accumulate, incipient heating sometimes occurred.

Some of the fires which took place in the Bullhurst seam developed with great rapidity, smoke being the first indication that combustion was in progress.   The following is a short description of one of the fires which occurred on October 1st, 1898, in this seam, at the Moss pits of the Harecastle collieries. It will be noticed from Fig. 3 (Plate V.) that drifting had only been carried on to a limited extent in this district. Smoke was found in the roadway, A. A dirt-stopping was built in this roadway, and the ventilation was cut off from the seat of the fire, so that the products of combustion might accumulate, and thus retard the process of combustion.   Wooden stoppings, 4 feet thick, had already been built in the intake and return roadways, B and C, with wagonways through them, so that, if the necessity arose, the district could have been sealed off in about 4 hours. It was decided to build the stoppings, before leaving them, of sufficient strength to withstand the blast from any explosion that might take place in the affected area; and, with this object, a pack-buttress was built, 12 feet thick, against and on the in-bye side of the wooden stoppings. At the same time, the opening through the wooden stoppings was being reduced, and another pack-buttress, also 12 feet thick, was built against and on the out-bye side of the wooden stoppings, a small opening being left through the whole of the stopping for the egress of the men working on the in-bye side of the stopping.

These operations proceeded simultaneously, and the small openings in the stoppings were finally closed at the same time. Observation-pipes were placed in the stoppings. These stoppings, over 27 feet thick, were completed, and the district sealed off, in about 10 hours. The stoppings were afterwards faced with another wooden stopping, 4 feet thick. During the time that the stoppings were being built, smoke was generated in large volumes from the fire, and came out along the roadway in a dense cloud, and at the time that the stoppings were closed, it had reached the points D, E and F (Fig. 3, Plate V.).

The hustle (carbonaceous shale) which is usually found associated with the Bullhurst seam, is often, no doubt, the chief factor in originating gob-fires in this seam; however, in the case just described no hustle was noticed in the coal in the vicinity of the fire.

As regards working gassy seams liable to spontaneous combustion, it is now generally admitted, at any rate in this district, that the best method to pursue is to divide the workings into small districts or panels, and, if possible, to drive out in the solid coal to the boundary before commencing drifting operations, to exclude air from the goaves, and to allow them to become charged with carburetted hydrogen so as to prevent oxidation. The panel-ribs should be left sufficiently strong to prevent the possibility of air passing through them. The roadways through these ribs should be as few as convenient, and they should have effectual stoppings built in them at suitable places in the rib or barrier. Of course, there are openings through the stoppings in the roadways which are used for working the district, and material should always be kept at hand ready for sealing them up promptly if the necessity arises. The writer believes that Messrs. Rigby & Company were the first to inaugurate this system of working, about 20 years ago, and it has proved very successful at their collieries.

If this method is adopted, and the emergency arises, a district can be closed with effectual stoppings in a minimum amount of time, and at most in a few hours, which is of paramount importance in working gassy mines liable to spontaneous combustion. The size that the panels should take can only be determined by experience, after taking into consideration the conditions under

which the seam has to be worked. It may be advisable in some
cases to place double temporary stoppings in the roadways, which
must be kept open for working the district; then, if a fire occurs,
both these stoppings can be closed, which would add to the
security of the mine. Obviously there would be less chance of
double stoppings being blown out by an explosion than single
ones, and it would take very little more time to complete double
than single stoppings.

It is desirable, where conditions will allow, to work by the
pillar-and-stall method; but conditions are such, that some
seams can only be worked to advantage on the longwall principle,
or some modified form of that system.

It is essential that pack-walls should be made solid, and all
timber extracted therefrom, so as to allow the roof to settle
uniformly on them. Of course, this is necessary for successful
working by the longwall method under any circumstances.

Where practicable, it might be of advantage, alternately
and systematically to change the position of the pack-walls,
and wastes at convenient distances, or systematically to put in
dirt-stoppings across the wastes at regular intervals of, say,
30 feet or so, thus making the goaf less susceptible to pass air-
currents and minimize the chance of oxidation in the gob.

Adverting to stoppings built of wooden blocks and mortar,
the writer has found them to be cheaper, less liable to crack,
more durable, more quickly erected, and also able to resist the
pressure from the superincumbent strata much better than
stoppings built of bricks-and-mortar. Of course, no one would
ever dream of building stoppings of wood in close proximity
to a fire, or where there was the least possibility of fire coming
into contact with them. Stoppings which are supported by a
good length of stowing are not often moved by an explosion,
but stoppings built of masonry only, unless they are of great
thickness, are easily moved. Ordinary stoppings, 9 inches thick
of brickwork, are sometimes blown out by a gust of air caused
by heavy falls of roof in the workings. Stoppings intended to
resist the lateral force of an explosion should not be built less
than 30 feet thick in a roadway, say, 4 or 5 feet square; and
in roadways having a larger sectional area they should be built
much thicker.

When a gob-fire does break out in a gassy mine, sometimes it is a very critical question to decide whether the stoppings should be worked at continuously, until they are strong enough to resist an explosion; or whether, after the stoppings have been closed and air cut off, the men should be brought out of the mine for a period, such period being fixed by a calculation as to the length of time that it will take for gas to accumulate and envelop the seat of combustion. Often such a calculation is conjectural, but sometimes it may be ascertained approximately. This point can only be decided after taking into consideration the special features connected with each case. The risk in some instances, may be too great to build continuously at the stoppings, but in most cases there will be less risk in working continuously at them until they are strong enough to resist an explosion, before leaving them. Seams in which there is any likelihood of fires occurring should be laid out from the commencement of the working, as if fires were expected; this will obviate many anxious moments, and very probably will be the means of preventing heavy losses from being incurred.

Undoubtedly, the safest course to take in dealing with a fire in a gassy mine is to exclude the air as speedily as possible from the affected area; and if this can be successfully accomplished, the oxygen in the pent-up area will after a time become exhausted, and the gases generated as the products of combustion will prevent combustion from being continued.

A short time since, a writer on this subject inferred that the only successful method of dealing with a gob-fire was to dig it out. This would be a suicidal policy to adopt in dealing with fires in a gassy mine; only under very favourable conditions, when the fire is confined to very narrow limits and easily accessible, would this course be justifiable. Unfortunately very few fires occur under these favourable conditions. Even in non-gassy mines, the least expensive and safest way of dealing with a gob-fire which is in any remote part of the working is, if possible, to build it off and exclude the air.

With regard to wax-walling and the use of clay-stoppings, as practised in some districts, this method must at all times be a source of great anxiety and trouble, in consequence of the cracking of the clay.

It is not often practicable to flood areas affected by gob-fire in gassy mines, the process being too slow. Inundation can, however, be successfully applied in non-gassy mines; but, when the fire is at a high point in the workings, it may be necessary to flood nearly the whole of the mine before the water reaches the fire.   Flooding should only be adopted as a last resource, as the unwatering process is often tedious, and very costly, in consequence of the damage done to the workings by the water.

In some cases, carbonic acid and other gases may be produced in large quantities, and directed to the seat of combustion.

Many serious fires have been caused by open lights, and when we reflect on this matter it is not at all surprising. How often do we find that open lights are frequently, through the carelessness of workmen, or for convenience, placed on timber in such a manner that the flame is actually in contact with the timber? It is very probable that fires of untraced origin have been caused in this way.   When these fires do occur they cause serious trouble, and are often only extinguished after an arduous struggle, in consequence of the large quantities of smoke usually generated by the burning timber and the absence of water near the seat of the conflagration.   Fires may be caused by open lights being carelessly handled near fodder in the stables.   Obviously these risks can only be entirely avoided by using enclosed lights; but the risk may be reduced to a minimum by strict supervision and careful instructions to workmen.

As a result of explosions of fire-damp and coal-dust, fires are often produced in the workings of a mine, and when an explosion does occur a large volume of air is usually passing in the mine, which may tend to fan up any inflammable material that may have been ignited by the flame from the explosion; and a recurrence of explosion may be the result until all the available oxygen has been consumed. When the supply of oxygen is not cut off by the action of the explosion, and explosions recur, then there is no alternative but to seal up the shafts hermetically or to flood the mine.   The closing of the shafts may be effected by a strong platform or scaffolding in them, plastic clay being thrown on the scaffold and finally covered with water, so as to ensure as far as possible that no air is passing into the workings. Even when these precautions are taken, air may be drawn past the scaffold through the brick-lining of the shaft, if the brickwork is not in good condition. In some cases it is not safe to put scaffolds in the shafts, and in these instances they may be closed by tumbling débris down them.

Underground boilers and furnaces often cause fires; but, at the present time, power being so easily transmitted, and there being so many suitable fans available, it is not necessary to introduce them into mines. Still, there are many in present use, and they may continue to be used for a long time to come.

A short description of the method of dealing with some fires which originated from these causes at the Clayhole pits, Far Green, Hanley, may be of interest. The writer is indebted to Mr. Thomas Roberts for the details readily placed at the writer's disposal, given in connection with these fires.

There were two shafts, one being used as an upcast, and the other as a downcast. These pits were in close proximity one to the other, only about 24 feet separating them. The heated gases from two boiler-fires and a furnace passed up the upcast shaft.

A dumb-drift rising to the upcast shaft at an inclination of about 1 in 3, and up which about 50,000 cubic feet of air passed, joined this shaft 90 feet above the boiler and furnace-flue, at a point where the Stony Eight-feet seam intersected the shaft. A fire broke out in this seam in the roof of the dumb-drift, a short distance away from the upcast shaft. To prevent the return-air, which at times became charged with fire-damp, from being brought into contact with the fire, two brick-walls were built above and below the seat of the conflagration, for the purpose of carrying air-tubes, through which it was intended to pass the ventilation. It was found, however, that this operation had the effect of retarding the ventilation to such an extent that the workings began to fill with gas. Consequently the pipes had to be taken out. The next step taken was to build a brick-archway along the dumb-drift; all the burning material was dug out from behind the brickwork; and the cavity was filled in with solid masonry. The fire extended all round the shafts, the burning coal was removed, and a cooling-chamber was left round the shafts, through which about 1,000 cubic feet of air was passed to carry away the heat as it was generated (Figs. 4 and 5, Plate V.).

The Rough Seven-feet, Ragman, Yard and Moss seams also took fire round the upcast shaft. The burning coal was removed, in each case, round the shaft, and masonry (3 feet thick) was built round the shaft, with sand, also 3 feet thick, intervening between the masonry and the coal. A plentiful supply of water was used in dealing with these fires from the downcast shaft.

The writer can speak from personal observation of the efficacy of the steps taken in these cases to prevent a recurrence of fires in these seams, as he had charge of these pits for about 5 years. No trouble was experienced during that time, and he believes that there has been no trouble at any of these places since the fires were dealt with some 20 years ago.

It seems scarcely necessary to mention that, to prevent a repetition of the experiences just described, the precautionary measures advisable are to insulate by masonry, sand, or cooling-chambers, whichever may be most suitable and convenient, all seams of coal in any shaft, before allowing heated gases to pass up them, and also the seams of coal in the neighbourhood of the boiler or furnace-houses.

As regards the probability of timber being set on fire by steam-pipes, possibly they never become sufficiently hot to ignite adjoining timber. If the pipes are covered with a suitable non-conducting material the likelihood of fire is remote, but if they are uncovered and in close proximity to timber, the timber may be charred to such an extent as to cause it readily to absorb oxygen, which may ultimately initiate a fire. Or, if the pipes are laid along a roadway in the coal, the increased temperature may cause spontaneous combustion. Coal has taken fire near to the exhaust from an underground engine, and probably the steam may have been the primary factor in producing spontaneous combustion.

Fires may be caused by sparks from machinery dropping on inflammable material.

The Report of the Departmental Committee appointed by the Home Secretary in October, 1903, to enquire into the use of electricity in mines, having recently been published, it would

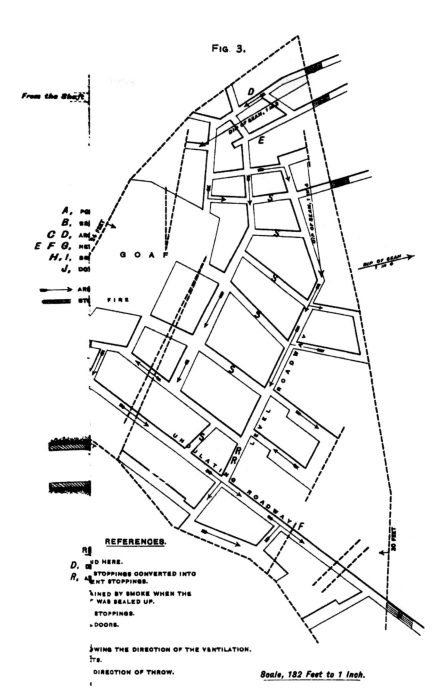

FIG. 3.

From the Shaft

A, PG
B, SR
C D, AR
E F G, NG
H, I, BR
J, DG

→ AR
━━━ ST

G O A F

F I R E

REFERENCES.

R

D, ND HERE.

R, A STOPPINGS CONVERTED INTO
NT STOPPINGS.

AINED BY SMOKE WHEN THE
WAS SEALED UP.

STOPPINGS.

DOORS.

WING THE DIRECTION OF THE VENTILATION.
TS.

DIRECTION OF THROW.

Scale, 182 Feet to 1 Inch.

be presumptuous on the writer's part to attempt to deal with this cause of fires. The report deals exhaustively with the dangers arising from the application of electricity in mines. It is, however, a matter for congratulation that the Committee's suggested Code of Rules is so liberally drawn up, as not to prevent the extension of the application of this most useful and convenient power in mines.

Many serious fires have been occasioned as the result of shot-firing; but, since the introduction of safety-explosives and shot-firing by electricity, they have been rarer in recent years.

Numerous devices and mechanical appliances have been invented, with the object of eliminating or rendering quite safe this necessary operation in mining, and if the best appliances are used and ordinary care is exercised by the persons carrying out these operations, the chance of flame being started is very remote indeed.

———

The PRESIDENT (Mr. A. M. Henshaw) said that there were few pits in North Staffordshire where spontaneous combustion had not been experienced at some time. In some parts of the district, it was almost a daily source of worry to those in charge of collieries.

Mr. JOHN CADMAN said that Mr. Lawton suggested that "thick seams . . are more friable, and usually there is greater rock-pressure on the coal," but he doubted this very much. He thought that Mr. Lawton had been somewhat unfortunate in quoting the particular theory that he had quoted, as it appeared to be one put forward some 5 or 6 years ago as a new theory, and one which brought forth considerable criticism. So far as the physical source of heat in the production of spontaneous combustion was concerned, one could not help noticing the enormous force exerted in compressing the gob; and also the high absorptive power of some coals, which not only generated heat by such absorption but as secondary agents were able to give up their absorbed oxygen, in order to produce chemical changes in the goaf-constituents.

With reference to the chemical source of heat, Mr. Lawton had mentioned pyrites, but he (Mr. Cadman) thought that pyrites played a much more important part than was generally supposed. It was an undisputed fact, as stated by Mr.

Lawton, that pyrites was known to oxidize, and in that oxidation, sulphuric acid was produced; and this, on the addition
of water, was capable of producing a considerable amount of
heat.   Again sulphuric acid, so formed, was capable of generating further heat by its action on other compounds (particularly
carbonates) in the goaf.   After or during a gob-fire, one frequently observed a deposition of sulphur along joints, etc., and
this might be produced by the formation of thiosulphates, or by
direct sublimation from the decomposition of pyrites.

Many fires in North Staffordshire had been traced to the
presence of moisture.   For example, an area of the Bullhurst
seam had been worked for a considerable time without any
appreciable signs of heating, but the introduction of moisture
had the effect of starting combustion.

The finding of matches underground was a very serious
matter, and managers should use every effort to prevent such
carelessness.   He (Mr. Cadman) believed that they were taken
down, in the majority of cases, by men ignorant of the fact and,
on finding them, the men had hidden them.   In many cases,
such matches were found behind timber, etc., but he (Mr.
Cadman) ventured to suggest that where one box was found
many others had not been discovered.   Matches, taken into a
mine, might become covered up in a dry gob, and it was not unreasonable to imagine the possibility of subsequent falls causing
the matches to ignite and initiate a gob-fire.

Mr. J. T. STOBBS suggested that the localities of the sections
should be stated in the paper.   The Great Row coal-seam was
stated to be 9½ feet thick, but it was not of that thickness over
the whole district.

Mr. T. E. STOREY said that, when he was manager of the
Leycett collieries, there were often gob-fires in the Ragman
seam.   He had worked the Bullhurst seam in the Kidsgrove
district, and he found that, where the seam was dry and free
from moisture, there were no gob-fires, but where moisture was
present gob-fires were started.

The PRESIDENT (Mr. A. M. Henshaw) proposed a vote of
thanks to Mr. Lawton for his paper.

The resolution was approved, and the discussion was
adjourned.

THE NORTH OF ENGLAND INSTITUTE OF MINING AND MECHANICAL ENGINEERS.

GENERAL MEETING,

HELD IN THE WOOD MEMORIAL HALL, NEWCASTLE-UPON-TYNE,
APRIL 9TH, 1904.

MR. W. O. WOOD, PRESIDENT, IN THE CHAIR.

The SECRETARY read the minutes of the last General Meeting, and reported the proceedings of the Council at their meetings on March 26th and that day.

The following gentlemen were elected, having been previously nominated:—

MEMBERS—

MR. JOHN BIBBY, Mine-manager, Harrietville, Victoria, Australia.

MR. HENRY MATTHEWS BULL, Manager, Bengal Coal Company, Limited, Rayhara, East Indian Railway, Palaman District, India.

MR. CECIL AUGUST ESMARCH, Electrical Engineer, 13, Westgate Road, Newcastle-upon-Tyne.

MR. STENER AUGUST FANGEN, Mining Engineer, Kroken, No. 7, Bergen, Norway.

MR. GATHORNE JOHN FISHER, Civil and Mining Engineer, Club Chambers, Pontypool, Monmouthshire.

MR. EDWARD GRIBBEN, Mine-manager, Harrietville, Victoria, Australia.

MR. THOMAS GRIFFITH, Agent and Director, Maes Gwyn, Cymmer, Porth, near Pontypridd, Glamorganshire.

MR. WILLIAM WALKER HOOD, Mining Engineer, Glyncornel, Llwynypia, Glamorganshire.

MR. RIENZI WALTON MACFARLANE, Mining Engineer, c/o Col. S. Beckett, 10, Treeland Road, Ealing, London, W.

MR. ROBERT MCINTOSH, Assistant Inspector of Mines, Dunedin, New Zealand.

MR. EDWARD DAVID MILES, Mining Engineer and Mine Superintendent, Charters Towers, Queensland, Australia.

MR. THOMAS THOMSON RANKIN, Principal of the Mining and Technical College, Wigan.

MR. PAUL SCHWARZ, Coke-oven Engineer, 4, South Avenue, Ryton-upon-Tyne.

MR. PETER B. STEVINSON, Engineer, 4, Warwick Street, Heaton, Newcastle-upon-Tyne.

MR. JAMES STEWART, Colliery Engineer, Garesfield Colliery, Rowlands Gill, Newcastle-upon-Tyne.

MR. WILLIAM STEWART, Colliery Manager, Milnthorp House, Sandal, Wakefield.

MR. ALFRED SIMEON TALLIS, Mining Engineer, The Rhyd, Tredegar, Monmouthshire.

MR. OTTO WITT, Engineer, Director of Altens Kobbergruber, Kaafjord, Finnmarken, Norway.

ASSOCIATE MEMBER—

MR. RAYMOND WILLIAM SCHUMACHER, P.O. Box 149, Johannesburg, Transvaal.

ASSOCIATES—

MR. THOMAS BURT, Overman, Engine Square, Washington Colliery, County Durham.

MR. JOHN COCKBURN, Under-manager, Fatfield Road, Washington, County Durham.

MR. JOHN JOSEPH MARSHALL, Back-overman, 14, Fairford Terrace, Stanley, R.S.O., County Durham.

MR. WILLIAM EDWARD SWAN, Mechanical Engineer, Washington Colliery, County Durham.

SUBSCRIBER—

MESSRS. MAVOR & COULSON, LIMITED, Electrical Engineers, 47, Broad Street, Mile-end, Glasgow.

DISCUSSION OF MR. J. M. MACLAREN'S PAPER ON "THE OCCURRENCE OF GOLD IN GREAT BRITAIN AND IRELAND."[*]

Mr. R. McLAREN (H.M. Inspector of Mines) wrote that Mr. Maclaren's paper was a valuable contribution to the *Transactions,* and as part of it related to the East of Scotland mines-inspection district, he would offer a few observations.

The writer states that "the gold of the Leadhills area is found in the streams, into which it has been washed from a gravelly clay, locally known as 'till,' which lies disposed on the slopes of the hills,"[†] and while this is so, it is perhaps of more importance to know that, in all the veins where galena has been discovered, gold has also been found. In all, 25 veins have been worked more or less in the area, and metal has been found in most. At the present time, there are two principal veins worked for lead-ore: the workings on the Brown vein have attained a depth of 1,320 feet, and on the Brow vein, to about half that depth. An analysis of the ore recently gave, in addition to a

[*] *Trans. Inst. M.E.,* 1902, vol. xxv., page 435.

[†] *Ibid.,* page 469.

large percentage of lead, about 4 ounces of silver. On the Lowther hills, traces of gold were found in stone, and a mine for prospecting was driven; but, failing to find either gold or lead, it was abandoned. It is the opinion of experts in the district, founded on close observation, that gold is to be found on the outcrop or "top" of the lead-veins, and that the gold is washed off and carried to the streams, where it is deposited. All the streams in the district have at some time yielded fine gold. The veins now being worked yield no trace of gold, which indicates that as depth is attained the gold-content disappears.

In many respects, Sutherland resembles Leadhills, both being in the same geological formation. Gold is found in the streams, and it is apparently brought from the surrounding hills. If further search was made it would probably be found that the gold was produced from the outcrop of some mineral-veins. These veins, like those in Leadhills, may consist of lead-ore, and quite possibly of sufficient richness to be payable. It does not appear that any special effort has been made to thoroughly prospect the hills near Kildonan and Gordon-bush to ascertain what, if any, minerals they contain.

Mr. H. M. CADELL wrote that the occurrence of gold in Scotland and especially among the crystalline schists of the Highlands, had frequently been an interesting subject of discussion, and like others who had visited gold-bearing countries he (Mr. Cadell) had often prospected on a small scale at home, where rocks are found similar to those of our colonial gold-fields. If no gold-bearing reefs had been discovered in the Highlands it was not because they had not been looked for, and, he (Mr. Cadell) feared that the prospects of success in this direction were not promising. He was well acquainted with the Kildonan "gold-field" and had washed small nuggets from the gravel there as an amusement, but as a matter of business he would say "Don't" to anyone bent on further digging at the Bail n'Oir. One reason why the Scottish glens were not likely to contain much auriferous alluvium was this:—During the Glacial Period the whole country was scoured clean, and all the rotted rock which might have contained the oxidized caps of gold-veins, as well as all the ancient Tertiary alluvial deposit in the valleys, was swept away by the vast ice-sheets

and glaciers, under which the country was deeply buried for
many thousands of years.   When the ice retreated it left the
land for the most part plastered over with a mass of drift
or boulder-clay, made up of rocks, soil and detritus, squeezed
and mixed together tumultuously and without any relation to
the specific gravities of the respective constituent materials.
If therefore any gold was derived from the ground-up rocks in
the boulder-clay, it must have been finely disseminated through
the mass, and could not be profitably separated from it.   Since
the Glacial Period, the streams and rivers had been busy washing
away the Glacial deposits and, to some extent, re-arranging
them in the valleys, according to their size and weight.   There
had, however, not yet elapsed nearly enough time for all the
Glacial drift to be thus washed down, sifted out and turned
into alluvium, and therefore it was useless even to expect to
find any very extensive alluvial gold-deposits north of the
Tweed.   No doubt gold occurred in Scotland as well as in other
countries, but for some unknown geological reason all the
evidence hitherto tended to prove that it was too finely disseminated to be worth working on a practical scale.   He (Mr.
Cadell) thought that Mr. Maclaren's paper contained an interesting and useful summary of our knowledge of an attractive
subject, but at present the interest seemed rather of an academic
and historical than of a practical nature.

Mr. J. T. Robson (H.M. Inspector of Mines, Swansea) wrote
that since the period referred to by Mr. Maclaren (1889 to 1891),
when the Ogofau gold-mine was worked by the South Wales
Gold-mining Company, one more attempt had been made to
work the mine.   This was in 1898, but the attempt was soon
given up; and in the early part of the following year there
was no one at the mine, except a caretaker.   No return of
mineral was made, and, probably, no gold was obtained from the
small quantity of quartz worked during 1898.   Since then, he
had not heard of any further attempt being made to work the
mine.   The Ogofau mine was the only one in South Wales from
which gold had been obtained during the past 17 years.

Prof. Grenville A. J. Cole (Royal College of Science for
Ireland, Dublin) wrote that, considering the occurrence of gold

in geyser-pipes and in peridotites, rocks that were presumably of deep-seated origin, he suspected that its presence in quartz-veins in British rocks indicated a deeper source than the immediately-surrounding country.

The nomenclature of the Silurian strata differed in Plates XVII. and XIX.; in the former it was that of the older Cambridge school, and in the latter that peculiar to Prof. E. Hull. The rocks of Rhobell-fawr (Plate XVII.) shown as of Bala age, were once classed as of Arenig age, but they were now known to lie on a Tremadoc horizon.

Mr. G. H. KINAHAN (Dublin) wrote that Mr. Maclaren ignored the northern portion of the Gold-mine valley district (Plate VI.); that was the area drained by the northern tributaries of the Darragh water or Aughrim river, the Macreddin stream and the Ow. The first stream was successfully worked for gold, by streaming in its head-waters, while there was a length of over 3 miles of untried deep diluvium between the workings and its junction with the Darragh water. In the gravel of the Ow, gold had been proved; while there was most successful streaming in its tributary, the Mucklagh brook. Mr. F. Acheson, after he had prospected the East Wicklow gold-field, considered from his previous Australian experience that the Ow valley was the best place to begin his adventure. He therefore diverted the Ow stream into the Roddenagh wood, by making a dam and a canal. He then, on the colonial system, in the lot laid out, had all the overbaring above the " pay-dirt " removed, intending to make one washing of the auriferous black sand. The fates, however, were against him: he had everything prepared to begin washing, when a " plump " or water-spout fell on Lugnaquillia, thus forming a spate in the river Ow, that broke down the dam and obliterated the working. From the Mucklagh brook to the Darragh water there were 6 miles of the bed of the Ow left unworked. Mr. Acheson subsequently did more prospecting than actual mining: he sank a shaft on a green tuffose bed that was similar in appearance to some of the Australian gold-bearers. Here again, the fates were against him, as a plump fell on Croghan Kinshelagh that flooded the pit. About this time, Mr. Acheson had perfected a theory, on the same principle as that followed by gold-seekers in Brazil,

in connection with the leads, which he calculated ought to
join and bunch; to find this bonanza, he devoted his explora-
tory efforts. Consequently he did very little at legitimate
streaming.

It was a mistake to believe that Mr. Thomas Weaver stopped
his works down stream, because the overbaring became heavy,
and the leads poor.   He and his colleagues were ordered
to take away all gold down to a limited depth, so that the
natives could not get it; and none of his workings were more
than 12 or 15 feet deep.   He went down the Coolbawn stream
until the overbaring became heavy; then he began below where
the drift-head was light, and worked up stream, till he stopped
for a similar reason, leaving betweeen the two workings about
3 miles of unworked ground.   In the Gold-mine river, about
¼ mile below and north-east of Ballinasilloge ford, there were
two leads, that to the south-east being shallow and the other
deep.   Mr. Weaver followed the first, leaving the second to be
partially worked by the Crockford and Carysfort Mining
Companies, and it was here that Mr. Crockford lifted a nugget
weighing 11 ounces.   Between the lowest works on this stream
there was over 1 mile of deep ground before it joined the
Darragh water.   From the already mentioned Coolbawn stream
to the river Ovoca, at Woodenbridge, there were over 7 miles of
unworked ground of the Darragh water, although at Ballycoog-
steps and Ballintemple auriferous gravel had been proved: there
should also be "dry gulches" and "shelf"-placers.*   While
these remained unexplored, it seemed rash to state that the gold
of East Wicklow was probably exhausted.

The mines of the Ovoca valley, also, should not be overlooked.   To the east of the river were the auriferous lodes at
Kilmacoo and Connary, that drain into the Ovoca by the
Shroughmore burn; next to them were the Magpie and Yellow
Bottom mines of Eastern Cronebane, both also auriferous.
Farther west were West Cronebane and Tigroney; in these, also
in Ballygahin and Ballymurtagh, to the west of the Ovoca,
the main or south lodes had long since lost their backs and
gossans by denudation; these, it might be suggested, were

* "The Possibility of Gold being found in County Wicklow," by Mr.
G. H. Kinahan, *Scientific Proceedings of the Royal Dublin Society*, 1883, new
series, vol. iv., page 39.

more or less auriferous because the Ballymurtagh lode was
in part at least so.   Where did the denuded portions go?
Probably southward with the ice; and partly to form the present
diluvium in the Ovoca valley.   We therefore had this detritus
and the drainage of the mines to generate gold in a river-bed over
8 miles long, from Shroughmore to the sea.   Gold had already
been found a short distance south of the Ovoca railway-station,
in the gravel of the reach left dry, when the river was diverted
during the construction of the Dublin, Wicklow and Wexford
railway.   Here it might be suggested that this now dry portion
of the old river-bed seemed a legitimate field for a syndicate to
spend a few hundred pounds in exploration-work.

As to the genesis of the gold, in the Gold-mine valley district,
it appeared nearly impossible that it could have come from the
East and West Ovoca mines; and he (Mr. Kinahan) would look
for the sources nearer afield.

In his (Mr. Kinahan's) paper, just now referred to, it was
suggested that if the gold of the Gold-mine valley had an existing " mother-rock " or reef, it ought to be found somewhere near
the upward limits of the gold in each stream; that was, in
Ballinasilloge to the north-west, and Knockmiller, Monaglogh,
and Mongan to the south-east.   It however, seemed improbable
that the source was in a reef, especially a quartz-vein; although
the startling statement had recently been made, that a quartz-
vein had been found that returned 8 ounces of gold to the ton.
Such a statement must be received with great caution, as the
vein was said to be in a most conspicuous place, that was, at the
mouth of Mr. Weaver's tunnel.   Now Mr. Weaver and his
colleagues, also Mr. Mallet, the agents of the Carysfort and
Crockford Mining Companies, and various other men of experience, had carefully examined all the quartz of the valley
and its neighbourhood; it seemed therefore a sort of miracle,
if this vein should have escaped them.   Little bits of quartz, on
which the gold seemed to have grown, had been found, but
fragments of quartz containing gold had not come to his
knowledge.

Mr. Maclaren suggested that " the present auriferous deposit, in fact, represents the concentrates of a pyritous lode,
which has suffered degradation."*   For this idea, there ap-

* *Trans. Inst. M.E.*, 1903, vol. xxv., page 497.

peared to be facts that would tend to such a conclusion. On the
ridge, where the Macreddin stream and the Mucklagh brook had
their source, there was a large ramp or bed of gossan: this would
seem to indicate the presence of a pyritous lode, which, however,
had not been found, although an open-cast was made across the
accumulation.   A trace of gold was said to have been found
in this gossan.   In the upper portion of the Coolbawn stream,
there was no black auriferous sand, but when it crossed the slope
below the Moneyteigue lode (iron and pyrites) the black sand
appeared with both gold and tin.   In the upper portion of the
Gold-mine valley stream no gold occurred, but when one came
to the shed from the Ballinasilloge and Ballycoog pyritous
lodes the gold and tin immediately appeared.   In the upper portion of the Monaglogh and Killahurler streams, there was no
gold; but lower down were the Lyra placers.   In connection
with these streams, no mineral-lode had been recorded; although in the valley of the adjoining townland of Mongan,
there was an excessive ramp of gossan strongly suggesting a
pyritous lode.   In the valley below Lyra, there was little or
no drift, but to the north-east in a dry gulch-placer, there was
large gold similar to that at Lyra; while farther north, at Clonwilliam slate-quarry, while making the mill-pond, some gold
was turned up.   It should be mentioned that some men who
had worked in California, Australia and the Transvaal believed
that a mass of green tuffose rock, similar to that on which Mr.
Acheson put down his shaft, was a gold-bearer.

The above-mentioned lodes and mineral accumulations might
supply the gold and black sand; but where did the tin come
from? As the granite of Croghan was of the same age as the
newer granite of Cornwall and Devon, it had been suggested
that the granite might have some connection with the tin: but
against this was the fact that the tin disappeared as the granite
was approached.   In the diggings, where there was tin, large gold
was found, but if the tin disappeared the gold also went or
became small (eyesills).

There was another feature in the area that might require consideration.   Mr. Acheson's drives in connection with the "Red
Hole," and Mr. Weaver's cut farther south-west at the road,
appeared to suggest that originally valleys were denuded in the
boulder-clay, much wider than the present ones; and that these

were filled by a sub-glacial accumulation. Mr. William Mallet suggested that the filling was due to " mud débâcles," the slush formed during the final melting of the ice and snow, that slid into and down the valleys. That such movements did take place was proved by the bending over of the tops of the slates and other rocks. This, by the American geologists, was called the " Champlain epoch " or " the time of the floods proceeding from the melting ice." These, as in Ireland, covered the slopes with rearranged drift.* This newer drift contained more or less angular shingle, while the black sand and accompanying stuff had in them rounded fragments. The nugget and grain-gold were usually abraded, but some were frosted as if gold had grown on them, while most of the eyesills seemed to have grown like the moss and dendritic gold in some of the Californian placers. It might, therefore, be suggested that the abovementioned pyritous lodes were denuded during the Glacial Period, and that the gold-sands and their adjuncts were washings from the portion of the Glacial Drift that was removed when the valleys were originally excavated. Such a theory, however, could not be accepted without due consideration. There was very little evidence as to the direction in which the ice was moving. To the north of the Darragh water, north and south of Knocknamohill house, the striæ pointed to the south-east, and south of the Gold-mine valley about 1 mile east-north-east of Slievefoore they pointed in the same direction, but at the summit of the hill they ran nearly eastward. The south-easterly direction would answer for the mineral detritus in the Macreddin and the Gold-mine valley areas, but not for the Coolbawn valley. In the latter case, although there were no proofs, it was possible that the line of hills or spur extending north-eastward from Croghan Kinshelagh had diverted the trend of the ice-flow to a south-westerly direction, in which case the mineral-detritus from the Moneyteigue lode would have been carried into the Coolbawn valley. If this glacial theory was correct, why was the gold not more widely distributed? While, if the gold-sands were adjuncts of the " Champlain epoch " why was the gold-sand below the recent drift, and not scattered through it?

In county Meath, there was said to be a quartz-lode carrying

---

* *Report of the Geological Exploration of the Fortieth Parallel*, 1878, vol. i., " Systematic Geology," by Mr. Clarence King, page 459.

gold, at the contact of Carboniferous Limestone and the Ordovician rocks. A specimen obtained by him (Mr. Kinahan) had
across it an auriferous streak, ⅜ inch wide, having in it a pure
vein from a string to ¼ inch wide. The gold was peculiar, carrying with it grains of blende, a very rare associate of gold.
Prof. Dana only mentioned two localities. As circumstances
had prevented the place from being visited and the find guaranteed by a competent authority, the locality for the present would
be left unnamed.

Mr. Maclaren ignored the West Wicklow gold-mines, in the
eastern plain of the Liffey, recorded in the *Annals of the Four
Masters*, also the auriferous lodes of Cork; one need not therefore say more about them, except that there seemed to be a
future for south-west Cork.

If the gold of the Gold-mine valley came from the detritus
of the mineral-lodes in the Carysfort mining setts, there ought
to be in the tract of country north-west of the valley, between it
and the mineral-lodes, gold-placers besides those in Ballintemple
stream and in the north-western branch of the Gold-mine valley
stream. It should be mentioned that in the gossan of the lodes
at Ballinasilloge, a trace of gold was detected. The approximate
position of the lodes of the Carysfort mining setts and the
known gold-placers are shewn in Fig. 1 (Plate VI.). According to the conclusions arrived at by Mr. J. M. Maclaren, the
gold-ores of the British Isles have their genesis in the older
rocks. In regard to the gold of Eastern Wicklow, he (Mr.
Kinahan) had come to a similar conclusion. The mineral-
lodes, for reasons given elsewhere, are probably of either Carboniferous or Devonian age, the minerals being leached out
of the associated Ordovician rocks. In south-west Cork, however, the auriferous lodes were post-Devonian, the ores being
leached out of the associated Devonian rocks of the country.
If there were an auriferous quartz-lode in county Meath, the
lode ought to be post-Carboniferous, but the gold might have
come from the Ordovician rocks, the lode being at the contact
of both.

Prof. J. P. O'REILLY (Dublin) wrote that, in matters of
mining, there was, however, always the speculative or prospective side, and perhaps sufficient justice was not always done to

The Institution of Mining Engineers.
Transactions 1903-1904.

Vol. XXVII, Plate VI.

To illustrate Mr. G. H. Kinahan's "Notes on the Occurrence of Gold in Wicklow."

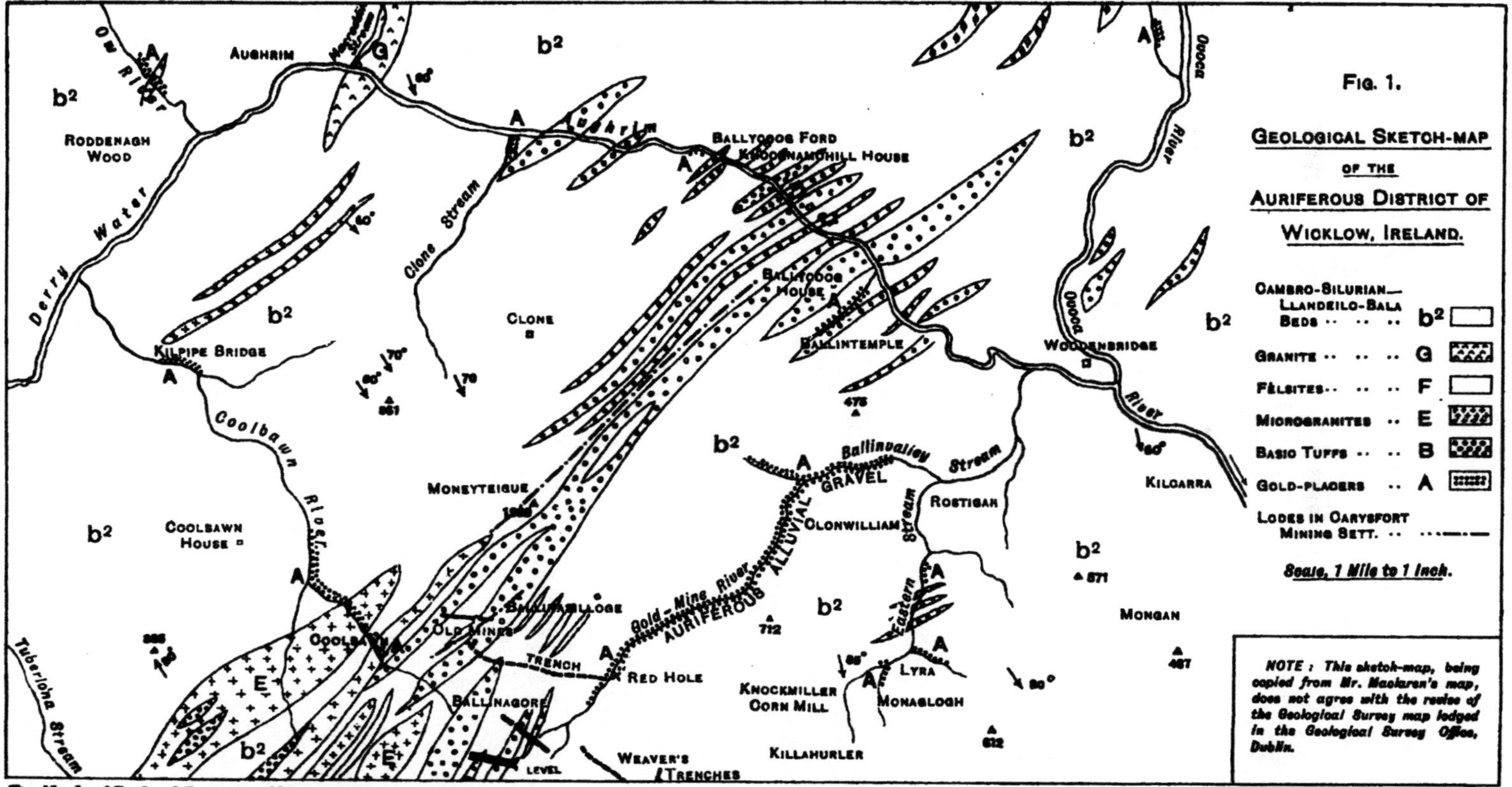
Fig. 1.
Geological Sketch-map of the Auriferous District of Wicklow, Ireland.
Cambro-Silurian—Llandeilo-Bala Beds b2
Granite .. .. .. G
Felsites .. .. .. F
Microgranites .. E
Basic Tuffs .. .. B
Gold-placers .. A
Lodes in Carysfort Mining Sett. ..
Scale, 1 Mile to 1 Inch.
NOTE: This sketch-map, being copied from Mr. Maclaren's map, does not agree with the revise of the Geological Survey map lodged in the Geological Survey Office, Dublin.
Aughrim
Roddenagh Wood
Derry Water
Kilpipe Bridge
Coolbawn River
Coolbawn House
Clone Stream
Clone
Moneyteigue
Ballycoog Ford
Knocknamohill House
Ballycoog House
Ballintemple
Woodenbridge
Ovoca River
Kilcarra
Mongan
Rostigah
Clonwilliam
Ballinvalley Stream
Alluvial Gravel
Auriferous Gold-Mine River
Red Hole
Trench
Old Mines
Ballinagore
Level
Weaver's Trenches
Knockmiller Corn Mill
Killahurler
Lyra
Monaglogh
Eastern Stream
Tuberlona Stream

The North of England Institute of Mining & Mechanical Engineers.
Transactions. 1903-1904.

Andw. Reid & Compy. Ltd. Newcastle upon Tyne.

Vol. LIV., Plate IV.

the " prospectors," the hardy adventurers of " perfervid imagination," the successors of the " Argonauts," the " Cossacks " of
science and discovery; and speaking on their behalf he (Prof.
O'Reilly) would wish to point out that there was still something
to be said in favour of further search for gold in Great Britain,
since, carefully examined and compared with other countries,
the rocks and their associated minerals point in that way.
Nobody had attempted to classify and enumerate the quartz-
reefs and lodes of this country, still less to give some analyses
or assays of them, and yet they were very abundant, as might be
learned by the perusal of the *Memoirs* of the Geological Survey.
Again, the association of diorites with gold was, he thought,
one of the things that might be considered as certainly frequent
if not constant. He was much struck on reading Sir Richard
Griffith's opinion on the lead-veins at the contact of the granite
with the Carboniferous formation in the counties of Dublin and
Wicklow; Sir R. Griffith said that " the whole of the true veins
which have been discovered in the county Wicklow are close to,
or within a short distance of, the junction of the granite and the
stratified rock.   The number of these veins is probably immense,
as there is no instance of any precipice along the eastern line of
junction, in which veins containing lead have not been discovered,
and we may reasonably infer that veins are equally numerous
in those parts, where the rock is not exposed to view. . . . .
From this fact, it may be concluded that in the whole line of
the granite-boundary we may expect to find lead-mines."*
Sir. R. Griffith was an experienced engineer and geologist, and
subsequently drew up the excellent and useful geological map
of Ireland known by his name; and he became famous
as head of the Land-valuation Department.   Now, in the
vicinity of Dublin, at Dalkey, not only is there an old and
abandoned lead-mine, but ribs of galena occur in the granite
in such a manner as to lead him (Prof. O'Reilly) to presume that
the gut or narrow strait between Dalkey Island and the shore,
so marked in its direction, really contains the main lode: but,
of course, under almost impossible conditions as regards exploration and working. Of course, argentiferous galena is, in
its way, a gold-ore, hence the interest of the whole question.

* *Fourth Report on Irish Bogs*, pages 172 and 173.

DISCUSSION OF MR. T. ADAMSON'S PAPER ON "WORKING A THICK COAL-SEAM IN BENGAL, INDIA."[*]

Mr. GEORGE A. STONIER (Chief Inspector of Mines in India) wrote that he agreed with the remarks of Mr. R. R. Simpson and Mr. James Grundy[†] on the methods of working at Giridih described in Mr. Adamson's paper. The security of the miners under the systems of working and the very careful supervision in vogue at these collieries was shown by Table I., which gave statistics for the collieries of the East Indian Railway Company and the coal-fields of Bengal, during the past three years.

TABLE I.—DEATHS BY ACCIDENTS IN INDIA.[*]

Collieries.	Years.	No. of Deaths by Accidents.	No. of Persons employed daily per Death	Annual Output per Death.	Output per Annum.
				Tons.	Tons.
East Indian Railway Company	1900	2	3,672	264,187	528,374
	1901	3	2,626	187,056	561,169
	1902	2	3,950	306,894	613,789
Bengal Coal-fields	1900	41	1,776	121,426	4,978,492
	1901	42	1,896	135,806	5,703,876
	1902	44	1,893	142,484	6,269,294

* The details of the five Bengal coal-fields are given in Mr. Stonier's reports for the years 1901 and 1902.

While in Australia, he (Mr. Stonier) made the following notes on the method of working a thick seam at the Australian Agricultural Company's Colliery, at Newcastle, New South Wales. The Bore-hole seam is 17 to 20 feet thick, contains 15 feet of coal, and has a dip of about 1 in 20, which varies in direction. The coal is bituminous and fairly hard. The cleat is well developed, and runs in one direction throughout the mine. Faults are uncommon, and dykes are few in number. The amount of cover varies from 230 to 320 feet. Naked lights are used by the workmen. The system of working the seam is divisible into two parts:—(a) The formation of the pillars; and (b) the extraction of the pillars.

(a) In the first lift (Fig. 7, Plate VII.), the headings, 12 feet wide, are driven more or less parallel to the cleat of the coal; and

* *Trans. Inst. M.E.*, 1903, vol. xxv., pages 10, 192 and 396; vol. xxvi., page 19; and vol. xxvii., page 10.

† *Ibid.*, vol. xxv., pages 192 and 399.

bords, 18 feet wide, are opened out at right angles to the cleat: leaving ribs, 36 feet wide, which are made into pillars, 210 feet long, by cut-throughs, 12 feet wide. The holing is made in the bottom-section of the big tops, immediately above the morgan (a dark bituminous shale), and the big tops, 4 inches coal, and the little tops (or in all 7½ feet of coal and 9 inches of shale) are excavated. The roof, formed by the parting, 1 inch of sandy shale and overlying coal, is generally sound and requires very little timber. The coal is loaded into tubs, and the shale is left temporarily in the bords and headings; but before commencing the second lift, this shale is stowed in bords from which the bottom coal has been won. The air-current is passed to the working-faces along passages, bratticed from the bords in the usual manner.

The miner receives the district hewing-rate of 4s. 2d. per ton for round coal, with an additional 1d. per ton for each inch of thickness of band excavated and ½d. per ton for small coal: the latter passes through screens, with bars ¾ inch apart, and amounts to 20 or 25 per cent. of the total amount of coal won.

In the second lift, the bottom coal is won in the bords and headings. Beneath the coal worked in the first lift, there are 2 feet of dark bituminous shale, covering 3 feet of coal: the whole thickness of 5 feet is excavated. As the face of the heading advances, the excavated coal is loaded into tubs, and the shale is packed on the upper side, leaving space for the rails on the lower side of the road: this space is ultimately filled with shale from the bords on the rise side, which are worked in the same manner, but the rails are placed in the centre of the bord (Figs. 1, 2 and 3, Plate VII.).

In the third lift, the top band (4 feet 2 inches thick) is worked in the bords for their whole length, except 15 feet back from the roads, and the roof of shale is allowed to fall. The shale from the preceding lifts generally fills the bord to within 6 feet of the top band, so that the latter is within easy reach.

(b) Only the upper thickness of 11 feet 2 inches is won from the pillars, and no attempt is made to mine the bottom coal, 3 feet 5 inches thick, which amounts to 8½ per cent. of the total quantity in the seam. A commencement is made 15 feet from a heading, and the pillar (210 feet long and 36 feet wide) is worked at right angles to the cleat for a width of 18 feet and a length of 90 feet, the rails being kept close up to the face and the coal loaded directly into the tubs. Round props, 4 inches in diameter, are set 1½ feet apart to support the roof. The remaining width, 18 feet, is worked back to the 15 feet stump at the heading, and the timber is withdrawn. The other half of the pillar is worked from the adjacent heading in a similar manner (Figs. 4, 5 and 6, Plate VII.).

The final operation is the removal of the stumps, 15 feet wide (one on each side of the heading), and the top coal along the headings and in the bords adjacent to the headings.

It will be seen that an attempt is being made to win 91 per cent. of the total amount in the seam, but the expenses are high compared with the costs in Indian collieries.

DISCUSSION OF MR. W. LECK'S PAPER ON "AMBU-LANCE-INSTRUCTION AT MINES."*

Mr. George A. Stonier (Chief Inspector of Mines in India) wrote that in India, although hospitals and medical attendance were provided in all the mining-fields, instruction in first-aid had only been introduced at the East Indian Railway collieries in Giridih, Bengal presidency, where 8,000 persons were daily employed. The medical staff cost £472 per year, and consisted of one assistant-surgeon (a native) and five native doctors, who were housed at different places, and were available at short notice for sickness and accident-cases. They also instructed the deputy overmen and sirdars (corresponding to back-overmen and deputies in England) in first-aid and held a weekly drill, so that the acquired knowledge was not forgotten. The special rules of the collieries referring to first-aid were as follow : —

28.—*Duties of Sirdars.*—Shall have a fair knowledge of first-aid to the injured, and shall for that purpose attend drill and instruction in this matter as ordered by the managers.

29. – Shall, in case of accident, send at once for the nearest European, the nearest native doctor, and the ambulance and accident-boxes.

161.—*Duties of Assistant-surgeon.*—Shall inspect the accident-boxes and ambulance-appliance, whether placed above or underground, frequently, and report the state of them to the superintendent, and see that the native doctors carry out the duties laid down for them.

* *Trans. Inst. M.E.*, 1903, vol. xxv., page 354 ; and vol. xxvi., pages 154 and 258.

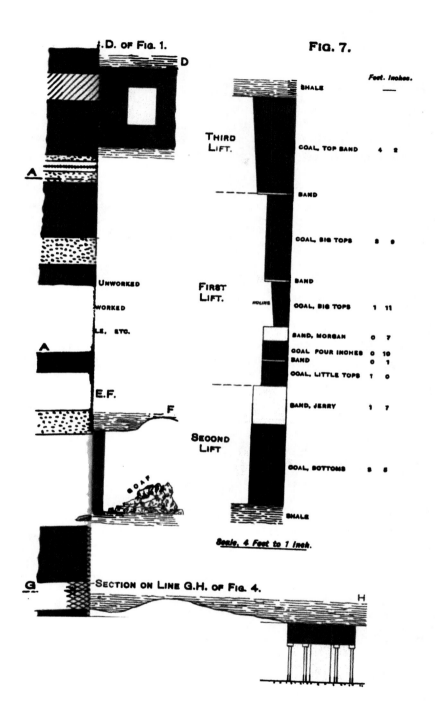

C.D. OF FIG. 1.

FIG. 7.

| | Feet. Inches. |

SHALE

THIRD LIFT.

COAL, TOP BAND 4 2

BAND

COAL, BIG TOPS 2 9

BAND

FIRST LIFT.

COAL, BIG TOPS 1 11

BAND, MORGAN 0 7

COAL, FOUR INCHES 0 10

BAND 0 1

COAL, LITTLE TOPS 1 0

E.F.

BAND, JERRY 1 7

SECOND LIFT

COAL, BOTTOMS 2 5

SHALE

Scale, 4 Feet to 1 Inch.

UNWORKED

WORKED

LE. ETC.

GOAF

SECTION ON LINE G.H. OF FIG. 4.

The North

162.—Shall see that sirdars are properly instructed in first-aid, and kept
efficient in the same.

165.—*Duties of Native Doctors.*—Shall promptly attend all accident-cases.
underground or on the surface, as the case may be, and all cases of serious illness,

167.—Shall instruct the sirdars in first-aid, and drill them regularly.

Accident-boxes are kept on the surface and underground,
and they contain:—4 ounces of carbolic oil, 4 ounces of ferric-
perchloride liquor, 1 ounce of chloroform, 1 ounce of opie tinc-
ture, 2 ounces of ammonia carbonas, 2 ounces of boric acid,
1 ounce of iodoform, 2 skeins of ligature (hempen and silk),

FIG. 1.—ASSISTANT-SURGEON TESTING THE RESULT OF AN
IMPROVISED TOURNIQUET ON THE FEMORAL ARTERY.

2 suture-needles, 2 bundles of tape, 2 dozen safety-pins, 2 ounces
of lint, 12 feet of mulmul (muslin), 2 perchloride towels, 1 piece
of soap, 4 ounces of emp. resina, 3 feet of plaster-cloth, 2 sponges,
1 minim measure, 1 pound of perchloride cotton-wool, 1 tour-
niquet-screw, and bandages (one, 6 inches wide; three, 4 inches;
six, 3 inches; twelve, 2½ inches; four, 1½ inches and three
double-headed).   In a separate bundle are the following

splints:—1 pair of clines legwood, 3 pairs of common arm,
1 liston-wood.

The course of action in cases of accident is as follows:—The
injured man is at once moved to a safe place near by, and first-
aid is rendered by a sirdar or deputy-overman (both of whom are
natives) with bandages improvised from turbans (*pagri*) and
pick-handles; while three messengers are bringing (1) an accident-box, (2) a native doctor, and (3) the nearest European. The
native doctor, on his arrival, puts on fresh bandages, if neces-

Fig. 2.—Sirdar and Deputy-overman setting a Fractured
Leg with Pick-handles and a Turban.

sary, from the accident-box, and makes the patient comfortable.
He then superintends his removal on a light iron-stretcher to
the surface and the hospital, where he hands the patient over
to the assistant-surgeon, who treats the case surgically.

Fig. 1 shows a sirdar stopping bleeding at the femoral artery
by a tourniquet improvised from a turban (*pagri*) and a stone.
The assistant-surgeon may be seen examining the pulse of the

ankle, to ascertain whether the bandage would stop the bleeding. Fig. 2 shows a sirdar and deputy-overman setting a broken leg

FIG. 3.—PATIENT BEING PLACED UPON AN AMBULANCE.

with pick-handles and turbans.   Fig. 3 shows the patient being placed on a stretcher under the charge of the assistant-surgeon.

———

Mr. W. C. MOUNTAIN read the following " Notes on Electric Power applied to Winding in Main Shafts ":—

NOTES ON ELECTRIC POWER APPLIED TO WINDING IN MAIN SHAFTS.

By W. C. MOUNTAIN.

In the following short paper, the author will attempt to describe some typical Continental and English winding-gears, in order to give an idea of what is being done in the way of applying electrical power to this class of work.

Very little in the way of electric winding has been done in this country, and considerable doubt appears to exist as to the advantage of the use of such gear. The author, however, is quite certain that it will pay colliery-owners to look carefully into this matter; and, while he does not wish to advocate the indiscriminate application of electricity for everything about a pit, and recognizes that every case will have to be considered on its merits, still he thinks that a description of some of the principal German winding-gears will be of interest to colliery-owners in this country.

The firm with which the author is connected has installed a few electric winding-gears in this country, and, while they cannot compare with those on the Continent in size, they have nevertheless been very successful.   Later on in the paper, a description of such a gear, fitted by the author's firm at a colliery of the Shelton Iron, Steel and Coal Company, Limited, and also at the Heckmondwike colliery, is given.

The author will begin by describing the two principal examples of electrical winding to be seen in Germany, namely, those at the Zollern II. colliery, Merklinde, and at the Preussen II. colliery, near Dortmund.   One can easily understand that there are advantages in the way of smooth running, starting and stopping, etc., to be obtained by the use of electricity, but the fact that they are economical to operate is remarkable.   This, however, is shown by the figures given further on in the paper.

ZOLLERN II. COLLIERY.

Introduction.—Zollern II. colliery is one of the latest, and unquestionably the finest, in Germany. It has been equipped on the most modern lines, the surface-arrangements are very complete, and the buildings are all fine examples of architecture of their kind.

The offices are situated in the centre of the colliery-yard, and are very elaborately arranged. Bath-rooms are provided for the officials, and special bathrooms in white marble with nickel and oxidized-silver fittings, for the directors.

An ambulance and first-aid room, situated near the entrance-gates, is equipped with all necessary instruments and material for performing a complete operation. As the "worm" is prevalent in Westphalia, a special building is provided for patients suffering from this complaint. All workmen, showing symptoms of this disease, are isolated from their fellow-men, and must use the special rooms provided for them.

Bathrooms for the use of the men are compulsory, and a splendid bathroom has been provided, in which 1,500 men can bath at one time. A partitioned portion for boys under 16 years of age, is capable of accommodating 500. Everything seems to have been provided for the comfort of the men, and the heating arrangements are most elaborate and carefully thought out. Adjoining the bathroom is a waiting-room, where the miners wait their turns to descend, and it is used as a dining-room by the surface-men.

The fitting-shop contains a horizontal planing-machine, three lathes, two drilling-machines, and a pipe-screwing machine; all these are driven from one motor. The blacksmith's shop contains 6 or 8 fires, a compressed-air forge-hammer (supplied with air from the main air-compressor situated in the power-house), and a motor-driven blower. All the necessary joinery-work for the colliery is carried out in the joiner's shop, containing a band-saw, a circular saw, and a planing-machine, all motor-driven.

Electric Power-house.—The power-house is a splendid steel-and-brick building, about 330 feet long and 80 feet wide, and it has a domed glass roof. The building and foundations cost about £12,500 (250,000 marks).

Engines.—Two sets of generating machinery are installed at present, and another similar set is on order. Each set consists of a triple-expansion horizontal engine with 4 cylinders, coupled tandem, of the following dimensions:—The high-pressure cylinder is 31·50 inches (800 millimetres), the intermediate, 39·37 inches (1,000 millimetres); and the two low-pressure, 43·31 inches (1,100 millimetres) in diameter respectively, with a stroke of 47·24 inches (1,200 millimetres). The piston-speed is 705 feet (215 metres) or 90 revolutions per minute. The pressure of the steam, at the engine, with full load, is 147 pounds per square inch or 10 atmospheres. The engines are made by the Maschinenbau Aktien-Gesellschaft.

Generators.—Each engine has, mounted on its crank-shaft, a continuous-current 16 pole generator producing 1,100 kilowatts at 525 volts and 2,100 ampères, when running at 90 revolutions per minute. These generators supply current to the whole of the machinery at the colliery, and one is sufficient to keep everything going, including electric winder, washery, fans, air-compressors, workshops and pumps. The average load on one generator is 750 kilowatts.

The observed readings, when running at 88 revolutions per minute, during the writer's visit on March 2nd, 1904, were:—

	Atmospheres.	Pounds per Square Inch.
Engine: Steam-pressures on cylinders—		
High pressure	6·5	96·0
Intermediate pressure	2·9	42·5
Low pressure	1·4	20·5
Steam-jacket	4·0	59·0
Vacuum	0·8	11·8

Generator: The average load read on the meters was 520 volts, 1,030 to 1,400 ampères, or 550 to 730 kilowatts. These readings, however, rise, at times, to 1,500 ampères.

Main Switchboard.—The switchboard is composed entirely of marble, the frame being of brown, and the switch-panels of white, marble. It is arranged as a double board, that is, two generator-panels and a battery-panel are placed in front, while the distributing-panels, situated in a room behind the main panels, are mounted in an iron frame. The two generator-panels are each fitted with a double-pole switch, a double-pole circuit-breaker, a field-rheostat, a voltmeter and an ammeter.

The battery-panel is of the usual type, with charging and discharging switches and meters. Three spare panels are provided for future extensions. All the connections, behind the boards and to the battery, are solid round copper, making an excellent fireproof arrangement.

Lighting-plant.—A motor-generator is provided for lighting: a motor of 500 volts and a generator of 110 volts.

Battery of Accumulators.—This battery consists of 250 Tudor cells, and was originally installed in connection with the electric winding-gear, but it was not found quite satisfactory. A motor-generator has taken its place. The cells are of the following capacity:—5,000 ampère-hours, 500 volts, and 250 cells. The battery, now used as a standby, is capable of running the whole plant, including the winder, for one hour, or of supplying 500 ampères for 10 hours. This enables the main engine to be shut down for a considerable time at week-ends.

The author was informed, however, in view of the experience gained on this plant, that a battery will not be installed for future work of this kind.

Battery-switches.—A most elaborate arrangement of battery-switches has been installed to work the winding-engine. These, however, can be entirely dispensed with since the motor-generator has been installed.

Charging Booster.—A booster, consisting of a motor and two generators, is provided for charging the battery.

Electric Winding-engine.—This is the largest electric winding-engine in existence, and was exhibited in the Düsseldorf Exhibition of 1902. It was constructed by the Bergwerksverein Friedrich-Wilhelmshütte, Mülheim-on-the-Ruhr, and the electric machinery by Messrs. Siemens and Halske, Berlin (Fig. 1). The drum, of the Kœpe type, is 19·69 feet (6 metres) in diameter. The rope, of round stranded steel-wire, is 1·97 inches (50 millimetres) in diameter. At present, the engine is winding from the depth of 984 feet (300 metres).

Two electric motors are fitted, one on each side of the drum. The shaft runs in two bearings only, on the outer ends of the motors. This shaft is 20·47 inches (520 millimetres) in diameter, 23·62 inches (600 millimetres) long in the journals; and 25·98 inches (660 millimetres) in diameter in the drum-boss. The distance between the centres of the bearings is 18·04 feet (5·5 metres). The distance between the centres of the motors is 8·20 feet (2·5 metres).

FIG. 1.—ELECTRIC WINDING-GEAR FOR ZOLLERN II. COLLIERY.

At present the maximum winding speed is 32·81 feet (10 metres) per second, but any speed can be obtained from as low as 1 foot to as high as 64 feet per second. When the lower seams are reached, the winding depth will be about 3,600 feet.

At present, the engine winds about 40 loads per hour, each load being about 5 tons (5,000 kilogrammes) of coal. Each cage weighs 4 tons (4,000 kilogrammes), and 6 tubs weigh 2·4 tons (2,400 kilogrammes). At an average of 40 winds per hour and 16 hours winding, the output from this machine is about 3,200 tons per day from one shaft.

There is also a small steam winding-engine upon this shaft, used only for repairs, or for winding men.

At the upcast pit, another electric winder will be erected to draw about 1,000 tons per day.

Motors.—Two continuous-current motors, each of the following capacity, are mounted one on each side of the drum :—705 normal horsepower, 500 volts, and 51 to 64 revolutions per minute. The field-magnets of these motors are separately excited from the main generators, and the armatures are coupled directly without switches to the generator-armature of the motor-generator (Fig. 2).

FIG. 2.—MOTOR-SIDE OF ELECTRIC WINDING-GEAR FOR ZOLLERN II. COLLIERY.

Controlling Switches.—The whole of the starting, stopping, speed-regulation and reversing is done by one switch operating a resistance placed in the shunt-field of the generator of the motor-generator. By this means, only currents of small magnitude are dealt with, consequent burning of contacts is avoided, and heavy and expensive resistances are not necessary. Practically no energy is wasted in resistance, when running the winder at reduced speeds, as it is only necessary to insert the resistance in the motor-generator field and to reduce the voltage supplied to the winder-motors as required. A small amount of energy is returned to the motor-generator from the winding-motors during retardation, and this is usefully applied

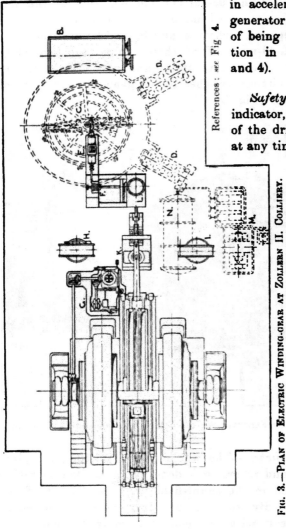

References : see Fig. 4.

Fig. 3.—Plan of Electric Winding-gear at Zollern II. Colliery.

in accelerating the motor-generator flywheel instead of being consumed by friction in braking (Figs. 3 and 4).

Safety-devices.—A depth indicator, fitted in front of the driver's stand, shows at any time the exact position of the hauling cage in the shaft. It works in connection with a safety device which prevents the maximum speed from being exceeded, and ensures automatically the required retardation when the hauling cage is nearing the pit-mouth. Should the lifting speed exceed at any time the maximum speed for which the engine is designed, the brake would immediately act automatically, and the electric current would be cut off from the motors. A slowing-down device turns the starting-gear lever gradually back, and thereby switches off the electric energy automatically when the cage approaches the surface, should the driver fail to operate the lever in time. This slowing-down device is an ingenious arrangement in the form of a wedge, which is gradually inserted behind a bell-crank lever connected to the starting gear. This was seen in operation, and worked in a perfectly satisfactory manner.

Brake.—The brake is operated by compressed air supplied by the main compressors, and, when these are standing, a small motor-driven air-compressor, of 12 horse-power, supplies the necessary air for working the brake.

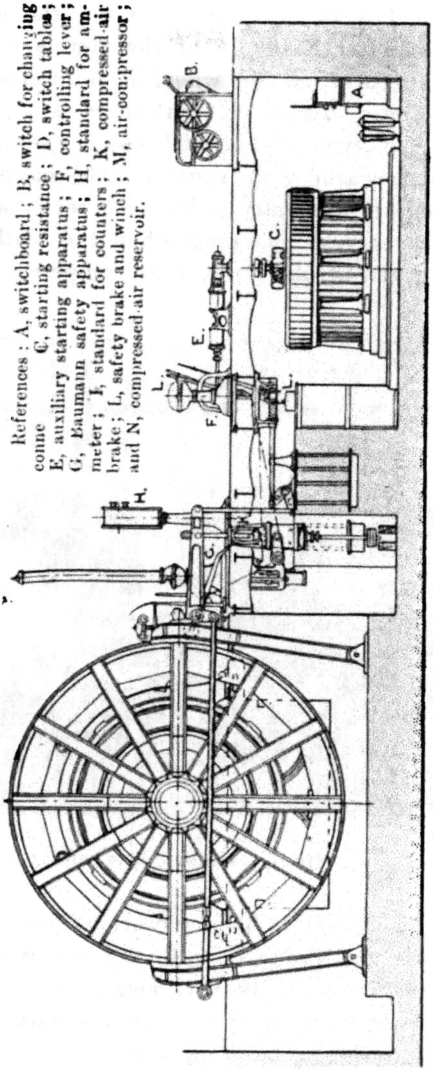

Motor-generator. — The motor-generator is the Ilgner arrangement, for applying the energy stored in a heavy revolving flywheel to provide the starting effort to accelerate the large winding-engine. It consists of a continuous-current motor and generator coupled together through a heavy flywheel (Fig. 5). The parts are as follows:—

(*a*) Motor-side : 500 volts, 300 effective horsepower and 300 to 345 revolutions per minute. The magnets are shunt-wound, and an automatic resistance of a most ingenious type placed in series with them, by adding to or cutting out the resistance, maintains the motor at a practically constant speed.

(*b*) Between the motor and the generator, a steel flywheel is mounted, and the shaft runs in two bearings, the armature of both the motor and the generator being overhung. The fly-

Fig. 4.—Side-elevation of Electric Winding-gear at Zollern II. Colliery.

References : A, switchboard ; B, switch for changing connexion ; C, starting resistance ; D, switch table ; E, auxiliary starting apparatus ; F, controlling lever ; G, Baumann safety apparatus ; H, standard for ammeter ; I, standard for counters ; K, compressed-air brake ; L, safety brake and winch ; M, air-compressor ; and N, compressed air reservoir.

wheel is 12·5 feet in diameter, the width on the rim is 2 feet
7 inches, the depth of the rim is 1 foot 1 inch, the width of
the centre of the wheel is 6 inches, and the total weight is
45 tons. The peripheral speed is 14,000 feet per minute.

(c) Dynamo-side: This is a special generator, with a double
commutator and special field-magnet system, having intermediate auxiliary poles with series-winding; and they effectu-
ally prevent sparking when the load comes on.   The dynamo
is rated at 550 volts, 1,800 ampères, 1,000 kilowatts, when run-

FIG. 5. — MOTOR-GENERATOR AT ZOLLERN II. COLLIERY.

ning at 300 to 345 revolutions per minute. The main magnet-
coils are shunt-wound, and separately excited from the main
generators.   By this means, any desired voltage is obtained
between zero and the full pressure of 550 volts. Through the
regulation of the strength of these coils, the voltage supplied to
the winding motors is regulated, and all speed-regulation on the
winders is obtained by this means. The electric currents handled
are therefore very small, and little or no energy is wasted in
resistances.

The following readings taken from the meters in connection with the winder and motor-generator were obtained during
ordinary working conditions: —

(a) Motor-side: input from steam-generators, the average of several readings.

	Ampères.	Volts.	kilowatts
Motor starting	600	520	300
At the end of 10 minutes, when the motor has reached its full speed	100	520	52
Winder starting, the current rises steadily to	100-450	520	235
Winder running, and after reaching the top speed, the current slowly drops ...	450-300	520	160
Winder slowing down, until it stops altogether	300-250	520	130
After the winder has been standing for 2½ or 3 minutes	100	520	52

In a general way, the winding-engine makes 40 winds per hour, and as it takes about 45 seconds to wind, it is standing for 45 seconds; therefore, the flywheel never reaches its full speed; and, consequently the average input into the motor-side varies between 250 and 450 ampères at 520 volts, or 130 to 235 kilowatts, or a variation in power of little over 100 kilowatts on the steam-generators. For many minutes, the variation was between 350 and 450 ampères, or about 70 horsepower, while the winding-engine stopped and started many times. This clearly showed the efficiency of the flywheel for providing the necessary starting effort. The variation in voltage on the steam-generator was less than 5 per cent.

(b) Generator-side of motor-generator: Winder, standing 0 ampères, starting 1,500 ampères, running at top speed 1,000 ampères, when stopping, current dropped to 0 ampères, and reverses, momentary, to 500 ampères to *nil*. The reverse current of 500 ampères, given by the winder-motors acting as generators, assists to accelerate the flywheel of the motor-generator. The time to wind varies from 43 to 45 seconds, and as the depth at present is 984 feet (300 metres), the average speed is about 24½ feet (7·5 metres) per second, and the maximum speed is about 30 to 33 feet (9 to 10 metres) per second. The average of several acceleration-readings was:—

	Ampères.	Seconds.
Winder, standing	0	0
,, starting	1,500	0
,, reached top speed	1,000	14
,, travelling at top speed	1,000	14
,, current cut off	0	0
,, current reverses	-500	0
,, comes to rest	0	17

On combining these observations, we can see at a glance the difference in the current supplied by and absorbed in the motor-generator during one wind of 45 seconds, as follows:—

	Motor-side or Primary side. Ampères.	Dynamo-side or Winder-side. Ampères.	Time. Seconds.
Winder, standing	100	0	0
,, starting	100-450	1,500	0
,, reached top speed ...	450-300	1,000	14
,, running at top speed ...	300	1,000	14
,, slowing down	300-230	0	0
,, current reverses ...	0	-500	17

Air-compressors.—In the same power-house are two air-compressors used for supplying air to machines placed underground. One of these has been running for some time, while the other is being erected. These compressors are driven by electric motors mounted directly on the crank-shaft of the compressor. The cranks are placed at an angle of 180 degrees, and the connecting-rod is fitted direct on to the trunk-gudgeon. The motor is of the following capacity:—500 volts, 670 ampères, 75½ to 130 revolutions per minute and 450 electric horsepower. The arrangement seemed extremely neat and compact.

Ventilating Fans.—Two fans are direct-coupled to their motor; and each fan is 23 feet in diameter. At 190 revolutions per minute, the air discharged is 176,600 cubic feet (5,000 cubic metres) per minute, and the water-gauge is 6·69 inches (170 millimetres). The motors generate 250 horsepower, at 500 volts and 435 ampères.

Coal-washery.—The coal-washery is capable of washing 1,000 tons of coal per day, and sorting it into five sizes. The whole washery is driven by two motors, each of the following size:— 142 ampères, 500 volts, and at 600 revolutions per minute each generates 95 horsepower.

The centrifugal pumps are driven by two motors of the same size as those described above, namely, 95 horsepower.

Boilers.—The boiler-room contains 6 double-drum Babcock-and-Wilcox boilers divided into 3 batteries of 2 each. The steam-pressure at the boilers is 206 pounds per square inch or 14 atmospheres and it is lowered through a reducing-valve to

10 atmospheres, or 147 pounds per square inch for use in the engines. The feed-pumps are driven by a motor of 22 horsepower.

During the writer's visit, 3 boilers only were at work, and duff-coal appeared to be used.

The following readings were taken from the various meters on the switchboard during ordinary working hours, and as the whole plant at the colliery is operated by electric motors, it is interesting to note the actual power going out and to compare it with the amount of steam which could be expected to pass through a cylinder, 31·5 inches in diameter and 47·24 inches stroke, running at 90 revolutions per minute.

	Ampères.	Volts.	Electric Horsepower.
Coal-washery	200	520	140
Condenser	24	520	16
Ventilating fan	100	520	70
Air-compressor	430	520	300
Workshops	10	520	7
Motor-generator of winder	300	520*	240
Totals	1,064	520	773

* Average of readings.

Consequently, the entire colliery, including the winding-engine, requires only an average load of 773 horsepower, when winding from 1,000 to 1,500 tons per day from a depth of 984 feet (300 metres). The coal-consumption to do the whole work at the colliery can be calculated from the dimensions of the engines and the horsepower shown by the meters, and it is only 2 to 2½ per cent. of the output.

PREUSSEN II. COLLIERY.

The Preussen II. colliery, one of the show-mines in Germany, has been laid out on elaborate lines. The chief point of interest, however, is the large electric winding-engine driven by a three-phase motor direct-coupled to the drum-shaft.

Power-house.—The generating plant is situated in a large power-house, adjoining the boiler-room; the air-compressors, ventilating fans and condensing plant are also situated in this power-house.

Engines.—The electric generating plant comprises three direct-coupled, slow-speed, cross-compound horizontal engines and three-phase generators. The high-pressure cylinder is 22·44 inches (570 millimetres) and the low-pressure, 37·40 inches (950 millimetres) in diameter respectively, with a stroke of 43·31 inches (1,100 millimetres). The steam-pressure is 125 pounds per square inch, or 8·5 atmospheres. When running at 94 revolutions per minute, the engine produces 750 horsepower when condensing, and 650 horsepower when non-condensing. These engines exhaust into a central condenser, which acts for the whole plant of the colliery. The engines were supplied by the Sächsische Maschinenfabrik Aktien-Gesellschaft (Fig. 6).

FIG. 6.—POWER-HOUSE AT PREUSSEN II. COLLIERY.

Generators.—The 550 kilowatts generators produce 2,100 volts and 150 ampères per phase, at 25 cycles per second when running at 94 revolutions per minute. These generators, supplied by the Allgemeine Elektricitäts-Gesellschaft, are of the flywheel-type, that is, the magnet-poles are mounted direct on to the flywheel. The armature-case is of the Allgemeine Elektricitäts-Gesellschaft braced type.

These three generators (two of which are necessary to keep the winding-engine running) supply current to the main switch-board, situated on a raised platform immediately behind the engine.

Exciters.—Two exciters are provided, one being driven by a steam-engine by belt on to a multipolar continuous-current generator. This is only used when all the alternators have been shut down, as the other exciter, being driven by a three-phase motor, cannot be started until an alternator is running up to its pressure. The motor-driven exciter, a three-phase motor of the slip-ring type, is coupled direct to a four-pole continuous-current generator, and provides the exciting current for the three alternators.

Switchboard.—The switchboard is of the usual Continental high-voltage type, with distance-controlled high-voltage switches. The arrangement is neat and simple: three main generator-panels being provided and an exciter-panel. It is mounted on an ornamental iron frame. The lighting panels are situated underneath the power-board platform, and are coupled to the steam-driven exciter-set, which also provides power for the lighting.

The following readings were taken on January 4th, 1904, when the winding-engine, drawing coal, came into operation:—

Winder Standing. Volts.	Winder Starting and Running. Volts.	Drop in Volts.	Ampères.	Apparent Kilowatts.
2,300	1,600	700	450	1,260
2,300	Motor holding the cage	—	50	200
2,300	1,700	600	300	880
—	2,000	Winder running	160	550
2,300	Winder standing	—	0	0
2,300	2,000	Winder moving the cage	130	450
2,300	1,400	900	200	480
2,400	Winder standing	—	0	0
2,400	1,600	800	—	450
—	2,000	—		100

The following readings were taken on February 29th, 1904, with the winder moving slowly, with the men in the pit:—

Volts.	Ampères.	Kilowatts.
2,300	0	Winder standing
1,900	170	200
2,300	0	Winder stopped
1,900	170	200

The speed of winding was 6½ feet (2 metres) per second. These readings repeated themselves whenever the cage was moved.

Air-compressors.—There are three horizontal two-stage cross-compound air-compressors, supplied by Messrs. Schuchtermann und Kremer.

Electric Winder.—The motor is coupled directly through its starting switches to the three-phase power-mains, no motor-generator arrangement, similar to that at Zollern II. colliery, being used. This arrangement, although simple, does not appear to be very satisfactory, as, while on a former visit, the engine was working on full load, two large alternators each of 550 kilowatts were required to keep the winder going, owing to the enormous drop of pressure caused by switching on the

FIG. 7.— ELECTRIC WINDING-GEAR AT PREUSSEN II. COLLIERY.

winding-motor (Fig. 7). This is only what might be expected, as the power-factor of a large three-phase motor starting with a dead load such as a winder must do, has a very bad effect on the pressure-regulation of the alternator supplying the current. On a subsequent visit to this pit, the winding-engine was working very slowly, as workmen were fixing cables in the shaft. One generator was running, but the load was practically nothing, seeing that the cages and ropes are balanced, and the motor had little to do except overcome the friction. The drop in pressure, was, however, very great, namely, from 2,300 to 1,800 volts, and the current was about 170 ampères. The power was about 200 kilowatts. An automatic regulator is about to be fitted, with a view of keeping the pressure more constant.

The three-phase motor, direct-coupled to the Kœpe drum, develops 950 to 1,500 brake-horsepower, at 53·5 revolutions per minute, 2,000 volts and 345 ampères (Fig. 8). The pit is 1,800

FIG. 8.—MOTOR-SIDE OF ELECTRIC WINDING-GEAR AT
PREUSSEN II. COLLIERY.

feet (560 metres) deep. The winder makes 30 revolutions per wind, made in 60 to 65 seconds. The winding speed is 52 feet (16 metres) per second when drawing coal, and 20 feet (6 metres) per second when drawing men. The cage carries 6 trams, containing 3 tons (3,000 kilogrammes) of coal.

Starting Switch for Winding-gear.—Fig. 9 shows the arrangement of the controlling-gear. The current from the mains flows through the high-tension safety-fuses and the safety cut-out or circuit-breaker, then through the change-over switch for reversing the motors, and straight to the stator of the motor.

These are all the switches necessary in the high-tension circuit.
The rotor of the motor, at a maximum pressure of only 300 volts,
has three slip-rings, from which its current is led to the special

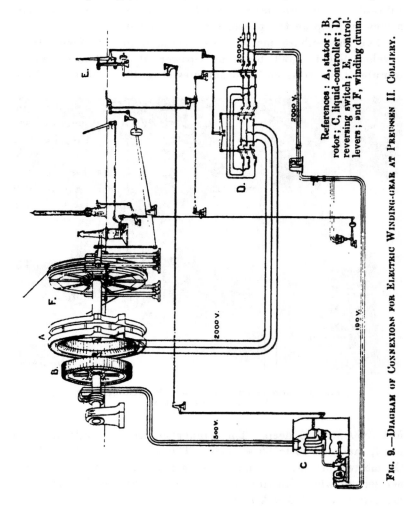

References : A, stator ; B,
rotor ; C, liquid-controller ; D,
reversing switch ; E, control-
levers ; and F, winding drum.

FIG. 9.—DIAGRAM OF CONNEXION FOR ELECTRIC WINDING-GEAR AT PREUSSEN II. COLLIERY.

liquid-starting rheostat (Fig. 10).   This rheostat is a well
thought-out arrangement, and is similar to that fitted on the
high-speed car which the Allgemeine Elektricitäts-Gesellschaft
recently succeeded in driving at a speed of 112 miles per hour.
Each of the three phases of the rotor-circuit is led to electrode-
plates, insulated and suspended in a tank in which circulates a

solution of soda. Instead of the plates being dipped down into the solution as in the liquid-starters made by the author's firm, the solution is made to rise up the plates, the water being pumped into the tank by a small centrifugal pump, which is kept continually running by a small motor. This pump delivers from a tank placed below the resistance-tank. When the liquid rises to its full height in the resistance-tank, it overflows into that placed below. A valve, fitted in the bottom of the resistance-tank, is worked by the driver's controlling-lever. If the valve

FIG. 10.—LIQUID-CONTROLLER FOR ELECTRIC WINDING-GEAR AT PREUSSEN II. COLLIERY.

is set full open, the liquid does not rise up the plates; if shut, the liquid rises upward at a regular steady rate, depending upon the rate of delivery of the fluid by the pump and independent of the driver, thus preventing him from switching on the current too quickly. The liquid can also be maintained at intermediate positions, according as the valve is set more or less open, thus giving intermediate speeds of the motor.

As compared with the Ilgner system first described, this system, although very successful, gives rather large voltage-drops on the main generators. The heavy starting efforts come directly on them, and although the average load is not great, still two generators of 550 kilowatts have to be kept running to supply this gear, together with the rest of the electric gear at the colliery, in order to keep the voltage-drop from interfering with the rest of the motors at the pit.

FIG. 11.—ELECTRIC WINDING-GEAR AT ARNIM COLLIERY.

Boilers.—There is a battery of 9 double Cornish boilers, with water-tube boilers set at the back of the tubes, forming a steam-generator of a very efficient type. The pressure varies from 8 to 10 atmospheres.

Fitting and Smith's Shops.—These shops contain many good tools, all driven by motors.

Offices and Bathrooms.—The offices and bathrooms are well arranged, but not on the same lavish scale as at Zollern II. colliery.

Having seen both the winding-gears described in operation, the writer is on the whole inclined to think that direct-

·current motors are more suitable for main winding-gears, but it is impossible to lay down hard-and-fast rules, and every case must, of course, be carefully considered on its merits.

WINDING-GEARS.

Fig. 11 shews a winding-gear fitted at the Arnim colliery, Zwickau. It is driven by· a three-phase motor of a maximum output of about 80 horsepower, with 500 volts at the terminals. The electric arrangements were designed and

FIG. 12.—UNDERGROUND ELECTRIC WINDING-GEAR AT CONSOLIDATED SALT-MINE, WESTEREGELN.

supplied by Messrs. Siemens and Halske. This winding-gear is remarkable for the fact that it is not, as usual, placed at the ·side of the shaft, but directly above it, so that the hoisting ropes drive in a perpendicular direction from the winding-·drum. The principal advantage of this arrangement lies in the smaller wear and tear of the rope, since it does not pass over ·a second pulley, and is therefore only bent once.

Fig. 12 shews a winding-gear fitted at the Consolidated ·salt-mine, Westeregeln. It is driven by a continuous-

difficulty was experienced in starting the motor or on arrival at bank; and whether any accidents or other difficulties had been experienced.

Mr. W. C. MOUNTAIN replied that the electric winder had been running at the Zollern II. colliery, he believed, for only a few months, and the winder at the Preussen II. colliery was also practically new. He believed that there were about 30 electric winders in use in Germany. At Essen, an electric winding-plant had been installed, the power being derived from a public supply-company.

The PRESIDENT (Mr. W. O. Wood) moved that a vote of thanks be accorded to Mr. Mountain for his paper.

Mr. J. G. WEEKS seconded the resolution which was cordially approved.

Fig. 14 shews a winding-gear fitted at one of the pits belonging to the Shelton Iron, Steel and Coal Company, Limited, and a similar gear has been erected at the Brook Pit, Heckmondwike. These electric winding-installations have now been at work for some time, and are giving every satisfaction.

At the Heckmondwike colliery, a large fault or throw occurs about 3,000 feet away from the pit-bottom. This fault throws down the coal-seam a vertical depth of 180 feet below its original level. To win the coal-field on the dip-side of the fault, it was necessary to drive either a large drift or sink a staple. If it

FIG. 14.—ELECTRIC WINDING-GEAR FOR THE SHELTON IRON, STEEL
AND COAL COMPANY, LIMITED.

had been decided to drive a drift, it would have been necessary to put down a powerful haulage-plant, at least equal to the size of the present winding-plant. The driving of such a drift, with a gradient of 1 in 6, would have involved 1,080 feet of drifting, and the cost would have been about £1,960. The sinking of the staple cost about £3 10s. per foot for 180 feet, or a nett saving of £1,360 against the drift-scheme. It also took nine months less time to win the coal by the staple than it would have done by the drift.

The shaft is 180 feet deep, and, with the present arrangements, the winding-gear is capable of raising from 150 to 200 tons per day of 10 hours. The winding drum is 3 feet in diameter, and a wind is made in 40 to 45 seconds, equal to a speed of, say, 300 feet per minute. The electric motor, driving the drum, is of the series-wound continuous-current type, capable of working up to 40 effective horsepower, when running at a speed of about 500 revolutions per minute. The motor is fitted with slotted-drum armature, with a specially large commutator fitted with carbon-block brushes, self-oiling bearings, and is of the usual type that the author's firm have designed for this class of work. A machine-cut pinion of forged steel is fitted on the end of the spindle, and gears into a spur-wheel on the countershaft. The first-motion gearing is machine-cut forged steel, and the second-motion gearing is machine-moulded cast-iron. A powerful foot-brake is provided for the drum, operated by a foot-lever. As a further protection, this gear is fitted with an electric brake, very similar to those on the German gears already described. This brake is fitted on the motor-shaft, and comes into operation as soon as the current is switched off, and thereby sustains the load. An indicator is also fitted on the winding-gear to shew the position of the cage in the shaft.

This winder is operated by means of a liquid controlling-switch, arranged for reversing and speed-regulation. The plates are arranged to dip into the liquid in an opposite manner to those of the gear at Preussen II. colliery, in which the liquid rises up the plates. This winder could have been fitted with a three-phase motor of the same power: in which case a similar type of liquid starter would have been employed.

The author hopes that this paper will be of general interest to the members, and afford some idea of the extent to which electric winding-gears are in use in this country and on the Continent. The large winding-gears, which the author saw in Germany, struck him as being put down somewhat on an extravagant scale, and while he would not advocate such elaborate and expensive gears in this country, he quite recognizes (as before stated) that every case must be considered on its own merits. He cannot, however, help thinking that a study of this problem, as it has been treated electrically, would well repay colliery-engineers in Great Britain.

The PRESIDENT (Mr. W. O. Wood) said that the members were indebted to Mr. Mountain for his excellent and instructive paper. He thought that less elaborate and less expensive arrangements might have been found quite as satisfactory, but German certainly set an example to British engineers worthy of being followed in several respects. The economy resulting from electric winders was high, but he would have expected to find that the direct application of steam would have been more economical than through an intermediary power.

Mr. T. E. FORSTER said that one naturally looked at the economical side of the question, and he was afraid that they were rather wedded in this country to commercial enterprise; but, if by spending money, they could save money, it was worth while for everyone to consider the matter. He would like to know whether water was pumped at the Zollern II. colliery, as in addition to winding and ventilating machinery, the paper only referred to the power required for driving the air-compressors and a coal-washery. It was stated that duff-coal was used; perhaps the writer could state what was the nature of this coal, as some people might include nuts and small coal in that category. The coal-consumption appeared to be low, although he believed that there were collieries in this country and in this district with as low, or a lower, consumption. He asked whether by applying electric power a larger output could be obtained from the same shaft; and remarked that possibly, at the German mines, they were not drawing so much mineral as at some of the winding shafts in the Midland district.

Mr. A. L. STEAVENSON said that the question of electric winding was one which might be considered, either from the point of view of economy of power or cost. They began with a given quantity of steam, and if they chose to have a first-class winding-engine (compound, condensing, and fitted with various arrangements which tended to reduce the consumption of steam), it was difficult to see how they were going to turn the steam into electricity, and again electricity into power, with more economy than they could directly use the steam in a first-class winding-engine. Take the case of a motor of, say, 350 horse-power applied to a first-class winding-engine; how much less coal would be required when using electric power as an intermediary? So far as the machinery described was considered as a working machine, it seemed to be perfect, and the work done was, they were given to understand, of a very high-class character; but, if it was an advantage to turn steam into electric power, where were they going to stop? They might use the electric power to compress air, and then use the compressed air as a source of power. He asked, therefore, why it was better to use electric power, rather than to use steam direct, if in both cases a first-class engine was used. His impression was that all subsidiary engines, of which there were generally eight or ten at a colliery, might be driven by electric power by cable owing to its great economy as compared with steam by steam-pipes; but in the case of the large engines employed for winding and pumping, he was inclined to doubt whether the intervention of electric power would prove economical.

Mr. T. W. Benson, referring to the application of electric power for winding up a staple, a long way in-bye, where there was a large downcast fault, said that in cases such as this, where steam could not be possibly used, the use of electric power would be a great advantage. Even with compressed air, there was a great waste of power, owing to the leakage of pipes, etc.

Mr. J. G. Weeks, referring to the statement that the consumption of coal was 2½ per cent. of the output at the Zollern II. colliery, asked whether much water was pumped; whether there was an extensive system of haulage; and finally, whether it was a new colliery, where there was less necessity to use much power. He supposed that the 2½ per cent. did not include coals supplied to workmen or for like purposes, and that it only included the coal-consumption at engines or boilers for power and electric lighting.

Mr. W. C. Mountain replied that the Zollern II. was an absolutely new colliery, at present winding from a seam at a depth of 984 feet and that the ultimate depth would be about 3,000 feet. There were no pumps just now. The maximum speed of winding was limited to 64 feet per second, and he mentioned that figure in his paper with a view of inviting criticism. He found that, in Germany, winding at deep pits was done at a much slower speed than usual at deep pits in this country. At one of the principal collieries, provided with a high-class steam winding-engine, drawing from a depth of 1,500 feet, a wind was made in about 1 minute, and the changing of the tubs required 1¼ minutes, or a total of 2¼ minutes per wind, although 25 men were employed at the pit-head to handle the output of 1,100 tons per day. He was informed that £500,000 had been spent on the total equipment of the Zollern II. colliery, and of this amount £125,000 had been spent on machinery in the power-house; and the utmost production was 3,000 tons in 16 hours, from a depth of 984 feet. This power-house had stained-glass windows and a domed glass-roof, and the floor was covered with tesselated tiles.

In discussing the economy of electric winding, people were apt to compare high-class compound or triple expansion engines producing electricity with antiquated steam "eaters," but to strike a true comparison it was necessary to consider the use of compound condensing winding-engines and to compare the results with compound or triple expansion engines producing electricity. Where steam was applied to winding-engines, within 100 or 150 feet of the boilers, he (Mr. Mountain) was doubtful whether any economy could be obtained by the intervention of electric power particularly when three-phase machinery was used, as at Preussen colliery. But a cheap source of power, such as the waste-heat from coke-ovens or blast-furnaces, might be used to generate electric current at a high pressure, at a central station, and thence it could be supplied to adjacent collieries. In such a case, electric power for winding, undoubtedly, had useful and economical applications. At Zollern II. colliery, the motor constantly absorbed only 300 horsepower, and the winder was taking a varying power from it, varying from nil up to 1,500 horsepower: consequently, if the constant steady load of 300 horsepower could be generated economically, the engine would do the work in an economical manner. It was the only economical system of electric winding that he had seen in Germany.

The coal used for firing would probably pass through about a mesh of ⅜ inch.

Mr. HENRY WHITE asked how long the electric winders had been in use at the Preussen and Zollern collieries; whether any difficulty was experienced in starting the motor or on arrival at bank; and whether any accidents or other difficulties had been experienced.

Mr. W. C. MOUNTAIN replied that the electric winder had been running at the Zollern II. colliery, he believed, for only a few months, and the winder at the Preussen II. colliery was also practically new. He believed that there were about 30 electric winders in use in Germany. At Essen, an electric winding-plant had been installed, the power being derived from a public supply-company.

The PRESIDENT (Mr. W. O. Wood) moved that a vote of thanks be accorded to Mr. Mountain for his paper.

Mr. J. G. WEEKS seconded the resolution which was cordially approved.

SOME SILVER-BEARING VEINS OF MEXICO.
(Concluded.)*

By EDWARD HALSE.

REAL OF SULTEPEC, STATE OF MEXICO.

According to La Croix,[†] the historian, the first mines worked by the Spanish *conquistadores*, who were guided, no doubt, by the surface-workings of the Aztecs, were those of Sultepec, Tasco, Tlapujahua and Pachuca. The Spanish workings are said to date from 1522, or two years only after the conquest. Alexander von Humboldt, in his essay on New Spain, states that from 1785 to 1789 there was put into the Royal Treasury, from the mines of Tasco, Zacualpan and Sultepec, silver to the amount of 7,806,472½ ounces troy, a quantity exceeded only by the classical districts of Guanajuato, Catorce and Zacatecas.

Sultepec was formerly called " Provincia de la Plata," and it has been recorded officially by José Vincente Cosío,[‡] a miner of the district, that in 1802-1804, and 1806-1808, or in all about 4½ years, 1,568 bars were remitted to the city of Mexico from the Real of Sultepec, weighing 1,512,274½ ounces troy, or an average of 336,061 ounces per annum. Of these bars only one was from smelted ore, the remainder having been reduced by the *patio* process.

Towards the close of the eighteenth century, Juan López de Saavedra acquired a large fortune in these mines; he organized the " Compañia Minera Saavedra y Socios," which is said to have extracted the large quantity of silver noted above. Saavedra died in 1806; a few years later, or in 1810, the long War of Independence was commenced, most of the mines were then practically deserted, being left to the tender mercies of

---

* *Trans. Inst. M.E.*, 1900, vol. xviii., page 370; 1901, vol. xxi., page 198; 1902, vol. xxiii., page 243; and 1902, vol. xxiv., page 41.

† *Historia de Mexico*, pages 41 and 146.

‡ *Diario del Imperio*, No. 440, Mexico, 1866.

the *buscones*,[*] and it is only within recent years that attempts have been made to work the principal veins on anything like an adequate scale.

The country-rock consists of argillaceous, aluminous and micaceous schists of Lower Cretaceous age, dipping flatly to the south, and striking a few degrees north of east, or obliquely to the *cordilleras* which trend north-west and south-east.   To the north, east and west, recent eruptive rocks predominate, consisting of granite, andesite and basalt, but in a southerly direction the belt of slate is traceable for many miles.   As pointed out by Mr. Santiago Ramirez,[†] the schistose area is in fact surrounded by a ring of igneous rocks in the shape of a horseshoe, Temascaltepec and Zacualpan being at its north-western and south-eastern extremities respectively, while Sultepec lies in the centre.

The schists to the east rest on a spur of the great extinct volcano Nevado de Toluca or Xinantecatl (altitude 15,898 feet) the mass of which is built up of hypersthene-andesites;[‡] and the Toluca valley; and the surface, for some considerable distance to the west, is covered by andesitic tuffs and breccias.

A few miles north-west of the town of Sultepec, near the Indian village of Santiago Texcaltitlan, are two small extinct volcanoes, with their craters well-preserved, resting on Cretaceous limestone, which is much broken and metamorphosed. These are regarded by Mexican geologists as satellites of the large volcano.

Here and there, patches of limestone are seen lying on the schists.   The former is sometimes overlain by Tertiary

---

[*] "Those who search for ore in abandoned mines, or who give notice of it in order to obtain a reward."—J. F. Gamboa, *Comentarios á las Ordenanzas de Mineria*, Madrid, 1761. It is derived from the verb *buscar*, to search. Closely related are the *rebotalleros*, who restrict their attentions to mine-dumps. By the term *gambucino* is generally understood one who searches for ore generally on his own account; he is in fact the *buscador* of the Spanish, the *faizquiero* of the Portuguese, and the *prospector* of the Anglo-Saxon race. Very nearly allied are the Spanish-American terms *barequeros* (*mazamorreros*, *burbuseros* or *gurguros*, searchers of alluvial), *catvadores* (or *cateros*, Mexican *pintistas*), *rumbeadores* (searchers of veins), and *monteadores* (searchers of mines in forests and mountains). The *fossicker* of the United States, the *hatter* of New Zealand, and the *pig-rooter* of Australia, are cousins-german of the above.

[†] *Noticia de la Riqueza Minera de Mexico*, 1884, page 502.

[‡] "Bosquejo Geológico de México," *Boletín del Instituto Geológico de México*, Nos. 4, 5 and 6, page 162.

andesites, which have partly altered it, as in the instance quoted above. Dykes and masses of basalt occur here and there, of more recent date than the andesites.

The schists are traversed by metalliferous veins, as well as by quartzose and felspathic igneous dykes, mostly andesitic, which often form one wall of the principal veins for considerable distances. When the dykes leave the lodes, the latter are usually barren, or of very low-grade value.* Hence, although the dykes are probably of more recent age than the principal fractures, the filling (as regards the silver-contents) is undoubtedly related to them.

The veins may be grouped as follows:—(1) North-west to south-east veins, dipping usually north-eastward 60 degrees. These are the principal veins, and the only ones that have been worked to any extent hitherto. (2) East-and-west veins, dip variable, (3) north-and-south veins, cutting across (1) somewhat obliquely, dip generally eastward.

The powerful outcrop of one of the principal veins of the district is traceable for upwards of ¾ mile; it then throws off a large branch to the east, while the main vein or leader can be traced for some distance farther south. This vein is on an average upwards of 60 feet thick; the schistose country is often highly metamorphosed near the vein, being changed in places into quartz-rock. The alteration has doubtless been effected by the dyke which generally accompanies the vein, as well as by the thermal waters emanating from the latter.

The ores of the vein consist of iron-pyrites (*bronce*), marcasite, argentite, ruby-silver (pyrargyrite and occasionally proustite), stephanite (rare), miargyrite ($Ag_2S.Sb_2S_3$, rare) and blende. The veinstone consists of white crystalline quartz, amethystine quartz, calcite, dolomite, brown spar, fluorspar and heavy spar (rare). The non-silver-bearing heavy minerals are copper-pyrites, stibnite (rare) and mispickel. Near the surface, native silver in fine threads and leaves, associated with gypsum, horn-silver and argentite, is sometimes disseminated in oxides and

* *Distrito y Real de Minas de Sultepec*, Mexico, 1888, page 14, an official report by Mr. Bartolomé Teodoro Villanueva. In Real del Monte, Hidalgo, the Moran vein (strike east and west, dip northwards) accompanies the dyke known as San Esteban, but the vein is, on the whole, of low-grade value.

hydroxides of iron, copper and manganese, constituting what Mexican miners term *ixtajales*,[*] an earthy variety of gossan.

Sometimes the structure of the veins is brecciated, but more often the ores and gangue are arranged in layers parallel to the walls; the banded structure, however, is not symmetrical, for one or more bands (*fajas* or *cintas*) of mineral may be succeeded by more or less barren layers (*bancos*) of country-rock; sometimes ore and matrix occupy the whole width of the vein, or ore, matrix and country-rock may be so intermixed that the filling is confused, and no definite structure can be made out.

According to Mr. Villanueva,[†] wide cavities occur here and there in the veins, filled with nodules of iron-pyrites, blende and galena mixed with quartz, which are coated with calcite or siderite (*espato ferrífero*). These may be compared to the cockade-, sphere- or ring-ores of some European regions, although in the latter, fragments of earthy veinstone or country-rock generally form the nucleus, the surrounding layers being composed of metalliferous minerals. The same author states[‡] that when the gangue consists largely of fluorspar, mixed with brown spar and limonite, galena is abundant and contains a high percentage of silver; whereas an increase of quartz in the vein is followed by an increase in blende and a diminution in galena, the percentage of silver in the latter decreasing also. It would appear, from the above, that open spaces existed here and there in the vein, which have subsequently been partly or wholly filled by ores, gangue and rock; but there is no doubt that a considerable portion of the filling has been produced by replacement or substitution.

The walls of the principal veins of the district are without selvages, and are frequently ill-defined. The barren filling consists usually of schist, generally altered, sometimes being highly quartzose (quartz-rock), and sometimes decomposed to a white clay (kaolin).

The ores, as a rule, are adapted to smelting, and unsuitable for treatment by the *patio* process. The Spaniards treated

* *Informe Provisional del Ingeniero Bartolomé T. Villanueva* . . . Mexico, 1888: compare *Trans. Inst. M.E.*, 1901, vol. xxi., page 206.

† *Distrito y Real de Minas de Sultepec*, page 12.

‡ *Op. cit.*, page 14.

only the docile ores, leaving the leady and pyritic ores untouched. They may be divided into:—(1) *Metales secos* (German *Dürrerze*), dry ores, or those with insufficient lead for smelting purposes. In these ores are included the silver-minerals, together with oxide of iron, etc., and a certain proportion of gold. The richer ores are smelted with (2) and (3), while the poorer ores are generally first roasted in kilns; (2) *ligas* or *metales plomosos, de fundición* or *de fuego*, leady or smelting ores, or those containing a good proportion of galena. The latter mineral is usually of coarse grain and poor in silver, and is known in the district as *soroche;* * (3) *metales piritosos* or *de quema*, pyritic or roasting ores, or those chiefly consisting of iron-pyrites and blende,† which are usually roasted before being smelted with (1) and (2). These ores are generally silver-bearing, and the iron-pyrites often contains gold as well.

The veins of the Malacate mine, one of the most important in the district, are shown in Fig. 1 (Plate VIII.). The relative positions and directions of these are, however, approximate only. The San Pascual lode, which forms a part of the master-lode already mentioned, strikes north 48½ degrees west to south 48½ degrees east. This and the Concepción, which is probably an accompanying quartzose dyke, strongly metalliferous in places, dip north-eastward 60 degrees for a vertical depth of 195 feet, then the lodes become much flatter, the dip being north-eastward 35 degrees for a further vertical depth of 120 feet, below which both lodes appear to resume their normal dip (north-eastward 60 degrees). The average combined thickness is about 50 feet.

Judging by the old workings, the ore has a tendency to pitch to the south-east. The pitch of the ore-bodies was most probably determined by the dip of the schistose country, which is southerly. The longitudinal section of the mine-workings shows two ore-shoots. The western one pitches south-eastward

* In Zacatecas, *soroche plomoso* is carbonate of lead, and *soroche reluciente* is argentiferous galena. According to Mr. J. B. Guim (*Nuevo Diccionario de la Lengua Castellana*, second edition, Paris, 1864), *soroche* in Spain is "mineral reluciente y quebradizo que tiene alguna plata, y puede reducirse á la clase de los negrillos." The same author describes *negrillo* as a miner's term for a variety of black native silver.

† As much blende as possible must be got rid of by hand-picking, and by concentration, as it is harmful in the smelting process.

30 degrees, and was worked by the Spaniards for a length of 70 feet (measured across the pitch) and to a depth of 140 feet (measured along the pitch). The second ore-shoot, 110 feet to the east (measured horizontally) pitches south-eastward 38½ degrees, and was worked by them for a length of 50 feet and a depth of 350 feet. This depth is much greater, if we include some old workings on a higher parallel ridge.

The writer has already pointed out[*] how necessary a knowledge of such local laws of ore-deposition is, in order to develop the veins in a rational manner. In estimating the probable reserves of ore, too, its occurrence in shoots must clearly be taken into account, and the unproductive ground between them must be eliminated from any estimate of the probable ore-reserves in depth.

The general structure of the San Pascual vein is seen in Fig. 2 (Plate VIII.). The iron-pyrites in the veins is usually fine-grained; the matrix consists of quartz, calcite (often in geodes), heavy spar and kaolinized country-rock. The silver-ores proper usually occur on the foot-wall. The bands, layers or streaks are frequently separated from each other by "horses" of country-rock (*caballetes* or *tabiques*), or by practically barren vein-filling; and sometimes the ores (leady with streaks of dry silver-ores) are continuous from wall to wall for a thickness of 30 feet and upwards. As a rule, the ore-streaks are of good workable size, varying from 2 to 4 feet and upwards in thickness.

The Concepción lode runs for some distance alongside the last in yellow schistose country-rock, and then branches off from it. The metalliferous contents are much the same, but the vein-stone, where the lode is ore-bearing, contains very little quartz, and much crystalline calcite, as well as some fluorspar. The ores are dry and leady. As the lode contains a large amount of calcite, the filling is probably of more recent date than that of San Pascual. La Concepción is probably a subsidiary fracture which has been formed on the lying or foot-wall side of the San Pascual vein; a few other smaller and less-defined parallel fractures occur to the south. The above may be compared to the branches that occur on the foot-wall side of the large veins of

---

[*] *Trans. Inst. M.E.*, 1901, vol. xxi., page 209.

-the region of Taviches,[*] Oaxaca, as well as to those occurring in Zacatecas,[†] and in the *veta madre* of Guanajuato.[‡]

The two veins, as cut by the Santa Helena tunnel to the west, have the structure shown in Fig. 3 (Plate VIII.). The pay-streak, *a*, occurs on what appears to be the foot-wall of the San Pascual vein; La Concepción is represented by a thick band of hard mineralized quartz-rock of low-grade value, *d*: in fact, at this point, the dyke (if it be such) is practically non-metalliferous.

The Capulin vein runs nearly east and west, but it would appear to be an eastern branch of the San Pascual lode, for it has not been traced to the west of the latter. It has a southerly dip at and near the surface; but, where it forms a junction with the San Pascual and Concepción, it follows the general northerly dip of these lodes. The structure is shown in Fig. 4 (Plate VIII.). In this instance, the high-grade ore, *a*, occurs on the hanging-wall. The foliated iron-pyrites, *b*, is sometimes stained with copper. In one place on the hanging-wall, there is a band of oxide of iron bearing free gold and no silver. This vein is characterized by having little or no galena and blende.

Prof. Edmund Fuchs[§] describes Capulin as formed of three or four veins close together, which are sometimes united by oblique veinlets or by a pyritic impregnation. These so-called "veins" are probably identical with what are termed, in this paper, bands, layers or streaks.

The Providencia is a parallel vein, some distance to the north of San Pascual: but the dip is south-westerly, and the vein is non-auriferous. The thickness varies up to 14 feet, the best ore being from 4 to 6 feet of iron pyrites in coarse cubical crystals, which carries up to 80 ounces of silver per ton. It also contains some silver-ores (native and ruby-silver), galena and blende: the gangue is quartz and crystalline calcite. The country-rock is a blackish schist stained with copper.

Just above the Guadalupe river, to the north-west of the Malacate ridge, a powerful vein is seen to strike straight into

[*] *Trans. Inst. M.E.*, 1900, vol. xviii., page 384.
[†] *Ibid.*, 1902, vol. xxiv., pages 42 and 43.　[‡] *Ibid.*, 1902, vol. xxiv., page 51.
[§] *Traité des Gîtes Minéraux et Métallifères*, Paris, 1893, pages 826 and 827.

the hill, or south 70 degrees east, dipping north-eastward 63 degrees, and traceable for a height of 500 feet. The proved thickness in one place is 20 feet, and the lode is said to be spotted throughout with argentite.* On the opposite bank of the river is an escarpment of columnar porphyritic rock, which no doubt cuts the vein off in that north-westerly direction.

One of the north-and-south cross-veins (French *croiseurs*), known as the Veta Nueva, is described by Prof. Fuchs† as attaining a thickness of 16½ feet. It contains galena, cerussite and iron-pyrites, all silver-bearing. Its greatest richness is concentrated towards its intersection with other veins.

Several *vetas ocultas* or "blind veins" (having no visible outcrops‡) have been met with in the underground workings. One of these trends north 60 degrees west to south 60 degrees east, and dips north-eastward 45 degrees. The structure is represented in Fig. 5 (Plate VIII.). The rib of quartzose ore, *a*, on the hanging-wall is 18 inches thick; but the pay-streak proper consists of only 3 inches of native silver and argentite. The rib on the foot-wall, *b*, is composed of 18 inches of iron-pyrites (gold- and silver-bearing) with quartz and calcite. The two streaks are separated by a band of country-rock, *c*, *in situ*, 25 feet thick, bespattered throughout with sulphides. Some of the so-called "blind veins" are doubtless nothing but local impregnations of the schistose country-rock.

The Socabon Grande, 600 feet above the Guadalupe river, and about 500 feet below the lowest workings of the Malacate mine, was driven into the hill by the Spaniards in a southerly direction for a length of upwards of 1,738 feet. The tunnel followed a cross-vein (north and south) for some distance, and several subsidiary veins were cut, but it falls a long way short of the principal veins. It is an interesting fact that the water from the upper workings here makes its way into the adit through a natural fissure in the rock.

Although rich patches of ore occur here and there in the veins, the contents, as a rule, are not of high-grade value, the

---

* The workings were under water at the time of the writer's visit.

† *Op. cit.*

‡ Known to the Mexicans as *filones sin crestones*.

average amount of silver in the ore-bodies not exceeding probably from 20 to 30 ounces of silver per ton. The gold is very irregularly distributed in the ores, but averages from 3 to 4 dwts. per ton.

The great drawback to the district is the want of a railway. A time is perhaps not far distant when Sultepec will be connected with the narrow-gauge National Railway at Toluca, which lies about 50 miles to the north-west. In this case, coke will replace the charcoal used in the water-jacket furnaces, and the district will doubtless become an important smelting centre.

There is a group of silver-bearing veins immediately to the south-east of the town of La Unión, which may be considered apart. The general strike is north-west and south-east; the veins are comparatively thin, averaging from $2\frac{1}{2}$ to 5 feet in thickness. Native silver is fairly abundant in these veins, and, according to Mr. Villanueva,[*] the average value of the ores ranges from 39 to $242\frac{1}{2}$ ounces troy per short ton of 2,000 pounds.

In the same district are parallel veins carrying argentiferous galena; these are much wider, being from 16 to 20 feet in thickness. One vein, indeed, known as the "Salon," from the enormous cavities which have resulted from the former workings, has a minimum thickness of 39 feet, and carries galena in large isolated masses mixed with iron-pyrites and blende. These ores yield on an average 20 per cent. of lead, and from 39 to $48\frac{1}{2}$ ounces of silver per short ton.[†]

Farther north-east is another group of silver-bearing veins proper, which, like the first, are thinner than those bearing lead. The ore consists of double sulphides of silver, antimony and iron, sometimes accompanied by blende. The average silver-contents vary from $48\frac{1}{2}$ to 97 ounces per short ton.[‡]

PACHUCA, STATE OF HIDALGO.

As Pachuca can be reached in a few hours by rail from the city of Mexico, it is probably more visited by foreigners than any other mining-camp in the Republic. It is famous historically, for the workings date from 1522; and in 1557 Bartolomé de Medina invented here the well-known *patio* or-

---

[*] *Op. cit.*, pages 29 to 35.   [†] *Op. cit.*, page 35.   [‡] *Op. cit.*, page 37.

·cold-amalgamation process.   It is famous, too, for the depth and
modern equipment of its mines.   Indeed a few years ago,
Pachuca was the most important silver-producer in Mexico, but
in December, 1895, a disastrous inundation took place in the
principal workings of the Vizcaina vein, which has considerably
retarded development.   Large pumping-plants have been installed in order to drain the flooded workings, and it is to be
hoped that the district will in the near future regain its old
supremacy.

So much has been written about Pachuca, notably by Messrs.
Humboldt, Burkart, H. G. Ward, L. De Launay, von Groddeck
and P. Laur, and quite recently by Messrs. J. G. Aguilera and
E. Ordoñez,* that it would be a superfluous task to describe
the district in anything like detail.   Consequently, the writer
will give only a general sketch of Pachuca, and point out the
main characteristics of the veins.

The Pachuca range of mountains, a branch of the eastern
·chain of the Sierra Madre, which tends approximately north-
west and south-east, the two districts of Pachuca and Real del
Monte being on its western and eastern slopes respectively, is
built up of Tertiary eruptive rocks, andesites, dacites, rhyolites
·and basalts, which have been erupted in the order named.   At
Regla, north-east of Real del Monte, columnar basalt rests
directly upon altered and greatly contorted beds of calcareous
sandstones, argillaceous and marly slates, with some layers of
limestone.   These beds are regarded as of Upper Cretaceous age.

The country-rock is essentially an altered pyroxene-andesite,
which was known to the older writers as "metalliferous porphyry."   Frequently the andesite is highly quartzose, forming
dacite; and, here and there, it contains much amphibole, be-
·coming hornblende-andesite.   The normal type is a green,
highly porphyritic compact rock with no particular structure;
the country-rock (*panino*), however, is generally dark-coloured,
·and arranged in thin layers or sheets (*lajas*), which are more or
less vertical, and usually parallel to the general direction of the
veins.   Near the latter, the rock is nearly always bespattered
with small crystals of iron-pyrites, which bears little or no

---

* "El Mineral de Pachuca," *Boletín del Instituto Geológico de México*, Nos.
7, 8 and 9, 1897.   The writer is largely indebted to this memoir for the facts
recorded in the text of his paper.

silver; while the same mineral in the veins is usually in crystalline aggregates, is nearly always silver-bearing, and sometimes to a considerable extent. For instance, according to Mr. C. B. Dahlgren* the iron-pyrites of the San Pedro mine was found to contain as much as 495 ounces of silver per ton.

The rock is frequently highly altered to some considerable distance from the veins, silicification, kaolinization, and other metasomatic processes often masking its real nature. According to Mr. Leopoldo Salazar† it is difficult to meet with unchanged rock, even 130 feet from the Vizcaina or mother-lode of the district, owing to the number of small veins that run parallel with it. The pyroxene is frequently altered into chlorite and epidote, the felspars (oligoclase and labradorite) into calcite, epidote and quartz, as well as into kaolinite (*arcilla*). Mr. Waldemar Lindgren says "on the whole, the similarity of this district to the Comstock, so far as the alteration is concerned, is very striking; and there is little doubt that the two deposits owe their origin to extremely similar solutions."‡

The veins of Pachuca are referred to one east-and-west system of fractures only.§ There are secondary veins branching off from, and sometimes diagonally uniting, the east-and-west parallel veins at angles seldom exceeding 30 degrees (Fig. 7, Plate VIII.). The veins are remarkable for their persistence in length. The Vizcaina or mother-lode can be traced for about 10 miles, and other veins are visible for considerable distances, but the thickness seldom exceeds 23 feet. According to Mr. Dahlgren, the Veta Vizcaina has a thickness of 16 feet, but the average thickness of all the principal veins would probably not exceed from 6 to 8 feet. The veins are pretty regular as regards strike, but they are frequently split up into branches, which unite again farther on. These branches may occur on either wall of the main fracture, and sometimes form, as seen in plan, arcs of great radius: this is particularly noteworthy in Los Analcos (Fig. 7, Plate VIII.).

* *Minas Historicas de la Republica Mexicana*, Mexico, 1887, page 200.

† *Estudio de la Veta Vizcaina en la parte que se explota en las minas San Rafael y Anexas*, Mexico, 1895, page 11.

‡ "Metasomatic Processes in Fissure-veins," *Trans. Am. Inst. M.E.*, 1900, vol. xxx., page 650.

§ In Real del Monte, newer north-and-south veins do occur, and are termed *trasversales*, to distinguish them from the east-and-west veins known as *legitimas*.

Some of the veins have a distinct outcrop of quartz, either more or less pure, or carrying pyrites, and its derivatives limonite and earthy peroxide of iron, as well as pyrolusite. In some veins no outcrop is visible, and they are only traceable by bands of highly silicified or kaolinized andesite, carrying some thin veins of quartz, occasionally alternating with calcite.

The general dip of the veins is southerly,* but very irregular, varying from 35 degrees to vertical. Changes in the dip are frequent, much more so than is the case in Real del Monte. The average dip of the principal veins is about 66 degrees. Among the exceptions may be mentioned the Cristo vein, which is vertical at and near the surface; the dip then changes to north 63 degrees. In the San Rafael mine, the Veta Vizcaina dips generally southward $72\frac{1}{2}$ to 87 degrees, but in places it is vertical, or dips northward up to 79 degrees. Mr. Salazar† points out that the more vertical portions of the lode carry the riches. This empirical rule has notable exceptions, even in Pachuca. Thus the Veta de los Analcos has one remarkable inflection, the dip changing suddenly from 38 to $84\frac{1}{2}$ degrees; the rich zone occurred in the flat portion of the vein above the steep bend. Again in the Bartolomé Medina vein the dip changes abruptly from 35 to 53 degrees, and the riches were found in the upper flatter portion of the vein.

The ores consist of iron-pyrites, fine-grained galena and argentite, together with some copper-pyrites, blende, stephanite ($5Ag_2S.Sb_2S_3$) and polybasite ($9Ag_2S.Sb_2S_3$). Of these latter minerals, copper-pyrites is somewhat scarce, and generally appears only in depth. Blende is infrequent, having a tendency to increase in depth, while stephanite and polybasite sometimes occur in depth below the *negros*, or simple sulphides, as a sub-zone associated with galena, iron-pyrites and some argentite. The iron-pyrites of the veins is always silver-bearing, galena less so, while the blende and copper-pyrites only occasionally carry silver.

The matrix consists largely of milky quartz with a greasy

* This is the rule also of the east-and-west veins of Real del Monte; the Moran vein is one exception, the dip being northerly.

† *Op. cit.*, page 17. This applies also to the veins of Real del Monte.

lustre, or it is bluish or chalcedonic, or in the form of agate, or amethystine. Calcite is not common, it has a tendency to diminish in depth, and, as in most districts, it appears to be the last mineral formed. Siderite is occasionally found, as well as dolomite—the latter usually crystallized on amethystine quartz. Manganese as rhodonite (silicate) and sometimes crystallized in geodes as rhodochroisite (carbonate) occurs not infrequently at various depths below the oxidized zone. Thus in the San Rafael mine, at a depth of 820 feet, the gangue (quartz and some calcite) is accompanied by rhodonite,[*] and its variety bustamantite. In the Santa Gertrudis mine, barytes has been met with at a depth of 500 feet, associated with quartz and calcite.

The rare or accidental minerals, most of which are of secondary origin, are comprised in the following list:— Arragonite (El Rosario mine), valencianite (a variety of orthoclase), apophyllite (Guadalupe and El Rosario), and its variety xonotlite ($4CaSiO_3 + H_2O$, Guadalupe), mountain-cork (a variety of asbestos, El Bordo), alabandite (MnS, El Rosario), massicot (PbO), vanadinite (a vanadate of lead, San Antonio and El Puerco), and pyromorphite (a phosphate of lead, Manzano). Native copper occurs in small quantities in the San Rafael mine at a depth of 1,148 feet. It has also been found in some of the upper levels of the same mine, together with the silicate (chrysocolla) and carbonate (malachite). When the ore has a greenish stain due to copper it is said to be *avardenillado*, and this is regarded as a favourable sign.

Quite remarkable is the absence of ruby-silver. Mr. E. Ordoñez says that it has never been found in the Pachuca veins,[†] but in the catalogue of minerals compiled by Mr. Aguilera[‡] dark ruby-silver or pyrargyrite ($3Ag_2S.Sb_2S_3$) is given as having been found in no less than 21 mines of the district, and is also recorded as having been found in Real del Monte. Light ruby-silver or proustite ($3Ag_2S.As_2S_3$) appears to be entirely absent in

[*] In the Santa Cruz vein of Real del Monte, which runs north and south and dips eastward, rose-red rhodonite frequently forms part of the veinstone in the bottom levels (depth 1,568 feet in 1895). The principal ore-body has a length of 656 feet, and is from 12 to 20 feet wide.

[†] "The Mining District of Pachuca, Mexico," *Trans. Am. Inst. M.E.*, 1902, vol. xxxii., page 238, and also the memoir already quoted.

[‡] "Catálogo sistemático y geográfico de las especies mineralógicas de la República Mexicana," *Boletin del Instituto Geológico de México*, 1898, No. 11, page 62.

both districts—at least, so far, there is no authentic record of its. occurrence.

In Pachuca, as in most of the mining districts of Mexico, an oxidized zone (*colorados*) occurs from the surface down to water-level, and is succeeded by the sulphide-zone (*negros*). Frequently, psilomelane, pyrolusite and wad predominate in the gossan, giving it a black, porous, cinder-like appearance—hence the term *quemazones* (from Spanish *quemar*, to burn), which is applied to this variety of *colorados*. The silver is found in this zone in the native state, or as chloride or chloro-bromide, or as simple sulphide (argentite). Gold usually accompanies the oxidized zone when limonite is abundant. The sulphide-zone is known as *pinta azul*, for the quartzose matrix is frequently of a grey-blue colour, owing to finely disseminated lead and various silver-ores as sulphides. In some veins, as, for instance, La Vizcaina and Santa Gertrudis, the oxidized zone is almost entirely absent, and the *bonanzas* of these mines have commenced only at depths ranging from 350 to 500 feet.* The depth to which the oxidized zone extends varies considerably in different mines, and even in the same mine. In the Santa Teresa mine, it reaches a depth of 70 feet only, while in the Bordo mine oxidized ores go down to a depth of over 984 feet. The two zones frequently overlap—in other words, the *colorados* and *negros* are often found side by side at the same level, the former, as will be seen when considering the structure of the veins, being usually confined to one or both walls.† The sulphide-zone has been workd to depths ranging from 1,000 to 1,300 feet, and, in rare instances, to a little over 1,500 feet, below which is a practically barren zone, characterized by galena and blende bearing little or no silver, in a quartz-matrix.

Judging by the structure, and the presence of selvages on the walls, the veins appear to be of the class known as fissure-veins. Faults are small and few in number in the district, but that considerable movement has taken place in the walls is

* The bonanzas of the Santa Inés vein of Real del Monte commenced at a depth of 328 feet (300 metres).

† Compare *Trans. Inst. M.E.*, 1900, vol. xviii., page 372, Plate XVIII., Fig. 2. In this instance the gouge of clay on the hanging-wall seems to have prevented oxidation on that side. In Real del Monte, an instance occurs of oxidized ores occupying the centre of the vein along which it has been re-opened, the sulphide-ores being confined to the walls. "El Real del Monte," *Boletin del Instituto Geológico de México*, No. 12, page 19, Fig. 3.

evident—for frequently the quartz and ore next to them are crushed to fine powder, locally called *lamas,** and forming selvages or *guardas*† which are frequently very rich. Sometimes on the hanging-wall there is a flucan of polished clay with slickensides. The structure of the veins is often banded, the layers of quartz sometimes having geodes, indicating that that mineral has crystallized out in open spaces in the veins: but the banded structure is seldom symmetrical,‡ for it has been interfered with by later re-openings and replacements.

The structure of the Maravillas vein, in one place, at a depth of 328 feet, is as follows:—Hanging-wall, oxide of iron and clay, calcite with quartz, sulphide-ore with a little quartz, amethystine quartz, foot-wall impregnated with ore (sulphides). At a short distance from this section, the amethystine quartz was absent.

In the Cal y Canto vein, there is:—Hanging-wall, quartz with ferruginous oxides and black sulphides, " horse " or band of country-rock, rib of pure quartz, black sulphides and quartz, calcite and fragments of country-rock, quartz with ferruginous oxides and black sulphides, foot-wall. As the writer states in another paper:—"The vein appears to have been re-opened along the centre, and what has the appearance of brecciated structure may in reality be an instance of the partial replacement of the country-rock; for, as a band of ore separates it from the foot-wall, it is difficult to conceive how it could have been produced by movement along that wall without leaving evidences of crushing in the band of ore also."§ Where the same vein is impoverished, at a depth of 722 feet, it consists of several parallel veinlets of sterile white quartz coursing through country-rock (hypersthene-andesite). These are enclosed by walls having selvages of crushed quartz and clay. Sometimes, the filling con-

* In the *patio* process, *lama* is slimes or ore ground into paste by the *tahonas* or *arrastres*. Crushed ore and gangue is still more noticeable in the veins of Real del Monte, and has resulted from reopening (termed *abra* in the district).

† " Se ven las guardas marcando la linea de separacion entre el cuerpo de la veta y la roca en que arma, fortemente impregnado de mineral " (Mr. Santiago Ramirez). *Guardas de arcilla* are salbands, flucans or selvages.

‡ One instance of normal or symmetrical banded structure occurs in the Cabrera vein of Real del Monte, *Boletin del Instituto Geológico de México*, No. 12, page 22, Fig. 7 ; and also *Trans. Fed. Inst. M.E.*, 1891-1892, vol. iii., Plate LX., Fig. 2. These examples are quoted, as symmetrical structure is rare in Mexico.

§ " Notes on the Structure of Ore-bearing Veins in Mexico," *Trans. Am. Inst. M.E.*, 1902, vol. xxxii., page 301.

sists of quartz and argentiferous sulphides (*azogues* of the miners), separated into bands by veinlets of pure or barren quartz, more hyaline than that of the rest of the filling. Sometimes, the bands of pure quartz are very irregular in shape; then no banded structure has resulted.

Very rarely, the banded structure has a symmetrical appearance; but even then it is very doubtful whether it has been built up by the different layers having been deposited *pari passu* from opposite walls: for the ribs of pure quartz, which help to make up the banded structure, have most probably been formed at a later date than the rest of the filling. The lodes, in fact, would appear in many instances to have been re-fractured in planes parallel to the walls, giving access to siliceous solutions. From these, quartz has been deposited, partly by crystallization in the open spaces resulting from the new fractures, and partly by replacement of the older filling. The veins sometimes have a brecciated structure. In Real del Monte, that is of far more frequent occurrence.

The Santa Gertrudis mine, which had a depth of 1,200 feet in 1895, has been sunk on a vein which runs a few degrees north of east, and dips southward 70 to 80 degrees. As the writer saw the vein in the *bonanza* portion in that year, the filling—often 20 feet thick—was very soft, the ore occurring in bands (*cintas*) or small bunches (*en boleo*), the matrix being quartz of the variety known as "sugar-spar." Two main shoots have been found, which have a tendency to pitch eastward. On the hanging-wall there is usually a layer several inches thick of highly polished blue clay with slickensides, termed "cab"[*] by the Cornish miners of the district. This wall is very treacherous, for near the lode on that side the rock is strongly kaolinized. The whole of the lode is taken out and timbered up by square sets, spilling or fore-poling sometimes having to be resorted to in stoping and driving. The levels are supported by masonry.

---

[*] In Cornwall, *cab* is (1) chalcedonic quartz: or (2) altered granite forming one and sometimes both walls of a tin-bearing lode. According to Sir Clement Le Neve Foster it consists at the Lovell mine of quartz and mica, gilbertite, chlorite, iron-pyrites, copper-pyrites, and a little schorl (common black tourmaline), from 6 to 12 inches thick. In West Cornwall, *cab* is the name of a horny gall on the hand caused by friction (Miss Courtney). When the side of a lode is as hard as flint, Cousin Jack says that it is "cabby."

A remarkable feature of the Pachuca veins is the orientation of the payable ore-bodies. Messrs. Aguilera and Ordoñez[*] point out that, although the veins course in a general east-and-west direction, the ore-bearing zones are directed north-west and south-east; and that, not infrequently, where one mine has an ore-shoot, the adjoining one will have a barren zone opposite it, so that the *bonanzas* of the different veins almost present an alternating position. This is directly contrary to the empiric rule known among miners as "ore-against-ore" (French *vis-à-vis*). The rule, however, is by no means a general one, for many exceptions occur.

Fig. 6 (Plate VIII.) is a theoretical diagram in which an explanation of this peculiarity is suggested. It is wellknown that in most districts a band of country of moderate hardness and usually of a distinct colour, and favourable for the deposition of ore (known to the Germans as *Erzträger*), is sooner or later succeeded by a band of very hard, or, on the other hand, of very soft country in which the veins are more or less sterile. In Pachuca, it is probable that the alternating bands are directed in a north-west to south-east direction, or parallel to the Pachuca range; consequently, two veins, A and B, which are a sufficient distance apart, will have their ore-bodies alternating *a*, *b*, *c* and *d*, as in the diagram. The two ore-bodies *b* and *d* of the vein A are opposite the sterile ground *f* and *h* of the lode B; while the two ore-bodies *a* and *c* of the lode B are opposite the sterile ground *e* and *g* of the lode A. We may suppose a third vein, C, contiguous to A. In this instance the ore-bodies will be practically opposite each other, or *vis-à-vis*.

Many of the ore-bodies are extremely irregular, having no definite shape and no determined pitch. A few of the largest, however, appear to have a general tendency to pitch to the eastward.[†] In the Cristo and San Juan Analco mines, this is fairly well-marked, but less so in the case of the Maravillas and Santa Gertrudis mines. In the Bordo mine, the ore is in small, narrow,

---

[*] *Boletín del Instituto Geológico de México*, Nos. 7, 8 and 9, page 73. In Mr. Ordoñez's paper, published in *Trans. Am. Inst. M.E.*, vol. xxxii., page 239, the payable zone is said to be oriented from north-east to south-west— this must be a misprint.

[†] This also appears to be the case with the Veta Vizcaina in Real del Monte. The pitch of the ore-shoots of the principal north-and-south veins (Santa Inés and Santa Brigida) seems to be southerly.

nearly vertical shoots (*clavos*). In the San Rafael y Anexas mines on the Vizcaina or mother-lode, the ore-shoot, worked from the 354½ feet (108 metres) level to a depth in 1895 of 1,230 feet, appears in inclined columns which pitch westward at angles varying from 57½ to 80 degrees, yet these columns may be regarded as forming one long course of ore (1,500 feet measured horizontally) which pitches eastward, or in the opposite direction.[*]

Frequently the ore occurs in small pockets (*boleos*) or in little bunches (*ojos*), and sometimes it is scattered throughout the vein in isolated spots known as *moscas*.[†]

It has not yet been definitely proved here, or indeed anywhere in the Republic, whether a third ore-bearing zone occurs below the impoverished zone. This is a problem that can only be solved by future development.

Mr. Ordoñez points out that in the San Rafael and Maravillas mines, " beyond an impoverished portion, there has immediately appeared native copper with polybasic mineral, blende and notably rich ores (naturally rebellious), showing what would be the mineral composition of the ores of the third zone."[‡] The presence of native copper looks as if the above complex ores had resulted from secondary enrichment, due to surface-waters, which have made their way down the lodes, or possibly to ascending thermal solutions. The fact of native silver having been found in another vein at a depth of 820 feet, and of chloro-bromide of silver having occurred in depth in the Santa Gertrudis mine, associated with galena and polybasite, seems to point to the same conclusion. Again, according to Mr. S. F. Emmons,[§] " in the San Juan region of Colorado, there is a very strong belief held by some of the more experienced and thoughtful mining engineers, that the rich silver-minerals in those veins, such as ruby- and brittle-silver [stephanite], are the result of secondary enrichment by descending waters." . . . At

---

[*] See the longitudinal section of the mine prepared by Mr. L. Salazar (*op. cit.*, page 20). In that given by Messrs. Aguilera and Ordoñez, the inclined columns pitching westward are not shown, the whole being coloured red, but the general easterly pitch can be made out.

[†] Compare *Trans. Inst. M.E.*, 1900, vol. xviii., page 379, for the *metal mosqueado* of the Taviches district, Oaxaca (hornblende-andesite).

[‡] "The Mining District of Pachuca, Mexico," *Trans. Am. Inst. M.E.*, 1902, vol. xxxii., page 240.

[§] "The Secondary Enrichment of Ore-deposits," *Trans. Am. Inst. M.E.*, 1900, vol. xxx., page 195.

Neihart, Montana, according to Mr. W. H. Weed, "the primary minerals are galena, blende and pyrite; the secondary sulphides are polybasite, ruby-silver, more rarely a pure transparent blende. Under the microscope galena is seen altering to a spongy polybasite. Polybasite and pyrargyrite are seen as crystalline aggregates and crusts on all other minerals, but in no cases coated by other minerals."[*]

In this connection we may cite Messrs. Aguilera and Ordoñez:—" Pronto se unen á estos sulfuros simples la estefanita y la polybasita íntimamente mezclados con la galena y sulfuros de plata, que se aislan en pequeñas cristales tabulosos en pequeñas geodas."[†]

It is undoubtedly of good augury for the future development of Pachuca that the veins, although impoverished, are well-formed, and retain their general structure at the greatest depth yet reached (1,653 feet); and that, whatever the later or secondary alterations may have been, the primary filling or enrichment shows evidence of having been produced by thermal and therefore ascending solutions. So far as we know, the above paragraph will apply to other classical districts of the Republic, worked for long periods, such as Real del Monte, Catorce (San Luis Potosí), Zacatecas and Guanajuato, where the greatest depths reached are 1,568, 1,600, 1,700 and 1,968 feet respectively.

CORRECTIONS AND ADDITIONS.

Region of Taviches, Oaxaca:—*Trans. Inst. M.E.*, 1900, vol. xviii., page 379, 2 lines from the bottom *for* "an amygdaloidal structure," *read* "a brecciated structure."

Tasco, Guerrero:—*Trans. Inst. M.E.*, 1901, vol. xxi., page 198, line 15, *after* "modern eruptive rocks," *add* "the veins of Tasco, according to Mr. J. G. Aguilera are related to andesite and rhyolite."[†]

Santiago Papasquiero and San Dimas, Durango:—*Trans. Inst. M.E.*, 1902, vol. xxiii., page 252, line 10 from bottom, *for* "silitication" *read* "silicification"; and page 256, lines 7 and 8, *for* "(? in the same district)" *read* "in the same district."

Zacatecas and Guanajuato:—*Trans. Inst. M.E.*, 1902, vol. xxiv., page 42, line 18, *for* "Upper Tertiary (Pliocene)," *read* "Miocene or Lower Pliocene."

[*] *Trans. Am. Inst. M.E.*, 1900, vol xxx., page 196.
[†] *Op. cit.*, page 70.
[‡] "Distribution of the Mineral Deposits of Mexico," *Trans. Am. Inst. M.E.*, 1902, vol. xxxii., page 514.

The red conglomerate of Zacatecas is Miocene; that of Guanajuato, Lower Pliocene, or even perhaps Miocene. The rhyolite and rhyolitic tuffs of both districts are of Pliocene age.*

Page 47, line 12, *for* "grey trachyte," *read* "rhyolite."

Note ‡ on same page. The bufas, south of Guanajuato city, are formed of rhyolitic tuff.†

Page 50, line 3 *for* "an eruptive or prior date" *read* "an eruptive of prior date."

Page 50, line 17. According to Mr. Ordoñez ‡ the succession of Tertiary eruptives in Mexico, commencing with the oldest, is as follows : —

Granites	—	Granulites.
Diorites	—	Diabases.
Andesites	—	Dacites.
Rhyolites.		
Dacites	-	Andesites.
Basalts	—	Basaltic andesites.

Mr. W. H. Weed, of the U.S. Geological Survey, who recently made a journey across the Sierra Madre from Parral to the Gulf of California, gives the order of succession of the rocks met with as follows :—(1) Andesite ; (2) trachyte ; (3) granite ; (4) dacite ; (5) rhyolite ; and (6) basalt.§

In Zacatecas, there appear to be three systems of veins :—(1) North-west and south-east, dip southward, northerly dip exceptional, comprising most of the principal veins ; (2) east and west, dip southward, northerly dip exceptional ; and (3) north and south ; this appears to be a minor system not well developed.

The Cantera and Quebradilla veins, and the Veta Grande, are examples of the first. The Cantera vein has already been described in the text. The Quebradilla vein (south of Zacatecas city) has an average strike of north 40 degrees west, and dips southward 55 degrees. The average thickness is 36 feet. This vein unites with the Cantera below the city.

The Veta Grande, 4 miles north of the city, has an average strike of north 52 degrees west, and dips southward from 55 to 60 degrees. It has been traced for 7 miles, but the principal workings are about 4 miles north of Zacatecas city, and extend for a length of a little over 2¾ miles. The thickness varies from 3 to 90 feet, but the average may be taken at about 30 feet. The following minerals have been found in the Veta Grande :—Native silver, argentite (*azul plomillosa* or the *polvorilla* variety), copper-pyrites (*bronce chino*), iron-pyrites (*bronce dorado*), pyrargyrite, cerargyrite (*plata azul* or *plata cornea*, AgCl), cerussite (*soroche plomoso*), and galena with or without silver (*soroche reluciente* and *tescatete* respectively).

The principal east and west vein is the important one known as the Veta de San Bernabé or Descubridora, 3 miles north of Zacatecas city. Malanoche, mentioned in the text, is on this vein. The dip is to the south. The silver contains 3 parts per 1,000 of gold. This vein contains iron-pyrites resembling

---

* "Las Rhyolitas de Mexico," by Mr. Ezequiel Ordoñez, *Boletin del Instituto Geológico de México*, No. 14, pages 26.28.

† *Op. cit.*, Mr. Ordoñez, pages 21-26.

‡ *Boletin del Instituto Geológico de México*, No. 14, page 66.

§ "Notes of a Section across the Sierra Madre Occidental of Chihuahua and Sinaloa, Mexico," *Trans. Am. Inst. M.E.*, 1901, vol. xxxii., page 458.

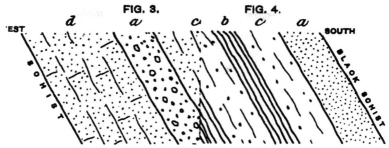

FIG. 3.   **FIG. 4.**

a, GALENA, IRON-PYRITES, BLENDE, SILVER-OVER-ORES, NATIVE GOLD AND SILVER, AND QUARTZ
b, NARROW STREAK OF ORE ON HANGING-WALL, IATED IRON-PYRITES
c, MINERALIZED COUNTRY-ROCK   MISTOSE COUNTRY-ROCK IMPREGNATED WITH QUARTZ,
d, MINERALIZED QUARTZ-ROCK   MISPICKEL AND IRON-PYRITES

Scale, 8 Feet to 1 Inch.   *Scale, 8 Feet to 1 Inch.*

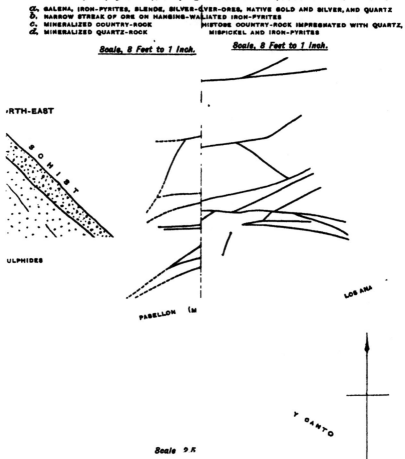

Scale 9.5

polished bronze in appearance, and called *bronze bruno*. It is highly argentiferous, containing up to 1,751 and 2,188 ounces of silver per short ton of 2,000 pounds. Ground in the *arrastres* it is known as *marmajas*.[*]

The silver-bearing veins of Zacatecas may contain, besides the ores mentioned in the text, stephanite (*azul acerado*), embolite (*plata verde* or chloro-bromide of silver), proustite (*rosicler claro*), polybasite and tetrahedrite or *fahlerz*.

The matrix is quartz, with some calcite, heavy spar, brown spar, and occasionally (as in the Cantera vein) fluorspar. Zinc-blende, magnetite, arsenical pyrites, malachite, azurite and stibnite (as, for instance, in the Quebradilla mine), are the heavy non-silver-bearing minerals.

Among the rare or accidental minerals found, nearly all of secondary origin, may be mentioned native sulphur, native bismuth (El Cristo and El Orito mines), bismutite (a hydrated carbonate of bismuth), cuprite (Cu_2O), bornite or variegated copper-pyrites ($3\ Cu_2S.Fe_2S_3$, from Refugio del Oro), bromyrite (AgBr from San Vicente and La Luz), iodyrite (AgI from Quebradilla), smithsonite (zinc carbonate from Albarradin), descloizite (vanadate of lead), and wulfenite (molybdate of lead).

The various deposits described in this series of papers occur in:—(1) recent eruptive rocks (Tertiary):—in Matapé (Sonora), in granite; in Taviches (Oaxaca), in hornblende-andesite; in Santiago Papasquiero and San Dimas (Durango), and in Zacatecas and Guanajuato, in andesites and rhyolites: in Pachuca (Hidalgo), in hypersthene-andesite; or (2) in sedimentary rocks with which recent eruptives are more or less closely related:—in limestone or schists of Cretaceous age in Santa Cruz de Alayá (Sinaloa), Tasco (Guerrero), and Sultepec (State of Mexico), and in schists of unknown age in Zacatecas and Guanajuato.

As already noted, at Taviches, Oaxaca, limestone has been found in one or two instances below the andesite. A belt of Jurassic limestone occurs 4 miles from the vein described near Matapé, Sonora; patches of Cretaceous limestone are found in the Zacatecas, Guanajuato and Sultepec regions, as well as in the vicinity of Pachuca, and the eruptive rocks of Durango probably cover Cretaceous limestone, as a belt of that rock runs in a general north-west to south-east direction immediately to the east of the eruptive masses. This general occurrence of a porous rock resembling limestone adjacent to, or not very far distant from, the silver-bearing veins described is noteworthy, and should be taken into account in considering the genesis of the deposits.

— —

The PRESIDENT (Mr. W. O. Wood) moved that a vote of thanks be accorded to Mr. Halse for his interesting paper.

Mr. T. E. FORSTER seconded the resolution, which was cordially approved.

———

[*] *Anales del Ministerio de Fomento*, vol. v., page 307.

THE NORTH OF ENGLAND INSTITUTE OF MINING AND MECHANICAL ENGINEERS.

GENERAL MEETING,
HELD IN THE WOOD MEMORIAL HALL, NEWCASTLE-UPON-TYNE,
JUNE 11TH, 1904.

MR. W. O. WOOD, PRESIDENT, IN THE CHAIR.

The SECRETARY read the minutes of the last General Meeting, and reported the proceedings of the Council at their meetings on May 28th and that day, and of the Council of The Institution of Mining Engineers.

The SECRETARY read the balloting list for the election of officers for the year 1904-1905.

The following gentlemen were elected, having been previously nominated :—

MEMBERS—

MR. VICTOR BUYERS COLLINS, Mine-surveyor, Lewis Street, Islington, *via* Newcastle, New South Wales.

MR. RICHARD STANLEY DAVIES, Mining Engineer, Trynant Hall, Llanhilleth, Monmouthshire.

MR. JOSEPH EDWIN GOWLAND, Mining Engineer, Luchana Mining Company, Limited, Apartado 45, Bilbao, Spain.

MR. JOHN LAIRD, Engineer, Alliance Jute-mills, Shamnagur, Bengal, India.

MR. CLEMENT ALFRED RITSON PEARSON, Engineer, The Old Hall, Denby, near Derby.

MR. WILLIAM PIERCY, Engineer, 32, Grainger Street West, Newcastle-upon-Tyne.

MR. JAMES SHEPHERD, Mechanical Engineer, Hulne Avenue, Tynemouth.

MR. GEORGE ALFRED STONIER, Chief Inspector of Mines in India, 6, Dacre's Lane, Calcutta, India.

MR. JOHN RICHARD WILLIAMS, Metallurgist, P.O. Box 149, Johannesburg, Transvaal.

ASSOCIATE MEMBERS—

MR. WILLIAM COURTENAY DAWES CRUTTENDEN, 5, Laurence Pountney Lane, Cannon Street, London, E.C.

MR. ISHMAEL MINNOW, 4, King's Road, Camborne, Cornwall.

DISCUSSION OF MR. T. ADAMSON'S PAPER ON "WORKING A THICK COAL-SEAM IN BENGAL, INDIA."[*]

Mr. THOMAS ADAMSON, referring to Mr. J. B. Atkinson's remarks on the scarcity of timber in India,[†] wrote that Mr. G. A. Stonier, chief inspector of mines in India, had stated that " in some cases, timbering is necessary, and it is as a rule in India well carried out. The rough village-carpenters become propping-mistries, and their work shows that they can be trained with considerable success.   The timber used in mines varies from second-hand railway-sleepers to round timber; excellent examples of the use of the latter can be seen in the Assam and Burma collieries where timbering is heavy.   One colliery[‡] in Bengal has used 30,000 props in a year."[§]   These props cost 1s. 4d. (1 rupee) each, so that the cost is about £2,000 a year for timber.

Round timber or sal-wood *(Gavœus Gaurus)* grows freely in the jungles on and surrounding the Giridih coal-field.   Timber, for use in the mines, is purchased close to the collieries, at a cost varying from 1s. to 1s. 4d. (12 annas to 1 rupee) per prop, measuring 7 to 8 inches in diameter and 22 feet in length. Second-hand railway-sleepers, used so extensively in the undergoing and in chocks, cost 1s. 4d. (1 rupee) each.

The statistics quoted by Mr. G. A. Stonier, chief inspector of mines in India, indicate that the method is relatively safe.

————

A " Memoir of the late William Cochrane," by Mr. C. A. Cochrane, was read as follows:—

.

[*] *Trans. Inst. M.E.*, 1903, vol. xxv., pages 10, 192 and 396 ; vol. xxvi., page 19; and vol. xxvii., pages 10 and 136.

[†] *Ibid.*, vol. xxvii., page 11.

[‡] The East Indian Railway collieries at Giridih.

[§] *Report of the Chief Inspector of Mines in India*, 1902, by Mr. Geo. A. Stonier, 1903, page 15.

MEMOIR OF THE LATE WILLIAM COCHRANE.

By CECIL A. COCHRANE.

William Cochrane was born on January 23rd, 1837, at Black-brook, near Dudley, being the second son of Alexander Brodie Cochrane, later of The Heath, Stourbridge, and of Sedgley Hall, Staffordshire, Deputy-Lieutenant for the county of Stafford and proprietor of the Woodside iron-works.

After being educated at a private school at Wilmslow in Cheshire, he proceeded to King's College, London, where he had a distinguished career. In addition to natural ability he possessed the power of application in a marked degree, and numerous prizes for mathematics, French, German and Divinity fell to him during the years 1852, 1853 and 1854, culminating in his election to an Associateship of the College in the latter year.

From King's College, it was intended that he should proceed to Cambridge with a view to reading for the Mathematical Tripos and afterwards taking up the Bar as a profession, but the breakdown of his father's health compelled him to give up a University career, a necessity he frequently regretted in after-life. What this decision must have meant to him may be understood from a letter written after his death by a distinguished contemporary of his at King's College. In it occurs the following passage :—" It (the news of his death) carried me back to the days when he and I sat in the same classes at King's College, and when I envied him and Charles* their infinite (as it seemed to me) power of work and mental vigour. We expected him to do great things at Cambridge, and were disappointed when business claimed him. The two brothers were, I think, the strongest ' heads ' we had at King's College."

From King's College, he went to his father's works and collieries in Staffordshire, and in 1857 came to the North of

* President of the Institution of Mechanical Engineers in 1889.

England to assist in the development of the coal-royalties in the possession of his father. After a brief residence in Darlington, he came to Newcastle-upon-Tyne in 1858 and was subsequently engaged in the sinking of Elswick colliery from the Low Main to the Brockwell seam, and the laying out and sinking of the Tursdale and New Brancepeth collieries. In addition to being connected at various periods of his lifetime with other mining properties (in South Wales and South Yorkshire), Mr. Cochrane was a director of, and consulting engineer to, the Blackwell Colliery Company, Limited, and the Nunnery Colliery Company, Limited, from their commencement.

At one time, he had an extensive practice as a consulting engineer, and his services were requisitioned in some very important mining cases and arbitrations.

It is perhaps, however, in connection with the mechanical ventilation of mines that his work as a member of the Institute will be best remembered. In conjunction with Mr. J. J. Atkinson, H.M. inspector of mines, and Mr. D. P. Morison, and later with Mr. Théophile Guibal, he conducted a prolonged and valuable series of experiments on ventilation, and embodied the results obtained in a number of papers read before the Institute. From the first a strong advocate of the use of centrifugal ventilators, he was instrumental in the introduction of the Guibal fan into this country in 1863, and on the death of Mr. Guibal in 1889 he read a brief notice of his life and work before the members of the Institute.

In 1875-1876, Mr. Cochrane co-operated with Prof. A. Freire-Marreco and Mr. D. P. Morison in investigating the conditions under which coal-dust would explode. A paper read in November, 1878, by Prof. Freire-Marreco, gave a complete account of the experiments carried out at the Elswick colliery and elsewhere.*

Mr. Cochrane was elected a member of the North of England Institute of Mining and Mechanical Engineers in 1859. In 1866, he was elected a member of Council, and in 1870 a Vice-President. He served the office of President in 1890 to 1891, and delivered a Presidential address in which he urged the importance of economy in the use of coal, having regard to

* " An Account of Some Recent Experiments on Coal-dust," *Trans. N.E. Inst.*, 1878, vol. xxviii., page 85.

the exhaustion of the thick seams and the increased difficulty and cost encountered in working thin seams at great depths.*

It is interesting to recall that Mr. Cochrane was Honorary secretary of the meeting of the Institute held at Manchester in 1865, and of the committee appointed to report upon the various Systems of Underground Haulage of Coal.   He took a keen interest in a scheme of technical education started by the Institute in 1869, and was a member of the committee appointed by the Institution to carry it through.   During the visit of the Institute to Douai in 1878 his mastery of the French language was of great service to the members.   He took a very active part in the entertainment of the Belgian engineers who visited Newcastle in 1892, and acted as interpreter during their stay.

He acted as one of the representatives of the North of England Institute of Mining and Mechanical Engineers upon the Council and the Finance Committee of The Institution of Mining Engineers from its formation in 1889.

In 1868, he was elected a member of the Institution of Mechanical Engineers,† and he was also for several years a member of the Institution of Civil Engineers.

From its inception he took a keen interest in the Durham College of Science, of which he was a member of Council; and in 1901 the University of Durham conferred upon him the honorary degree of M.Sc. in recognition of his scientific attainments and labours on its behalf.

He took a prominent part in the promotion and management of the successful Exhibition held at Newcastle in 1887, and

---

* The papers communicated to the *Transactions* include :—
"Description of the Guibal Ventilator at Elswick Colliery," *Trans. N.E. Inst.*, vol. xiv., page 73.
"The Harrison Cast-iron Steam-boiler," *Trans. N.E. Inst.*, vol. xvi., page 35.
"A Comparison of the Guibal and Lemielle Systems of Mechanical Ventilators," *Trans. N.E. Inst.*, vol. xvi., page 57.
"Remarks on the Guibal and Lemielle Systems of Ventilation," *Trans. N.E. Inst.*, vol. xviii., page 139.
"The Advantages of Centrifugal-action Machines for the Ventilation of Mines," *Trans. N.E. Inst.*, vol. xxvi., page 161.
"Obituary Notice of the late Théophile Guibal," *Trans. Inst. M.E.*, vol. i., page 79.
"Presidential Address," *Trans. Inst. M.E.*, vol. ii., page 181.
"A Duplex Arrangement of Centrifugal Ventilating Machines," *Trans. Inst. M.E.*, vol. ii., page 483.

† "The Various Systems of Ventilation of Mines," *Proceedings of the Institution of Mechanical Engineers*, 1869, page 137.

WILLIAM COCHRANE,

PRESIDENT OF THE NORTH OF ENGLAND INSTITUTE OF MINING AND MECHANICAL ENGINEERS, 1890-1891.

Born January 23rd, 1837, and died on November 25th, 1903.

(Presented by the North of England Institute of Mining and Mechanical Engineers).

devoted much time to the various committees of which he was a member.

No notice of Mr. Cochrane's life would be complete without a reference to the Sick Children's Hospital with which he was identified so closely. Himself one of the founders, he served the Hospital as honorary secretary for 25 years, and never ceased to work in its interest to the day of his death.

In 1898, he succeeded his brother Charles as chairman of Cochrane and Company, Limited, a position which he held at his death.

In 1859, he married Eliza, second daughter of William Blow Collis of Wollaston Hall, Stourbridge, who predeceased him. Mr. Cochrane died on November 25th, 1903, at his residence, Oakfield House, Gosforth, after a long illness, leaving two sons and one daughter.

———

Mr. A. L. STEAVENSON said that Mr. Cochrane joined the Institute a few years later than himself, and from that time until his death they were constantly in communication on matters which affected the interests of the Institute. On questions of ventilation, they took perhaps opposite sides for many years, Mr. Cochrane supporting the centrifugal fan, while he (Mr. Steavenson) thought, and still for some reasons thought, that the displacement-fan was the best: although, when they considered the large volumes of air which it was now necessary to deal with at some of the mines, he agreed that the use of the centrifugal fan was preferable. He felt great regret at now being present on the occasion of the reading of a memoir of their late friend.

Rev. Principal GURNEY said that the Durham College of Science owed a great debt to this Institute, for it was founded jointly by the Institute and the University of Durham, and there was no member who took a more active interest in the development of the College than the late Mr. Cochrane. It must be satisfactory to the Institute, as founders of the College, to have observed the great strides which that college had made; and the measure of success which it had achieved was very largely due to the unsparing labours and the wise advice of Mr. Cochrane. On behalf of his colleagues, the staff and members of the Council of the College, he deeply deplored his loss, which in some respects would be irreparable.

Mr. J. H. MERIVALE, in moving a vote of thanks to Mr. Cecil A. Cochrane, for his kindness in contributing a memoir of the late Mr. Cochrane, said that it would be incomplete without a reference to one of the strongest points in his character, namely, his great kindness of heart and liberality.

The PRESIDENT (Mr. W. O. Wood), in seconding the resolution, said that the members would heartily and sincerely endorse the remarks of previous speakers.

The resolution was approved.

THE LATE SIR CLEMENT LE NEVE FOSTER.

Prof. HENRY LOUIS moved a vote of condolence with the widow and family of the late Sir Clement Le Neve Foster. Although that gentleman had retired from the position of H.M. inspector of mines, he was, at the time of his death, professor at the Royal School of Mines.

Mr. J. H. MERIVALE seconded the proposal, which was adopted.

The PRESIDENT (Mr. W. O. Wood) read the following paper on " The Re-tubbing of the Middle Pit, Murton Colliery, 1903 " : —

THE RE-TUBBING OF THE MIDDLE PIT, MURTON COLLIERY, 1903.

By W. O. WOOD.

Introduction.—The middle pit, Murton colliery, is one of the three historic shafts sunk at that colliery through the Magnesian Limestone and underlaying sand during the years 1838 to 1842. A paper on "Murton Winning in the County of Durham" was contributed by the late Mr. Edward Potter to the Institute.[*] It contains a very complete and interesting account of the operations, from the breaking of the ground on February 19th, 1838, until the three pits reached the Hutton Seam in April, 1843, five years later. A few leading facts of interest may be quoted, as follows:—

The depth to the bottom of the Magnesian Limestone is 456 feet. The thickness of the sand at the base of the Magnesian Limestone was found to be 34 feet 6 inches at the east pit, 27 feet 8 inches at the middle pit, and 26 feet at the west pit.

The engines employed were estimated to develop 1,584 horse-power, and were supplied with steam from 39 boilers. The engines drove 27 setts of pumps, of which 9 delivered to bank, and the total quantity of water raised when sinking through the sand was 9,306 gallons per minute from a depth of 540 feet.

From April 16th to May 10th, 1841, 17 buckets and clacks were changed per day, the cost of grathing the buckets being £2 5s. 10½d. each for the 19 inches setts, and £1 17s. 6d. each for the 16 inches setts. The consumption of coal, when all the engines were employed, amounted to 1,000 chaldrons (2,650 tons) per fortnight.

The pressure of the water at the base of. the tubbing is upwards of 108 pounds per square inch.

[*] *Trans. N.E. Inst.*, 1856, vol. v., pages 43 to 61.

Many years ago, the west pit, used as an upcast, was entirely re-tubbed, and the east pit was re-tubbed in 1891.

Having regard to the age of the tubbing in the middle pit (1838), and to the fact that it was a little out of shape, although strengthened with internal strengthening-rings, it was decided to re-line this shaft also for a depth of 232 feet. The shaft is 13 feet 9 inches in diameter and is divided by a brattice into two drawing-shafts, the west side or polka pit drawing from the Main coal-seam, and the east side or middle pit drawing from the level of the Low Main coal-seam. The polka pit is a double-shift pit, drawing coals incessantly 20 hours per day; and the middle pit is a single-shift pit, drawing coals 10 hours per day, and occupied after coal-drawing hours in changing men, sending down material, etc.

The difficulty of re-lining this busy shaft, with four cages constantly ascending and descending, without stopping the pits appeared well-nigh insuperable. It was, however, accomplished without interfering with the coal-drawing in the least, or stopping either pit a single day; and a short description of the operation may be interesting, and prove of service to others similarly situated.

Figs. 1, 2 and 3 (Plate X.) indicate the general arrangements, showing the position of the engines, crabs, etc., employed, and the cradles, kibbles and ropes in the shaft.

Fig. 4 (Plate X.) shews the old tubbing, strengthening-rings, cages, timber, delivery-column, cables, pipes, and other obstacles. Fig. 5 (Plate X.) shews the shaft with the lining in position.

Figs. 6, etc. (Plate X.) shew the new tubbing, the bell-mouth, and special courses in detail.

A suitable bed for the new tubbing was selected at a depth of 96 feet below the lowermost wedging-crib.

Description of the Lining.—Very careful measurements were made at three points, and afterwards checked, of the shaft and fittings from the proposed position of the new wedging-cribs up to the point to which it was decided to carry the new tubbing, specially designed to suit the conditions. The metal used for the tubbing is of special quality and to specification

The brackets are all internal. The height of each segment is 2 feet, and there are ten segments in each round or ring, except in special cases (Figs. 7 and 8, Plate X.).

Method of forming the Rings of Tubbing.—The joints are machine-planed so as to make a perfect vertical joint, and then bolted firmly together into a complete ring. The ring, thus formed, is then placed on the face-plate of a special lathe, large enough to take in the ring, and both of the horizontal flanges are properly faced to gauge. Part of the tubbing was dealt with by machinery in operation for making the metal-lining of some of the London tube-railways and with the same accuracy.

The vertical and horizontal joints have three V-shaped grooves on each face, and are made metal to metal (as it was not considered advisable to use any of the usual jointing-material). There are 40 bolts, 1 inch in diameter, in the entire circle, and 4 bolts are used in each of the vertical joints.

Each ring is cast with bracketted grooves (Figs. 9 and 10, Plate X.) to receive the brattice, which divides the pit into two portions, and every third ring is cast with pockets (Figs. 11, 12 and 13, Plate X.) to receive the guide-buntons and stringing-planks. The individual segments were numbered round the ring, and each completed ring was marked from the lowermost crib upward, so as to come into its proper place: no alterations of the bolt-holes, etc., being required in fitting them together in the shaft.

The largest inside diameter of the old tubbing is 13 feet 9 inches, and that of the new tubbing is 12 feet 6 inches throughout. The space between the two tubbings is about 4 inches, and it is filled with a cement-concrete composed of 3 parts of shingle, sifted over a ¼ inch screen, and 1 part of best Portland cement. The cement-concrete was rammed solid, as each ring of tubbing was placed in position, each course of cement-concrete being left about 1 foot below the upper edge of each ring of tubbing; and if there should be any chance of a parting in the cement-concrete (which is hardly probable) it will not occur directly opposite an horizontal joint in the tubbing.

The lower courses of the tubbing were, of course, large enough to suit the crib-bed, A and B, made to receive it after the brick-walling had been taken out.   It consists of one bottom-crib, A, 13 feet 9 inches in inside diameter and 1 foot 5½ inches broad on the bed (Figs. 14, 15 and 16, Plate X.).   This was wedged in the usual way, and a second crib, B (Figs, 14, 17 and 18, Plate X.), 1 foot 8½ inches broad on the bed was laid upon it and wedged.   The upper crib, B, is made with a snug or lip, a, all round it, to which the first or foundation-course, C, of tubbing (Figs. 14, 19, 20 and 21, Plate X.) is bolted, both surfaces being machine-planed.

The foundation-course of the tubbing is narrowed in diameter and each of the four bell-courses, D, E, F and G (Fig. 14, Plate X.), are inclined at the same angle.

The top course reduces the inside diameter to 12 feet 6 inches —the finished size of the rest of the new tubbing.   The whole of this portion of tubbing is backed with cement-concrete.

There are two special courses of tubbing, the necessity for the first one, H (Figs. 6 and 22, Plate X.) arising at the top of the forty-fifth course of tubbing.   It is a combined wedging-crib and water-ring, placed opposite to a crib in the old tubbing.

The second special course, I, was necessarily used (Figs. 6, 23 and 24, Plate X.) to pass a wedging-crib, which has a strengthening-ring placed in front of it.   It was not deemed advisable to remove this ring, and a special course of tubbing was designed to pass the obstruction.   The body of this course extends to, and in a line with, the flanges of the tubbing; and the flanges of the course are only used at such positions in the shaft as to be quite clear of the passing cages.

Immediately above this course, I, is placed one course of the plain tubbing, and, behind it, is fixed a ring, K, of wrought-iron pipes, 2 inches in diameter, with a number of ⅜ inch holes drilled in it; no cement is placed between the ring, K, and the plug-holes in the tubbing (Figs. 23 and 24, Plate X.). This ring was placed at the top of the eighty-third course of tubbing, for the purpose of removing any water that might find its way through the old tubbing above, and was protected above (and below) with a layer of wood and connected to water-ring pipes, M, 3 inches in diameter.

The total number of 2 feet courses of plain tubbing used is 114, together with 2 special courses, and the work, including

the shifting of the rising main, occupied 723 hours, as follows:—
Taking-out walling and making ready the lowermost crib-bed,
96 hours; laying and wedging the two main cribs, 80 hours;
putting in the remainder of the bell-tubbing, 32 hours; replacing the kep-buntons and shifting the rising main, 64 hours;
putting on the straight courses of tubbing, and including the
special courses, 259 hours; and taking out the cage-timber,
getting the cradles, etc., and replacing the same in time for coal-
work, 192 hours.

The work was accomplished without the slightest accident
of any kind: being commenced on Good Friday, April 10th,
and finished on November 7th, 1903. These dates include
6 week-ends, when operations were entirely suspended from
various causes.

Method of Working while the Tubbing was being put in.—The
four coal-cages were left in the bottom of the pit, and the
down ropes were lashed to the side of the pit (Figs. 1, 2 and 3,
Plate X.).

Two cradles were made to suit each side of the pit, the
polka-pit-side cradle being attached to the horse-crab, N (Fig. 2,
Plate X.), and the middle-pit-side cradle to the crab-engine,
O. The west-pit jack-engine, P, "waited on" with a kibble
at the polka pit; and the east-pit jack-engine, Q, performed
the same duty for the middle pit.

None of the dividing brattice was removed during the operations. The ends were cut to suit the brackets designed for
its support in the new tubbing.

Four men worked on each cradle in each shaft.

The cement-concrete was properly gauged and mixed at
bank, previous to being sent down the pit.

The preparations for getting the pit ready, placing the ropes
over the pulleys, hanging the cradles, refixing the cage-timbers,
lifting out the cradles, and hanging the cages for coal-work,
occupied 8 hours for each spell of work at the new tubbing.

———

APPENDIX.—JOURNAL OF OPERATIONS IN RELINING THE TUBBING OF THE
POLKA AND MIDDLE PITS, MURTON COLLIERY, IN 1903.

			Hours.	Hours.
April	10.	Took out walling and prepared the crib-bed ...	72	
,,	18.	Do. do. do. ...	24	
			—	96

APPENDIX.—*Continued.*

			Hours.	Hours.
May	25.	Laid and wedged the lowermost crib	24	
,,	9.	Laid and wedged the second crib, and put on the first and second courses of tubbing ...	32	
,,	16.	Put in concrete backing	24	
,,	23.	Put on 4 courses and finished the bell-tubbing	32	112
,,	30.	Put on 2 courses of straight tubbing, and replaced the kep-buntons of the rising main with steel-girders		72
June	6.	No work done.		
,,	13.	Finished the kep-buntons, and shifted the rising main		24
,,	20.	Put on 7 courses of tubbing		32.
,,	27.	Put on 6 courses of tubbing		24
July	4.	Put on 7 courses of tubbing		23.
,,	11.	Put on 9 courses of tubbing		24
,,	18.	Put on 9 courses of tubbing, and cut out the strengthening ring in front of an old wedging-crib ...		22.
,,	25.	Took out an old water-ring, below the old wedging-crib, and put in a new one below the new crib, carrying the 3 inches pipes up to it. Put on 2 courses of ordinary tubbing, and one course, 1 foot 10 inches high, and finished level with the bottom of the old wedging-crib, making 48 courses to this point		24
August	1.	Put in additional collarings to the rising main, also a new wedging-crib and water-ring combined, H, in front of an old crib, wedged it, and put on 10 courses of tubbing		48.
,,	8.	Put on 4 courses of tubbing		16.
,,	15.	Put on 7 courses of tubbing		24
,,	27.	No work done.		
,,	29.	Put on 6 courses of tubbing		24
Septr.	5.	No work done.		
,,	12.	All the water was wedged off at the sand-crib. Put on 6 courses and one special course of tubbing, I, to pass the wedging-crib and rings without lessening the diameter of the shaft. Put on one course above the special course, making 83 courses in all to this point		32:
,.	19.	A circle of wrought-iron pipes, 2 inches in diameter, with ⅛ inch holes, was put in between the new and old tubbing, and the cement-concrete was left out below this ring and the plug-holes of the new tubbing immediately below. Above this circular piping, wooden sheeting was put in, and above the sheeting the space was filled to the top of the last segment put on the previous week-end with stiff cement-concrete. Part of the plugs of the last course of new tubbing were left out, and a quarter water-ring was put in below these holes on the south side of the shaft. The water-pipes, 3 inches in diameter, were then connected to this ring		30.

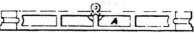

FIG. 15.—ELEVATION.

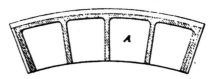

FIG. 16.—PLAN.

to 1 Inch.

FIG. 17.—ELEVATION.

FIG. 18.—PLAN.

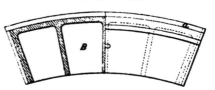

Inch.

APPENDIX.—*Continued.*

			Hours.
Septr.	26.	Put on 12 courses of tubbing. The strengthening-rings on the old tubbing terminated at the tenth course or ninety-third course from the bottom, making 95 courses in all	32
October	3.	Put on 8 courses of tubbing	28
,,	10.	Put on 4 courses of tubbing	13
,,	17.	No work done.	
,,	24.	No work done.	
,,	31.	No work done.	
Nov.	7.	Put on 7 courses of tubbing, topping the last with 3 courses of brick-and-cement to protect it, making 114 courses in all	23
		Total time	723

NOTE.—From the bottom of the lowest wedging-crib, it is 13 feet 7 inches to the top of another crib. From the bottom of this crib, it is 53 feet 10 inches to the bottom of the crib, I, at the top of the sand, where most of the water is coming out from the closing course below it. All the old tubbing, level with the ninety-third course of new tubbing, had strengthening-rings placed in front of it and wooden filleting placed between these rings. This filleting was taken out before the cement-concrete was put in.

———

The PRESIDENT (Mr. W. O. Wood) said that he was very much indebted to the resident manager of the colliery (Mr. S. Hare) and to the engineer (Mr. R. H. Oughton) for the way in which the work had been carried out. It was a long and arduous piece of work, to which it was necessary to devote Sunday after Sunday for several months. It had to be done very carefully, as the members might judge from the precision with which the tubbing was manufactured.

Mr. T. E. FORSTER said that papers of this kind, although they might not evoke much discussion, were of great value and assistance to others. He had pleasure in proposing a vote of thanks to the President for his paper.

Mr. C. C. LEACH, in seconding the resolution, asked why the tubbing was bolted together, instead of being wedged in the ordinary way.

Mr. T. E. FORSTER said that he had tried tubbing with inside flanges and using ordinary sheeting, but it was a most difficult matter to make it watertight.

Mr. A. L. Steavenson remarked that the paper was a record of the successful completion of one of the most difficult mining operations with which he had met, and it was noteworthy that it had been carried out without loss of life.

Mr. W. C. Blackett asked whether the old tubbing leaked; whether the new tubbing was subject to hydraulic pressure in consequence; and whether, in contemplating the removal of the pressure from the old to the new tubbing, consideration had been given to the possibility of any motion of the old tubbing.

The President (Mr. W. O. Wood) said that there were several reasons for the adoption of the method of tubbing described in the paper, but the principal of these was that he had previously tried the usual method of wedging the tubbing in the east pit shaft; and, if there had not been what might be described as a "Providential strike" at the time, they would never have accomplished the work, within a reasonable time. The work was done at week-ends and holidays; it took 8 hours to prepare the pit, and on ordinary days there was only an interval of 4 hours between the cessation and resumption of coal-drawing, as the pit drew coals for 20 out of 24 hours, so no work could be done during the week. There were a few leaks, amounting to about 40 gallons per minute, but he did not consider it advisable that they should be closed. One of the special segments of the new tubbing was provided to take away the water from one of the largest leaks in the old tubbing. The tubbing was subject to hydraulic pressure, just the same as the old, and he expected, by this time, that the new tubbing and the cement-concrete behind it was taking all the pressure; if there had been any opportunity for an accumulation of air behind the old tubbing, it might cause some movement, but the new tubbing strengthened it, and with the excellent cement-concrete lining, the water could have no effect on the new tubbing.

The vote of thanks was cordially approved.

———

The Secretary stated that Mr. Clarence R. Claghorn's prize had been awarded to Mr. H. W. G. Halbaum for the following essay on "The Action, Influence and Control of the Roof in Longwall Workings":—

THE ACTION, INFLUENCE AND CONTROL OF THE ROOF IN LONGWALL WORKINGS.

By H. W. G. HALBAUM.

The order of division adopted in this paper is that indicated by the above title. The observations herein contained are suggested partly by personal experience of longwall workings, and partly by mechanical and geological theory. Scientific terms, such as pressure, action, etc., are, for the most part, used in the same wide sense as that employed by practical men in common conversation. And the aim of the paper is to indicate broad principles, rather than to discuss the particular manner of their application to local cases.

I.--THE ACTION OF THE ROOF.

Every roof is subject to the law of gravitation, in virtue of which it is endowed with weight, pressure, and, in the actual longwall working, with motion, momentum and kinetic energy. Every roof is also subject to fracture, decay and disintegration. Every roof is, moreover, endowed with a certain amount of elasticity, either real or practical, for the miner will tell you that the roof bends before it breaks, while the geologist need only point to the anticlines, synclines and contorted strata of the earth's crust in order to prove that, from the practical standpoint at least, the rocks possess the ability to yield to pressure by bending.

(1) *The Pressure of the Roof.*—The case of the unworked strata is simple. Here we have a statical pressure, a pressure at rest. There is a vertical line of pressure evidently due to gravity. But there is also a horizontal line of pressure due to the lateral compression of the rocks. This also is primarily due to gravity, since all the rock-particles press to the centre of gravity; for it must be remembered that the various lines of gravity are not parallel but radial, and the convergence of

all the rock-masses to a common centre must necessarily set
up a lateral compression. The end-thrust thus created probably
sustains part of the vertical weight, and general experience of
working coal-mines shows that it is only after you have got
the first big fall in the goaf, and so relieved the lateral compression, that you get the "weight" on the coal below to the
fullest extent.

(a) *The Draw.*—The mine-working being got under way,
and the lateral compression of the rocks being relieved by the
first big fall in the goaf, it is reasonable to suppose that for
some distance beyond the coal-face, and above the solid coal,
the original lateral thrust of the overlying strata will find effect
in a lateral movement of such strata towards the face outward.
This movement in turn will develop incipient lines of fracture
above the solid coal, the roof begins to crush the coal by its
sheer weight, and the resistance of the artificial packs being
less than that of the solid coal, the strata begin to bend down
over the coal-face on to the packs behind. The effect of this
disturbance is observed belowground in the crushing of the
coal, and is often visible at the surface at a point many feet
in advance of the working face. This point lies on the shore-
line of the subsidence at the surface caused by the working
below, and the distance by which this point leads the line of
the working-face is technically known among mining surveyors
as the "draw." The draw, it will be understood, always lies
beyond the face-line, while the goaf and the subsidence proper
lie behind the face-line (Fig. 1, Plate XI., and List of Illustrations).

(b) *Oblique Direction of the Total Pressure.*—It appears,
therefore, that the action of the roof is not quite so simple as
might at first be supposed. We have, first, a force due to the
sheer weight of the strata, acting along the line of gravity. We
have, in the second place, a force due to the end-thrust of the
strata as their natural lateral compression is relieved by
successive falls into the excavation. Wherefore, in the third
place, we have the oblique resultant of these two component
forces. In cases of subsidences due to purely natural causes
acting in former geological ages, the oblique resultant has had

time to define its direction clearly, as seen in the hade-lines of various faults. In the case of a mine, however, the area of the subsidence is continually being enlarged; the fault-line is continually shifted forward as the working-face advances, and the incipient hade existing for the time being cannot be so clearly traced and defined. But it seems reasonable to infer that this incipient hade-line, and therefore, the resultant of the component forces which produce it, lie approximately along the hypothenuse of a right-angled triangle whose two sides are roughly equal to the depth of the seam and the length of the draw respectively. For every force produces its effects in the line of its own action.

(c) *Nature of the Pressure.*—In many cases, and for varying intervals of time, the pressure may remain practically statical. But " weightings " of the roof, and the gradual descent of the roof, are matters of common knowledge. Hence, the pressure, so called, may often be in reality the power due to kinetic energy and accumulated work. And from the practical standpoint, the pressures, so far as the most violent effects are concerned, may certainly be described as intermittent in their action.

II.—THE INFLUENCE OF THE ROOF.

The Influence on the Coal.—This sub-head is taken first in order, only because it is the first that would occur to a practical miner. The effects of the roof-action on the coal are to crush it more or less, and render it easier to be gotten by the miner. The general principle of longwall working from this particular point of view may be briefly explained.

The bord-and-pillar system and the longwall mode have this principal point of difference. In the bord-and-pillar system, the goaf being allowed to fall freely, the coal at the goaf-edge sustains the normal pressure due to the weight and the draw. The longwall system brings a greater pressure to bear upon the coal. The roof is not allowed to break off bodily at the face-line, but is projected therefrom like a cantilever over the face of work and on to the sinking heads of the packs behind. In the bord-and-pillar system the action of the roof is more nearly allied to that of a simple weight; in longwall, it is more nearly allied to the action of a powerful lever. We may exert a power

sufficient to bend this lever, and do well. Or we may guide the
power in such a manner that the lever snaps on its fulcrum and
becomes useless. Referring to Fig. 2 (Plate XI.), if the length
of the lever be AD, it will be seen that if it snaps (in the
practical sense) at B, the advance of the face will soon make the
relative position of the fracture occupy the site C, and the crushing effect of the lever on the coal will be reduced to zero. The
great object of longwall, therefore, is to secure the continuous
transmission of pressure to the coal-head, not only from the
roof above the coal itself, but also from the roof superposed
above the face of work and the packs behind.

The intensity of the effects produced by this leverage on
the coal will be modified by the nature of both the roof and
the floor. If both be very soft, the intensity of the effects will
be less. If both be very hard and strong, the effects on the
coal may be so violent as to crush it unduly and reduce its
market-value. This phase of the matter will be further considered at a later stage.

The Reaction on the Roof itself.—The influence of the roof-
action recoils in some measure on the roof itself. The nature
of the pressure, especially that portion of it more particularly
due to the draw has already been alluded to. The action on
the coal is distinctly intermittent, sometimes it is merely a
statical pressure, at other times the action is more truly defined
as kinetic power. The effects of this intermittent pressure or
power are those of a hammer at work, the roof corresponding to
the hammer, and the coal, etc., corresponding to the anvil. The
effects on the anvil have been referred to in the previous section.

Theory of the Reaction.—But according to a wellknown
physical law, acti n and reaction are equal and opposite. Hence,
the action of the hammer on the anvil is equal to the reaction
of the anvil on the hammer. The preservation of the roof in
longwall is a matter of importance, and it is by no means an
academical nicety to state the following practical proposition : —
In any longwall working, if the action of the roof be small,
the reaction on the roof itself will also be small; and conversely,
if the action be large, the reaction will also be large. In applying this principle to the deterioration of a roof, however, it is

necessary to remember that the deterioration of the roof will depend not on the total reaction, but upon the reaction per unit of mass. For example, if the roof be very thick, you may have a very large total action, and an equally large total reaction; yet the deterioration may be comparatively small, simply because, although the total reaction be large, the reaction per unit of mass is small.

But, according to the same theorem, action and reaction are not only equal in amount, but also opposite in direction. The oblique direction of the total action has already been shown. The line of the reaction, therefore, is simply the line of action with contrary motion. Hence the action of the roof, the reaction upon the roof, and the fractures produced in the roof by such reaction, all lie in the same straight line. Therefore, the great lines of fracture in the longwall roof are deflected from the true vertical line, and lean in the direction of the limit of the draw.

The foregoing seems to disclose an interesting and important principle of practical value. It clearly proceeds *a priori* from physical theory; it is proved *a posteriori* by geological phenomena and by longwall experience, and it explains mining phenomena which would otherwise be inexplicable. The obliquity of the line of fracture accounts for many things which could not be reconciled with any theory of vertical fractures.

Associated Phenomena of the Reaction.—Assume, as an example, that in a given case, the angle of the obliquity is 10 degrees. Then, referring to Fig. 3 (Plate XI.), the depth of the mine and the length of the draw are in the same ratio as the lines AB and AC; and the angle of the obliquity is CBA. The direction of the roof-action is CB; that of the reaction is BC, and the lines of fracture lie in, or parallel with, the same line.

Transmission of Pressure.—As the working-face advances, successive lines of fracture are produced as shown in Fig. 4 (Plate XI.). Taking any pair of them, it is seen that the section of roof between them reposes on the section beyond at an angle of 10 degrees. Thus, e reclines on d, d on c, c on b, b on a and a finally reclines on the solid roof at A. Hence, pressure is transmitted from e through d, c, and b to a, and thence to the solid roof at A and the coal below. It is evident that this transmission of pressure could not take place if the fractures were vertical, or if they were deflected to the opposite side of the vertical line.

"Weightings" of the Roof.—Also, each section, *a*, *b*, etc., must, by the nature of the case, slip down from its hanging-wall before it slips on its foot-wall. Thus, in the circumstances portrayed by Fig. 4 (Plate XL), section *a* will presently slip down on its hanging side as *b* did before it. But the hanging-wall of *a* is the foot-wall of *b*, and a similar relation obtains between each pair of adjacent sections. Hence when *a* slips from its hanging-wall, the foot-wall of *b* slips also. The slip of *b* in turn depresses the foot-wall of *c*, and so on throughout the entire series. Thus the slip of *a* is followed by the bodily slip of the whole roof over the face of work and the packs just behind, the result being known as a "weighting."

The Bending of the Roof.—Again. When the section *a* slips on its hanging side, the bodily motion of *b* follows, and so on to *e* and *f*. It is evident that the leverage of each section varies directly as its distance from A. The movement commencing at *a* continually gathers momentum as it proceeds. The fall of *b* must exceed that of *a*; the fall of *c* exceeds that of *b*, and so on. If we plotted to scale these different falls of the various sections, we should require to take these accelerations into account, and their magnitudes vary as the distances from A.

But the motion described is arrested, and the manner of this arrestment must also be considered. When section *a* slips on its hanging side, it is clear that section *b* would immediately slide right down to the floor if nothing occurred to prevent it. Now what prevents it from doing so? Only the fact that the combined weights of all the sections from *c* onward immediately force *b* against its foot-wall, and hold it there by lateral thrust. It is literally caught by a pair of self-acting clamps. Similarly, section *c* is arrested in turn by the lateral thrust due to the combined weights of all the sections from *d* onward, and so on throughout the series. Here, then, we have a thrust which comes into play at each successive motion originated at A, a thrust which arrests the motion of each section in turn, and a thrust the grip of which on each section varies inversely as the distance from A.

Then, since the moment varies as the distance from A, and since the power of arrestment varies as the reciprocal of this distance, it follows that during any given weighting, the travels of the various sections must vary as the squares of the distances from A. Plotting the various travels on these lines, the bases of the several sections arrange themselves in the newly-depressed roof-line along a parabolic curve the vertex of which is at A. Thus, the effect of all the various accelerations taken together is the bending, so-called, of the longwall roof. This result, much exaggerated of course, is shown in Fig. 5 (Plate XI.). The positive amounts of travel accomplished by the various sections in the ordinary practical case are too small to be appreciated by the senses, or we might not perhaps, in our mining idiom, have come to speak of them, or of their effects, in the way that we do.

The "Locking" of the Roof.—The miner often speaks of the roof "locking" itself. One may go into a longwall face of work and find all the timbers, or most of them, broken, or split, or sunk for inches into their soles and caps. And yet it often happens that if the whole of the timber, sound and otherwise, be removed, the entire roof remains in position. Why? The answer is obvious. The great roof-sections, by successive slips, have descended a few inches; the precipitation of the enormous weight on to the timber has smashed it to splinters, it may be; but the motion has been arrested for the time being by the lateral thrust above described, and the great body of strata remains securely gripped in the powerful jaws of its natural clamps. It appears, then, that a simple application of physical theory (the third law of motion, in fact,) easily and naturally explains many notable longwall phenomena.

The Mechanical Disintegration of the Roof-stratum.—By the roof-stratum is meant that stratum (or, it may be, those strata) lying immediately above the coal-head. From one point of view, the roof-stratum is a bridge that extends across the face of work. When the roof "bends," the strain is greatest at the roof-line, because the radius of curvature is least at the roof-line. In a previous section it was seen that this "bridge" withstands immense "weightings" and exhibits apparent "bendings" simply because the arrangement of its constituent

parts is such that the sudden "weighting" of the bridge has
the effect of automatically putting it into compression from
end to end, so that the bending of the bridge under its load
becomes also a bending which accentuates the curvature of
the arch.   In short, the stresses on the roof-stratum are variable in amount, variable in kind, and intermittent in their
application.   Such a combination of stresses, and such a manner
of their application cannot but exert a most destructive effect
on the roof-line.   The mechanical disintegration of the roof-
stratum is, therefore, much more rapid than that of the superposed strata.   Hence, while the upper strata remain practically
locked in position, the roof-stratum is liable to be broken into
fragments.

Efficiency of the Timber.—It is the office of the timber to
prevent the fragments of the roof-stratum from falling out
and causing injury to the workers.   The timber can never
support the weight above the bridge at the roof-line; it is merely
a system of centres, so to speak, the object of which is to maintain the constituent parts of the bridge in their relative positions.
The timber no more sustains the weight of the body of roof than
the copper-pins in the self-detaching hook sustain the weight
of the loaded cage.   The sole object of the timbering and its
perfect efficiency are secured if the fragments of the shattered
roof-stratum are maintained in their relative positions.

Mysterious Accidents from Falls of the Roof-stratum.—It
seems very probable that the theory of intermittent stresses,
compressive and tensional, set up in the roof-line especially,
accounts very easily for at least some of those mysterious
accidents which arise from the sudden fall of more or less
massive fragments from the roof-stratum.   Observe that in the
great majority of instances, the victim is injured or killed by
a falling fragment of the roof-stratum only, and not by a large
fall from the great roof-body.   A large proportion of these
accidents appear mysterious by reason of two notorious facts:
Firstly, they occur under an apparently good roof; and, secondly,
they occur soon after a satisfactory examination of the roof.
If we assume that the examination took place in the interval
between the action of two stresses of opposite kinds, the mystery

of the accident vanishes at once. It merely emphasizes the old lesson that we should trust no roof whatever, but timber all of them with the utmost promptitude.

The Influence of the Roof upon the Roads.—The effects produced on the roads are so wellknown that they require little treatment here. The pressure is often sufficient to cause the gobbing to act as a semi-fluid, transmitting much lateral force against its binding walls and thrusting them into the road. If the floor be moderately soft, pressure is again transmitted laterally from underneath the packs to the road-floor. These two lateral pressures, one from the foundation of each pack, meet in the middle of the floor, and finding their least line of resistance upward, burst the floor-stratum, which is said to " heave " or " creep," as it rises into the road-area. Such results are illustrated in Fig. 6 (Plate XI.), where ABCD is the original cross-section of an imaginary road, and EFGH the cross-section of the same road as subsequently reduced by the pressure. Other things being equal, such effects are more marked in the deeper mines, on the softer floors, and in the thicker seams.

III.—THE CONTROL OF THE ROOF IN LONGWALL WORKINGS.

The term, control of the roof, is herein construed with a good deal of elasticity. By the successful control of the roof is meant the obtaining of good results, whether they be obtained by aid of the roof-action, or in spite of it, or by simply running away from it. Such a definition of control may not be in strict harmony with the letter of etymology, but it harmonizes perfectly with the practical spirit of this paper.

Control by Direction of Working.—The direction of working is often defined with reference to the cardinal points. In this paper, however, it must be defined with reference only to two lines " graven in the rocks " by geological forces. One of these is the line of dip; the other is the line of cleavage. The angle at which these lines intersect each other varies in different localities. Hence in one locality the angle made by the direction of working with one line may accentuate the effects produced by the angle made with other line; whilst in another locality the angle made with one line may to some extent

counteract the effects due to the other angle referred to. It is
therefore necessary to consider the angle with dip and the angle
with cleavage in quite separate sections.

(a) *The Angle with the Line of Dip.*—Whether the working
advance to the dip or to the rise, the force due to the sheer
weight of the overlying strata will act in the vertical line.
But the useful effect of the lateral thrust will be different. For,
in working to the dip, the lateral thrust of the strata beyond
the face-line is resisted by all the weight of the strata behind;
whilst in working to the rise the lateral thrust of the strata
beyond the face-line is not merely unresisted, but is, in a very
positive sense, facilitated by the weight, which naturally tends
to slide downhill, and thus accentuate the draw at the surface
and exert a mangling action on the coal-head such as cannot be
obtained in working to the dip. On the other hand, the roof is
more liable to break off at the face-line in rise-workings, unless
the packs be kept well up to the face.

As for the influence of dip on the maintenance of the roads,
it may be remarked that roads along the line of strike are more
difficult to maintain than roads in the line of dip. For, owing
to the direction of the roof-action, the liability of the packwalls
to lateral displacement is greater when the road is in the line
of the strike, and such walls should therefore be tied with substantial cross-walls at frequent intervals.

Working to the rise also involves a larger expenditure of
timber than does working to the dip. For, whatever tends to
break off the roof at the face-line equally tends to throw the
weight on the timber and packs behind. And it is hardly
necessary to add that whatever impairs or reduces the efficiency
of the timbering must coincidently affect the standard of safety.
Hence, several pros and cons are to be considered before one
can, in any particular case, decide on the definite angle that
the direction of working should make with the line of dip.

(b) *The Angle with the Line of Cleavage.*—The important
effect of this angle is too well understood to need much description here. But it is necessary to consider this angle from the
standpoint of the roof as well as from that of the coal. It may
firstly be enunciated, as a cardinal principle, that the planes

of cleavage, incipient or pronounced, existing in the overlying roof-strata, strike in the same general direction as the planes of cleavage existing in the coal below. The importance of this principle, and the necessity of its acceptance, justify a reference, by way of proof, to the natural philosophy of the case. According to geological theory, the cleavage in the coal and in the roof-strata was produced by the action of the same force. Assuming that in a given case, the planes of cleavage are vertical, the theory is that some force, acting laterally, and at right angles to what are now planes of cleavage, was the cause of such cleavage being created in the strata. Such a lateral force is supplied by the shrinkage of the earth's crust, this in turn being due to the cooling of the planet through vast ages of time. This force, acting with immense energy on the particles of matter in the strata, and subjecting them to enormous lateral compression, obliged such particles so to rearrange themselves that their longer axes finally lay at right angles to the line of action of the compressing force. The planes of cleavage are thus defined as the planes in which the particles of matter now extend their longer axes. Men of science, assuming *a priori* the truth of this theory, and employing artificial agencies, have produced distinct planes of cleavage even in such an unpromising substance as plastic clay. The theory of cleavage propounded by the geologists cannot, therefore, be dismissed as a pure speculation. It is supported by such evidence as the nature of the case either permits or requires. And therefore, we may, so far as the normal case is concerned, conclude that the planes of cleavage in the coal below and in the roof above strike in the same general direction.

A little reflection will show that this theory of the cleavage, in conjunction with the theory of roof-action herein explained, sufficiently account for the wellknown fact that the coal, as a general experience, is more easily got by the miner when the direction of the working is at right angles with the direction of the cleavage. Several of the most notable roof-troubles frequently met with are explained, and their remedy, or remedies, indicated by the same principles. It may first be shown why the coal is more easily won in the working which advances directly across the cleavage.

Advancing directly across the cleavage, the coal is more easily got by the miner because : —

(1) The planes of cleavage in the overlying strata, running continuously along the face-line, and in planes continuously parallel thereto, afford, as lines of least resistance to fracture, their maximum assistance to the vertical component of the roof-action.

(2) Because the natural compression of the strata being greatest along the line of the force which produced the cleavage-planes, the relief of compression is greatest along that line when falls occur in the goaf behind. Hence the lateral thrust of the strata, due to the relief of the compression, the thrust which gives birth to the horizontal component of the roof-action —the thrust which produces the draw at the surface and "mangles" the coal below, is greatest in magnitude and richest in power when the direction of working is at right angles to the line of cleavage.

(3) Because both components being at their maximum value, the total roof-action is a maximum also.

(4) Because the compression of the coal itself is greatest in the line at right angles to the cleavage, and the coal is thus more likely (on its own account) to burst out from the face in such circumstances.

(5) And finally, whilst, in the circumstances supposed, the horizontal component of the roof-action is of maximum power, the resistance of the coal, from the standpoint of its structure, is at its minimum value. For the planes of cleavage divide the coal into a number of vertical slices. These run parallel with the face-line, and have no lateral support at that line. Hence the oblique roof-action attacks the slices on their sides, where they are least able to resist attack. See Fig. 7 (Plate XI.), where the arrow, A, may denote the direction of the roof-action, the vertical lines representing the planes of cleavage, and the horizontal lines the planes of stratification. If, however, we advance in the same line as the cleavage, all the vertical slices of coal receive mutual lateral support, the roof-action is less oblique, and it only strikes the slices in the direction of their lengths, in which direction their resistance is evidently much greater, as will be apparent on reference to Fig. 8. It is therefore small wonder that most mining men prefer, if possible, to advance their longwall working directly across the cleav-age.

In some cases, however, it is found undesirable to advance in this particular direction. The factors above-named may, in cases, develop more power than is desired, crushing the coal unduly, and producing an altogether abnormal proportion of slack, or pulverized coal. On the other hand, the development of excessive power may, even in this direction, overreach its object, and actually render the coal more difficult to win. This phase of the problem, however, is best considered along with the roof-trouble with which it is usually associated.

The roof-action set up in advancing at right angles to the cleavage may overreach its object in the following manner. Where the cleavage-planes are very fully developed, the disturbance may be so violent as literally to break off the roof at the face-line. In such an instance, the coal is at once relieved of much of the weight. Already disturbed by the pressure, then losing such pressure, and also, it may be, drained of its compressed gases, the coal becomes abnormally tough— the miner calls it "winded"—and correspondingly difficult to win. This toughness continues until the further advance of the face develops a new "weighting," which, of course, may expend its energy in the same violent manner as its predecessor, and with the same disastrous effects on the coal and on the timber along the face of work.

In any case, it will be evident that where the planes of cleavage are well developed in the rocks, the securing of the face of work will involve a larger cost in timber when advancing across the cleavage. For the vertical slices of roof-strata then run from end to end of the face of work. They are, therefore, very liable to give way at the first joint of stratification above the roof-line, and a very liberal expenditure of timber is required in order to maintain the standard of safety. When the direction of working, however, is in the same line as the cleavage, the situation is different. The vertical slices of roof-strata then merely bridge the shorter span across the face of work, being supported by the coal-head on one side and the packs on the other. They are not, therefore, so liable to give way at the first joint, or at the first plane of stratification above the roof-line. In the case described, it may, of course, be wise to think twice before finally adopting the favourite direction of working.

Control by Mode of Working.—It has already been noted that
the roof-action on a (presumably straight) longwall face, advancing directly across the cleavage, may, in cases, be so unduly
violent as repeatedly to break the roof clean off at the face-
line. If this cannot be remedied or prevented by ordinary
means, such as the institution of more efficient packing, etc.,
one radical remedy is to alter the direction of working and
advance at a lesser angle with the cleavage. Such a remedy is
usually effective, but it possesses its own share of disadvantages.
There is, however, another method of cure. Instead of advancing a single long line of face, we may, without altering the
direction of working, evade the surplus power of the roof-action
by installing a system of shorter working-faces which follow
each other at stated intervals. Such a system is often called the
" step-system," " longwall by steps," or " stepped longwall."
It is illustrated in plan by Fig. 9 (Plate XI.), and it has often
proved successful where a bad roof appeared to render the operation of the long, straight face a practical impossibility.

Referring to Fig. 9, each short working-face is 12 yards long,
and it leads the face next behind by the same distance. In
practice, the lengths of face and amounts of lead show some
considerable variation between one colliery and another. View-
ing the step-system as a weapon of roof-control, its efficiency
is easily accounted for. Taking as an example the case
illustrated by Fig. 9, it is seen that whilst each man
works on a " bord-ways " face of coal 12 yards in length,
he continually leaves on his right-hand side a " head-ways "
or " end-on " face of coal of similar length. The " bord-
ways " face is the working-face; and the headways-face, in
mining idiom, is simply the " wall-side." The roof-action,
however, takes no account of such distinctions, but is exerted
on both faces alike. If the bord-ways face existed alone, the
draw at the surface would be projected forward in a line at right
angles to this face. On the other hand, if the wall-side, or " end-
on " face alone existed, the draw at the surface would be projected in a line at right angles to the vertical plane of the wall-
side. But since both faces co-exist together, the respective draws
due to one and to the other become the rectangular components
of a resultant draw the direction of which is at right angles
to the vertical plane of what we may call the " mean face-line."

Thus, in Fig. 9, the mean face-line is on the line AB, and the direction of the resultant draw at right angles thereto is shown by the arrows, a, b, c and d. Now, the projection of this resultant draw, and the projection of the total roof-action on the horizontal plane, represent forces acting in the same line but in opposite directions. Hence, since the arrows a, b, c and d, show the direction of the draw, the arrows, x, y and z, show the direction of the roof-action in the horizontal plane. It will thus be noticed that whilst each man works on a face of coal which advances directly across the lines of cleavage, the action of the great body of roof is directed upon a mean face-line which crosses the cleavage at an angle of 45 degrees only. Hence, considering this system of "stepped longwall" purely as a weapon of roof-control, its efficiency arises from the fact that the roof is here controlled by the angle which the cleavage-line makes with the mean face-line; and not by the angle that it makes with the direction of working. And it is clear that the angle at which the mean face-line crosses the cleavage-line may be made as large or as small as we please, by merely varying, as the circumstances may dictate, the numerical ratio which obtains between the width of each face and the length of its lead. Similar principles apply in the cases of some other so-called modifications of true longwall.

Where a roof is very bad, and the seam of coal is some 10 or 12 inches thicker than the height of the tub, or mine-wagon, the system illustrated in Figs. 10 and 11 (Plate XII.) may sometimes be adopted with advantage. It is, in effect, the same case as that of Fig. 9 (Plate XI.), only that the faces of work may be still shorter and the leads still longer. The men work on the " bord-ways " faces of the "lifts," which may be from 6 to 9 yards wide. These lifts are advanced at right angles to the line of the pack-walls which sustain the gateways through the goaf. They are timbered by a double row of stout props along the tubway (which is carried close to the wall-side) and by temporary timber which is drawn out from behind the double row of props as the lift advances, leaving only enough of it next the face to protect the men—-the transmission of pressure from the goaf being thereby reduced. As the lift continually advances, the roof is thus allowed to fall freely behind the double row of props and behind the men. As soon as a lift crosses the line of the next gateway, it is worked from that gateway, the pack-walls being brought up to within a few feet of the wall-side. The double row of timber left behind is then drawn out and recovered. In the ideal arrangement, all the lifts reach their objective gateways simultaneously, each lift requiring the use of a fresh gateway just when the lift immediately preceding is done with it. The gateroads may be from 30 to 60 yards apart. The crosses show the method of timbering, so far as the propping is concerned. No brushing or ripping is required in the temporary tub-roads between the pairs of gateways.

Control by Rate of Advance.—In most longwall workings, the time-element has an important bearing on the control of the roof-action. The action of the roof does not come to a sudden stop when the men cease working at the close of the shift, neither is it immediately and appreciably accelerated on the recommencement of work next morning. In fact, it would appear that whilst the men continue working the venue of the action is being largely shifted forward; when they cease working, the action, from a practical point of view, concentrates its power upon the making of definite fractures at the face-line. Even in bord-and-pillar workings, we remove the "stook," or last remnant of the pillar, as rapidly as possible, knowing that a leisurely rate of working will probably result in the stook being smothered by the roof before the coal is secured. And in most mining systems, longwall or what not, if a stoppage occurs from any cause, such as a strike, we invariably find, on re-entering the mine, that, although the men have been on strike for so many weeks, the roof-action has been at work all the time, rendering necessary the execution of extensive repairs before any resumption of coal-getting can be made.

It is therefore evident that we may find a weapon of roof-control, not only in the direction of working, but also in the rate of working. Suppose, for instance, that the safety of the face of work demands its efficient timbering for at least 4 yards behind the face-line. Then, it is clear that each prop must stand for the same length of time as that required to advance the working 4 yards. Thus each prop must stand the pressure for a length of time which is measured inversely by the rate of advance. If the rate be doubled, the time of advance through

4 yards will be halved. Similarly, if the time occupied by the travel of the face through, say, 30 yards, be sufficient to so settle the roof-body that the cross-gate is then "worn-out," the result is that a new cross-gate must be made every 30 yards. But if the rate of advance be accelerated, the efficiency of each cross-gate will be increased. The increase of the rate of advance will allow the spacing of the cross-gates at longer intervals. It may not much affect the life of each individual cross-gate, but it will certainly affect the total number of cross-gates requiring to be made in each square mile of royalty. These are illustrations of the principle embodied in the rate at which the longwall face advances, and the principle itself can be studied from many standpoints, in addition to the two above occupied.

The rate of advance may be accelerated by putting more men on a given length of face. Or, what amounts to the same thing, it may be accelerated by freeing the nominal coal-getters from all duties beyond actual coal-getting—all other operations, such as ripping, brushing, packing, timber-drawing, etc., being entrusted to other classes of workmen, who should discharge these duties at night-time. The "stall-man," who is partly a hewer, partly a shifter, and partly a deputy, scarcely represents the system most favourable to rapid advance. In some cases, again, the rate of advance may be considerably augmented by the use of coal-cutting machinery. And, finally, in every system of longwall, the rate of advance will the more nearly approach the practical maximum for each system accordingly as the proportion of work paid for " by piece " exceeds the proportion paid for " by time."

Control by Arrangement of Pack-walls.—It is not necessary to refer to this item otherwise than briefly. Scarcely anything has been more fully studied by the average practical man than the proper arrangement and construction of his pack-walls; therefore scarcely anything is better understood by him, and it behoves one to be very careful not to say too much to him on the subject. The average practical man will therefore kindly understand that the following observations are made by way of quoting him, and not by way of teaching, or attempting to teach, him anything about the matter. Briefly, then, it is an

axiom of his that to keep the weight on the coal, you must
keep the packs well up to the face of the coal. Also, the greater
the width of the packs, and the closer they range in parallel
lines, the greater is the efficiency of the arrangement with
respect to the object named.   Cross-walls, properly tied into
the longitudinal walls of each pack, and at sufficiently frequent
intervals, are valuable by reason of their tying effect and their
greater or lesser efficiency to prevent the bulging of the
longitudinal walls. For obvious reasons, the width of the pack
should be greater where the height of the seam is greater; and
the minimum width, in seams less than 30 inches thick, should
not be taken as under 6 feet. In longwall workings, the packs
stand in the place of the pillars formed in the first working of
the bord-and-pillar system; and, to a large extent the same
rule applies in both cases, namely, the creep in the roads will
be less as the dimensions of the pack (or the pillar, as the case
may be) are greater; and *vice versa*. Where the roof-stratum
is very thick and strong, the packs should also be of very strong
and substantial construction and dimensions; for, in such a case,
the packs must sustain (at least for a time) the entire weight of
the roof-body, and offer a resistance sufficiently rigid to enable
the strong roof-stratum to break itself off finally in the goaf.
On the other hand, when the roof-stratum is fragile, and the
strata for a considerable distance above it fall freely into the
goaf behind the packs, the goaf speedily becomes choked, by
reason of the simple fact that any given weight of the fallen
rock cannot, from any practical standpoint, be compressed into
a space equal to that which the given weight of rock occupied
in situ. The result of this fact is that much of the pressure is
now transmitted through the fallen débris to the floor on which
it lies; the pressure on the artificial packs is thus greatly
relieved, and it is not necessary to build them of such substantial dimensions as those required where the roof is stronger.
It will be seen that, under the softer roof, it is not the mere
artificial pack of the longwall system which corresponds to the
" pillar " of the bord-and-pillar system.   In the case of the
stronger roof, without a doubt, the analogy is to some considerable degree along those lines. But, under the softer roof, the
dimensions of our analogous " pillar " are more truly defined
by the area of the parallelogram formed by two lines, namely :—

The distance separating each pair of cross-gates, and the distance separating each pair of ordinary gateroads. The artificial pack runs round this area as a double wall, but the great body of area enclosed in the parallelogram has been " packed " by natural means—by the law of falling bodies, in fact, and by the falling bodies themselves. The " pillar," however, is none the less efficient on that account. That is to say, the natural filling is probably quite as efficient as the artificial filling. Of course, neither the one nor the other can ever form a pillar as unyielding as a natural coal-pillar of equal dimensions. But both alike possess a measure of ability to delay the descent of the great roof-body; both tend to modify the magnitude of the vertical component of the total roof-action; and in so far as they do this, they coincidently tend to accentuate the obliquity of the total action—the preservation of which obliquity it is the chief object of roof-control to secure.

Where the roof is excessively strong, and the packed roads a considerable distance apart, it may become necessary to build butts, or stone pillars about 4 yards square, at intervals along the working face. Such a roof will hang a long time and for a considerable distance into the goaf; but when at last it does break itself off, it will probably carry all before it, unless the face of work be fortified in some such way. Such a system of packing is sometimes called the " checker " system, from its fancied resemblance to the board on which the game of draughts is played (see Fig. 12, Plate XII.). With such an unduly strong roof, however, it is questionable if pure longwall is the best system to adopt for winning the coal—unless, indeed, the seam contains a thick stone-band, or a number of thinner bands, sufficient to pack thoroughly a large proportion of the gob.

It may here be remarked that the successful working of longwall demands a roof only moderately strong. Extremes in either direction are baneful, notwithstanding the assertion so frequently made in the textbooks that "longwall working requires a very strong roof." As a matter of fact, a very strong roof is more difficult to control than a roof which is distinctly tender. And if the seam be thick, and free from dirt-partings, such a roof may render any system of pure longwall a wholly objectionable mode of working, so far as the winning of that particular seam is concerned. Of course, if the roof were

infinitely strong instead of being only very strong, it would
make all the difference; for, in that event, the roof would never
break away at all.

Control by Arrangement of Roads.—After a pure longwall
system has been opened up, it will often be found that the roads
proceeding in one particular direction are more difficult to maintain than roads which proceed in other directions. Sometimes
this is due to the particular angle (if any) made with the line
of cleavage; sometimes it arises from the angle that these roads
make with the line of dip; and sometimes, again, it may be
due to the angle made with the strike of abnormal lines of
reduced resistance to fracture in the roof-strata. In other
cases, indeed, the cause of the evil encountered by roads proceeding in one particular direction may be as obscure as the
effects themselves are obvious. Quite frequently, however, the
difficulty may be surmounted, or, to speak more accurately,
evaded, without entailing serious inconvenience. For it is
usually practicable to deflect the class of roads thus affected
into a different direction, along which the conditions may be
more favourable. In fact, in a pure longwall system there are
only one or two roads—the mother-gate and the main return airways—that we are bound to carry in one particular direction.
These must evidently follow the direction of working. All the
minor roads, however, may be deflected through considerable
angles.

For example, let the bearing of the main mother-gate be
zero or 0°. Suppose, then, that the obnoxious direction above
alluded to lies in the bearing of 90°. Then the arrangement
of roads shown in the skeleton plan (Fig. 13, Plate XII.) might
be adopted, for in this arrangement, no roads bear at an angle
of 90°. Suppose, however, that the obnoxious line lies in the
direction of working. Then the arrangement of Fig. 14 (Plate
XII.) may be adopted, and all the minor roads will make angles
with the undesirable line. And finally, if the obnoxious bearing be neither 0° nor 90°, but along some intermediate line, the
arrangement illustrated in Fig. 15 (Plate XII.) would probably
answer, for all the roads proceed either in the bearing the angle
of which is 0° or in that the angle of which is 90°. Of course
such a dodging of the roof-action is not control from the

etymological standpoint, but from the practical standpoint it amounts to precisely the same thing. The results obtained are those desired.

Sometimes, again, apart from its angular bearing, the position of a road may be ill-chosen. For example, it is a mistake to run a gob-road close to a solid wall of coal. The road-roof sinks on the pack at one side and on the other breaks off clean at the coal-edge, owing to the resistance of the solid coal *in situ* being much greater than that of the artificial pack. A gully is thus formed in the roof and runs from one end of the road to the other. Such a result is pictured in Fig. 16 (Plate XII.). The roads most apt to be thoughtlessly placed in such positions are, perhaps, the initial cross-gates and the gate-roads leading to the two extremities of the face of work. These may subsequently furnish large portions of the main return-airways, so that the oversight becomes doubly unfortunate, both from the standpoint of safety and from the standpoint of cost of maintenance.

The correct site for the road is shown in Fig. 17 (Plate XII.). Here, the road lies at least 4 or 5 yards from the solid coal, the intervening space being properly packed. The roof, of course, still breaks off at the coal-line, but no longer shatters the roof of the road, the resistance at each side of the road now being equalized.

Control by System of Timbering.—The general principles of mine-timbering apply over a much wider area than the long-wall workings with which alone this paper is concerned. Reference is therefore required only to such principles of timbering as require more especial study in longwall than in other mine-workings. And probably there is only one main principle which really comes within this category. This principle recognizes: (1) That the timbering must be capable of sustaining the weight of what has herein been called the roof-stratum. This refers to the strength of the timber. (2) That the timbering must be capable of maintaining the roof-stratum in position, notwithstanding the mechanical disintegration of that stratum. This means that the timbering must be sufficiently close. (3) That the timbering should readily yield (without breaking) to the irresistible pressure and motion of the great roof-body.

No. (2) needs no amplification. No. 1 states that the timber should be sufficiently strong, and No. 3 states that it should be sufficiently elastic. For the roof-stratum must be maintained in position, whilst the whole body of roof descends, it may be, through a distance of several inches. The ideal prop to answer this two-fold purpose would be a telescopic steel-prop in two or more sections—the sections sliding on each other against powerful steel-springs within—the springs being able to sustain a pressure of several tons. Such a prop would sustain the roof-stratum and also yield, without loss of efficiency, to the gradual descent of the body of roof. Such a prop is not yet forthcoming, however, and other means must be resorted to. Any ordinary pit-prop, standing by itself, will only yield to the irresistible pressure by breaking, after which, of course, it is not of further value in either of the directions indicated.

Since we cannot get a prop at once sufficiently strong and sufficiently elastic to answer our double object, we must fall back on a system of timberings. The only way known to the writer by which this double purpose may be attained with any degree of success is to employ tapered or " chamfered " props, resting on thick soles, and capped by substantial lids. The soles may further be bedded on a thin layer of dirt. As the roof slowly descends, the prop-head is forced through its lid, the foot is forced through the sole, the layer of mine-dirt is ground out, and the prop itself " burrs " or spreads out its fibres at either end. The roof may thus sink through several inches without breaking the prop itself, and without appreciably diminishing its efficiency as a support of the disintegrated roof-stratum. Such systems of timbering have recently been described and illustrated very fully in the *Transactions of the Institution of Mining Engineers*, to which enquirers are referred for further particulars.

As for the timbering of main roads, especially main return-airways, it is possible that the Gothic or " herring-bone " system of timbering a bad roof is not appreciated as fully as it deserves. In all main return-airways, a liberal height should be made in each section before the face leaves it far behind. A second set of rippings is undesirable. In fiery mines, the blasting in such return-airways is objectionable, and in any mine whatever, one does not want to carry the débris thus made a mile before finding a place wherein to stow it. The height should be made at first and once for all, the return-airway should be temporarily timbered, and then left until the strata have comparatively settled. By this time, the roof of the return-airway, in one section after another, will have assumed the form of a natural arch (Fig. 18, Plate XII.). Each of these sections should then be dealt with in the following manner:—Without shearing back the sides—an operation which always invites fresh falls along the centre-line—the section should be timbered as illustrated in Fig. 17. Firstly, the dirt which has crumbled from the roof is levelled down—the liberal height previously made doing away with the necessity of carrying it out of the return-airway. Secondly, the soles, a and a, being firmly embedded in and upon the dirt-stratum thus obtained, and secured in position by flat stones d, d, d . . . set up edgewise in the dirt-stratum, vertical props are placed thereon to receive the longitudinal timbers, b and b, each of these being 8 or 9 feet long. Another longitudinal timber, c, of equal length or a little more, is then raised and temporarily supported on a " middle " prop. The diagonal struts, connecting b and b with c, are then placed and carefully wedged in position. Each set of vertical and diagonal timbers must lie in the same vertical plane, and there should be at least three sets of these along each set of longitudinal timbers. Lagging sufficient to make solid the resistance in the various lines of thrust is then inserted, the middle prop supporting c is removed, and the erection of the next set commenced with. Such a method of timbering is hardly as expensive as it looks; for the vertical and diagonal timbers may frequently be cut from bigger timber which has been broken in the main haulage-road; besides which, the vertical props and the soles, a and a, may quite frequently be entirely dispensed with by the exercise of a little ingenuity. In drawing Fig. 18 (Plate XII.), it was assumed that the roof had fallen in such a manner that the diagonal struts required to incline at an angle of 45° from the horizontal. Other circumstances, of course, might require a much smaller inclination, say along b e, as shown in Fig. 18, by the lower and more faintly-outlined struts. But whatever the angle, the ultimate strength of such a system of timbering is obvious, and its great elasticity under pressure is not less manifest. Even though the first cost be a little heavy, the system is a cheap one in the end, for, to use the idiom of the engineer, it makes "an everlasting job."

Conclusion.—The paper is concluded, but the subject is very far from being exhausted. Had opportunity permitted, the writer would have very much liked to refer to one or two other matters. However, it is hoped that the paper, even in its present imperfect form, may furnish some little food for reflection on this very important subject of the action, influence and control of the roof in longwall workings.

———

LIST OF ILLUSTRATIONS (PLATES XI. AND XII.).

Fig. 1 illustrates the phenomena of subsidence and draw at the surface, as produced by mining operations below. The original profile of the surface is shown by the dotted line A B C, and the profile of the subsided surface by A H C. The line B E E₁ is in the vertical plane of the working-face, and G F F₁ is in the vertical plane of the solid coal-face left behind at the commencement of operations. A B is the draw on the surface beyond the working-face, and G C the draw extend. ing over the solid coal left in the rear. The oblique line C D D₁ is that referred to in the text as the incipient hade-line. As drawn in the figure, it takes no account of the disparity of the vertical and horizontal scales, the angle C D M being shown at something like a normal value. It correlates the mining subsi. dence with the natural subsidence associated with the normal fault, and it suggests that the great lines of fracture in the mine-roof must practically lie at the angle indicated by the incipient hade.

Figs. 2 to 4 illustrate roof-fractures, and the influence of their obliquity on leverage and transmission of pressure.

Fig. 5 illustrates the probable philosophy of a "bending" roof. The roof, originally on the line D E, bends by steps, *a, b, c, d*, etc., into the parabolic curve, D F. A B is the vertical plane of the working-face, and B C the direction of advance. The pack-walls are not indicated.

Fig. 6 illustrates the effect of the pressure of the roof on the gate-roads, the original cross-section A B C D being reduced by pressure to E F G H.

Figs. 7 and 8 illustrate the difference of the resistance offered by the coal to the oblique roof-action, according as the direction of working is across the planes of cleavage or in the same line.

Figs. 9 to 12 illustrate modifications, so-called, of longwall, the object of which is to evade the most violent effects of excessive roof-action, by directing such action on to a "mean face-line" which intersects the line of the actual working-face at a considerable angle.

Figs. 13 to 17 illustrate arrangements of roads designed to evade particularly obnoxious developments of the roof-action.

Fig. 18 illustrates the Gothic or herring-bone system of timbering a main return-air way, carried through the gob, and under a bad roof.

FIG. 6.--TRANSVERSE SECTIONS OF GATEWAY.

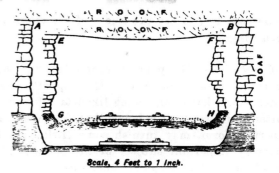

Scale, 4 Feet to 1 Inch.

FIG. 9.—PLAN OF GATEWAYS

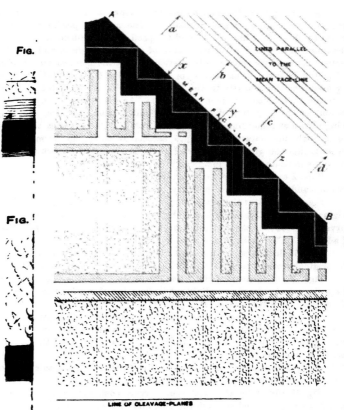

Scale, 100 Feet to 1 Inch.

The North

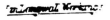

in the end, for, to use the idiom of the engineer, it ⟩
everlasting job."

Conclusion.—The paper is concluded, but the subje
far from being exhausted.   Had opportunity perm;
writer would have very much liked to refer to one or ʇ
matters.   However, it is hoped that the paper, even in it
imperfect form, may furnish some little food for refle
this very important subject of the action, influence a
trol of the roof in longwall workings.

LIST OF ILLUSTRATIONS (PLATES XI. AND XII.).

Fig. 1 illustrates the phenomena of subsidence and draw at the su
produced by mining operations below.   The original profile of the su
shown by the dotted line A B C, and the profile of the subsided surface by
The line B E E_1 is in the vertical plane of the working-face, and G F F_1 is
vertical plane of the solid coal-face left behind at the commencement of oper
A B is the draw on the surface beyond the working-face, and G C the draw e
ing over the solid coal left in the rear.   The oblique line C D D_1 is that re
to in the text as the incipient hade-line.   As drawn in the figure, it tak
account of the disparity of the vertical and horizontal scales, the angle (
being shown at something like a normal value.   It correlates the mining ʂ
dence with the natural subsidence associated with the normal fault, an
suggests that the great lines of fracture in the mine-roof must practically li
the angle indicated by the incipient hade.

Figs. 2 to 4 illustrate roof-fractures, and the influence of their obliquity
leverage and transmission of pressure.

Fig. 5 illustrates the probable philosophy of a "bending" roof.   The ro
originally on the line D E, bends by steps, *a*, *b*, *c*, *d*, etc., into the parabo
curve, D F.   A B is the vertical plane of the working-face, and B C the directi
of advance.   The pack-walls are not indicated.

Fig. 6 illustrates the effect of the pressure of the roof on the gate-road₂
the original cross-section A B C D being reduced by pressure to E F G H.

Figs. 7 and 8 illustrate the difference of the resistance offered by the coal to
the oblique roof-action, according as the direction of working is across the planes
of cleavage or in the same line.

Figs. 9 to 12 illustrate modifications, so-called, of longwall, the object of
which is to evade the most violent effects of excessive roof-action, by directing such
action on to a "mean face-line" which intersects the line of the actual working-
face at a considerable angle.

Figs. 13 to 17 illustrate arrangements of roads designed to evade particularly
obnoxious developments of the roof-action.

Fig. 18 illustrates the Gothic or herring-bone system of timbering a main
return-air way, carried through the gob, and under a bad roof.

Fig. 6.—Transverse Sections of Gateway.

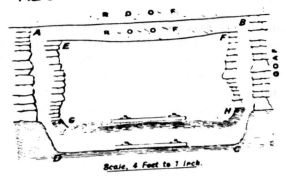

Scale, 4 Feet to 1 Inch.

Fig. 9.—Plan of Gateways

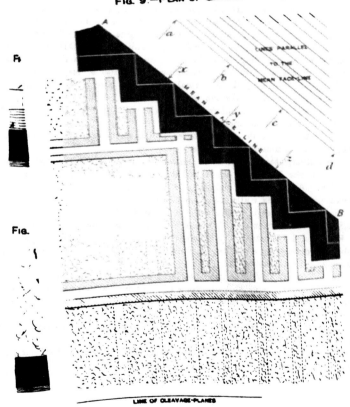

LINE OF CLEAVAGE-PLANES

Scale, 100 Feet to 1 Inch.

in the end, for, to use the idiom of the engineer, it makes "an everlasting job."

Conclusion.—The paper is concluded, but the subject is very far from being exhausted. Had opportunity permitted, the writer would have very much liked to refer to one or two other matters. However, it is hoped that the paper, even in its present imperfect form, may furnish some little food for reflection on this very important subject of the action, influence and control of the roof in longwall workings.

LIST OF ILLUSTRATIONS (PLATES XI. AND XII.).

Fig. 1 illustrates the phenomena of subsidence and draw at the surface, as produced by mining operations below. The original profile of the surface is shown by the dotted line A B C, and the profile of the subsided surface by A H C. The line B E E_1 is in the vertical plane of the working-face, and G F F_1 is in the vertical plane of the solid coal-face left behind at the commencement of operations. A B is the draw on the surface beyond the working-face, and G C the draw extending over the solid coal left in the rear. The oblique line C D D_1 is that referred to in the text as the incipient hade-line. As drawn in the figure, it takes no account of the disparity of the vertical and horizontal scales, the angle C D M being shown at something like a normal value. It correlates the mining subsidence with the natural subsidence associated with the normal fault, and it suggests that the great lines of fracture in the mine-roof must practically lie at the angle indicated by the incipient hade.

Figs. 2 to 4 illustrate roof-fractures, and the influence of their obliquity on leverage and transmission of pressure.

Fig. 5 illustrates the probable philosophy of a "bending" roof. The roof, originally on the line D E, bends by steps, *a*, *b*, *c*, *d*, etc., into the parabolic curve, D F. A B is the vertical plane of the working-face, and B C the direction of advance. The pack-walls are not indicated.

Fig. 6 illustrates the effect of the pressure of the roof on the gate-roads, the original cross-section A B C D being reduced by pressure to E F G H.

Figs. 7 and 8 illustrate the difference of the resistance offered by the coal to the oblique roof-action, according as the direction of working is across the planes of cleavage or in the same line.

Figs. 9 to 12 illustrate modifications, so-called, of longwall, the object of which is to evade the most violent effects of excessive roof-action, by directing such action on to a "mean face-line" which intersects the line of the actual working-face at a considerable angle.

Figs. 13 to 17 illustrate arrangements of roads designed to evade particularly obnoxious developments of the roof-action.

Fig. 18 illustrates the Gothic or herring-bone system of timbering a main return-air way, carried through the gob, and under a bad roof.

FIG. 6.—TRANSVERSE SECTIONS OF GATEWAY.

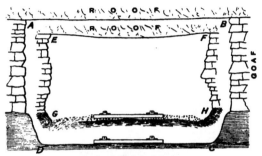

Scale, 4 Feet to 1 Inch.

FIG. 9.—PLAN OF GATEWAYS

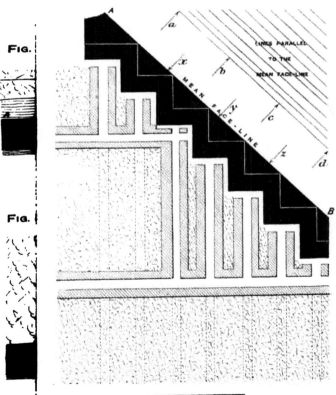

Fig. 12.—Plan of Gateways.

MEAN FACE-LINE

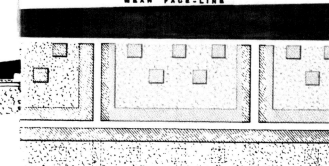

Scale, 100 Feet to 1 Inch.

Fig. 15 —Plan of Gateways.

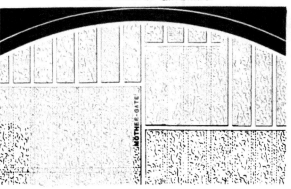

MOTHER-GATE

Scale, 200 Feet to 1 Inch.

Fig 18.— Section through a Goaf-road.

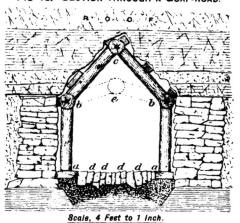

Scale, 4 Feet to 1 Inch.

DISCUSSION OF MR. T. E. PARRINGTON'S PAPER ON "THE ADOPTION OF A BALANCE-ROPE AT HYLTON COLLIERY."*

Mr. JOHN MORISON referred to the discussion which took place at the previous meeting on this subject, when the question was raised by Mr. A. L. Steavenson as to the effect of over-balancing a winding-rope, by some means which he did not suggest.† An interesting discussion arose, in which Mr. W. C. Blackett referred to a tapered rope which Mr. John Daglish had projected for balancing the Silksworth winding-engine in 1879. Mr. W. C. Blackett and Mr. T. E. Forster, who were both conversant with Mr. Daglish's idea, referred at some length to the matter; and in the course of that discussion, he (Mr. Morison) said that a parallel rope, of the same weight as a tapered rope, would have as good an effect and in a simpler form. Mr. Blackett had supplied him with some figures with regard to the proposed Silksworth rope, and from these figures, he (Mr. Morison) had constructed a diagram (Plate XIII.) showing the effect of the tapered balance-rope as proposed by Mr. Daglish; also the effect of a balance-rope of the same weight made in the ordinary way, and the effect of the load without any balance-rope. The comparison between the tapered rope and the parallel rope showed that the weight assisting the winding-engine at the start ceased its effect at about the fifth revolution; and conversely in retarding the load, it began to come into effect at the fifth stroke from the end of the wind. An ordinary rope of the same gross weight would commence at the beginning of the wind with the same load assisting the engine, and would carry its assistance through to the meeting of the cages, and at meetings the reverse effect began to take place. In the diagram (Plate XIII.), more assistance was rendered to the engine from the commencement of the wind to meetings, an object which he did not think would be attained by using a tapered rope. He (Mr. Morison) agreed that it would be desirable to make the balance-rope a little heavier than the winding-rope. The weight of the winding-rope in the case referred to was 8,004 pounds, and of the counterbalance-rope 9,908 pounds. There was not much difference in the weights, but he thought

* *Trans. Inst. M.E.*, 1903, vol. xxvi., page 294.   † *Ibid.*, page 298.

that the effect in winding would be improved by using a still heavier counterbalance-rope. He suggested that if a subject could be got, a diagram should be made in which the load put in motion by the engine should be plotted in a continuous line. He thought that it would be possible to construct such a diagram, and, coupled with the paper which had recently been read by Mr. Thacker on "The Dynamics of the Winding-engine," it would form a great advance in the knowledge of the operation of winding.

Mr. C. C. LEACH agreed with Mr. Morison that the balance-rope should be heavier than the winding-rope. There was great loss in the winding-engine, and probably the first two or three revolutions used three-fourths of the total volume of steam used in a wind. He did not see any advantage in using a tapered balance-rope, as it was more desirable to overbalance than to balance the load.

Mr. W. C. BLACKETT stated that the proposed tapered balance-rope was heavier than the winding-rope, but the point under discussion was where the weight should be placed in the rope. The tapered balance-rope had the advantage because, in the first few strokes, it helped the load. The Silksworth engine ran for a long time, the pit being deep, and once started, there was no need to continue the assistance to meetings, but towards the beginning of the fifth stroke from the end, the weight of the tapered end of the balance-rope was brought into use to help to stop the ascending cage. Any member who wished to look more closely into this matter should read Mr. Daglish's paper.* It was written prior to the suggestion of tapered balance-ropes, when counterbalancing of the load was secured by the use of staples or conical drums.

Mr. C. C. LEACH said that Mr. Blackett evidently desired to make the winding-engine run at a uniform pace, which could not be done.

Mr. BLACKETT replied that such was not his suggestion.

Mr. J. H. MERIVALE remarked that Mr. Blackett had simply spoken of the rope, but it was necessary to take the inertia and

* *Trans. N.E. Inst.*, 1876, vol. xxv., page 201.

: THE FACTORS GIVEN FOR
'NE.

REVOLUTIONS							REVOLUTIONS	
0	000	22,000	24,000	26,000	28,000	30,000	31,000	
START							START	

momentum into account.   Additional weight was required at
the beginning, not only on account of the weight of the rope,
but also on account of the inertia of the mass to be put into
motion.   The engine did not require the balance equally from
the top to the bottom, but a little extra weight at the beginning
of the wind to overcome the inertia, and a little extra weight
against the engine at the end of the wind to overcome the
momentum.

Mr. T. E. FORSTER said that he had made a diagram as well
as Mr. Morison.   He thought that theoretically Mr. Daglish's
tapered balance-rope was a good suggestion, but the proposed
rope was not made.   He (Mr. T. E. Forster) found that several
people used a balance-rope a little heavier than the winding-
rope.   It was an advantage, and he thought that if they made it
about 10 per cent. heavier in the case of an ordinary shallow
shaft, that would be about the best weight.   In a pit, 2,580
feet deep, he was astonished to see what the winding-engine
could do: the engine was not more than half the size of what
they would put in now, and the balance-rope was of a form that
he had not seen elsewhere: it was a flat one.   The manager told
him that he had found the advantage of using a flat rope was
that it was not so lively as a round one, and it did not twist
about in the shaft.

Mr. S. HARE said that at Murton colliery a flat balance-rope
had been in use for many years with very little trouble.   Once
or twice some few years ago it kinked, but not recently.   It was
used at the shaft where the men were changed.   Originally a
guiding-pulley was used, but it had been taken out.

Mr. JOHN MORISON said that the point in dispute between
Mr. Blackett and himself was not only as to whether the balance-
rope should be heavier, but whether or not the tapered balance-
rope was as good or better than the parallel balance-rope.   He
thought that the diagram (Plate XIII.) demonstrated that the
parallel balance-rope was better than the tapered rope, which
was so troublesome to design that it had never been made.   He
had seen many diagrams, and had many diagrams of winding-
engines, and they were not all shallow pits; but he had never
seen a diagram yet proving that it was advantageous to cut off

the assisting balance-rope at the fifth revolution, and he would like to see such a diagram. All the diagrams that he had seen showed that the engine required assistance from the start very nearly to meetings, and as soon as possible afterwards the balance-rope was used to retard the engine, and the further they carried it towards the meetings the better was the result. Every separate winding-engine would require a different arrangement of balance-rope. It would, to begin with, depend on the capacity of the engine for its load, and one might be economical where another was not; but he thought that he had demonstrated (at least to his own satisfaction, if not to Mr. Blackett's) that the parallel rope gave a better balance than any tapered rope that could be made.

Mr. W. C. BLACKETT said, with reference to the curved line, JNOK,* arranged by Mr. Daglish for the Silksworth engine, that it was at the fifth revolution that the assistance stopped. With the ordinary counterbalance, with chains hanging in a staple, the weight was removed at a certain part of the stroke to give the exact line which he had indicated, and the conical drum also was made to give that identical line. He was only putting forward what Mr. Daglish at that time considered would give the best possible result for the Silksworth winding-engine; and, if the members would follow out the subject, they would find that the engine was working economically after the fifth stroke. Mr. Morison's straight line was certainly against Mr. Daglish's theory and actual practice.

The PRESIDENT (Mr. W. O. Wood) said that he had had some experience with balance-ropes. About 30 years ago a winding-engine broke down, and by means of a balance-rope one cylinder was able to perform the work that the two had been doing. The tapered balance-rope was theoretically correct, but he thought that it would not answer in practice, as it would coil and kink. The flat balance-rope at Murton colliery worked beautifully for a year or two, and then, without any apparent reason, the rope would kink and twist and give a great deal of trouble. As Mr. Hare had said, there had not been the slightest trouble for the last two or three years, but it might go wrong any day.

---

* *Trans. N.E. Inst.*, 1876, vol. xxv., page 214, Plate LIV

WELDLESS STEEL PIT-PROPS.

The Sommer prop for mines is made of two tubes, one tele-scoped into the other, thus enabling the prop to be lengthened or shortened as required. The two tubes are held in position by means of a clamp, which can be screwed together so as to carry a load varying from 15 to 16 tons without giving way. The clamp is placed upon and attached to the outer tube in such a way that it cannot slide downward. The heel of the lower tube is either made flat or pointed, or fitted with a base-plate suitable for a soft bottom-stone. The upper tube can be supplied either with a flat or semicircular end to receive a small wooden

FIG. 1 —STEEL PIT-PROPS.

block, whenever its use is found necessary. The props can easily be placed in position, after the proper length has been adjusted, in a similar way to wooden props.

The principal aim being to use these steel props over and over again, a special device has been adopted, so as to recover them after the coal has been worked out. The clamp is then loosened by means of a spanner or ratchet, and the latter can, if necessary, be worked with an iron-hook from a distance, so as to avoid danger (Fig. 1).

It is claimed that the Sommer steel props will never bend, break or collapse by upheavals of the bottom, or subsidence of the roof. In any case, where the maximum pressure of 16 tons is exceeded the upper tube will adjust itself, by automatically telescoping into the lower tube, and will remain stationary and intact as soon as the pressure abates. It is a common experience that wooden props break one after the other, thus gradually weakening the resistance until the entire roof collapses. Sommer steel props, however, will never act in this way; and if one prop is overloaded it will telescope sufficiently until the pressure is met by the other props and equally distributed.

The Sommer props are made 4 to 6 feet long, with tubes 2½ and 3 inches in diameter; 6 to 9 feet long, with tubes 3 and 3¼ inches in diameter; 9 to 12 feet long, with tubes 3¼ and 3⅝ inches in diameter; and 12 to 18 feet long, with tubes 4⅛ and 4¼ inches in diameter. A steel prop, 7 feet long, weighs about 112 pounds.

DISCUSSION OF MR. J. J. MUIR'S PAPER ON "AN IMPROVED FORCED METHOD OF TREATMENT OF LOW-GRADE COPPER-ORES."*

Mr. H. LIPSON HANCOCK (Moonta Mines, South Australia) wrote that he was struck with the percentage of the ore termed "low-grade," which would here be considered good grade. However, Mr. Muir did good service to the members by giving them the benefit of his investigations, and deserved their appreciation. If the ore in the material considered to be typical of Lyell, in Tasmania, is finely disseminated throughout the gangue containing 48 per cent. of silica, the process outlined should have much to recommend it, although he was rather doubtful whether the operation could be completed for 14s. per ton. Even accepting the figures quoted, the extraction of copper from a 3 per cent. ore, with a good extraction, would require an expenditure of £28 per ton of copper to bring it up to the precipitate form, and a further £3 per ton to make it into good metal, or a total cost of £31 per ton. It appeared to him (Mr. Hancock) that too little attention, in a general way, was paid to the possibilities of economical cracking, sorting and

* Trans, Inst, M.E., 1903, vol, xxvi., page 40,

mechanical concentration, which with many Australian ores (where the specific gravity of the gangue was not more than 2·8 to 3), would often be found more easily applied, especially as there were so many minerals, which increased the quantity of acid required to dissolve the copper.   As an instance of this, at Wallaroo mines, 3 per cent. sulphide ore is treated by mechanical dressing and smelting, and produces copper at a cost of about £26 per ton, as against the cost of applying Mr. Muir's process at a cost of £31 per ton of copper.   If the ores at these mines were equal to an average of 4½ per cent. of copper, the Wallaroo mechanical processes combined with smelting would, he thought, be preferable from a commercial point of view to the process outlined in Mr. Muir's paper.

Mr. WALTER T. HOLBERTON (Copiapo, Chile) wrote that, although the principle on which Mr. Muir's method was based was by no means a new one, the paper was interesting; and if the process could be worked to be commercially successful, undoubtedly great benefit would be derived in localities where that class of ore, as described, was abundant.   The method had been attempted in Chile locally, under various modified forms, with a view of profitably treating the large deposits of very low-grade carbonate ores, that existed throughout that country, by the treatment with sulphurous acid obtained from low-grade pyritic ores, but hitherto it had not been a commercial success. It seemed to him (Mr. Holberton) that Mr. Muir's deductions were made more from experiments than from treatment on a working scale, and, he feared that gentleman would find great difficulty in using the sulphurous fumes derived from the selected portion of the ores, when roasted in a reverberatory furnace, because the fumes would be mixed with the products of combustion of the fuel.   When using a retort and compressed air, his expenditure would be high, but local conditions would have to be taken into consideration.

He (Mr. Muir) also stated that "in operation, it is advisable to have an excess of sulphurous acid (H_2SO_3) in the solution."[*] In actual practice he would find that it was not only advisable, but that it was absolutely necessary, because a certain amount of sulphite of copper was formed, which was not soluble in water but was so in a solution of sulphurous acid.

* *Trans. Inst. M.E.*, 1903, vol. xxvi., page 44.

He (Mr. Holberton) did not quite agree with Mr. Muir when he said that "it is also necessary . . . . to show a speed in the work that will compare with that obtained in a smelting operation."[*] He could not see that great speed in treatment was necessary, so long as the capital required for the plant was not too great, and that a good profit was the result of the treatment. The Rio Tinto Company probably treated larger quantities of this class of poor ores than any other such undertaking in the world, by a process which required several months to treat a given pile of ore and yet they paid big dividends.

He thought that Mr. Muir was certainly on the right track, and that his investigations would be watched with considerable interest by the copper community. It was to be hoped that he would favour the members with further details, when his method had had a continuous trial on a working scale.

Mr. LUKE WILLIAMS (Claremont, Tasmania) wrote that Mr. Muir estimated that the cost of treatment of crude schistose ores would be 14s. per ton, to this he would add from 3s. to 6s. per ton (according to the nature of the deposit) for mining charges, supervision and disposal of the products, the cost of treatment thus averaging about £1 per ton of crude ore. With copper at £50 per ton (and it was not safe to calculate on a higher value, when looking into the future of any new process) it would require 2 per cent. of copper in the crude ore to pay all expenses. He therefore thought that the cost of treatment must be reduced before the process could be extensively used, as the Mount Lyell mine had paid dividends by direct smelting of an ore containing only 2¼ per cent. of copper (with copper under £60 per ton), 2 ounces of silver and 1½ to 2 dwts. of gold per ton. He considered that Mr. Muir was deserving of great credit for the work that he had done up to the present with his laboratory-tests of his process, and he (Mr. Williams) would like to see it in operation at some mine working on a commercial scale.

Mr. G. D. VAN ARSDALE (New York City, U.S.A.), wrote that the process described by Mr. Muir, provided that he had overcome the mechanical and other difficulties hitherto attending the use of sulphur dioxide as a solvent, would seem to be adapted to

[*] *Trans. Inst. M.E.*, 1903, vol. xxvi., page 41.

the treatment of an ore of the analysis given, since it is claimed by those advocating similar methods that sulphur dioxide is not consumed to so great an extent as sulphuric acid in attacking soluble constituents, other than the copper, of an ore. Unless this particular ore contained them in a more or less insoluble state, the comparatively large amounts of lime, magnesia, etc., would practically prohibit leaching by ordinary methods. In certain cases, however, a choice of the proper strength of the solvent, is of importance, as certain magnesium-aluminium silicates are decomposed by stronger solvents in preference to the copper; while, with weaker solutions, the copper may be extracted without consuming too much of the reagent. Mr. Muir's conclusion as to the good extraction obtainable by most of the main processes proposed for copper-leaching, tended to show that probably most of these could be successfully applied under proper conditions, but that very few would prove economical under conditions of high cost of fuel, chemicals, including iron, etc. In considering a choice of processes, there are, besides costs and the chemical composition of the ore, sometimes other factors, difficult to determine by analysis or experiment on a small scale, that largely influence results. For example, experiments were made for the wet extraction of the copper from an ore of seemingly nearly identical composition with another, which was yielding good results on a large scale. The experiments did not succeed, although the reason for their failure could not be predicted from the chemical composition of the ore. He hoped that Mr. Muir would publish the results obtained on a large scale. It would be especially interesting to know whether he found in practice that the amount of sulphur contained in these ores was sufficient to effect the solution of their copper. Regarding the precipitation of the copper, it seems probable that Mr. Muir intends to make use of the reaction proposed in the Neill process (precipitation of the copper from solution as sulphite, by driving off the excess of sulphur dioxide) in addition to precipitation by iron, since otherwise the cost given by him for pig-iron would seem rather low, unless this was obtainable at a very low price. The leaching of copper, while entirely practicable, as shown by the large amounts annually extracted from the Spanish and Portuguese deposits and by the Henderson process, had not made the same progress as copper-smelting, possibly because of the need for a more universally applicable process; but doubtless there would be applications made in the future, since there exist large deposits of ore only amenable to a wet process, and if Mr. Muir's work proved to be a positive step towards this end, he was to be congratulated.

THE MINING INSTITUTE OF SCOTLAND.

GENERAL MEETING.
Held in the Hall of the Institute, Hamilton,
June 9th, 1904.

Mr. ROBERT THOMAS MOORE, President, in the Chair.

The minutes of the last General Meeting were read and confirmed.

The following gentlemen were elected:—

MEMBERS—
Mr. Stewart Chambers, Northfield Colliery, Prestonpans.
Mr. James Eadie, Eastfield, Harthill.
Mr. John McLuckie, Cross House, Larkhall.
Mr. John Robertson, Lorne Villa, Polmont Station.
Mr. James Wilkinson, Burngrange, Motherwell.
Mr. William Wilson, Cardenden Colliery, Cardenden.

ASSOCIATE MEMBER—
Mr. Frederick W. J. Bowie, 200, Glenpark Road, Glasgow.

DISCUSSION OF MR. J. R. WILSON'S PAPER ON A "PROPOSED METHOD OF SINKING THROUGH SOFT SURFACE."*

Mr. Robert Martin (Portobello) wrote that Mr. Wilson's paper would have been more interesting if it had been the description of an actual experience. The difficulties of "skin-friction" and keeping a cylinder in a "truly vertical position" are exaggerated by Mr. Wilson, and can be overcome by sinking the first cylinder for a portion of the distance only, and making it large enough to sink another or more inside it. As a matter of experience, boulders are no trouble to a cylinder where the sinking is in soft strata, such as silt or mud. But, if for example,

* *Trans. Inst. M.E.*, 1904, vol. xxvii., page 86.

the first 20 feet of the sinking is through boulder-clay, followed
by mud, boring is almost impossible. At Olive Bank, Mussel-
burgh, the borer sank a small pit through the boulder-clay,
after wasting a good deal of time in trying to bore one hole
through it. A bore-hole is as impossible in mud as in the sea. In
mud, silt or soft clay, the borer drives a tube into it and pumps
out the soft matter—in fact he is just sinking a small cylinder.
He must start with a tube large enough for the purpose, knowing that it will be forced only to a limited distance, so as to
enable him to insert smaller sizes, and leave him at the rock-
head with a tube large enough for boring into the rocks below.
The sinking of a series of such tubes, 100 feet into mud, silt
or soft clay so regularly that they afterwards could be wedged,
seems no easy task, and would probably cost far more than a
steel cylinder of ½ inch plate covering the same area. The tubes
would be at least 8 inches in diameter at the surface. The real
trouble, however, would only begin when the boring and tubing
were (let them suppose) satisfactorily finished. The tubes are
not watertight or mud-tight, till they are wedged, but how is
the wedging to be done if there be a constant inrush of water
and mud? Should the material to be sunk through be dry
enough and firm enough to allow of the removal of 18 or 9 or
even 4½ inches at a time, so as to admit of wedging and the
fixing of the barring thereto, what would happen when the
pressure due to a depth of 400 or 500 feet is encountered, and
a feeder of water and mud prevents the sinkers from even seeing
within a couple of feet or more of the lowest barring? The
constant inrush of water and silt will in all probability hinder
sinking, and if the water and silt be removed unequal sinking
of the ground around the shaft will take place. The woodwork
and tubing already in place will be distorted, and if the sinking
be persisted in long enough the shaft will be ruined.

Mr. T. H. MOTTRAM (Glasgow) remarked that Mr. Wilson
did not state the diameter of the malleable-iron rods, although
he gave their length. Mr. Wilson only suggested that his
system was feasible to a depth of 100 feet; but Mr. Martin, on
the other hand, referred to the difficulties that would be experienced at depths of 400 and 500 feet. Personally, he suggested
to Mr. Wilson the advisability of working from the surface with

a larger template than one only " slightly larger than the actual finished size of the shaft," as this would in a measure provide for the bore-holes getting out of plumb.

The PRESIDENT (Mr. R. T. Moore) said that by Mr. Wilson's system there would be considerable trouble in getting down the bore-holes, and most people would think of adopting Mr. Mottram's suggestion in regard to a larger shaft. Mr. Wilson's paper, however, even although the members did not agree with his proposals, gave opportunities for discussion on a very interesting subject.

The discussion was closed, and a vote of thanks was awarded to Mr. Wilson for his paper.

———

DISCUSSION OF MR. R. W. DRON'S PAPER ON "THE OCCURRENCE OF CALCAREOUS COAL IN A LANARKSHIRE COAL-FIELD."[*]

Mr. JAMES C. WEIR (Motherwell) wrote that he had had a similar experience in working the Ell, Pyotshaw and Main coal-seams. In certain areas, the coal is burnt or altered in tracts varying in area to 100 square yards or thereabouts. In the Ell coal-seam, a portion, 1 foot thick, next the pavement is burnt, but the remaining 6 feet is of a very good quality. The Pyotshaw seam, immediately under, 2 feet 3 inches thick, is much harder than the average seam, and some portions are unfit for market purposes. The change in the Main seam is erratic; in some places only a few inches next the roof or pavement is burnt or altered, and at other parts the alteration extends to the full height of the seam. The Splint seam, immediately underneath, has been thoroughly proved and is fully 5 feet high. It is of good quality, and there are no signs of burning or alteration of the coal. The Splint seam was proved previous to the opening out of the Main coal-seam, and it was expected that the latter seam would be of good average quality and thickness, but on cutting into the seam he found it next to an upthrow slip of 20 feet fully 3 feet 4 inches thick, but within a distance of 20 feet, the thickness of the coal decreased to 11 inches, and it was burnt and useless. The road was continued for a distance of 100 feet

* *Trans. Inst M.E.*, 1904, vol. xxvii., page 92.

without any alteration; a place was then turned to the left for a distance of 150 feet, and the coal at the face is 1 foot 11 inches thick, the upper 10 inches being burnt and useless. The depth from the surface at this point is about 300 feet, and so far as proved the seams below this depth are of good thickness and quality, showing no signs of burnt or calcareous coal. He (Mr. Weir) was of opinion that the coal was altered by the infiltration of lime, etc. from above, at intervals during the deposition of the coal; but he could not understand why the coal below that depth had escaped alteration. He had observed that the rocks above the altered coal are harder than those above the unaltered coal.

Mr. G. H. STANLEY (Durham College of Science) wrote that he had analysed a sample of the burnt coal, submitted by Mr. R. W. Dron, with the following percentage-results:—Volatile matter, 39·30 per cent.; fixed carbon, 22·72 per cent.; and ash, 37·98 per cent. The coal comprized sulphur, 0·375 per cent.; silica and insoluble matter, 1·30 per cent.; ferrous oxide, 4·73 per cent.; alumina, 8·66 per cent; lime, 20·02 per cent.; magnesia, 2·03 per cent.; soda, 0·45 per cent.; and potash, 0·09 per cent. The specific gravity is 2·06. Very little iron-pyrites is present, and practically all the iron exists as ferrous carbonate. The analysis indicates the probable presence of carbonates, as follows:—Ferrous carbonate, 7·62 per cent.; calcium carbonate, 35·75 per cent.; and magnesium carbonate, 4·22 per cent. Traces of manganese, and soluble sulphates and chlorides are also present.

The chief points of difference between this analysis and that printed in Mr. Dron's paper are the existence of the iron in the ferrous and not in the ferric condition, and the great difference in the percentage of magnesia.

Mr. R. W. DRON, remarked that the difference between the previous analysis and that made by Mr. G. H. Stanley was due to the material, which was very irregular in its chemical composition.

Mr. ANDREW WATSON (Glasgow) referred to an alteration in the coal, which had occurred in the Main seam in the Carmyle district, about 5 feet in thickness. In this seam, the bottom and not the top part of the coal was affected by " burning," to a thickness varying from 2 to 12 inches. This alteration was not caused by whin, nor was the coal in the neighbourhood of whin.

Mr. T. H. MOTTRAM (Glasgow) asked whether there were any joints or cleat-lines in the altered coal, and whether these were coated with carbonate of lime as is generally the case in the Main coal-seam ; and also whether there were any " wants " in the calcareous-coal district which Mr. Dron had described in his paper. It seemed probable that the faults described by Mr. Dron had some connection with the altered coal, although it was curious to note that in the Main coal-seam, the coal was clean in one place near one of the faults, but altered from top to bottom in a distance of 50 feet.

Water of springs percolating through strata is rarely free from iron, carbonate of lime, or other earthy ingredients ; and as some water has the power of dissolving calcareous rock over which it flows, it is possible that at the time when the dislocation of the strata took place, the calcareous matter found its way into empty spaces or interstices by infiltration or segregation from the surrounding rocks. The altered coal, occurring in some places at the bottom of the seam, rather points to the calcareous matter coming from below or to its being deposited contemporaneously with the formation of the coal. The discoloration of the roof at Rosehall colliery and elsewhere, referred to by Messrs. J. S. Dixon and J. Hamilton, appeared to be purely local. In Ayrshire, for instance, where the field is much cut up with whin-dykes and where there is a good deal of burnt coal, the red coloration—or red ground—is rarely seen.

For some time it was difficult to account for the burnt coal at Rosehall colliery, but the problem was ultimately solved by the discovery of a whin-dyke, the existence of which had been previously unknown.

Mr. WILLIAM SMITH (Dalmellington) said that they had come across some burnt coal in their colliery lately, and in driving a dook they encountered a number of whin boulders in the coal. The boulders were isolated, and they had no connection whatever with any outside dyke. He had often wondered how they came to be there, and whether they were deposited contemporaneously with the coal. The coal was burnt round the boulders for perhaps 8 or 9 inches, then there would be 1 foot or 2 feet of clean coal, and then another boulder. The boulders were of irregular shape, and varied in size from 1 foot to 4 or 5 feet in diameter. In their part of the field in Ayrshire, they found that the softer the coal was burnt the thicker was the whin; and, if the coal was burnt hard, the whin was generally not so thick.

Mr. R. W. DRON said that at the last discussion the President and Mr. Dixon took exception to his calling this "kingle' and "calcareous" rather than "burnt" coal; but he thought the discussion that day had justified him in giving a name to this species of altered coal. The particular alteration which he had described was so entirely distinct from any action of whin, that unless they gave it a separate and distinctive name there was apt to be confusion; and for that reason he was endeavouring to call it "kingle" or "calcareous" coal. There was absolutely no connection between such coal and coal affected by whin, either in the chemical composition or the physical state: the only resemblance being in the outward appearance. In the hand, one could easily recognize that the specific gravity was greater than that of common coal; and miners distinguished it readily by the weight. Mr. J. S. Dixon and other speakers had remarked that calcareous coal was rather common in the Hamilton district. He had hoped that some of the members who were acquainted with the occurrences at Rosehall and Carmyle would have given details similar to those that he had adduced, so that the subject could be studied and referred to at any time. Mr. H. Mungall had written to him (Mr. Dron) that "throughout the east country he had never met with this class of coal. There was plenty of burnt coal in Fife, but neither there nor in Edinburghshire had he (Mr. Mungall) met with this calcareous coal." The fact that the bottom and not the top part of the coal-seam was affected, had aroused the suggestion from one or two members that the coal must have been altered during its formation; but he (Mr. Dron) did not agree with that opinion. There was the strongest possible evidence of the fact that the occurrence took place after the formation of the coal: for instance, the coal on one side of the fault was of a calcareous composition, while on the other side of the fault, and iu the same seam, but at a greater depth, the coal was perfectly clean. In that case, apparently, the depth from the surface had been the greatest factor in the alteration. There could be no doubt whatever that the infiltration of the water with lime, magnesia, etc., was more intense along the line of faulting. Joints or cleat-lines containing calcite could be observed, the bands of calcite passing through both the clean coal and the altered coal. He had not referred to the occurrence of a red roof in the course of his paper; but to some extent a red coloration was apparent in the roof overlying the coal that he had described. This, however, did not prevail to the same degree as at some other collieries that had been mentioned, where the miners were completely covered with red dust. A few " wants " occurred where the coal had become entirely altered or " burnt." He thought that the points raised by Mr. Smith should not be touched upon in the present discussion, dealing with a different subject.

The discussion was closed, and a vote of thanks was awarded to Mr. R. W. Dron for his interesting paper.

THE CHAMPION COAL-CUTTER.

The members had an opportunity of examining the Champion coal-cutter in operation at the Wishaw Coal Company's Dalzell Colliery, Motherwell, in the afternoon.

Mr. R. W. DRON said that the Wishaw Coal Company, Limited, had used the Champion coal-cutting machine for the past three or four months. One machine was introduced experimentally, and they were so satisfied with the results that they were erecting plant with the intention of installing more. The practical results of the machine were satisfactory, although they did not come up to the estimates given by Dr. Simon in his paper.* They found, so far as they had gone, that a machine could deal with two places per day. These places were 11 feet wide, and the coal was undercut to a maximum depth of 5 feet, giving an undercutting, altogether, of 110 square feet; and in addition the machine also put in two shot-holes. The work of a shift of two men, who were controlling the machine, was

* *Trans. Inst. M.E.*, 1903, vol. xxvi., page 322.

represented by an undercut in two places, charging the shot-holes, firing them, and leaving the coal on the pavement. The coal was cut in a seam 4 feet 9 inches to 5 feet thick, and the amount produced was about 13 tons. He calculated that the cost of a ton of coal put on the pavement was 1s. 4½d., the actual cutting cost was 1s. 1d. per ton, 2d. per ton was allowed for powder, while the upkeep of the machine and fuel were reckoned at 1½d. per ton. The machine was employed in hard coal, nevertheless, he was satisfied that·the work was being done, as compared with ordinary pick-labour, at a reduction in cost of from 20 to 25 per cent. As to the time occupied, he (Mr. Dron) found that at Dalzell colliery, with 2 men on the machine, it took 3 hours and 20 minutes to cut a place 11 feet wide and 5 feet deep: and to bore two holes. Dr. Simon had expressed the opinion that a man and a boy could efficiently control the machine, but it had been his (Mr. Dron's) experience that this type of boy was not to be found in Scotland. It was necessary to explain, however, that of the 3 hours 20 minutes referred to, the actual time spent in cutting was 1 hour 50 minutes; and the balance of the time was spent in lifting the machine into position; and included stoppages, slight break-downs, and so on. He (Mr. Dron) had given a fair average of what the machine could do, and what they had been able to get out of the men in a shift of 8 hours from bank to bank; and he hoped to get better results out of these machines when they commenced working in more regular places. In the mean-time, the machines were being applied to "deficient places," where the coal was more costly and more difficult to work than in ordinary places. The machines had proved very useful in cutting through hitches and stonework of that kind, while they had proved advantageous in driving stone-mines. In the softer rock, the machine was not very beneficially employed, but in hard rock, it was certainly profitable to use the machine.

The coal-cutter had also been employed in a seam, 2 feet 4 inches thick, for the purpose of driving a dook, 30 feet wide. They found that 2 men took 8 hours to make that undercut, without being able to bore the holes. The principal difficulty was that the place was very wet, and the holing-dirt became clogged in the undercut. Even in this case, the machine was cheaper than pick-labour.

A good feature of the Champion machine, to his mind, was that the expenditure on repairs was very small. Then again, a workman of ordinary intelligence could learn to operate the machine successfully in the course of a week. The trials at Dalzell colliery had all been made in hard coal, which could not be worked by manual labour at a profitable rate per ton. A Morgan-Gardner electric heading-machine, weighing about 30 cwts., had been introduced for the purpose of cutting through this hard coal, but they found that the cost was about 6d. per ton more than with hand-labour, chiefly because the machine was unwieldly; and from 2 to 2½ hours were required to shift the machine from one place to another. People who were satisfied with a moderate result would be quite pleased with the Champion machine, as it enabled them to work fairly hard coal at miners' ordinary rates.

Dr. A. SIMON said that the Champion machine had been tried at the Cannock and Rugeley collieries, in ordinary strata, with good results. The rock was of moderate texture—not very hard. Some of the men had tendered to drive a place through a fault at the rate of £2 per unit; but one of the men, who appeared to know how the machine should be handled, tendered for the work at £1 per unit. He got the contract and did excellent work, making more than his ordinary wages. The Champion machine had been operated upon oil-shale, and the highest rate of cutting that he had been able to obtain was 16 square feet per hour. It would not pay at that speed, and if the machine were to be a profitable investment in oil-shale mines it would require to cut 25 or 30 square feet in the same time. The experiments, however, had not yet been concluded, and he did not despair of success. He attributed the failure of the first trial to the shape of the cutter, which required to be specially adapted for oil-shale. The trial was made at Pumpherston, and there the shale behaved like leather, and could hardly be penetrated. The machine could be used for boring shot-holes in shale, and he had bored to a depth of 6 feet in 12 or 15 minutes. Generally speaking, however, the machine had not, so far, proved successful in its operations in oil-shale.

At Aldridge colliery, near Walsall, the Champion machine had cut, in consecutive shifts of 8 hours, 360, 340 and 350 square feet in a fire-clay found on the top of the coal. In other cases, where the fire-clay was located underneath the coal, the rate of speed was in excess of cutting in the coal; but on the other hand it sometimes happened that the reverse was the case. Generally speaking, reasonable and good results could be got from the use of the machine in fire-clay, if no clogging occurred. To overcome any eventual clogging, a special device had been adopted by which a jet of water was made to play intermittently from the centre of the cutting-bit.

Mr. HENRY KING (Lanemark) said that the Champion machine had been in use at his colliery for about 2 months. It was first tried in the Eight-feet seam, comprizing a bottom coal about 3 feet thick, then about 2 feet of dirt, and finally the top coal.   The machine was set for holing in the dirt immediately on the top of the bottom coal : when going uphill and driving ahead, it was found that the men were kept busily employed in redding up the dirt.   After a week's trial, the machine was removed to another seam, consisting of 4¼ to 4½ feet of ordinary house-coal, with 9 inches of cannel coal. The holing in that seam (and the seam was still being worked) was on the top of the house-coal, lying immediately underneath the cannel-coal.   The place was being worked by stoop-and-room, and the men were holing two places in each 8 hours' shift. Each place was about 9 feet wide, and the men were holing to a depth of 6 to 8 feet.   Two men could easily move the machine about, when it was in a level-course working ; but in steep work-ings assistance was required.   There was more difficulty, to his mind, in shearing than holing with the Champion machine, because so much more dust was produced.   There was another difficulty in connection with shearing, namely, that when the drill missed the coal or stone and struck into empty space, it flew out and stuck.   Attempts had been made to square the corners of the cuts, but they were never squared in the true sense of squaring.

Mr. THOMAS THOMSON (Hamilton) said that, when he saw the Champion machine working some time ago, it certainly cut to an equal depth all round. The manager, Mr. McBride, said that if the machine could be allowed to run without squaring the corners, it would do one half more work. He (Mr. Thomson) was of opinion that there was no advantage in squaring the corners, because the coal could be blasted off in a semicircle as well as if the face were kept straight; but, in the case of a coal with good backs, etc., the results might be different. He was also of opinion that when the coal was holed, the machine did not require to be removed, and could bore holes straight forward and into the sides as required, which he thought would do as well as shifting the machine from side to side.

Mr. R. W. DRON said that the Wishaw Coal Company started to drive a mine by using the machine for channelling: they found that this system was not economical, being no cheaper than hand-labour, so they took it out again. When, however, they came into the hard rock the machine was put back into the mine; and when they had passed through the hard rock, the machine was once more withdrawn.

Mr. W. SMITH (Dalmellington) said that in the Wishaw Coal Company's pit that day the Champion machine ran for 30 minutes. The holing was 2 feet 3 inches under in the centre; at $1\frac{1}{2}$ feet from the centre, it was 1 foot 6 inches; and it was only 9 inches deep at the corners. The length of the work was $8\frac{1}{2}$ feet.

Mr. JAMES BARROWMAN said that there had been no attempt to cut out the corners in the course of the tests made that day.

The PRESIDENT (Mr. R. T. Moore) remarked that a miner in his ordinary working of a place did not square the corners every time.

Mr. HENRY KING said that very little attempt was made to square the corners at his colliery.

Dr. A. SIMON said that the squaring of the corners had been a difficulty with several users of the Champion machine. Mr. Smith had mentioned that, in connection with the operations at Motherwell that day, the holing was 2 feet 3 inches deep in the middle, and only 9 inches at the sides. The members were aware, however, that the longer the rod with which the cutting was made, the flatter would be the arc. Anyhow he could assure the members that there was no difficulty in squaring out the corners, if only the operators would take the trouble to think a little of what they were doing. Mr. H. King stated that one drawback to the machine was the tendency which it showed for sticking or stopping in front when missing the solid coal-face in striking. He, however, could provide Mr. King with a machine which did not stick in front, but it would not cut the same amount of square feet per day as the other: a novice preferred a machine which did not stop, though the skilled operator gave his preference to the machine which did most work and he could handle the machine so that it did not stop. Mr. King had also remarked that the Champion machine did not work so fast at shearing as at holing. That day, on their visit to the colliery at Motherwell, the members had seen the reverse: the holing took 35 minutes, and the shearing to a slightly greater depth, 3 feet 6 inches in a $5\frac{1}{2}$ feet seam, only occupied some 18 or 19 minutes.

Mr. R. W. DRON said that, in order to avoid a false impression getting abroad, it would be as well to add that the shearing had been made in a very soft coal and the holing in a hard coal.

Dr. A. SIMON, continuing, said that the holing was made in what was described as hard coal, close to the burnt coal; whereas the shearing had been performed through all the layers of the seam. In other mines, where soft layers alternated with hard, he found that the shearing required more skill and time than the holing. Many of the so-called drawbacks pointed out in the course of the discussion, were attributable to the fact that the men in charge of the machine had not yet acquired a sufficient acquaintance with it; and usually most of the above-mentioned drawbacks vanished as soon as a rate for piece-work could be arranged.

A vote of thanks was awarded to the Wishaw Coal Company, Limited, and to Dr. Simon.

———

Mr. D. JACKSON read the following paper on "Natural Coke in Douglas Colliery, Lanarkshire":—

NATURAL COKE IN DOUGLAS COLLIERY, LANARKSHIRE.

By DOUGLAS JACKSON.

In Douglas Colliery, Ponfeigh, an area of burnt coal has been found in the Big Drum seam. This seam, at a depth of 569 feet from the surface, is 6 feet 4 inches in thickness, and is worked by the stoop-and-room system. In a section of the workings, going towards a fault of 85 feet down to the east, and when about 100 feet or thereby from the fault, the coal gradually became more and more carbonized, and after driving 25 feet, the coal changed into a natural coke, the thickness of the seam being reduced to 5 feet 6 inches. The burnt coal has been proved to extend in a direction parallel with the fault to a distance of 240 feet, and apparently extends still further in that direction.

The same seam of coal on the downcast side of the fault has been worked and found in its normal condition.

The following analyses have been made of the coal, in its natural state and when burnt:—

	Natural state. Per cent.	Burnt coal. Per cent.
Volatile matter	34·00	9·75
Sulphur	1·76	2·85
Water	6·52	—
Fixed carbon	52·86	81·13
Ash	4·86	6·27

Where a stone-mine has been cut through the fault from the lower seam, the "vees" of the fault is well defined, and shews that the strata have slipped forward about 6 feet, as well as downward 85 feet. The intervening space of 6 feet is filled with unstratified débris. On the stone-mine, just referred to, being continued in the same straight line for 210 feet, another large fault was met with, 284 feet down to the south-east. The two faults join together at a point about 300 feet north of where the burnt coal has been proved, and it has occurred to the writer that the burning of this coal may have been produced by great friction when the faulting took place.

This is the only case of the kind that has come before the writer's notice. The burnt part of the seam, so far as can be seen at present, extends to about 1 acre, and has the appearance of having been rapidly cooled down by submergence in water, or by the rent of the fault becoming quickly filled up, thus excluding atmospheric air from the heat or fire.

————

The PRESIDENT (Mr. R. T. Moore) said that he remembered seeing a similar formation in the Denny district. The first indication of burning was the finding of what the miners described as "coal-tatties" in the seam. These were round lumps of coal, exactly shaped like potatoes. They burnt well, and it was only by their shape that they appeared to be different from the ordinary seam of coal. After this, a width of about 20 feet of natural coke was found and gradually the seam became a regular burnt coal.

NORTH STAFFORDSHIRE INSTITUTE OF MINING AND MECHANICAL ENGINEERS.

GENERAL MEETING,
HELD AT STOKE-UPON-TRENT, MAY 9TH, 1904.

MR. E. B. WAIN, IN THE CHAIR.

The minutes of the last General Meeting were read and confirmed.

The following gentlemen, having been previously nominated, were elected :—

HONORARY MEMBER—
MR. B. WOODWORTH, 21, Hamilton Road, Normacot.

MEMBER —
MR. H. GRAINGER, Zululand Collieries, Somkele, Zululand.

ASSOCIATE — ;
MR. W. G. SALT, Crackley, Silverdale.

Mr. WILLIAM LOCKETT read the following "Record of the Failure of a Locked-coil Winding-rope":—

RECORD OF THE FAILURE OF A LOCKED-COIL WINDING-ROPE.

By WILLIAM LOCKETT.

The locked-coil rope under consideration was made of patent steel, and constructed as shewn in Fig. 1 (Plate XIV.). There were 33 recess-shaped wires in the outer ring of the rope, 27 wedge-shaped wires in the second ring, 20 wedge-shaped wires in the third ring, 11 round wires in the fourth ring, 6 round wires in the fifth ring and the central core is a round wire, $\frac{1}{8}$ inch in diameter. The total length of the rope was 1,460 feet; it was $1\frac{1}{4}$ inches in diameter, and weighed $52\frac{1}{2}$ cwts. The breaking-strain was about 70 tons, and the working-load was from 0·125 to 0·10 of the breaking-strain.

This rope was used for winding coal in double-decked cages in a downcast shaft, and wound upon a drum, 12 feet 10 inches in diameter. The average working-load was about 5 tons $11\frac{3}{4}$ cwts. The time of winding and decking was 50 seconds. The steam-pressure varied from 45 to 55 pounds per square inch. The rope was what is termed the " low rope," being that wound on the under side of the drum.

The rope raised 287,273 tons during a period of 1 year 9 months and 12 days; and for comparison with this performance it may be stated that ordinary ropes working under identical conditions have had a life of 3 years.

The place, in the rope, where the failure occurred was about 1,073 feet from the cap or attachment to the tackler-chains of the cage.

The broken wires are indicated by the letters A, B, C, D, E, F, G and H (Fig. 8). The rupture is mostly observed at G in the outer casing of wires; and of the six broken wires, four of these appear in the outer and two in the inner ring.

Two possible causes may be suggested for the breakage of the wires :—

(1) Their brittle nature. The straight fracture-line of the wires favours this notion, yet when a portion of the remaining wire, removed from near the point of fracture, was bent backward and forward it was broken with difficulty.   Still, this opinion is further supported by the breakage of the wires inside the cap.

(2) Owing to the peculiar construction of the rope, the writer is of opinion that more stress may be thrown upon the outer layer or ring of the rope than upon the inner rings, inasmuch as it is quite feasible to strip off the outer ring by hand

Fig. 8. – View of the Ruptured Wires.

with the aid of plyers, and to leave the inner part sound and complete.   Consequently, the suggestion arises that the mutual grip of the strands, which is found in ropes of ordinary construction, has no equivalent in locked-coil ropes.   Furthermore, when the outer ring is stripped off the rope, a certain amount of rust or corrosion is seen to exist upon the inner rings.   This is not found in ropes of the ordinary construction, and the writer is of opinion that this result is owing to want of lubrication, as it is impossible to lubricate the inner strands or rings of this class of rope.

(3) It is also suggested that a possible weakness is created during the process of the wires being drawn out or thinned to form the grooved or recess-shaped wires, in order that these wires may interlock one with the other. These wires cannot be of strength equal to the usual round wires, owing to unequal expansion and contraction and to the different rate of cooling in different portions of the section during manufacture.

The examination of this rope seems to show that the failure might have occurred suddenly, although there is no positive evidence to prove the fact. The ordinary daily examination was made on the previous Saturday, in accordance with the Coal-mines Regulation Act, and it proved satisfactory. The writer's attention was directed to the broken wires on the following Monday afternoon; he gave instructions that no men were to travel on this rope, and it was changed on the following day.

The photograph (Fig. 8) shews that six broken wires were found at one place, and five others singly, hence arose the special danger. Emphasis must be laid upon the necessity for special examination, not only in accordance with the General Rules of the Coal-mines Regulation Act, but also by the makers before the ropes are delivered at the mine. Further, it might be argued that all ropes used in connection with winding at collieries should be subjected to a Government test, and that a written guarantee of the working-load of each rope should be given to the user by the maker.

The writer may also mention a suggestion, made by Mr. W. N. Atkinson, H.M. inspector of mines for the Staffordshire district, which is carried out at the colliery under his charge, that all ropes should be capped at least once in each six months.

Another failure, occurring to a rope-capping under the writer's charge, is shewn in Fig. 9. Fig. 1 (Plate XIV.) explains the section of the wires, and the cap of rope is shewn in Figs. 2, 3 and 4 (Plate XIV.).

This failure seems to give support to the suggested cause of the unusual brittleness of the recess-shaped wires. It was reported that the copper-wire wrapping used in the capping of ropes had become loose upon a locked-coil rope, resulting in the slipping of the cap. A new cap was put on at once, although the capping in question had only been working about a month.

A close inspection of the wires inside the old cap was made, with the following startling results:—Thirty recess-shaped wires, twenty wedged-shaped, and six round wires were broken. These wires fell loosely away, as soon as the iron shoe or cap was liberated. The writer has had several of these broken wires tested, and the breaking-strain of each class of wire is recorded in Table I.

TABLE I.—MEASUREMENTS OF THE AREA OF THE SECTION AND BREAKING-STRAIN OF THE SEVERAL WIRES.

Description of Wires.	No. of Wires in a Rope.	Area of Section. Square Inch.	Breaking-strain per Square Inch. Tons.	No. of Broken Wires.	Percentage of Broken Wires.
Recess-shaped	33	0·0098	75·1	30	91·00
Wedge-shaped	47	0·0100	67·0	20	42·55
Round	17	0·1100	85·5	6	35·30

FIG. 9.—VIEW OF THE RUPTURED WIRES.

The breakage of the wires may be attributed to the following causes:—

(1) Brittleness of the wires. It was found that some of the wires snapped off very easily, while others were broken with great difficulty.

(2) Hammering back upon the feathers or iron (Figs. 3 and 4, Plate XIV.) upon which the wires are bent backward.

The ends of the feathers, in the writer's opinion, may have formed a sharp or cutting edge. The feathers (Fig. 4) were placed close together, instead of being set slightly apart (Fig. 7) and this, in the writer's opinion, would create a tendency for the cap to slip upon the feathers. The point of fracture of the wires is shewn at A (Fig. 2). When the feathers were replaced, the radius of the curve at the ends was lengthened, and all sharp edges were filed away.

The writer submitted specimens of these feathers, one to shew the original form (Figs. 3 and 4), and another to shew the improved form (Figs. 6 and 7).

(3) The construction of the recess-shaped wires may be unsuited to their being bent backwards, owing to the groove or recess formed in them.

The writer, after careful thought as to the failures recorded in this paper, is of opinion that the load, in the course of working, is not evenly distributed in the several rings of wires of the locked-coil rope. He had observed, in one case, that the outer ring was loose upon the inner rings, leading to a stretching or lengthening of one set of wires and to a consequent relative slipping of the outer ring of the recess-shaped wires upon the inner rings of the rope.

There is an absence between the rings of a locked-coil rope of that unity which is found in ropes of ordinary construction, in which every strand is twisted round the remaining five or six strands. The pull in that case is more uniform, and each strand is forced to bear its proper proportion of the weight to which the rope is subjected.

While locked-coil ropes possess certain advantages over ropes of ordinary construction, for instance, in the sinking of shafts, because they do not spin or twist; and while they form good cage-conductors, on account of the smooth surface for the cage-slippers to run upon, locked-coil ropes have also the advantage that they do not stretch to the same extent as ordinary ropes. A locked-coil rope will stretch about 1 foot, in course of working, whereas a rope of ordinary construction will stretch about 15 feet. A locked-coil rope offers great difficulty in examination, inasmuch as the inner strands or rings cannot be inspected and the sudden rupture of the wires, without

FIG. 2.—SECTION OF CAPPING OF ROPE.   FIG. 5.—SECTION OF CAPPING OF ROPE.

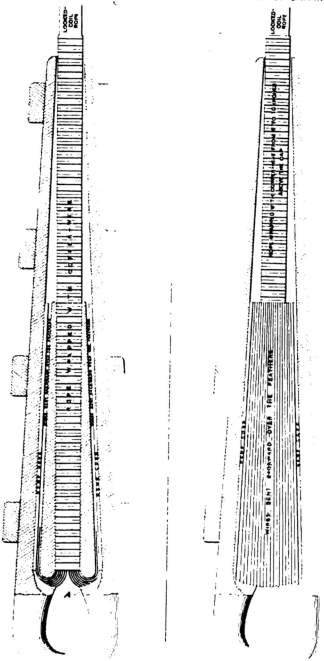

sufficient warning, introduces a grave danger; whereas, in ordinary twisted ropes, the wires, each coming in turn to the outside of the rope, afford every facility for a thorough and satisfactory inspection.

The writer does not desire to find fault with rope-makers, as the subject of his paper is strictly confined to a comparison of the different methods of construction, and the information is given entirely without prejudice.

———

Mr. JOHN GREGORY said that Mr. Lockett's paper raised the important question as to whether locked-coil ropes were as reliable in use as those of ordinary construction. Their chief advantage appeared to be that they were less liable to spin than the older forms, but against this one advantage there were several disadvantages. It was certainly very difficult, if not impossible, to get grease into the inside of the rope, and the rust that Mr. Lockett found on the inner rings was almost certain to follow. Again, it was impossible to ascertain the condition of the inner wires: the outside wires could be inspected, but the only means of forming an estimate of the condition of the inner wires was by re-capping, and then the extreme end of the rope only had to serve as an indication of the condition of the whole. A still further objection was the difficulty of detecting broken wires in the outer ring. The ordinary method of passing the whole length of the winding-rope slowly through the hand was not applicable here, as the interlocking of the outer wires would in many cases prevent any broken wires from protruding, and if there was a thick coating of external grease they would probably escape detection.

Mr. J. T. STOBBS said that he had examined the rope and the capping referred to in Mr. Lockett's paper. He thought that in working locked-coil ropes there was a differential lengthening in the several rings of wires. He did not get the figures of percentage; but whether it meant that the different shape of wires did not support one another like wires in an ordinary rope, or whether the inner or ordinary rope lengthened more than the outside wires, and extra weight was thereby thrown upon the outside wires that did not lengthen, were important questions for the consideration of the users of locked-coil ropes. The difficulty seemed to be to get the wires to support one another in the same way as in the ordinary type of rope. It was evident that they did not do so, because Fig. 8 showed, after the wire had snapped, the amount of lengthening of the inner portion of the rope due to loading.

Mr. G. E. LAWTON said that he had used two locked-coil ropes at winding-shafts about 720 feet deep, drawing comparatively light loads. The special caps, supplied by the makers of the rope, were 15 inches long, and after they had been working for three months it was noticed that the rope was drawing from the cap. On being opened, it was found that all the wires, which passed over the small hoop or thimble, were more or less broken. New caps were made, 3½ feet in length; three rings were put round them, copper-wire was used instead of a thimble, and then the ropes worked satisfactorily, and had the usual length of life as winding-ropes. Afterwards, they were used as haulage-ropes. The lubricant could not penetrate to the inner portion of the rope, and there was a possibility of internal corrosion.

Mr. T. PEACOCK said that, twenty years ago, it was stated that it would be impossible for internal corrosion to take place in locked-coil ropes, because, owing to the casing outside, moisture could not penetrate the rope,* but that statement had been proved to be wrong. Possibly, the reason of the outer layers breaking more frequently than the inner layers might be accounted for in the manufacture. It might be that the maker put more stress on the outer layer than on the inner layers in the manufacture. He became connected with a colliery, some years ago, where he found that the ropes were capped in the old-fashioned way. Within a few weeks he was requested to look at one of those ropes, as the capping was giving way, and about one-third of the wires were broken. He had all the caps altered, and had the wires doubled back in the opposite direction to the lay of the rope, and tucked under the strands behind the small hoop or feather.† Since then, he never had had another experience of a rope giving way, although they had been severely tested.

* *Transactions of the Midland Counties Institution of Engineers*, 1885, vol. xiv., page 263.

† *Ibid.*, 1885, vol. xiii., page 395.

Mr. G. A. MITCHESON said, with respect to the use of locked-coil ropes, that he had used a number for different purposes, but he was afraid that he was not competent to pronounce an opinion as to whether the locked-coil rope was a safe rope or not. With a heavy load and a deep shaft with wire-guides, he was inclined to think that the ordinary type of rope was not so efficient as a locked-coil rope. The danger of corrosion was common to both sorts of ropes, and he thought that there was really less danger from corrosion in a locked-coil than in a rope of ordinary construction. There was a great difference between handling ropes for shallow depths and light loads, and handling heavy ropes for big loads and great depths. He did not know anything about the failure of a locked-coil rope at the cap such as that mentioned. The capping of a big locked-coil rope was a difficult operation, because they had not only the difficulties of the actual capping, but they had to be careful not to loosen the outside jacket of the rope; and he did not know whether it was wise to risk more than they could help. He had used locked-coil ropes in a sinking pit, and it was thought advisable to use them for winding; but it was decided that they should be tested so as to see what their breaking-strain would be. They were sent to a testing-house and the results were very low. The maker said that the ropes were all right, and they proved to be all right. He had known several cases of failure of locked-coil ropes, owing to the outside jackets having become slack, but after the makers had put on a new jacket they had been worked satisfactorily. He supposed that locked-coil ropes were more difficult to make than others, and that there was less known about them.

Mr. J. GREGORY remarked that the tests recorded in the paper showed the advantage of using round wires over shaped wires.

The CHAIRMAN (Mr. E. B. Wain) said that the inner wire was the one that received the least strain when at work. He thought that there were few, even of the best plough steel ropes, that would not harden considerably when in use, and that might account for the higher tensile strain of the inner coil of wire, when tested after the failure. The rope seemed to be a big one for the duty that it had to perform. One detail mentioned by Mr. Lockett was the exact point of fracture, which, he said, occurred where the rope would receive a gradual strain, when it became necessary to steady the cage to the bottom.   There had recently been a lamentable accident in Yorkshire, showing the danger in connection with sudden retardation instead of the steady action of the brake being used.   If a rope was subjected to unusual strain at any particular point, naturally, in re-capping, they would shift the point where the severe strain was received, and therefore it seemed that it was desirable for capping to take place at fairly frequent intervals.

Mr. JOHN CADMAN moved a vote of thanks to Mr. Lockett for his paper.

Mr. G. A. MITCHESON, in seconding the vote of thanks, said that, in Mr. Lockett's case, the failure occurred at the particular part where retardation would begin.   He asked Mr. Lockett whether he had any idea at what speed the load was travelling when retardation began, and what sort of brake was used. A suggestion had been made as to re-capping, and altering the position of the rope where the strain of retardation would be first felt.   A far better plan than re-capping was to make pro-vision inside the drum, and carry several coils rather than loosen out the rope.

The resolution was cordially approved.

———

DISCUSSION OF DR. J. S. HALDANE'S PAPER ON "MINERS' ANÆMIA, OR ANKYLOSTOMIASIS,"[*] AND MR. F. W. GRAY'S PAPER ON "ANKYLO-STOMIASIS: THE WORM-DISEASE IN MINES."[†]

Mr. JOHN CADMAN said that he had investigated about 20 per cent. of the collieries in the North Staffordshire district, and at the present moment there was no sign of the disease in them and no ova had been found in the samples of fæces taken from mines, although he had found the ova of some nematoids, so like the ova of *Ankylostoma*, that it had given rise to considerable anxiety. He (Mr. Cadman) described some of the features in connec-

[*] *Trans. Inst. M.E.*, 1903, vol. xxv., page 643.

[†] *Ibid.*, 1903, vol. xxvi., page 183.

tion with the disease at Dolcoath mine, and said that it was
evident that every man in that mine was suffering from the
disease and the miners' capacity for labour had been seriously
affected. There were many mines in North Staffordshire where
the temperature and moisture-conditions were such that the
loathsome disease could be propagated, and it was for the
managers of collieries to be watchful, and do whatever could
be done to prevent an outbreak of the disease. The great point
was to keep men, who came from districts where the disease
prevailed, away from the mines. They must not be allowed to
enter the mines of this district, unless it could be proved that
they were free from the disease. If the foremen and managers
would enforce a rule that the men should pay proper regard to
cleanliness in the pits and before entering the pits, the disease
would not be so likely to occur. In the Levant mine, Cornwall, which extended under the sea, the temperature was high,
and the conditions seemed to be favourable for the spread of
the disease. The disease, however, was not prevalent, owing
to the fact that the water contained salt in solution, which
seemed to kill the larva as soon as it was hatched. He (Mr.
Cadman) exhibited some ova and larvæ under the microscope.

Mr. J. T. STOBBS said that the life-history of the worm
referred to in both papers was a matter for biologists, who
should look after it, and state (1) how to keep the disease out
of the mines; or (2) if the germ got into the mines, how to deal
with it and stamp it out. It seemed, looking at the thing
generally, that the disease had spread most rapidly in ladder-
pits or ladder-mines; that was where the hand and the foot
followed one another in travelling or going from place to place.
Mr. Cadman had mentioned that salt-water destroyed the larvæ
in the Cornish mines. There was salt in the water in some
of the mines in the North Staffordshire district, and it should
have a destructive effect and prevent the propagation there of
the worms.

Mr. G. A. MITCHESON remarked that it would be too late to
deal with the disease when it had once come among them.
It appeared to him that, with such a migratory working population as they had, any attempt on the part of the officials to
deal with it in any particular place would be of very little value

if the same precautions were not taken all round. Personally, he was not prepared to give an opinion as to whether the time had yet come for the Home Secretary to deal with it; but when the time did arrive, it certainly was a matter which ought to be dealt with collectively. It would not be fair for the officials at one colliery to put themselves to expense and exercise great care to prevent the evil, and for the officials of an adjoining colliery to pay no attention to the matter. It was a matter in which the officials of every colliery ought to receive the loyal assistance of every working-miner in the district: a few black sheep might do serious mischief. He had thought that in this district circumstances were not favourable to the propagation of ankylostomiasis, but Mr. Cadman now told the members that they were.

Mr. JOHN CADMAN stated that the ova hatched at a temperature of 60° Fahr., and even as low as 58°. As long as the fæces remained in a plastic condition, the ova could hatch. Men with the disease in them might go about, and not know it; they might show no outward symptoms, and yet, carrying in their bodies a few worms, they might infect a large district.

Mr. F. W. GRAY (Birdwell, near Barnsley), wrote that Dr. Boycott's remarks on the detrimental effect of salt-water upon young *Ankylostoma*-larvæ were confirmed by the statements of continental doctors. Dr. Previtera stated as the outcome of his experiments, that the disinfection of the mine-floor and the pit-water can be completely secured by the use of a saturated solution of common salt;[*] and among the many disinfectants that had been tried in Westphalia, salt had been as efficacious as any. But the utter futility of attempting to disinfect the whole of a large mine is now tacitly admitted on the continent, and all efforts are being directed to the most obvious means of combating the disease, that is, to prevent the pollution of the mines by human excreta. It is needless to comment on the desirability of this, and Dr. A. E. Boycott's remarks on the prevalence of intestinal parasites, other than *Ankylostoma*, among miners are very significant.[†] The rapidity with which salt acts upon the newly-hatched larvæ suggests its use in underground closets.

---

[*] *Rassegna Mineraria*, 1903, vol. xix., page 115.
[†] *Report on the Diagnosis of Ankylostoma-infection, etc.*, 1904, page 4.

DISCUSSION OF MR. F. H. WYNNE'S PAPER ON "NATIVE METHODS OF SMELTING AND MANUFACTURING IRON IN JABALPUR, CENTRAL PROVINCES, INDIA."*

Mr. R. R. SIMPSON (Geological Survey of India) wrote that Mr. Wynne spoke of the native Indian iron-industry as having prevailed throughout centuries with little or no attempt at improvement,† but surely the reverse was the case. It was known that the famous blades of Damascus were made from Indian crucible-steel known as *wootz*. It had been stated‡ that there were reasons for believing that *wootz* was exported to the west 2,000 years ago, while some years ago it was in considerable demand by cutlers in England. Again, that art was surely distinctly decadent in which castings of a few pounds were made, as compared with those of several tons in weight such as the famous memorial-pillar of Kutub, near Delhi. This pillar, by the way, Mr. Wynne declared to be 50 feet in length, but‖ General Cunningham stated that it was 23 feet 8 inches long and varied in diameter from 12 inches near the top to 16·4 inches near the base, while the actual base was bulbous and measured 2 feet 4 inches in diameter. The total weight was said to be more than 6 tons, and analysis had shown that the pillar consisted of pure, malleable iron without any alloy.

In speaking of the *bhatti-wallas* or smelters, Mr. Wynne, somewhat mistakenly, depreciated their intelligence as compared with that of the natives of other castes, and remarked that "only those who are not clever enough to earn a livelihood as smiths, become *bhatti-wallas*."§ Now so far as he (Mr. Simpson) was aware, the two trades were quite distinct, and it would be a rare occurrence for a smelter to become a smith or *vice versa*. It should be remembered that in no country was it easy for a skilled artizan to change his trade; whilst in India, owing to the rigid caste-rules, it was almost an impossibility. On the question of fluxes, it was interesting to note that, on the authority of Dr.

---

* *Trans. Inst. M.E.*, 1904, vol. xxvi., page 231.

† *Ibid.*, page 231.

‡ *A Manual of the Geology of India*, part. iii., "Economic Geology," by Mr. V. Ball, page 340.

‖ *Ibid.*, page 339.

§ *Trans. Inst. M.E.*, 1904, vol. xxvi., page 232.

Verchere, limestone was added to the charge in Waziristan, whilst in another locality (Tendukhera, Nerbada valley) the ore itself was stated to be calcareous. In Kathiawar, a type of reverberatory furnace was said to be used.[*]

As the question of the geological age of Indian ores had been raised, it might be remarked that, up to the present time, the most extensively-worked deposits were the clay-ironstones found in the Raniganj coal-field. There ores occurred in bands and lenticular masses in the Ironstone-shales group of the Damuda division of the Gondwana system. The Ironstone-shales group was intermediate between the Raniganj and Barakar groups, in which the bulk of the coal of India occurred. It was, however, only in the Raniganj coal-field that the iron-ores had been found to be worth working. They were largely used by the company manufacturing iron at Barakar, but of late years the latter had drawn part of its supplies from the transition-rocks of Manbhum and Singhbhum in the province of Chota Nagpur. During the course of a brief visit to these localities, he (Mr. Simpson) observed both magnetite- and hæmatite-deposits, the latter being most important. The ores occurred in lenticles, a few feet long, but he (Mr. Simpson) was unable to trace any definite vein or band of iron-bearing rocks. The country was hilly and covered by talus, and the ores which were being mined were almost all obtained from these degraded deposits, which were, however, doubtless originally derived from the lenticular deposits in the ancient rocks. The rich ores of Jabalpur, according to Mr. Hacket, as quoted by Dr. T. Oldham,[†] occurred " in the Lora group of the Bijawar Series," which belonged to that vast system of unfossiliferous rocks which Indian geologists had called " transition," and considered to be of Archaean age.

The production of iron in India, during 1902,[‡] was 80,869 tons, and of this amount, 76,056 tons was manufactured in the blast-furnaces of the Bengal Iron and Steel Company, near Barakar. The remainder, therefore, or 4,813 tons must be the total amount produced by native smelters, there being no other

[*] *A Manual of the Geology of India*, part. iii., " Economic Geology," by Mr. V. Ball, page 340.

[†] *Records of the Geological Survey of India*, vol. v., page 9.

[‡] *Statistics of Mineral Production in India*, 1902.

European firm engaged in iron-manufacture. When it was mentioned that the imports of iron and steel* into India during the same year amounted to 3,612,088 tons it could be understood of what small dimensions, compared with the market, was the Indian iron-industry, and still more of what trifling importance was the native iron-industry. It was, however, worthy of mention that pig-iron, manufactured by a British company at Porto Novo in Southern India, was, at one time, largely exported to Great Britain and converted into steel for use in building the Menai and Britannia tubular bridges, and for the manufacture of railway-tyres.†

One of the principal causes of the decline of the native industry was the difficulty of obtaining charcoal, due to the stricter conservation of the once extensive, but now greatly depleted forests. The great bar to the development of the iron-industry in India, on European lines, was the large amount of inorganic matter in Indian coals. In the coals of the Central Provinces this was most pronounced, and they were only occasionally of coking quality. To establish an iron-industry in Jabalpur, which appeared to be one of the most promising localities, would necessitate the importation of fuel from Bengal. As the railway-freight on coal carried from the Bengal coal-fields to Jabalpur was 15 rupees (£1) per ton, this would entail a heavy charge on the finished product.

Mr. F. H. WYNNE wrote that he thanked Mr. R. R. Simpson for replying, in part, to Mr. Stobbs' question as to the geological age of the beds in which the iron-ores mentioned occur. It was difficult to state, without further evidence, whether the iron-ore was a product of alteration, as the beds have, hitherto, only been worked in close proximity to the outcrop. He (Mr. Wynne) still contended that the decadent state of the industry was chiefly due to the introduction of iron manufactured more cheaply by European methods, with the consequent reduction of the demand for the native product. The Kutub pillar, near Delhi, was now generally recognised to have been made by the successive weldings together of small bars made from such blooms of crude-iron as he had described in his paper. He

* Statistics of Imports into India, 1902.

† A Manual of the Geology of India, part iii., " Economic Geology," by Mr. V. Ball, page 350.

thought that it was decidedly erroneous to speak of it as a casting of several tons weight, especially seeing that the natives have not, and never had, much knowledge of the methods of dealing with iron in the molten state. His (Mr. Wynne's) remark that "only those [*lohars*], who are not clever enough to earn a livelihood as smiths, become *bhatti-wallas*," was based upon competent authority, and it was further borne out by an instance which had come under his own observation; but he did not wish to suggest that the reverse had happened or that it was likely to happen.

THE SOUTH STAFFORDSHIRE AND EAST WORCESTER-SHIRE INSTITUTE OF MINING ENGINEERS.

GENERAL MEETING,
HELD AT THE UNIVERSITY, BIRMINGHAM, APRIL 18TH, 1904.

PROF. R. A. S. REDMAYNE, VICE-PRESIDENT, IN THE CHAIR.

The minutes of the last General Meeting and of Council Meetings were read and confirmed.

The following gentlemen were elected:—

MEMBERS—
Mr. J. T. BROWNE, Griff Colliery, Nuneaton.
Mr. E. F. MELLY, Griff Colliery, Nuneaton.
Mr. W. NOWELL, Haunchwood Colliery, Nuneaton.

DISCUSSION OF DR. A. SIMON'S PAPER ON "COAL-CUTTING MACHINERY OF THE PERCUSSION-TYPE."*

The CHAIRMAN (Prof. Redmayne) observed that coal-cutting machinery of the percussion-type was hardly likely to compete successfully with machines of the rotary type in longwall working. The Champion machine should prove invaluable, however, in working steep seams, driving headings in stall and bord-and-pillar working, etc., and in many places where it was impracticable to use the rotary type of machine.

Dr. A. SIMON said that, while admitting that a rotary long-wall machine, machine for machine, did more work than the Champion machine, it must be borne in mind that the rotary machine was a much more powerful unit than a Champion machine; but it was only such a unit at the expense of the consumption of very much more power than the Champion machine, and of being a very much more complicated machine alto-

* *Trans. Inst. M.E.*, 1903, vol. xxvi., page 322.

THE MIDLAND COUNTIES INSTITUTION OF ENGINEERS.

EXCURSION MEETING,
APRIL 26TH, 1904.

The members visited the works of the Derwent Valley Water Board, near Bamford, Derbyshire. The sites of the Howden and Derwent dams were inspected, together with the stone-quarries at Grindleford.

DERWENT VALLEY WATER-SCHEME.

The Derwent Valley water-scheme is for the supply of water to the towns of Leicester, Derby, Sheffield and Nottingham, and to the counties of Derby and Nottingham. The daily supply to each authority will be approximately as follows:—Leicester, 9,800,000 gallons; Derby, 6,800,000 gallons; Sheffield, 6,800,000 gallons; Nottingham, 3,900,000 gallons; Derbyshire, 5,000,000 gallons; Nottinghamshire, until 1930, 1,000,000 gallons; a total of 33,300,000 gallons. The scheme includes the construction of:—(1) Five reservoirs for impounding the waters of the rivers Derwent and Ashop; (2) about 100 miles of aqueduct for distributing the water to various authorities; (3) about 20 acres of filter-beds at Bamford; and (4) a service-reservoir at Ambergate.

The water-scheme divides the works into three sections, each capable of yielding about 11,000,000 gallons per day. The cut-and-cover and tunnel work will be constructed to carry the full quantity, but only one pipe-line will be carried out for each section. The first section involves the construction of the Howden and Derwent reservoirs; the second, the Hagglee reservoir; and the third, the Ashopton and Bamford reservoirs.

The average rainfall is about 49 inches.

th the Champion machine had been
re feet in a shift of about 8 hours.
vas 350 blows per minute, but the
essure of the compressed air work-
w pressure of 25 pounds per square
'ows, doing good work in coal of
er pressure of 75 or 85 pounds
that was hardly ever exceeded.
depended upon (1) the pressure
s of the coal; and (3) the skill
kill to work the machine was
the fact that at the Hindley
machine was worked by two
2s· per day. With ordinary
5 to 50 square feet per hour.
90 square feet an hour, and
shire, with an air-pressure
30 to 35 square feet per
le occupied in removing
If that was taken into
60 square feet an hour;
'et an hour; and with
. 20 to 25 square feet.
ine was cutting 45 to

'ing machinery was
uld be fairly and
hich were difficult
·as easily remov-
·nmand of space.
·r this type of
orked by long-
and. It could
uld compare
·orking; for
·iving head-

·red to be
whether

thought that it was decidedly erroneoi
ing of several tons weight, especially se
not, and never had, much knowledge
ing with iron in the molten state. His (J
" only those [*lohars*], who are not clever
hood as smiths, become *bhatti-wallas*,"
petent authority, and it was further bo
which had come under his own observatio
to suggest that the reverse had happened
happen.

so that a rotary
much air as the
~hine was gener-
upion machine.
rd to the two
vorking cost.
cost of cut-

JEFFREY

y.
Rhondda.
? inches.

: PAPER ON "COAL-
ñ THE PERCUSSION-

: bserved that coal-cutting
· % hardly likely to compete
: rtary type in longwall w.:
: ·aid prove invaluable, h·
· ·nving headings in stall
·al in many places where
·ñ trp of machine.

: While admitting that a ro
: machine, did more wor'
: be borne in mind the
: :% powerful unit than
· ·! nch a unit at the exp
: more pwer than the (1
: m·: x·y complie?
L, 1893. w. w. :
·q 322.

·IN. ʹ·ʹ

LAST WORCESTER-
ʹ ENGINEERS

ʹ·,

· April 19TH, 1904.

··ODENT, IN THE CHAIR.

l Meeting and of Council

elected:—

eaton.
ton.
, Nuneaton.

ION'S PAPER ON "COAL-
OF THE PERCUSSION-

mayne) observed that coal-cutting
type was hardly likely to compete
f the rotary type in longwall work-
ine should prove invaluable, how-
ms, driving headings in stall and
c., and in many places where it was
tary type of machine.

t, while admitting that a rotary long-
or machine, did more work than the
nust be borne in mind that the rotary
more powerful unit than a Champion
ly such a unit at the expense of the con-
more power than the Champion machine.
much more complicated machine alto-

003, vol. xxvi., page 322.

h the Champion machine had been
·e feet in a shift of about 8 hours.
as 350 blows per minute, but the
ssure of the compressed air work-
· pressure of 25 pounds per square
ws, doing good work in coal of
r pressure of 75 or 85 pounds
hat was hardly ever exceeded.
epended upon (1) the pressure
of the coal; and (3) the skill
ill to work the machine was
ιe fact that at the Hindley
ιachine was worked by two
ι. per day. With ordinary
to 50 square feet per hour.
) square feet an hour, and
ιire, with an air-pressure
ł0 to 35 square feet per
· occupied in removing
If that was taken into
0 square feet an hour;
·t an hour; and with
20 to 25 square feet.
ne was cutting 45 to

ing machinery was
·ld be fairly and
ιich were difficult
as easily remov-
ιmand of space.
·r this type of
ιrked by long-
and. It could
ιuld compare
·orking; for
·iving head-

ιred to be
· whether

a chapel on Sundays, and a large recreation-hall and reading-room, containing billiard-tables, etc., at one end, and a stage at the other end. There is also a water-supply, sewerage-system, and public baths. The population is between 600 and 700.

Railway.—The railway, about 7 miles long, connects with the Midland railway about ½ mile west of Bamford Station, runs up the valley on the western side of the river Derwent, and terminates at the Howden dam.

Fig. 4.—Trench for the Derwent Dam.

Aqueduct.—The main aqueduct extends from the Howden to the Ambergate reservoir, a distance of 30 miles; and will consist of the following sections :—Tunnels, 6¼ feet in diameter, 4 miles; cut-and-cover, 6¼ feet in diameter, 8½ miles; and pipes, 45 inches in diameter, 17½ miles. The aqueduct from Ambergate to Leicester will be constructed by the Board as far as the river Trent, at Sawley, and the remainder by the Leicester Corporation; and from Little Eaton, on the above line, Derby will take its share of water. In addition, another line of pipes, 8 miles in length, will be laid from the Ambergate reservoir towards Nottingham, ending at the county-boundary at Langley Mill, where the Nottingham pipes will be connected.

All the water, except the Sheffield supply, will be filtered at Bamford.

Quarry.—The quarry for supplying stone to the dams is situated at Grindleford, on the Dore and Chinley branch of the Midland railway, about 4 miles east of Bamford. The stone is from the Rivelin-bed of the Millstone Grit Series. About 1,000,000 tons of stone will be required for the Howden and Derwent dams. A connection, between the quarry and the Midland railway, is formed by means of a self-acting incline, on which the full trucks of stone haul the empty trucks from the sidings at the bottom.

The trucks are conveyed by the Midland Railway Company between Grindleford station and the sidings at Bamford.

THE MIDLAND COUNTIES INSTITUTION OF ENGINEERS.

GENERAL MEETING,
HELD AT UNIVERSITY COLLEGE, NOTTINGHAM, JUNE 25TH, 1904.

MR. W. B. M. JACKSON, PRESIDENT, IN THE CHAIR.

DEATH OF MR. WILLIAM WILDE.

The PRESIDENT (Mr. W. B. M. Jackson) moved a vote of condolence with the widow and family of the late Mr. William Wilde, a member of the Institution from the commencement and also for many years a member of the Council. Mr. Wilde was very highly respected, and his loss was severely felt. He (Mr. Jackson) was sure that they would all join in conveying to his widow and family the sense which they had of his loss.

The resolution of condolence was passed in silence.

The SECRETARY announced the election of the following gentlemen :—

MEMBERS—

Mr. THOMAS WILLIAM ALLEN, Colliery Manager, Birch Coppice Colliery, Polesworth, Tamworth.

Mr. THOMAS R. GAINSFORD, Colliery Proprietor, Woodthorpe Hall, Sheffield.

Mr. WILLIAM CRADOCK KNIGHT, Electrical Engineer, 14 and 15, Rodgers Chambers, Norfolk Street, Sheffield.

Mr. EDGAR LINES, Civil Engineer, Union Offices, Chesterfield.

Mr. HARRY JOHN SUTHERLAND MACKAY, Mechanical Engineer, 53, Deansgate Arcade, Manchester.

Mr. FRANK HOWARD POCHIN, Coal-owner, Fern Bank, Narborough, Leicestershire.

Mr. THOMAS RYAN, General Manager, Buxton Lime Firms Company, Limited, Woodlands, Buxton.

Mr. C. A. SLATER, Electrical and Mechanical Engineer, Market Place Buildings, Sheffield.

Mr. THOMAS WILLIAM WARD, Mechanical Engineer, Riverdale Road, Sheffield.

Mr. Walter W. White, Engineer, c/o Simplex Coke-oven Company, Temple Bar House, London.

Mr. William John Wilkinson, Colliery Manager, The Pilsley Coal Company, Pilsley, near Chesterfield.

ASSOCIATES—

Mr. Walter Bell, Under Manager, Clay Cross Collieries, Chesterfield.

Mr. Samuel Clarke, Colliery Engineer, Shirebrook Colliery, Mansfield.

Mr. John Roger Harvey, Under Manager, Moor Lane, Ockbrook, Derby.

Mr. William Hudson, Under Manager, 17, Prospect Drive, Shirebrook.

Mr. George Sellers, Colliery Deputy, 24, Church Drive, Shirebrook.

Mr. William Sword, Mining Surveyor, Hall's Collieries, Swadlincote, Burton-on-Trent.

Mr. George William Thomley, Colliery Deputy, 4, Central Drive, Shirebrook.

Mr. Ernest Wheldon, Surveyor, Shipley, Derby.

STUDENT—

Mr. John Howard Moorby, Mining Student, The Villas, Creswell, Mansfield.

———

The SECRETARY announced the nomination of officers for the ensuing year.

———

Mr. W. Hay read the following paper on "The Three-phase Electric Haulage-plant at Shirebrook Colliery, Mansfield":—

THE THREE-PHASE ELECTRIC HAULAGE-PLANT
AT SHIREBROOK COLLIERY, MANSFIELD.

By WILLIAM HAY.

As there are few three-phase electric haulage-installations at work in this country, the writer thought that a description of the plant at Shirebrook colliery might be of interest to the members.

It is not, at many collieries, an easy matter to ascertain the efficiency of the electric haulage-plant, owing to the practice prevailing at collieries of taking the steam required for the plant from the main steam-pipes attached to a range of boilers; and also to the fact that, frequently, an electric power-plant, originally intended for haulage, has eventually its mains tapped and the power used for driving pumps, coal-cutting machines, etc. Under such circumstances, it is very difficult to eliminate accurately the power absorbed by the coal-cutting machines, pumps, etc., and tests made under such conditions are always of a doubtful character.

The boiler, working the plant about to be described, was put down by the writer, and its connections were so arranged, that it could either be coupled up to the main steam-pipes, or isolated from the main range and connected to the electric power-plant direct.   It is a Lancashire boiler, 8 feet in diameter and 30 feet in length, with two plain flues 3 feet 2 inches in diameter, and fitted with a Proctor mechanical stoker.

There is a considerable length of steam-pipes from the boiler to the engine, and in order to obtain moderately-dry steam, a Bolton downtake superheater is placed at the back end of the boiler in the downtake-flue.   The superheater consists of a top steam-chamber connected to a series of tubes forming the heating-surface.   There are two equal sections of tubes through which the steam passes in succession, being dried in the first section of tubes and superheated in the second. The superheating varies from 100° to 140° Fahr. above the temperature of saturated steam, and this materially reduces the initial condensation, prevents condensation in the pipes, and enables the engines to receive a supply of comparatively-dry steam. The steam, on leaving the superheater, passes through 248 feet of pipes, 6 inches in diameter, to the engine. The relative position of the boiler, superheater, steam-pipes, etc., is shown in the plan (Fig. 8, Plate XVI.).

The alternator is driven by a tandem compound engine, having two high-pressure cylinders, 10 inches in diameter, two low-pressure cylinders, 16½ inches in diameter and 32 inches stroke, and fitted with Meyer expansion-gear, running at 95 revolutions per minute. The power is transmitted from the fly-wheel, 12 feet in diameter, by means of a double leather-belt, 19 inches wide, to the countershaft, and by a belt, 15 inches wide, to the generator. The Westinghouse three-phase ten-pole generator, giving 75 kilowatts at 440 volts and 7,200 alternations, is a belt-driven, composite-wound machine. The lower half of the field and the supports of the bearings constitute a single casting, which ensures accurate centreing of the armature. The pole-pieces are made of soft laminated steel, and the bearings are self-lubricating and self-aligning. The armature-coils lie in slots, and are held in position by retaining wedges driven into notches, near the top of the slots, parallel to the shaft. Band-wires are not used, and the core and windings are well ventilated. The machine runs at 720 revolutions per minute, and is driven by a belt-pulley, 25 inches in diameter and 16 inches wide on the face. The field-excitation is derived from a separate exciter, consisting of a 1·5 kilowatts, 125 volts, compound-wound, four-pole generator running at 1,300 revolutions per minute; and it is belt-driven from the counter-shaft (Figs. 2, 3 and 4, Plate XV.).

The generator-switchboard consists of polished-marble panels, 2 inches thick with ½ inch bevelled edges, mounted on a rigid iron-frame of L-section. There is a clear space of 2 feet between the bottom of the marble and the floor, and a space of 3 feet from the switchboard to the wall of the engine-house. All wires and omnibus-bars are thoroughly insulated, and attention has been given to spacing the leads, of opposite potential, as far apart as practicable.   The fittings consist of a voltmeter, three ammeters, a three-pole automatic quick-break switch and a rheostat.   The instruments for the direct-current exciter are attached to a separate panel on this switchboard, and consist of a voltmeter, ammeter, knife-switch and rheostat.   Both rheostats are attached to the back of the switchboard, with the face-plates only on the front, and are independently operated by concentric hand-wheels on the front of the switchboard.   Figs. 2, 3 and 4 (Plate XV.) show the arrangement of the engine, alternator and switchboard.

A three-core cable of 19 tinned copper-wires of No. 16 standard wire-gauge, 2,160 feet long, conveys the current from the generator to the motor underground.   The cable is run down the shaft in steel-tubes, and is supported at the surface on insulated wooden clamps, and at intervals there are suitable junction-boxes in the shaft.

The motor underground is a Westinghouse three-phase machine giving 100 horsepower, at 400 volts and 7,200 alternations, when running at 580 revolutions per minute. The primary element, which is directly magnetized by the currents supplied from the power-circuit, is stationary; and the secondary element, in which low-potential currents are induced by the action of the primary, revolves. The mechanical construction is very simple. The stator is connected to the main circuits, and the rotor has no electric connection with any external circuit; there are, therefore, no electric contacts or adjustments, and, in consequence, there is an entire absence of sparking. The rotor consists of laminations built on the shaft, and having no coils or wire wound thereon, but simply copper-bars inserted in the periphery. The ends of these copper-bars, where they project at each side over the laminations, are all connected and short-circuited by means of phosphor-bronze rings, cast with vanes on them, for ventilating purposes. The only wearing parts are the shaft and the journal-boxes. That the friction is very slight has been amply proved by the fact that this machine has been working 18 hours per day for the last 2½ years, and beyond oiling and cleaning has required no further attention. The bearings were overhauled a short time ago, and showed very slight signs of wear.

The position of the motor in relation to the hauling-ropes, and other particulars are given in Figs. 5, 6 and 7 (Plate XVI.).

The switchboard belowground is similar to the one on the surface, with the addition of an auto-starter, consisting of a double-throw switch mounted on a cast-iron box, containing two auto-transformers. When the switch is thrown into one position, the auto-transformers are connected across the circuits, and deliver a low-pressure current to the motor. By throwing the switch into the opposite position, the auto-transformers are cut out, and the motor is coupled direct to the main circuit.

The generator and exciter are guaranteed to withstand, without injurious heating, an overload of from 50 to 75 per cent. temporarily, or of 25 per cent. for a period of 2 hours; and the motor will stand an overload of 100 per cent. temporarily, or of 25 per cent. for periods varying from 4 to 6 hours.

Figs. 5, 6 and 7 (Plate XVI.) are end and side elevations, and a plan of the motor and hauling-gear. The plant was set to work in August, 1901, and has worked on an average 18 hours per day and 5 days per week since that time; and the cost of repairs for the 2 years and 8 months running is *nil*. During the last 2 years and 8 months, the amount of coal brought from the workings to the shaft by this motor is 1,042,606 tons, through an average distance of about 1,890 feet. The power is transmitted from the motor to the hauling-gear by means of a belt, 13 inches wide. The rope-driving pulleys are 6 feet in diameter, slightly tapered on the face, curving a little towards the outer edges, and are thrown in and out of gear by means of a friction-clutch of the ordinary wedge and expanding-ring type.

The speed of the haulage-ropes is 2 miles per hour. The ropes are endless, and the tightening arrangement consists of a balance-bogey, carrying a pulley 5 feet in diameter, working on an inclined plane; and there is sufficient spare rope to admit of resplicing when necessary. The tubs are run in trains of 6 trams, attached to the ropes by means of Smallman clips; and also in trains of 14 tubs where the roadways are single. Owing to the tender nature of the roof, the roadways are, at times, a little rough, especially as the roads approach the face (Fig. 1, Plate XV.). The haulage-ropes are ⅞ inch in diameter, the north district rope being 10,500 feet long and the south district rope 7,680 feet long.

A six hours' test was made with this plant on May 6th, 1904. As the engines are arranged to drive a continuous-current machine of 75 kilowatts, in addition to the alternator herein described, this dynamo was thrown out of gear, and the driving belt removed during the time that the test was being made. Four indicators were attached to the cylinders and diagrams (Figs. 9 to 14, Plate XVI.) were taken every 5 minutes during the time of the test, and the results are recorded in Table I.

TABLE I.—INDICATED HORSEPOWER OF THE ENGINE AND THE DISTRIBUTION OF THE POWER.

					Horsepower	Per cent.
Engine and countershaft	...	...	...	...	30·74	29·6
Alternator and exciter	...	...	...	...	10·18	9·8
Cables, motor and hauling-gear	...	...	...	...	21·04	20·3
Ropes	...	...	...	...	8·87	8·5
Load, including coal and tubs	...	...	...	...	33·07	31·8
Totals		...	...	...	103·90	100·0

The coal used at the boiler was carefully weighed. The water was measured in a 200 gallons tank, and pumped by means of a separate pump direct into the boiler. The water- and steam-connections were disconnected from the range of boilers. The results of the trial of the boiler are recorded in Table II.

TABLE II.—RESULTS OF THE TRIAL OF THE BOILER.

Quality of coal used	Nutty slack.
Coal burnt per hour	529 pounds.
Water evaporated per hour	395 gallons.
Water evaporated per pound of coal used	7·46 pounds.
Average temperature of the feed-water	120° Fahr.
Equivalent evaporation of water from and at 212° Fahr.	8·36 pounds.
Average temperature of the gases in the side-flues ...	535° Fahr.
Average temperature of the steam in the superheater ...	455° Fahr.
Average pressure of the steam per square inch	112 pounds.

In calculating the efficiency of the plant, it should be borne in mind that the engine is capable of developing 200 horsepower; and, as the power used in this test is only 104 horsepower, the power absorbed by the engines and countershaft should be divided *pro rata*.

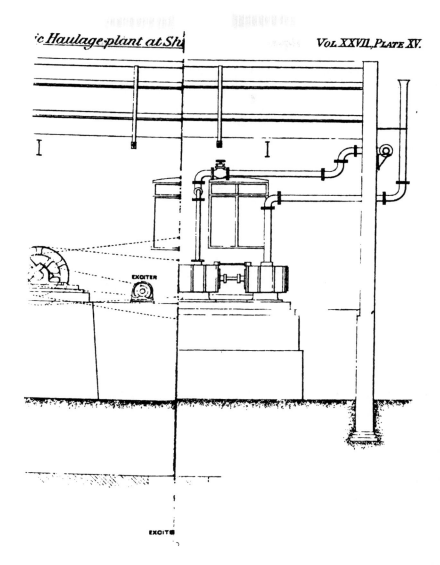

EXCITER

EXCITE

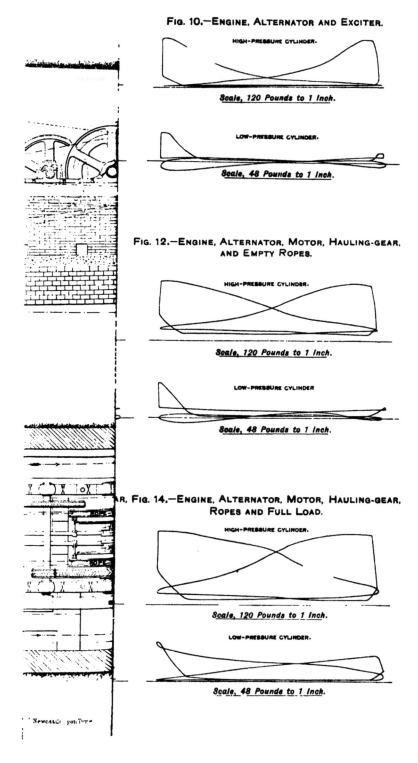

Fig. 10.—Engine, Alternator and Exciter.

HIGH-PRESSURE CYLINDER.

Scale, 120 Pounds to 1 Inch.

LOW-PRESSURE CYLINDER.

Scale, 48 Pounds to 1 Inch.

Fig. 12.—Engine, Alternator, Motor, Hauling-gear,
and Empty Ropes.

HIGH-PRESSURE CYLINDER.

Scale, 120 Pounds to 1 Inch.

LOW-PRESSURE CYLINDER

Scale, 48 Pounds to 1 Inch.

R, Fig. 14.—Engine, Alternator, Motor, Hauling-gear,
Ropes and Full Load.

HIGH-PRESSURE CYLINDER.

Scale, 120 Pounds to 1 Inch.

LOW-PRESSURE CYLINDER.

Scale, 48 Pounds to 1 Inch.

Newcastle-upon-Tyne

The mechanical efficiency of the engine, *plus* the power used to overcome the friction of the countershaft and belts is 0·8. The efficiency of the plant, as now working, is 40·3 per cent.; but, if working at full load, the efficiency would be very much increased. The electrical efficiency is [(41·94 × 100) ÷ 73·16 =] 57·3 per cent. The steam-consumption is 38 pounds per indicated horsepower.

The quantity of coal delivered to the pit-bottom per hour during the test was 115·2 tons, and the wages-cost was 0·76d. per ton.

———

Mr. Arnold Lupton (Leeds) said that he knew the plant, and it worked well. He regarded it as a useful example of three-phase electric power-transmission for underground haulage. The compound engine, working with high-pressure superheated steam, every care being taken to ensure economy, had a steam-consumption of 38 pounds per horsepower per hour: three times as much as the minimum possible consumption of steam. The consumption was high, when compared with the practice of marine engineers, but low, when compared with the common practice of mining engineers. He thought that a compound engine, with this proportion of high- and low-pressure cylinders and this steam-pressure, was not an economical steam-user, unless a condenser was added to it.

The President (Mr. W. B. M. Jackson) asked whether Mr. Hay had found the superheater to give good results. No doubt Mr. Hay would have obtained better results, if his engine had been doing more work; but, as he pointed out in his paper, it had not been doing full work. The feed-water did not seem to attain a very high temperature: it was only 120° Fahr. He asked whether the wages-cost, which was put at 0·76d. per ton, covered the entire cost in connection with the haulage, or only a portion of it.

Mr. Roslyn Holiday (Ackton Hall) noticed that the generator was driven by means of a countershaft. It seemed uneconomical to have two belts; and it would be interesting to know what power was absorbed by the countershaft and the large belt from the engine. He agreed with Mr. Lupton that a steam-consumption of 38 pounds per horsepower per hour was rather high, although it must be admitted that, compared with many colliery-engines, it was low. He erected a steam-turbine a short time ago, and was gratified to find that the steam-consumption was only 39 pounds per kilowatt per hour, as compared with 38 pounds per horsepower per hour at Shirebrook colliery.

Mr. J. W. FRYAR (Sherwood) asked whether Mr. Hay could say what was exactly the amount of superheat at the engine. He (Mr. Fryar) was using superheaters and had experienced some difficulty in carrying the superheat from the heater to the engine.

Mr. J. H. W. LAVERICK (Pye Hill) asked whether the arrangement of driving from a countershaft was thoroughly satisfactory. He thought that the power absorbed in the engine and countershaft (30·74 horsepower) was rather high; and asked whether it would not be possible, with a direct-driven machine, to reduce it to 10 or 15 per cent. He (Mr. Laverick) asked what arrangement had been adopted for heating the feed-water. If the water was good enough for an economizer, he had no doubt that a temperature of 280° to 300° Fahr. could be attained, as was the case at one of the collieries at Riddings. As to the question of superheating, he had not ascertained what amount of superheat reached the engines, but he knew that the exhaust-steam was as nearly dry as it could be—whereas it used to be very wet.

Mr. W. HAY, replying to the discussion, said that he did not feel very highly satisfied with the steam-consumption of 38 pounds per horsepower per hour. He was alive to the fact that there was room for considerable improvement, although, compared with that of many collieries, it was perhaps good; and if a condenser was added, better results would be obtained. One factor which contributed very largely to the result was the long range of steam-pipes. There was a considerable loss in these pipes, and if the engine was worked at its full power, a better result would be attained, inasmuch as they would benefit from any condensation which took place; also, if the leakage from the joints were deducted, the results would appear more favourable. The feed-water was heated by means of the exhaust-steam from the winding-engine, and the average temperature was about 180° Fahr., but, in this test, the feed-water was obtained from another source, and the temperature was much lower, namely, 120° Fahr.   The cost included all men and boys, including the motor at the bottom and corporals, but pony-drivers and engine-men were excluded.   The results would be better if the engine and dynamo were directly coupled.   The question as to the superheater had been answered by Mr. Laverick.   The depth of the shaft was 1,605 feet.   The junction-boxes were placed 300 feet apart, and there had not been the slightest trouble with them.   The power required to start the underground motor was great, and as the machine would not start under a load, it was started empty.

The PRESIDENT (Mr. W. B. M. Jackson), in moving a vote of thanks to Mr. Hay, said that the members were glad to know that satisfactory work was being obtained on a fairly large scale with three-phase haulage-plant.

Mr. J. H. W. LAVERICK seconded the proposal, which was agreed to.

———

Mr. W. MAURICE read the following "Comments on the Proposed 'Rules for the Installation and Use of Electricity in Mines'":—

COMMENTS ON THE PROPOSED "RULES FOR THE INSTALLATION AND USE OF ELECTRICITY IN MINES."

By WM. MAURICE, Assoc. M.I.E.E.

Introduction.—A Departmental Committee was appointed by H.M. Home Secretary on October 16th, 1902, " to enquire into the use of electricity in Coal and Metalliferous Mines, and the dangers attending it; and to report what measures should be adopted in the interests of safety by the establishment of Special Rules or otherwise."

The Committee commenced to take evidence on November 26th following, and concluded their enquiry on May 6th, 1903. The minutes of evidence and a report on the question at issue have recently been presented.*

The preliminary pages of the *Report* are occupied with useful introductory matter. The Committee then submit four general principles, which they think should govern the employment of electricity in mines. These are :—

(1) The electric plant should always be treated as a source of potential danger.

(2) The plant, in the first instance, should be of thoroughly good quality, and so designed as to insure immunity from danger by shock or fire ; and periodical tests should be made to see that this state of efficiency is being maintained.

(3) All electrical apparatus should be under the charge of competent persons.

(4) All electrical apparatus, which may be used when there is a possibility of danger arising from the presence of gas, should be so enclosed as to prevent such gas being fired by sparking of the apparatus ; when any machine is working, every precaution should be taken to detect the existence of danger, and, on the presence of gas being noticed, such machines should be immediately stopped.†

It must be supposed that these are the general principles upon which the conclusions of the Committee are based. Every engineer will probably accept the three somewhat obvious

---

* *Report of the Departmental Committee on the Use of Electricity in Mines,* 1904, [Cd. 1916] ; and *Minutes of Evidence taken before the Departmental Committee on the Use of Electricity in Mines; with Appendices and Index,* 1904, [Cd. 1917].

† *Report*, page 10.

" principles " (Nos. 1, 2 and 3), although as regards the third it cannot be held that every electric plant requires the constant attention of an electrician. The first part of the fourth " principle " enunciates that electric apparatus in possibly gassy mines should be enclosed. This is an arbitrary " principle," and therefore open to criticism, and further remarks on this question will appear in a subsequent section.

The *Report* then proceeds to group the various applications of electricity to mining work into the nine following divisions :— (1) Generating stations; (2) cables, switches, fuses, etc.; (3) stationary motors; (4) portable motors for coal-cutters, drills, etc.; (5) electric locomotives; (6) electric lighting; (7) shot-firing; (8) signalling; and (9) electric re-lighting of safety-lamps.* Each of these sections is briefly reviewed, and the general conclusions are then put into definite shape through the medium of " a set of rules, which we [the Committee] think might, with advantage, be introduced into all mines."† In order to differentiate between fiery and non-fiery mines, the Committee have " indicated in our rules that certain additional precautions should be taken in those parts of a mine which are required to be worked with safety-lamps."‡ The ultimate responsibility for installing and using electricity in dangerous places, however, remains entirely with both colliery-owners and managers. There is no absolute prohibition; it is, however, pointed out that H.M. inspectors of mines have powers under section 42 of the Coal-mines Regulation Act, 1887, to deal with any such special dangers as are not expressly guarded against in the rules submitted.

As precautionary measures must necessarily vary according to the pressure of supply, the Committee have divided electrical installations into four groups, according to the classification of pressures, recently settled by the Board of Trade, namely :— Low, under 250 volts; medium, from 251 to 650 volts; high, from 651 to 3,000 volts; and extra high, any pressure above 3,000 volts.§

* *Report of the Departmental Committee on the Use of Electricity in Mines,* 1904, [Cd. 1916], page 10.

† *Ibid.,* page 20.   ‡ *Ibid.,* page 19.

§ *Ibid.,* pages 11 and 23.

The *Minutes of Evidence* and the *Report* together form an exceedingly valuable contribution to the literature of electric mining-practice, and merit most careful attention. The definitive conclusions do not, however, lend themselves to universal acceptance, and consequently the writer will submit some detailed comments.

There can be no question as to the practical value of rules for the regulation of the details of industries. It is, however, permissible to doubt whether they should not as far as possible be left in the hands of the industries concerned. There seems to be a tendency, which one cannot help but deplore, to follow the lead of other nations. The British race may or may not have lagged behind, but at any rate it can only progress by thinking and acting in its own hereditary way.

The writer considers that the interests of safety can be met in a simpler way than by the multiplication of Special Rules under the Coal-mines Regulation Act, and submits that it is the legitimate function of a representative body of mining engineers to draft such precautions as may be necessary, and to modify them from time to time in order to meet changed conditions.

The writer offers the following comments and notes on some of the proposed definitions and rules.

Definitions.—Arbitrary classification of pressures, such as those contained in the preliminary section of definitions, have already been established by the Board of Trade. It was not clear that the existing rules had served any useful purpose, and it did not seem desirable that their application should be extended to mines. The inevitable tendency of such regulations was to limit the range of ideas and to establish standard methods of obtaining certain results.

Section I. General.—1 and 2. It will doubtless be conceded, as a mere matter of common sense, that generally all machinery should be "sufficient in size and power for the work it may be called upon to do," and that it should be installed in suitable places; but it is difficult to determine why it should be deemed necessary to give an official expression to this factor. Machinery may, when put down, be capable of transmitting double the power required, and yet eventually become overloaded. In such an event, the management would appear to be the proper authority to decide whether to continue running an overload or to put in a larger machine.   If a breakdown, necessarily or even probably, involved risk to life, it would be desirable that the temptation to run the risk should be brought under official control.   The fact remains, however, that generally there is no such risk.

3. The third rule covers the ground as expressed in Nos. 1 and 2 rules.   It is a far-reaching rule and does not properly fall within the scope of official administration.

4. This is a comprehensive rule, and may cause great practical inconvenience without any corresponding addition to general safety.*   Commutators, for example, are more likely to give trouble if " effectively covered " than if left open.   The knowledge that it is possible to receive a shock is in itself the best preventative of accidents.   As to the use of electric apparatus under No. 8 General Rule of the Coal-mines Regulation Act, 1887, it is doubtful whether the enclosure of motors is really a precaution. The danger of brush-sparks igniting mine-gas is conjectural and remote.   The enclosure of a motor frequently necessitates the placing of terminals and contact-pieces in awkward positions; and if a motor is made troublesome to examine, it is fairly certain that some day or other the examination will be " scamped."   Some of the parts will in consequence become hot, perhaps burn off, and more risk and damage will be incurred than would be reasonably likely to develop with an open-type and easily accessible machine.

5. (a) (1) and (2).†   Expert opinion is not unanimous on the question of the earthing of motor-frames; and in mines there might be serious practical difficulties in the way of carrying out this rule.   In the case of a coal-cutter, running on skids on a dry bed, unless it were considered to " earth " itself by making a considerable area of contact with the floor, the earthing would simply be an additional complication.   Again, it would surely not be considered necessary to earth an electric drill every time that it is set up.   Someone would be sure to trip over the earth-wire, knock down the drill, and cause general confusion.

---

* This rule may be compared with one of the *General Rules recommended for Wiring for the Supply of Electrical Energy : The Institution of Electrical Engineers, 1903,* as follows :—" 17. Switches and fuses, not in an engine-room or compartment specially arranged for the purpose, must be covered."

† Compare with Nos. 6, 9, 10, 11 and 12 rules of the Institution of Electrical Engineers, 1903.

5.   (3) and (4). It is not usual to provide a separate switch
    each arc-lamp, where a number are connected in series,
    the lamps must necessarily be either all on or all off at
    given time. Moreover, there are conditions in mines under
    this rule might cause more danger than it is designed to
    and in many cases it would cause a great deal of incon-
    and complication. Would it be necessary to carry a
    switch and also a starting-switch for a portable coal-
    In this case, the double-pole switch, in its cast-iron
    would be nearly as large, if not larger than the motor
    then the switch-box would have to be earthed.
    with a coal-cutter, the provision of a double-pole switch
    simply means the addition of another weak

    cutter probably shakes more, when running, than
    of machinery to which an electric motor has
    Every bolt and nut on it is a source of possible
    every bolt and nut that forms part of an electric
    an addition to the risks; no matter how care-
    be, they will come loose; and then inevitably,
    the usual concomitants of bad contact will ensue.
    such conditions as this, where the addition of
    for safety actually increase the number of
    perfect system would be to have a motor at
    and a dynamo and switchboard at the other
    necessities, however, usually demand the
    switches, etc., but the fact should not be lost
    extra device attached to current-carrying
    possible point of breakdown.

    would appear to prohibit the use of
    bank or siding, although it is noted that
    Section III., rule 14) permitted.

    appears to mean that the switches, fuses,
    switch- or distributing-board must be

    that no higher pressure than 650 volts
    surface.

    object in making 75 brake-horse-
    of a high-pressure motor or trans-

be
the
seen
The
range
certain

Section
as a mere
should be
called upon to
places: but it is
necessary to give
may, when put down
required, and yet

7. (2) This rule is too severe and would not effectively secure safety, because many instances are known of " earthing " having failed to do what was expected of it (Section I., rule 5 (*a*) (1)).

7. (3) This rule seems to allow no such alternative to armoured cables as is provided for on a medium-pressure supply (Section I., rule 5 (*a*) (2)). Incidently, it may be observed that many objections may be urged against the use of armoured cables.

7. (4) The writer suggests that confusion might arise through an electrical " danger board " being mistaken for one placed at the entrance to a fenced-off working-place or district of the mine. The suggestion that cables should be coated with vermilion paint is not a sufficiently practical one. The danger of high-pressure cables ought, of course, to be indicated, but there does not appear to be any reason why a rule should be formulated to that end.

7. (5) It is possible that much practical inconvenience might arise from the operation of this rule. For all practical purposes, it should be omitted, and, from the point of view of safety would probably never be missed.

8. It is customary to carry out the test required by this proposed rule.

9.* It would be a very difficult matter to keep up a large colliery-installation with 1/10,000th of the maximum-supply current as the greatest permissible leakage, and might add substantially to the cost of maintenance. It is desirable, of course, that the management should recognise a certain leakage-limit, because leakage-records are of great practical value. It would, however, be absolutely impossible to officially enforce the maintenance of any standard, owing to the extremely variable conditions of electric work in mines.

10. It is not usually considered the best practice to carry out tests of the nature indicated in this rule. It is somewhat analogous to that of straining metal to its elastic limit, and the test may permanently injure the material tested.

11.† The useful 2 lamps leakage-indicator could not be used under this rule, since it provides no readings. In the case of

* Compare with Nos. 78 and 79 rules of the Institution of Electrical Engineers, 1903.

† Compare with No. 80 rule of the Institution of Electrical Engineers, 1903.

5. (*a*) (3) and (4). It is not usual to provide a separate switch for each arc-lamp, where a number are connected in series, as all the lamps must necessarily be either all on or all off at any given time. Moreover, there are conditions in mines under which this rule might cause more danger than it is designed to prevent, and in many cases it would cause a great deal of inconvenience and complication. Would it be necessary to carry a double-pole switch and also a starting-switch for a portable coal-drill? In this case, the double-pole switch, in its cast-iron case, would be nearly as large, if not larger than the motor itself: and then the switch-box would have to be earthed. Further, with a coal-cutter, the provision of a double-pole switch on the machine simply means the addition of another weak place.

A coal-cutter probably shakes more, when running, than any other piece of machinery to which an electric motor has been applied. Every bolt and nut on it is a source of possible breakdown, and every bolt and nut that forms part of an electric circuit constitutes an addition to the risks; no matter how careful a driver may be, they will come loose; and then inevitably, heat and fire, the usual concomitants of bad contact will ensue. There are many such conditions as this, where the addition of devices intended for safety actually increase the number of risks. The ideally perfect system would be to have a motor at one end of a cable, and a dynamo and switchboard at the other end of it. Practical necessities, however, usually demand the addition of local switches, etc., but the fact should not be lost sight of, that every extra device attached to current-carrying mains constitutes a possible point of breakdown.

5. (*b*) (1) This rule would appear to prohibit the use of aerial cables on a pit-bank or siding, although it is noted that they are elsewhere (Section III., rule 14) permitted.

5. (*b*) (2) This rule appears to mean that the switches, fuses, etc., on every main switch- or distributing-board must be enclosed.

6. This rule implies that no higher pressure than 650 volts is permissible at the coal-face.

7. (1) There is no apparent object in making 75 brake-horsepower the minimum rating of a high-pressure motor or transformer.

7. (2) This rule is too severe and would not effectively secure safety, because many instances are known of " earthing " having failed to do what was expected of it (Section I., rule 5 (a) (1)).

7. (3) This rule seems to allow no such alternative to armoured cables as is provided for on a medium-pressure supply (Section I., rule 5 (a) (2)). Incidently, it may be observed that many objections may be urged against the use of armoured cables.

7. (4) The writer suggests that confusion might arise through an electrical " danger board " being mistaken for one placed at the entrance to a fenced-off working-place or district of the mine. The suggestion that cables should be coated with vermilion paint is not a sufficiently practical one. The danger of high-pressure cables ought, of course, to be indicated, but there does not appear to be any reason why a rule should be formulated to that end.

7. (5) It is possible that much practical inconvenience might arise from the operation of this rule. For all practical purposes, it should be omitted, and, from the point of view of safety would probably never be missed.

8. It is customary to carry out the test required by this proposed rule.

9.* It would be a very difficult matter to keep up a large colliery-installation with 1/10,000th of the maximum-supply current as the greatest permissible leakage, and might add substantially to the cost of maintenance. It is desirable, of course, that the management should recognise a certain leakage-limit, because leakage-records are of great practical value. It would, however, be absolutely impossible to officially enforce the maintenance of any standard, owing to the extremely variable conditions of electric work in mines.

10. It is not usually considered the best practice to carry out tests of the nature indicated in this rule. It is somewhat analogous to that of straining metal to its elastic limit, and the test may permanently injure the material tested.

11.† The useful 2 lamps leakage-indicator could not be used under this rule, since it provides no readings. In the case of

* Compare with Nos. 78 and 79 rules of the Institution of Electrical Engineers, 1903.

† Compare with No. 80 rule of the Institution of Electrical Engineers, 1903.

5. (*a*) (3) and (4). It is not usual to provide a separate switch for each arc-lamp, where a number are connected in series, as all the lamps must necessarily be either all on or all off at any given time. Moreover, there are conditions in mines under which this rule might cause more danger than it is designed to prevent, and in many cases it would cause a great deal of inconvenience and complication. Would it be necessary to carry a double-pole switch and also a starting-switch for a portable coaldrill? In this case, the double-pole switch, in its cast-iron case, would be nearly as large, if not larger than the motor itself; and then the switch-box would have to be earthed. Further, with a coal-cutter, the provision of a double-pole switch on the machine simply means the addition of another weak place.

A coal-cutter probably shakes more, when running, than any other piece of machinery to which an electric motor has been applied. Every bolt and nut on it is a source of possible breakdown, and every bolt and nut that forms part of an electric circuit constitutes an addition to the risks; no matter how careful a driver may be, they will come loose; and then inevitably, heat and fire, the usual concomitants of bad contact will ensue. There are many such conditions as this, where the addition of devices intended for safety actually increase the number of risks. The ideally perfect system would be to have a motor at one end of a cable, and a dynamo and switchboard at the other end of it. Practical necessities, however, usually demand the addition of local switches, etc., but the fact should not be lost sight of, that every extra device attached to current-carrying mains constitutes a possible point of breakdown.

5. (*b*) (1) This rule would appear to prohibit the use of aerial cables on a pit-bank or siding, although it is noted that they are elsewhere (Section III., rule 14) permitted.

5. (*b*) (2) This rule appears to mean that the switches, fuses, etc., on every main switch- or distributing-board must be enclosed.

6. This rule implies that no higher pressure than 650 volts is permissible at the coal-face.

7. (1) There is no apparent object in making 75 brake-horse-power the minimum rating of a high-pressure motor or transformer.

7. (2) This rule is too severe and would not effectively secure safety, because many instances are known of " earthing " having failed to do what was expected of it (Section I., rule 5 (a) (1)).

7. (3) This rule seems to allow no such alternative to armoured cables as is provided for on a medium-pressure supply (Section I., rule 5 (a) (2)). Incidently, it may be observed that many objections may be urged against the use of armoured cables.

7. (4) The writer suggests that confusion might arise through an electrical " danger board " being mistaken for one placed at the entrance to a fenced-off working-place or district of the mine. The suggestion that cables should be coated with vermilion paint is not a sufficiently practical one. The danger of high-pressure cables ought, of course, to be indicated, but there does not appear to be any reason why a rule should be formulated to that end.

7. (5) It is possible that much practical inconvenience might arise from the operation of this rule. For all practical purposes, it should be omitted, and, from the point of view of safety would probably never be missed.

8. It is customary to carry out the test required by this proposed rule.

9.* It would be a very difficult matter to keep up a large colliery-installation with 1/10,000th of the maximum-supply current as the greatest permissible leakage, and might add substantially to the cost of maintenance. It is desirable, of course, that the management should recognise a certain leakage-limit, because leakage-records are of great practical value. It would, however, be absolutely impossible to officially enforce the maintenance of any standard, owing to the extremely variable conditions of electric work in mines.

10. It is not usually considered the best practice to carry out tests of the nature indicated in this rule. It is somewhat analogous to that of straining metal to its elastic limit, and the test may permanently injure the material tested.

11.† The useful 2 lamps leakage-indicator could not be used under this rule, since it provides no readings. In the case of

* Compare with Nos. 78 and 79 rules of the Institution of Electrical Engineers, 1903.

† Compare with No. 80 rule of the Institution of Electrical Engineers, 1903.

5. (*a*) (3) and (4). It is not usual to provide a separate switch for each arc-lamp, where a number are connected in series, as all the lamps must necessarily be either all on or all off at any given time. Moreover, there are conditions in mines under which this rule might cause more danger than it is designed to prevent, and in many cases it would cause a great deal of inconvenience and complication. Would it be necessary to carry a double-pole switch and also a starting-switch for a portable coal-drill? In this case, the double-pole switch, in its cast-iron case. would be nearly as large, if not larger than the motor itself; and then the switch-box would have to be earthed. Further, with a coal-cutter, the provision of a double-pole switch on the machine simply means the addition of another weak place.

A coal-cutter probably shakes more, when running, than any other piece of machinery to which an electric motor has been applied. Every bolt and nut on it is a source of possible breakdown, and every bolt and nut that forms part of an electric circuit constitutes an addition to the risks; no matter how careful a driver may be, they will come loose; and then inevitably, heat and fire, the usual concomitants of bad contact will ensue. There are many such conditions as this, where the addition of devices intended for safety actually increase the number of risks. The ideally perfect system would be to have a motor at one end of a cable, and a dynamo and switchboard at the other end of it. Practical necessities, however, usually demand the addition of local switches, etc., but the fact should not be lost sight of, that every extra device attached to current-carrying mains constitutes a possible point of breakdown.

5. (*b*) (1) This rule would appear to prohibit the use of aerial cables on a pit-bank or siding, although it is noted that they are elsewhere (Section III., rule 14) permitted.

5. (*b*) (2) This rule appears to mean that the switches, fuses, etc., on every main switch- or distributing-board must be enclosed.

6. This rule implies that no higher pressure than 650 volts is permissible at the coal-face.

7. (1) There is no apparent object in making 75 brake-horse-power the minimum rating of a high-pressure motor or transformer.

7. (2) This rule is too severe and would not effectively secure safety, because many instances are known of " earthing " having failed to do what was expected of it (Section I., rule 5 (a) (1)).

7. (3) This rule seems to allow no such alternative to armoured cables as is provided for on a medium-pressure supply (Section I., rule 5 (a) (2)).   Incidently, it may be observed that many objections may be urged against the use of armoured cables.

7. (4) The writer suggests that confusion might arise through an electrical " danger board " being mistaken for one placed at the entrance to a fenced-off working-place or district of the mine. The suggestion that cables should be coated with vermilion paint is not a sufficiently practical one. The danger of high-pressure cables ought, of course, to be indicated, but there does not appear to be any reason why a rule should be formulated to that end.

7. (5) It is possible that much practical inconvenience might arise from the operation of this rule. For all practical purposes, it should be omitted, and, from the point of view of safety would probably never be missed.

8. It is customary to carry out the test required by this proposed rule.

9.* It would be a very difficult matter to keep up a large colliery-installation with 1/10,000th of the maximum-supply current as the greatest permissible leakage, and might add substantially to the cost of maintenance. It is desirable, of course, that the management should recognise a certain leakage-limit, because leakage-records are of great practical value. It would, however, be absolutely impossible to officially enforce the maintenance of any standard, owing to the extremely variable conditions of electric work in mines.

10. It is not usually considered the best practice to carry out tests of the nature indicated in this rule. It is somewhat analogous to that of straining metal to its elastic limit, and the test may permanently injure the material tested.

11.† The useful 2 lamps leakage-indicator could not be used under this rule, since it provides no readings. In the case of

* Compare with Nos. 78 and 79 rules of the Institution of Electrical Engineers, 1903.

† Compare with No. 80 rule of the Institution of Electrical Engineers, 1903.

5. (*a*) (3) and (4). It is not usual to provide a separate switch for each arc-lamp, where a number are connected in series, as all the lamps must necessarily be either all on or all off at any given time.   Moreover, there are conditions in mines under which this rule might cause more danger than it is designed to prevent, and in many cases it would cause a great deal of inconvenience and complication.   Would it be necessary to carry a double-pole switch and also a starting-switch for a portable coal-drill?   In this case, the double-pole switch, in its cast-iron case, would be nearly as large, if not larger than the motor itself; and then the switch-box would have to be earthed. Further, with a coal-cutter, the provision of a double-pole switch on the machine simply means the addition of another weak place.

A coal-cutter probably shakes more, when running, than any other piece of machinery to which an electric motor has been applied.   Every bolt and nut on it is a source of possible breakdown, and every bolt and nut that forms part of an electric circuit constitutes an addition to the risks; no matter how careful a driver may be, they will come loose; and then inevitably, heat and fire, the usual concomitants of bad contact will ensue. There are many such conditions as this, where the addition of devices intended for safety actually increase the number of risks.   The ideally perfect system would be to have a motor at one end of a cable, and a dynamo and switchboard at the other end of it.   Practical necessities, however, usually demand the addition of local switches, etc., but the fact should not be lost sight of, that every extra device attached to current-carrying mains constitutes a possible point of breakdown.

5. (*b*) (1) This rule would appear to prohibit the use of aerial cables on a pit-bank or siding, although it is noted that they are elsewhere (Section III., rule 14) permitted.

5. (*b*) (2) This rule appears to mean that the switches, fuses, etc., on every main switch- or distributing-board must be enclosed.

6. This rule implies that no higher pressure than 650 volts is permissible at the coal-face.

7. (1) There is no apparent object in making 75 brake-horse-power the minimum rating of a high-pressure motor or transformer.

7. (2) This rule is too severe and would not effectively secure safety, because many instances are known of " earthing " having failed to do what was expected of it (Section I., rule 5 (a) (1)).

7. (3) This rule seems to allow no such alternative to armoured cables as is provided for on a medium-pressure supply (Section I., rule 5 (a) (2)). Incidently, it may be observed that many objections may be urged against the use of armoured cables.

7. (4) The writer suggests that confusion might arise through an electrical "danger board " being mistaken for one placed at the entrance to a fenced-off working-place or district of the mine. The suggestion that cables should be coated with vermilion paint is not a sufficiently practical one. The danger of high-pressure cables ought, of course, to be indicated, but there does not appear to be any reason why a rule should be formulated to that end.

7. (5) It is possible that much practical inconvenience might arise from the operation of this rule. For all practical purposes, it should be omitted, and, from the point of view of safety would probably never be missed.

8. It is customary to carry out the test required by this proposed rule.

9.* It would be a very difficult matter to keep up a large colliery-installation with 1/10,000th of the maximum-supply current as the greatest permissible leakage, and might add substantially to the cost of maintenance. It is desirable, of course, that the management should recognise a certain leakage-limit, because leakage-records are of great practical value. It would, however, be absolutely impossible to officially enforce the maintenance of any standard, owing to the extremely variable conditions of electric work in mines.

10. It is not usually considered the best practice to carry out tests of the nature indicated in this rule. It is somewhat analogous to that of straining metal to its elastic limit, and the test may permanently injure the material tested.

11.† The useful 2 lamps leakage-indicator could not be used under this rule, since it provides no readings. In the case of

* Compare with Nos. 78 and 79 rules of the Institution of Electrical Engineers, 1903.
† Compare with No. 80 rule of the Institution of Electrical Engineers, 1903.

a mine having a number of transformer-stations, it would be an additional complication, and an unnecessary refinement, to place a megohmmeter at each station.

12 and 13.* The suggestions contained in these proposed rules are current practice.

14.† It may easily occur that conductors are safer when bunched together than when placed separately.

15.‡ An automatic cut-out would presumably be recognised as an effective substitute for a switch and a fuse. To enact that every circuit carrying more than 3 ampères must have a separate double-pole switch and fuse might have the effect of so multiplying the number of connections as to materially increase the risks and render a high standard of insulation-resistance an absolute impossibility.

16. This rule might be eliminated, since it would be generally acted upon and as a safety precaution is covered by other rules.

17. It is useful to have a supply of sand in a high-tension station, but the rule should be of local rather than of general enactment. In the great majority of electric plants at mines, a competent person could put out an incipient fire in less time than it would take another man to get hold of a bucket. Workmen should not be led to believe that they will be allowed to throw sand on an electric machine every time that they may see a flame.

18.§ Circumstances might arise under which it would be impossible to carry out this rule, and, in any case, it is unnecessarily restrictive. It might be necessary, for example, to repair a brush-holder or to put a new set of brushes upon a dynamo or motor whilst the machine was running. To a man, who knows what he is doing, there is absolutely no danger. In the case of some types of starting-switches, small adjustments and repairs can be made more advantageously when the current is on than when it is off.

19. This, in itself, is a good rule but it is scarcely worth while to give it legal recognition. The practical value of the rule may be readily deduced by anyone who knows the British workman.

* Compare with Nos. 56, 57, 58, 59 and 60 rules of the Institution of Electrical Engineers, 1903.

† Compare with No. 4 rule of the Institution of Electrical Engineers, 1903.

‡ Compare with Nos. 7 and 8 rules of the Institution of Electrical Engineers, 1903.

§ Compare with No. 81 rule of the Institution of Electrical Engineers, 1903.

20. The inspection of electrical apparatus, etc., is comprised under the existing rules as to the inspection of machinery.

21. Unless this rule be greatly modified, it might be necessary to keep a man underground when there was absolutely nothing for him to do.

It is difficult to make any comment on these proposed rules (20 and 21), without knowing how the expression "competent person" is to be interpreted.

22. With reference to this and other of the proposed rules, it is to be noted that there are no emergency clauses. Nobody is permitted to do anything, unless he has been duly authorised! Men are converted into machines and action, born of independent thought, becomes a breach of the law.

23 and 24. These requirements are already covered by the Special Rules in force in the various mines-inspection districts.

25. It is desirable that the treatment for electric shock should be made generally known. The printed instructions should be brief, lucid and in large type, otherwise they will either not be understood or will not be read.

26. It might be somewhat of a hardship to insist on the provision of telephonic communication at every colliery. Extensive underground telephone-systems have not always been found to be wholly advantageous.

27. It is probable that the subject-matter of this rule lies outside of the functions of the Departmental Committee. The application of electricity is only a branch of general engineering practice and should not logically be placed under official control to a greater extent than similar applications of air- or steam- or water-power.

28.* This rule requires that a wiring-plan of the entire installation shall be prepared and corrected to date. The writer suggests that it would be sufficient to indicate, say, by special colouring on the mine-plans, the position of all fixed transformers or motors, and the roads along which main cables are conveyed.

29. It is usual to say that storage batteries shall be placed in well-ventilated rooms, but it is rather premature as a rule for mines.

* Compare with Nos. 73 and 74 rules of the Institution of Electrical Engineers, 1903.

Section II. Generating-stations and Machine-rooms.—2. The provision of a switch-house, at the pit-mouth, would, in many cases, be costly and entirely superfluous.

3. The provision of a rail in front of the switchboard would be likely to hinder the attendant from carrying out his duties. A platform sometimes gives an air of finish to a switchboard, but it is usually more inconvenient than useful. The writer has known a man to trip over the edge of a platform, and fall with his hand upon a switch.

4. The clauses of this rule are arbitrary and introduce unnecessary restrictions.

5. It is the general practice to make the provisions required by this proposed rule, and it seems scarcely necessary to formulate them.

6. The ammeters would be an additional expense, without contributing towards the prevention of accidents. It would be advantageous to know precisely what is intended by the expression "feeder-circuit." The use of the term appears to be extended beyond its technical meaning.

7. The practical necessity for this rule is not apparent, and it will entail additional complications, with additional risk of breakdown.

8. The use of lightning arresters in connection with overhead circuits is so much a part of current practice as to render the rule unnecessary.

9. It is not clear why unenclosed fuses should be placed so near the floor. There are many open-type fuses available that will "blow" without danger to anyone who may be standing near; and to place a fuse in an inconvenient situation is the readiest way of ensuring that someone will render it inoperative.

10. Switchboards to comply with this rule would require to be specially constructed, and large stocks of well-designed and standard switchboards will become useless. It seems probable that this rule is intended to refer to high-pressure installations.

11. This rule is not desirable, because it stereotypes a method of carrying out practical details, and hampers the work of expert designers.

12. This rule is similar to Section I., rule 5 (*a*) (1).

13. The first portion of this rule is presumably not intended to apply to switchboards.

The detail suggested by the second portion of this rule should be left to the designers of dynamos. It is difficult to recognize its value as a safety-precaution; and a competent dynamo-attendant ought to, and would, handle the brushes if it became necessary to do so during a run. So far as adjustment is required, modern machines will stand very wide variations of the load without alteration of the "lead" of the brushes. The rule, moreover, cannot conveniently be applied to all motors.

14. This rule may be compared with Section I., rule 19.

15. The writer, as one who has spent many days and nights in charge of generators, considers that a fence around the commutator-end of a machine is an obstacle deliberately placed in the way of the person who is authorized, and therefore presumably competent, to attend to it.

16. A similar rule is already in force at every colliery.

Section III. Cables.—1. See Section I., rules 27 and 28.

2, 3 and 4.* These rules are essentially part of established practice.

5. If this rule includes shaft-cables, it would seem that the only permitted alternative to a lead-covered armoured cable is a cable enclosed in a water-tight metal-tube; or possibly this rule is intended to exclude the use of armoured, but not of lead-covered cables, or of cables insulated, braided and with or without leaden sheathing and bedded in grooved wooden casing. It is very desirable that the intention of this rule should be made perfectly clear. An experience of 14 years has served to convince the writer that underground is the wrong place for cables placed in " water-tight metal-tubes."

6.† This excellent rule might with advantage be more fully recognized. It is, however, almost invariably set aside, as a detail of no particular importance, even by fairly competent and experienced wiremen. It can scarcely, however, be considered to have any appreciable bearing on the safety of a mine.

7.‡ This proposed rule is a part of current practice.

* Compare with Nos. 20, 23, 24, 25, 26, 27 and 29 rules of the Institution of Electrical Engineers, 1903.

† Compare with No. 30 rule of the Institution of Electrical Engineers, 1903.

‡ Compare with No. 21 rule of the Institution of Electrical Engineers, 1903.

8. The protection of cables from overheating is sufficiently covered by other rules.

9. Essentially this rule embodies current practice.   As a safety rule, it is not necessary having in view the all embracing rule (Section I., rule 3).   In any case, it is not desirable that such definite limitations should be imposed.

10. This is a sound practical rule, but it is too trivial for official recognition.

11. The writer infers that " joint-box " will permit the use of any kind of clamp that will serve to make a mechanically and electrically perfect joint.

12.* In this rule also it seems desirable that " joint-box " should be considered to include all satisfactory kinds of connections; otherwise the rule is liable to cause considerable inconvenience without conducing in any way to general safety.

13. The expression " their equivalent " in this rule, appears to be destroyed by the subsequent sentence, which determines one specified type of connector.   There are many other and simpler ways of making a joint than the one permitted by this rule.   Moreover, it would frequently be desirable to make permanent joints in a district of a mine worked with safety-lamps.

14. The writer has already referred to the use of aerial cables in Section I., rule 5, (b) (1).

15. The first clause appears to prohibit the use of any system of bare conductors in insulated channels.

16. It has been stated that this rule was drafted so as to permit of the use of bare cables in shafts; but it would be useful to learn if such is the case.

17 (b). It does not appear desirable or necessary that the minimum distance between cables and tubs should be exactly defined.

17 (c). The writer would like to know whether the use of a metallic fastening with an insulated support would be prohibited, and if so, why?

17 (e). If trailing cables were always carried in front (as recommended and practised by the writer) instead of behind the machine, there would be no particular necessity for special treatment, since the cable is not liable to be knocked about to anything like the same extent.   The cable lies on a clean road,

* Compare with No. 37 rule of the Institution of Electrical Engineers, 1903.

out of everyone's way, and (excepting falls) is scarcely touched except when laid hold of to be pulled forward. It may be observed, incidentally, that a " specially flexible " cable almost ceases to possess that advantage when heavily braided and armoured.

17 (*f*).* This rule provides a double-pole switch and a terminal-box at the points where flexible conductors join the main cables. The switch, being enclosed (Section IV., rule 4), may presumably be combined with or in itself constitute a terminalbox.

Section IV.   Switches, Fuses and Cut-outs.—1.† " Form an arc " in this rule is obviously a slip of the pen and is intended to mean " maintain an arc."

(2) To insist that the covers of fuses must be made of " rigid metal lined with insulating incombustible material " seems to be an unduly severe limitation of type.   If, for example, an enclosed fuse were mounted in a cast-iron box, it would surely not be considered necessary to line the box with " insulating incombustible material."

3. The requirement that " they shall be stamped or marked with the current for which they are intended to be used " limits the type of fuse to two or three patterns not usually found in mines.   If it is desired that fuses shall always be the protective devices—the safety-valves, as it were—which they are intended to be, that end is best secured by simplifying fuse-boxes in every possible way.   A simple copper-wire, fastened under two screw-heads, at a distance apart proportionate to the pressure of supply, is better than any " fancy " fuse for pressures under or about 500 volts.   It is impossible to insert the wrong fuse without breaking the previous rule or Section I., rule 24.

6. Oil-break fuses are better adapted for alternating than for continuous-current supply.   In continuous-current working, especially with a motor or other inductive load, the sudden rupture caused by an oil-break produces an extreme rise of pressure, which may do a considerable amount of damage.   It may be observed that a fuse in oil will carry more than double its current in " air " without blowing.   There is also the further

* Compare with No. 54 rule of the Institution of Electrical Engineers, 1903.
† Compare with No. 45 rule of the Institution of Electrical Engineers, 1903.

possibility of oil being ignited. This rule might be considerably modified, since it introduces the speculative fourth general principle of the Departmental Committee.

Section V. Motors.—1. Generally, it is desirable that motors should be capable of being disconnected by means of a local double-pole switch. In mines, however, there are frequent cases where the presence of such an appliance is more likely to cause a breakdown than to prevent an accident. The reader may refer to the note on Section I., rule 5 (*a*) (3) and (4).

2. An ammeter is an extremely useful instrument in the hands of an expert, but to the type of workman ordinarily employed in motor-driving its indications would convey no particular meaning. Moreover, a colliery fitting-shop might, for example, have 6 or more motors, attached to various tools; and it is difficult to conceive that any practical object will be served by inserting an ammeter in each motor-circuit. The fuse or cut-out is available to give an indication of any overload; and the precise amount of current absorbed at any given moment is only a matter of academic interest.

3. The maximum permissible rise of temperature might safely be raised to, say, 100° Fahr., without any increased risk. It ought, however, to rest with the management to decide how machinery should be worked.

4. This rule is a detail of current practice.

5. In view of the fact that the reasons for making this proviso are almost wholly a matter of speculation, the rule, if inserted, should be greatly modified. Although there is a certain amount of theory in favour of some such rule, it is probable that the practical effect would be merely to exchange one possible danger for another.

6. The writer has not observed any tendency to ignore this precaution, except amongst drivers of coal-cutters. The use of tramway-controllers and geared starters for coal-cutters is now developing so rapidly that, in a short time, it will be almost impossible even intentionally to break this rule. Moreover, a rule which cannot by any possibility be generally enforced is of doubtful value.

7. The writer is strongly of opinion that it is a mistake to enclose anything which may, with reasonable safety, be left

·open; and it is probably safer, in the end, to risk giving a man
·a shock (on low- or medium-pressure circuits) than to enclose
·everything, and inevitably—sooner or later—have a blaze. (It
must not of course be inferred that one would risk causing a
man actual injury: the idea in the writer's mind is that a rule
which tends to make an individual forget the necessity for
·caution is in itself a dangerous one.)

8. This rule is already essentially in force, as No. 7 General
Rule of the Coal-mines Regulation Act, 1887.

9. The cutters may become jammed in many different ways,
for example, it is not at all unusual for a machineman's shovel
·to be removed from his hand and carried under the holing.
If this proposed rule must be obeyed, an hour or more might be
wasted in removing the coal from the cutting-wheel. The
simple operation of reversing the motor usually results in the
·ejection of such fragments of shovel as remain, and the machine
is then free to advance. Further, there does not seem to be
any valid reason why a motor should not start on an " increased
·load " if this load be within its safe working-range. Even if
the extra load should prove to be excessive, the fuse or automatic
·cut-out would operate and prevent any damage from being done.

10. Noting that the average period of work of coal-cutters does
·not amount to half of the shift, it would appear to be somewhat
·premature to arrange for artificial stops.

11. The fact that enclosed motors generally require attention
·several times a day, renders this rule somewhat unnecessary.

12. This rule might with advantage be eliminated, for the
·abovenamed reason.

13. As a rule, a first-class workman will not perform a piece
·of work that he can make the machine do. If a cable has to be
·pulled along, it does not particularly matter whether the machine
or the man does the pulling. The only essential point of the
rule is that the strain of the pull should not fall on the terminals.

14. This rule would be better made the subject of local
bye-laws. It is sometimes a great convenience to be able to
·execute small adjustments or repairs without switching off the
current. Noting that the said repairs must be carried out by a
competent person it is difficult to see why he should be so
·hemmed in with restrictions.

15. This rule is unnecessary and might be harmful. The

majority of incipient faults can be set right in less time than it might take to give notice that the machine must be stopped. This is another specimen of the kind of rule, which in the writer's view tends to lessen the individual sense of responsibility and leads towards the development of an inferior type of workman.

16. This rule is covered by local Special Rules relating to the use of sprags when holing.

Section VI. Electric Locomotives.—This section does not appear to be required.

Section VII. Electric Lighting.—1. It is the universal practice to carry out the provisions of this rule.

2. This rule would seem to indicate that an arc-lamp may not be used without a globe or even without a netted globe. The chance of a piece of hot carbon falling upon a person and causing him injury is too remote a contingency to justify an official reduction of from 10 to 30 per cent. in the available amount of light produced per lamp.

5. This rule appears to prohibit the practice of hanging lamps, coats, etc., on electrical mains, and the use of pendent fittings.

6. The writer suggests that it presumably would not be held that No. 8 General Rule of the Coal-mines Regulation Act, 1887, applies wherever safety-lamps are in use. The wiring restrictions are unduly severe, and apparently contradictory to the fourth rule of this section. The enclosure of lamps in strong glass fittings is of doubtful value. In the case of an exposed lamp, the probability is that a blow would completely smash it and allow no sufficient time-interval for a possible ignition of gas. A similar accident to a rigidly-cased lamp does sometimes create conditions appropriate to the starting of a fire and so proves a source of danger instead of a means of prevention. As the probabilities are fairly even either way, the most desirable plan would be to omit the rule.

7. This rule is provided for in Section I., rule 24.

Section VIII. Shot-firing.—1. This rule enacting that high-tension magneto-generators shall be enclosed in flame-tight cases is unnecessary, and appears to be based upon an exaggerated sense of danger. If required, it would be easy to produce a sparkless machine, as to the interior mechanism.   In fact, there are many such machines in use and there is no reason whatever why they should be made "flame-tight."

2 and 3. These are useful practical rules.

4. There is an excellent exploder, which does not generate· sufficient current to ignite a fuse until the armature attains a· given speed. The attainment of this speed is indicated by a hammer, which raps loudly and continuously upon a plate within. the box. It would not be necessary to remove the handle from. such a machine to prevent an accidental explosion. Moreover, if it is desired to keep the number of rules as low as is consistent with safety, why insert this one? If the machine is not connected to the shot-firing cable, what does it matter whether it. carries a handle or not?

5. If the firing-cells are mounted in a wooden case, and the· external connections are covered by separate leathern flaps. (buttoned down if desired), there does not appear to be any need for the departure from simplicity proposed by this rule.

6, 7 and 8. These are useful rules. As regards the last rule,. contact with signalling wires is a greater source of accidental shot-firing than contact with lighting or power cables, since the latter would ordinarily be insulated.

Section IX. Signalling.—1. This precaution is covered many times over by other rules.

There does not appear to be any particular necessity for any of the rules enacted in this section. The clause as to the 10 volts limit, and use of relays, is entirely unnecessary. If an ignition of gas is ever caused by signalling apparatus in the ordinary working of a mine, it is more than likely that relays will be at the bottom of the mischief.

Section X. Electric Relighting of Safety-lamps.—1. The first sentence is useful and practical. If, however, a man has to be provided to handle lamps brought to every automatic re-lighter,. the economic advantages to be derived from the use of such an appliance would disappear.

2. This rule would be a useful one, if sparks really mattered in places where a manager would use this type of apparatus.

Section XI. Exemptions and Miscellaneous.—1. It will be noted that the rules are intended to be retrospective or otherwise, at the option of H.M. inspectors of mines, subject to any prescribed conditions.

2. An inspector has power under this rule to grant any exemption, which he deems necessary or desirable.

Conclusion.—In conclusion, the writer submits that although the publications of the Departmental Committee on the Use of Electricity in Mines are of great educational value, the actual carrying into effect of the proposed Special Rules should be delayed until the necessity for some of them and the probable operation of others is more clearly seen. It is not desirable that the Home Office should rush into the issue of numerous Special Rules when it is apparent that the necessity for most of them is rapidly passing away.

Considerably more than half of the provisions will be found embodied in any elementary text-book. The writer fails to see any more reason why such rules should be made law than there is for legalising the memoranda contained in Sir Guilford Molesworth's *Pocket-book of Engineering Formulæ.* If it were enacted that neglect to take those precautions, which are commonly accepted as essential to the safe working of electrical machinery and apparatus, be an offence under the Coal-mines Regulation Acts, detailed rules to meet the situation might very properly be drafted by a committee of experts, as already suggested. The step would be more in accordance with racial characteristics, and any suggestions which might be proposed would become absorbed in everyday practice, without giving rise to a feeling that the British mining industry is by way of being " regulated " to death.

———

Mr. A. H. STOKES (H.M. Inspector of Mines) said that he should not have intervened so early in the discussion, had he not been anxious to explain one or two points, and to refer to some remarks by Mr. Maurice upon which further information or explanation was desirable. Mr. Maurice said that " it cannot be held that every electric plant requires the constant attention of an electrician."* The rule did not necessitate " the con-

* *Trans. Inst. M.E.*, 1904, vol. xxvii., page 291.

stant attention of an electrician," it only required that all electric apparatus, when in use, should be under the charge of some "competent person." In the next three lines, Mr. Maurice referred to the arbitrary principle that "electric apparatus in possibly gassy mines should be enclosed."[*] Surely it was not an arbitrary principle to enclose naked lights in mines where gas was likely to occur; and, if so, then arbitrary principles were required for safety. A little further on, Mr. Maurice remarks that "there can be no question as to the practical value of rules for the regulation of the details of industries. It is, however, permissible to doubt whether they should not as far as possible be left in the hands of the industries concerned."[†] He (Mr. Stokes) would like to enquire whether Mr. Maurice had ever met with, or could cite a single instance in which a mining or an electrical industry had voluntarily introduced special rules for regulating the details of their works.

A little further on, Mr. Maurice said that "the knowledge that it is possible to receive a shock is in itself the best preventative of accidents;"[‡] but was that so? Had the knowledge that falls of roof were likely to occur abolished accidents and loss of life from such cause? And, even recently, timbering rules had been established to prevent such accidents. Everybody knew that it was possible to incur an accident from a fall of roof, and that such accidents frequently happen, but this knowledge did not prevent the accidents from happening, and therefore the "best preventative" was found in good rules strictly enforced. Mr. Maurice was doubtful whether the enclosure of motors was really a precaution, adding that "the danger of brush-sparks igniting mine-gas is conjectural and remote:"[§] equally so was a match in a miner's pocket, but they would not allow the miner to have it. If it were a safety-lamp mine, they did not discuss the question whether the miner was likely to strike it or not—they simply would not permit him to possess the match. Had there not been explosions of fire-damp from electric sparks? He did not think that it would tax the members' memory much to call to mind an explosion from the sparking of electric cables. If members would read the evi-

* *Trans. Inst. M.E.*, 1904, vol. xxvii., page 291.
† *Ibid.*, page 292.   ‡ *Ibid.*, page 293.   § *Ibid.*, page 293.

dence given before the Departmental Committee on the Use of
Electricity in Mines, they would find interesting evidence of an
explosion of fire-damp in a coal-mine due to electricity.   Mr.
Maurice went on to say that "if a motor is made troublesome
to examine, it is fairly certain that some day or other the
examination will be 'scamped',"* but were not safety-lamps
troublesome to examine, and were they scamped?   At some
pits, 1,000 lamps were issued daily, and a careful examination
was or ought to be made of every lamp.   He would, however,
read an extract from Mr. Maurice's evidence before the Committee.   Question 4,891 was:—"You have had sparking at
your electrical machines?" And his answer was:—"Yes, everybody has sparking more or less.   I do not attach the slightest
importance to sparks in an atmosphere, which does not show a
'cap.'"   There was that beautiful "cap" again—they had
heard about it for years.   But in this case the cap did not fit.
No one would object to a naked light in an atmosphere always
free from gas; but a manager's idea was to anticipate danger,
and in mines where gas was occasionally found the men often
used safety-lamps as a precautionary measure of safety, and
equally so in such mines the parts of motors likely to spark
should be enclosed.   A few sentences further on, Mr. Maurice
said that "in the case of a coal-cutter, running on skids on a
dry bed, unless it were considered to 'earth' itself by making a
considerable area of contact with the floor, the earthing would
simply be an additional complication."†   Well if that kind of
earth did not suit, there was no reason whatever why a separate
wire should not be used to convey the current back; and by
that means, there would be a perfect "earth."

Mr. ARNOLD LUPTON asked what voltage would be allowed
on the return cable.

Mr. A. H. STOKES replied that any voltage desired could be
used, provided the return cable were big enough.   Further, Mr.
Maurice said that "someone would be sure to trip over the
earth-wire, knock down the drill, and cause general confusion."‡
They would no more be liable to trip over the earth-wire than
over the cable itself; and besides the earth-wire could be put

* *Trans. Inst. M.E.*, 1904, vol. xxvii., page 293.
† *Ibid.*, page 293.   ‡ *Ibid.*, page 293.

into the power-cable. There must be a cable of some sort, and if an additional wire for a return or earth be put into it, then one cable only was required. Mr. Maurice asked "Would it be necessary to carry a double-pole switch and also a starting-switch for a portable coal-drill?"[*] There would be a switch at the gate-end, and then wherever the drill was and wherever it was taken, a trailing cable would be carried along with it. Mr. Maurice proceeded to say that "a coal-cutter probably shakes more, when running, than any other piece of machinery to which an electric motor has been applied. Every bolt and nut on it is a source of possible breakdown, and every bolt and nut that forms part of an electric circuit constitutes an addition to the risks; no matter how careful a driver may be, they will come loose; and then inevitably, heat and fire, the usual concomitants of bad contact will ensue."[†] Imagine every bolt and nut coming loose! What a case Mr. Maurice had made for the necessity of frequent careful examination of such machinery! So serious a description as that given by Mr. Maurice he (Mr. Stokes) contended showed the absolute necessity for stringent rules as to coal-cutters. Rule 5 (*b*) (1) appeared, according to Mr. Maurice, "to prohibit the use of aerial cables on a pit-bank or siding;" but if the members looked at Section III., rule 14, they would find the words:—"It shall be permissible to use on the surface bare overhead wires under the following conditions;" and surely that was clear enough. Mr. Maurice suggested that "confusion might arise through an electrical ' danger board ' being mistaken for one placed at the entrance to a fenced-off working-place or district of the mine."[‡] The electrical "danger board" was altogether a different thing from a board fencing off a dangerous part of a mine. On rule 10, Mr. Maurice's comment was that "it is not usually considered the best practice to carry out tests of the nature indicated in this rule. It is somewhat analogous to that of straining metal to its elastic limit, and the test may permanently injure the material tested."[§] Did not Mr. Maurice know that similar tests were made on boilers, cables, chains, anchors and tubes? And was it to be said that such tests permanently injured the boilers, the cables or the tubes? As to rules 20 and 21, Mr. Maurice said

[*] *Trans. Inst. M.E.*, 1904, vol. xxvii., page 294.
[†] *Ibid.*, page 294.   [‡] *Ibid.*, page 295.   [§] *Ibid.*, page 295.

that "it is difficult to make any comment on these proposed rules, without knowing how the expression 'competent person' is to be interpreted."[*] Mr. Maurice was a colliery manager: had he not yet found out the meaning of "competent person," which had been used as a statutory expression under the Coal-mines Regulation Acts and the Special Rules for the past seventeen years? Mr. Maurice, himself, held a manager's certificate as a "competent person," and yet he did not know what a "competent person" meant! Mr. Maurice's comment on rule 26 was that "it might be somewhat of a hardship to insist on the provision of telephonic communication at every colliery," but he (Mr. Stokes) did not know why. He should like to see telephones used more than they were in mines, for they proved very useful. It paid the management to know what was going on in various parts of the mine; it helped in regulating the supply of tubs and other work; and it was especially useful in cases of accident. Rule 28 suggested a wiring plan, and why not? Were there not, at most mines, young men learning the profession, who would derive advantage from having to make a plan of the mine, and putting upon it the details of the electrical installation? It would be very good training for them. The mine-plan should not be complicated by different colourings for electrical purposes: paper was cheap, and the time of a young student would be well employed in making a separate plan of the electrical installation.

On Section II., rule 3, Mr. Maurice remarked that "the provision of a rail in front of the switchboard would be likely to hinder the attendant from carrying out his duties. A platform sometimes gives an air of finish to a switchboard, but it is usually more inconvenient than useful. The writer has known a man to trip over the edge of a platform, and fall with his hand upon a switch."[†] Mr. Maurice shows by his own remarks the necessity for the rail being placed there: it showed its value. Mr. Maurice thought that "it is not clear why unenclosed fuses should be placed so near the floor,"[‡] but it was quite clear to him (Mr. Stokes). One day, a workman stooping to do work under an unenclosed fuse, got something hot down his neck, causing him to make an involuntary grasp at the wires. He

* *Trans. Inst M.E.*, 1904, vol. xxvii., page 297.
† *Ibid.*, page 298.   ‡ *Ibid.*, page 298.

then came to the conclusion that it would be better for the hot
material to fall on the floor than down a man's neck. "A fence
around the commutator-end of a machine" was regarded by
Mr. Maurice as "an obstacle deliberately placed in the way of
the person who is authorized, and therefore presumably competent, to attend to it;"* but it was just as much an obstacle
to prevent people from getting inadvertently into the machine :
a consideration not without importance, in view of the Workmen's Compensation Act.

Mr. Maurice thought that "to insist that the covers of fuses
must be made of 'rigid metal lined with insulating incombustible material' seems to be an unduly severe limitation of
type. If, for example, an enclosed fuse were mounted in a cast-
iron box, it would surely not be considered necessary to line
the box with 'insulating incombustible material;'"† but why
not? If the box-cover became live metal it would be dangerous; and there had been a case in which an arc was set up,
and the side of the box was burnt through. Mr. Maurice contended that the type of fuse would practically be limited "to
two or three patterns not usually found in mines,"‡ but there
was no limit to the type of fuse. Any shape could be used, so
long as it was marked with its standard capacity. A serious
accident had occurred through a man doing what Mr. Maurice
suggested a little further on: "a simple copper-wire, fastened
under two screw-heads." The man had not another piece of
"simple copper-wire," so he used a piece of bell-wire, and the
members could guess the result.

There were other points in the paper which he should have
liked, had time permitted, to deal with, in which he thought that
Mr. Maurice had taken a wrong view, probably through a mis-
understanding of the meaning of the proposed Special Rules.
In conclusion, Mr. Maurice expressed the conviction that precautions such as these under consideration should emanate from
a recognized professional institution. He (Mr. Stokes) did
not know whether Mr. Maurice meant The Institution of Mining
Engineers or the Institution of Electrical Engineers. If the
latter, he disagreed with Mr. Maurice, unless they went into

---

* *Trans. Inst.. M.E.*, 1904, vol. xxvii., page 299.
† *Ibid.*, page 301. ‡ *Ibid.*, page 301.

the mines and saw for themselves the conditions under which the work was carried on. The proposed Special Rules had been drafted by a Committee representing law, electricity, coal-owners, managers, workmen and inspectors of mines. The evidence was taken from gentlemen representing all sides of the question; Mr. Maurice, himself, gave evidence, and had every opportunity for placing his views before the Committee.

Mr. JONATHAN PIGGFORD (Teversal) wrote that he agreed with the definitions of the *Report* as to the meaning of low, medium, high and extra high pressures, and generally with Section I.; but he thought that in a dry and dusty mine it might be difficult to earth efficiently the frames of motors (rule 5 (*a*) (1)). Rule 13 provides that " conductors must radiate from distributing centres, and in large systems from those centres to sub-centres. At these centres, suitable switchboards shall be fixed, by which the pressure can be cut off from any circuit." This raises the important question as to who is to decide concerning the large systems which require sub-central switchboards : this ought, and must be definitely stated. Rule 14 is not quite clear, and if it applies to two or three core-cables, it is altogether too stringent and must be modified. Rule 21 is too stringent, for it would be quite possible to have no electricity in a mine whatsoever, except one motor of 200 brake-horsepower, and it would certainly be absurd to have a man on duty belowground just on account of this one machine.

Section II. generally is very stringent. Where the generating house is situated on the surface, no reason is perceptible why there should be Special Rules differing from those in force for ordinary power-houses. Rule 7 would be very difficult to carry out, as it practically means the division and isolation of the omnibus-bars behind the switchboard going to the various feeder-switches. Rule 9, requiring that automatic cut-outs must be placed at least 6½ feet above the flooring, should depend upon the type of cut-out. If some of the types now in use were placed at that distance from the floor, it would be very difficult to throw them in, and there is really no necessity for altering this, so long as sufficient space is left clear in front of the switchboard. Rule 10 is unnecessarily severe, and it should only apply to pressures above medium pressure (650 volts), otherwise it practically means that for any circuit above 250 volts, a high-tension type of switchboard must be installed similar to that at present supplied for 1,000 and 2,000 volts. Tramway switchboards are allowed to have live material in front; and, as these are about the same voltage (that is medium), there should be, and in fact is, no necessity to have a high-tension type of board at a colliery generating-station, where none but authorized persons are allowed to touch the boards.

Section III., rule 17 specifies that trailing cables for portable machines should be protected with either galvanized wire, or armouring, etc. In his opinion, wire-armouring for trailing cables is a source of danger, as it is apt to get damaged and the insulation is punctured, causing an earth throughout the system. Probably stout braiding, hose-pipe or leather would form a much safer protection.

Section V., rule 2: it seems quite unnecessary that every motor should be provided with an ammeter. In a three-phase system, particularly, an ammeter does not convey a true meaning, as the true power depends on the power-factor; but, even with a direct-current system, in many cases an ammeter would be quite unnecessary.

These are some of the main points which strike one as being too stringent, but there are several other matters which are stricter than is absolutely necessary, and it is highly desirable that they should be modified.

Mr. ROSLYN HOLIDAY (Ackton Hall) thought that it would have been better if Mr. Maurice had dealt with a few general principles. As an instance of what the proposed Special Rules would lead to if strictly carried out, he might mention that, if a manager wished to put an electric bell in his office, he would first have to notify H.M. inspector of mines, then he must get a certificate from the makers that the wire for the bell had been tested at 2,000 volts for 1 hour; and after erection he must test the whole circuit at 200 volts. He would be prohibited from using any staples to fasten the wire, he must show the wires on the plan, and no repairs must be done to the bell-circuit until the battery is switched off the bell-wire. He thought that the proposed Special Rules wanted reconstruction before they went further: for, as they stood, there was no exemption in the case of such things as bell-wires. He did not know why it had been thought necessary to make such stringent rules for surface-plant. Where there was gas in mines, arbitrary conditions were needed: Special Rules were made for those who would not take care, and were necessary where dangers existed under-ground. But why should Special Rules be made for surface-works, which could do no more injury than similar installations at iron- or steel-works, or railway-sidings which were not governed by Special Rules, unless they voluntarily chose to follow those of the Institution of Electrical Engineers? He had used armoured cables in his pits, and after some time he had to take them out again. Whenever the air was moist and hot, the insulation deteriorated, especially if it were damaged by a fall of rock; smouldering began, and, working its way along, a fire was soon started. When the proposed Special Rules were pub-lished, noticing that armoured cables were so strongly advo-cated, he thought that he would make another trial in a district where the conditions which he had described prevailed, and within a week the same thing happened again, and the cable took fire. He had stripped off all the armouring, and he was hoping that the Committee would reconsider their decision and look more favourably upon the use of unarmoured cables.

Mr. ARNOLD LUPTON (Leeds) said that he was much obliged to Mr. Maurice for his paper, and to Mr. Stokes for his very instructive and interesting criticisms upon it. Personally, he greatly objected to too many rules, which fettered development and invention. When people were tied down to almost innumerable rules, which they must carry out, they did not give that play to their inventive faculties which otherwise they would be stimulated to do: they learnt to rely upon rules only, and then they had accidents.

With regard to what had been said about "earthing," it was of no use talking about earthing motors, unless they stated what was the voltage to be allowed on the earthing cable, and if they had to go to the shaft-top or bottom to get a good earth. When they knew that, they could calculate the size of cable which would have to be made, in order to earth the motor-frame in case of a leak of any given fraction of the total current in the power-supply.

Mr. W. MAURICE wrote that it was a pleasure to have Mr. A. H. Stokes' criticisms and elucidations of some of the moot points in the proposed rules. The expression "competent person" was susceptible of elastic interpretation. Under the Coal-mines Regulation Acts, it had served the purpose for which it was drafted, and he was glad to have the assurance of Mr. Stokes that the proposed new "competent person" would fall into the same category.

He (Mr. Maurice) thought that Mr. Stokes was rather begging the question when he compared the enclosure of naked lights with the similar treatment of electrical apparatus. Why was it proposed to treat electric apparatus on an equality with a naked light as a potential source of danger? He (Mr. Maurice) was bound to confess that he did not know, as it was only by the simultaneous presentation of a group of abnormal conditions that the electric ignition of mine-gas could be effected. There had never been any loss of life in a mine, due to an electric ignition of gas, nor had there ever been a mine-explosion due to sparking on any electrical machine or switch; and the enquiry of the Departmental Committee on the Use of Electricity in Mines had not elicited any more information on the subject of the spark-ignition of explosive-mixtures than was known many years ago. If everything that was luminous was a light, and every naked light was set down as a source of danger, it would be logical to prohibit the presence underground of fire-flies and glow-worms. A good deal of experimental research had been conducted, which was not without a certain value, but of conclusive evidence carrying sufficient weight to justify arbitrary action there was practically none.

Mr. Stokes had asked whether he (Mr. Maurice) could cite any instance of a mining or electrical industry voluntarily introducing special rules. He knew of no other industry so extensively controlled by laws and regulations as that of mining. Yet almost every colliery had its own bye-laws, designed to meet contingencies not officially provided for. Many collieries had their own electrical rules already, and since the question had received so much publicity the necessity for compulsory action had almost passed away.

The operations of the electrical industry were entirely self-controlled in matters of detail. It had extensive rules, based on equally extensive experience, but they were not legally compulsory. It was true that the electrical industry was controlled by electric-lighting Acts and Board of Trade regulations. It was twenty-two years since the first electric-lighting Act became law. Yet, even with so long a perspective through which to take a calm view of its operation, it was still held to have done the most serious injury to a promising industry. It was not to be inferred that he (Mr. Maurice) was opposed to rules of any kind. A few broad-based rules, designed to ensure the recognition of certain fundamental scientific laws, would be extremely valuable. The present proposals, which would give to a great number of exceedingly trivial rules all the force of a legalized act, were strangely reminiscent of that legislation known as Gilbertian.

Mr. A. H. Stokes had compared brush-sparks with matches. The flame from a match might quite fairly be compared with certain heating effects of the electric current, so far as was referable to gas-ignition. He (Mr. Maurice) dissented from the view that spark and flame should be considered as interchangeable expressions. The ignition of fire-damp, which Mr. A. H. Stokes cited as having been caused by "electric sparks" was stated to have been due to "a clear case of short-circuiting, and consequent melting or fuzing away of the copper wires* by the electric current."[†] The writer does not see that it would have been prevented by any electrical rules formulated under the Departmental Committee's fourth general principle.[‡]

As to "earthing" machines, etc., he (Mr. Maurice) failed to see any evidence in the *Minutes of Evidence*, etc., indicating its necessity. "Earthing" from the point of view of an electrical engineer was, under certain circumstances, quite the right thing to do. It was academically sound, and had even been known not to fail in practice—for a short time. The question was undoubtedly one which ought to be decided by an expert, after careful consideration of local conditions.

It was contended that he (Mr. Maurice) had made out a strong case for the enactment of stringent rules applying to coal-cutters; but all the rules in the world would not stop a nut

* Coal-cutter cables.

† *Mines and Quarries: Reports . . . for the Midland District . . . for the Year 1901*, [Cd. 1062—1] 1902, page 15.

‡ *Report*, page 10.

from coming loose. Frequent, indeed continuous watchfulness was so essential a part of a machine-driver's work that it was difficult to conceive what object could be served by making such watchfulness the subject of rules.

Then, Mr. Stokes referred to a certain electrical test as analogous to tests made on boilers, cables, chains, etc. He (Mr. Maurice) imagined it was common knowledge that a great many boilers, cables, chains, etc., had been permanently injured by improper testing.

Telephones certainly did enable " the management to know what was going on in various parts of the mine;" and they were equally available for preventing that most desirable consummation.

He (Mr. Maurice) did not think that it was sound legislation to change a world-wide system of disposing fuses, simply because one man had had his neck burnt. Mr. Stokes further contended that there was no limit to the type of fuse, but everything depended upon the interpretation of the clause (Section IV., rule 3): if a fuse might be stamped anywhere on its base or cover or terminals, instead of on the fuse itself (as was only done with one design), then of course any make of fuse would be available.

In Mr. Stokes' concluding remarks, reference was again made to his (Mr. Maurice's) contention that such safety rules as were under consideration should emanate from a professional body. There should be no difficulty in forming, say, a committee of mining engineers to formulate a set of electrical mining rules and to recommend their general adoption. Many of the rules, now proposed, would doubtless find acceptance if put forward in this way. Being adopted voluntarily, they would possess rigidity or elasticity at the discretion of individual mine-managers. But, if given the force of law, they would tend to stereotype methods of work, cripple imagination, destroy individuality and initiative, and add one more influence towards the making of a nation of automatons.

He (Mr. Maurice) gave his evidence before the Departmental Committee on the Use of Electricity in Mines at extremely short notice, and without any information as to the lines of enquiry. It was not, however, the Departmental Committee, but their conclusions which were the subject of criticism.

From his analysis of the evidence it would appear that out of 56 witnesses, 35 had had no practical experience of electrical coal-cutting, 22 professed to have no personal acquaintance with electricity in mines, and 7 witnesses stated that they had not been underground for an average period exceeding 12 years. He was glad to notice that the opinions expressed by Mr. Jno. Piggford, Mr. Roslyn Holiday and Prof. A. Lupton were substantially in accord with his own views.

The PRESIDENT (Mr. W. B. M. Jackson) moved a vote of thanks to Mr. Maurice for his paper.

Mr. ARNOLD LUPTON seconded the resolution, which was cordially approved.

———

Mr. A. B. HEWITT's paper on " An Improved Roller-journal for Haulage-ropes " was taken as read as follows : —

AN IMPROVED ROLLER-JOURNAL FOR HAULAGE-ROPES.

By A. B. HEWITT.

This roller-journal is made of cast iron, with two recesses, A and B, in either of which the spindle of the roller may be placed (Figs. 1 and 2). These recesses are practically covered, so as to prevent dust and dirt from getting to the spindle, and also to prevent the roller from being knocked out by the haulage-clips of passing tubs. It will be recognized that there are two journals in one, for nearly the same first cost.

Ordinary tub-grease is used as the lubricant, whereas in the old type oil was used, at twice the cost of tub-grease : the oil being poured on the dirt surrounding the spindle, and not upon the spindle itself.

The old type of roller-journal requires oiling four times a day, as the oil runs off freely, and at each oiling ¾ ounce is used ; and with oil at 10s. 6d. per cwt., the cost is 4s. 2·6d. per roller per annum. The new type of roller-journal requires 3 ounces of grease for two bearings every 12 days ; and with grease at 5s. 3d. per cwt., the cost is 2·1d. per roller per annum. The saving is thus at least 4s. per roller per annum. So great a saving in lubricant and labour is a large item in the reduction of colliery-costs.

It will be seen from Fig. 1 that it is practically impossible for the roller to be forced out of the journals, and so to throw the tubs off the rails. The spindle of the roller is placed in the recess nearest to the travelling tubs, and in case a clip strikes the roller, the spindle is forced into the opposite recess, in the direction shown by the arrow in Fig. 1. The shock due to the clip striking the roller is greatly reduced, by its giving way to the blow, and the spindle falling into position again in the opposite recess. If placed on an incline, and the loaded tubs are travelling to the rise, the spindle of the roller is placed in the lower recess, so that if the roller be struck by a clip, the spindle will pass into the upper one and *vice versa*, if placed on the empty side. The journal is specially adapted for use on steep gradients, as the roller will not come out, even if inclined at an angle of 90 degrees. When one recess is worn out, the journal is reversed, and it has consequently twice the life of any other type.

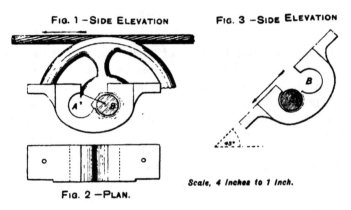

FIG. 1 – SIDE ELEVATION   FIG. 3 – SIDE ELEVATION

FIG. 2 – PLAN.   Scale, 4 Inches to 1 Inch.

This journal has been in general use at Messrs. J. and N. Nadin's Stanton colliery, near Burton-on-Trent, for the past 2½ years. Previous to this journal being introduced, the rope on the engine-plane had a life of 12 months, but with the new journal, the present rope has been working 24 months. The engine-plane at Stanton colliery is 4,500 feet long, and there are 200 rollers. Formerly, a man, paid 5s. 6d. per day, did nothing else but oil the rollers four times a day. He would travel 3½ miles per day and carry (200 rollers by 3 ounces equals) 600 ounces or 37½ pounds of oil. The lubrication, with the new type of journal, is done once every 12 days or 20 times per year. The rollers under the old system were not oiled efficiently, because the oil ran off after the first few revolutions. The man who attends to the greasing of the rollers has now 220 spare days per year to relay rails, etc., and help coal-turning in many ways.

The cost at Stanton colliery had been as follows:—

(*a*) Old type of roller-journal: the cost of oil for 200 rollers, at 3 ounces per day for 240 days, or, say, 80 cwts. at 10s. 6d. per cwt., was £42 per annum; the cost of labour, 1 man for 240 days at 5s. 6d. per day, was £66 per annum; and the total cost was £108 per annum.

(b) New type of roller-journal: the cost of tub-grease for 200 rollers, at 3 ounces every 12 days or 20 times per annum, or, say, 6½ cwts. at 5s. 3d. per cwt., is £1 14s. 2d.; the cost of labour, 1 man for 20 days at 5s. 6d. per day, is £5 10s.; and the total cost is £7 4s. 2d. per annum.

The saving per annum is therefore £100 15s. 10d., for lubricant and labour; and, inclusive of a new rope, it is £200 15s. 10d.

Not a single accident has been known to occur on this engine-plane through a roller being forced out, since the new type of journal has been adopted; but with the old journal as many as three such accidents were experienced per day.

THE INSTITUTION OF MINING ENGINEERS.

GENERAL MEETING,

HELD IN THE ROOMS OF THE GEOLOGICAL SOCIETY, BURLINGTON HOUSE, LONDON, JUNE 2ND, 1904.

MR. J. C. CADMAN, PRESIDENT, IN THE CHAIR.

PRIZES.

The SECRETARY reported that the Council had awarded prizes of books to the writers of the following papers, which had been printed in volumes xxiv. and xxv. of the *Transactions*:—

"The Fernie Explosion." By Mr. W. Blakemore.

"Sinking by the Freezing Method at Washington, County Durham." By Mr. Mark Ford.

"Miners' Anæmia, or Ankylostomiasis." By Dr. J. S. Haldane.

"The Marl-slate and Yellow Sands of Northumberland and Durham." By Prof. G. A. Lebour.

"The Occurrence of Gold in Great Britain and Ireland." By Mr. J. Malcolm Maclaren.

"The Probability of finding Workable Seams of Coal in the Carboniferous Limestone or Bernician Formation, beneath the Regular Coal-measures of Northumberland and Durham, with an account of a recent Deep Boring made, in Chopwell Woods, below the Brockwell Seam." By Mr. J. B. Simpson.

Mr. J. C. CADMAN delivered the following "Presidential Address":—

PRESIDENTIAL ADDRESS.

By J. C. CADMAN.

In taking the chair as your President, I feel sure that you will appreciate my feelings of doubt in being able to fulfil the duties in the same efficient manner as did so many of the eminent men who have in previous years occupied this chair; however, with the help of the Council, I will endeavour to conduct the business in such a way that the usefulness of our Institution shall not suffer. Allow me again to thank you sincerely for the very great honour which you conferred upon me when electing me to the position of President.

In the first place, I must allude to the serious loss that the Institution and the mining world in general have sustained in the deaths of Mr. Henry Aitken and Sir Clement Le Neve Foster. It was only at the Council Meeting in February last, that Sir Clement Le Neve Foster was unanimously elected an Honorary Member of the Institution. Sir Clement Le Neve Foster deserves more than a passing word: he held a very high position, was a great authority in all mining matters, and contributed largely to mining literature. He was ever ready to give help and advice, and it was whilst assisting at the Snaefell calamity that he was overcome by carbon monoxide, the effects of which ultimately contributed to his death. A letter of sympathy to his widow and family had been forwarded on your behalf.

The amalgamation of local institutions had now reached its fifteenth year; such a federation had been frequently suggested by members of the various institutes, its birth, however, took place after the late Mr. Theophilus Wood Bunning read his paper in 1887 before the North of England Institute of Mining and Mechanical Engineers, and it was circulated amongst all the mining societies of Great Britain. After several meetings presided over by Sir Lowthian Bell, the inaugural general meeting was held on January 22nd, 1890, under the presidency of the late Mr. John Marley. The number of members at that time was 1,239; and to-day it stands at 2,688, including a considerable number of colonial and foreign members. The Institution comprises the following:—The North of England Institute of Mining and Mechanical Engineers; the Midland Counties Institution of Engineers; the Midland Institute of Mining, Civil and Mechanical Engineers; the Mining Institute of Scotland; the North Staffordshire Institute of Mining and Mechanical Engineers; and the South Staffordshire and East Worcestershire Institute of Mining Engineers.

No less than 79 papers have been read and discussed during the past year, and notes of 120 papers from colonial and foreign societies are also printed in the *Transactions*. These figures show that the Institution has been attended with success.

I am pleased to note and welcome the valuable addition to our ranks of the Manchester Geological and Mining Society. It is to be hoped that the remaining societies who have not yet joined the federation, will ere long see their way to do so. I particularly refer to the South Wales Institute of Engineers, and the Institution of Mining and Metallurgy, who are both doing admirable work. These additions to our ranks would not only increase the value of our *Transactions*, but would enable their papers to become of national use.

The progress of the Institution has advanced with great rapidity, the *Transactions* have increased in number and value, not only by papers read by home members but by papers from colonial and foreign members, so that the time is not far distant when the original proposals laid down by the late Mr. Theophilus Wood Bunning appear likely to become, not only desirable but essential, namely, the further enlargement and formation of an Imperial Institution of Mining Engineers with its headquarters in London. I look forward to the time when our Institution will be to the mining world what the Iron and Steel Institute is to the iron and steel industry. I also see no reason why the *Transactions* should not be issued to the members with the same regularity and on the same lines as in the case of the institution referred to.

Two reports of the Royal Commission on Coal-supplies, together with minutes of evidence, have been issued, and it is with interest that we await the final report of the Royal Commissioners. It will in all probability show that there is an increased quantity of available coal remaining to be got, compared with that stated in the report of the last Royal Commission, 33 years ago, on account of the considerable extensions which have been discovered in many of the coal-fields. I would like here to refer to the valuable work done by the Geological Survey in this direction, and I am sure that the publication of the Geological Survey maps of the coal-fields on the scale of 6 inches to the mile will be welcomed by all mining engineers, particularly those working on the borders of the various coal-fields.

Although the work of the Royal Commission is very necessary and valuable, there is one important point which must not be lost sight of, namely, the cost of production of coal from seams lying at great depths, of small thickness, and of steep inclination; for I feel convinced that a time will arrive' (although it may be yet far distant) when coal will be imported into this country, at a less cost, from existing and unproved fields abroad.

In order to delay this period, a more economic use of fuel is desirable; however, much is now being done to increase the value and useful effect of small coal by the adoption of extensive washing-plants. As an example, by the sorting, sizing and washing to a minute degree, I have known cases where the value of small coal has been increased by 200 per cent., with a corresponding increase in the demand; and where possible, with a further refinement by the use of bye-product coking plants, this profit may be still further increased. This is a serious question at collieries working soft bituminous coals, yielding large percentages of small, and the necessity of the equipment of such a plant is very obvious.

In my own district of North Staffordshire, very few, if any, attempts have been made to effect a saving in fuel in the manufacture of pottery, since the early days, when it became the staple trade of the district; and practically the same construction of ovens and methods of firing are still in vogue, with the

result that the loss of heat and the consumption of fuel continues
to be very great. Several attempts have been made, in a more
or less crude fashion, to use gas as a fuel, but generally they
have proved unsuccessful. I have no doubt, however, when
Mond producer-gas or other gas, is available and properly
applied, that successful results will follow; and I understand
that gas has been satisfactorily applied to similar purposes
in Germany.

These remarks, with reference to the manufacture of china
and earthenware, will also apply to the manufacture of other
articles, where heat is essential; and gas-firing has so many
advantages that its economic usefulness must receive that
serious treatment at the hands of manufacturers which it
deserves.

The Report and Minutes of Evidence of the Departmental
Committee on the use of Electricity in Mines, recently issued,
fully shows the importance to which electricity has already
attained as a motive power in mines, and the very large part
that it is likely to play in the mines of the future. It is pleasing to note that the Committee did not recommend any restrictions that might be likely to hamper the development of a source
of power, which bids fair to become an essential feature in the
economy of mining operations. The part that electrical power
will play in the future, owing to its adaptability and efficiency
as compared with other means of transmission of power, will
no doubt be revolutionary. Without wishing to be prophetic,
I anticipate the adoption of central power-stations, comprising
gas-engines, driven by producer-gas, and working generators,
from which electric power can be distributed throughout a
colliery, doing away with steam-boilers and engines and removing from the landscape the gigantic chimney-stacks—the land-
mark of every extensive works.

The country is now awakening to the necessity of more
technical and scientific education, to enable her to compete with
other nations. I note, with pleasure, that great advancement
has been made during the last few years in this direction, by
the opening of colleges and technical schools; and the arrangement of classes by County Councils and other bodies in nearly

all the mining centres of Great Britain. All such have received considerable support and encouragement from this Institution. This is, however, a work, which is more or less of a national character, and much more help should be given by the Government: it might even be suggested that a moiety of the tax levied upon the export of coal might very fittingly be allotted for this object.

With the everyday increase in the necessity of technical education for mining engineers, it is with pleasure that I refer to the Act, recently brought into force, allowing two years of the time spent at a recognized college to count in the statutory period required by the Coal-mines Regulation Acts in qualifying mine-managers for the certificate of competency. Whilst it is most imperative that time spent in obtaining practical experience should take up the major portion of the education of a mining engineer, it is also absolutely necessary that the student of to-day should receive a thorough technical training in the sciences applied to mining, because the engineer of the future will be required to undertake and develop problems (owing to increased depth and extension of workings) of such a nature, that he will be totally unfit to design and carry them out, unless he has received an extensive scientific education in conjunction with practical work.

At the present moment, our miners and, consequently, the mining industry of this country are threatened with a formidable and loathsome disease. I refer to ankylostomiasis, but thanks to the energetic action of Dr. J. S. Haldane and Dr. A. E. Boycott, who discovered the existence of the disease in Cornwall, in bringing before us the necessity of sanitation in our mines, an outbreak may yet be prevented. Should ankylostomiasis find its way into the coal-mines of this country, the result would be most serious to the workmen and the coal-industry. The question of sanitation is one which hitherto has received very little, if any attention, and the time is now ripe for some systematic steps to be taken. Valuable reports have recently been issued by the Home Secretary, containing further informa-as to the character of the disease, its mode of detection and treatment, together with hints and suggestions to be adopted for stamping it out. It is also necessary for local medical men, whose work lies amongst the mines, to get well posted up in the symptoms, so as to be able at once to diagnose the disease and so prevent a similar out-break to that now prevalent in Cornwall.

The quantity of coal mined in the United Kingdom still continues to increase, the returns for 1903 showing an output of 230,323,391 tons, an increase of 3,238,520 tons as compared with the output for 1902. The total number of persons employed was 842,066, an increase of 17,275 as compared with 1902. The statistics for 1903 show that, although there was a considerable increase over the whole of the United Kingdom, there was, nevertheless, a considerable decrease in the two midland mines-inspection districts. This decrease is probably due to two causes, (1) to the fact that the whole of the coal has to be carried by rail; and, therefore, owing to the high rate of carriage, it cannot compete with other districts having the benefit of the cheaper rates of shipment; and (2) to the somewhat higher cost of production over these areas. A few years ago, the question of railway-rates was discussed by the midland coal-owners with the railway-companies, at a joint meeting held in London; the falling off in the quantity of coal delivered to London and the south was pointed out and suggestions made that the comparative rate between railborne coal and seaborne coal should be revised, so as to place each district on about the same basis. The railway-companies, however, refused to entertain any such revision, with the result that a still further relative decrease has occurred.

The total deaths due to fatal accidents for 1903 was 1,067, showing an increase of 43 compared with 1902: these figures include 566 deaths from falls of roof and side, being an increase of 114. What had brought about this increase? During this period, the rules for systematic timbering have been in force, and a reduction in the number of accidents was naturally anticipated. No doubt, the heavy rainfall experienced during last year had been an agent in some mines in producing falls, by finding its way into fissures and joints, and bringing about disintegration of the roof and sides, thus leaving them more susceptible to falls.

While mining engineers and managers are doing all they possibly can to reduce the number of accidents, by the adoption of Special Rules and safety-appliances, they are receiving, in some cases, very poor encouragement from the administrators of the law. The way in which some local magistrates override the provisions of the Coal-mines Regulation Acts had recently been brought under notice; I refer to a case in which a magistrate stated that "to search a man for matches below-ground was tantamount to highway robbery" and dismissed the case, although the search actually resulted in the discovery of matches on twenty-six men. In another case, in which a miner was prosecuted by a manager for a breach of the Special Rules with reference to timbering, the magistrate in dismissing the charge stated that "the collier is the best judge where spraggs ought to be used." Such decisions cannot be allowed to stand, otherwise the enforcement of the Coal-mines Regulation Acts and Special Rules would become an impossibility. It is to be hoped that the proper authorities will note such cases.

In conclusion, the records show that the Institution continues to progress and its sphere of usefulness to extend. The members should not, however, stop to think of limits, until the Institution has spread its wings to the extent of an Imperial Institution, that may be of utility to the whole of the mining community over which the British flag flies.

———

Mr. M. H. MILLS (Mansfield), in proposing a vote of thanks to the President for his address, hoped that some of the remarks would take root and prove of advantage in the future. Report had said that we had come to the time when more technical knowledge was required; that Germany was going ahead and so on; but only a few weeks ago some of the principal coal-owners of Germany came over to England, and he had the pleasure of taking them to see some of our largest collieries. These gentlemen came, no doubt, because they had much to learn from us.

Mr. J. G. WEEKS (Newcastle-upon-Tyne), in seconding the vote of thanks, felt sure that the President would forgive him for stating that the proposal to provide for the future education of mining engineers in this country from the proceeds of an export-tax on coal was not one which he must expect to be gratefully received by every one present. If it was necessary for the country to be taxed for any such doubtful purpose, it should be on a very much wider basis than that of export-coal. Why should a tax not be imposed on the production, not only of all coal, but also of corn, iron, steam-shipping and everything else? He failed to see why the unfortunate colliery-owner who was an exporter of coal should be alone called upon to bear this unfair burden.

The PRESIDENT, in acknowledging the vote of thanks, said that he did not for a moment suggest that any further tax or additional charge should be put upon export-coal; but, so long as the present duty was imposed, he did not see why a moiety of it should not be allocated for the advancement of mining, since from the mining industry it was derived.

————

The members then divided into two sections for the reading and discussion of papers. The President (Mr. J. C. Cadman) presided over one section and Mr. H. C. Peake (past-President) over the second section, in the rooms of the Royal Astronomical Society, Burlington House.

————

Mr. G. A. GREENER's paper on "The Coal-fields of the Faröe Islands" was read as follows:—

THE COAL-FIELDS OF THE FARÖE ISLANDS.

By G. A. GREENER.

Introduction.—The Faröe Islands are a group of 23 islands belonging to Denmark, situated in the North Atlantic Ocean about 180 miles from the extreme northern end of Scotland and on the direct steamer-route to Iceland (Fig. 1, Plate XVII.).

A very small area of the land is cultivated, that surrounding

FIG. 4.—THE WESTERN COAST, NEAR QVALBÖ.

the fjords, for very little soil covers the rocks on the hills. The inhabitants live chiefly by fishing and sheep-farming, and spin most of the wool into clothing with the old-fashioned spinning-wheel. At the present time, the number of inhabitants on the island of Suderö is 3,100.

Coal has been found on four of the islands, namely:— Myggenaes, Gaasholm, Vaagö and Suderö. The deposits are not extensive, except those in the last-named island.

The Faröe Islands are of volcanic origin.   About 10 miles west-south-west of Suderö is the Faröe bank, a large level plateau covered by about 50 to 70 fathoms of water.   On the eastern side of this bank, about 8 miles south 62 degrees west of Suderö, there is a great depression over 200 fathoms deep, and between the plateau and the island are vertical rocks covered by 80 to 100 fathoms of water.   This appears to have been the site of the volcano, and its ejected lava formed the basalt-rock, which is the seat or foundation of the coal-formation.   The ejection of lava from this volcano ceased, the

FIG. 5.—THE WESTERN COAST, NEAR FAMOYEN.

basalt was covered with clay, and plants grew upon this clay, where their remains are now found in the form of coal.   Later, this volcano resumed its activity, and the whole of the islands were covered by immense sheets of lava.

It is probable that this group of islands, at one time, formed one stretch of land, and that by denuding agencies, including inland ice, similar to that found in Greenland, and the action of water on the softer parts of the strata, the fjords, valleys, etc., and the islands have all been carved into their present form. Suderö is the southernmost of the group, and lies between

61 degrees 25 minutes and 61 degrees 42 minutes of north latitude, and about 7 degrees west of Greenwich. This island is about 20 miles long and 7 miles broad in its widest part (Fig. 2, Plate XVII.).

The climate is very mild, owing to the influence of the Gulf Stream, snow seldom lies for more than a few days, and the fjords are never icebound, being open all the year round.

On the western coast, the mountains are very precipitous, rising sheer out of the water to heights of over 1,000 feet, and presenting cliff-scenery of the grandest description (Figs. 4, 5 and 6).

FIG. 6.--THE WESTERN COAST, NEAR SUMBÖ.

On the eastern side, the coast is somewhat similar, but it is indented with several deep inlets or fjords, making excellent harbours. Trangisvaag fjord (Fig. 7) is a splendid deep-water haven, with good anchorage, and being surrounded by high mountains, affords a shelter for vessels, which can hardly be surpassed. Approximately, it is 4 miles long by 1 mile broad, and the 3 fathoms line lies within a short distance of the shore. The coal-seams occur in the mountains surrounding this fjord and that of Qvalbö, in the north of the island.

Coal-field.—The area of the coal-field is shewn on the map of Suderö (Fig. 2, Plate XVII.). There is one workable seam extending over an area of more than 6,000 acres of this coal-field, and there is evidence of the occurrence of other seams. Table I. contains representative sections of the seam at present being worked, and the positions are indicated by the respective numbers on the map (Fig. 2, Plate XVII.).

The coal-seam lying south and north of Trangisvaag fjord is one and the same; the intervening space forming the fjord having been washed away by the action of land-ice and water.

TABLE I.— REPRESENTATIVE SECTIONS OF THE COAL-SEAM.

1.—AXELN MINE, TRANGISVAAG, 350 feet above sea-level.

Roof:

	Ft. In.	Ft. In.
Brown clay	—	—

Seam:

	Ft. In.	Ft. In.
COAL, bright ...	1 9	
Brown clay	0 2	
COAL	1 5	
Clay	0 9	
COAL, bright ...	0 10	
Shale	0 1	
COAL	1 3	
		6 3

Thill:

Fire-clay	6 0	
		6 0

2.—VEST-I-SKAAR, about 400 feet above sea-level.

Roof:

	Ft. In.	Ft. In.
Brown clay	20 0	
COAL and clay, mixed	3 0	
		23 0

Seam:

COAL, bright ..	1 11	
Brown clay	1 0	
COAL	1 5	
White clay ...	0 4	
COAL, bright ..	1 0	
COAL, mixed, bright and dull	2 0	
White clay	0 3	
COAL, bright ...	1 6	
		9 5

Thill:

White clay	—	—

3.—NORD-I-WEST-I-SKAAR, about 400 feet above sea-level.*

Roof:

Brown clay	12 0	
		12 0

	Ft. In.	Ft In
Seam:		
COAL	0 6	
Brown clay ...	2 0	
COAL, bright ...	0 8	
Brown clay	9 0	
COAL, bright ...	1 0	
Brown clay ...	2 2	
COAL, bright ..	2 6	
White clay ...	0 2	
COAL, bright ...	0 5	
Brown clay ...	0 8	
COAL	0 9	
		19 10

Thill:

White clay ...	2 0	
Basalt	30 0	
COAL, bright ...	1 6	
White clay	3 0	
Basalt, down to sea-level	330 0	
		366 6

4.—QVANHAVE, about 20 feet above sea-level.

Roof:

Brown clay	—	

Seam:

COAL, bright ..	1 3	
Brown clay	0 4	
COAL, with thin bands of clay ...	2 6	
COAL, bright ..	0 6	
COAL	0 6	
Brown clay ...	0 8	
COAL, bright ...	1 0	
		(

Thill:

White clay	—	—

* This section was taken with extreme difficulty: the writer was lowered over the cliff by a rope held by men at the top, being preceded by a Faröeman, a noted mountain-climber, and by Mr. S. Lind, overman at the mine. The first seam was reached about 650 feet from the top.

TABLE I,—*Continued.*

5.—FRODEBÖ, at sea-level.†

	Ft. In.	Ft In.
Roof:		
Brown clay	10 0	
		10 0
Seam:		
COAL	1 3	
COAL, bright	0 11	
		2 2
Thill:		
Brown clay	5 0	
		5 0

6.—STUGGESENDE, at the outcrop, about 340 feet above sea-level.

	Ft. In.	Ft In.
Roof:		
Clay with ironstone-balls	10 0	
Light-brown clay	3 0	
Light-grey clay	0 7	
		13 7
Seam:		
COAL	0 8	
COAL, bright	0 7	
White clay	0 3	
COAL, bright	0 6	
		2 0
Thill:		
White clay	—	—

7.—KING'S MINE, ORNEFJELD, SOUTH OF TRANGISVAAG FJORD, about 740 feet above sea-level.‡

	Ft. In	Ft. In.
Roof:		
Clay with woody *coal*	0 6	
		0 6
Seam:		
COAL, bright	0 6	
Clay	0 2	
COAL	1 6	
Brown clay	0 1	
COAL	0 8	
		2 11
Thill:		
White clay	—	—

8.—RANGABOTN MINE, about 820 feet above sea-level.§

	Ft. In.	Ft. In.
Roof:		
Brown clay	20 0	
		20 0
Seam:		
COAL, bright	2 6	
Shale	0 2	
COAL	0 6	
Brown clay	0 3	
COAL	1 3	
Thill:		
White clay	—	—

The section (1) at the Axeln mine is taken from the seam now being worked by the present holders of the concession, and is 350 feet above sea-level. The seam dips from 3 to 4 degrees in a north-easterly direction, and underlying it is a bed of fire-clay, 6 feet thick.

From Stuggesende (6) to the western coast (2), the coal-seam runs fairly level, but from Stuggesende (6) to the eastern coast (5), it changes its level and falls towards Frodebö (5), as shewn in Fig. 3 (Plate XVII.). The rock underlying the coal is basalt, in regular columnar masses (Fig. 8).

Overlying the coal, the rock is rather similar, but it does not occur in so regular a manner (Fig. 9) and is a dolerite. It offers less resistance to water and the action of the atmosphere, and seems to be of a somewhat porous nature.

On the southernmost point of the island of Suderö, copper is found in its native form, as thin layers embedded in a hard

† This section can hardly be said to be a true one, as the strata at this point are distorted and fallen, owing to the action of the sea.

‡ This section was taken about 90 feet from the open-cast.

§ This section was taken about 90 feet from the open-cast or outcrop.

test of the same outcrop coal yielded:—Fixed carbon, 62·82 per
cent.; volatile matter, 31·37 per cent.; and ash, 5·81 per cent.
A later analysis shews that the coal is a lignite of Tertiary
age containing 50 per cent. of fixed carbon, 35 per cent. of
volatile matter and 15 per cent. of ash. The following are the
calorific values:—Ordinary coal, 9·35 pounds of water; and
bright coal, 9·60 pounds of water per pound of coal.

FIG. 9.—BLAA-FOS.

Markets. — A
considerable de-
mand for this coal
is already assured
locally; and, no
doubt, Suderö will
become a local
centre for coaling
for the large fleet
of British steam-
trawlers, and Nor-
wegian steam fish-
ing - boats and
whalers, the Faröe
Islands being one
of their centres of
call. The coal
may also be
shipped to Den-
mark, Norway,
Sweden and Ice-
land, and as a set-off
against the higher freight, which may have to be paid from
the Faröe Islands to European ports, it should be remembered
that, as the islands belong to Denmark, Faröe coal will be
admitted into that country free of the import-duty of about
1s. 5d. per ton; and it will also escape the British export-duty
of 1s. per ton. Denmark imports more than 1½ million tons of
coal annually.

The photographs reproduced in Figs. 4, 5 and 6, were taken
from a small rowing-boat, at different points, on the western

coast of Suderö. The Trangisvaag fjord (Fig. 7) is a splendid harbour, and there is a natural breakwater at the entrance. The surrounding hills form a further shelter for the protection of vessels and during the severe storms, which are prevalent, many gunboats and fishing-trawlers make use of the harbour. The basalt formation, underlying the coal-seams, is shewn very clearly in Fig. 8, and the dolerite formation, above the coal-seams, in its broken and irregular formation, is shewn in Fig. 9. This view was taken at the Blaa-fos, near the Blaa-fos mine.

APPENDIX.—HISTORY OF COAL-MINING IN THE ISLAND OF SUDERÖ.

In 1673, Lucas Debe mentioned that coal existed at one place, difficult of access. In 1709, Commander Yule from Denmark, while making a map of the Faröe Islands, found coal on the southern shore of Qvalbö fjord. In 1723, the Danish Government proposed to examine the Faröe coal-formation, but the project fell through.

In 1733, a company was formed with the same object, and an Englishman, who was sent to Suderö, discovered coal under the Ornefjeld, near Ordevig. A shaft, 6 feet square and 18 feet deep, was sunk on the plateau, but this attempt was abandoned. The site of this shaft can still be seen, but it has fallen in.

The question was again discussed in 1756, but the Governor at that time did not consider himself capable of judging as to the quality of the coal; there was no one who understood the method of working it; and, in his opinion, it would delay the Government ships too long to go to Suderö and load coal there.

In 1777, a mines-directorate was formed, to undertake and carry out thorough investigations. Dr. Henckel was put in charge, and he found coal on Myggenaes, Gaasholm, Vaago and Suderö. Together with a miner named Kuster, he tried the coal at these places, but found that only the coal of the Ornefjeld, near Ordevig, was workable. He sent samples of the coal to Copenhagen, and they proved to be good and free from sulphur. In 1778, Dr. Henckel was again sent with two workmen from Kongsberg, and he continued until 1779 searching for a seam favourable for regular working. In addition to these two men, he employed several Faröemen, who caused him much trouble by their independent ways and dislike of regular work. He drove eight drifts, 6 to 8 feet wide and 6 to 8 feet high, at Olafesende, north of Famien. A drift was driven on the north-western side of the Ornefjeld, near Ordevig, and a number of trials were made in the bed of a rivulet on the northern side of the Trangisvaag valley. Trials made in many other parts of Sudero proved unsuccessful. In 1780, Dr. Henckel returned home, and left the two Kongsberg workmen behind. They continued the mine in the Ornefjeld, but, as the coal-seam was very irregular and difficult of access, it was ultimately abandoned. They next began a drift in the Praesterfjeld, in the Qvalbo valley, and here work was continued until 1797.

In 1800, Mr. Landt gave the following information respecting the Qvalbö mines:—"Here are fifteen winnings which rise to the south-west. The breadth of these winnings is at first 2 to 2½ alen [4 to 5 feet] but they widen out to 5 to 7 alen [10 to 14 feet]. This last breadth proves to be too great. They are from 2½ to 2½ alen [4½ feet to 5 feet] high." He records the lengths of several of the mines, of which many were closed and fallen in.   He states that "the reason why work goes so slowly is that the mines are so low, and a man cannot use his whole strength on his pick.   The daily product of two men rarely exceeds 3 tondes [about ¼ ton] of clean coal ; but as the pick-shafts have up to now always been of firwood, because shafts of other wood are not provided, and these shafts are continually breaking during work, the work is often stopped while new shafts are being fitted ; if this difficulty is to be overcome, better shafts of birch, beech or ash must be procured ; or the eye of the picks must be made larger, and iron strengtheners put on the shafts.   If some men were to work in the coal and were paid 8 to 12 öre [1d. to 1½d.] per tonde and other men putting at 33 öre [about 4d.] the day, it would well repay the employers."   He added that "the coal is brought down by horses and carts, and they can bring down a daily quantity of 9 tondes [1½ tons].   One can calculate, therefore, that when work is being carried on partly by Norwegians, who must have 48 öre [6½d.] per day, and partly by natives who should have 32 öre [4½d.] : one tonde of coal, at the strand, will cost 33 to 40 öre [4d. to 5½d.] without taking into consideration wear-and-tear, repairs, stores and pay to officials.   If the coal-mines in Faröe are to go successfully, and on a large scale, 50 workmen are necessary ; but where the 50 men are to come from is a serious question, not easy to answer owing to the foolishness and mischievousness of the Faroemen."

In 1798, the mines were transferred to the Welfare Company, who, however' abandoned them in 1804.

From 1804 to 1827, coal was only worked in very small quantities for household use by the inhabitants of Qvalbo.

In 1827, work was recommenced by the Government in the Qvalbo mines, with disappointing results.   In 1829, 2,206 tondes were sold in Copenhagen at 14 marks a tonde [£1 16s. 3d. per British ton], but later the coal was sold for only 4 marks [about 10s. per British ton].   In 1839, work was stopped, leaving 8,000 tondes in Suderö, as at such low prices the sales would not repay the transport-expenses.

In 1840, the Peninsular Steam Navigation Company were refused the concessions by the Danish Government.   Since then and until recently, work had been carried on at the Qvalbo mines by the inhabitants for household purposes.

At the present time, the Qvalbo mines are worked by the inhabitants in a very primitive fashion.   The mines resemble rabbit-burrows, being driven in a zigzag and irregular manner : sometimes 3 feet and at other times 14 to 15 feet wide, and one mine is frequently "holed" into another.   Some of the mines have been driven over 300 feet from the outcrop, without any timber being used, and it is more by good luck than good management that the roof has not fallen in.

The pick used has only one point, which is very seldom sharpened.   Light is obtained from home-spun wool, hanging out of the corner of a square tin, filled with whale-oil, which gives off a most noxious stench.   The coal is carried from the mines on the backs of both men and women in wooden boxes or *lobes :* a broad piece of canvas being attached to the *lob* and passed across the forehead.   Some of the men can thus carry 250 pounds down the hill-side.

———

Mr. M. H. MILLS (Mansfield) asked whether the lavas had any injurious effects on the coal found in the Faröe islands.

To illustrate Mr G. A. Greener's Paper on "The Coal-fields of the Faröe Islands."

FIG. 1.—MAP OF WESTERN EUROPE.

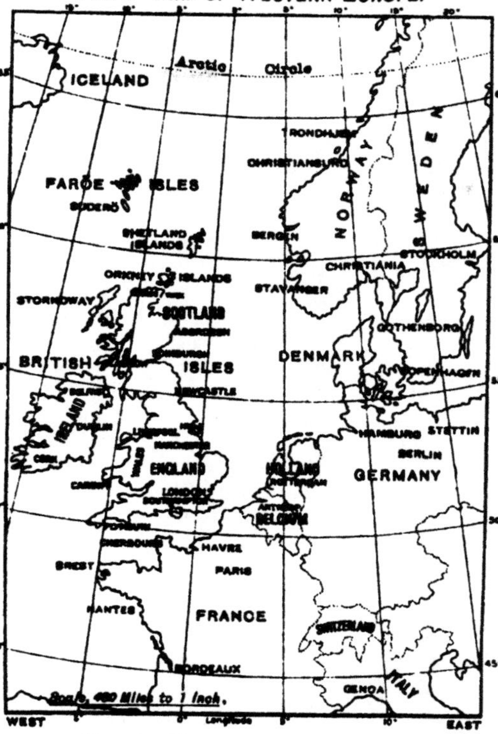

FIG. 2.—MAP OF SUDERÖ.

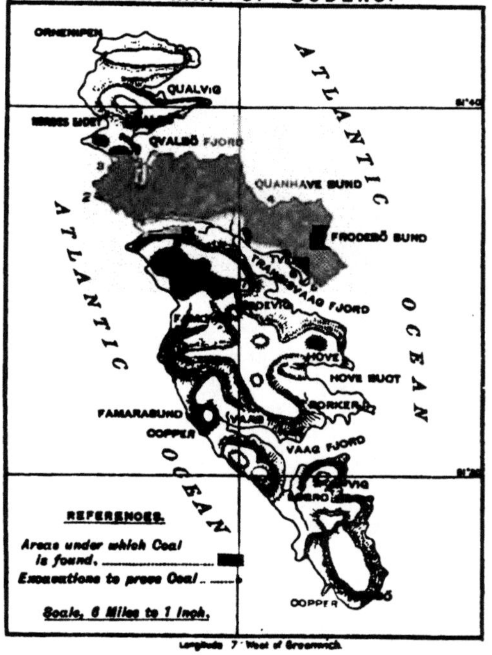

FIG. 3.—SECTION OF COAL-FIELD.

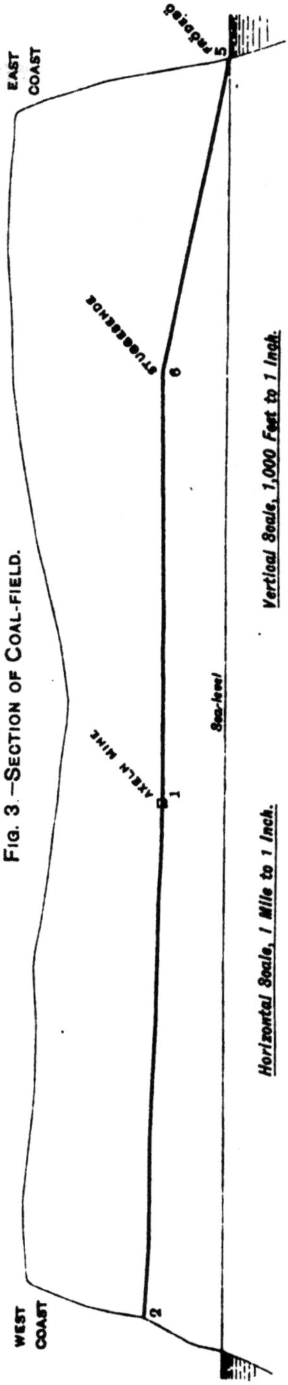

Mr. HENRY RICHARDSON (Westminster) asked whether any difficulty was experienced with regard to the ignition of the coal. He was told by a gentleman who had visited the mines, when they were only worked on a very small scale, that he had tested the coal and found that it only ignited on admixture with other coal which he had on board his vessel. Was it likely that this was due to the samples being taken from near the outcrop, or was it due to the lavas? He had pleasure in proposing a vote of thanks to Mr. Greener for his paper.

Prof. HENRY LOUIS (Newcastle-upon-Tyne), in seconding the vote of thanks, said that 20 per cent. of oxygen in the coal struck him as being so utterly abnormal that he would like to know how the analysis was made, and who was responsible for it. If the analysis was correct the stuff was not coal, whatever else it might be. In all his experience, the only coal anything like that of this district was on the Lofoten islands (off the Norwegian coast) where a good deal had been said as to the discovery of workable seams of coal, but on trials being made they proved to be thin seams, so patchy and unevenly distributed that economical working was impossible. He did not suggest anything of the kind in connection with the district described in the paper, but he would like to know what area of coal had been proved, not only by bore-holes but by continuous working. Also, could the writer say what was the geological age of the coal-seams?

The REV. J. M. MELLO (Warwick) wrote that the coal, a lignite of Miocene age, enclosed amongst the doleritic rocks of the islands, was of course wellknown, and had been described by various writers, a good many years ago. Amongst others by Roth in " The Faröe Islands and their Coal-beds." Messrs. Beghin and Ch. Mène also gave an analysis of the Suderö coal,[*] and from its composition arrived at the conclusion that it is of Tertiary age, as did Prof. Mohr who described it as of Miocene age. It is somewhat remarkable that a lignite of so comparatively recent origin should have become even partly converted into anthracite. In No. 3 section, Mr. Greener showed a seam of coal, 18 inches thick, not only resting on a great thickness of

---

[*] *Comptes-rendus hebdomadaires des Séances de l'Académie des Sciences*, 1875, vol. lxxx., page 1404.

basalt, but also overlain by 30 feet of basalt. He (Mr. Mello) asked whether the igneous rock was in direct contact with the coal; and if so, what had been the effect upon the coal? One would naturally suppose that, if an overflow of molten lava had taken place, it could not have done so without producing a marked effect upon the underlying seam, and he (Mr. Mello) would like to know whether this had been observed.

Mr. M. WALTON BROWN (Newcastle-upon-Tyne) wrote that the geology and mineralogy of the Faröe Islands had been described by Mr. J. Durocher.[*]

The vote of thanks was cordially approved.

Mr. G. A. GREENER, replying to the discussion, said that the effect of the lava was very distinct; in several places, the coal was burnt. The ignition of the coal was the opposite of that to which Mr. Richardson had referred: the mineral was too quickly ignited, if anything, and that was why an admixture had to be made. As regards the area of coal worked, there were two drifts, respectively 550 and 480 feet long; to the left four places were set away, averaging about 100 feet in length, while those on the right were 80 feet long. He was not prepared to state the geological age of the Faröe coal-field.

———

Mr. W. T. SAUNDERS' paper on "Tin-mining in the Straits Settlements," was read as follows:—

———

[*] "Recherches sur les Roches et les Minéraux des Îles Féröe," *Annales des Mines*, series 3, 1841, vol. xix., page 547; "Notice Géologique sur les Îles Féroe," *Ibid.*, series 4, 1844, vol. vi., page 437.

TIN-MINING IN THE STRAITS SETTLEMENTS.

By W. T. SAUNDERS.

About 60 per cent. of the world's supply of tin comes from the Straits Settlements.

Alluvial Deposits.—For years past, most of the tin has been obtained from alluvial deposits; and, while there appears to be a never-failing quantity of this for treatment, much attention is now being given to lode-mining.

Alluvial mining is carried on almost entirely on the western side of the peninsula, although large bodies of alluvial tin are known to exist on the eastern side. Some of these eastern areas in the state of Pahang have recently been leased from the Government for the purpose of extensive working.

Lodes.—Lode-mining is in operation on the eastern side. The principal mines belong to the Pahang Corporation, Limited; the Pahang Kabang, Limited; the Royal Johore; and the Bundy.

The most important mines are wrought by the Pahang Corporation, Limited, and the Pahang Kabang, Limited. These mines are situated about 40 miles up the river Kuantan from the port of Kuantan. Transport is carried on by steamer from Singapore to the port of Kuantan—a distance of about 225 miles—and thence to the mines by boats worked by Malays. The river being shallow, with numerous rapids in the upper reaches near the mines, this transport constitutes a difficult and somewhat costly business, as the boats are necessarily small (not carrying more than 5 tons of cargo) and taking about 4 days to reach the mines. The freight from Singapore (whence all stores, machinery, food, etc., are shipped) to Kuantan costs about 14s. per ton and boatage thence to the mines, costs about 17s. per ton.

At present, the principal workings of the Pahang Corporation, Limited, are confined to four lodes, known respectively as Willink's, Nicholson's, Bell's north and Bell's south, worked from Willink's shaft, sunk to a vertical depth of about 350 feet below adit-level.

Several other lodes have been developed, notably Pollock's and Jeram-Batang. A vertical shaft, about 520 feet deep, has been sunk on the former lode, which has returned a large amount of valuable stone, but it is not being worked at present, partly on account of the inefficiency of the pumping- and winding-machinery (which is being remodelled), and partly because the Willink's shaft is in a position to produce as much stone as the mill is capable of crushing.

The Jeram-Batang lode is worked from a vertical shaft about 300 feet in depth, but is at present idle pending the completion of a tramway to facilitate transport to the mill.

The principal workings of the Pahang Kabang, Limited, are on Fraser's lode, present supplies being drawn from above adit-level. A shaft is now being sunk on this lode for deep developments, and another shaft is being sunk on the Myah lode.

The formation is slate, the tin-bearing ground traversing the country-rock in an east-and-west direction. The lodes have a varying dip, sometimes to the north and sometimes to the south. They vary greatly in thickness and, for the most part, are undefined by any distinct walls.

The concession of the Pahang Corporation, Limited, comprizes an area of about 500 square miles, and is leased from the British Government, who exact a royalty of 5 per cent. on the tin worked.

The stone, from both mines, is crushed in the same mill. The mineral is conveyed from the mines to the mill by a tramway of 2 feet gauge, worked by two small locomotives.

The mill comprizes a stamp-battery with 60 heads of 850 pounds stamps, vanner-tables, buddles, grinding-pans, furnaces, etc. It is capable of crushing about 3,500 tons monthly. The stone varies greatly in value, the Pahang Corporation, Limited, running from 2 to 2¾ per cent., and the Pahang Kabang, Limited, bearing rather less to the ton.

The tin oxide carries from 71 to 72 per cent. of metal, which is a very high value compared with Cornwall, where tin-ore is rarely dressed higher than 65 per cent.

Labour.—The labour is practically all Chinese. The native Malay, with very few exceptions, is absolutely useless for any work other than boating and procuring timber suitable for mining and structural purposes. The cost and the difficulty of obtaining timber has greatly handicapped the development of the mines.

The Chinese are procured from agents in Singapore for £5 to £6 per man, and are brought to the mines under an agreement to work 300 days. In compliance with their agreement, they are supplied with food and clothes, and sufficient money to purchase tobacco or opium. At the end of this agreement, they, in nearly every instance, continue in the service of the company as free men, and not infrequently (for them) build up big fortunes. As free labourers, they start at a wage equivalent to about 8d. per day and advance as they become proficient as fitters, engine-drivers, miners, tin-dressers, etc., to £3 to £5 per month, and on contract-work frequently make considerably more. Some of the expert fitters are paid from £6 to £8 per month.

Whenever practicable, all work is let by contract, the management dealing with the contractors who engage their own gangs of labour and purchase their own tools, explosives, etc., from the company's stores, which are well-stocked with all necessary materials. The contract-price for driving levels runs from £4 to £7 per fathom (according to the nature of the rock). When the driving is done by machine-drills, the task is done by day-work, under constant European supervision.

Fuel.—The fuel used for the boilers and for the roasting-furnaces, etc., is wood; and light tramways are run into the jungle for conveying it to the works. The wood-cutting is done by contract by Chinese. The wood is delivered cut into lengths of 4 feet, and split to a suitable size. The contractors are paid from 10s. to 12s. per cord or stack, measuring 8 feet by 4 feet by 4 feet. The erection of bridges, tramways, transport, etc., brings the total cost of the wood to about 16s. per cord. The wood is soft, and has very poor calorific properties: 1 ton of fair-quality Welsh coal, being equivalent to $2\frac{1}{2}$ to 3 cords of wood. There are large areas of virgin-forest or jungle within (comparatively speaking) easy distance of the mines; but, owing to

sickness amongst the coolies engaged on this work, and the
frequent damage to the tramroads caused by the heavy rains
and floods (particularly during the wet season), the maintenance
of an ample supply of fuel is very difficult, and latterly the
wood has been supplemented by coal. The writer found, despite
the high cost of transport from Singapore to the mines, that
coal is not a more expensive fuel than wood; but, under existing conditions, it is not possible to transport up the river a
sufficient quantity to supply the mines entirely with this fuel.

Costs.—The working-costs of these mines necessarily
fluctuate, owing to the varying rate of exchange of the dollar,
in which coinage all wages are paid and working cost-sheets
made out.

The cost per ton of crushed stone may be taken as an
average. with the dollar at about 1s. 9d., as follows:—

	Dollars.	£	s.	d.
Development	2·00	0	3	6
Mining	4·00	0	7	0
Dressing	2·28	0	4	0
General expenses, including royalties, tin-charges, maintenance and repairs, office and all other expenses	4·00	0	7	0
Totals ..	12·28	1	1	6

The production of 1 ton of oxide, in which form the ore is
sent from the mines, costs from £45 to £60, but it, of course.
varies greatly owing to the variable percentage of oxide in the
stone treated.

The produce of these and nearly all other tin-mines in the
Straits is sent for smelting into metal to the Straits Trading
Company, Singapore, who are the largest tin-smelters in the
world, producing over 100 tons of metal per day.

Chinese Labour.—The fact that Chinese labour is shortly to
be introduced into South Africa may make a few remarks on
the characteristics and treatment of Chinese of some interest
to those connected with mining in that country.

The ordinary Chinese coolie, coming straight from China,
being unaccustomed to the class of work (and method of doing
it) to which he is put under European supervision, has to learn
everything from the very beginning, with the disadvantage of

being unable to understand any European language and the Europeans being equally unable to understand Chinese, for very few Europeans can speak the Chinese language.   There are seven distinct dialects of Chinese and several derivatives from each of these ; and in a batch of 50 or so coolies from China, it will frequently be found that there are several dialects among them and that they cannot converse among themselves.

In the Straits Settlements, the language universally spoken is Malay, and it is easily and quickly acquired by both Europeans and Chinese.   It is very noticeable in the Straits Settlements, that when gangs of Chinese are talking with one another they frequently use the Malay language.

Like most (if not all) other races, the Chinaman is not a good workman when working by the day, in fact he is very indifferent ; but, when working on contract or paid by results, he is an excellent workman and capable of performing a fair day's work.   He is, however, very apt, and quickly acquires the method of doing work according to European ideas.

A good trait of the Chinese labourer is that when justly punished he bears no illwill.   It may be mentioned, however, that whatever punishment a European might inflict upon him for an offence, it is infinitesimal when compared with the punishment meted out to him in his own country.

From a European point of view, the Chinese are, perhaps, decidedly cruel amongst themselves, but one must bear in mind the fact that they are a people of a different race, and as regards nerves, temperament, etc., it must be remembered that they are very differently constituted from Europeans.   The most satisfactory punishment for wrong-doing is financial, that is, to stop a portion of their pay or to fine them.   They feel a fine more severely than any other form of punishment.

The usual food of the Chinese is rice, salt and fresh fish, pork, poultry (especially ducks), and dried fruits and vegetables from their own country.   Unlike Mohammedans and many of the Indian races, their religion does not prohibit them from eating any food which takes their fancy, and they quickly accustom themselves to the vegetables and meat produced in the country of their adoption.

The system usually adopted of feeding coolies is to allow the *mandor* (equivalent to a European foreman) a certain

amount of food and money to feed the gang in his *kongsi* (or
building in which they are housed). Should these remarks be
read by anyone employing Chinese labour and not accustomed
to it, the writer would strongly advise that a sharp eye be kept
upon the feeding of the coolies. Fear of the mandor and
inability to express themselves properly in a European language
may for a considerable time prevent the fact of their being
underfed and the mandors making money on the food-allow-
ance from coming under the notice of the Europeans in charge.

In the Straits Settlements, where most of the work is done
by contract, the contractor takes over his coolies from the company paying for them whatever they have cost. This is the best
system for everyone concerned, as the contractor is alive to the
fact that a badly-fed coolie cannot work efficiently, and also, if he
falls sick or dies, it is a considerable loss to the contractor personally. In the case of coolies employed at odd work (and in a mine
or works a large number are usually so employed) a sharp eye
should be kept on the food, etc., supplied to them, as the Chinese
mandor is indifferent to their condition of health, he suffers no
loss if they die, and, in nearly every case, he will make as much as
he can out of the coolies unless carefully watched.

The principal ailments of coolies in tropical and hot countries
are malaria, acute diarrhœa, dysentery, and beri-beri. The
latter is a disease peculiar to Asiatic races (Europeans
very rarely contract it), and although the medical profession
have devoted considerable attention to this disease of late years
they have not so far discovered a satisfactory cure. The symptoms are much the same as dropsy, that is, the patient swells
abnormally (usually in the legs). Beri-beri carries off thousands
of Asiatics yearly, and, speaking generally, once contracted it
invariably proves fatal, unless the patient is removed to the
coast, where he can be treated with salt-water baths, the best
treatment now known. The diarrhœa is usually the result
of the hot climate and indiscreet feeding on such things as fat
pork, etc., a somewhat dangerous diet in a hot country.

Another ailment which causes trouble is mosquito-bites :
the coolies scratch these with nails not over clean and their blood
being poor the wounds fester (on their legs more particularly) and
frequently necessitate their being sent to hospital for treatment.
A good deal of sickness may be prevented by the European staff,

if they see that all wounds are dressed with iodoform or some other disinfectant, and by giving to the coolies occasional doses of salts and quinine.

In conclusion, the writer thinks that it would be money well spent if the Chinese were assisted with money-grants in the building of a joss-house (a house of religion). A few tom-toms and musical instruments should be kept in each kongsi, affording them a means of recreation when the day's work is done, keeping them more or less contented and happy and to some extent preventing them from seeking other and undesirable forms of amusement.

———

Prof. H. LOUIS (Newcastle-upon-Tyne) said that he had not much criticism to offer on the paper, as, on the whole, he was thoroughly in agreement with the writer. He had had considerable experience with the Chinese, who were excellent miners and good, careful and steady workers underground. He had also had the opportunity of seeing a good deal of mining done by Chinese, when quite free from European supervision : he had been consulting engineer to a Chinese company, and could testify to the high standard of their work. Perhaps their timbering was a little more carelessly done than would be allowed by inspectors of mines in this country, but the men were by no means unintelligent.

His experience of tin-mining was almost wholly of alluvial mining, but he did not agree that alluvial mining was carried on almost entirely on the western side of the Settlements. Some of the largest alluvial tin-fields with which he was acquainted lay on the eastern side of the axis, and no doubt there were very large areas of tin on that side. It should be mentioned that the lode-tin produced by the Pahang Corporation, Limited, was very inferior to the alluvial tin, and gave much more trouble in smelting. He had pleasure in proposing a vote of thanks to Mr. Saunders for his paper.

Mr. N. SAMWELL, in seconding the vote of thanks, endorsed the remarks of Prof. Louis, and said that the metal extracted from the ore would generally be found to average 62 to 65 per cent., and not 71 or 72 per cent. as mentioned by the writer.

The vote of thanks was cordially approved.

Mr. W. T. SAUNDERS wrote that, as Prof. Louis had observed, the tin of the Pahang Corporation, Limited, was subject to a discount on account of its containing a small percentage of copper; but he might mention that in a conversation with the manager of the Straits Trading Company, that gentleman explained that they now got on very much better with tin-ores containing copper than formerly. He asked for a sample of one of the lodes containing about 1½ per cent. of tin and 1½ per cent. of copper, which up to that time they had refused to purchase at anything like a reasonable price, as they were unable to separate properly the tin from the copper at the mines.

The figures, 71 to 72 per cent. of metal in the oxide, were quite correct, so far as the mines of the Pahang Corporation, Limited, and the Pahang Kabang, Limited, were concerned, and he (Mr. Saunders) believed that most of the mines exporting alluvial tin sent it down dressed to a higher value than this.

—— -

Prof. HANS HOEFER's paper on "Underground Temperatures, especially in Coal-mines" was read as follows:—-

UNDERGROUND TEMPERATURES, ESPECIALLY IN COAL-MINES.

By Prof. HANS HOEFER.

I.—INTRODUCTION.

A considerable portion of the progress made in exact scientific research is due to the perfecting of methods of observation and of the aids used for this purpose, and especially is this true of instruments.

Geology, like other natural sciences, works as the sculptor does. He conceives an idea from some impression received or observation made, carries it out at first in the rough, and gradually perfects his work with more and more delicate tools. His ideal continually inspires him to work towards this perfection. This has been and still is the case with geothermic studies.

Schapelmann,[*] mining superintendent of the Herrengrund mine, Hungary, first pointed out in 1664 that in his mine the lower down one went the warmer it became. Gensanne, in 1740, took the first temperature-measurements in the mines of Alsace. Alexander von Humboldt, towards the end of the eighteenth century, conceived the idea of investigating this matter more closely and so became the father of geothermy. Since that time, both scientists and engineers have laboured at the perfecting of this idea, the fruits of which are a benefit not only to the scientific world but to the public in general. It is superfluous to mention here the great value of geothermic experiments in judging of the workable amount of coal in a basin, for the natural ventilation of the mines as well as for the long driving of an Alpine tunnel.

Possibly the practical significance of these studies in relation to thermal springs (which are for the greater part medicinal) is

---

[*] *Mundus Subterraneus*, II., by Kirchner.

less in Great Britain than in Austria, where there are for example, the international treasure Karlsbad, not to mention Teplitz and Gastein and other hot-springs of world-wide renown. We have to secure them from being disturbed and to fix a preserve for them from subterranean workings, especially mining, by means of which complete security shall be assured to the springs while at the same time mining operations shall not be unnecessarily restricted. The solution of this problem necessitates the study of geothermy.

In the coal-mines of Karlsbad and Teplitz in Bohemia, very warm water was found, so that a connection with the medicinal springs and possible interference with them was feared. In both cases the writer was commissioned by the Government to draw up a report on these occurrences, and was therefore obliged to examine most carefully the geothermic conditions of the North-west Bohemian brown-coal or lignite-basin. The most important result of general interest in these investigations was the surprising fact that these seams of brown-coal possess an unusually high degree of natural warmth.

The question now arose whether this warmth really belonged to the coal-seam, or whether it might not be connected with the volcanic phenomena of this district, which, having begun with the eruption of basalts, phonolites and tephrites in Tertiary time, are now gradually approaching their extinction in the hot-springs of Karlsbad and Teplitz, and in the acid springs of Bilin, Krondorf, Gieshübel and Franzensbad. Austrian geologists summarized all these volcanic phenomena under the term " Bohemian thermal fissures." The problem is an interesting one, and its solution is equally important for medicinal springs and for mining, as well as full of responsibility for experts.

II.—THE BRÜX BASIN OF NORTH-WESTERN BOHEMIA.

The temperature-measurements taken by Prof. Puluj[*] in a bore-hole, penetrating the Bilin waters, near Teplitz, which are especially rich in carbonic acid, are of great importance in the solution of this problem. Close by is situated the curiously formed Borschen, a hill which consists of phonolite and basalt.

---

[*] *Zeitschrift des Oesterreichischen Ingenieur und Architekten-Vereines*, 1890, page 98.

In the Emeran mine, near Bilin, a trachytic phonolite has broken through the coal-seam and changed the brown-coal into anthracite. One is, therefore, in the midst of a volcanic district.

The bore-hole, between the Borschen and the Emeran colliery, outside the coal-field, and close to the renowned Bilin waters, was sunk through gneiss, which consistently showed the geothermic grade of 1° Cent. for a depth of 32·07 metres (105·2 feet). Even if the depth of this bore-hole is not very great, we must take into account the exactitude of the measurements, their number and the uniformity of the results deducted from them, on which account a high value must be set upon them. Appendix I. records the details of Prof. Puluj's observations and calculations.

The gneiss, the subsoil of the Tertiary coal-field, shows the normal geothermic grade, although volcanic phenomena are present on all sides; these cannot therefore be a necessary condition to the abnormally high temperatures of the coal-seams to be noted further on.

The important question here to be decided, might still seem to justify the second supposition, that perhaps the primitive rocks under the coal-seam have a higher temperature in consequence of the thermal springs and communicate this to the seam. On this subject, there is unfortunately only one direct observation; but the facts to be stated below will show that this hypothesis is untenable.

This observation was taken in the district of the Nelson and Gisela shaft, near Dux, which lies from 4 to 5 miles westward of the Teplitz thermal springs. These shafts, among others, have been undesirably affected by the influx of hot water since 1879. Here bore-holes were made in the sandstone immediately below the thill of the coal-seam. The experiment showed that a rise of 1° Cent. was attained for depths of 11·7, 12·7 and 12 metres (38·4, 41·7 and 39·4 feet); on the other hand in the neighbouring extension-shaft, in the bore-hole made in the primitive rock (gneiss?) the thermal gradient was 16·3 metres (53·5 feet), a proof that the primitive rock is, comparatively speaking, cooler than the immediate thill of a coal-seam, and thus the second hypothesis falls through, while the theory that coal itself is a source of heat, embedded in the rocks is confirmed.

Decisively in favour of the last theory were the measurements taken at the Alexander shaft* near those before mentioned. They were taken by Mr. Woschachlik during the sinking of the shaft, by means of thermometers placed on rods 1·5 metres (4·9 feet) long, in bore-holes of corresponding length, immediately after being made. The results were as follows:—

Depths.		Temperatures.		Formation.
Metres.	Feet.	Degrees Cent.	Degrees Fahr.	
100	328	15	59	
200	656	20	68	} Clay-slate
300	984	25	77	
332	1,089	31·0 to 31·2	87·8 to 88·16	Coal-seam

In the clay-slate, the temperature regularly increased by 5° Cent. for a depth of 100 metres (328 feet), so that the geothermic grade here is 20 metres (65·6 feet); but in the last 32 metres (105 feet) the heat increased by 6·2° Cent. so that the geothermic gradient for the lowest part of the shaft could only be reckoned as 5·2 metres (17·1 feet). This value will be further decreased if one takes into account that from 300 metres (984 feet) downward the temperature will have increased, not suddenly, but gradually.

Thus, on approaching the coal-seam, a rapid rise of temperature was proved so that the source of heat must be sought in the coal-seam; all the more so as, some time after, on boring the seam in all directions, starting from the Alexander shaft, a temperature of 31·5° Cent. was everywhere found in the freshly bored coal, but no indication of thermal fissures.

Hence, the abnormally high temperatures in the neighbourhood of Dux and Brüx cannot possibly be accounted for by the influence of thermal springs—on the contrary, the seam itself contains a large amount of natural heat.

* Prof. F. Schwackhöfer, in *Die Kohlen Oesterreich- Ungarns und Preussisch Schlesien* (The Coal of Austro-Hungary and Prussian Silesia), pages 184 and 185, stated that the coal of the Alexander shaft consists of:—

Nut-coal.			I.				II.	
Carbon	...	...	62·71	76·37	...	...	54·22	74·76
Hydrogen	...	...	4·40	5·36	...	..	3·73	5·15
Oxygen	...	...	14·24	17·35	...	...	13·72	18·90
Nitrogen	...	...	0·76	0·92	...	...	0·86	1·19
Water	...	...	16·73	excluded	...	...	22·96	excluded
Ash	...	...	...	1·16	do.	...	...	4·51	do.
Combustible sulphur	0·46			...	...	1·19	—

The miners in the brown-coal district of north-western Bohemia, when sinking a shaft, prophesy that they will soon strike a seam, as soon as they feel a rapid rise of temperature.

But the observations in the Alexander shaft also prove that the clay-slate is a bad conductor of heat, as the temperature, on approaching the seam, rose so very quickly. In any case, the even distribution of the heat and the comparatively low level of the clay-slate also deserve consideration.

As almost the same temperature of the coal, at almost the same depth as the Alexander shaft, is found in the Bruch shafts, which lie rather more to the west; one may assume this to be a regularly recurring local phenomenon, which demands a satisfactory explanation.

It is universally accepted that coal has been formed from plants, which have successively become peat, lignite and brown coal, and have, respectively, become black coal and anthracite. This metamorphosis is called carbonization. By means of it, the carbon is, comparatively speaking, accumulated by more hydrogen and oxygen being separated. The products of this disintegration are water, carbon dioxide and marsh-gas (methane or CH_4) all of which, in fact, issue from coal-seams. It has been proved in north-western Bohemia that the gas issuing from brown coal contains a comparatively large amount of carbonic acid—certainly more than that contained in black coal. The presence of carbon dioxide (CO_2) is a proof that an oxidation of carbon, a so-called silent combustion, goes on during which heat is developed. In the formation of methane (CH_4) too, great quantities of heat are given off. We have then here already two important sources of heat in coal-seams, and the task now before us is to judge of it as to quantity.

Mr. F. Toldt, a friend of the writer's, undertook this calculation, and found that, in the case of the formation of 1 kilogramme of brown coal from cellulose, 4,048 calories were liberated, but in the transformation of brown coal into black coal only 1,407 calories. His calculation showed further that, in the case of the first change, 46·4 per cent. by volume of methane (CH_4) and 52·4 per cent. by volume of carbon dioxide (CO_2); in the case of the latter, however, 71·2 per cent. of methane (CH_4) and 26·3 per cent. of carbon dioxide (CO_2) are liberated. An analysis of the gas of the coal-seam at the Pluto shaft, one of the Bruch shafts (near Brüx) shewed 54·4 per cent. by volume of methane (CH_4) and 45·6 per cent. by volume of carbon dioxide (CO_2). If one takes into consideration that the process of carbonization has already advanced some way from the plant-cellulose, and that therefore the carbon dioxide must already be diminishing, Mr. Toldt's theory will be fully confirmed as regards the analysis of this gas.

The fact that during the metamorphism of plant-fibre into brown coal about 3 times as much heat is liberated as during that of the latter into black coal, deserves serious consideration. the more so, as the period of formation of the brown coal, extending as far back as the Oligocene period, is short compared with that of the black coal, which for the greater part goes back as far as the Carboniferous period. From this we conclude that :--

(1) The process of carbonization does not always develop an equal amount of heat, that at the brown-coal stage greater quantities of heat are developed, and that therefore brown coal-seams must possess more natural heat than black coal-seams.

(2) As the heat liberated in black coal took a very long time to develop, therefore it would be communicated more equally to the neighbouring rocks, and so is less perceptible in the seams. Such a seam may be compared to an oven, in which a fire, formerly kindled, only remains as a glow on the point of extinction, while at the brown-coal stage it may be compared to an overheated oven.

The great natural heat of the north-west Bohemian brown coal-seam makes itself felt, much to the inconvenience of the miner, not only because the work done is less efficient, but also because during the process of carbonization, a large quantity of methane or fire-damp is liberated. This necessitates the introduction of a large amount of fresh air, which, further, facilitates the oxidation or silent combustion of the already hot brown coal, so that fires in the mines may easily occur. In north-western Bohemia, therefore, one is always exposed to the danger of a fire-damp explosion on the one hand and a fire in the mine on the other. The task of the mining engineer is to steer successfully between these difficulties.

In the Brüx basin, they are only working one seam about 30 metres (100 feet) in thickness.

III.—THE FALKENAU BASIN.

In the Falkenau district, to the west of Brüx, at the foot of the Erzgebirge, in the north-western corner of Bohemia, are three seams that are workable and two that are only partially so.

The lowest, situated immediately over the Oligocene sandstone, called the Josefi seam,* yields a very good brown coal and is covered by light, thickly bedded clay, with iron-pyrites. It is from 3 to 4·5 metres (10 to 15 feet) thick. Separated by intervening strata, there are two seams above this called the Josefi seams I. and II. This last in some places joins the real Josefi III. seam, and then reaches a thickness of 7 metres (23 feet). Seam I. can only be worked to a slight extent. The higher strata of clay are green, red or brown, and contain clay-ironstone in lumps and small blocks. The clays, before-mentioned, are looked upon as basaltic tuff. Above this lies the Agnes seam, containing valuable gas-coal, which can only, however, be worked in the southern and western districts of the basin. It is covered with basaltic tuff, in places also by basalt, or by sand and sandy clays. After the cessation of volcanic activity the Antoni or lignite-seam, was deposited, 20 to 30 metres (66 to 100 feet) in thickness. This yields good lignitic brown coal and is covered by foliated clay-slate (Cypris slate).

While the Josefi and Agnes seams have been subjected to violent convulsions during their formation, the lignite-seam has been less disturbed, though it is interrupted by several faults. The convulsions, which began during the Miocene period, therefore must have continued, but gradually diminished in intensity.

---

* Two analyses of the coal of the Josefi seam of the Falkenau basin are as follow :—

I. From the Maria pit, new deep working. Analysed by Prof. Schwackhofer.
II. From the Friedrich-Anna pit. Analysed by Dr. H. Bunde and Dr. P. Eitner.

			I.				II.	
Carbon	...	...	43·16	73·64	...	...	50·30	78·03
Hydrogen	...	...	3·12	5·33	...	...	4·18	6·46
Oxygen	...	...	11·88	20·26 ⎫	...	...	10·04	15·51
Nitrogen	...	...	0·45	0·77 ⎭				
Water	...	...	37·12	excluded	...	...	26·45	excluded
Ash	...	...	4·27	do.	...	...	8·83	do.
Combustible sulphur			2·79	do.	...	...	2·73	do.

The writer's object, therefore, was here also to acquaint himself with the thermal conditions of the seams. An opportunity soon offered, as an influx of hot water took place in the Maria II. shaft of the Britannia mine, east of Falkenau, which might possibly have affected the Karlsbad springs lying 10 miles to the east. The above-named shaft was sunk through clay-slate to a depth of 86 metres (282 feet), then through the lignite-seam of 25 metres (82 feet) in thickness, passed on through variously coloured clays (basaltic stage) and reached a depth of 170 metres (558 feet). In the sump, a bore-hole was made, which after penetrating a distance of 12 metres (39 feet), therefore 182 metres (597 feet) in all deep, struck the Josefi III. seam.

This method of advance-boring is customary in north-western Bohemia, in order to let out gradually the water contained in the seam. Here, too, immediately after the opening of the Josefi III. seam, water spurted out of the bore-hole at first (June, 1898), 100 litres (22 gallons) per minute; then (September 16th, 1898), 864 litres (190 gallons) per minute, and finally (November 8th, 1898) 1,052 litres (231 gallons) per minute. The temperature of the water was 30·8° Cent. (87·44° Fahr.). After due consideration as to whether this was thermal or seam-water, the government at last gave permission for the further sinking of the shaft, which was carried to a depth of 185·55 metres (608·8 feet) and after being carried through the Josefi seam soon penetrated the underlying Oligocene sandstone. The latter was found to be dry, all the hot water had spurted and welled from the small cracks of the seam: the temperature in the northerly drift being 32·0° to 30·2° Cent. (89·6° to 86·36° Fahr.)[*] and in the lower part of the seam only 28° Cent. (82·4° Fahr.). The hot water was therefore distinctly seam-water and, fortunately, the writer was relieved of all fear as to the safety of the Karlsbad springs.

In the Josefi seam, which becomes very shallow towards the north, a cross-cut was made towards the roof and thence a gallery was carried along. Here too, hot water ran and spurted from all the cracks of the coal-seam, its temperature varying between 29° and 32·5° Cent. (84·2° and 90·5° Fahr.). We may assume

---

[*] The temperature-measurements taken in the Josefi seam of the Falkenau basin are chiefly official. Doubtful values have everywhere been omitted.

an average of 31·0° Cent. (87·8° Fahr.). Bore-holes drilled for the purpose of taking temperature were immediately filled with water, so that the thermometer could only be used in the latter. The average temperature of air on the earth-surface is here 9° Cent. (48·2° Fahr.), from which we may calculate the grade at [(185·55−20)÷(31−9)=] 7·5 metres (24·6 feet), if we assume that the neutral stratum is 20 metres (65·6 feet) below the surface.

The Britannia Mining Company, about 377 feet north-west of the Maria shaft, has sunk the air-shaft V at 419·88 metres (1,377·6 feet) above sea-level and struck the same strata and seams as before, as well as the underlying Oligocene sandstone. The latter was here, too, perfectly dry and firm; the water issued from the Josefi III. seam in all directions—on the eastern side only in very small quantities. In four different directions and at four different heights, bore-holes 2 metres (6·6 feet) long were drilled in the coal, which all turned out to be wet. The following table shows that the thermometer registered constant temperatures:—

Side of shaft			Depth below surface.		Temperature.	
			Metres.	Feet.	Degrees Cent.	Degrees Fahr.
North	...	...	178·95	587·1	31·62	88·92
East	...	...	179·01	587·3	31·12	88·02
South	...	...	178·85	586·8	31·10	87·98
West	...	...	178·95	587·1	31·10	87·98
Averages...	...		178·94	587·1	31·23	88·21

Taking the temperature of the air at 9° Cent. (48·2° Fahr.) and the depth of the neutral stratum at 20 metres (65·6 feet), the geothermic grade comes out at 7·15 metres (23·5 feet), almost as low, therefore, as in the Maria II. shaft.

Temperature-measurements were also taken above the seam in the clay-roof, three weeks after the making of a bore-hole 1·9 metres (6·2 feet) deep and the temperature of the air was found to be 29° to 33° Cent. (84·2° to 91·4° Fahr.) at a depth of 164·7 metres (540·4 feet). Therefore, between this place and the Josefi III. seam the geothermic grade is 6·38 metres (20·9 feet). As this temperature was only ascertained three weeks after the sinking of the part of the shaft in question, this statement is not altogether free from objection; although the great depth of the bore-hole (1·9 metres or 6·2 feet) allows one to suppose that the influence of the air upon the mineral had not

yet penetrated so deeply.   Later, on the same day, the temperature of the air was registered as 16° Cent. (60·8° Fahr.) at a
depth of 132 metres (433 feet) when the shaft had been exposed
to the air.

In air-shaft V, the temperature was registered at 30·30°
Cent. (86·54° Fahr.) in clay, at a depth of 176·55 metres (579·2
feet), that is, only 2·59 metres (8·5 feet) from the Josefi seam;
the temperature therefore rose up to the middle measurements in the seam by 0·93° Cent. (1·67° Fahr.) so that near the
seam the grade sinks down to 2·79 metres (9·2 feet), an incon-
trovertible proof that the seam is an important source of heat.

During the time of the first excitement about a possible
danger to Karlsbad from the Maria shaft, several temperatures
were registered in the Britannia colliery.   In No. II. bore-hole
of the Commission (No. 42 of the mining company), which was
begun at 452·21 metres (1,483·7 feet) above sea-level, 700 metres
(2,297 feet) north-east of the Maria II. shaft, hot water was
found during the sinking of the shaft at a depth of 21·10 metres
(69·2 feet) which showed a temperature of 11° Cent. (51·8°
Fahr.).   While the measurements given below were being
taken, water was found at a depth of 75 metres (246 feet).   It
showed the following temperatures:—At a depth of 139·4 metres
(457·4 feet) in the most inclined part of the lignite-seam, 22·5°
Cent. (72·5° Fahr.) at +2·8° Cent. (37·04° Fahr.) temperature of
the air.   At a depth of 212·3 metres (696·5 feet), 31·6° Cent.
(88·88° Fahr.) at + 4·75° Cent. (40·55° Fahr.) air-temperature.
These two measurements were taken in the writer's presence;
over and above this, official observations were made at a depth
of 218·35 metres (716·4 feet) the temperature was 32° Cent.
(89·6° Fahr.) and at the bottom of the bore-hole, at a depth of
225·95 metres (741·3 feet) a temperature of 34·4° Cent. (93·92°
Fahr.) was found (Fig. 1).   As a result of the last measurement,
the grade in relation to the temperature at a depth of 21·1 metres
(69·2 feet) is 8·75 metres (28·7 feet); and at a depth of 139·4
metres (457·4 feet) is 7·27 metres (23·8 feet).   If one compares
the temperature at 139·4 metres (457·4 feet) with that at 212·3
metres (696·5 feet) one gets a gradient of 6·68 metres (21·9 feet).

The Government Board of Mines in Falkenau also registered
temperatures in other parts of this basin.   In the recently
worked Josefi seam, the following results were recorded:—

In the Dionys and Laurenzi pit, at Zieditz, 3½ miles west of the Maria shaft, in the so-called Haselbach basin, on July 9th, 1902, 21·5° Cent. (70·7° Fahr.) were registered in the extreme north of the place, while 20·0° Cent. (68° Fahr.) were registered in the west; on March 27th, the water of the seam showed temperatures of 19·8°, 20·8°, 20·8° and 20·5° Cent. (67·64°, 69·44°, 69·44° and 68·9° Fahr.), therefore an average temperature of 20·53° Cent. (68·95° Fahr.), while the water dropping from the roof in a drift showed 17·5° Cent. (63·5° Fahr.). The seam-water has, therefore, the mean temperature of the coal. The overlying strata of the northern part were 140 metres (459 feet) thick. The geothermic gradient is here, therefore, [(140 − 20) ÷ (21 − 9) =] 10 metres (32·8 feet).

In the Mathias shaft, at Zwodau (2½ miles north-west of the Maria shaft), in the so-called Neusattel basin, seam-water was found at a depth of 85 metres (279 feet) with a temperature of 14·5° to 15·1° Cent. (58·10° to 59·18° Fahr.). The grade is

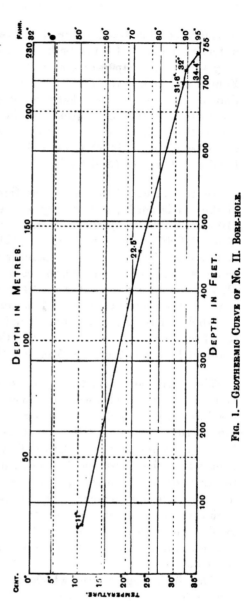

FIG. 1.—GEOTHERMIC CURVE OF NO. II. BORE-HOLE.

therefore 11·20 metres (36·7 feet). It should be noticed that the water, which several times broke out from the thill of sandstone, only showed a temperature of 11° to 12° Cent. (51·8° to 53·6° Fahr.), and therefore only 3·3° Cent. (5·94° Fahr.) cooler than that of the seam, a proof that the hot water does not come from the thill. In the same mine, in the Agnes seam, which lies nearer the surface, seam-water was found in quantities, having a temperature of 14·5° Cent. (58·1° Fahr.) and the grade is fixed at 8·4 metres (27·6 feet).

In the Friedrich-Anna pit, at Grasseth (1 mile south-east of the Maria shaft), in a freshly worked cross-cut at a depth of 110 metres (361 feet), the drop-water in the Josefi seam was 16° Cent. (60·8° Fahr.), whence we obtain the gradient of 13 metres (42·6 feet). In a dry bore-hole made from the surface, water was found in the Josefi seam at a depth of 78 metres (256 feet) with a temperature of 24·4° Cent. (75·92° Fahr.): the gradient therefore amounts to 4 metres (13·1 feet). In another bore-hole, close by, water was found at a depth of 87 metres (285 feet), the temperature was 22·5° Cent. (72·5° Fahr.), and the gradient here was therefore 5 metres (16·4 feet).

The writer accounts for these unusually low grades by the fact that the bore-holes were put down in the southern district of the Josefi seam, which does not come near the surface but underground runs against an Oligocene sandstone-ridge. On the other hand, the seam in the northern district comes to the surface, from which the water trickles down, flows through the whole of the heat-giving Josefi seam, and reaches the southern district by means of artesian pressure. This is also a warning not to set too high a value on artesian well-water for the purpose of fixing the geothermic gradient. On the other hand, it is clear that water trickling in, up to a certain point, must lower the temperature of the seam.

The official report, from which the temperature-measurements were taken, records a succession of observations made in the sandstone-water of the thill of the Josefi seam, which prove that its temperature is lower than that of the seam-water.

From the Reichenau district (west of the Maria shaft), another very interesting observation has come to the writer's notice. The Josefi seam, which here divides and has in both roof and thill light-coloured basaltic clay, was bored to at a depth of 74·19 metres (243·4 feet) below the surface (401·82 metres or 1,318·3 feet above sea-level). An artesian spring of hot-water at 21° Cent. (69·8° Fahr.) immediately came up. The grade therefore is 4·5 metres (14·8 feet). As this bore-hole was likewise put down in the south side of the basin, the writer takes tor granted the same circumstances as in the bore-holes of the Friedrich-Anna mine, which indeed shew almost the same low grades.

TABLE I.—FALKENAU BASIN.

Mines	Height of the Surface above Sea-level		Depths.		Gradients.	
	Metres.	Feet.	Metres.	Feet	Metres.	Feet.
In Mines						
Maria II. shaft	—	—	0·0 to 185·6	0·0 to 608·9	7·50	24·6
Air-shaft V. ..	419·90	1,377·6	0·0 ,, 178·9	0·0 ,, 586·9	7·15	23·5
Do. do. ...	,,	,,	164·7 ,, 178·7	540·4 ,, 586·3	6·38	20·9
Do. do. ..	,,	,,	176·4 ,, 178·7	578·7 ,, 586·3	2·79	9·2
Dionys and Laurenzi pit	457·00	1,499·4	0·0 ,, 140·0	0·0 ,, 459·3	10·00	32·8
Mathias shaft ...	421·00	1,381·3	0·0 ,, 85·0	0·0 ,, 278·9	11·20	36·7
Friedrich - Anna mine	—	—	0·0 ,, 110·0	0·0 ,, 360·9	13·00	42·6
In Bore-holes—						
Britannia Borehole II. ...	452·20	1,483·6	21·1 ,, 226·0	69·2 ,, 741·5	8·75	28·7
Do. do. ...	,,	,,	21·1 ,, 139·4	69·2 ,, 457·4	7·27	23·8
Do. do. ...	,,	,,	139·4 ,, 226·0	457·4 ,, 741·5	6·68	21·9
Friedrich-Anna	439·50	1,442·0	0·0 ,, 78·0	0·0 ,, 255·9	4·00	13·1
Do	450·00	1,476·4	0·0 ,, 87·0	0·0 ,, 285·4	5·00	16·4
Reichenau ...	401·82	1,318·3	0·0 ,, 74·2	0·0 ,, 243·4	4·50	14·8

Table I. contains the gradients found for the Josefi seam of the Falkenau basin. It will be seen that the grade diminishes, as the distance from the Josefi seam decreases. In a mine, the grade increases in proportion to the proximity of the seam to the surface, because of the water which collects from the latter. One cannot strike an average of the values given, because they extend over too wide a range (2·79 metres to 13 metres or 9·2 feet to 42·6 feet); the amount of heat is also too much influenced by the circulation of the water. If we summarize the foregoing conclusions, the fact remains (so often confirmed that we may consider it to be established) that certain coal-seams of the Lower Miocene strata in the east as well as in the west of north-western Bohemia, contain an unusual degree of natural heat on being first opened up, which can only be accounted for by a certain stage in the process of carbonization and by the low-conducting power of the neighbouring clay formation.

This fact, which was of scientific as well as of practical interest, caused the Austrian Minister of Agriculture to institute a similar inquiry in all Austrian mines—as well in brown as in black coal, and especially in those places where shafts or cross-cuts had been driven towards a seam. Good maximum thermometers, mounted in brass, with large mercury-bulbs and degrees marked in 0·1 were used for the measurements.

The directions issued by the Minister of Agriculture and the Government Board of Mines are as follows:—

Work undertaken: Temperature in . Shaft, Adit-level or Crosscut of ... . Mine.

No. of Bore-hole.	Day and Hour of Observation.	Depth of Bore-hole.		Geological Section of former and present Bore-hole.	Line of Least Resistance of the Bore-hole.	Is the Bore-hole dry or wet?	Interval between		Temperature in Degrees Cent.		Observations.
		From the Surface.	Normal to the Surface.				Boring and the Thermometer Readings.	The preceding Thermometer Reading.	In the Bore-hole.	In the Air.	
		Metres.	Metres.		Centimetres.		Days. Hrs.	Days. Hrs.	Degs.	Degs.	

EXPLANATIONS.

(a) Geothermic observations are to be made when a shaft has presumably reached a depth of 100 metres (328 feet) or more, or, if an adit-level or cross-cut attains about that perpendicular depth. These measurements are to be taken with suitably prepared instruments in bore-holes, which should be as dry as possible, and must have at least 2 metres (6·6 feet) of line of least resistance, and are usually to be made in the gallery or drift, with the greatest exactitude practicable.

(b) In choosing the position of the bore-hole, which should always be in the driest face or wall, care should be taken that it lies as far as possible from all local disturbances of temperature, especially from issuing water (springs), and that the bore-hole is not influenced, as regards temperature, by external causes, such as dripping or inflowing water.

(c) The thermometer to be used shall be a maximum thermometer, it shall have a large mercury-bulb, protected by a strong perforated metal-case. The scale of the thermometer (extending from 0° to 45° Cent., or 32° to 113° Fahr.) shall be graduated to fifths of a degree Cent. At the upper end, the thermometer shall be provided with a chain, so that it may be more easily withdrawn from the bore-hole. The thermometer must further be calibrated, or have its accuracy otherwise tested.

(d) The observations shall be commenced immediately after the sinking or driving of the particular part of the shaft or adit-level, where they are to be taken, so that the face or wall still retains its original temperature. The maximum thermometer must have its index lowered in the mercurial column by immersion in cold water before inserting it into the bore-hole; and immediately after removal, and during the reading, the bulb must also be immersed in cold water. The thermometer must be pushed in, until its bulb attains the end of the bore-hole, and the mouth of the hole is to be lightly closed with a plug, for example wooden plug wrapped with greased tow.

(e) The thermometer should be inserted about 24 hours after the boring of the hole and is to remain in the sealed bore-hole for 2 days, after which the first reading should be made. The reading should be repeated daily until the temperature remains constant.

(f) The first observation should be taken where the shaft or adit-level attains the perpendicular depth of 25 metres (82 feet) from the surface, and thence forward at vertical intervals of 50 metres (164 feet).

(g) If, on account of old workings or of certain guiding strata, the point where the seam will be struck is at least approximately known, five measurements of temperature shall be taken—at vertical distances of 40, 30, 20, 10 and 5 metres (131, 98, 66, 33 and 16 feet) before reaching the seam. The temperature of the seam shall be ascertained at 1 metre (3 feet) from the roof, in the middle of the seam and 1 metre (3 feet) above the thill.

(h) If the seam has been passed through, the same measurements are to be taken as on approaching it—at vertical distances of 5, 10, 20, 30, 40 and 50 metres (16, 33, 66, 98, 131 and 164 feet), and besides this, at 100 and 150 metres (328 and 492 feet). . . . If a lower seam is expected, the same rules apply as were given on approaching the first seam.

(i) When a seam is reached, an average sample of the coal should be taken, in the ordinary way, and shall be subjected, as soon as possible, to an elementary analysis. The result of the analysis shall be entered as an appendix to the observations of temperature.

(k) When the seam has been reached, further observations should be taken, immediately after the main-road has been driven, at distances of 100 metres (328 feet), in the manner above described. The position of each hole relative to the roof or floor should, if possible, be noted.

As will be seen from this scheme and its explanation these directions are intended to eliminate sources of error as far as possible and to investigate the secondary influences which disturb the normal distribution of heat in the earth.

———

APPENDIX I.

Prof. J. Puluj´used a telethermometer,[*] which, before and after being used, was compared with a normal thermometer at Bilin. The temperature of the air, over the water-level, was $+ 3\cdot2°$ Cent. ($37\cdot76°$ Fahr.). The measurements were taken, while work in the bore-hole was suspended for 48 hours. The temperature in the cross-cut was always taken in five places, that is at four points situated at right angles to each other in the circumference, and one in the

———

[*] "Ein Telethermometer," *Sitzungsberichte der Kaiserlichen Akademie der Wissenschaften*, Wien, Abtheilung IIa., 1890, vol. xcviii., page 1502.

middle. The five measurements taken in the same cross-cut agreed. On January 2nd and 3rd, 1890, 4 tours of inspection were completed, the differences were insignificant and the averages were calculated.

At depths of 6 to 20 metres (19·7 to 65·6 feet) 10 observations were made, and the constant temperature was found at 20 metres (65·6 feet).

Taking the formula, $T = a + b(H - 30)$ as a basis, in which T represents the temperature at H, the vertical depth below the surface, and a and b constants, the law was deduced (calculating by the method of the least squares) by which the temperature increases with depth. The result is as follows :—

$$T = 11·45908° + 0·031182(H - 30),$$

with a probable mean error in single observations of $\pm 0·06°$ Cent. and that of the result of $\pm 0·02°$ Cent.

Depth from Surface.		Observed Temperature.		Depth from Surface.		Observed Temperature.		Calculated Temperature.		Differences between observed and calculated Temperature.	
Metres.	Feet.	Degrees Cent.	Degrees Fahr.	Metres.	Feet.	Degrees Cent.	Degrees Fahr.	Degrees Cent.	Degrees Fahr.	Degrees Cent.	Degrees Fahr.
6·0	19·7	6·4	43·5	30	98·4	11·4	52·52	11·46	52·63	− 0·06	− 0·11
10·0	32·8	6·5	43·7	40	131·2	11·8	53·24	11·77	53·19	+ 0·03	+ 0·05
11·0	36·1	6·6	43·9	50	164·0	12·0	53·60	12·08	53·74	− 0·08	− 0·14
12·0	39·4	6·7	44·1	60	196·9	12·4	54·32	12·39	54·30	+ 0·01	− 0·02
12·5	41·0	7·3	45·1	70	229·7	12·9	55·22	12·71	54·88	+ 0·19	+ 0·34
13·0	42·7	7·9	46·2	80	262·5	13·1	55·58	13·02	55·44	+ 0·08	+ 0·14
13·5	44·3	8·8	47·8	90	295·3	13·3	55·94	13·33	55·99	− 0·03	− 0·05
14·0	45·9	9·4	48·9	100	328·1	13·5	56·30	13·64	56·55	− 0·14	− 0·25
15·0	49·2	9·8	49·6	110	360·9	13·9	57·02	13·95	57·11	− 0·05	− 0·09
20·0	65·6	10·4	50·7	120	393·7	14·3	57·74	14·27	57·69	+ 0·03	+ 0·05
25·0	82·0	11·1	52·0	130	426·5	14·6	58·28	14·58	58·24	+ 0·02	+ 0·04

APPENDIX II.—SECTION OF No. II. BORE-HOLE (No. 42) AT BRITANNIA COLLIERY.

No.	Description of Strata.				Thickness of Strata.		Depth from Surface.	
					Metres.	Feet.	Metres.	Feet.
1	Vegetable soil	...	..	...	0·15	0·5	0·15	0·5
2	Loam, with gravel	...	...	...	1·60	5·2	1·75	5·7
3	Grey impure clay	..	...	...	1·30	4·3	3·05	10·0
4	Yellowish-grey clay-slate		...	...	14·55	47·7	17·60	57·7
5	Marly limestone	...	...	...	0·10	0·3	17·70	58·0
6	Yellowish-grey clay-slate		...	..	3·25	10·7	20·95	68·7
7	Marly limestone	...	...	...	0·15	0·5	21·10	69·2
8	Yellowish-grey clay-slate		...	...	24·45	80·2	45·55	149·4
9	Marly limestone	...	...	...	0·10	0·3	45·65	149·7
10	Yellowish-grey impure clay		...	...	0·55	1·8	46·20	151·5
11	Marly limestone	...	...	...	0·15	0·5	46·35	152·0
12	Blue impure clay	...	...	...	2·50	8·2	48·85	160·2
13	Marly limestone	...	...	...	0·20	0·7	49·05	160·9
14	Blue impure clay	...	...	...	9·60	31·5	58·65	192·4
15	Marly limestone	..	...	...	0·10	0·3	58·75	192·7
16	Greyish-blue impure clay		...	...	0·75	2·5	59·50	195·2
17	Marly limestone	...	...	...	0·25	0·8	59·75	196·0
18	Greyish-blue impure clay		...	...	14·55	47·7	74·30	243·7
19	Marly limestone	...	...	...	0·35	1·2	74·65	244·9
20	Greyish-blue impure clay		...	...	0·85	2·8	75·50	247·7
21	Marly limestone	...	...	...	0·30	0·9	75·80	248·6
22	Greyish-blue impure clay		...	...	18·80	61·8	94·60	310·4
23	Marly limestone	..	...	...	0·30	0·9	94·90	311·3
24	Greyish-blue impure clay		...	...	2·30	7·5	97·20	318·8
25	Marly limestone	..	...	...	0·10	0·4	97·30	319·2
26	Greyish-blue impure clay		...	...	16·70	54·8	114·00	374·0
27	Brown impure clay	...	...	...	0·10	0·3	114·10	374·3
28	Greyish-blue impure clay		...	...	24·30	79·8	138·40	454·1

APPENDIX II.—*Continued.*

No.	Description of Strata.	Thickness of Strata.				Depth from Surface.	
		Metres.	Feet.	Metres	Feet.	Metres.	Feet.
29	**LIGNITE**	18·65	61·2			157·05	515·3
30	**COAL**, with impure clay	0·40	1·3			157·45	516·6
31	**COAL**	1·45	4·7			158·90	521·3
32	**COAL**, with impure clay	2·40	7·9			161·30	529·2
33	**COAL**	12·10	39·7			173·40	568·9
34	Greyish - brown clay, with sandy *coal* ...	0·80	2·6			174·20	571·5
35	**COAL**	2·10	6·9			176·30	578·4
				37·90	124·3		
36	Greyish-brown clay, with *coal*			7·30	24·0	183·60	602·4
37	Sand with clay			0·40	1·3	184·00	603·7
38	Greyish-brown impure clay			0·85	2·7	184·85	606·4
39	Marly limestone			0·90	3·0	185·75	609·4
40	Sand with impure clay			0·50	1·6	186·25	611·0
41	Brown clay, with *coal*			0·95	3·1	187·20	614·1
42	Grey sandy clay			4·30	14·1	191·50	628·2
43	**COAL**			0·80	2·7	192·30	630·9
44	Light-grey clay			0·95	3·1	193·25	634·0
45	Marly limestone			0·15	0·5	193·40	634·5
46	Grey impure clay			4·90	16·1	198·30	650·6
47	Bluish-grey clay			4·00	13·1	202·30	663·7
48	Grey clay, with *coal*			1·65	5·4	203·95	669·1
49	**COAL**			0·65	2·1	204·60	671·2
50	Grey clay, with *coal*			2·95	9·7	207·55	680·9
51	Sandstone			0·45	1·5	208·00	682·4
52	Grey clay, with *coal* and pyrites ...			2·35	7·7	210·35	690·1
53	Grey impure clay			1·35	4·4	211·70	694·5
54	Grey impure clay, interstratified with white clay			1·55	5·1	213·25	699·6
55	Sandstone			0·30	1·0	213·55	700·6
56	Grey sandy clay			0·80	2·6	214·35	703·2
57	Brown clay, with *coal*			1·90	6·2	216·25	709·4
58	Grey clay, with pyrites			2·10	7·0	218·35	716·4
59	**COAL**			0·40	1·2	218·75	717·6
60	Grey impure clay			0·95	3·1	219·70	720·7
61	Grey clay, with *coal*			1·55	5·1	221·25	725·8
62	Clay, with mica			4·70	15·5	225·95	741·3

Mr. B. H. BROUGH felt sure that most of the members believed that hot springs would certainly have an effect on the thermal gradient, and he considered that, if in translating this paper, the gradient had been expressed not in metres per degree Cent. but in feet per degree Fahr., the members would have been better able to compare the results given by the author with the elaborate collection of underground temperatures tabulated year after year by Prof. J. D. Everett in his reports on underground temperatures to the British Association for the Advancement of Science. As far as he (Mr. Brough) could make out, at Teplitz, the measurements gave exactly the normal reading which one would expect, that of 33 metres per degree centigrade. Prof. Hoefer had certainly made out a good case in favour of his hypothesis of the internal heat of the brown coal-seams, but it seemed to him (Mr. Brough) that there was another side to the question. It would have enabled the members to form an impartial judgment, if they could have had a paper from the school of thought representing the authorities of the Teplitz and Karlsbad hot springs. For some years past, there had been a good deal of bickering between the authorities who had charge of the hot springs and the coal-miners, it being alleged that the collieries were tapping the Karlsbad spring, which had been such a source of wealth to the district; while the representatives of the coal-miners said that the hot water found in the collieries had nothing whatever to do with the hot springs. Prof. Hoefer represented the views of the coal-miners, but a paper was published* in the *Transactions* of the Austrian Geological Survey, giving the view held by the hot-spring authorities. It was rather premature to form an absolute opinion upon the subject, but he (Mr. Brough) would await with great interest the result of the elaborate enquiry which the Austrian Government were going to make into the temperatures of the brown coal-seams.

Mr. H. HALL (H.M. Inspector of Mines) said that, so far as he could make out from the paper, the temperatures rose much more quickly than they did in this country. The temperatures given in the paper were much higher at given depths than was considered to be the case in Great Britain. Recently, he (Mr. Hall) had made some experiments in newly opened seams, at a depth of 2,400 feet. The experiments were carefully made, the thermometer was kept buried in the bore-hole for 6 or 7 hours, and care was taken that the warmth of the lamp, afterwards, did not come into contact with it. The readings were carefully taken, and they bore out entirely the old value, at any rate in Lancashire, that a rise of 1 degree Fahr. for 60 feet of depth was almost absolutely correct. He was disposed to think that Prof. Hoefer, somehow or other in his experiments, had confused the heat due to chemical action in the coal and the heat due to depth; and, possibly, water, coming from greater

* "Studien über unterirdische Wasserbewegung," by Dr. Franz E. Suess, *Jahrbuch der Kaiserlich-Königlichen Geologischen Reichsanstalt*, 1898, vol. xlviii., pages 425-516.

depths than where he was experimenting, might have, to some extent, accounted for the high temperatures recorded in his paper. He (Mr. Hall) believed that it had been generally held that the increase of temperature with depth was due to the internal heat of the earth and the permeability or otherwise of the rocks through which it had to pass, but he thought that the correctness of this view was by no means certain. At the mine, where he obtained these high temperatures, the upper seams had all been worked out, affording, he thought, a more ready exit for the heat below. He (Mr. Hall) thought that Prof. Hoefer's idea of coal-seams being hotter than the surrounding strata, could have been readily tested by taking temperatures at depths where there were no coal-seams, or in an area where there was no coal. From his own experiments, he did not think that the coal-seams were hotter than the surrounding strata. He had several bore-holes made in the strata above the seams as well as in the coal itself, and the readings corresponded.

Prof. H. Louis (Newcastle-upon-Tyne) said that if they compared a bore-hole put down through hard rock by means of a percussive drill working wet, with a bore-hole in a soft rock made by a rotary drill working dry, the amount of heat would be considerably higher in the case of the rotary drill. Taking the temperature once every 24 hours was not sufficient to overcome this enhanced reading, and by the time that the mechanically-evolved heat might have passed away the heat produced by oxidation might have taken its place. Dr. P. Phillips Bedson's experiments had proved that many coals were capable of absorbing oxygen directly, and there was no difficulty in imagining that coal, if it was assumed that the seam was permeable, might be slowly oxidized; and that would account for a rise of temperature in such coal-seams. He (Prof. Louis) asked whether Prof. Hoefer had considered the possibility that heat derived from chemical action might be the original source of the heat in those seams. If the existing coal-seams were remnants of a much larger body that had been compressed, the compression must have been the source of heat at the period of compression; but whether that heat had been dissipated since that time was of course another question.

Mr. J. GERRARD (H.M. Inspector of Mines) said that it seemed impossible to lay down one uniform rate of increase of the natural temperature of the strata for all parts of the world; and the actual ratio must be ascertained at the mine that was being worked. The gradient might range from 40 to 80 feet, and they could not lay down a law for 40 feet which would apply to 80 feet, or *vice versa*. It had been suggested, and he believed with some force, that the inclination of the seams, the volume of the strata, the presence of water, and so on, might play a part in connection with temperature. The rate of increase of the temperature in the steeper measures in the neighbourhood of Manchester was lower than where the measures were flatter.

Prof. H. HOEFER wrote that the sole object of his paper was to make the members of the Institution of Mining Engineers acquainted with the present state of his investigations in regard to temperature-conditions in the brown-coal district of North-western Bohemia; while he reserved for a future occasion the project of extending his researches to other coalfields in the Austrian Empire. He had already pointed out that the conditions of temperature in " black-coal " districts, reasoning from theoretical premises alone, must be entirely different from those prevailing in North-western Bohemia; and, in point of fact, they were so. A provisional examination of a fairly abundant mass of material from the " black-coal " districts of his own country, placed at his disposal through the kind offices of the Imperial and Royal Ministry of Agriculture, confirmed this view, since for those mines the temperature-gradient worked out almost exactly at 60 feet per degree Fahrenheit. This, of course, agreed with the results obtained from the majority of English coal-mines, and with the recent determinations of Mr. H. Hall.

No one disputes the fact, confirmed as it is by many observations, both official and unofficial, that the seams of the North-west Bohemian coal-field possess an unusually high temperature, which by no means corresponds to their depth below the surface. The important problem is to find the source of this heat. Is it a characteristic property of the seam itself, or is it of external origin, derived, say, from an upwelling thermal spring? At the sittings of the Commission for the Protection of the Karlsbad springs, the two mining experts and the three sworn official geologists, including (besides the writer) Prof.

Dr. G. Laube of Prague University and Prof. Dr. V. Uhlig of Vienna University, after long-continued investigation of all the evidence, came unanimously to the conclusion that the extraordinarily high temperature of the pit-waters was due to the heat of the coal-seams, and not to any such thermal spring as that suggested above. The latter hypothesis, however, was maintained by a member of the Austrian Geological Survey and municipal councillor for Karlsbad, but he was unable to adduce any convincing proof in support of his contention. Engineer J. Knett, geologist to the city of Karlsbad, abstained from enunciating any opinion as to the source of the high temperature of the seams. (This is mentioned, in order to supplement Mr. B. H. Brough's remarks.)

That the source of the high temperature resides within the seams themselves is proved, beyond the possibility of doubt, by the extraordinarily rapid rise in temperature as the seams are approached. If the source (say, a thermal spring) lay below the seams, the ascertained facts would be inexplicable, since the overlying clay-slate is a better conductor of heat than the coal. On the other hand, the gneiss, which forms the basement-rock of the coal-bearing Tertiaries, is found to show, close to the junction-line, the normal temperature-gradient.

That rock-metamorphism can give rise to great evolution of heat has been proved in detail, and mathematically at that, by the chief engineer and geologist of the St. Gothard tunnel, Dr. F. Stapff; and he attributed the low temperature-gradient beneath the Andermatt plateau to the weathering of the rocks there. It is, moreover, wellknown that the high temperature of the Great Comstock lode was traced to the same cause.

Now, the Bohemian brown coal, too, is in process of metamorphism, as shown by the abundant evolution of gases and vapours from it, and great quantities of heat must be thereby set at liberty, *teste* the concordant calculations of such distinguished chemists as Dr. F. Fischer of Goettingen, Baron von Jueptner of Vienna, and Dr. F. Toldt of Graz. The writer did not reproduce these calculations in the paper under discussion, as he had already published them in detail elsewhere.*

* "Die Wärmeverhältnisse im Kohle führenden Gebirge," *Oesterreichische Zeitschrift für Berg- und Hüttenwesen*, 1901, vol. xlix., pages 249, 267 and 286 ; and " Les Conditions calorifiques des Terraines à Combustibles," *Revue Universelle des Mines*, etc., 1904, vol. vi., page 159.

Since this metamorphosis has been and is associated with actual loss of substance, it may be naturally inferred that the consequent compression of the seams gives rise to further evolution of heat. This new source of heat within the coal itself was apparently not taken into account by the chemists, when making the aforesaid calculations.

In the publications quoted above, the writer has further shown that other brown coal-basins, situated far away from thermal springs or from eruptive rocks of comparatively recent date, such, for example, as Fohnsdorf in Styria (20·46 metres or 67·1 feet per degree Cent.) and Monte Massi in the Tuscan Maremma (12·5 metres or 41·0 feet per degree Cent.) are also conspicuous for low gradients. The coal in these basins has reached the same stage of metamorphosis as the Bohemian brown coal. Bituminous shales, too, as in the borings at Neuffen in Würtemberg (11·1 metres or 36·4 feet per degree Cent.) and Sulz-unter-dem-Walde in Alsace (gradient variable according to the depth and the nature of the rocks) are characterized by low geothermic gradients.

The PRESIDENT (Mr. J. C. Cadman) moved a hearty vote of thanks to Prof. Hoefer for his interesting paper, and remarked that contributions from foreign engineers were always welcome.

The REV. G. M. CAPELL seconded the resolution, which was cordially approved.

Mr. ADOLPHUS F. EOLL's paper on " The Hammer-Fennel Tachymeter-theodolite " was read as follows:—

THE HAMMER-FENNEL TACHYMETER-THEODOLITE.

By ADOLPHUS F. EOLL.

The Hammer-Fennel tacheometer or the auto-reducing tacheometer, as it more adequately might be designated, permits by one sight both the distance and the difference of altitude

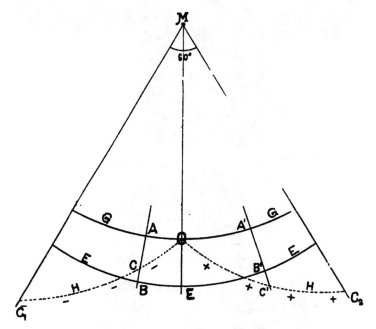

FIG. 1.—CURVES FOR DISTANCE AND ALTITUDE.

(with $D - \text{Constant} = 100$, and $h - \text{Constant} = 20$) to be practically read from a vertically held staff—thus reducing tacheometric operations to the simplicity of ordinary levelling—thereby affording a great saving of time and labour, and accordingly rendering general tacheometry more productive without any relative sacrifice in accuracy.

The characteristic feature of this instrument is that, resting upon a trunnion of the Porro - telescope, and close to an aperture in its tube is a diagram of certain curves (on glass, similar to a magic-lantern slide), which can be viewed at the same time as the tacheometer-staff, by means of an extra optical system pro-

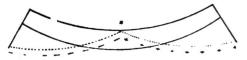

FIG. 2.—CURVES FOR DISTANCE AND ALTITUDE.

FIG. 3.—THE AUTO-REDUCING TACHEOMETER.

vided within the tube of the telescope. Figs. 1 and 2 show the curves for distance and altitude; Figs. 3 and 14 (Plate XVIII.), the position of the diagram on its carrier beside the telescope; Fig. 15 (Plate XVIII.), the same in section and its protective glass-plate, together with the required optical arrangements, which include a simple prism, an achromatic lens and a double prism; while Fig. 5 is a view of the diagram and part of the staff, as seen through the telescope for a gradient of about + 7 degrees at a distance of 14 metres (45·9 feet).

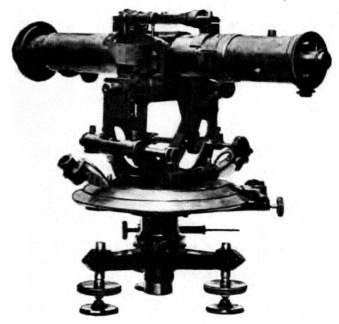

FIG. 4.—THE AUTO-REDUCING TACHEOMETER.

All other parts, except the telescope and the diagram-carrier, are similar to those of an ordinary theodolite, so that, whether or not a compass be included, or a vertical arc for occasional purposes, the kind of levels, their number and arrangement, the system of tripod or how the instrument should be attached to the tripod-head, etc., are all matters of individual choice. The telescope can also be adapted for use on the plane-table. In the eyepiece of the telescope, the vertical edge of the double prism makes a vertical spider-line unnecessary; but, in Fig. 5, it will be noticed that the horizontal line, NN, is somewhat above the centre of the field of view, for the better utilization of the latter.

When the Hammer-Fennel tacheometer is properly adjusted, and the line of collimation is horizontal, a plane passing through the latter at right angles to the diagram will touch the arc, GG, at 0, but, since the centre of the arc, GG, lies in the centre of the horizontal axis, it will touch at some other point of GG for any degree of inclination or decline of the telescope. For this reason also, when, to the eye, the diagram appears to pass across its field of view (the left half of the circular field of view), the arc, GG, continually appears to touch the line, NN (zero-line), at the point of intersection with the vertical line, LL.

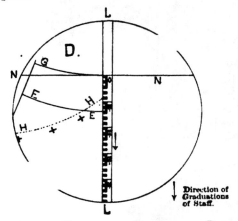

FIG. 5.—VIEW OF THE DIAGRAM AND PART
OF THE STAFF.

Distances are obtained, after the zero-mark of the vertical staff is sighted, by multiplying the reading taken at the point indicated by the curve, EE, by 100; and the difference of altitude, without any further manipulation, after multiplying by 20 the value indicated by the curve, HH, and noting the sign + or − below the curve. The zero on the staff corresponds to the height of the instrument when set up, or rather of the diagram above the ground (approximately 1·40 metres or 4·59 feet), so that the value read on the curve, HH, requires simply to be added algebraically to the given altitude of the station; this, perhaps together with the multiplication by 20, constitutes all the office-work necessary for that otherwise more or less tedious part of tacheometric calculations.

The diagrams are made by careful photographic reduction from the original design (scale, 20 to 1), which was made with all precision on paper affixed to a plate of the best glass—care being also taken to avoid any detrimental influence through changes of temperature during transport.

In using or rectifying the Hammer-Fennel tacheometer, no knowledge of the construction of the diagram-curves is required;

but, as many engineers would not be satisfied with only a super-ficial description of the appearance and qualities of the instrument, without being able to judge of the principles theoretically involved, it is desirable to give the following explanation. In this connection, however, the writer would just refer to the usual formulæ for tacheometric reductions:—

$$E = c + kl^1 \text{ or } = kl^1 \text{ for Porro-telescopes} \qquad . \qquad (1)$$
$$e = E \cos^2 a \qquad . \qquad . \qquad . \qquad . \qquad . \qquad . \qquad (2)$$
$$h = e \tan a = E \tfrac{1}{2}\sin 2a \text{ (Fig. 7)} \qquad . \qquad . \qquad . \qquad (3)$$

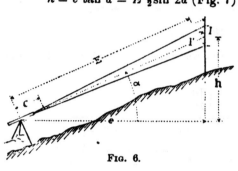

FIG. 6.

To obtain e and h, graphic or graphic-mechanical contrivances (by Messrs. Goulier, Gillman, etc.) or slide-rules of varied arrangements and pretensions (most commonly the Eschmann-Wild rule), or numerical tables (Prof. W. Jordan's and others) are used.

In order to facilitate that wearisome part of tacheometry which is expressed by the formula (3), various attempts at instrumental improvement have been made during the last decades, amongst others by Messrs. Goulier, Peaucellier-Wagner, Sanguet, Schrader, Ziegler-Hager, Eckhold, Roncagli-Urbani and Nasso. As a particular class of instrument, the projective tacheometers or tachygraphometers (by Messrs. Wagner-Fennel, Kreuter, etc.) may be mentioned, possessing in place of a vertical arc a so-called projector, bearing a vertical and a horizontal scale by which distances and altitude can be read, but, except in the case of the Puller-Breithaupt instrument, each requiring the staff to be held obliquely.

The constants chosen for the Hammer-Fennel tacheometer in pre-ference to others, are as above-stated, C_1 is 100, for distance: and C_2 is 20, for altitude. The objective of the Porro-telescope has two lenses at a permanent distance (in contrast to the Huyghens system), with a focus f equal to 350·0 millimetres (achromatic chief lens) and f_1 equal to 220·0 millimetres; the distance, m, between the lenses, chosen as large as possible so as to ensure a large view, is 340·0 millimetres.

The focus of the equivalent lens is therefore :—

$$F = \frac{ff_1}{f + f_1 - m} = 334 \cdot 78 \text{ millimetres (see Fig. 7). (1)}$$

The additive-constant of the instrument becomes zero, by taking

$$b = \frac{f(m - f_1)}{f + f_1 - m} = 183 \text{ millimetres approximately. (2)}$$

As, however, the objective-tube is shifted for varied distances, the additive-constant does not always remain at zero. but, as it never exceeds a few centimetres, it may be totally neglected. Further, in case the distance, OE, between the two curves were not

$$a = \frac{F}{100} = 3 \cdot 348 \text{ millimetres} \quad . \quad . \quad . \quad . \quad (3)$$

it would not signify, because the diagram can be adjusted to make up for it.

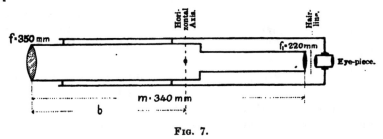

FIG. 7.

The dimensions that follow refer to the original diagram, drawn to the scale of 20 to 1.

The zero-arc, GG, was constructed, for an angle of 60 degrees to a radius of 600·0 millimetres, by co-ordinates to every 2 degrees of angle, thus simultaneously furnishing zero-points from which to fix, at the radial distance, the points of the other curves. The values were computed to two decimals of the millimetre, and doubly checked: the Breithaupt scale and offset being used in plotting. To fix the points of the curves, EE and HH, radii were drawn through the 2 degrees points of the arc GG; the distance between G and E at 0 being :—

$$20a = 66 \cdot 955 \text{ millimetres} \quad . \quad . \quad . \quad . \quad . \quad (4)$$

The constant 10 could not be chosen, as otherwise the field of view would allow but small angles of inclination, and thus limit the scope of the instrument; the constant 50 would not ensure sufficient accuracy in the altitudes; while 25 would be less convenient as a multiple than 20.

The radial distance for other points of the E curve from the zero-arc are, apparently, for the enlarged diagram

$$a^1 = 66 \cdot 955 \cos^2 a \text{ millimetres or } a^1 = a \cos^2 a, \quad . \quad (5)$$

as would follow from the equations :—e is equal to $kl \cos^2 a$, and e is equal to $C_1 l_1$, in which l represents the distance read off the vertical staff between the upper and lower cross-lines of an ordinary tacheometer; a, the vertical angle to the central cross-line; while l_1 represents the part read at the line, EE, after sighting zero with the tachy-tacheometer.

For points of the H curve, the radial distance results from the equations :—

$$h = kl \tfrac{1}{2}\sin 2a = e \tan a = 100 l_1 \tan a,$$

and $\quad h = C_2 l_2$,

where l_2 denotes the reading to the curve, HH, so that in the diagram for an angle a

$$h^1 = 20 l_2 = 100 l_1 \tan a;$$

and because $\quad \dfrac{h^1}{a^1} = \dfrac{l_2}{l_1}$,

therefore, $\quad h^1 = 5 a^1 \tan a, \text{ or } = \tfrac{5}{2} \times 66 \cdot 955 \sin 2a \quad . \quad . \quad . \quad (6)$
These formulæ (5 and 6) give the first series of values, for a^1 and h^1, on the large diagram (see Table I.).

TABLE I.—INITIAL VALUES FOR THE DIAGRAM TO A SCALE OF 20 TO 1.

a	a'	h'	a	a'	h'
Degrees.	Millimetres.	Millimetres.	Degrees.	Millimetres.	Millimetres.
− 30	50·25	145·09	∓ 0	66·955	0·00
− 28	52·23	138·88	+ 2	66·88	11·68
− 26	54·12	131·98	+ 4	66·63	23·30
− 24	55·90	124·44	+ 6	66·22	34·80
− 22	57·58	116·32	+ 8	65·66	46·14
− 20	59·14	107·63	+ 10	64·94	57·25
− 18	60·57	98·40	+ 12	64·06	68·08
− 16	61·88	88·72	+ 14	63·05	78·60
− 14	63·05	78·60	+ 16	61·88	88·72
− 12	64·06	68·08	+ 18	60·57	98·40
− 10	64·94	57·25	+ 20	59·14	107·63
− 8	65·66	46·14	+ 22	57·58	116·32
− 6	66·22	34·80	+ 24	55·90	124·44
− 4	66·63	23·30	+ 26	54·12	131·98
− 2	66·88	11·68	+ 28	52·23	138·88
∓ 0	66·955	0·00	+ 30	50·25	145·09

These simple and symmetrical values require, however, a certain modification owing to the fact that the line of observation does not coincide with the optical axis of the telescope. In order to utilize more fully the field of view, the zero-line, NN, is fixed somewhat above the centre. To obtain the corrected a^1, let a be first considered as the vertical angle to the line of observation, through the horizontal cross-line, NN, directed towards M (Fig. 8), the zero-mark being 1·40 metres above the ground; and l, as the reading on the staff in the case of the usual central horizontal cross-line at a distance from NN:—

$$a = \frac{F}{100} = 3\cdot348 \text{ millimetres, so that if } \beta - \frac{1}{100}s = 0^\circ\ 34'\ 22\cdot65''$$

represents the diastimometric angle to the length of the staff or l, then:—

$$e = kl \cos^2\left(a + \frac{\beta}{2}\right) = 100\ l \cos^2\left(a + \frac{\beta}{2}\right), \quad . \quad . \quad . \quad (7)$$

and if l_1 be the reading actually necessary to obtain e directly by multiplication by 100, then

$$e = C_1\ l_1 = 100\ l_1. \quad . \quad . \quad . \quad . \quad . \quad . \quad . \quad . \quad . \quad (8)$$

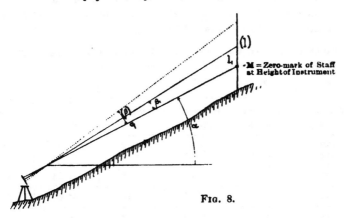

FIG. 8.

Therefore, e being the horizontal distance, and β_1 the micrometric angle variable with a, by combining the equations (7 and 8):—

$$l_1 = l \cos^2\left(a + \frac{\beta}{2}\right), \text{ or } \frac{l_1}{l} = \cos^2\left(a + \frac{\beta}{2}\right) \quad . \quad . \quad . \quad (9)$$

If provisionally, q denote the inclined distance from the horizontal axis to M, then:—

$$l = q\ \frac{\sin\beta}{\cos(a+\beta)}, \text{ and } l_1 = q\frac{\sin\beta_1}{\cos(a+\beta_1)}; \quad . \quad . \quad . \quad (10)$$

by comparing the equations (9 and 10) :—

$$\frac{l_1}{l} = \cos^2\left(a+\frac{\beta}{2}\right) = \frac{\sin \beta_1 \cos (a+\beta)}{\sin \beta \cos (a+\beta_1)}, \quad \dots \dots (11)$$

therefore with β equal to $0° 34' 22.65''$, the micrometric angle, β_1, corresponding to any positive angle a would be obtained by :—

$$\frac{\sin \beta_1}{\cos (a+\beta_1)} = \cos^2\left(a+\frac{\beta}{2}\right)\frac{\sin \beta}{\cos (a+\beta)}, \quad \dots (12)$$

and by proportion :—

$$a^1 : a = \tan \beta_1 : \tan \beta. \quad \dots \dots \dots (13)$$

For negative angles, by an analogous consideration, but taking a as absolute, the formula is :—

$$\frac{\sin \beta_1}{\cos (a-\beta_1)} = \cos^2\left(a-\frac{\beta}{2}\right)\frac{\sin \beta}{\cos (a-\beta)}. \quad \dots (12^1)$$

The formula (13) applies to both cases.

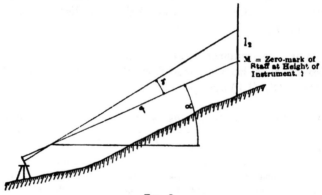

M = Zero-mark of Staff at Height of Instrument. ?

FIG. 9.

After calculating the value of β_1, the distances a^1 were found by the equations :—

$$a^1 = a\,\frac{\tan \beta_1}{\tan \beta}, \text{ and } \log a^1 = \log\left(\frac{a}{\tan \beta}\right) + \log \tan \beta_1,$$

or $\log a^1 = 3.82577 + \log \tan \beta_1$ for a scale of 20 to 1 in millimetres. (14)

Some twelve results, thus computed, together with those in Table I., sufficed to fix the E curve.

In the computation of the values for the curves of altitude, a is first regarded as positive, and if l_2 denotes the length of the staff from zero, M, read to the curve, HH, then :—

$$C_2\, l_2 = 100\, l_1 \tan a, \text{ and } l_2 = \frac{100}{C_2}l_1 \tan a \quad \dots \quad (15)$$

or, with l_1 applied to the equation (10) and C_2 equal to 20 :—

$$l_2 = 5 \tan a \; q \; \frac{\sin \beta_1}{\cos (a + \beta_1)}. \qquad \ldots \ldots \quad (16)$$

If γ denote the angle corresponding to l_2 (Fig. 9), then

$$l_2 = q \; \frac{\sin \gamma}{\cos (a + \gamma)}. \qquad \ldots \ldots \ldots \quad (17)$$

and, combining the equations (16 and 17), γ is deduced from :—

$$\frac{\sin \gamma}{\cos (a + \gamma)} = 5 \tan a \; \frac{\sin \beta_1}{\cos (a + \beta_1)}, \qquad \ldots \ldots \quad (18)$$

and the radial ordinates of the diagram:—

$$h^1 = a^1 \frac{\tan \gamma}{\tan \beta_1} \qquad \ldots \ldots \ldots \quad (19)$$

Similarly, for angles of declination and with a taken absolutely:

$$\frac{\sin \gamma}{\cos (a - \gamma)} = 5 \tan a \; \frac{\sin \beta_1}{\cos (a - \beta_1)} \qquad \ldots \ldots \quad (18^1)$$

By means of twelve ordinates, thus reckoned, and by adequately correcting the initial values, the definite radial ordinates were obtained (compare Tables I. and II.).

It must be remarked that minimal values of no practical significance have been omitted from the formulæ (12), (13), (12^1), (13^1), (14), (18), (18^1) and (19).

TABLE II.—CORRECTED VALUES FOR THE DIAGRAM TO A SCALE OF 20 TO 1.

a	a'	h'	a	a'	h'
Degrees.	Millimetres.	Millimetres.	Degrees.	Millimetres.	Millimetres.
− 30	50·43	146·80	∓ 0	66·955	0·00
− 28	52·42	140·33	+ 2	66·86	11·68
− 26	54·30	133·19	+ 4	66·59	23·29
− 24	56·09	125·43	+ 6	66·15	34·78
− 22	57·76	117·13	+ 8	65·57	46·09
− 20	59·32	108·26	+ 10	64·83	57·16
− 18	60·74	98·86	+ 12	63·93	67·93
− 16	62·04	89·04	+ 14	62·90	78·37
− 14	63·19	78·82	+ 16	61·71	88·38
− 12	64·19	68·23	+ 18	60·38	97·92
− 10	65·05	57·34	+ 20	58·94	106·98
− 8	65·75	46·19	+ 22	57·37	115·47
− 6	66·29	34·82	+ 24	55·67	123·37
− 4	66·67	23·31	+ 26	53·88	130·66
− 2	66·90	11·68	+ 28	52·00	137·28
∓ 0	66·955	0·00	+ 30	50·00	143·17

In the adjustment of the Hammer-Fennel tacheometer, certain requirements common to all theodolites hold good, such as :—The alidade-level must be perpendicular to the vertical axis, the line of collimation (or plane of collimation) must be perpendicular to the horizontal axis, and the horizontal axis must be perpendicular to the vertical axis.

Further adjustments are :—

(1) The axis of the level above the telescope must be made parallel with the line of observation (the latter not being coincident with the optical axis); (2) in case the constant C (and with it C_2), does not appear sufficiently correct, the achromatic lens between the two prisms must be shifted appropriately; (3) the diagram must be adequately set, so that the NN line (zero-line) appears to touch the zero-arc for any inclination of the telescope; and when the line of observation is horizontal, the curve of altitude should indicate zero while the edge of the double prism simultaneously bisects the image of the two asterisks $\vdots$ (Fig. 2). For inclinations, $\pm$ 30 degrees, the border-lines will then agree with the prism-edge. Thanks to check-screws, and by the general arrangement or action of the adjusting screws, re-adjustment is seldom required and proves to be not especially intricate.

When at work with the Hammer-Fennel tacheometer, it is simply needful, besides the ordinary reading of the horizontal limb or compass, to sight the zero-point of the vertical staff, and (without any further manipulation) to take the two readings :—l_1. where indicated by the E curve; and l_2, where indicated by the H curve; noting the sign + or − below the curve; whereupon the staff-holder can be sent to another station. It is thus quite possible to employ two or more staves alternately, since one sight and a reading off a vertical arc, as required with ordinary tacheometers (besides a further staff-reading with some), are saved and time gained.

As to the staff employed, it may again be noticed that the zero lies at the height of the diagram above the ground, which varies between 1·35 and 1·42 metres, according to circumstances; but in general tacheometry, for which the Hammer-Fennel tacheometer is primarily intended, it can be neglected.

When the staff is 4·00 metres (13·12 feet) in length, distances may equal 260 metres (850 feet), and where it is 4·40 metres (14·44 feet), distances may attain 300 metres (980 feet).

The graduation of the staff can, of course, be very much varied, the one used in the first trials had centimetre block-marks (Fig. 10) so that 1 millimetre could be read. An inaccuracy of 1 millimetre would give 2 centimetres of error of altitude or 1 decimetre in distances, but in order to read milli-metres the length of the sights is limited to 120 or 150 metres (400 to 500 feet). Another staff had scale-marks 1 centimetre apart and larger decimetre-numbers (Fig. 11). So-called tacheo-meter-staves in Germany are generally furnished with ½ deci-metre or 1 decimetre lines on one side of the staff (Fig. 12) and on the other side chequered centimetre-marks as on the ordinary levelling-stave. Fig. 13 is a type that the writer designed some years ago for the surveyors' school, after experience in pre-liminary surveys for railway-purposes, and it was found clearly readable for the most extended sights.

Sometimes it becomes necessary to level with a tacheo-meter, in which case, if a particular graduation be not provided on the back of the staff, the front may be complemented by centimetre-marks numbered downward from the zero-point; the readings must then be used with inverted signs.

It is advantageous to have the downward graduations painted in another colour, say, red, in which case all readings to red numbers refer to points so and so much higher than the station, or *vice versa*; and a back-sight to red or black numbers has to be subtracted from or added to the height of the bench-mark or given altitude to furnish the height of the station, while a fore-sight or intermediate sight to red or black numbers must be added to or subtracted from the height of the station.

Experimental work to which the Hammer-Fennel tacheometer, in its first form with $C_1 = 98$ and $C_2 = 24\cdot8$ and the diagram not yet adjustable, was applied in October, 1900, gave results which would be satisfactory for some purposes. Owing to the lines in the diagram being too thick and the edge of the double prism not sufficiently nice, the average error was for distances 0·5 metre and for altitudes 0·15 metre (maximum, 0·40 metre) on points well checked by other measurement and double levelling. The distances varied between 30 and 250 metres with inclina-

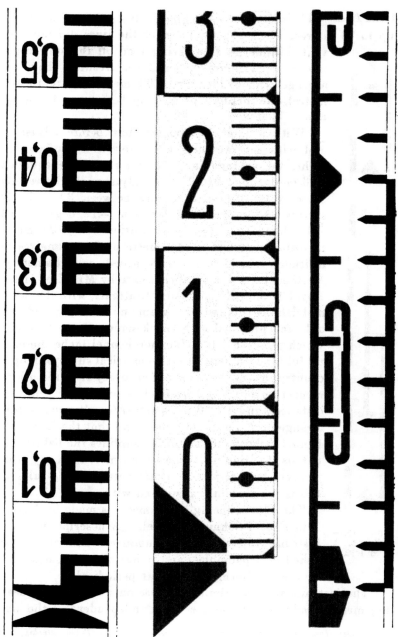

FIG. 10.—SCALE, 1 TO 3·55.   FIG. 11.—SCALE, 1 TO 2·40.

FIG. 12.—
SCALE, 1 TO 4·55.

tions from −14 degrees to −4 degrees. With shorter distances up to 36 metres, where the thickness of the lines did not prevent millimetres from being read off the staff bearing centimetre-graduations, the results show an average error of altitude of 0·1 metre (maximum 0·2 metre), the inclinations varying between +4 degrees and +14 degrees.

With a later instrument, but before adjusting and with several parts not screwed up sufficiently tight, the average inaccuracy for altitude equalled 0·23 metre; but, after proper adjustment, the results show, with distances varying between 30 and 260 metres and inclinations between −12 degrees and +12 degrees, an average variation of ±0·23 metre in distance and of ±0·06 metre or 6 centimetres in altitude, which may be regarded as satisfactory. Further trials of more recent date by Mr. Heer of the Royal Technical University, Stuttgart, with the latest and definite type (with diagram and other parts adjustable and provided with check-screws) furnish results which prove that the Hammer-Fennel tacheometer is useful for such general tacheometrical surveys in open country, where the degree of accuracy required allows a mean error of ±2 decimetres or 0·2 per cent. for distances, and of ±6 centimetres for altitudes within a range of 260 or 300 metres; and that the Hammer-Fennel tacheometer not only simplifies the field-work, but also facilitates the office-work more than any other tacheometer known. Tacheometrical slide-rules or tables can be totally dispensed with.

The auto-reducing tacheometer is made by Messrs. Otto Fennel Söhne, in Cassel. A description of the instrument, by Dr. E. Hammer,* has been published as a pamphlet, and it has also been described in various German technical periodicals.

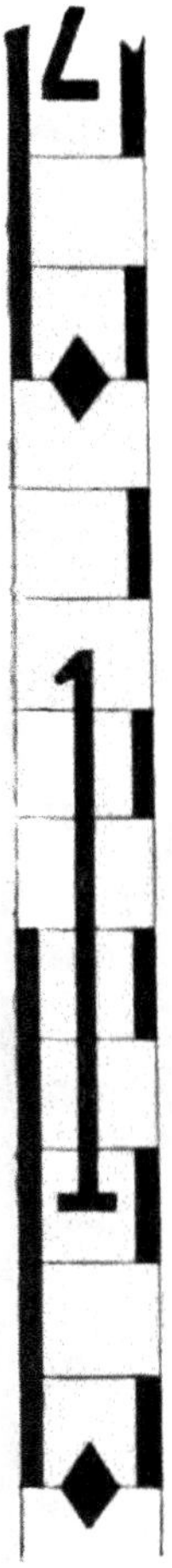

The instrument is not limited for use only with staves bearing metric graduations: the use being independent of the unit

FIG. 13.—
SCALE, 1
TO 1·12.

* Der Hammer-Fennel'sche Tachymeter-theodolit und die Tachymeterkippregel zur unmittelbaren Lattenablesung von Horizontal Distanz und Höhenunterschied D. R. Patent, Nr. 122,901, Stuttgart, 1901.

FIG. 15.—CROSS-SECTION ON LINE AB
OF FIG. 14.

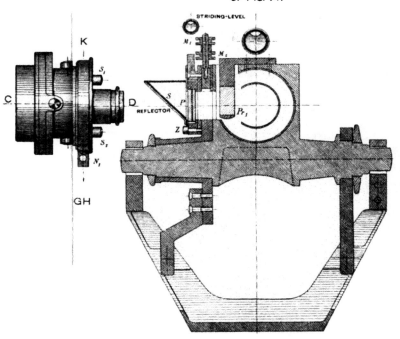

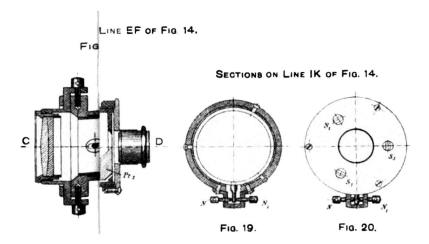

LINE EF OF FIG. 14.

FIG

SECTIONS ON LINE IK OF FIG. 14.

FIG. 19.

FIG. 20.

of measurement; and this, in addition to its good qualities alluded to above, allows the writer to predict with some degree of assurance that the Hammer-Fennel tacheometer will be largely used in the future.

———

The PRESIDENT (Mr. J. C. Cadman) moved a vote of thanks to Mr. Eoll for his interesting paper.

Mr. M. WALTON BROWN seconded the resolution, which was cordially approved.

———

Mr. SYDNEY F. WALKER read the following paper on " The Report of the Departmental Committee on the use of Electricity in Mines ":—

THE REPORT OF THE DEPARTMENTAL COMMITTEE ON THE USE OF ELECTRICITY IN MINES.

By S. F. WALKER.

INTRODUCTION.

Those who have watched the progress of the enquiry that was held in the early part of last year, into the use of electricity in coal and metalliferous mines, must have been struck with the completeness with which the enquiry was conducted, and the thoroughly practical nature of the investigation. Everyone who had any information to give that would throw any light upon the subject was welcome, from the working miner who had run an electrically-driven coal-cutter, and the miner's agent, keenly alive to the interests of his *clientèle*, to the professor who explained the principles upon which polyphase-machinery worked. The burden of every question asked by the members of the committee was, " what is the practical effect of so and so? " The experiments which were carried out also by Mr. W. E. Garforth, Mr. W. C. Mountain and others, on the effect of gaseous mixtures and enclosed motors, for the information of the committee, are of immense value.*

The thorough nature of the enquiry is especially creditable to the committee, as owing to unfortunate circumstances, they received very little assistance from the electrical member.

THE REPORT OF THE COMMITTEE.

The *Report* of the committee breathes the same liberal spirit as that in which the enquiry was carried out. The committee fully recognize the great usefulness of electricity in the present, and its still greater probable usefulness in the future. They

---

* *Report of the Departmental Committee on the Use of Electricity in Mines,* 1904, [Cd. 1916]; and *Minutes of Evidence taken before the Departmental Committee on the Use of Electricity in Mines ; with Appendices and Index,* 1904, [Cd. 1917].

have recognized, in a liberal spirit that, while electricity may
be dangerous, under certain conditions, it is not so dangerous
per se as other agents that are habitually used in mining operations, and that the saving of life and health to be obtained,
as well as the economy of working, by its aid, more than counterbalance any risk that is incurred by its use.   Some valuable
hints, as to working, are also given in the *Report*, such as that
trailing cables for coal-cutting machines and shot-firing cables,
shall be connected to the coal-cutting machine, and to the fuses
respectively first, and to the supply-service last.   Simple rules
of this kind may save many lives and much property.

A perusal of the *Report* leaves one with a feeling that the
fear which was so commonly expressed, when the enquiry was
first instituted, that the development of the use of electricity
in coal-mines would be restricted, was unfounded.

THE PROPOSED SPECIAL RULES.

In the proposed Special Rules, however, the liberal sentiments
expressed in the *Report*, and the practical line taken in the
enquiry appear to have been given a second place.   The fears
which were expressed at the commencement of the enquiry
will have a very substantial foundation, if the proposed Special
Rules become law, and we shall have once more to thank a
Government, which never helps British industries, for giving
American and German rivals additional chances in their favour.

The first thing that strikes an observer is the great length
of the proposed rules.   Considering the present number of the
General and Special Rules, it would have been thought that
a few simple rules, covering the main lines upon which electrical
development will proceed, would have been sufficient for the
purpose in hand, namely:—The protection of life and property.
A few simple rules, on these lines, would be of great service to
the colliery-manager.   A number of rules, such as those
suggested, can only be a great hindrance, and must lead to the
throttling of the development of the use of electricity.   The
committee themselves appear to have had this feeling, as in the
concluding paragraphs of their *Report*, they apologize for the
length of the rules, which they fear will cause alarm; but they
say that they are unable to deal with the subject effectively in
any other way.   The present writer respectfully begs to differ

from this view, and he ventures to submit a few simple rules, which he suggests will meet the objects that the committee had in view.

The writer suggests also that, in this case, the rules that are adopted should be specially adapted to the conditions of mining, and that it would be wise to break away from departmental tradition on the one hand, and to avoid a slavish imitation of methods in other branches of industry where electricity is employed. Grave mistakes have been made in the development of electricity in the lighting of towns, and in power-distribution, and the success which has attended those industries, not always financially great, has been achieved in spite of these grave errors. If they are to be carried into mining work, the results may be disastrous.

The Special Rules proposed by the committee appear to consist very largely of what is merely advice, which it is almost impossible to enforce, such as that the materials are to be of the best. Who is to decide what is the best, and with the advances that are so continuous, how are we to know what is the best at any particular moment? As advice, the fact that there is only from 10 to 20 per cent. of difference in cost between the best and the worst, is excellent, but in the form of a Special Rule, the same thing becomes absurd.

The division of installations into low, medium, high and extra-high pressures, is good for the Board of Trade, for work that they have to supervise; though British electrical engineers complain bitterly of the way in which the industry has been cramped by Parliamentary action and Board of Trade supervision. But for mining work, the division renders the work of the colliery-manager additionally and needlessly puzzling. Many of the rules for medium pressures should be applied equally to low pressures, if the maximum security is to be obtained; and again, many of the rules for medium pressures are repeated for high pressures, as though they were applicable only to those. Why should the colliery-manager be puzzled in this way? He wants help, not obstruction.

Again, in many of the rules, terms are used which will be unfamiliar to mining engineers, and are really inapplicable, such as " feeders." The term is one that was introduced many years ago by Mr. Thomas A. Edison, to describe cables which did perform the office of feeding a net-work in a town-lighting service, and it has since been applied generally to town-cables which are connected with the main switchboard, but there is no reason for its use in mining work. Such an use only leads to difficulties in the interpretation of the rules. The term " live metal," although applicable, should hardly have been used without explanation. Many of the Special Rules also are very vague, thus fuses are to go " on short circuit," or with 100 per cent. increase of current. A rule of this kind can only be misleading. If a fuse goes with 100 per cent. of increase of current, it must go on " short circuit," unless the conditions are very special, and such as would hardly require the protection of a fuse.

It is common experience that it is very difficult indeed to provide for all cases, in matters of this kind, and it is particularly difficult where, as with electricity, knowledge and experience are advancing rapidly, while points such as " earth " are by no means thoroughly understood. On the other hand, it is perfectly practicable to lay down a few general rules which shall cover all contingencies, and at the same time allow for development. In the Special Rules given below, which the present writer ventures to suggest, he submits that the requirements of the committee are all provided for fully, although the method of meeting their requirements is left open. There is plenty of room, within the four corners of the Coal-mines Regulation Acts, and the General and Special Rules, for enforcing all the requirements of the committee, on the lines that have hitherto ruled in mining administration. It should be noted also that many of the Special Rules proposed by the committee are provided for by either the General or Special Rules already in force, such as those referring to damage, fencing machinery, and others.

Rationale of the Suggested Special Rules.

In the suggested Special Rules, the idea has been to follow the spirit of the Coal-mines Regulation Acts, and to make use of the machinery provided in them, so far as it is available, and thus to avoid the necessity of further burdening the already heavily-weighted colliery-manager. The idea has also been to develop the principle of responsibility, in place of attempting to provide leading strings.   The writer's view is that, if you enact that certain perfectly-practicable things are to be done, and you hold a given official responsible for their being done efficiently, all experience shows that they will be done. While, if you endeavour to provide for every contingency, with the idea that you are assisting those who have to carry out the rules, you bar out the best class of men, those who are not afraid to accept responsibility, and you provide innumerable loop-holes for the devolution of responsibility. Meanwhile you have not really added to the safety of the apparatus.

As the writer understands the Coal-mines Regulation Acts, H.M. inspector of mines for the district is the guiding spirit, and the judge of first instance; the colliery-manager is the responsible official, and he is assisted in matters requiring special knowledge of mechanical engineering, by a properly qualified mechanical engineer. The idea is that the same spirit shall rule with electrical apparatus. H.M. inspectors of mines must already have acquired a considerable amount of experience in the matter, especially those who have had mines under their charge, in which accidents have occurred, and as times goes on, their experience will naturally increase. It will follow, as it has in other matters connected with mining work, that what H.M. inspector of mines wishes, will be done, unless there are special reasons against it; such as that the expense would be very heavy, while the advantage would be small, or when there is a grave difference of opinion as to its advisability. Cases will arise similar to those that have arisen in connection with the use of safety-lamps in certain mines, and there will probably be a great many such instances as the use of electricity in coal-mines advances.

In these cases, however they may arise, the machinery already existing in the Coal-mines Regulation Acts is quite sufficient for all purposes. The manager of each colliery, the writer suggests, should be advised by an electrical engineer, whose qualifications and remuneration should vary with the size of the plant, and of the colliery, just as he is now advised by a mechanical engineer for all mechanical matters.

The committee have indicated in the proposed Special Rules that electrical engineers must be employed, but they have hesitated to enforce their view in the only manner that is likely to work satisfactorily. The electrical engineer at the colliery must be responsible for all the electrical apparatus at the colliery, to the colliery-manager, just as the mechanical engineer is, and must be supreme in his own department, subject to the over-riding authority of the manager, and of H.M. inspector of mines.

As indicated above, the qualifications of the proposed electrical engineer will vary with the size of the plant and with the size of the colliery. One important qualification that he must have is familiarity with mines. In the writer's view, probably, for large firms, the best man to be found would be a mining engineer who had qualified in electricity and mechanics. What is wanted for the work is, not a knowledge of the latest theory of electricity, but an intimate acquaintance with mining conditions, and the ability to instruct his men in the way of quickly putting little things right. In electrical work there are so many little matters which may stop a plant from working, that a man with the requisite knowledge, not necessarily deep, can quickly put right.

The present mechanical engineers at collieries also, who qualify in electricity, should very well fill the positions that will be open, if the suggestion is carried out, and in any case before very long.

The writer confesses that he would place very little faith in electrical engineers, pure and simple, unless they have had considerable experience in mining work. Electrical students, fresh from technical colleges, should be rigorously avoided, at any rate until they have had opportunities of acquiring sound experience under men who are thoroughly conversant with mining, and with electricity.

For small mines, and small plants, men who correspond to the engine-wrights at small collieries would be suitable, and they might be subjected to the occasional supervision of the electrical engineer of one of the large firms.

THE SUGGESTED SPECIAL RULES.

(1) Pressures, exceeding 3,000 volts, shall not be used anywhere in or about a mine, and pressures above 650 volts shall not be used outside the main intake airways, or chambers supplied directly from the main intake airways without the permission in writing of H.M. inspector of mines in charge of the district.

(2) All machine-carcases, cases enclosing switches and fuses, and any other metals (excluding nails and screws) used to protect, or to support conductors belonging to the electric service, but not in conductive connection with the service, shall be connected to earth, except in cases specially exempted by H.M. inspector of mines in charge of the district.

(3) No part of the conductive system of an electric service at a mine shall be connected to earth, except by special permission of H.M. inspector of mines in charge of the district.

(4) All electric apparatus in use at a mine shall be so fixed, and maintained that in the ordinary working of the mine:—(a) No part of the apparatus shall become unduly heated; (b) no arc shall be formed in any part of any apparatus; and (c) it shall not be possible for those using or engaged in the use of the apparatus to sustain a shock, while carrying out their duties in a proper manner.

(5) All cables and machinery shall be tested not less than once a week for insulation, with a current the pressure of which is not less than that of the working current, the tests being recorded in a book provided for the purpose, and the insulation of the whole of the apparatus must be maintained in such a condition that the leakage-current at any instant is not greater than 0·0001 of the maximum total current.

(6) All connections between service-cables and generators, motors, transformers and lamps shall be properly controlled by switches, and protected by fuses or electro-magnetic apparatus, in such a manner that No. 4 Special Rule is fully complied with, and in those portions of mines where No. 8 General Rule applies, all switches, fuses and junctions between cables and any places where the circuit is made and broken, shall be enclosed in gas-and-flame-tight boxes. Where trailing cables are used, in connection with machinery in motion, or for shot-firing, a connection shall not be made with the supply-service, until a connection has been made with the motor, or fuses, and everything is arranged; and no electric motor shall be worked in an explosive atmosphere.

EXPLANATION OF THE SUGGESTED SPECIAL RULES.

(1) The writer is of opinion that the first suggested Special Rule enacts all that is necessary in regard to the use of pressures, and it is specially drawn so as not to cramp the development of the use of electricity in mines. Probably it will be some time before it becomes economically advantageous to use pressures exceeding 3,000 volts; while on the other hand it is not difficult to imagine some engineer who may have been appointed to a colliery with an imperfect knowledge of engineering using higher pressures without necessity and without economy. There is nothing in the committee's proposed Special Rules to prevent him from doing so, provided that he uses motors of 75 horsepower. The provision that only 650 volts shall be used outside the main intake airways also ensures safety, while the provision that H.M. inspector of mines may permit higher pressures in both cases, allows of expansion with increased knowledge and experience. The writer is also of opinion that the provisions necessary to ensure safety from shock above a pressure of 250 volts are equally necessary below it. There have been deaths, in towns, with 200 volts, and less, under conditions apparently very much more favourable to the victims than will often exist in a mine. The time-factor has an important bearing on the subject; even 100 volts will kill if time be allowed, and it has been shown that 5,000 volts may not kill if the victim's skin is burnt quickly so as to block the passage of the current through him.

(2) The second rule suggested by the writer embodies the views of the committee repeatedly expressed in their *Report,* and in their proposed Special Rules. On the other hand, it is open to grave objections in many cases, and the idea is that these cases should be represented to H.M. inspector of mines.

(3) The third rule also embodies the views of the committee.

(4) The fourth rule comprizes the different instructions which run through the Special Rules proposed by the committee, but it differs from the method of the committee in that, while strictly enforcing arrangements that ensure safety, it refrains from attempting to say how they shall be carried out. The

provision that no part of an electric apparatus shall be unduly
heated, is, it will be seen, very far-reaching. To ensure that
this is so, the engineer must follow certain rules with regard
to the density of current in the cables, the insulation of
the cables, the size and arrangement of the switches, fuses and
junctions; and in the size and insulation of the conductors in
the generators, motors, etc. He must also provide proper
measuring instruments. In fact, he has to design and maintain
the apparatus as well as the state of knowledge at the time
allows. The writer has purposely refrained from specifying
that a certain rise of temperature must not be exceeded, as in
his view this would hardly meet the case of the very variable
conditions of mine-working, varying air-currents, varying loads,
varying time of working, etc. He is of opinion, as already
indicated, that H.M. inspector of mines would be the guide.
If the inspector saw a motor that in his opinion was working
too hot, he would call for it to be altered. In the matter of
density of current in cables, economy in transmission will generally rule a low density, but again the permissible densities vary
considerably with the conditions. In a powerful air-current,
cool from the surface, a comparatively high density may easily
be quite safe, while the density allowed by the Special Rules
proposed by the committee might be unsafe in others. Personal
observation must be the guide in these matters.

(4) The second section (b) of the fourth rule is really a part
of the first section (a), but it is placed by itself, so as to direct
especial attention to it. The rules for guarding against arcing
in electrical apparatus are now well-known, and the test would
be in practice. A provision that fillets shall be fitted or that
a certain distance shall exist between certain parts will not be
sufficient. An arc can be made over a considerable distance with
the aid of a deposit of copper or mine-dust. Again, the idea is
to show what should not be done, and to throw upon the colliery-
manager, and his electrical adviser, how it shall be arranged.
H.M. inspector of mines in charge of the district should be able
to give valuable advice, as he would see different forms of
apparatus, and what is of far more value, different forms in
which the same trouble may arise.

The third section (c) of the suggested fourth rule again

embodies all that runs through the *Report* of the committee and their Special Rules, but it also throws upon the colliery-officials the onus of providing efficient means of carrying out the rule.

(5) The fifth suggested rule provides all that the committee have aimed at, and the writer is of opinion that, if the provisions are carried out, there is no necessity for an "earth" indicating device. It is very doubtful, indeed, whether it is wise to adopt the "earth" indicating device. It might very possibly introduce complications into the system, and if careful tests are taken periodically, and the measuring instruments carefully watched, ample warning will be given of any coming trouble. In all of these cases, it is common experience that the smallest number of "accidents" occur where the man in charge constantly watches his apparatus, and keeps in close touch with it, and this he would be able to do by means of the insulation-tests, and the measuring instruments. It will be noticed that no mention is made in the suggested Special Rules as to how the cables are to be insulated, of the insulation-resistance, whether naked cables shall be allowed, nor of how cables are to be tested. The writer is of opinion that all of the above requirements are provided for. Naked cables would be perfectly permissible in situations where there would be no liability to shock, or to accidental connexion between them, and where the insulation can be maintained. How the cables are to be insulated, and what tests are made on them before they are fixed, may also be left to the manager and his electrical adviser. Naturally, the manager and his adviser, and the manufacturers of cables, will observe those tests which will ensure the cable having the longest life, so far as insulation is concerned. It will also be observed that the case of arcing, or discharging between cables, is provided for in the fourth suggested Special Rule.

The writer may observe, however, *en passant*, that he views with considerable disfavour the Special Rule proposed by the committee, providing that cables shall be tested with higher voltages than those to which they will be subjected in work He is of opinion that by so doing the insulation of the cable is strained, and deprived of a portion of its life as to insulation-resistance, and resistance to sparking. Test a sample-piece as

much as you like, in whatever way you like, but do not strain
the cable itself.

Objection has been taken to the quantity of the leakage-
current allowed. This is a matter that can be adjusted, but in
the writer's opinion, the higher the standard that is fixed, and
the lower the maximum leakage-current allowed, the better will
the apparatus work as a whole.

(6) The sixth suggested Special Rule is really provided for
in the other rules, but it has been thought desirable to add it,
in order to emphasize the points previously mentioned. Again,
it will be noticed that this Special Rule, together with the others,
provides all that is required in the matter of the control of
motors, lamps, transformers, etc., but without specifying the
method. Arcing, for instance, in a starting switch and overheating in a starting switch are equally forbidden, although the
means of ensuring that they shall not occur are left to the judgment of the responsible officials. The possible reversal of a
generator when running in parallel is also forbidden, without
stating that a reverse automatic cut-out is to be used. It would
probably be wise, for some time to come, not to run generators
in parallel, in colliery electrical services, but where they are,
reversal would lead to arcing, and to heating, and is therefore
forbidden. The last portion of the sixth suggested Special
Rule provides for the case of fiery mines, and it also provides
for the case of motors working in fiery mines, the commutator
and the slip-rings being places where the circuit is broken.

The Special Rule of the committee providing that the electric
motor working a coal-cutting machine shall only be run for a
certain time, and shall be stopped when gas is present, is provided for by the suggested Special Rules against undue heating, and the clause of the sixth suggested Special Rule, for-
bidding electric motors from working in an explosive atmosphere.
The question of repairs and the disconnection of the service,
while they are being done, is provided for by the rule against
shock.

The writer believes that in the suggested Special Rules, he
has provided for all contingencies, on the lines indicated, the

responsibility of the manager and the guidance of H. M. inspector of mines. He wishes to say, however, that the suggested Special Rules are put forward tentatively, with a view to the discussion of the whole subject, and if the main lines on which they are drawn are approved, they can be added to, subtracted from, or altered as may be desired.

EARTH.

There is one important matter which the writer thinks it well to discuss. One of the few points on which he disagrees with the committee, in their *Report*, is the way in which "earth" is dealt with. Cables carrying currents above 650 volts pressure, are to be enclosed in armour or in metal pipes, the armour or the pipes being earthed. Also, all cases, etc., as described in the second suggested Special Rule, are to be earthed.

The writer is very strongly of opinion that cables should only be armoured in certain special cases, as in that of a three-phase cable in a shaft, and that the wisdom of earthing everything is doubtful. His view is that "earth" should be kept out of, and away from any electrical service, everywhere, but especially in collieries. He would permit "earth" in special cases, but he would rather not use it.

The questions of what is earth? and how do you earth? are almost like the celebrated question asked by Pontius Pilate. In a coal-mine, earth is the coal-seam, the metals that are connected with it, and any body of water such as a river, with which the metals may be connected. Both water and coal are poor conductors, but both acquire considerable conductive ability when present in large masses, as in a coal-seam, or the water of a large river. To obtain connection with each, large surfaces must take part in the contact, and, with water, this is comparatively easy. A large metal-plate dipping into the water makes good connection with earth. With the coal-seam it is not so easy, as the surfaces will not cling together as will those of water and an immersed metal-plate. With both also, electro-chemical actions, due to the currents which pass to "earth," often alter the whole problem. In both cases, earth-connections are frequently made when not wanted, yet are difficult to make where they are wanted. It is for these reasons that the writer has suggested that the matter shall be left to the judgment of H.M. inspectors of mines.

THE ADDITIONAL WORK TO BE PLACED UPON H.M. INSPECTORS OF MINES.

It may be objected that, by the suggested Special Rules too much responsibility is thrown upon H.M. inspectors of mines. The reply is that no more is thrown upon them than they should bear, in accordance with the spirit of the Coal-mines Regulation Acts, and the developments that have taken place in electrical matters. These Acts especially provide that the Secretary of State may appoint as many inspectors of mines as he may think fit, and presumably he may give electrical advice to any of H.M. inspectors of mines when needed. But, as already pointed out, it is the colliery-manager on whom responsibility would fall, and he would be able to use his judgment, instead of being tied by Special Rules which, even if correct to-day, in the rapid advance of electricity, may become obsolete at an early date.

To take a simple instance, under the suggested Special Rules, if a fire occurs in an electric-generator house, the colliery-manager will be responsible, no matter what the cause; but under the Special Rules proposed by the committee, if he could show that he had carried out the Special Rules, his responsibility would cease. Surely the suggested Special Rule makes far more for safety than the second rule of the Departmental Committee, while it at the same time allows the manager to study the factor which British governments so persistently ignore, but which American and German governments so persistently study, the interest of the colliery-owner. The colliery-manager has also, in addition to the suggested electrical engineer, many engineers able and willing to advise him on every question that may arise.

———

Mr. T. LINDSAY GALLOWAY (Glasgow) asked whether the maximum current of 650 volts mentioned by Mr. Walker in his paper was the virtual or actual voltage.

Mr. W. C. MOUNTAIN (Newcastle-upon-Tyne) said that, whilst appreciating what Mr. Walker had done in endeavouring to reduce the number of rules to six, he (Mr. Mountain) thought that it would be far better to have rules which would be a guide to engineers as to what they should do, and would enable them to erect installations in such a way as would protect life.

and limb; but the rules suggested by Mr. Walker, he thought, did not do this. Although the proposed Special Rules were too long for many engineers, he did not think that they would prove so objectionable as some of them might think at the present time; at the same time, the proposed Special Rules might require some amendment, as some of them, if carried out, would mean the expenditure of much money without any resultant increased safety. He (Mr. Mountain) thought that the proposed regulation as to the fencing of gearing, would prove a source of danger and not of safety, and if anyone wished to cut off the current he would probably fall over the handrails.

He (Mr. Mountain) had suggested, at a recent meeting of the Institution of Electrical Engineers, that three-phase motors up to 1,000 volts should be of 20 horsepower, up to 2,000 of 30 horsepower, and up to 4,000 volts of 50 horsepower; but he thought that they might go up to 4,000 volts with motors of 30 horsepower. Respecting the rating of motors, he (Mr. Mountain) did not think it wise to make any hard-and-fast rules on the subject, and the question of temperatures would therefore have to be carefully revised.

The question of earthing was also under consideration. Mr. Walker had drawn attention to a very important matter, as there seemed to be a little confusion in people's minds as to what earthing meant. Of course the earthing of the motor-frames and switch-cases was a different matter, but he thought, on the whole, that all the rules respecting earthing of wires, etc., were unnecessary, and meant additional expense without additional safety.

With respect to the use of the words " competent person " in Section 1, rule 21, he thought that the rule, as it now stood, was a fair one, and did not bind anyone to employ a highly paid electrician, who would probably not be of much use on the job.

The Departmental Committee had suggested that cables should be tested for ½ hour with a voltage equal to the working voltage, which seemed to him to be all that was necessary, and any higher voltage would only strain the cable. If the members would carefully read the proposed Special Rules they would see that the whole question had been left almost entirely to the colliery-manager, with regard to the armouring of cables, as it was not necessary to armour them if they were fixed properly (Section 3, rule 5).

He (Mr. Mountain) believed that, after careful consideration by all parties concerned, the proposed Special Rules would be amended where necessary, and if this were done it seemed to him better to have a code of Special Rules for the installation and use of electricity in mines rather than a few rules which left one in doubt on many points.

Mr. C. C. LEACH (Seghill Colliery) considered that H.M. inspectors of mines should not be empowered to fix any limit as regards pressure; if he (Mr. Leach) desired to double the pressure, an inspector of mines should not interfere, as he (Mr. Leach) took all responsibility.

Mr. H. RICHARDSON HEWITT (H.M. Inspector of Mines) alluding to the last speaker's remark, said that, if an accident happened, an inspector of mines was called in. Mr. Walker, while complaining of the great length of the proposed Special Rules, had drafted six suggested Special Rules, and then added two or three pages of explanations, making them nearly as long as the proposed Special Rules that he condemned. Mr. Walker failed to recognize that there were very few collieries where the whole of the proposed Special Rules would be required; and those rules applying to the requirements of the particular colliery only would be adopted, so that the rules were not so formidable as they seemed. He did not think that the proposed Special Rules respecting the use of electric locomotives would be required in any colliery. With reference to the protection of gearing, etc., and the liability of a man falling over the fencing, he (Mr. Hewitt) thought that the fencing should be erected so that a man in his normal condition could not possibly fall over it. Mr. Walker suggested that an inspector of mines, when he saw "a motor . . . . working too hot" would interfere and stop its working. An inspector of mines did not devote all his time to one colliery, possibly each mine was visited twice a year, and, therefore, he was not able to see everything that took place. Mr. Walker stated that he disagreed with the manner in which "earth" was dealt with by the Departmental Committee, and then, at great length, explained what the committee did mean.

Mr. HERBERT A. JONES (Bradford) said that he did not think that the proposed Special Rule was necessary, requiring the current, when it was switched off under full load, not to form an arc. Then, with regard to the question of leakage, they might have a transformer which was only 10 feet or it might be 10 miles away, and the leakage should not be the same in each case. With regard to motors, which had to run for six hours at their normal load, it was very difficult to find out what that load was in the case of coal-cutting. He hoped that the Departmental Committee would look into these and other points from the point of view of the manufacturer. There were plenty of difficulties put in their way already, and a series of rules, such as these, were enough to frighten manufacturers of electrical machinery. There was, in fact, at present a scare in certain districts and a desire not to order electric plant until colliery-managers saw whether the proposed Special Rules were to be made compulsory; if they could be reduced to ten or twelve rules, covering the general principles, he thought that it would be better for both the mining industry and the electrical industry.

Mr. G. S. CORLETT (Wigan) said that Mr. S. F. Walker had tried to deal with what seemed to him, and what had always seemed to him since the enquiry had commenced, an extremely difficult situation. He confessed that he had always thought that it was too difficult for an ordinary man to draft a set of rules, which would be applicable to all possible conditions of British mining, and yet would be sufficiently rigid to ensure the main object of the rules being attained, namely, that the work throughout should be good. The Departmental Committee had drafted Special Rules, and if a colliery-manager did not do everything that the rules required, unless he were exempted by an inspector of mines, he would be liable to prosecution. Mr. Walker had substituted six rules, and he (Mr. Corlett) considered that it was not an unfair paraphrase of these six rules to condense them into one, namely, "No electricity shall be used in a mine, except by special permission of the inspector of mines in charge of the district."

He (Mr. Corlett) argued, from the point of view of an electrical engineer, that they should know a little in advance what sort of plant was going to be satisfactory, not only to the colliery-manager but to the man who was going to pass it. It seemed perfectly absurd to him, if a particular colliery wished to instal electric plant, that a contractor or would-be contractor should worry an inspector of mines and ascertain whether he approved of the proposed installation.

With regard to the question of limiting the pressure to 3,000 volts, power-companies had pressures of 5,000 volts, and if a colliery wanted to purchase a small amount of current from one of these companies it would have to be transformed down to a lower pressure for use in the mine; and it would be necessary for land to be bought outside the area owned by the electric company, in order that transformers might be erected.

In South Wales, there were many square miles where it was impossible to get satisfactory earthing, and the same could be said of other districts. He (Mr. Corlett) believed that any one who had practical experience in the matter would agree with him in stating that a bad earth was worse than no earth at all. There should not, in his opinion, be a rigid rule requiring all machine-carcases, etc., to be earthed. In many collieries the armouring disappeared entirely in about three months, from dampness and other causes, and it was therefore impracticable to depend upon it for the purpose of earthing.

Mr. ROSLYN HOLIDAY (Ackton Hall Colliery) said that he had found several cases, where the proposed Special Rules if enacted as they now stood, meant that the collieries would be closed, as they could only be worked economically with coal-cutters. Further, if a colliery-manager wanted a bell put into his office, he must advise the inspector of mines that he was going to put in an electric installation; he would have to put in a cable to stand a pressure of 2,000 volts; when the bell was installed, he must make and keep a plan of the installation; and, he supposed, submit it to the inspector of mines. Further, nobody must touch that bell without first switching off the current. It was difficult in reading the proposed Special Rules to see how they would apply in all cases, but, in some instances, it had been found that they were really impracticable.

In his second suggested Special Rule, Mr. Walker proposed that all machines-carcases, cases enclosing switches and fuses, etc., should be connected to earth, but he (Mr. Holiday) agreed with a previous speaker that this was a matter of great difficulty, especially in South Wales, where it was almost impossible to get any connection with earth at all. Then in the proposed Special Rules, it was required that the armouring should be continuous as well as earthed. If a section were switched off at the end of, say, a haulage-plane, in order to do some repairs, and the armouring was continuous there; but if in some other part of the district there was a fault which forced the armouring, and the other armouring became earthed, the men might receive a severe shock.

It was impossible to carry out Mr. Walker's proposed Special Rule (4b) that "No arc shall be formed in any part of any apparatus." The rule should read that "No arc shall be sustained," etc.

The 0·0001 of the maximum total current was infinitely too high in many cases occurring in practical experience. The question was one of the most difficult to settle, and he did not know how it was to be solved. There was one colliery where they were working in a very wet seam, with everything insulated in a first-class manner, and yet the leakage came up to 0·4 ampère in 3,000 feet. The installation was well supervised; the pressure was 400 volts; and the insulation was 3,000 megohms indiarubber. This was therefore one of the places that would be closed if this rule came into force. They might, of course, obtain exemption from the inspector of mines.

He thought that Mr. Walker's proposed Special Rules placed too much responsibility upon the inspectors of mines. It was not fair that an inspector of mines should be required to give exemption from some of the rules; the inspector did not know all the conditions, and might say to himself: "If I grant this exemption and an accident happens, the manager can say that he held my exemption." It was absurd to say that the leakage should not exceed so much of the maximum current; they might as well expect a gas company to make a similar rule : there was more leakage on a circuit 5 miles long than on a circuit 1 mile long.

The last of Mr. Walker's suggested Special Rules requiring that "No electric motor shall be worked in an explosive atmosphere," he (Mr. Holiday) considered unnecessary. If an atmosphere were explosive, gas would be burning on the flame of the lamp, and this was forbidden by the General Rules. In such a case, the miners cleared out, and the coal-cutter would be left unworked.

Mr. C. C. LEACH (Seghill) protested that the Home Secretary should not force colliery-managers to consult the inspectors of mines, in order to settle points which they (colliery-managers) had hitherto settled for themselves, and still were able to do.

Mr. A. H. STOKES (H.M. Inspector of Mines), wrote that it was pleasing to note that Mr. S. F. Walker admitted the thorough and practical nature of the inquiry, and this inquiry being the basis of the rules, it would naturally be thought that a Committee, who had made such an investigation, would also be thorough and practical in any code of rules that they might suggest. If the proposed Special Rules had not been of a thorough and detailed character, the Departmental Committee would have been charged with having made bad use of the material that they had laboriously gathered. It might be granted that the rules were numerous, but the use of electricity in mines was modern, and its application and control so little understood, that it would have been a grave error not to have provided rules to meet, as far as possible, all the dangers brought to the notice of the Committee. The rules were prepared principally for those having the care and control of the electric installation, and not for the ordinary miner. If the standard fixed by the rules only prevented the introduction of shoddy work, they would accomplish much towards the prevention of accidents.

The author stated that "Grave mistakes have been made in the development of electricity," and he (Mr. Stokes) might add that the gravest mistake had been in not having controlling rules to secure good work and safety. In another part of the paper, complaint was made of the use of technical terms, but such terms were the recognized terms used by electricians. The term "live metal" was objected to, but could a more expressive term be found for metal charged with electricity? Was not "live metal" a better-understood term than "machine-carcases," used by Mr. Walker in his second suggested Special Rule?

The most interesting part of the paper was where the author proposed to deal with the whole question by six Special Rules, but it only required a cursory glance at such rules to see that not only could the proverbial carriage and pair be driven through them, but a tyro in mining could drive an elephant and a pullman car with ease through any of the six rules. The first of the author's rules contained an interesting conundrum, namely:— Where did a main intake airway end? The second rule was a curious example of the style of an author who complained of the use of the term " live metal," but himself used such terms as " conductive connections " and " electric service." The fourth rule was ambiguity itself: what was " unduly heated "? and " it shall not be possible for those using or engaged in the use of the apparatus to sustain a shock, while carrying out their duties in a proper manner?" If a person did get a shock, it would be *prima-facie* evidence that the person was not " carrying out his duties in a proper manner."

The first three rules in a large measure depended upon " the special exemption by H.M. inspector of mines." He did not think that it would be difficult for the inspector of mines to make one rule which would meet all conditions of electrical work; but he was afraid that neither owner, agent, manager, nor workmen would be willing to accept the same. The gist of the author's rules is that H.M. inspector of mines should become an electrical consulting engineer, and that rules are quite unnecessary.

It was not necessary to follow the author further through either his rules, or the remarks thereon. Special Rules were of no use unless made statutory; and, if statutory, then they should express clearly and in detail the object of the rule, so that those having the responsibility of carrying them out should have no difficulty in understanding the extent of their application, and those having to enforce them should equally know the limit of their power to have them observed.

It will be known to those who have read the evidence taken by the Departmental Committee, that the question " Do you believe in good work?" was put to experts over and over again; and of course everyone replied " yes," but when asked how they would secure " good work " a discreet silence was observed, and the question was not once satisfactorily answered. Clearly then, the only way to secure good work was to make a code of Special Rules which would necessitate good work so as to comply with the standard laid down by such rules.

Mr. S. F. WALKER, replying to the discussion, said that, with regard to the question of leakage, which he had expected would be taken up, he might say that he agreed with the gentleman who said that it should vary with the distance. He had simply taken the 0·0001 from the *Report* of the Departmental Committee. He had in his paper taken the *Report* and endeavoured to get behind the minds of the Committee, and had put into his suggested Special Rules what he considered that the Committee had intended to convey. The proper way to treat the question of leakage would, of course, be to take the mileage estimate, an ampère per mile or 1,000 feet, or so on. With respect to the question of the arcing of switches, they all knew that every switch made an arc : the word should have been " sustained," which would make the rule intelligible.

He considered that the greatest justification for the remarks that he had made in his paper, if justification were wanted, was found in the discussion that had taken place. Nobody was satisfied with the proposed Special Rules; everybody hoped that they would be modified; but nobody could say how they should be amended. Mr. G. S. Corlett had suggested that they should be condensed into one rule; and the members could not have a better one than the one he had suggested.

The reason why he had put the inspector of mines in so prominent a position was because, so far as he (Mr. Walker) understood the Coal-mines Regulation Acts, he held that position with regard to everything about a mine. Mr. Leach had referred to the adoption of extra pressure, and was promptly taken up by an inspector of mines, who said that if an accident occurred, the inspector of mines would soon be there. He believed that it would be better for an inspector of mines to give advice, as at present, rather than to make him an overriding authority. Take the question of earth for instance, there were two extremes, where one could not get earth and where one got it whether one wanted it or not. In such a case, the inspector of mines might be consulted as an adviser. If a boiler was erected and the inspector of mines objected, the manager took the responsibility. The manager can explain to the inspector of mines that he cannot get earth, and there is no question of exemption if the inspector of mines does not see his way to accept the responsibility. The colliery-manager had had the responsibility in the past; let him retain it in the future if they liked. Inspectors of mines gathered experience, and should be able and were always willing to give advice and assistance. In any trouble, the manager should communicate with the inspector of mines, and the manager would surely be in a much stronger position, if he consulted the inspector of mines.

Some one had suggested that there was no necessity for limiting the pressure to 3,000 volts, and with this he agreed. He had not, however, suggested it; the rule was that the pressure was limited to 3,000 volts, unless the inspector of mines said that a higher voltage might be used. Surely it was a good reason for going to 5,000 volts, if a supply could be obtained at that pressure; and if the Board of Trade allowed that pressure to be supplied, he thought that the inspector of mines, as a reasonable man, would sanction its use.

The CHAIRMAN (Mr. H. C. Peake) said that the most cordial thanks of the members were due to Mr. S. F. Walker for his paper and for the trouble he had taken in writing it. The paper had been rendered more valuable by the discussion which had taken place, and he thought that very few could go away without having learned a great deal from it. He (Mr. Peake) had not much to say about the Special Rules, but his feeling was to condense them into one rule: " No electricity shall be used in a mine," as that seemed to be the ultimate effect of the proposed Special Rules. He had pleasure in proposing a vote of thanks to Mr. Walker for his paper.

Mr. C. C. LEACH seconded the resolution, which was cordially approved.

———

Mr. ROSLYN HOLIDAY read the following paper on " A Comparison of Three-phase and Continuous Currents for Mining Purposes ":—

A COMPARISON OF THREE-PHASE AND CONTINUOUS CURRENTS FOR MINING PURPOSES.

By ROSLYN HOLIDAY.

Not many years ago the use of electricity was new to mining engineers generally, and any papers on the subject prepared for this Institution dealt at length on the differences between series and shunt-machines, and whether a motor could possibly be applied to many of the operations connected with mining. Things have so changed now, that sometimes the meetings of this Institution might almost be mistaken for meetings of the Institution of Electrical Engineers. The conditions under which electricity is used about a mine are so very different from its other applications, and the amount of experience in this special field is comparatively so limited when contrasted with compressed air for instance, that it is in the best interests of mining that there should be as much exchange of opinions and experience as possible, and this is the author's apology for another paper dealing with electricity.

Three-phase Current.—The question which now always presents itself to those about to apply electricity to one or more purposes in their mines, is, which system shall be used: continuous or three-phase current. As the use of three-phase currents in mines is only recent, it may not be out of place to say a little about this system. Three-phase currents are alternating, each of the three currents varying between a maximum and a minimum of pressure and current. Each complete rise and fall is called a period or phase, and the number of these per second is spoken of as the periodicity or frequency.

In order to generate these currents, a dynamo is used, generally spoken of as an alternator, but the armature-coils, in which the current is generated, are usually stationary, and the magnets and magnet-coils revolve, surrounded by the armature-coils.

When generating currents at high pressures, this is a great advantage, as the coils carrying the high-pressure currents are stationary, and not subject to mechanical vibration, while the revolving coils only carry a low-tension current used for magnetizing the magnets. The alternating three-phase current cannot be used direct to magnetize the magnets, so that in most cases a separate small continuous-current dynamo is used to supply the current for this purpose.

The mains are connected direct with the armature-coils, without the intervention of rubbing contacts, or any liability of sparking. For the transmission of three-phase currents, three wires must be used, and switches, fuses, etc., must be of the three-pole type. One of the most valuable features of the three-phase system is the facility with which the pressure can be transformed up or down, without the intervention of an apparatus containing any moving parts. Where power has to be transmitted for a considerable distance, this can be done on relatively small cables at high pressure; and at the various points of distribution it can be transformed down to a pressure suitable for the various purposes to which it is to be applied. Three-phase currents can be applied to the running of motors and the lighting of lamps, either arc or incandescent, but they cannot be used to charge accumulators direct.

The special features about the three-phase system are the motors. These consist of a circular stationary case (wound with three series of coils), termed the stator, and a revolving drum (either wound with three series of coils or covered with a squirrel-cage formation of copper-bars) termed the rotor. The high-pressure current is supplied direct to the stator-coils, the rotor need have no connection from outside, as any current flowing there is generated by induction, and is at a very low pressure.

Up to 30 horsepower, these motors can be started without the use of any resistances or other complications. The writer has used a number of these for driving coal-cutting machines, and as far as he knows, he was the first in this country to use three-phase current for this purpose. With larger sizes of motors, there are two principal systems of starting in use at present. One is by means of a special form of switch and small transformer combined, termed an auto-transformer, in the main circuit. In the other system, there is a three-pole switch in the main circuit, and on the rotor-shaft there are three insulated brass-rings, connected to each of the three series of rotor-coils. From these rings, the current is led to an arrangement of variable resistances. For motors up to 30 horsepower, the first-mentioned method of starting cannot be beaten for absolute simplicity, and this is essential for motors used to drive coal-cutting machines. These motors can also be reversed from full speed in one direction to full speed in the opposite direction, without shutting off the current. This facility of reversal is of great use in coal-cutters of the disc-type, as a disc jammed by a piece of cubical coal wedged corner-wise, which would necessitate great power to crush the piece of coal, can be liberated instantly by reversing the motor for a few seconds.

Three-phase motors cannot, by any simple means, be made to run continuously at various speeds. The speed does not vary with the voltage of the current, but depends approximately on the periodicity of the current supplied, and the number of pairs of poles on the motor; thus the speed of a six-pole motor per minute with a current frequency of 50 per second is found thus:—The number of periods per minute divided by the number of the pairs of poles equals the number of revolutions per minute or $(50 \times 60 \div 3 =)$ 1,000 revolutions per minute.

When this condition obtains, the motor is said to be running in step with the driving current. If the motor be sufficiently overloaded, it is pulled out of step and comes to rest. The writer finds that, if this takes place when coal-cutting, there is a very considerable increase in the flow of current, but that it is not such as almost immediately to burn the motor in the event of the fuses not blowing. In fact, even if the attendant is engaged fixing timber behind the machine when it is pulled up, by the time he has got to it and switched off, there is no per-ceptible rise in temperature of the motor. This fact is due to the motor, even when stationary, having by induction a choking-effect on the current and so preventing it from rising to dangerous proportions. This feature of three-phase motors is found of the greatest benefit in coal-cutting, for should the cables be short-circuited, the fuses would be instantly blown, but they are not blown every time that the cutter is jammed;

in fact, the writer, in his experience, has not known of a single instance of a fuse being blown through this cause. As a consequence, he has placed the fuses on the surface only, thus removing from the pit one of the possible sources of danger.

One disadvantage of the three-phase system is, that it is necessary to use three wires for the motors; but, as, later on, the various points for or against the two systems are compared, this will receive its due place and weight.

In fitting up a colliery with plant on the three-phase system, it is only necessary to have one type of generating-plant, which is equally available either for lighting or power-supply. The current is generated at a suitable pressure for transmission to the motors, and by means of static transformers it is transformed down to a pressure suitable for lighting.

Continuous Current.—Let us now consider some of the points connected with the continuous-current system. In the first place, 1,000 volts is about the practical limit at which continuous current can be generated by one dynamo, and then it can only be transformed down by means of a rotary transformer. As a matter of fact, the current generated in the armature-coils of the dynamo, is alternating, and it is only by means of the commutator that we obtain it in the form of a continuous current.

The commutator consists of a number of segments which have to be insulated from each other and from the spindle. This insulation is liable to give way, owing to vibration, oil, moisture, or expansion and contraction due to varying temperatures. The brushes rest on the commutator to collect the current, and sparking may take place there, owing either to electrical causes and vibration, or both combined. The sparking burns away the surface of the commutator, and, in the pit it is a source of danger, as continuous-current motors must be fitted with a commutator. A continuous-current dynamo can be run as a motor or *vice versa*.

In order to start a continuous-current motor, it is necessary to place some form of resistance in the circuit, otherwise, if the current were switched on direct, the armature would soon be burnt up. When the armature is stationary, it has no effect in checking the flow of the current in the same way that a three-phase motor has: it therefore follows that, if while running, the armature by means of an excessive load is brought to rest, it will be burnt out in a few seconds unless the fuses blow. In coal-cutting with continuous-current motors, the men find it very annoying to have constantly to be replacing fuses, so that, unless the supervision is strict, they insert larger and larger fuses until the armature is burnt out. A letter of complaint then goes to the manufacturers, which, however, does not state that the men have been trying to make a motor of 25 horsepower develop over 100 horsepower. In the evidence given recently before the Departmental Committee on the Use of Electricity in Mines, it was stated on behalf of a company insuring electrical plant at collieries, that the rate of breakdown of colliery-motors and dynamos was 30 per cent. and of these, 85·7 per cent. occurred to armatures and commutators and were principally due to over-loading.* With the three-phase motor, on the contrary, if the load is very excessive, the motor is brought to rest, and even if the load is not sufficient to do this, at any rate a continuous overload is soon shown by the motor-case getting warm, as the wires are near to the surface of it all round. For the continuous-current system, two wires only are necessary, which means that the risk of stoppage or leakage through damage to cables is apparently reduced by one-third, as compared with the three-phase system. In practice this is hardly borne out, for when there is a bad fall of roof, as frequently as not, both wires are damaged in a continuous-current system, and the presence of a third wire would not stop the motors any more completely if it were broken. It is not as if each single wire were run along a separate road, the addition then of another wire would add to the liabilities to stoppage which were not common to the other wires. The writer has found after several years' experience, that there is no more difficulty in maintaining a three-phase circuit in good order than a continuous-current circuit.

Comparison.—In drawing a direct comparison between the two systems under consideration, it is found that it almost consists in stating the defects in the continuous-current system,

---

* *Minutes of Evidence taken before the Departmental Committee on the Use of Electricity in Mines* [Cd. 1917], 1904, page 168, question 5,948.

and then on the other hand stating that these do not exist in the three-phase system. With the continuous-current system, the high-pressure current is on the moving parts of dynamos and motors; there are commutators with their wear-and-tear, sparking and liability to breakdown; starting resistances are a necessity for motors. Transformers must be of the rotary type. With the three-phase system there is no high-pressure current on the moving parts of motors or dynamos; there are no commutators, therefore no sparking; starting resistances can be dispensed with; the pressure can be transformed up or down by means of static transformers. Continuous-current series-wound motors start better against a load than three-phase motors; but, as has been shown, the armatures are very liable to being burnt out when used for driving coal-cutting machines.

With the exception of charging storage-batteries, three-phase currents can be used for any purpose to which continuous currents can be applied. For efficiency, simplicity and reliability, the writer is strongly in favour of the three-phase system for use in mines. The writer has an alternator coupled direct to a steam-turbine of 200 horsepower, and it has a small continuous-current dynamo, as exciter, on the same spindle. This has run almost night and day since 1897 with absolutely no repairs, and the only part which to-day shows any signs of wear and tear, is the commutator of the exciter.

In a few years, mining engineers will have considerable practical experience, as well as manufacturers' figures, in regard to the use of three-phase plant with which to form a decision on its merits or demerits. In the meantime the writer hopes that this short paper, based on experience in the mine, may be of some use to the members.

———

Mr. W. C. MOUNTAIN (Newcastle-upon-Tyne) said that he had read Mr. Holiday's paper carefully and with appreciation, as coming from a practical man, but he did not think that it was yet demonstrated that the continuous current, which had proved a good servant in the past, would not be a good servant in the future. Mr. Holiday had made certain comparisons between three-phase and continuous currents, to the detriment of the latter; though in a few cases there were objections to the use of continuous current—not to the system, but in supposing that it could be one of universal application without taking into consideration the purpose for which it was to be used. For instance, with three-phase single generators of small size there was a difficulty in making a generator with an electromotive force capable of maintaining the pressure: an ordinary specification provided for a drop of 15 per cent. on full load. Then again they had to remember that on a long line of cable the outer-phase current, which was always present in three-phase motors, meant that, although the current did not take much power to develop, there was a considerable drop on the line; and if there was a large drop in pressure, it had a serious effect upon the efficiency of the motor, and with a drop from 500 to 250 volts on the motor, only one-quarter of the starting effort was obtained. The constant speed, at which three-phase motors had to be run to secure anything like economy, was also another objection to their use; for instance, it was absolutely necessary that a fan should be run in accordance with the air-supply required. At a colliery where there was a mixed installation of pumps, haulage-gears, etc., and it was necessary to vary the speed of these gears, there were difficulties with the three-phase current, and these difficulties could be surmounted to a certain extent with continuous current. He believed that Mr. Holiday was the only colliery-manager in England, except Mr. W. E. Garforth, who had been able to introduce a successful three-phase coal-cutter; and he would be pleased to learn what work these three-phase coal-cutters had been doing with three-phase current. He believed, however, that there would be a great advantage in their use if three-phase coal-cutters could be made really efficient and satisfactory, and commutators dispensed with, but so far, he did not think that the problem had been solved. Much money had been spent in experimenting in this direction, though hitherto no successful machine had been worked by three-phase current.

The electric winding-engine at Zollern II. colliery was the finest engine of the kind in Germany, in regard to work and results, and it was worked entirely by continuous current.[*] The motor-generator, a 300 horsepower motor driving a generator of 1,000 kilowatts capacity, with a flywheel weighing 45 tons, was

---

[*] *Trans. Inst. M.E.*, 1904, vol. xxvii., page 143.

placed between the main supply and the winding-gear, the result being that the load on the main generators supplying current to the winding-engine was practically constant, only varying by about 70 horsepower, the flywheel taking up and giving out the balance of the power required. Of course, this system could be utilized by using a three-phase instead of a continuous-current motor on the motor-generator, but continuous current was used on the winding-engine motors. At Preussen II. colliery, when the three-phase winding-gear was working, the electromotive force, which was about 2,300 volts when the winding-gear was standing, fell to about 1,400 volts; and when the winding-gear was started two compound engines with generators of 550 kilowatts were required to supply current to the winding-gear.* Either of these engines, coupled direct to the winding-gear, would have been ample to have done the work; and this was not economical working. Electricity was no doubt a subject that was coming well to the front, and each case had to be considered on its own merits; but, so far as he (Mr. Mountain) had seen, there could be no doubt that continuous motors on winding-gears, with motor-generators driven by either three-phase current or continuous current, were the most economical.

Three-phase current was of advantage for long-distance work, where power had to be transmitted over great distances, but where all the machinery was situated within reasonable distance of the generator he did not think that there was any economy in adopting the three-phase system. He did not, however, desire to disparage either system, and both were good servants when properly used.

Mr. W. A. CHAMEN said that Mr. Holiday had not considered anything but what were known as induction-motors for three-phase work, and no doubt this was a wise line to take. He could quite understand that alternating-current motors would not be so suitable for colliery-work. With induction-motors the clearances were essentially very fine, that is, compared with those of continuous-current motors; and he asked Mr. Holiday, supposing that the bearings wore at all, whether trouble was likely to arise. He also asked Mr. Holiday whether he had had

* *Trans. Inst. M.E.*, 1904, vol. xxvii., page 153.

sufficiently long experience with the use of three-phase induction motors in driving coal-cutters or other colliery-machines to say whether they were going to give serious trouble or not.

Mr. W. C. MOUNTAIN said that he had some induction-motors of 55 horsepower running and the clearance was only a shade under a millimetre. One of these motors stopped, and an examination showed that it was heating: they could not understand this, as the motor was very carefully looked after, but ultimately a film of fine dust was discovered between the rotor and the stator.

Mr. HERBERT A. JONES asked whether Mr. Holiday could give the members any comparative costs with regard to equipment in the case of continuous and three-phase motors; and whether he had had any experience of slip-ring motors. Mr. Mountain had very properly remarked that the drop which took place with three-phase motors might have a serious result. Mr. Holiday had made some remarks, with regard to fuses being inserted of a strength greater than should be employed by men working coal-cutters driven with continuous current, which damaged the machines. As a manufacturer of mining machinery, if such were the case, he felt inclined to go home and destroy all his continuous-current machinery, but he had had no experience of the kind suggested by the author. Mr. Holiday had stated that starting resistances were not necessary for three-phase motors; but he (Mr. Jones) thought that the author meant motors of a certain type, as starting resistances were just as valuable in three-phase motors as in motors worked with continuous current.

Mr. T. LINDSAY GALLOWAY (Glasgow) said that, having recently had occasion to erect an electric plant at Campbeltown colliery, and after making enquiry as to cost and other matters, he had come to the conclusion that the three-phase system was more applicable for pumping, on account of its simplicity; the cost being almost the same for both systems. The plant consisted of a generator of about 100 horsepower driving a motor of 80 horsepower for pumping and a lighting installation. The plant had not given the slightest trouble, and he would not change it in favour of the continuous-current system. There were cases, however, where the three-phase current was not applicable, and he thought that the members would make a mistake if they were to devote their attention entirely to one system. He might add that the starting, in the case he had mentioned, was done by slip-rings, and the pump had a bye-pass, so that it did not start at full load. He thought that winding could be done just as easily with three-phase current as with continuous current.

Mr. SYDNEY F. WALKER (London) said that during the past 30 years alternating current had gone through a succession of changes which was really like its own current; at one time it was the only thing to use, and then continuous current was in favour. He (Mr. Walker) would like to remind the members, however, that three-phase was merely a stage on the way to single-phase current. Single-phase current had never succeeded in the way that they had hoped it would and as he thought it might in time; it was making its way, however. There were substantial propositions for driving motor-carriages on railways by the single-phase current at 3,000 volts, and transforming it down by single-phase motors. He thought that fact, however, need not trouble any one who had adopted three-phase plant; because it merely meant that, if the single-phase apparatus established itself, they would have simply to do away with one of their three cores and rewind the plant to suit. Three-phase plant was slightly cheaper as a rule in first cost than continuous-current plant, and simpler in many respects; and if single-phase could be properly established, it would be enormously simpler still than three-phase, because there would not only be two cables instead of three but merely a single set of stator-coils, and so on; in fact the whole plant would be much simpler.

With regard to Mr. Holiday's paper, and the mention made therein about not burning coal-cutting machines, that had struck him for some time as a very important feature: for, supposing that a coal-cutting machine was jammed, with a continuous-current motor an enormous current would be passing through the armature-coils, with a consequent burn in a short time. In a coal-mine, it was requisite that the plant should be kept running, and this was an important point to be remembered in connection with the use of three-phase current. The question of having three cables was not, he thought, a serious one, because he believed that in nearly every instance the three cables were made up into one, and it was laid as one. No mistake could be made in connecting the motor, as there were three ends to be joined on to three terminals.

The question of electric winding was, of course, a much more serious matter, but he thought that the members would agree with him that electric winding was only in its infancy. The difficulties in connection with three-phase motors as applied to electric winding, together with the problem of the variation of speed, would, when seriously tackled, be worked out successfully. Electrical engineers had never yet failed in working out problems of this kind, and if they had failed in mining work it was because they did not understand the conditions of mining, with which they were called upon to deal, as they were not mining engineers.

Mr. G. S. CORLETT (Wigan) said that he entirely agreed with the final conclusion of Mr. Holiday's paper that, generally speaking, three-phase current was most suitable for mining work; but with some of the other remarks he was not in accord. With three-phase current there would only be one generator or one series of generators at the same pressure, at the generating station for all the power. He had tried squirrel-cage three-phase motors, at a colliery, up to 15 or 20 horsepower; and he had often found in practice that, at starting, there was a large variation in the pressure, and that these variations were altogether irrespective of engine-governing. About six years ago, he had erected three-phase motors of various sizes, from 4 to 35 horsepower, and they had had rather a rough-and-tumble experience, being constantly shifted about and doing nearly as much work as there were hours in the week. Up to the present time, he had only had one case where there had been contact between the stator and the rotor, and then it was discovered before any damage was done. He had simply adjusted the bearings and set it to work again, the clearance being $\frac{1}{10}$ inch.

Mr. Holiday stated that, in order to start a continuous-current motor, it was necessary to place some form of resistance in the circuit, otherwise, if the current were switched on direct, the armature would soon be burnt up. Some allusion had been made to the bad starting effect of three-phase motors, but he (Mr. Corlett) was using such motors in everyday work for winding, main-and-tail rope haulage, and endless-rope haulage, and in practice he had had not the least trouble in starting the load. In main-and-tail rope haulage, with perhaps as many as 40 tubs on a train, they could be readily started and the load set into motion without any trouble. It seemed a mistake to lay down any hard-and-fast law with reference to mining work—whether it should be done with three-phase or continuous current—although three-phase seemed to be coming very much to the front at present.

Mr. W. H. PATCHELL (London) said that the chief point in Mr. Holiday's paper seemed to be that induction-motors did not blow their fuses; but, on the other hand, they could maintain their voltage. He had lately seen a winding-engine worked off a normal pressure of about 2,000 volts three-phase. As soon as the winding-engine was started, the pressure fell from 2,300 volts to the lowest reading on the voltmeter scale, 1,600 volts, and stopped there while the motor was running; and, as soon as the winding stopped, the pressure rose again to the normal. Such an effect might indefinitely postpone the blowing of the fuses, but it could hardly be called satisfactory working.

Mr. F. W. HURD (Bothwell) wrote that Mr. Holiday's description of the various merits of the three-phase, as compared with the direct-current, system of electric power, as applied to mining work, was of considerable interest. There was one feature of three-phase motor work which Mr. Holiday had not touched upon, and that was the excessive current required to start the motor, unless slip-rings and an external resistance were used. On a coal-cutter, the use of slip-rings and resistances was scarcely permissible, and in actual practice the current required to start the motor was from two to three times the normal working-current. This meant that the engine and dynamo must be larger and stronger in proportion, and therefore more expensive, for a three-phase plant than for a direct-current plant capable of performing the same work. Again, the combination of a three-phase plant, for lighting and power, was not satisfactory, owing to the great variation in voltage when motors were thrown in and out. A separate plant for lighting should be installed if satisfactory results were desired. He (Mr. Hurd) could corroborate the claim that Mr. Holiday was the first to apply a three-phase motor to a Hurd bar-machine about 4½ years ago. The application was entirely successful, and since that time a considerable number of machines had been so fitted. These machines had, without exception, given entire satisfaction, and there could be no doubt that a three-phase motor coal-cutter was a more simple and a more reliable tool than a direct-current motor-machine. Where considerable areas had to be served, three-phase current generated at a high voltage and reduced near the point where the power was required undoubtedly had the advantage; but, for an ordinary colliery, a direct-current plant at a pressure of 500 volts, with good concentric wiring on the earthed return-system, was safer, had more elasticity, and was more easily applied to the various power-requirements of pumping, hauling and coal-cutting.

Mr. H. RICHARDSON HEWITT (H.M. Inspector of Mines) said that Mr. Holiday was doing two things with the three-phase current, which others had up to the present time been unable to do. He was working coal-cutters successfully, and as all fuses were placed on the surface, he had removed a serious danger from the mine. He (Mr. Hewitt) would like Mr. Holiday to give full details as to how he was able to do these things, and, also, what was the loss of current caused by using static transformers.

Mr. H. M. HOBART (London) wrote that electricity was now widely employed in mining work, and whether the best results were to be obtained by continuous current or by polyphase current, was a matter to be decided in each special case. In many instances, a polyphase supply was alone available, and the question arose whether the advantages of the continuous current were sufficiently great to justify the expense of employing rotary transformers or motor-generators, in order to provide it. If the periodicity was low—say 25 cycles per second—polyphase motors for the coal-cutters might be designed with diameters nearly or quite as small as for continuous-current motors; and, so far as related to such cases, a transformation to continuous current was the less necessary. A periodicity of 50 cycles per second, necessitated a motor of larger diameter, and even then the motor would be less satisfactory in most respects. Hence there would be greater reason for employing a motor-generator and obtaining continuous current. Even then, however, it was not justifiable to draw a sweeping conclusion, as the nature and location of the coal and numerous other circumstances required to be considered. Excellent results had been obtained with polyphase current, even at 50 cycles, in many mines. The same was, however, also perfectly true of the continuous-current system. Mr. Holiday's disparaging references to the continuous-current motor were decidedly misleading, as many firms were building excellent motors of this type, which frequently had many advantages over the polyphase type, especially for mining work. Thus, a continuous-current motor for operating a coal-cutter, would for a given output and speed, have a smaller diameter, a much deeper air-gap, a higher efficiency, a lower temperature-rise, and would not so greatly interfere with good regulation on the lighting circuits. In many and probably most cases where a generating plant was installed directly at the mine, continuous current was preferable. The polyphase motor was necessarily of larger diameter for a given output and speed; and unless the air-gap was very small, the power-factor would be low and the wattless current would be high. This necessitated the provision of heavier transmission-cables, and interfered with good regulation on the lighting circuits. When suitably designed, the squirrel-cage induction-motor would have a fair torque at starting, but the accompanying rush of current was far larger than was required by a continuous-current motor for the same starting torque; and unless much greater outlay for cables of large cross-section was made, there would—during such starting periods—be a heavy line-drop. The conditions as to dimensions and performance became worse the higher the periodicity employed. With decreasing periodicities, the polyphase squirrel-cage motor gradually approached the continuous-current motor in its dimensions, and in the quality of its performance. Both continuous-current and polyphase motors were, however, capable of doing the work thoroughly satisfactorily, and Mr. Holiday made a great mistake in handicapping the work of introducing electrical operation into mines, by encouraging a feeling that only one system was available. This was the greater pity since the system favoured by him, while often excellent and almost always practicable, was in many, if not the majority of cases, inferior. Both systems should be clearly recognized as worthy of consideration, and an impartial consideration of the conditions of each case should determine the selection.

Mr. PERCY C. GREAVES (Wakefield) wrote that, as a mining engineer, it seemed very questionable even to consider the advisability of carrying a higher pressure into a mine than 750 volts, because the current was not likely to be used at a greater distance than 3 miles from the pit-bottom save in exceptional cases, and if large enough cables were used there should not be a loss of more than 10 per cent. in transmission. He would certainly hesitate very much before allowing a greater pressure than 750 volts in a mine. It must not be forgotten that it was practically impossible to keep the same standard of insulation in a mine as it was on the surface, as in tramway work, and therefore one hesitated to adopt the use of very high pressures. He agreed with Mr. Holiday's contention that three cables were not more likely to be injured than two, provided that they were three separate cables; but the polyphase system was often run by means of one cable containing three wires, and, in that case, the danger might be greater. He thought that too much was made of the sparking of the brushes on continuous-current motors, as the danger existed much more in the cable than in the motor, and a spark was produced every time that a switch was used, whether the current was polyphase or continuous. He asked Mr. Holiday to say something regarding the danger of shock from the two systems. He had always understood that the shock of alternating current was more dangerous than continuous current, owing to the muscles of the persons suffering from shock being contracted. Continuous current had been most successful up to the present time, when used in connection with coal-cutters; but it was necessary that the supervision should be strict to prevent men replacing fuses by larger wire : the men would soon find out when the coal-cutter was not working properly, and switch off the current instead of blowing the fuse.

Mr. SAM MAVOR (Glasgow) wrote that Mr. Holiday had the authority of experience, and his claims for the three-phase system were no doubt fully justified in the conditions under which his own plant was working; but his comprehensive recommendation of three-phase plant for all descriptions of work at collieries, irrespective of the nature of the power-requirements, distance of transmission, etc., was, in his opinion, unwarranted. The three-phase system for colliery work presented many attractive features both to manufacturers and users, but it was also subject to important limitations. It was agreed, for large colliery power-plants, that the three-phase system had, at present, no rival; but in a considerable proportion of the collieries in this country, especially in the North of England and in Scotland, the power-requirements were moderate or small. In cases where the power required was under, say, 200 horsepower, it was by no means a foregone conclusion that the three-phase system should be adopted; and there were many cases, where the power-requirements were larger than 200 horsepower, in which direct current would be more suitable. The magnificently-engineered three-phase installations at some of the Westphalian and Belgian collieries, while they might be admirably adapted to the conditions there, should not be allowed to set the fashion for general practice in this country, where the conditions were often so widely different. The circumstances of each case required special consideration, and while the balance of advantages would sometimes be in favour of three-phase, he (Mr. Mavor) believed that, in a large proportion of the smaller collieries, the balance of advantages was in favour of direct current. The simplicity of starting three-phase induction-motors, as described by Mr. Holiday, could only be realized in cases where the generating plant was large relatively to the motor, otherwise the starting of the motor by these simple means would involve stoppage of all the other motors in the pit. The three-phase motor could not be used for driving single or main-and-tail haulages without resort to slip-ring rotors with rubbing contacts, and resistance switches; and these corresponded to the commutators and starting switches of direct-current motors. The starting torque of the three-phase type was much less than that of the direct-current series-wound motor; for even a moderate starting torque the line-current was high, and the low overload capacity was a positive disadvantage. The choice of speeds of rotation with three-phase motors was strictly limited, and the means of obtaining variable speeds were wasteful. It was claimed that the three-phase current was equally suitable for power and for lighting, but. the periodicity best suited for power was unsuited for lighting, especially for arc lamps. A compromise might be effected by adopting a periodicity with which both lighting and power could be supplied, but it would be effected at the expense of some of the useful characteristics of the motors. The three-phase system required that three conductors should be led to every switch, fuse, motor and such like appliances. Where large powers were being used this was a matter of little relative importance, but in the case of small collieries and for small motors it became a matter of considerable importance. Every switch and fuse was a possible source of trouble, and the number of such apparatus should be reduced to a minimum. A three-phase generator, as Mr. Holiday pointed out, required the use of a direct-current generator to excite its magnets, and in small installations this complication was also relatively important. The commutator was still a link and in the three-phase plant it was an added link in the chain. In plants of large dimensions for general colliery purposes, the disadvantages which applied to the use of the three-phase system for small collieries became of less relative importance and the advantages gained in transmission were overwhelming. Still, a usual adjunct of a large three-phase plant was a motor-generator used to supply direct current for duties which could not be satisfactorily performed with three-phase current. There was no duty at a colliery that could be performed by a three-phase motor which could not be done as well by a direct-current motor; and many of the duties were much better and more efficiently performed by the direct-current motor. But the question as between three-phase and direct current for colliery work was rather one of transmission than of the application of power. It was to the advantages gained in transmission that the three-phase system in large plants owed its unquestioned superiority.

A colliery-manager, who had under his control direct-current electric power-plant for pumping, hauling, coal-cutting, etc. (some of which had been in regular use for ten or twelve years) at a considerable number of collieries, recently consulted him (Mr. Mavor) as to the relative advantages of direct current and three-phase current for a new plant. The chief advantage claimed for the three-phase system, namely, the abolition of commutators and brushes, had no weight with this colliery manager, as his experience of the direct-current plant had been so entirely satisfactory; and he decided that the limitations of three-phase as applied to plants of 80 to 150 horsepower rendered it less suitable for his general purposes, wherefore he adhered to the direct-current system.   The error of ignoring recent developments in the design of direct-current machinery occurred too frequently when direct current was being compared with three-phase plant.

It was generally admitted that in fiery mines the motors of coal-cutters required most careful protection.   As a manufacturer of both direct and three-phase current coal-cutters, he (Mr. Mavor) had no hesitation in saying that both types, when properly enclosed, were equally safe.   Such danger as existed from sparking was associated with the cables, and this danger was the same in both cases.   Mr. Holiday deserved the credit of demonstrating the practicability (which had been doubted) of driving coal-cutters by three-phase motors, and there would no doubt be a large demand for coal-cutters of this type for collieries equipped with three-phase plant and also where power could be purchased from supply-companies.

Mr. Holiday's paper would serve a useful purpose, if advantage was taken of the opportunity that it afforded of discussing a matter which was of lively present interest to many colliery managers.   Three-phase plant had in its proper sphere a wide field of application at collieries, but an unqualified advocacy of the system for all colliery-work in this country should be deprecated.

Mr. Roslyn Holiday, replying to the discussion, said that Mr. Mountain had stated that continuous-current motors had been old and good servants in the past, and should not be tabooed. All he could say was that he had had rather a peculiar experience with a continuous-current plant, and having to put coal-cutters into a seam in which there was gas, and knowing that the three-phase current had the advantage of not involving the risk of sparking, he had put in a converter and made experiments. The great electrical engineers told him that three-phase was useless, and that the Morgan-Gardner Electric Company, after spending thousands of pounds, had given up all idea of running machinery with three-phase current.   He had bought motors, discs and brackets, etc., from the manufacturers, and built coal-cutting machines, and after many changes he had at last been successful. He knew that three-phase plant was not as efficient as it ought to be, but with coal-cutters and machinery in a pit, the great necessity was reliability, and the loss was only the extra amount of coal burned: they might have highly efficient machinery, but if it broke down, a large amount of money would be lost. In a seam 5½ feet high, 250 feet per shift was cut with four machines. Mr. Garforth was also using coal-cutting machines of three-phase type, as were also Messrs. Mavor and Coulson. In regard to haulage, he did not maintain the necessity of short-circuiting, but in coal-cutting it was used on account of its simplicity. With haulage machinery, slip-rings or auto-transformers were used, as it was really better to do so. Referring to the clearance of induction-motors, it was true that the clearance was very small, but there had not been any trouble whatever. A stoppage occurred with one of the motors, the machine moved round easily by hand, but when the current was turned on it would not move at all, although they had ascertained that it was perfectly free. He placed his hand on the spindle, and it was then found that the current, when turned on, lifted it a little and so jammed the machine, but as soon as the current was stopped it was perfectly free: that was, of course, no fault of the machine. The periodicity was 42. He could not remember exactly the cost of the cables, but their life was rather longer than when used with continuous current. He had had pumps running with slip-rings for some time, and had never had any trouble with them. The drop of voltage was considerable at the coal-face, but he had had no difficulty with the starting of coal-cutters. Where the slip-ring motor was used, resistances must be employed.

The Chairman (Mr. H. C. Peake) said that he had great pleasure in moving a vote of thanks to Mr. Holiday for his excellent paper, which had been very instructive.

Mr. T. L. Galloway seconded the resolution, which was cordially approved.

Mr. Beverley S. Randolph's "Comparison of Electric and Compressed-air Locomotives in American Mines" was read as follows:—

COMPARISON OF ELECTRIC AND COMPRESSED-AIR LOCOMOTIVES IN AMERICAN MINES.

By BEVERLEY S. RANDOLPH.

In .approaching this subject, it must be borne in mind that there are fashions in machinery as there are fashions in dress, and just now electricity is fashionable in America. To the average man unfamiliar with the facts, electricity appears as something magical; producing results without the expense and trouble incident to old fashioned means. Careful attention to the statements of generally well-informed men will often develop the fact that it is even conceived as a source of power rather than as a means of transmission.

From a careful study of the subject, the writer believes that the advisability of either form of haulage is entirely a question of the conditions which obtain in each case. The pneumatic locomotive (Fig. 1) is large and cumbersome, and therefore not so well adapted to low seams as the electric locomotive, except where its safety in the presence of fire-damp may be held to counterbalance this disadvantage. The delays incident to charging are not so great as would at first appear, since, even in the best-managed establishments, there is more or less lost time at terminals which may be utilized for charging. Again, any unusual amount of work between charging-stations, such as may be due to badly lubricated cars or assisting in replacing derailed cars, is liable to exhaust the supply of air in the .tanks and necessitate a run to the charging-station and back before the train can be brought forward.

On the other hand, electricity has never been successfully applied to the gathering of cars from the rooms or working-places, owing to the expense involved in wiring each place and the difficulty in passing round short turns without displacing the trolley. A compressed-air locomotive is very successful for

FIG. 1.—MAIN HAULING LOCOMOTIVE: LENGTH, 20 FEET; EXTREME WIDTH, 5 FEET 8 INCHES; EXTREME HEIGHT, 5 FEET 7 INCHES; AND WEIGHT, 30,000 POUNDS.

this purpose. The small gathering locomotive (Fig. 2) has the same dimensions as the mine-cars in the mine in which it was designed to work, and it can, at any time, travel on any road traversed by these cars.

The machinery used with compressed air so closely resembles that used with steam, that mechanics familiar with the one have little to learn in managing the other.   Steam is so generally understood that men competent to manage pneumatic plants are easily obtained, while experts in electricity are scarce.

FIG. 2.—GATHERING LOCOMOTIVE: LENGTH, 10 FEET; EXTREME WIDTH, 3 FEET 10 INCHES; EXTREME HEIGHT, 5 FEET 1 INCH; AND WEIGHT, 8,000 POUNDS.

In the matter of cost, Mr. A. de Gennes had stated that the cost of an electric installation was only one quarter that of a pneumatic plant.* This is by no means the first time that this statement had been made, and its vitality in the face of easily ascertained facts was an interesting psychological phenomenon.†

As a matter of fact, the electric plant is usually the more expensive, and we have not far to go to discover the reason. In the construction of the locomotive, the same weight of metal must of necessity be used, and as there is little difference in the labour involved the cost is practically the same.

Comparing the pipe and the transmission-wire, the latter is cheaper for short distances, but, as its cost increases as the

* Annales des Mines, 1900, vol. xviii., page 244.
† Trans. Inst. M.E., 1903, vol. xxv., pages 532, 538 and 543.

square of the distance while the cost of the pipe increases directly as the distance, a point is soon reached where the cost is equal; and, beyond this, the difference grows rapidly against the electric plant.   In fact, one prominent establishment, in the State of Pennsylvania, is reported to have reached a point beyond which the cost of the conductors becomes so great that it will be necessary to remodel the entire plant on a basis of higher voltage. It is the usual practice in pneumatic plants to adjust the diameter of the pipe with a view to its storage-capacity, rather than simple transmission, in order that the locomotives may be charged promptly at a pressure near the normal; and consequently additions can be often made with a smaller size of pipe.

The greatest difference of cost is found in the generation of the power.   In the electric plant, this feature must be sufficient to meet the greatest demand that can be brought upon it at any one time, although the remainder of the run may require only a small part of this maximum demand.   Not only does this involve a large first-cost, but no such plant can be operated economically, if it runs the greater part of its time much below its capacity, especially when it must be kept in constant readiness to respond to this maximum demand at any time without notice. In the pneumatic system, the elasticity of the air provides for this excessive demand, and distributes the work over a considerable time.   This not only admits of the installation of a smaller plant, but it allows of its adjustment to an even rate of energy-production, which is very conducive to economy of operation. Consequently, the electric plant must be proportioned solely to meet the maximum demand, while the pneumatic plant is proportioned to meet the average demand; and it is not difficult to imagine a case where this average demand will be only a small fraction of the maximum demand.

These disadvantages appear to balance the somewhat lower efficiency of the pneumatic system, due to the loss of heat in compression, so that there is little difference in the cost of operation.

During the past year, the writer had an opportunity of comparing the actual cost of plants in successful operation: the maximum haul in each case being about 8,000 feet.

A. Compressed-air plant operating two main hauling loco-motives, each weighing 30,000 pounds, and five gathering loco-motives, each weighing 8,000 pounds, a total weight of 100,000 pounds. The cost was as follows:—

	Dollars.	£
A three-stage air compressor and a compound steam-engine	5,300	1,104
5,600 feet of pipes, 5 inches in diameter ...	5,600	1,167
3,100 ,, ,, 2½ ,, ,, ...	1,700	354
1,000 ,, ,, 1½ ,, ,, ...	300	63
Two locomotives, each weighing 30,000 pounds	6,000	1,250
Five ,, ,, 8,000 ,,	10,000	2,083
Two boilers, each of 80 horsepower	1,000	208
Installation	4,000	833
Totals	33,900	7,062

B. Electric plant of four locomotives, each weighing 26,000 pounds, a total weight of 104,000 pounds. The cost was as follows:—

	Dollars.	£
Generator, producing 225 kilowatts	3,900	812
Engine, 500 horsepower	4,800	1,000
Boilers, 600 horsepower	4,400	917
Foundations, piping, etc.	1,500	312
Wiring	8,000	1,667
Four locomotives, each weighing 26,000 pounds	9,500	1,979
Totals	32,100	6,687

C. Electric plant, comprising two locomotives, each weighing 26,000 pounds, a total weight of 52,000 pounds. The cost was as follows:—

	Dollars	£
Generators, etc.	24,000	5,000

The A plant was installed, and is being operated under the direction of the writer. The costs for the B and C plants were supplied by the superintendent now in charge of their operation: the B plant was installed under his direction, and the C plant by his predecessor. There is no reason to suppose that these values are to any extent misleading, and as the gentleman is a pronounced electrical enthusiast, they are certainly not exaggerated.

The draft of a locomotive is a function of its weight, and it forms a convenient unit for comparison. The costs of the three plants per 1,000 pounds of locomotive operated, on this basis, are as follows:—A, compressed air, £70·62 or 339 dollars; B, electric, £64·29 or 309 dollars; and C, electric, £96·15 or 461 dollars.

The B plant possesses an advantage in this comparison, due to the fact that it works against a gradient in hauling outbye only; and when hauling inbye all the gradients are descending. There are two parallel roads, and two locomotives are going inbye and two are coming outbye all the time; and the generator is never loaded with more than two locomotives. If worked on undulating gradients, as in the case of the other two plants, where all the locomotives might be on maximum gradients at the same time, it could not be relied on to drive more than three locomotives, with a total weight of 78,000 pounds. The plant, with three locomotives, would have cost £6,192 or 29,725 dollars.

An inspection of the cost given above for the A plant, will show that its locomotives cost much more than those of the other two; but this is largely due to the number of small gathering units. If this plant had been equipped with four locomotives, each weighing 25,000 pounds, they would not have cost more than £583¼ or 2,800 dollars each (or £2,333 or 11,200 dollars for the total weight of 100,000 pounds). The plant would then have cost £6,062 or 29,100 dollars.

Amending the calculations on the basis of these revised locomotive-weights, the costs of the three plants per 1,000 pounds of locomotives operated are as follows:—A, compressed air, £60·62 or 291 dollars; B, electric, £79·38 or 381 dollars; and C, electric, £96·15 or 461 dollars.

Upon the question of cost of repairs, there is a considerable difference of statement and opinion. This is doubtless due to the fact that this item is always largely influenced by the *personnel* of the operating force. A few years ago, the balance was largely against the electric system, but owing to later improvements, the fruits of earlier experience, the advantage is now probably with this system.

The prices given for various items in this paper are based on market conditions obtaining several years ago, when these plants were installed, and they will probably be found to be 10 to 15 per cent. higher than present-day prices.

Mr. SYDNEY F. WALKER (London) said that he would not be suspected of undue leaning against electricity, but he thought that it would be better if compressed-air engineers would tackle the work properly, in the same way as electrical engineers had tackled theirs, and they would then make a better show than they did. One thing essential was for them to use higher pressures, and if they did so compressed air would then be able to make a much better show with regard to running costs. If air were used at higher pressures, although more heat would be generated per pound of air, a pound of air would do more work. The two powers, electricity and compressed air, ran on parallel lines. When air was compressed heat was generated, and the same thing occurred in electricity; and the best thing that could be done in both cases was to remove this heat as soon as generated. Electricity had a great advantage over compressed air in the matter of transmission. No attempt at present, so far as he knew, had been made to use the air expansively. One part of the problem to be solved was the avoidance of the enormous leakage that accompanied the use of air at high pressures. In America, one heard of pressures of 1,000 pounds per square inch being transmitted inbye, and if this could be done in America it ought to be accomplished in Great Britain and elsewhere. Mr. Randolph stated that electric locomotives had never been used for gathering tubs, as there was a difficulty in getting the trolley to go round the corners; but he (Mr. Walker) thought that one hardly needed to be reminded that when electric tramcars in streets were first introduced there had been the same difficulty, and it had been overcome. It was quite correct to say that the cost of the electrical transmission-line varied as the square of the distance; but this was only true if the loss in the transmission-line was kept within the same limits. In the case of compressed air, the same thing occurred and to keep down the loss the size of the pipe must be increased. The costs given by the author were misleading, and he (Mr. Walker) might point out that the load-factor which he claimed for a compressed-air plant applied very largely to an electric plant. Where a number of locomotives or coal-cutting machines were at work, there was always a certain number only at work at one time; and, therefore, if any one wanted to cut matters fine, they could easily do so by erecting a plant capable of supplying the mean demand only. Mr. Randolph also said that an electric plant would not run economically, unless it were run at the maximum load. Of course, any plant ran more economically and efficiently, with the least charge for conversion, when it was doing all the work that it could; but one of the features of electricity was that the efficiency did not fall as the load decreased in the same proportion as it did in the case of other power-plants, so that the electrical plant had the advantage under such conditions. The electrical plant, mentioned by Mr. Randolph, was evidently erected as an experimental one some years ago, and was not put in so economically as it would have been at the present time. It was, therefore, hardly fair to judge the two plants in the way that he had done. He (Mr. Walker) thought that if the whole question was gone into where electric locomotives were used—although unfortunately there were not many coal-mines in the United Kingdom where they could be used—where they could be employed safely and where sparking at the trolley did not matter—it would be found that electric locomotives were more economical in first cost and in running cost than compressed-air locomotives.

Mr. HENRY HALL (H.M. Inspector of Mines) said that he was glad to hear from Mr. Walker that there appeared to be a future for compressed air. Electricity was a profession, and had some very clever men at the back of it, but no engineer in particular had taken up compressed air, and it had, therefore, to fight its own battle, whilst electricity had been backed tooth and nail by very able people. From the point of view of safety, compressed air had a great advantage over electricity; the members had only to read the proposed Special Rules to realize the dangers resulting from the use of electricity.

Mr. C. C. LEACH said that it seemed hardly possible to keep a compressed-air plant at the same state of efficiency as an electrical plant.

The REV. G. M. CAPELL (Passenham) said that the electric locomotives in use at the Marles colliery, Pas-de-Calais, were small, about 7 feet long, and they ran on a narrow gauge belowground. They hauled a load of about 15 tons each, and were managed generally by a small boy. The same kind of locomotive was in use at the Grand Hornu colliery in Belgium for conveying coal aboveground from the various centres to the cleaning-plant, and the boys in charge, generally about 13 or 14 years of age, worked the locomotives with the greatest ease and intelligence. He suggested that electricity might be used to work an air-compressor belowground at a distance inbye, and thus drive on compressed air for use in the gassy and dangerous portions of the mines.

Mr. ROSLYN HOLIDAY said that he had taken tenders for a plant to work coal-cutters, one driven by electricity and one by compressed air, and he had found that a combined plant to run two coal-cutters could be obtained for the same money as would run four coal-cutters in the ordinary way.

Mr. H. RICHARDSON HEWITT (H.M. Inspector of Mines) wrote that he quite agreed with Mr. Randolph that the conditions appertaining to each case must control the type of locomotive to be used, and further that electric locomotives would be prohibited in fiery mines. If electric locomotives were ever used in this country, it would be necessary to provide a separate travelling road for men and horses, owing to the proximity of the trolley-wires carrying the current; and he presumed that the driver was protected by the covering of his cab, or was placed in such a position that he was clear of the trolley-wires. Mr. Randolph gave examples of power-generation, and it would be interesting if he would give the tonnage drawn by each plant described, the distance travelled by the locomotives of each type, and the average cost of upkeep.

Mr. A. S. E. ACKERMANN (London) asked Mr. Randolph to explain why the cost of copper-conductors increased as the square of the distance, while the cost of piping for compressed air was only directly as the distance. It was also astonishing to find that the estimate for the air-power plant was for only 160 horsepower, while that for the electric plant was for 600 horsepower: such disproportion certainly needed some further explanation. No mention had been made of the very much greater rapidity with which electric conductors could be fixed than lines of pipes, nor of the fact that the fixing was simpler and cheaper. There was also no corrosion in the case of the copper-mains, whereas, in

some mines, the water was particularly corrosive and caused considerable damage to the air-mains. He could not help thinking
that the vibration caused by the want of balance of pneumatic
locomotives would cause considerably more wear-and-tear of the
track than electric locomotives. With regard to the difficulty of
the maximum demand in the case of an electric installation, it
was quite possible to use a storage-battery, which would permit
of a considerably smaller electric generating-plant, much in the
same way as air-storage vessels permit of a smaller air-com-
pressing plant being used than would otherwise be necessary.

Mr. HENRY DAVIS (Derby) wrote that it would be instructive
to be able to compare the weights, effective horsepower or draw-
bar pull, etc., of the electric and compressed-air locomotives
referred to by Mr. Beverley S. Randolph, particularly of the
gathering locomotive weighing 8,000 pounds, and to see whether
this would compare in draw-bar pull with the electric locomotives
made by the Jeffrey Manufacturing Company of a similar weight.
So far as one could gather from Mr. Randolph's paper the comparative figures were as follows:—

Description.	Weight of Locomotive.	Draw-bar pull.	Speed per Hour.	Minimum Gauge.	Width over all, with Minimum Gauge.	Height over all, excluding the Trolley.	Length over all, excluding the Buffers.
	Pounds	Pounds	Miles.	Inches.	Inches.	Inches.	Inches
Air-power loco-motive ...	8,000	Not	stated	Not stated	46	61	120
Jeffrey electric locomotive	8,000	1,000	6 to 10	36	55	39	93

The cost of the two engines would probably approximate
one to the other, but Mr. Randolph would not wish to give the
impression that air-power could be conveyed any considerable
distance with the same efficiency as electric power. It would also
be interesting to hear whether Mr. Randolph could state whether
the prominent establishment in the State of Pennsylvania to
which he referred, had discontinued the use of electrical engines
in favour of air-power, or whether, as would appear, a higher
pressure for the electric transmission had been adopted.

Mr. J. F. LEE (Glapwell Colliery, Chesterfield) wrote that it
would interest the members and be of value to them if Mr.

Randolph would give the cost per horsepower for transmission of compressed air and electric power in the plants that he described in his paper. There were various opinions as to whether the. first cost or capital-outlay for an air-compressor or an electric installation was the highest; but he thought that a matter of more importance was the cost of transmission after the plant had been installed. Whether the power was used for locomotives,. rope-haulage, coal-cutting, pumping or any other underground work required to be done, after the question of safety had been considered, the chief point was to adopt the power that would give the greatest efficiency and consequently the least cost of· transmission. The dangers of electric power for underground work had been greatly reduced by the introduction of induction-motors, with switches, fuses, and all parts where sparking was likely to occur, properly covered by suitable oil or carefully pro-. tected in some other way; and, with cables laid in a groove in the floor of the mine in clay or other substance so as to protect them from water and corrosion, the danger from falls of roof and· damage by haulage-trains would be entirely obviated. With a sensitive governor attached to the steam-engine, generating the· electric power, the current required could be easily adapted to the work in the same way as the elasticity controlled the supply of compressed air. From his (Mr. Lee's) observations there was no doubt that the cost of electrical transmission was very much less than that of compressed-air power in the ordinary way as used in British mines. He (Mr. Lee) had made trials of the efficiency at a single-stage ordinary water-jacketed air-compressor, with the outside of the cylinders kept cool by water-circulation. Indicator-diagrams were taken simultaneously at both the air-compressor and the engines doing the work underground. The temperature of the compressed air was taken as it left the compressor, and also just before it passed the throttle-valve at the motor, so as to note the loss due to fall in temperature; and pressure-gauge readings were taken at the same points to ascertain the loss due to friction in the pipes. The following results were obtained:--Loss by the fall in temperature, 54·85 per cent.; loss due to friction, as shown by the loss of pressure, 5·42 per cent.; loss not accounted for, but which may be ascribed to piston and valve-clearance, leakages, etc., 27·04 per cent.; leaving an efficiency at the engine, in actual work, of only 12·69 per cent.

This compressed-air plant had recently been replaced by an electric installation, and although he had not, up to the present time, been able to obtain the losses in detail for electric-power transmission, he had been able to satisfy himself that the cost of the latter was very much less. The electric plant was doing the work of the air-compressor, together with considerable additional hauling and pumping; and further, the pumping as well as the hauling, had all been done during the day, thus knocking off entirely the working of the power-plant at night.

Mr. HENRY LAWRENCE (Newcastle-upon-Tyne) wrote that locomotives worked by compressed air were introduced into the mines of the Earl of Durham by Messrs. W. Lishman and W. Young about 1885, and one was exhibited by the Grange Iron Company in the hauling-ground of the Newcastle Exhibition in 1887, and formed part of the exhibits of several systems of underground haulage. The compressed air, formed in three stages by a vertical engine on the site, was delivered directly into the receiver or boiler of the engine, at a pressure of 400 to 600 pounds per square inch. To prevent the attendant applying too great a pressure, more than was necessary for the ordinary working of the load, a reducing valve was introduced limiting the maximum pressure necessary for working the load. There was no difficulty in working the engines underground, except that they required better roads and heavier rails than were generally used. The air-pipes carried the air into the workings, and the refilling of the receivers was very simply effected and occupied about 1 minute.

The engines illustrated in Mr. Randolph's paper could not be applied in Great Britain, on account of their great weight, length and height, unless suitable working-places, curves and rails were prepared for their use. These changes might easily be incurred in opening out a new mine, but the greatest drawback to their use in this country was the production of sparks when the wheels of the engine skidded at various gradients and at starting.

This sparking prevented the use of compressed-air locomotives in South Wales; and added to the danger from sparks was the upheaval of the floor of the mine, displacing the rails that formed the way, etc. Experiments were made by the writer (Mr. Lawrence) and others during two nights in one of the mines in South Wales, and during the time occupied in running inbye, about 1½ miles, the roads were all right; but on the return journey, the rails were shifted in several places, and had to be re-adjusted before the locomotive could pass. Further experiments were tried as to the danger of sparks, and it was proved that they would ignite gas.

Although it appeared from Mr. Randolph's paper that a compressed-air plant was more expensive than an electric plant, on account of the loss of heat by compression, he (Mr. Lawrence) preferred the use of compressed-air locomotives to those worked by electricity, because compressed air in mines, especially deep ones, was of service by the exhaust air aiding the ventilation and there being no necessity for the use of return pipes; while in an electrical plant, a return cable would be required and there would be a great danger of the cables being displaced by falls from the roof, etc.

The compressed-air engines, as worked in mines in Durham, were small, those used by the putters in collecting the tubs to the main roads weighing only 1,350 to 1,575 pounds each, and the larger ones not more than 4,500 pounds each, against the 8,000 and 30,000 pounds locomotives illustrated in Mr. Randolph's paper.

In conclusion, he (Mr. Lawrence) might mention that, some years ago, some small compressed-air locomotives were made for use in tunnelling abroad, and to save piping the air into the tunnel a series of tenders were made, containing as much compressed air as would work for a length of 1 mile, and an easy method of coupling to the engine one or more of these tenders was applied.

The CHAIRMAN (Mr. H. C. Peake), in moving a vote of thanks to Mr. Randolph for his paper, said that he was rather surprised at the costs given for electricity. The writer had stated that for a short distance electricity would be cheaper, but as the distance increased the cost would increase so greatly that compressed air would be the cheapest in the end. He (Mr. Peake), however, thought that the increase of cost in both would be in the same ratio. In the case of compressed air, larger pipes would have to be used, and for electricity larger conductors would have to be employed, to convey the power from the point where it was produced to the point where it was used. He agreed with Mr. Walker that compressed air would have come more to the front had it not been for electricity, which had taken the lead because it could be more easily conveyed to any distance where it was wanted. He asked the author to supply the cost of working the locomotives, the quantity of the loads, and the distance they were hauled.

The REV. G. M. CAPELL seconded the resolution, which was cordially approved.

———

Mr. C. C. LEACH moved a vote of thanks to Mr. H. C. Peake for his services as chairman of their meeting on that day.

Mr. M. WALTON BROWN seconded the resolution, which was cordially approved.

THE INSTITUTION OF MINING ENGINEERS.

GENERAL MEETING,

HELD IN THE ROOMS OF THE GEOLOGICAL SOCIETY, BURLINGTON HOUSE, LONDON, JUNE 3RD, 1904.

MR. J. C. CADMAN, PRESIDENT, IN THE CHAIR.

Mr. G. A. Mitchell (past-President) presided over the second section of the members in the rooms of the Royal Astronomical Society, Burlington House.

DISCUSSION OF MR. S. L. THACKER'S PAPER ON "THE DYNAMICS OF THE WINDING-ENGINE."*

Mr. S. L. THACKER wrote that the following corrections should be made in his paper:—

Page 447, line 4, "$r\sqrt{2}$" should read "$r + \sqrt{2}$."

Page 448, line 10, the equation should read

$$\text{``} \frac{r_l}{r_s} = \frac{l + 2l_c + 2l_r}{l + 2l_c} \qquad . \qquad . \qquad . \qquad . \qquad . \qquad . \qquad (15) \text{''}$$

Page 449, line 2, the equation should read

$$\text{``} L = (l + l_e + l_r) - \frac{r_l}{r_s}(l_c), \qquad . \qquad . \qquad . \qquad . \qquad (20) \text{''}$$

Page 450, line 8, the equation should read

$$\text{``} M \times 2\pi n_t + \frac{W_1 v^2}{2g} + \frac{W_2 v_1^2}{2g} + \frac{W_d v_g^2}{2g} \qquad . \qquad . \qquad . \qquad (36) \text{''}$$

Page 464, line 27, "19 feet" should read "24 feet."

Mr. T. LINDSAY GALLOWAY (Glasgow) said that the author did not appear to have taken into consideration in his paper the old-fashioned way of a slow-moving counterbalance similar to that used in the North of England for many years. In the case of a balance-rope, the weight moved at the same rate as the cage.

* *Trans. Inst. M.E.*, 1903, vol. xxvi., page 445.

In the case of electrical winding, which was now attracting attention, there was a comparatively small mass moving at a great velocity. He would suggest that Mr. Thacker might include in his reply to the discussion the case of a heavy counter-balance moving at a low velocity. It had certain advantages, in so far that the kinetic energy which was developed in the counterbalance was the product of the square of the velocity multiplied by the mass, so that if the mass were increased a less velocity could be used, and consequently the kinetic energy which was required to start the mass into motion would be much less. He did not suggest that the old form of counterbalance was superior to the balance-rope, which was simple and commended itself to the mining engineer at once; but he merely suggested as a matter of interest that the author might include it as another of the various cases which he had analysed. So far as he (Mr. Galloway) was aware, most mining engineers had come to the conclusion that the scroll-drum was a clumsy and heavy appliance, and he thought that this had been proved by the calculations contained in Mr. Thacker's paper. In that case, there was a large mass moving at a high velocity, and the author had clearly shown that the handiness of the winding was very much curtailed by its use.

Mr FRANK R. SIMPSON (Blaydon-upon-Tyne) said that he had noticed in Belgium that flat tapered hemp-ropes were still in use, at a pit about 3,600 feet deep. These ropes were probably used owing to the perfect balance attained with them.

———

Mr. VICTOR WATTEYNE's paper on "The Purpose and Present State of the First Experiments on Safety-lamps and Explosives carried out at the Frameries Experimental Station, Belgium" was read as follows:—

THE PURPOSE AND PRESENT STATE OF THE FIRST EXPERIMENTS ON SAFETY-LAMPS AND EXPLOSIVES CARRIED OUT AT THE FRAMERIES EXPERIMENTAL STATION, BELGIUM.*

By V. WATTEYNE,

Ingénieur en Chef, Directeur des Mines et Directeur du Service des Accidents Miniers et du Grisou.

Introduction.

It is hardly possible as yet to record publicly the results of the experiments on safety-lamps and explosives undertaken by the Belgian Department of Mining Accidents and Fire-damp, under the writer's direction, at the Government experimental station at Frameries. These experiments are, as a matter of fact, incomplete at present; and, in so delicate a matter and in the case of a methodical experimental investigation of this kind, it is important to suspend judgment until every single experiment has been repeated over and over again, under all the conditions likely to be encountered in mining practice. The publication of partial results, inadequately checked, and open possibly to revision in the course of later experiments, might be attended with the grave disadvantage of propagating mistaken conclusions, highly detrimental to the security of those who work in mines.

Nor does the writer think it necessary to recapitulate once more the details as to the structure, installation, etc., of this experimental station, and to state what is the nature of the experiments there carried out, for full descriptions and ample details have already been set forth in previous publications.†

---

* Translated by Mr. L. L. Belinfante, M.Sc.

† "Emploi des Explosifs dans les Mines de Houille de Belgique en 1901," *Annales des Mines de Belgique*, 1902, vol. vii., page 993; and "Le Siége d'Expériences de l'Administration des Mines à Frameries," *ibid.*, vol. ix., page 149.

But the members will perhaps be interested to know what is the actual state of the question that we have set ourselves in Belgium to solve by experiment, that is, the question of safety-lamps and explosives.

I.—SAFETY-LAMPS.

The lighting of fiery mines is regulated in Belgium by the Royal Decree of April 28th, 1884. Among other provisions, this decree includes the following:—

43.—The use of safety-lamps fed with pure vegetable oil is obligatory in fiery mines.

44.—. . . The Mueseler type of lamp . . . shall be alone used, to the exclusion of all other lighting-appliances, in mines of the second and third class . . .

49.—Lamps extinguished in the mines shall be sent back, either to bank, or to the immediate neighbourhood of the downcast shaft . . .

At the time when the decree of 1884 was drafted, these provisions were undoubtedly justified. The Mueseler lamp was then, indeed, of all those known, and especially of all those experimented with in Belgium,* by far the best when exposed to air-currents of great velocity. On the other hand, mineral oils (although the matter had not been gone into very deeply) were considered too dangerous to permit of their introduction into mines; and the possibility of a practical method of internal ignition (relighting), absolutely free from risk, had not been foreseen.

But much progress has been accomplished since then: lamps guaranteeing a degree of safety superior to that of the Mueseler lamp have been invented; essentially volatile oils, such as benzine, have come into current use in other countries, yielding a more intense and more constant illumination, without introducing any fresh danger into the mine; and internal ignition by sure and convenient methods has become general in several mining districts.

Since then, the formal provisions of our regulations prevented our miners from taking advantage of the progress which had been accomplished, it became necessary to proceed to a revision

* It may be observed that the experiments, whence the Committee, who drafted the regulations of 1884, drew their inspiration, dated as far back as 1870.1873.

of the aforesaid regulations. In order to carry out that revision, however, it was imperative, first of all, to demonstrate beyond cavil that the advantages to be gained from the adoption of the new methods would not be counterbalanced or perchance more than counterbalanced, by special risks inherent in the said new methods.

The three main points, to which reference has been made, are as follows:—

(1) *The Use of Benzine.*—Assuredly lamps, which are well made and supplied with this fluid, yield incomparably-better light than lamps fed with vegetable oil, especially if one compares the lamps after they have been in use for several hours in dusty workings. Now it is known, and the British Royal Commission on Accidents in Mines long ago drew special attention to the fact, that good light is an essential element of safety in mines and a protection against innumerable accidents which have resulted in the sacrifice of many lives.

Does not, however, the use of this highly-inflammable fluid diminish the degree of safety of the lamps in which it is burnt, by the overheating which it brings about and the inflammable atmosphere which is formed within the lamp itself?

(2) *The Suppression of the Exclusive Monopoly of the Mueseler Safety-lamp.*—Are there really other lamps in existence possessing the same degree of safety as, or a greater degree of safety than, the Mueseler lamp?

(3) *Internal Ignition (relighting).*—The chief advantage of internal ignition, from the point of view of safety alone, is that it cuts off from the workman any temptation to open his lamp for the purpose of relighting it when in the workings. Resort is too often had to this reprehensible and most dangerous proceeding, despite special locks, careful inspection, and rigid fines, because the workman is anxious to avoid the delay, and consequent pecuniary loss, which is involved in replacing an extinguished lamp or in sending it up to bank. Moreover, in the case of serious disaster, the workmen who are fleeing from an explosion which has perhaps extinguished all the lamps, if they are provided with the means of again obtaining light will effect

their escape with a greater assurance of safety than if they have to stumble about in pitch darkness, over falls of rock and among perils of all kinds.

Does not, however, the working of the igniter (or relighting appliance) present some danger, in the always possible event of its utilization in the midst of an atmosphere charged with fire-damp?

Experiments made in other countries had already thrown some light on these various points. Yet, before altering our own regulations, we held that it was necessary to repeat those experiments in Belgium, and, if possible, to make them more complete and more conclusive by taking into account the diverse conditions met with in Belgian mines.

Members are perhaps aware that, for the purpose of our experiments, we are in possession of a natural source of fire-damp of a high degree of purity, unexcelled in mining practice. Thus we are enabled to experiment with genuine pit-gas, which the writer regards as an essential factor in arriving at really conclusive results.

The apparatus used allows of the subjection of the lamps to the action of gaseous currents of any desired composition and velocity up to as much as 60 and 66 feet, or 18 and 20 metres per second.   Nor are these currents horizontal only, they are upward or downward, oblique or vertical.   Moreover, the densest atmosphere known to exist in deep mines has been reproduced; and in Belgium there are several mines 3,000 or even 4,000 feet deep.

With the view of simulating all the diverse conditions met with in practice, experiments have been conducted not only in gas-laden atmospheres, but also in dusty atmospheres, wherein the proportions of dust and fire-damp were varied as desired, and with lamps which had been kept for several hours in the underground workings.

From the point of view of relighting, the extreme case has been simulated, wherein the lamp-gauze before extinction has become red-hot in an atmosphere charged with fire-damp, and the workman ignorant or reckless of the danger, has immediately set the igniter at work without even waiting for the gauze to cool down.

As the writer remarked at the outset, the Frameries experiments have not yet been brought to an end. Nevertheless, those that have been carried out in regard to lamps have reached a sufficiently-advanced stage to permit of the following brief synopsis of some of the conclusions to which they lead:—

(1) The use of benzine does not diminish in any notable degree the safety of the lamps in which it is burnt.

(2) Several shielded or bonneted lamps, especially the Marsaut lamp and the Wolf benzine-lamp are shewn to be considerably safer than the Mueseler lamp. The last-named, which shews a perfect capacity of resistance in horizontal currents, rapidly breaks down in upward currents.

(3) The internal igniter, if made of phosphorized paste and worked by friction only, may be often used without danger in the midst of atmospheres charged to extreme with fire-damp. This does not, however, apply to every type of lamp, especially not to the Mueseler, where the working of the igniter causes the flame to pass into the upper gauze.*

II.—EXPLOSIVES.

In Belgium, the use of explosives in mines is regulated by the Royal Decree of December 13th, 1895. The provisions of this decree are very stringent in regard to the use of explosives generally. Thus, for instance, the use of any explosive whatsoever is prohibited:—

(1) For breaking down masses of coal;

(2) For breaking down masses of rock in workings ventilated by a downward current;

(3) In places where one may expect to come upon ancient workings;

(4) For driving roads: in the upper return-airway, in all galleries save the intake-airway, or in all working places, according as they belong to mines of class 2*a*, of class 2*b* or of class 3;

(5) In all mines of class 2*b*, for driving all galleries driven somewhat aside from the normal air-current;

(6) In the same class of mines, for all workings ventilated by a downward current;

(7) In cross-measure stone-drifts approaching a seam, which spontaneously gives off fire-damp, etc.

If, also, the different conditions with which the use of explosives is hedged round, even when allowable, be taken into

---

* In all these observations with regard to lamps the writer must be understood to exclude the electric lamp. That subject must needs be dealt with, specially, and it has not entered into the experiments herein described.

consideration, it will be admitted that the Belgian regulations are the most prohibitive that are enforced in any country. On the other hand, no specification is made in regard to the particular explosives that may be used, with the exception of blackpowder which is absolutely forbidden in many cases; and the most dangerous shattering explosives are allowed on the same footing as the safest (or, if you will, the least dangerous) safety-explosives.

This gap has been in part made good by Ministerial ordinances in the manner now to be described. In consequence of the researches which Mr. Lucien Denoël and the writer had conducted in regard to safety-explosives (some of the results of which were laid before the International Congress of Mining and Metallurgy at Paris in 1900),[*] a list of explosives was drawn up, consisting of all such as, for theoretical reasons or from analogy with explosives really submitted to experiment, appeared to fulfil most nearly the appellation of safety-explosives from among all those in use in Belgium.

The list is recapitulated here, merely as a matter of record, for it may be stated at once that our experiments, incomplete as they still are, have conclusively shown that a number of these explosives by no means deserve the appellation of safety-explosives. We had, indeed, made the most formal reservations in regard to this matter, and had demonstrated, with the help of arguments which it is needless to recapitulate here but which the members will find in the paper just quoted, that the problem of safety-explosives could be solved by experiment alone. These reservations have proved to be amply justified. However that may be, the list is given on the opposite page.

Now, it so happens that, in consequence of the very stringency of the provisions regulating the use of explosives, frequent exceptions have to be allowed. The chief inspectors of mines in granting such exemptions, make them dependent on conditions of more or less stringency, according to the extent of the risk which they believe is incurred, one of these conditions being very often that safety-explosives alone should be used. On October 27th, 1900, a Ministerial circular declared this condition to be obligatory, and defined provisionally as safety-

---

[*] " Les Explosifs dans les Mines de Houille de Belgique," *Bulletin de la Société de l'Industrie Minérale*, 1900, third series, vol. xiv., page 59.

explosives those enumerated in the foregoing list.   In this way, the gap in the regulations, to which the writer has already referred, was made good.

LIST OF SAFETY-EXPLOSIVES.

Name of Explosive.	Composition.	Per Cent.	Name of Explosive.	Composition.	Per Cent.
Antigrisou Favier No. II.	Ammonium nitrate	80·90	Flammivore.	Ammonium nitrate	85·00
	Dinitronaphthaline	11·70		Ammonium sulphate	5·00
	Ammonium chloride	7·40		Collodion cotton ...	10·00
Favier No. IV.	Ammonium nitrate	95·50	Dahmenite A.	Ammonium nitrate	91·30
	Dinitronaphthaline	4·50		Naphthaline ...	6·50
Matagne grisoutite and Baelen antigrisouteuse dynamite.	Nitroglycerine ...	44·00		Potassium bichromate ..	2·20
	Cellulose	12·00	Nitroferrite No. I.	Ammonium nitrate	93·50
	Magnesium sulphate	44·00		Potassium ferrocyanide	2·00
Densite D.	Ammonium nitrate	81·40		Crystallized sugar ..	2·50
	Strontium nitrate ...	10·40		Trinitronaphthaline	2·00
	Trinitrotoluol ...	8·20	Cornil white-powder.	Ammonium nitrate	90·00
Densite E.	Ammonium nitrate	82·74		Ammonium chloride	5·00
	Strontium nitrate	11·42		Trinitronaphthaline	3·00
	Trinitrotoluol ...	5·84		Sulphur	2·00
Arendonck antigrisou.	Nitroglycerine ...	27·00	Fractorite.	Ammonium nitrate	90·00
	Guncotton	1·00		Resin	4·00
	Ammonium nitrate	72·00		Dextrin	4·00
Ammoniated gelignite.	Nitroglycerine ...	29·30		Potassium bichromate	2·00
	Collodion-cotton ...	0·70	Lebeau or Casteau No I.	Ammonium nitrate	90·00
	Ammonium nitrate	70·00		Nitro-dextrin ...	10·00
Antigrisouteuse forcite No. I.	Nitroglycerine ...	29·40	Baelenite.	Ammonium nitrate	85·00
	Nitrated cotton ...	0·60		Trinitrotoluol ...	15·00
	Ammonium nitrate	70·00	Velterine No. 2.	Ammonium nitrate	93·00
Ammoniated gelatine A or No. 2.	Nitroglycerine ...	30·00		Ammonium trinitrocresylate ...	7·00
	Nitrocellulose ...	1·00	Wallonite.	Ammonium nitrate	90·00
	Ammonium nitrate	69·00		Nitrated tar ...	10·00
Safety-dynamite.	Nitroglycerine	24·00	Westphalite.	Ammonium nitrate	91·00
	Nitrated cotton	1·00		Potassium nitrate	5·00
	Ammonium nitrate	75·00		Resin	4·00

But now, that we have at our disposal an experimental station allowing of a close personal study of safety-explosives; now, that we are enabled to determine their practical value by fixing the maximum charge (that is, the highest charge that can be detonated without igniting the most dangerous mixture known of fire-damp and dust), it will be possible to remodel the regulations with greater effect, and bring them abreast of the most recent progress.

The manner in which the experiments are conducted has been dealt with elsewhere, and it is needless to recur to it here.

It will suffice to mention that they are carried out in a gallery of the same dimensions as that usual in a mine, with genuine pit-gas, and with increasing charges of explosives which are detonated without stemming in an atmosphere charged with fire-damp, accompanied or unaccompanied by dust. It need scarcely be added that the power of the explosives is also determined by the leaden block or Trauzl method. It is proposed to repeat some of the experiments, at all events, with stemming, so as to simulate as closely as possible the conditions which obtain in practice. Also, to carry out experiments in smaller or partly-obstructed galleries; and finally, to settle exactly the influence of dust on the limitation of the maximum charge, etc. The results of all these experiments will be published at a later date. The sole object of this brief paper is, meanwhile, to convey some idea of the actual state of the experiments and of the aims to which they are directed.

———

Mr. W. N. ATKINSON (H.M. Inspector of Mines) wrote that the experiments at Frameries on safety-lamps and explosives were likely to yield valuable results. The Belgian regulations with reference to safety-lamps seemed very conservative. The exclusive use of one defined type of lamp and one kind of oil must have tended to check improvements. Safety-lamps had arrived at a state of security such that the most important feature now appeared to be an increase of lighting-power. The use of essentially volatile oils, such as benzine, could hardly be said to have introduced no fresh danger, in view of the number of lamp-rooms, which had been burnt down where such oils were in use. He would direct Mr. Watteyne's attention to petroleum as an illuminant; it was safer than benzine or colzaline, and when used with a suitable wick and burner gave a better light, nor did it heat the lamp so much. With reference to internal igniters, the danger appeared to be, not so much that a red-hot lamp might be relighted, as that a damaged lamp might be relighted. The lamp might be damaged by the same cause as that which extinguished it, and if it was capable of being relighted by the workman in the dark the risk seemed very appreciable. It was questionable, therefore, whether the benefit of being able to relight a lamp at once after an explosion warranted such risk being taken. The temptation to workmen, to unlock their lamps in order to relight them, could be provided against by the use of proper locks. The use of electric relighters was not referred to. These were valuable for lighting a large number of lamps on the surface, and, when of proper construction and used under proper regulations, did not appear to be unduly dangerous underground.

The Belgian regulations of December 13th, 1895, as to the purposes for which and the places where explosives might be used were undoubtedly stringent—many would say unduly stringent—in view of the necessity for "frequent exceptions." The line now adopted of regulating more efficiently the kind of explosives to be used, will probably result in great benefit without increasing the danger. The fixing of the maximum safe-charge of different explosives would be especially valuable.

Mr. HERBERT PERKIN (Leeds) said that the adoption of the Wolf safety-lamp, which was lit by an internal igniter, in this country would be a step backward, because every miner could not be trusted with an igniter in his lamp. Further, when the glass was broken, the igniter was practically equivalent to the presence of a box of matches in the pit. One advantage claimed was in the case of an explosive mixture extinguishing a lamp, but it was a very rare occurrence for all the lamps in a district to be put out. It was pointed out in the paper that the chief advantage of this appliance was that it cut off from the workman any temptation to open his lamp for the purpose of relighting it when in the workings, but in his own opinion this temptation did not exist in this country to anything like the same extent that it did 15 or 20 years ago. In a longwall-face, the other workmen would act as detectives, and if one man attempted to open his lamp he would doubtless be prevented by his fellows. He thought that there could not be a better system than that in vogue of having lighting-stations in the charge of an experienced man, who would not relight any lamp without examining it.

Mr. H. R. HEWITT (H.M. Inspector of Mines) said that he could not agree with the writer that the introduction of benzine into a mine did not entail additional danger. It was a highly volatile oil and gave off a great amount of vapour after the lamp was extinguished, while the lamp was still hot, and even at ordinary temperatures. In the Midland mines-inspection district, there was a Special Rule providing that no appliance used for striking a light should be carried by any person, and to place a self-igniting safety-lamp in the hands of the collier would be a breach of that Special Rule. Very few benzine-lamps were used in this country, but there was an equally dangerous, highly volatile oil called " colzaline." He agreed that the use of benzine and oils of that class made the lamp very hot, and a considerable ventilating current, which was sometimes not present in the workings, was required for the purpose of cooling the lamp and lessening the results of the vapour becoming explosive in the lamp. In consequence of the wick-tube or sucker fitting badly, or having been badly put into the casing, explosions had occurred inside the lamp owing to the vapour rising too rapidly. In many mines, the electric method of relighting lamps had been adopted, and he thought that it was a step in the right direction, so long as the electric relighting-stations were placed in the intake-airways; but he thought that it would be dangerous to place the means of relighting a lamp in the hands of a workman. Every safety-lamp, after being extinguished, should be properly examined by a competent person, even if that competent person were only a road-contractor or a day-man on the roads. He had found that in far-off districts, 2 or 3 miles from the pit-bottom, when the only relighting-station was on the pit-bank (as they had in North Derbyshire in some cases) the percentage of extinguished safety-lamps during the shift (although boys were provided to carry them to the pit-bottom, but there was 2 or 3 hours' delay in getting a light) was much below that at collieries where the lamp-station was placed conveniently for the men. In the case of explosions occurring, it would only add to the danger to strike a light inside the lamp, and thereby probably cause a further explosion. It was satisfactory to know that these experiments had been made during the past few years with a supply of natural fire-damp. He believed that fire-damp was less easily ignited than manufactured lighting gas, and that experiments were better conducted with the natural supply; and he agreed that tests made under such natural conditions would be very conclusive. Mr. Watteyne stated that " the internal igniter, if made of phosphorized paste and worked by friction only, may be often used without danger in the midst of atmospheres charged to extreme with fire-damp." He (Mr. Hewitt) could not see, if the atmosphere were charged to extreme with fire-damp, how the men could relight a lamp without again igniting the gas: perhaps the writer would give some explanation of this paragraph. The part of the paper devoted to explosives stated that the Belgians had been making experiments in a similar manner to those made at Woolwich, and, as a consequence of those experiments, their rules, regarding the use of explosives generally, were far more stringent than our own, and possibly those in charge of the Home Office testing-station would take notice of the fact. The Belgians fixed a maximum charge for each of their permitted explosives, and, probably, a catastrophe at some future time might be prevented, if we adopted a maximum charge in each case. The leaden block or Trauzl method was known as the best and was used in the Belgian tests, and he would look forward with great interest for the results of those further experiments, which the author promised to communicate to the Institution.

Prof. HENRY LOUIS (Newcastle-upon-Tyne) stated that the experiments on explosives conducted by the North of England Institute of Mining and Mechanical Engineers, at Hebburn, were made with natural gas, so that this was not the first time that these conditions had been observed. Mining engineers, interested in coal-mining, owed a deep debt of gratitude to the Belgian Government for the excellent experimental work which they were doing, and this country (which produced so much coal) should be ashamed to be indebted to a country which was so small a producer as Belgium for the pecuniary outlay which was involved in such experimental work. In this country, the Royal Commission (he believed) recommended a mixture of vegetable and mineral oils, and he supposed that mixture was used in the large majority of our collieries. It would be very interesting indeed to have some experimental details from Mr. Watteyne, comparing the relative safety of the two. He did not think that it followed from what was said in the paper that the Belgians were by any means likely to be lax in the use of safety-lamps; and it was obviously very valuable to know from experimental data what was the margin of safety. It did not follow that, because under certain conditions, they showed the Wolf safety-lamp to be safe, they would rush in and legalize its use under all possible conditions; but it was useful to know that under certain conditions the lamp could be used, and he only hoped that they would have many more papers like this, recording the results of experiments which were immensely valuable to them all.

Mr. A. M. HENSHAW (North Staffordshire) said that the cause of the last explosion at Talk-o'-the-hill colliery was for some time shrouded in doubt, and experiments were necessary for ascertaining the initial cause. Amongst other possible causes there was an idea that one of the electric bells had short-circuited during the week-end when the accident happened, and at a time when the district, where the bell was placed, had become charged with an explosive mixture. It was found by experiments that, although it was possible with an ordinary electric bell, with five Leclanché cells, to cause an explosion of ordinary manufactured illuminating gas by ordinary sparking, it was impossible (so far as the experiments had gone) to cause an explosion of ordinary pit-gas by the same sparking.

The PRESIDENT (Mr. J. C. Cadman) moved a vote of thanks to Mr. Watteyne for his valuable paper.

Mr. H. R. HEWITT seconded the resolution, which was cordially approved.

———

Mr. A. R. SAWYER's paper on " The Transvaal Kromdraai Conglomerates " was read as follows :—

THE TRANSVAAL KROMDRAAI CONGLOMERATES.

By A. R. SAWYER.

"Igneous material . . . . . . occurs locally at the surface and intrudes into the conglomerates and other beds constituting the hill behind Jeppe's Township. . . . . . The rock is a quartz-amygdaloid lava of an intermediate character, containing porphyritic crystals of a dirty-white felspar. This rock also fills up a large portion of the valley east of Doornfontein."[*]   So wrote Mr. Walcot Gibson, in a paper on "The Geology of the Gold-bearing and Associated Rocks of the Southern Transvaal" read before the Geological Society of London on April 6th, 1892.

Fig. 1 (Plate XIX.) shows the extension of this sheet. It is based on a report made by the writer in 1899, in which it is stated that:—"The sheet exhibits foliation, with a southerly dip. It consists in part of amygdaloidal diabase, the amygdules being filled with quartz. It is also in part porphyritic, the larger felspar-crystals (orthoclase) showing twinning. In other parts, the diabase shows neither of these characteristics. So far, this sheet resembles to some extent the amygdaloidal diabase-sheet of Klipriversberg and other parts of South Africa. It exhibits, north of Rietfontein farm, special peculiarities; it contains pebbles and blocks, angular and rounded, of talcose quartzite, diabase, etc., thus resembling the Dwyka Conglomerate of the Cape Colony."

Fig. 3 (Plate XIX.) is a section along the line AB of Fig. 2 (Plate XIX.), taken from a report made in 1895. It shows the relative position of the Kromdraai Conglomerates (on the Kromdraai farm, No. 493, north of Krugersdorp) to the overlying Black Reef Series, and further shows the Black Reef Series unconformably overlying the Kromdraai Conglomerates. The

---

[*] *The Quarterly Journal of the Geological Society of London*, 1892, vol. xlviii., page 427.

writer described these Conglomerates in 1896,[*] as follows:—
" A large hilly region occurs, which consists of schists and slates
containing pebbles and blocks of all sizes.   The pebbles and
blocks consist of quartz, quartzite, conglomerate and breccia.
The schists, being softer than the pebbles and blocks, which they
contain, have weathered more readily, leaving an accumulation
of débris on the tops and slopes of the hills.   Amongst these,
numerous pieces of banket occur, leading a casual observer to
infer that they must have broken off from a reef in the immediate
vicinity.   In two instances, blocks of auriferous banket occur
in the schists on the top of one of these hills, which simulate
the outcrop of a reef.   In each case, prospectors misled by their
appearance sank in them, only to find the schists a few feet
beneath."   A similar case occurred at Rietfontein farm, No.
286 (Fig. 1, Plate XIX.).   What appeared to prospectors to be an
outcrop of conglomerate or quartzite with a vertical dip, proved
on being exposed, to be a large piece of " float " (Fig. 4,
Plate XIX.).

From these descriptions it will be apparent that there is a
strong resemblance between the Kromdraai Conglomerates and
parts of the sheet in the Bezuidenhout valley, and it is, the writer
thinks, admitted that they form part of the same series.

Dr. G. A. F. Molengraaff, in his able treatise on the " Géologie
de la République Sud-Africaine du Transvaal,"[†] gives a section
through Kromdraai, and shows the Conglomerate also with the
Black Reef Series lying unconformably upon it.   He calls these
Conglomerates: " Conglomérat schisteux du Système primaire."[‡]
This section does not (to the writer's mind) sufficiently indicate
the importance of this conglomerate-mass, which reaches to a
greater height than that shown by the scope of Dr. Molengraaff's
section.   The writer has given the height to which it reaches
in Fig. 3 (Plate XIX.).

Dr. G. A. F. Molengraaff and Dr. F. H. Hatch have recently
directed attention to deposits which occur in the Orange River
Colony, Ventersdorp and elsewhere, in detailed and interesting

[*] *Transactions of the Geological Society of South Africa*, 1896, vol. ii.,
page 16.

[†] *Bulletin de la Société Géologique de France*, 1901, series 4, vol. i., page 13.

[‡] *Ibid.*, page 34.

papers entitled respectively :—" Preliminary Note on a hitherto unrecognized Formation underlying the Black Reef Series."[*] and " Note on an unusual Basal Development of the Black Reef Series in the Orange River Colony."[†] Dr. Hatch has since written a further paper on the subject entitled " The Boulder Beds of Ventersdorp, Transvaal."[‡] Both of these writers in these papers appear to admit that these newly-described beds or masses resemble, or are identical with, the Kromdraai Conglomerates. Dr. F. H. Hatch states that no name has hitherto been given to this series, and suggests that it should bear the name of " Ventersdorp Beds," because of its particularly large development there.[§] But, as the writer has shown, the Kromdraai Conglomerates are also particularly well developed; and they have been known and described for so many years that he thinks it unfortunate that a new name should now be given to them.   " Dwyka Conglomerate " is used, although the terms " Dwyka Breccia " or " Dwyka Boulder Bed " would (in many cases) be much more appropriate.

Dr. G. S. Corstorphine in a paper on " The Volcanic Series underlying the Black Reef " appears to consider the " conglomerate " which he saw at Rietfontein farm, No. 286 (Fig. 1, Plate XIX.) to be the same as " the extensive conglomerate-outcrops on Kromdraai."[‖]

Mr. David Draper has informed the writer that, in 1890, he and Dr. G. A. F. Molengraaff found a series of beds containing boulders of Hospital Hill slate, conglomerate, diabase and quartz, lying almost horizontal, and enclosed in a matrix, which weathered into a gritty sandstone, on the Zendelingsfontein and Schietfontein farms beyond Klerksdorp.[¶] He further found similar beds occurring with amygdaloidal diabase on the Vaalbank farm beyond Ventersdorp, and noticed that they dipped directly under the Black Reef.[**] Dr. Molengraaff subsequently

* *Transactions of the Geological Society of South Africa,* 1903, vol. vi., page 68.
  † *Ibid.,* page 69.　　　　　　　　　　‡ *Ibid.,* page 95.
  § *Ibid.,* page 96.　　　　　　　　　　‖ *Ibid.,* page 99.
  ¶ " Beiträg zur Geologie der Umgegend der Goldfelder auf dem Hoogeveld in der Südafrikanischen Republik," by Dr. G. A. F. Molengraaff, *Neues Jahrbuch für Mineralogie, Geologie und Palaentologie,* 1894, vol. ix., page 239.
  ** *Transactions of the Geological Society of South Africa,* 1896, vol. i., page 45.

accompanied him to these farms, and carefully studied this discovery.

The question arises, as to where all the detrital matter came from. This the writer has indicated in the section which accompanies his paper on "The Northern Extension of the Witwatersrand Gold-field" read in 1889.[*] He there indicated the raising of the Witwatersrand Beds above the basement granite, and showed by dotted lines how the beds may have lain before denudation; the granite, itself, having been pushed up by earth-movements, thereby raising the Witwatersrand Beds.[†] The amount of material removed by denudation must have been almost incalculable, so that it is no wonder that, here and there, we still find attenuated remains of what must at one time have been very extensive deposits. Where the pebbles, blocks and large pieces of conglomerate occur, these were probably dropped not far from the original beds, sheets or dykes from which they were derived, in the form of talus-slopes with probably steep inclinations, originally.

In a country such as this, where enormous changes of temperature occur within 24 hours, large angular pieces of rock are readily thrown off and this will have been the case in a greater degree when the escarpments were so very much higher than they are now. Large rock-slides probably also occurred.

As pointed out, in a paper read in July, 1903, the lava, which now forms the amygdaloidal diabase-sheet so well represented at Klipriversberg, was not deposited in a day. On the contrary, there was a long interval, during which "the Witwatersrand Beds had already in part been tilted and broken before the final deposition of this lava upon it."[‡] It is not, therefore, astonishing to find the detrital remains formed by the denudation of the Witwatersrand Beds in close proximity to amygdaloidal diabase, and in some instances seeming to form part of it. The detrital accumulations formed during the overflow of the lava which is now classified under the name of "Klipriversberg

---

[*] *Transactions of the North Staffordshire Institute of Mining and Mechanical Engineers*, 1889, vol. x., Fig. 1, Plate X. *bis*, page 132.

[†] *Ibid.*, vol. x., page 132.

[‡] *Transactions of the Geological Society of South Africa*, 1903, vol. vi., page 49.

Amygdaloid" belong therefore to the same period. Further detritus, no doubt, accumulated after the last flow of this lava and before the deposition of the Black Reef Series, and the Kromdraai Conglomerates at Kromdraai are probably some of these. It will be very interesting to hear Dr. G. A. F. Molengraaff's further remarks on these peculiar deposits, which he foreshadows in his " Preliminary Note."[*] He and Dr. F. H. Hatch have recently thrown a good deal of light on them.

Amygdaloidal diabase occurs below the Black Reef Series on the Rietfontein farm, No. 286 (Fig. 1, Plate XIX.), and like that at the Orion Gold-mining Company, south of Johannesburg, its foot-wall consists of a whitish clay-slate, which is probably decomposed diabase.

The writer has noticed similar conglomerates and breccias in Manica land, and he has described them in two papers[†] read before the members some years ago. He there stated that:—" These conglomerates and breccias are very thick in places. They resemble in part the well-known Dwyka Conglomerate."[‡] " They are mostly sheared and contain quartz, quartzite, and occasionally fragments of sandstone, all these occurring as angular, subangular and rounded pebbles. Some of the pebbles are composed of banded quartzite, . . . and others of decomposed basic [igneous] rocks and felsite."[§] Fig. 2 (Plate XI.)[||] in the first of these two papers shows " chlorite-schist, with included pebbles of banded quartzite, the pebbles being flat, rounded, and up to 10 inches long, occurs on the top of the Hog's Back,"[¶] one of the highest ridges, and its foliation is vertical.

Fig. 1 (Plate XIX.) is interesting, in that it shows the turn of the Witwatersrand Beds northwards round the granite-mass, which lies between Johannesburg and Pretoria. In a report made by the writer in 1895[**] on farms situated to the north of

* *Transactions of the Geological Society of South Africa*, 1903, vol. vi., page 68.

† " The Portuguese Manica Gold-field," *Trans. Inst. M.E.*, 1900, vol. xix., page 265 ; and " Further Remarks on the Portuguese Manica Gold-field," *ibid.*, 1903, vol. xxv., page 637.

‡ *Ibid.*, 1900, vol. xix., page 275.

§ *Ibid.*, page 275. || *Ibid.*, page 278.   ¶ *Ibid.*, page 275.

** *South African Mines*, Johannesburg, 1903, vol. i., No. 13.

this granite-mass, south of Pretoria, he stated with regard to certain reefs that "they bear great resemblance to some found in the Witwatersrand formation." "The discovery of the northern slope of the Witwatersrand formation, for such he believes it to be, is of immense importance ; and, as it undoubtedly underlies the Black Reef Series, and the Dolomitic Limestone formation, its nature should be ascertained by boring."

The PRESIDENT (Mr. J. C. Cadman) moved a vote of thanks to Mr. A. R. Sawyer for his interesting paper.

Mr. M. H. MILLS seconded the resolution, which was cordially approved.

Mr. G. F. MONCKTON's paper on the " Cinnabar-bearing Rocks of British Columbia " was read as follows: -

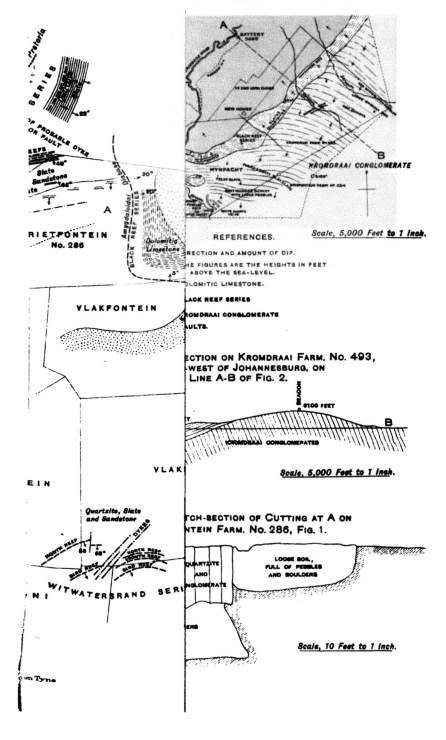

KROMDRAAI CONGLOMERATE

Scale, 5,000 Feet to 1 Inch.

REFERENCES.

...RECTION AND AMOUNT OF DIP.

...HE FIGURES ARE THE HEIGHTS IN FEET
...ABOVE THE SEA-LEVEL.

...OLOMITIC LIMESTONE.

...LACK REEF SERIES

...ROMDRAAI CONGLOMERATE

...ULTS.

...ECTION ON KROMDRAAI FARM, No. 493,
...-WEST OF JOHANNESBURG, ON
LINE A-B OF FIG. 2.

6100 FEET

KROMDRAAI CONGLOMERATES

Scale, 5,000 Feet to 1 Inch.

...TCH-SECTION OF CUTTING AT A ON
...NTEIN FARM. No. 286, FIG. 1.

LOOSE SOIL,
FULL OF PEBBLES
AND BOULDERS

QUARTZITE
AND
NGLOMERATE

Scale, 10 Feet to 1 Inch.

SERIES

Pretoria

OF PROBABLE DYKE
OR FAULT

EEFS

Slate
Sandstone

A

RIETFONTEIN
No. 286

Amygdaloidal Dykes
BLACK REEF SERIES

Dolomitic Limestone

VLAKFONTEIN

VLAK

EIN

Quartzite, Slate
and Sandstone

DYKES

NORTH REEF

WITWATERSRAND SERI

on Tyne

CINNABAR-BEARING ROCKS OF BRITISH COLUMBIA.

By G. F. MONCKTON.

The quicksilver-bearing zone of British Columbia may be said to be 2 miles wide, and has been traced for a length of 30 miles. Its trend is north and south. Isolated occurrences of the metal are known 8 miles east and west of this belt. The quicksilver occurs as sulphide of mercury, generally known as cinnabar. The belt crosses Kamloops Lake, 3 miles above the lower end of it (Fig. 1, Plate XX.).

The region is geologically much disturbed, and the formation of the remarkable rent in which the lake lies was ascribed by the late Dr. G. M. Dawson to faulting, due to volcanic phenomena. It is a difficult matter to arrange the different dykes, beds of ash, and sedimentary deposits in their proper order, and Dr. Dawson stated of one of his sections of this area that it was " little more than diagrammatic and largely hypothetical."[*] At this point, where the more important of the quicksilver-bearing rocks should approach the lake, and be visible in the bluffs, their places are taken by intrusive basalt, porphyry and andesite, and it is not until a point is reached some distance back from the lake, that the main rocks of the hills can be seen. The whole country bears out the comprehensive description of the puzzled miner who said that it was " hove, busted and swung around, and then hoisted and broke and shook." It was perhaps in compensation for its hard usage, that Nature has endowed it with one of the finest climates and some of the most magnificent scenery in the world.

The writer has had the advantage of studying this district intimately after some £15,000 had been spent on underground work, and a considerable sum expended on cutting trails round the steep sides of the hills. These exposed many rocks hitherto invisible, opportunities which were denied to the writer of the

* "Report on the Area of the Kamloops Map-sheet, British Columbia," *Geological Survey of Canada, Annual Report*, 1894, new series, vol. vii., report B, page 162.

Canadian Survey reports. The sections, appended, are believed to represent correctly the position of the strata behind the precipitous face of the mountains immediately above the lake. Much, however, yet remains to be discovered, and alterations will very likely have to be made in the sections to some extent (Figs. 2 and 3, Plate XX.).

The quicksilver-zone begins at the eastern edge of the Nicola rocks, which are of Triassic age. This series is composed of volcanic rocks, with some limestones, conglomerates and argillites; and a limestone (No. 1) which lies at a considerable depth in the series, crops out on the eastern side of a synclinal fold on the north side of Kamloops Lake. This bed contains some cinnabar, but it is not yet of importance. Where it crops out on the western side of the same syncline, in an Indian reservation, it is also known to contain cinnabar. About ½ mile east of the outcrop of No. 1 bed is dolomite (No. 2) carrying cinnabar. This stratum is 300 feet thick, and appears to underlie No. 1 bed, from which it is separated by intrusive rocks. This is followed by volcanic materials, and the basal conglomerates of the Coldwater or Oligocene series, which are in turn overlain by a massive bed of dolomite (No. 3) with which are intercalated some thin black shales and grey argillites resembling arkose. This dolomite is about 300 feet thick, and is overlain by sheets of basalt and porphyry and some conglomerates, which latter swell out higher up the mountain. These are followed by another bed of dolomite (No. 4), and beyond a fault, No. 3 bed re-appears and is the principal carrier of cinnabar. Beyond this, the bed-rock on the line of the section is concealed for ½ mile by surface-soil, but judging from outcrops farther north this space is filled by volcanic ash and lava. This brings us to the Tranquille Tertiary ash-beds at Copper Creek, which limit the cinnabar-bearing zone on the east. The basal conglomerates of the Oligocene lie unconformably upon the Nicola series, and dip eastward. They vary greatly in width, as such beds often do (Fig. 2, Plate XX.).

On the south side of the lake, the only difference appears to be that the thickness of the volcanic accumulations between No. 3 and No. 4 beds is not so great, and some tuffs of the Nicola series underlying No. 2 bed come to the surface. Cinnabar is found in all the dolomites, and in veins of dolomite cutting through decomposed volcanic rocks, also in volcanic ash and conglomerate, and at one point in granite, but only the dolomites have produced workable quantities at present. Judging by patches of drift which have been found, the first, second and fourth dolomites will eventually take a prominent place.

In reference to the accompanying section (Fig. 2, Plate XX.), it may be mentioned that there is some question as to whether the bed of dolomite immediately west of Copper Creek is No. 3 or No. 4, but exposures at Hardie mountain point to it being No. 3 bed.

From his use of the word "zone" in reference to these dolomites, Dr. G. M. Dawson was apparently under the impression that they were not sedimentary deposits. But when he examined the district, they were largely covered by surface-soil, and since that time some 4,000 feet of drifting has been done on No. 3 bed alone. This and surface-prospecting have convinced the writer that these are all ordinary beds of limestone *in situ*, which have been altered to some extent and faulted. They are traceable to the south of the lake, 10 miles to the Toonkwa Lake, and northward 10 miles to Criss Creek, where they pass under later flows of basalt, and emerge again 6 miles farther north on the Deadman river, where they are again covered up.

It is not easy to trace each separate bed for this distance, as they are frequently traversed along their strike by dykes, and overlain by sheets of lava, but the series as a whole can be followed with ease, and future prospecting will render it possible to define them at all points.

It is important to note that the cinnabar always occurs in the dolomite near porphyry, thus on the Briar mine (Rosebush of Dr. Dawson) the porphyry is from 10 to 50 feet away; on a claim south of Kamloops Lake, it is at the contact; on the Toonkwa, about 30 feet distant. On Sabiston Creek, it occurs in a seam of dolomite separated from the main bed by porphyry. At one point on Criss Creek, it is only 5 feet from porphyry; at another on the contact; and, in a third case, it is in a narrow seam of dolomite, between porphyry and conglomerate. This would tend to show that the deposition of the cinnabar was due to the heat generated by volcanic action, at a period subsequent to the deposition of the dolomite. In many cases, notably at Hardie mountain, the fissures of old thermal springs may be seen. One may, therefore, gather that the formation of these deposits took place over a long period of cooling, subsequent to the era of volcanic action which resulted in the accumulation of 5,000 feet of strata in the Tertiary period.

More work has been done on the property owned by the British Columbia Cinnabar-mines Company, at the mouth of Copper Creek than on any other. About £10,000 in all has been expended on this mine. The first work resulted in the extraction of about 150 tons, from which 114 flasks worth about £900 were obtained. The management then decided to mine low-grade ore and to treat it on a large scale, and for this purpose a 25 tons Granza coarse ore-furnace was built. Several hundred tons were treated in this, but the results not being satisfactory the works were closed down in 1897. Of this plant it may be said that it consisted of a shaft-furnace, in which the ore, charged through a vertical shaft, sifted gradually down over a series of inclined shelves. The mercury reduced from the ore during its descent was carried off in fumes to the condensers. The pipes through which it passed were earthenware. The first two condensers were iron, and were succeeded by a brick one, and this by eight earthenware condensers, the lower part of which was wood. The mercury could be drawn off by doors at the bottom of the condensers, and water was kept pouring on the condensers by means of spray-pipes. Various causes are assigned for the failure of this plant. It is usual in furnaces of this kind to provide a fan, to draw the fumes through the condensers, but this was omitted. It appears that the ore fed was of very low-grade, and it was fed too large, some pieces being nearly a foot long. The fuel available was not good, being chiefly pitch-pine, which has a tendency to evolve sudden heat instead of an even temperature. The supply of water was not sufficient, especially in view of the poor quality of the fuel. The ore, previously treated successfully, was passed through a modified Bavarian furnace, in which two cast-iron retorts were built in an arched brick-furnace. The ore, being broken small and mixed with charcoal, was placed in iron-pans and inserted in the retorts. The quicksilver passed through iron-pipes into condensers under water. The charges were withdrawn at intervals, and replaced by others. The labour-cost was large on this plant, and the total expense amounted to £1 per ton. This would be a small matter where the ore could be picked to a high grade, as this indeed could. It was unfortunate for the district that this plant was dismantled, as it would have enabled prospectors to treat their ores on a working scale. The ore, at present showing on this mine, consists of three high-grade streaks on the Briar mine, and some lower-grade material to the west of it. Most of this has been uncovered since the stoppage in 1897. There are also some less important and untested outcrops. It is a pity that Dr. Dawson's advice to trench across the veins was not adopted, as it might have resulted in finding more ore. The cinnabar is contained in quartz-veins and associated with antimonite. Bornite and hæmatite are occasionally found.

The mines of the Hardie Mountain Mines, Limited, were developed in part by a local company, and recently by an English company, which spent £4,000. This development was apparently being done with the intention of preparing for work on a large scale later on, and but little of it was done in ore. There is one high-grade streak on this property, which appears promising; but as a rule the ore at this point is of low grade, and would need to be handled on a considerable scale. Cinnabar is found in conglomerate and tuff here. No ore from this mine has yet been treated on a working scale.

Some work has been done on other claims on this mountain, and on Criss Creek a good deal of prospecting took place some years ago with fair results. Cinnabar can be found in the bed of the stream where it is crossed by the dolomites. The mineral has also been found on the slopes of the mountain, where the dolomites appear on Deadman river, but not as yet *in situ*. At the Toonkwa, 10 miles south of Kamloops Lake, cinnabar is found in the dolomite, not in veins of quartz, and it is sometimes accompanied by antimonite. Sometimes it occurs at this point in high-grade streaks, but much of it is found in a zone several feet wide through which it is disseminated. Some of the solid streaks, from the Briar and Toonkwa mines, are 2 inches thick. As a seam of solid cinnabar 2 inches thick may seem small, the writer may point out that if the ore were to be

picked to represent a vein 1 foot thick, an inch of cinnabar
would represent £14 a ton on the vein of 12 inches, taking an
average price. This affords a basis for comparing it with other
minerals. In this district, the cost of extracting such ore
would be £1 4s. per ton, and the cost of treatment would vary
from 5s. to 15s. per ton, according to the size of the furnace.
Quicksilver being one of the most difficult metals to follow,
a further allowance of, say, £5 should be made for development.

In reference to this difficulty of following the ore it may not
be amiss to quote Prof. S. B. Christy, who had special facilities
for studying the New Almaden, the most famous of Californian
quicksilver-mines. " The quicksilver-deposits of California are
characterized by great and persistent irregularity, which makes
the mining of these ores much more difficult than that of other
metals. New Almaden is a striking example of this irregularity.
It has often occurred in the history of this mine that there was
no ore or scarcely any in sight. Very frequently large bodies
of ore would almost completely run out, and there would be
visible in the face of the works only a slight coloration of the
vein-matter to indicate that there was any ore left in that particular place, and by following this little string of ore very
carefully it might lead to a large deposit." The New Almaden
was £320,000 in debt in 1870, and on the point of being
abandoned, but in the following twenty years it produced over
£3,000,000.

Besides the occurrences mentioned, cinnabar has been found
in many places in the district, some of which are a little outside
the main quicksilver-bearing zone, but these are at present of no
importance, although its occurrence in a bed of dolomite north
of Cherry Creek, 10 miles east of Copper Creek, may be said to
offer possibilities of other large deposits, as this dolomite extends
many miles south of Kamloops Lake. The fact that the higher
part of the mountains is much covered with lava-flows and deep
surface-soil is a stumbling block to the prospector. The liberal
conditions of British Columbian mining laws have also militated
against the development of this form of mining, as they have
enabled parties to hold large areas without working them, in
hopes of re-selling at high prices. This hope, however, has been

To illuring Rocks of British Columbia."

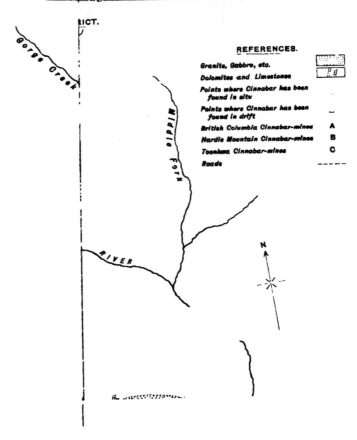

REFERENCES.

Granite, Gabbro, etc.	
Dolomites and Limestones	
Points where Cinnabar has been found in situ	
Points where Cinnabar has been found in drift	
British Columbia Cinnabar-mines	A
Hardie Mountain Cinnabar-mines	B
Toonkwa Cinnabar-mines	C
Roads	

falsified, and the only holders now are miners who are slowly developing their properties. There is plenty of room in the district for prospectors, both to work abandoned claims and to search for new outcrops of ore. It may be said that the area alone of the dolomites which are exposed is not less than 10 square miles, hardly any of which has been tested. The great need of the district at present is a furnace to give working tests. The only alternative is to ship ore to Great Britain, but this is only practicable with very high-grade ores.

For information regarding some parts of the district the writer is indebted to Mr. A. J. Colquhoun, who has made a special study of cinnabar and its occurrence.

———

Mr. W. D. VERSCHOYLE (Seattle) wrote that Mr. Monckton's paper gave the impression that the camp was in a somewhat depressed condition, and this appeared to be due either to want of capital or want of ore. If developments up to the present time had resulted in the finding of any considerable tonnage of even low-grade ore, or had shown that there was a probability that such ore would be found, it would appear that there was an opportunity for a strong company to take hold of the best showings, get to work with a diamond-drill, and eventually erect a central reduction-works, if favourable results were obtained. The topographical and general information required as a basis for even advocating such a course were not, however, contained in the paper; and whilst a notice of the occurrence of the ore was interesting, a consideration of the commercial aspect must always be still more so to the mining engineer. Therefore, a little more information as to the probable extent of the deposits and their topographical distribution would be desirable, and, in conjunction with data as to the best means of testing them, would afford a basis upon which to make an approximate estimation of the commercial importance of the discovery.

Mr. J. D. KENDALL (London) wrote that he had traversed the area described by Mr. Monckton, and had examined such of the workings of the Rosebush mine as were open at the time (1898), but he was unable to obtain any clear idea of the geological structure of the ground in the neighbourhood of the mine. The rocks, near the surface, were very much decomposed and covered by detrital and other matter; and many of the drifts had collapsed, so that it was not easy to determine the true character of the different rocks. Moreover, it was quite impossible to ascertain their relations to one another, except at one or two points. At the inner end of the workings, the rocks were less changed than near the surface, and consequently it was possible to learn something there (although not much) as to the nature of this cinnabar-deposit. The section (Fig. 4, Plate XXI.) was observed at the furthest point from the day reached by the Rose-bush drifts. The direction of the contact between the ore and the decomposed volcanic rock, shown in Fig. 4, was magnetic north and south, the dip being westerly at an angle of about 75 degrees. The thin irregular strings of cinnabar were more or less parallel to the contact. Figs. 5 and 6 (Plate XXI.) were views of a piece of ore from this point, Fig. 5 shewing the east-and-west face as it occurred in the mine, and Fig. 6, the north-and-south face.

When seen under the microscope, the ore was more complex in structure than might be supposed from its macroscopic aspect, as would be seen from Fig. 7 (Plate XXI.). The ragged contact of the decomposed volcanic rock and the calcite was suggestive of replacement.

From an economic standpoint, these deposits were not interesting, as would be understood when the work done and the amount of mercury obtained were considered. Drifting and sinking or raising had been done on, and in the neighbourhood of, the deposits to the following extent:—

In 1896, on the Rosebush mine, there was 750 feet of drifting and a raise of 50 feet to the surface. In 1897, on the Yellow Jacket claim, 800 feet of drifting, two raises of 80 and 100 feet respectively, and 1,200 feet of boring. In 1897, on the Columbia claim, 145 feet of drifting. In 1898, on the Almaden claim, 60 feet of drifting. The totals included 1,755 feet of drifting, 230 feet of raising and sinking, and 1,200 feet of boring.

In 1896, two retorts were built to treat high-grade ore, but they were only operated for a few weeks, until the ore of that grade was exhausted. Then a furnace for treating low-grade ore was erected, and starting about March, 1897, it ran only for a short time, owing to defective design. Fig. 8 (Plate XXI.) shows the details of this furnace.

"To[...]ing Rocks of British Columbia"

7.—Cinnabar-ore. Magnified 20 Diameters.

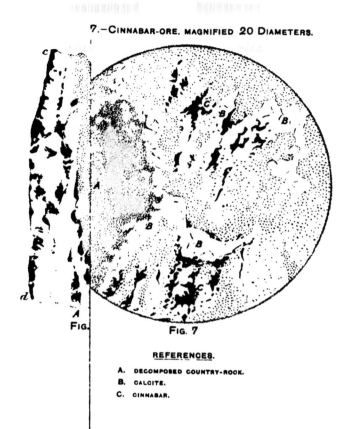

Fig. 7

Fig.

REFERENCES.

A. DECOMPOSED COUNTRY-ROCK.
B. CALCITE.
C. CINNABAR.

Fig. 4.—Elevation of Furnace for Treating Cinnabar-ore.

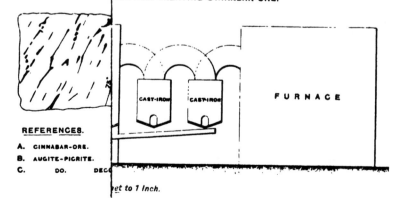

CAST-IRON   CAST-IRON   FURNACE

REFERENCES.

A. CINNABAR-ORE.
B. AUGITE-PICRITE.
C.   DO.   DEC[...]

[...]et to 1 Inch.

The amount of mercury, obtained from the properties noticed in the government reports, was 100 flasks extracted from high-grade ore in 1896.

He (Mr. Kendall) assayed an average sample of low-grade ore from one of the claims in March, 1896, and it yielded only 0·3 per cent. of mercury.

The PRESIDENT (Mr. J. C. Cadman) moved a vote of thanks to Mr. G. F. Monckton for his interesting paper.

Mr. A. M. HENSHAW seconded the resolution, which was cordially approved.

DISCUSSION OF DR. J. S. HALDANE'S PAPER ON "MINERS' ANÆMIA OR ANKYLOSTOMIASIS."*

Mr. R. R. SIMPSON (Geological Survey of India) wrote that it has frequently been asserted, no doubt with truth, that ankylostomiasis is a tropical disease. Mr. F. W. Gray stated that it is "essentially a dirt-disease, and no doubt had its origin in uncleanly habits in tropical countries."† To one acquainted with Oriental habits, the facility with which ova or larvæ might be transferred by means of the finger-nails and mouth into the stomach is apparent. Taking part in the general interest which the subject had aroused among mining engineers, he (Mr. Simpson) some time ago, made enquiries as to the prevalence of ankylostomiasis in Indian coal-mines.

In Bengal, whence comes the bulk of Indian coal, the disease appears to be unknown. Dr. J. W. C. O'Connor, chief medical officer of the Indian Mining Association, considers this to be due to the following causes:—(1) The salubrious climate; (2) the non-prevalence of malaria, hence the coolies are generally of strong physique; (3) the absence of the habit of clay-eating among the coolies; and (4) free labour-conditions. To these, he (Mr. Simpson) would add (5) the almost universal dryness of the mines.

In the Central Provinces, at the Mohpani collieries, no cases of ankylostomiasis have been diagnosed. In these collieries, for

---

* *Trans. Inst. M.E.*, 1903, vol. xxv., page 643; and vol. xxvii., pages 11, 36, 47 and 262.

† *Ibid.*, 1903, vol. xxvi., page 200.

many years, an organized gang of *mehtars* or sweepers have been employed in removing excrement from the workings and in sprinkling disinfectants on the places of deposit. Latrines are not provided: the coolies relieving themselves where they have occasion.

At the Dandot collieries of the North-western Railway Company in the Punjab Salt-range, the disease has not been recognized, but its presence may be suspected; for Mr. James Grundy, chief inspector of mines in India, in a report on the mines, published in 1896, stated that "those who have worked in the mines for some time have become pale in colour, and complain of being weak and anæmic."* From the hospital returns, for 1894-1895, given in the same report, the anæmic cases are 12 per cent. of the total number of cases treated.

He (Mr. Simpson) had no information regarding ankylostomiasis in the Makum coal-mines, Assam; but it is wellknown to be extremely prevalent in that province. Dr. J. W. C. O'Connor, who had studied the disease in the tea-gardens, stated that it was the chief cause of mortality among the labour-forces. The tea-districts have a heavy rainfall, and the atmosphere is a damp one. Malaria is rife, and doubtless the effects of its repeated attacks weaken the constitution, and predispose its victims to become an easy subject for the reception of the *Ankylostoma*-larvæ.

Mr. J. CADMAN (H.M. Inspector of Mines) said that he had the good fortune to be associated with Dr. A. E. Boycott in a considerable number of his investigations in regard to this matter, and, having taken considerable interest in the physique of men in other districts, he was interested in investigating the disease in Dolcoath. The men to an inexperienced eye seemed feeble, but their appearance in many cases was not sufficiently evident to him to say that they had got the disease; but, on seeing them at work, it was evident that they were in an indifferent state of health. There were many mines in North Staffordshire, having a high gradient, a high temperature, and more or less water, where the disease if it once got in would soon become prevalent. There were two points that required consideration: (1) the sanita-

* *Report on the Inspection of the Dandot and Pidh Mines, Salt Range, near Khewrah, Jhelum District, Punjab*, by Mr. James Grundy, Calcutta, 1896, page 47,

tion of the mines, where infection had not occurred; and (2) it was equally important, to watch that soldiers, militia, aliens and men who had worked abroad, did not enter the mine until they had been examined.

Mr. HENRY HALL (H.M. Inspector of Mines) said that scientific men had told the members all about the worm-disease, and they now recommended that sanitation should be attempted in the pits; but it was difficult to adopt sanitary measures, where there were neither sewers nor water. They said that the best remedy was dryness and that the larvæ speedily died when dry. Now if the fæces had to be dealt with in bulk, this salutary dryness would be absent, and the precaution would defeat its own object. In the recent report of Dr. A. E. Boycott it was pointed out that $2\frac{1}{2}$ per cent. of salt in water was fatal to the larvæ, here apparently was a simple remedy and one that would also mitigate the danger of coal-dust. At the Levant mine, in Cornwall, where the circumstances were favourable for the propagation of the disease, Dr. Boycott stated that he could not find a single case.

Mr. A. M. HENSHAW (Stoke-upon-Trent) thought that some ordinary sanitary precautions should be adopted in mines. He had come to the conclusion that, in the North Staffordshire district, there were several collieries where the disease, if it once got a footing, would become seriously prevalent. There were seams of coal worked under ideal conditions for the propagation of the disease; deep, warm and with a humid atmosphere owing to the proximity of water-bearing strata over the coal.

Mr. THOMAS DOUGLAS (Darlington) suggested that it might be made incumbent on the men to sprinkle some chemical powder over the fæces, and so prevent the occurrence of objectionable conditions. He did not think that it would be desirable to provide conveniences in the mine. Properly constructed closets might be placed near the mine, their use before descending into the mine would be a wise precaution, and only in very exceptional cases need evacuations be made underground. Aliens should be examined by a qualified medical man, who understood the disease, and they should be certified as being free from disease before being allowed to enter the mine.

Mr. J. GERRARD (H.M. Inspector of Mines) said that on Saturday last he visited the Shamrock collieries, and had an opportunity of seeing the interesting arrangements adopted in Westphalia.   Over 16,000 samples had been examined since November, 1902; and the actual expenditure, exclusive of loss of work, in dealing with this dreadful disease amounted to ½d. per ton of coal raised.   The percentage of sufferers from the worm-disease had been reduced from 40 to 0·4.   The members could imagine the difficulty of working a colliery, with 40 per cent. of the workers unable to work, and laid idle every fortnight for three days, whilst they were being treated at home with remedies, the wages of the workmen for the three days being paid by the coal-owners.   At another colliery, 70 per cent. of the workers had suffered from this disease.   Mr. G. A. Meyer said that the original idea of the spread of this disease being due to watering the mines, as required by the Government, hardly met the case; and that other conditions, such as temperature, and the high inclination of the seams, had all played their part in the spread of the disease.   Mr. Meyer had carefully noted the position in the mine of every worker who had suffered from the disease, and the greatest number of sufferers were employed in the highly-inclined portions of the seams. Although there were no ladderways, as in metalliferous mines, owing to the system of stowing the goaf, the workmen had to crawl on their hands and knees over the débris, and the fæces deposited there, until the pail-system was adopted.

Mr. S. F. WALKER (London) suggested that one of the best disinfectants was coal-dust; and possibly it might account for the immunity of British mines from worm-disease.

Mr. C. C. ELLISON (Barnsley) said that it was better to go on the principle that prevention was better than cure.   If they did that forthwith they could go on as they were; but if they continued to talk for some years before adopting the principle of examination, they would be put to the enormous expense which it had been necessary to incur on the Continent.

Dr. J. S. HALDANE (Oxford), replying to the discussion, said that the remarks made by Mr. R. R. Simpson as to the conditions prevailing in India were very interesting, particularly as to the absence of this disease in the collieries of Bengal and the Central provinces, but the fact that the collieries there were dry was certainly a determining factor. Ankylostomiasis was a disease common in the population nearly all over India; and it certainly must have got into the collieries, but he took it that the collieries were dry and therefore that the infection did not spread. If they allowed fæces to be deposited they might become dry, some man might tread in them and afterwards step into some wet place and infect a puddle, where the ova would easily develop. Even with the fæces deposited in dry places, where the ova would never develop, there was always that danger. The same objection, to a certain extent, applied to Mr. Douglas' suggestion that the men should sprinkle some disinfectant over the fæces; supposing that could be carried out, one must remember that a disinfectant would not kill all the ova; and a man might easily step upon the fæces and carry the germs to some wet place, and in that way the mine would become infected. The proper course was to remove the fæces: if they could be disposed of in the goaf so much the better, but they should not be left where the men could tread upon them, particularly in wet parts of the mine, and also where it was rather warm. In some mines, about 50 per cent. of the men made a regular practice of relieving themselves underground; and it was monstrous that this should be so. He hoped that measures would be agreed upon all over the mining districts, for devising common action for dealing with the disease in the way of prevention. He should certainly advise that all men should be examined who came to a mine from Cornwall and Italy, or from any other country abroad, including men coming from India and South Africa, or any tropical country. A sample of the fæces should be sent to some laboratory, as the local doctor was usually unable to undertake the examination; that must be made by someone who had done it before, or was skilled in microscopic work. With regard to the Shamrock collieries, although 40 per cent. of the men were infected with the disease, a much smaller percentage were appreciably affected by it in health, and only 1 or 2 per cent. were unable to work. They had been extraordinarily successful in eradicating the disease in Westphalia, but the expense had been enormous.

. Dr. A. E. BOYCOTT (London), with reference to the Levant mine, in Cornwall, where the disease should have been prevalent,

but was not, said that sea-water, leaking into the mine, killed
the larvæ.   It seemed an easy way of stopping them, but he
understood that there was considerable difficulty in the way of
using salt-water in a mine, owing to its effect on pumping
machinery, etc. His own case was a good illustration of the
danger which existed, for he had no symptoms whatever, and
yet he had had the disease for 4 or 5 months. He had been
making experiments on larvæ in daylight, and his hands never
touched his mouth, under any circumstances, unless they were
first washed. Long experience in the laboratory rendered this
precaution almost mechanical, but notwithstanding every precaution he was infected, and that was an interesting point,
because it showed how difficult it was to avoid infection.
Another point of interest with regard to his own case was that
it was found by examining his own blood, and no traces whatever were discovered in the fæces until some 35 or 36 days after
he had found by the blood-test that he was infected. His blood
continued to give the well-marked change until about a month
ago; and only on two occasions had he found eggs in the stools.
His case illustrated the advantage of examining the blood; it
was much quicker, and would show in cases where the ordinary
method failed.

Mr. J. CUNNINGHAM BOWIE, M.B., C.M., D.P.H. (Cardiff),
wrote that so far as the coal-fields of the United Kingdom were
concerned, they were practically free from the presence of the
disease, and consequently it appeared that attention should be
concentrated largely upon measures of prevention.   Some of
the suggestions which had already been put forward in this
direction—no doubt with the best intentions—seemed to be
lacking in the possibility of practical application in this country.
Perhaps he (Mr. Bowie) could best explain his meaning by
setting forth the proposals which in his judgment would be the
more effectual for combating the introduction of the disease, and
at the same time for preventing the spread of it should it at any
time become more or less endemic.

There were two phases which were worthy of consideration on
this branch of the subject.   First, as to the measures which
could be adopted in the mine.   It was conceded that the disease
was spread solely by the pollution of the ground with infected

fæces and by the colliers coming into contact therewith. It should therefore be obligatory on the part of the miners to defecate in old and disused stalls, which were in process of being filled with rubbish: this plan being preferable to a system of pails, for the following reasons:—(1) It would avoid the risk arising from the probability, if not certainty, of the contents of the buckets being spilt over the roads and cages in the process of removal to the surface; and (2) it would render unnecessary the use of strong-smelling antiseptics, such as izal, in the colliery, the objection to which is obvious. If the plan that he (Mr. Bowie) advocated were adopted, the fæces could be sprinkled with lime daily before being covered up by the rubbish, thus destroying any ova which might be contained in such fæces.

Then, with regard to the watering of the mine, he (Mr. Bowie) was decidedly opposed to a suggestion which had been put forward for discontinuing this. In the South Wales coal-field with which he was more intimately acquainted than that of any other district in the United Kingdom, it would certainly be preferable to risk the introduction of ankylostomiasis than to stop watering. He need scarcely point out the extreme danger which would be occasioned to the lives of the colliers by the increased liability to explosions, particularly in dusty mines, but there was also the additional risk of anthracosis or miners' phthisis, being brought about by the inhalation of the dust. It was well-known that the *Ankylostoma*-larvæ—the real source of infection —required moisture, and it was for this reason no doubt that the discontinuance of watering had been advocated; but, to meet this difficulty, he recommended that in the process of watering a solution of salt, with the specific gravity of sea-water, or even stronger, should be used. This could be done either by first sprinkling the roads with an ascertained quantity of sodium chloride, or by adding sufficient salt to the water to be used in the ordinary process of watering. The result would be, that not only would the larvæ be destroyed, but the possibility of the ova being hatched would be prevented. A further beneficial effect would be the reduction of the temperature of those portions of the colliery thus dealt with, while it must also be remembered that the salt-solution would lessen the frequency of watering as it would tend to prolong the period of moisture on the roads. He (Mr. Bowie) was assured by Mr. Fred. A. Gray,

H.M. inspector of mines for the Cardiff district, that his proposals were practicable and could be carried out without difficulty. He might add that the salt would produce no deleterious effect on the workmen, and it would yield no smell, points worthy of consideration as compared with the use of antiseptics, some of the more powerful of which (apart from the disagreeableness of their odour) were of an exceedingly poisonous nature.

Assuming that a case of ankylostomiasis was introduced into a mine in which this system was carried out it would be practically impossible for the disease to become endemic as neither the ova nor the larvæ would be able to develop under such conditions.

Now, as to the second phase of the subject—the hygienic surroundings of the collier's home, a matter of more importance than might at first sight appear. Every miner's dwelling should be provided with a combined bath, range and boiler in the scullery, such as was recommended by Mr. W. Thompson, of Richmond.* This could be fitted into any cottage for the sum of £8 to £10, and would obviate the proposal to erect the very questionable system of douche-baths at the pit-head, as was done on the Continent,—for these last-named have been shewn to really become in themselves centres of infection. Besides, it was extremely doubtful whether the British collier would ever submit to this washing being a part of the condition of his employment, and for this he could hardly be blamed. The bath which he had suggested would admit of the miner's working clothes being thoroughly washed, and here again it would be possible to disinfect them by a salt solution being added to the water, thus again limiting the possibility of infection either to members of the household or fellow-workmen. Of course there were other directions in which the sanitation of the collier's home could be improved, but it would be out of place to dilate upon them here.

As to the treatment of the disease, Dr. J. S. Haldane and other authorities recommended either thymol or extract of male-fern. Having for some years past used in his (Mr. Bowie's) practice the first-named drug, he considered that it was more efficacious than male-fern extract, and whilst agreeing that large doses were necessary in the initial treatment of the disease, he was of opinion that much benefit would accrue by adminis-

* *Housing Handbook*, page 211.

tering small doses subsequently and for a somewhat prolonged period. He was convinced that the best method of administering thymol in small doses was by dissolving it in almond-oil.

Dr. T. EUSTACE HILL (Medical Officer of Health for the County Council of Durham) wrote that he had prepared, in April last, a special report upon the disease, containing suggestions for its repression, which concluded as follows:—

I would point out that, although with the exception of one case in Scotland, the disease has not been detected in any coal-mine in this country, the danger of our warmer mines becoming infected is a very real one. The invasion by the disease is very slow and insidious, for in Belgium and Germany, although isolated cases were detected as far back as 1884, ten years elapsed before any general prevalence was noted; and in Cornwall the disease existed for nearly ten years before its presence was detected. It is therefore quite possible, as stated by Dr. Haldane, that at the present moment a similar slow spread of infection is occurring among certain English collieries, and that infection may already be widely scattered among them.

The general prevalence of the disease in Cornwall will certainly for some years be a menace to other mining districts; and, moreover, the recent practice of watering the mines for the laying of coal-dust greatly favours the development of the disease in infected mines by providing the necessary moisture, as the marked increase in the number of cases in Westphalia, immediately following the introduction of government regulations as to the watering of coal-dust, appears to conclusively prove.

In Dr. Haldane's opinion, there are strong reasons for making compulsory regulations for preventing the spread of ankylostomiasis, at any rate in mines with a temperature exceeding 70° Fahr.; but the necessity for such compulsory regulations will largely depend on whether the mine-owners and the men actively co-operate in carrying out the precautions which are so essential for the prevention of ankylostomiasis in the warmer pits of this country.

Dr. J. W. HEMBROUGH (Medical Officer of Health for the County Council of Northumberland) wrote that he had reported to the County Council on April 6th, 1904, as to the danger of the disease being imported into the coal-mines of this county, although owing to the efficient ventilation which was very generally provided, the conditions did not appear favourable to the development of *Ankylostoma*-eggs into encapsulated larvæ. He believed that a large number of miners would be averse to any attempt to ensure cleanly habits in the pit-workings, and unless it was made possible for these to be adopted, the present disgusting habit of every man defecating wherever he liked must of necessity continue. Dr. J. S. Haldane stated that it was unusual for miners anywhere to drink underground water. His (Dr. Hembrough's) experience did not corroborate this statement:

In this county it was not uncommon for a miner to drink underground water when at work, and many thousands were entirely dependent upon water pumped from the pit for their domestic supply. In his report, he had made the following suggestions for the prevention of the spread of the disease : —

(1) The exclusion of men, known to be infected, from underground workings. (2) The prevention of the entrance of ripe larvæ into the body either by the mouth or skin. This can only be accomplished by (3) the prevention of pollution of underground workings by human excrement. This would of itself be sufficient to prevent the spread of the disease, as in the absence of such pollution eggs could not develop into encapsulated larvæ. No precautions are at present taken in this direction ; it is not customary to provide ash-closets near the pit's mouth ; no underground receptacles are provided ; fæcal matter is deposited anywhere and everywhere, and is carried by the men's boots in all directions ; tools, clothes and timbering are constantly liable to pollution, and it is probable that the absence of conditions favourable to the development of ova into encapsulated worms, that is, a sufficiently high temperature and moist ground, has alone protected innumerable mines from infection. (4) In order that it may be possible for workers in mines to avoid the present filthy and so general habit of defecating in any portion of the underground workings, it appears absolutely necessary that a sufficient number of ash-closets or other conveniences should be provided aboveground near the shaft, and when the workings are at a long distance from the shaft, watertight metal tubs should be supplied underground. The thorough disinfection of these receptacles is simple, as the newly hatched larvæ are easily killed, and so their development to the encapsulated stage (which is the infective stage) prevented.

The co-operation of the workers in mines is of course absolutely necessary, if cleanliness is to be secured, and doubtless some difficulty will be experienced in securing such co-operation ; but unless measures are adopted which make cleanly habits possible, it is useless to lay any stress upon their great importance.

(5) It seems desirable that in all lectures in any way associated with mining, some instruction should be given on the above disease, the precautions which are necessary in order to prevent its spread, and the dangers which may result from uncleanly habits amongst persons employed in mines. (6) It also seems important that arrangements should be made for the microscopical examination of fæces and blood, with a view to the early detection of the disease in any mine in which its existence is suspected.

DISCUSSION OF MR. FRANCIS FOX'S PAPER ON "RAPID TUNNELLING."[*]

Mr. FRANCIS FOX (London) wrote that on page 404, the seventh and sixth lines from the foot of the page should read " from numerous shafts and break-ups, was 1 foot 6 inches per day per ' face ' of excavated and brick-arched tunnel."

Mr. F. C. SWALLOW (Nuneaton) wrote that the value of Mr. Francis Fox's paper would have been enhanced, if some refer-

[*] *Trans. Inst. M.E.*, 1903. vol. xxvi., page 403.

ence and figures had been submitted as to the cost per foot obtained by the adoption of the Brandt rock-drill in tunnelling. As pointed out in the introduction of the paper, however, if suitable machinery could be applied whereby rapid driving was obtained, the loss of interest on capital expended was correspondingly reduced. The question of rapid and cheap tunnelling or rock-heading, was at times a matter of great moment to colliery-managers, and a statement of the cost per foot (either lineal of a stated width or cubic) obtained over a length of say 1,000 feet, driven in the Simplon tunnel, would be of value to those interested in the subject. It was apparent, however, that the particular circumstances of the case must (to a great extent) decide as to what system of driving a long rock-heading, where speed and first cost were of prime importance, should be employed. Although no statement was given as to the outlay on the plant required for the system employed in driving the Simplon tunnel, it appeared, however, that the method would be too costly for the majority of mine-tunnels in this country, unless a considerable length of rock-heading was to be driven. It would be interesting, therefore, if the writer of the paper would say what would be the approximate minimum length of a tunnel or rock-heading to be driven, where speed of driving and initial outlay were questions of great moment, in which the system advocated would prove a financial gain, if adopted.

The nature and angle of dip of the strata and other circumstances of a colliery, together with the direction of the rock-heading relative to the dip of the strata, would largely determine the method to be employed. In the ordinary Coal-measure strata of this country it was questionable whether any mechanical system of heading compared so favourably as did the application of hand-drilling machines, such as the ratchet hand-drilling machine now in general use in collieries; and blasting by one of the medium high explosives, ignited by ordinary tape safety-fuses. For heading through fairly hard Coal-measure strata, rock-binds, fire-clays, etc., this latter system was most commonly employed in this country and with good results. For comparatively long-distance mine-tunnels, exceeding 1,000 or 1,500 feet in length, pneumatic or electrically-driven rock-drills for drilling the shot-holes were advantageously used under certain conditions, and when systematic water-spraying was employed during drilling operations in order to lay the dust, which was occasioned by the working of the rock-drill, together with the provision of adequate ventilation, the danger of phthisis among the workmen was wholly avoided.

The following methods might be cited in the driving of rock-headings, the dimensions of which are 9 feet by 6 feet within the supports :—

(1) In Coal-measure binds, fire-clays and shales requiring careful timbering to support the roof and sides of the heading, the employment of ratchet hand-drilling machines and blasting with an approved medium high explosive, together with the exercise of judgment in selecting the best position for the shots so as not to disturb the timber-supports and to take the best advantage of the lines of break or cleavage and dip or rise of strata, would be found to yield most satisfactory results both as regards speed and cost of driving.

(2) In fairly hard and dry rock-measures, the employment of air-driven rock-drills, if compressed air is available at the point of heading, would result in economy of labour for drilling the shot-holes, again employing careful judgment in selecting the most advantageous position for the holes. The air-pressure for working the rock-drills should not be less than 60 pounds per square inch.

(3) In hard rock, the same system as advocated in the second method, would be applicable, except that the air-pressure should be higher, up to 80 pounds per square inch. To obtain this pressure at the drills, a stage-air compressor, placed as near to the point of heading as possible, was recommended.

Simultaneous electric blasting was not so satisfactory or efficient as ordinary tape-fuse ignition of the shots, and in the latter case the shots should be so arranged as to be fired in succession: the fuses being of unequal lengths, so as to ignite the charges a few seconds one after the other. By this means, the strata were more thoroughly loosened and less damage was done to the sides and roof of the heading, and the timber and other supports, farther back, were not so liable to be disturbed. By the adoption of electric simultaneous blasting, the surrounding strata were often much shattered and the supports blown out, wherefore the time gained in blasting operations was more than lost in resetting timber and securing the face of the heading.

The following statement recording the cost of a rock-heading which he (Mr. Swallow) was in course of driving might be of interest:—

(A) The rock-heading was being driven so as to be 9 feet by 6 feet between the supports, which consisted of brick side-walls, 14 inches thick, and steel-girders for the roof (second-hand tram-rails, weighing about 30 pounds per foot): the heading rising at a gradient of 1 in 100.

The strata dipped at an angle of about 1 in 5 in the direction in which the heading was being driven—the latter passing through the measures. The strata consisted of shales, binds, fire-clays, wet sandstones, and thin seams of coal and ironstone: these strata varying in thickness measured at right angles to the dip, from a few inches to about 12 feet.

(B) The length of the heading was 1,350 feet.

(C) The system employed in driving the heading through the softer measures was described in the first method; and where the strata met with were hard, and it would prove advantageous, the second method was employed, except that the air-pressure for working the rock-drills did not exceed 45 pounds per square inch at the drill—this being the working-pressure available at the colliery.

(D) The costs per foot were as follows:—(a) Average labour-cost for excavating a face about 11 feet by 7 feet, including small tools, lights, explosives, building brick side-wall, 14 inches thick, placing girders, etc., £1·025. (b) Average cost of supervision, general labour both underground and at the surface, including tool-sharpening, tramming, onsetting, banking, winding, tipping débris, pumping, sundry labour, etc., £0·807. (c) Average cost for certain and other rents, rates, taxes, insurances, etc., £0·281. (d) Average cost of the outlay on rock-drills and other tools, £0·230. (e) Average cost of the amount of boiler-fuel used for steam-raising, both for winding, driving air-compressors, pumping, etc.; also bricks, mortar, girders, timber, tubs, pipes, air-troughs, sundry stores, oils, etc., £1·238. Total cost per foot, £3·581.

———

Mr. JOHN H. MERIVALE's paper on "The Prevention of Accidents in Winding" was read as follows:—

THE PREVENTION OF ACCIDENTS IN WINDING.

By J. H. MERIVALE, M.A.

The objects of the automatic gear, about to be described, the invention of Mr. T. Campbell Futers, engineer, Broomhill Colliery, Northumberland, are:—(1) The prevention of accidents, by substituting a simple mechanical contrivance for human agency. (2) Lessened cost, as a highly-skilled brakesman is not required to take charge of the winding-engine. In fact, if necessary, it can be worked by the banksman, and the brakesman may be dispensed with altogether. (3) Economy of steam.

The gear automatically controls the steam-valve (or in the case of an electric winding-engine, the starting-switch), the brake and the signals of the winding-engine in such a way that over-winding is impossible, and the engine cannot be started until the signals have been given by both the banksman and the onsetter.

So far, it has been practically tried in its simplest form only, that is, without the automatic brake and signalling apparatus, at Broomhill colliery, upon a hoist, which has been in constant and successful use since November last. The general arrangement of the hoist is shewn in Fig. 1 (Plate XXII.); while Fig. 2 shews the arrangement of the gear as applied to the steam-winch working the hoist. This engine is an ordinary ship's steam-winch, with a ratio of gearing of 13·6 to 1 between the engine-shaft and the rope-barrel; and it is not so suitable for the application of the controlling-gear, as a direct-acting engine, such as an ordinary winding-engine would be. The automatic gear is applied to the ordinary link-reversing motion, and takes the place of the reversing-lever. It consists of a hand-wheel, A, mounted on a screwed spindle, B, working through the screwed boss of a bracket, C, and fitted telescopically into a spindle, D, carrying a mitre-wheel, E. The reversing weigh-bar has an arm fixed to it and this is attached to a vertical screwed spindle, G, which is keyed to a mitre-wheel, F, by means of a sliding feather, and is screwed into a mitre-wheel nut, H, which has a fixed position vertically on the bracket, C. This nut, H, gears with another mitre-wheel, K, attached to a spindle, J, on the other end of which is keyed a worm-wheel, L, gearing into a worm, M, fixed upon the crank-shaft of the engine.

The action of the appliance is as follows:—On the hand-wheel, A, being turned, the motion is transmitted to the vertical spindle, G, by means of the mitre-wheels, and the mitre-wheel nut, H, being held stationary, the vertical spindle, G, rises or falls according to the direction in which it is turned; and, being attached to the reversing weigh-bar, it raises or lowers the links, which start the engine, steam being always on. Immediately the engine starts, the worm, M, on the crank-shaft, acting through the worm-wheel and mitre-wheel, turns the mitre-wheel nut, H, and so long as the mitre-wheel nut and vertical spindle, G, turn in the same direction, and at the same rate, no up-or-down motion of the vertical spindle, G, will take place; and the reversing links will remain stationary. On the hand-wheel, A, being stopped, the vertical spindle, G, stops also, but the mitre-wheel nut continues to be turned by the engine, and raises or lowers the vertical spindle, G, and consequently the reversing links, until they are brought to their mid-position, and then the engine stops. The hand-wheel spindle, B, being screwed and provided with stops, it follows that the number of revolutions is limited to the distance between the stops and to the pitch of the screw. By properly proportioning the number of revolutions of the hand-wheel, A, to the number of revolutions of the engine required for the distance of the wind, there can be no over-winding, as the hand-wheel, A, cannot be turned further, once it comes against the stop, and then to move the engine the hand-wheel, A, must be turned in the opposite direction, and this reverses the engine. Further than this, the hand-wheel, A, must be kept in motion or the engine stops, and consequently should the engineman be seized with any sudden illness, so as to prevent him from turning the hand-wheel, when drawing men, no accident to the men could occur, as the engine and cages would simply come to a standstill. Finally, as the engine is absolutely controlled both in speed,

starting, stopping and reversing, by the one hand-wheel, which
can only be turned a fixed number of times in either direction,
little or no skill or training is required of the engineman, and,
in cases of emergency, the engine could be handled by anyone
with perfect safety.

The above is a description of the apparatus in its simplest
form, as it may now be seen running successfully at Broomhill
colliery. It may, however, evidently be modified in various
ways to suit special circumstances, some of which the author
will now describe.

Figs. 3 and 4 (Plate XXII.) is a diagrammatic sketch of
an arrangement as applied to a direct-acting winding-engine,
in which the gear also automatically controls a steam-brake.
It will be seen that the vertical spindle, G, is extended, and,
works a special steam-brake valve, by which the steam is
admitted, from the centre of the valve, through ports to the
top or bottom of the steam-brake cylinder for the purpose of
applying or releasing the brake. When the hand-wheel, A, is
turned so as to raise or lower the reversing link it also puts the
steam-brake valve into position for releasing the steam-brake;
and, when the engine automatically raises or lowers the reversing links to stop the engine, it simultaneously turns the steam-
brake valve so as to apply the brake, which of course may be
proportioned so as to give any amount of braking power.

It is evident that the gear driven by hand, as above described,
would not be suitable for winding coals from great depths: the
continuous turning of the hand-wheel becoming then too tedious.
The apparatus can be so arranged as to be thrown altogether out
of use when drawing coals by the addition of two couplings,
one on the reversing-lever weigh-bar and the other on the
spindle, J, connecting the worm-wheel, L, and the mitre-
wheel, K. Or a small motor might be used for this purpose,
worked by a simple reversing starting-switch: the motor driving
the hand-wheel spindle, B, through a small friction-clutch,
which would be put out of gear when drawing men, and the
hand-wheel, A, then worked by hand.

In the case of heavy engines, where it is necessary to have
steam-reversing apparatus, the motor could be sufficiently

powerful to take the place of this apparatus, and would be controlled with a starting and speed-regulating switch. In either case, the motor being arranged to work with a friction-clutch, the clutch would slip in case the motor continued to turn after the horizontal spindle, B, was screwed up to the stops; or by means of an automatic quick-break switch (to be presently described), the motor may be made to stop after a certain number of revolutions corresponding to the number of revolutions of the hand-wheel spindle, B, but the actual number of revolutions of the engine would be absolutely fixed by the stops.

Where it is necessary to change decks of cages, movable stops are provided in addition to the fixed stops, which allow the cage to be run to the first deck; the movable stop is then lifted out by means of a small lever, shewn in Figs. 5 and 6 (Plate XXII.), and the thickness of the movable stop is such as, when raised, to permit of the cage travelling the height of the deck. The engine is then reversed and the stops dropped into position again. Supposing, however, that these stops were left out, the engine could only run far enough to bring the lower deck on a level with the flat-sheets.

The apparatus can also be arranged so as to be worked by the banksman, the winding-engineman being altogether dispensed with. Fig. 7 (Plate XXII.) shews the electrical connections for this purpose. A and B are two ordinary starting-switches, with a small holding-on magnet and no-voltage release. The motor is series wound. C is an automatic quick-break switch, worked by the engine or motor—in this case by the motor. Figs. 8 and 9 (Plate XXII.) shews the proposed arrangement of the switch. It will be seen that movable contacts are arranged upon a barrel, which has two positions: one in which the circuit is closed for the motor working in one direction, and the other for the reverse; and it is always in one or other of these two positions. Following the diagram (Fig. 7) of connections and assuming that the automatic switch is in position to make contact, as shewn by the continuous lines, and the switch, D, at the pit-bottom is closed; if the starting handle on the switch, A, is pushed over, the motor at once starts the winding-engine, so as to lower the cage on the side nearest to the switch, and the handle is kept over by the holding-on magnet Immediately the motor revolves, it turns the worm-quadrant (Fig. 8), and this, at the end of the wind, brings the spring into the position shewn by the dotted lines, and being then extended, the force is sufficient to turn the switch-barrel quickly over: breaking the electrical connection on that side, and making connection on the other. Simultaneously, the switch, D, at the pit-bottom is opened automatically by the cage on reaching the pit-bottom. At once, the electrical connection is broken, the current is cut off from the motor, and the handle on the switch, A, flies back to the off-position, and it is then impossible to start the engine by this switch as the electrical connection is broken (*a*) by the automatic switch, C, and (*b*) by the switch, D, at the pit-bottom. Moreover, the banksman cannot start the engine in the reverse direction until the onsetter closes the switch, E. It may also be mentioned that, should any of the electrical connections break or the current fail, during winding, the engine would simply come to a stop.

This arrangement, however, (the banksman to act also as brakesman), could not very well be applied to cases where it is necessary to change the decks of the cages, unless the bottom-deck could be changed first, when the starting-switch could be used to lower the cage, in order to change the top-deck.

Another arrangement has been devised so that the brakesman cannot start the engine before receiving the signal from both the banksman and the onsetter; and accidents due to the brakesman starting to wind before he has received the proper signals are not uncommon under present conditions. Fig. 10 (Plate XXII.) illustrates the electrical connections of an arrangement in which the brakesman is retained, but by means of an electrically-controlled apparatus, he cannot start the winding-engine until the circuit is completed by both the banksman and the onsetter, and moreover should either the banksman or the onsetter require to stop the engine after it had started, either of them could do so without the co-operation of the brakesman. A is a starting and regulating switch, controlling the motor in the engine-house, and worked by the brakesman ; B and C are two switches at the top and bottom of the pit respectively, and arranged so that they are automatically thrown out by the cages,

FIG. 6.

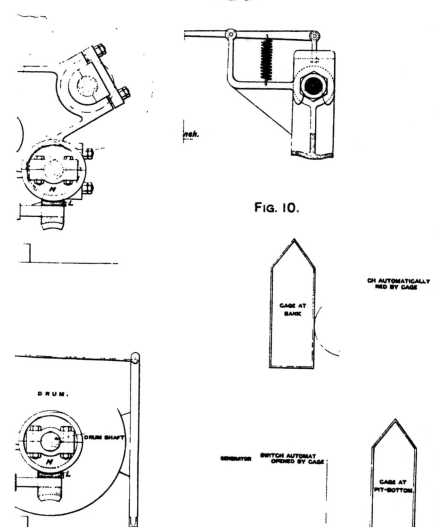

FIG. 10.

DRUM.

DRUM SHAFT

CAGE AT
BANK

CH AUTOMATICALLY
NED BY CAGE

GENERATOR   SWITCH AUTOMAT
OPENED BY CAGE

CAGE AT
PIT-BOTTOM.

and that the circuit is broken both at the top and bottom of the pit. Before the engineman, therefore, can start the winding-engine, both the banksman and the onsetter must close the switches, B and C, and keep them closed until the cages have moved, as they are so arranged that they will not remain closed so long as the cages are at the top and bottom of the pit respectively. At the same time, as the circuit is completed, a bell may be arranged to ring at the various points, namely, in the engine-house, on the bank-top, and at the pit-bottom, so that each one is aware that the engine may start.

The advantages claimed for this gear are as follows:—(1) The prevention of accidents; (2) simplicity and ease in working the winding-engine, a skilled winding-engineman not being required; and (3) economy in steam-consumption, as the steam being always on, the pipes and valve-chests are always kept at one temperature; thereby reducing losses due to condensation, and throttling at the steam-valve.

Where electrical power is in use for winding, the gear would be arranged to act upon the starting and regulating switches, controlling the motor or motors working the winding-drum, in much the same way as upon the reversing-link mechanism of steam-engines.

————

Mr. G. E. COKE (Nottingham) said it seemed to him that the Futers gear attempted to deal in a complicated way with what was already dealt with by the ordinary system of preventing overwinding. At Sherwood colliery, there was an automatic system by which it was impossible to overwind: as soon as the cage rose above a fixed level, the steam was cut off and the brake put on. He did not think that anyone would feel desirous of going down a shaft, without a skilled man in charge of the engine; certainly he would not do so himself. The arrangement was no doubt an ingenious one, but he very much doubted whether it would be applicable for a winding-shaft; of course, that remained to be seen.

Mr. C. C. LEACH (Seghill) asked what was the speed of winding with the Futers gear, and how quickly the cage ceased to move when the engine was left to itself to come to rest.

Mr. T. E. FORSTER (Newcastle-upon-Tyne) said he gathered

that practically the apparatus shut off the steam, and applied a brake. At present, the arrangement had only been applied to a small winch with a small drum; and when put on to a large engine with a big drum, the momentum required should be sufficient to shut off the steam and apply the brake, so as to bring the cage exactly to its place to an inch. At many collieries, especially in the North of England, where the heapstead and pulley-frames were not of sufficient height for safety-hooks, and where they had to run to a few inches, he did not think that it would be safe to dispense with a skilled man under such circumstances. There might be something in what Mr. Merivale had said, that it might make a man who was not a properly qualified engineman able to work a winding-engine, but he thought that there would be many cases where it could not be done. It was, however, a start in the right direction, as they wanted something of the kind to help them when the enginemen were troublesome, and possibly the appliance might be of assistance under such circumstances.

Mr. S. F. WALKER (London) asked the author to give further information with respect to the automatic switches at the bottom and top of the pit. Switches placed in such positions would require careful supervision, and to work them automatically was, to his mind, putting a little more strain on the electric apparatus than he would care to see.

Mr. C. C. LEACH said that he had the points mentioned by Mr. Forster in his mind. He had been drawn up the shaft at Sherwood colliery at a speed of 4,000 feet a minute, and he would not have been there, unless a qualified engineman had been in charge of the engine.

Mr. W. C. MOUNTAIN (Newcastle-upon-Tyne) said that, at the Zollern II. colliery in Germany, a switch was provided for cutting off the current at the top and at the bottom of the wind; it was attached to the indicator. The momentum of the drum might be easily taken up by means of a magnetic clutch or brake, which would operate as soon as the current was cut off. He had already applied it successfully to two winding-gears.

Mr. HERBERT A. JONES (Bradford) believed with Mr. Walker that the automatic switches proposed for the pit-bottom for electric winding would be found unreliable. He did not agree with

the remarks at the end of the paper, in which it was stated that
the apparatus could be applied to electric winding, that is, if
the author meant main-shaft winding, because in his experience,
automatic devices were largely matters which worked only in
works or laboratories, and had a knack of failing at a critical
time in practical working; and at a pressure of 500 volts he
believed that it would be futile to try main-shaft winding,
with automatic controls at the top and the bottom of the pit.
Messrs. Siemens & Halske seemed to have devised, so far, the
best arrangement for winding: they made use of the energy
stored in the flywheel and obtained a fairly even load-factor.
In their arrangement, the whole of the control was effected in
the enginehouse by means of what was practically the equivalent of a shunt-regulating switch; and, in these, no main circuits were broken and sparking was entirely eliminated.

Mr. T. LINDSAY GALLOWAY (Glasgow) said that the arrangement resembled a steam steering-gear, which was devised on the
principle that, so long as the wheel was kept turning, the helm
moved over, and the steam was automatically shut off.   For
small hoists for steamers, for raising ashes, etc., the arrangement might work efficiently; but at high velocities it would not
work so well; and he believed with other speakers that a simpler
and less costly arrangement could be devised.

Mr. C. C. LEACH said that there would be a very high consumption of steam when using the arrangement, as the full pressure of the steam would be on the engine all the time.

Mr. F. S. PANTON (Sunderland) wrote that the arrangement
described in Mr. Merivale's paper was a very ingenious contrivance, and would no doubt work on the small hoisting-engine
tried at Broomhill colliery; but the arrangement would, he
thought, be found too slow in its action if applied to a large
winding-engine. A very similar arrangement was applied to
the winding-engine of the second pit at Silksworth colliery,
when it was built in 1868. It was worked by means of a spiral
worm on the main shaft of the engine, and it was used to reverse
the engine and to put on the steam-brake in the case of over-
winding; but it was found in practice to be too slow in its
action, with the result that the cage reached the pulleys before
the engine was stopped.

Mr. HENRY LAWRENCE (Newcastle-upon-Tyne) wrote that Mr. T. Campbell Futers' invention, as fitted to a steam-winch for the prevention of overwinding, was very ingenious, and he had no doubt that it would work successfully as applied to a hoist. From his (Mr. Lawrence's) experience in the manufacture of machinery as fitted to winding-engines for any depth of pit now in use, he was of opinion that the apparatus must be perfectly automatic and instantaneous in its action and always ready.   An apparatus, made by the Grange Iron Company, possesses these qualifications; and in 1886, two winding-engines fitted with this apparatus were manufactured for the Australian Agricultural Company, Limited, for their mines in Australia.   These engines were completed and tried under steam in the fitting-shops, and they proved satisfactory to all concerned.

With this apparatus, when the cage arrives at the danger-point, that is after passing a few feet above the flat-sheets, it comes into contact with a lever or tappet, which lifts a wedge-lock and allows the rod and double piston-valve, to which it is attached, to move at the velocity of the steam-pressure.   It simultaneously opens the port giving the full boiler pressure to the steam brake-cylinder, shuts off the steam from the steam-engine cylinders and opens a port, which allows all four ends of the steam-engine cylinders to have free vent to the atmosphere, and thus the ascending cage is instantly brought to rest.   For further safety, the cages are fitted with steel rack-teeth which are held there with falls, the ordinary kep-levers at the top of the shaft, for ordinary working, being attached as usual. The drum is fitted with strong brakes on each rim, that on one rim being applied on the contrary side to the other, so that whichever way the winding-engine may be running, there will be equal brake-power.   The brake-straps are made in four parts, and thus should one break, only one-fourth of the brake-power will be lost.   The foot-brake and steam-brake can be used, as usual, without interfering with the automatic brake-gear.   The cage is purposely overwound, every day, to ascertain that all is in working order: the gear and engine being again ready for use in 1 to 3 minutes after the overwind has taken place.

He (Mr. Lawrence) might mention that this apparatus was attached to a winding-engine at the Lingdale mines,* belonging

* *Trans. N.E. Inst.*, 1880, vol. xxix., page 188.

to Messrs. Pease and Partners, Limited, in 1886, and it was still working efficiently.

Mr. T. C. FUTERS (Broomhill collieries) said that a question had been raised with regard to the speed of winding. At Broomhill colliery, a very high speed was not attained, as the arrangement was applied to a little hoist used for lifting material from the ground to the flat-sheets, and a great speed was not required. He did not see, however, in the case of a winding-engine running even at a speed of 4,000 feet a minute, why the apparatus should not be applied, as it acted directly on the reversing links. The question of momentum was probably one of greater importance, and so far nothing had been done in that direction, but he thought that it would be met by a proper proportioning of the gear on the spindle and on the hand-wheel. Where there was a little variation of the steam-pressure, it might not be possible to come to the same dead point every time, by merely letting the hand-wheel run against the stop. One thing, however, was absolutely certain, and that was that the engine must stop, for supposing that sufficient momentum was stored in the moving masses, and that the engine kept turning after the links had reached the mid-position, it would automatically reverse and gradually put steam against the engine. The question of momentum he must leave over until he had had more experience with the use of the gear.

Another question had been raised with respect to the automatic switches. Figs. 7, 8 and 9 (Plate XXII.) showed an auxiliary arrangement to the gear. The gear could be applied successfully to single-deck cages, and there would be no mechanical difficulty with respect to the switches at the pit-bottom, as they might be enclosed and so kept thoroughly clean.

There were several appliances for the prevention of over-winding, but all more or less depended to a certain extent upon the presence of a skilled man, and the majority depended upon coming into action after the engine was out of control. Some of them were controlled by governors: if the speed attained a certain limit, the apparatus came into action by means of trip-gear. Others, like the apparatus described by Mr. H. Lawrence, were put into action by the cage, when it was raised above the flat-sheet level, and after an overwind the gear had usually to be re-set. These appliances were very complicated; but his arrangement was absolutely certain, and although at first sight, it appeared to be complex, it was not really so; and, moreover, it was not liable to get out of order; it did not depend for its action upon the speed of the engine rising above a certain limit, or the cage being raised into a dangerous position above the flat-sheet level, but was in operation all the time. There were some electric winding-engines in Germany, but, owing to the high initial expenditure, it was unlikely that they would displace steam-engines in this country, except in cases where power could be cheaply purchased from an electric-power company. In any case, however, the gear could be applied to the controlling mechanism, irrespective of the current or voltage.

The apparatus was very similar to a steam steering-gear, as by turning the hand-wheel in one direction the engine ran one way, and *vice versa*. The arrangement, of course, was adopted, not with the view of doing away with a skilled engineman, but for preventing mistakes in winding. It made the winding-engine absolutely controllable in a very simple manner by the use of a hand-wheel; and he had no hesitation in stating that any one, who had never even seen an engine before, could draw men with absolute safety.

He (Mr. Futers) did not claim much in the way of steam-economy, but he could not agree with Mr. Leach, as a winding-engineman usually throttled the steam at the stop-valve, which was far from being economical. With this gear, however, once the engine attained its speed, the engineman could easily link-up to any required point of cut-off and thus work the steam expansively; and keeping the pipes and valve-chests full of high-pressure steam would not, he ventured to think, waste so much steam as re-filling them at the beginning of every wind.

The CHAIRMAN (Mr. George A. Mitchell, Glasgow) in moving a vote of thanks to Mr. Merivale for his paper, said that Mr. Futers, who was present, was the inventor of the apparatus. Such papers were always interesting, and although the apparatus described seemed to be complicated, possibly the inventor might be able to simplify it and thus also make it a success.

Mr. C. C. LEACH seconded the resolution, which was cordially approved.

———

Dr. P. DVORKOVITZ read the following paper on " Petroleum, and its Use for Illumination, Lubricating and Fuel-purposes."

PETROLEUM, AND ITS USE FOR ILLUMINATION, LUBRICATING AND FUEL-PURPOSES.

By Dr. P. DVORKOVITZ.

Introduction.—Petroleum represents a liquid which in most cases is either fluid or semi-fluid, and it consists chiefly of a chemical combination of carbon and hydrogen in which carbon is the more prominent factor, being present to an extent varying from 81 to 87 per cent., and hydrogen to the extent of 10 to 15 per cent. In addition, however, to the abovementioned chemical combinations, there are some oils which contain a small percentage of nitrogen, oxygen or sulphur.

Regarding the formation of petroleum, many speculative theories have been expounded, but all of them can be divided into two classes: the organic and the inorganic. The carbon groups of the organic rocks comprize three distinct kinds:— (1) Organic substances of indefinite composition, stored in the earth and transformed into coal and mineral-oils; (2) graphite; and (3) diamonds. From the pure-wood stage there is every gradation from graphite to peat, lignite and anthracite; and allied to this group is asphaltum, which in turn passes into mineral-oils, and hence to indefinite carbonaceous compounds stored in the rocks. Carbon is widely disseminated through many of the stratified rocks, especially in shales and limestones: the blackness of which is often due to hydrocarbons derived from the remains of animals or plants contained in the rocks. The inorganic theory is presented by Prof. D. Mendelejeff, the celebrated Russian chemist, who considers that petroleum-hydrocarbons are the result of molten iron carbide in the centre of the earth coming into contact with water. Thereupon the carbon of the iron enters into a combination with the hydrogen of the water, and various hydrocarbon groups are formed.

The organic theories can be easily divided into two groups, namely :—(1) The formation of petroleum from various deposits of organic matter, especially derived from animals and fishes: this view has found great support among various investigators in Europe and America; and (2) the theory that petroleum is of vegetable origin, and derived from the decomposition of plants. This latter theory appeals to the writer, for the reason that in his investigations of peat-deposits in Ireland and also on the Continent, and in the distillation of coal at a low temperature, he has found a large percentage of hydrocarbons, very similar in character to petroleum.

Petroleum has been known from time immemorial, and it is on record that petroleum-gases were used as a fuel for fires for religious purposes before the commencement of the Christian era. However, it was not until the sixteenth century that the Persians used liquid petroleum for lighting purposes. About that time, in the now great oil-producing district of Baku, the Persians dug wells, and, after collecting the petroleum, used it for illuminating purposes in the crude state.

At the end of the seventeenth century, Le Bel in Alsace, then belonging to France, started mining for petroleum, by sinking a shaft. This remarkable way of obtaining petroleum held sway for a time, but it was discontinued about the middle of the nineteenth century; although the shaft, at which the experiments were carried out, remained in existence until the end of the past century, when it fell into decay.

The discovery made by Col. — Drake in 1854, that it was possible to bore for petroleum, instead of digging wells, gave an impetus to the industry which, since that time, had developed by leaps and bounds. Suffice to say that, since the first boring was made, no less than three or four hundred thousand wells have been bored in all parts of the world, and that the total production for the past twelve months of the world's oil-fields has reached 30,000,000 tons.

Drilling for Petroleum.—Various ingenious systems of boring have from time to time been employed in various parts of the world; but, for the present purpose, it will be sufficient to describe the systems of boring introduced into the Russian oil-fields, some of the most important in the world.

Innumerable arrangements for boring have been invented, yet the old free-fall system continues to be the most favoured, and holds its own against all comers. Modified, improved and developed, as it undoubtedly is, the underlying principle still remains, from which there has been little change, and from which there is little likelihood of immediate alteration. Roughly made, badly fitted, and carelessly kept, as are the boring-machines, they nevertheless perform their work, although slowly. The systems of boring adopted in the Russian oil-fields may be well divided under three headings: percussion, rotary and hydraulic. The percussion-system can again be subdivided into:—(a) The Russian free-fall system; (b) the American rope system; (c) the wire-rope system; (d) the Canadian; (e) the Galician; and (f) the Mather-and-Platt flat rope, the last-named, however, being very little used now. Among rotary drills, the diamond and the calyx drills may be mentioned; while, under the hydraulic processes, the Chapman spindle-top rig is placed, a system in which water is made to perform much of the work.

The Russian free-fall system is an ingenious appliance, by which the chisel in conjunction with the sinker-bar is raised to a certain height, and is then allowed to fall freely again. The internal sliding portion of the free fall is connected to the rods, and attaches itself automatically to a projection on the body of the instrument when the lowest point is reached. By this arrangement, the boring-chisel, sinker-bar, etc., can be lifted and released at each oscillation of the walking-beam, and from 20 to 40 powerful blows per minute can be delivered to the rock, which is pulverized into a fine powder, suitable for removal in buckets or other clearing tools.

The character of the strata in the Baku oil-district is such that boring cannot proceed for any great distance without caving-in taking place; consequently, the American rope system cannot be applied without some modification. The chief advantage gained by the adoption of the rope principle is the rapidity with which the tools can be raised or lowered, compared with rods, and the diminished risk through the absence of so many feet of iron-rods, and numerous screwed joints, which necessarily means that there is less breakage to be overcome.

In the wire-rope system, the tools are much the same as in

the ordinary Russian system, but the chief peculiarity is the ingenious free-fall contrivance placed immediately above the sinker-bar, which automatically imparts a rotary motion to the drill during the operation of boring. In this system, ordinary tools can be used without alteration. The Canadian pole-tool is almost identical with the common American rope-rig, the only difference being that pine rods are used instead of a Manilla rope. The Galician system is a modification of the Canadian arrangement, but in place of wooden rods, very light iron drilling-rods with screwed ends are used.

The diamond drill is the best known rotary drill, and the boring is executed by a steel-crown studded with black diamonds, in the presence of a powerful flush of water. In this system, hollow rods are continued to the surface, and rotated by a geared frame. This system holds sway in a number of oil-fields, but so far as the Russian oil-fields are concerned, the strata have been found unsuitable for the diamond drill. The calyx drill, one of the latest introductions, has been well received since its adoption. This drill is after the type of the diamond drill, the peculiar features lying in the construction of the cutter, which takes the place of the diamond-crown. This is made of a collar of extra-quality steel, formed into a number of teeth of special shape, which are impelled to seize hold of the stratum by the application of weight from the surface. Above the core-barrel, and attached to the hollow rods, is the calyx, a long cylindrical vessel of the same diameter as the core-barrel; and, in addition to guiding the bit, it receives the chippings and particles resulting from the action of the cutter. Water is forced through the hollow rods by a pump, and emerges at the mouth of the bit at a high speed.

Of hydraulic boring-apparatus, the Chapman spindle-top rig has been and is most widely employed in America, where it has proved very successful in passing through beds of quicksand. The surface-apparatus consists of a heavy geared frame, and revolving hollow drill-rods, through which a powerful flush of water is forced during work. The chisel has two expanding wings, which loosen the strata, and the water conveys the disintegrated material to the surface, but if a hard formation is encountered, then another class of drill must, necessarily, be used.

The tubing of the wells is especially rendered necessary in Russia, on account of the special nature of the strata. Owing to this circumstance, the Baku oil-wells have been made large in diameter, and, in turn, this has led to the manufacture of special lining tubes or casing. The usual form of tube employed consists of rectangular sheets of iron rolled into a circular shape and riveted along the lapped joints as well as round the sockets at either end. The general length of these tubes is between 4 and 5 feet, this length having by experience been found to meet the requirements most satisfactorily. Much greater lengths have been suggested, but it has been found that inasmuch as the joints of the tubes increase their strength, it was advisable that greater lengths should not be adopted.

The tubing of the wells by no means completes the work which has to be carried out before petroleum is got. One vital question is that of shutting off water, and in the past, numerous wells have been ruined by a neglect of this matter. In order to accomplish successfully the exclusion of water, it is the usual practice to cut away and remove a number of the tubes, and into this space, some impermeable substance, such as sand or cement, is inserted. The infilling is performed by pouring a well-stirred mixture of sand and water through a funnel and pipe into the space, where it gradually settles and forms a water-tight dam, and if the work be placed in the hands of a reliable workman, the water will be effectively kept out of the well.

Petroleum.—The crude oil is a mixture of various hydro-carbons of widely different boiling points, specific gravities and viscosities, yet all the oils produced in the world may be classified in four groups. These are (1) oils containing a small percentage of petroleum-spirit, that is, light petroleum which distils up to 125° Cent.; a considerable quantity of illuminating-oil which generally boils over between 125° and 300° Cent.; a large percentage of lubricating-oil; and a small percentage of pitch; (2) oils containing a large percentage of benzine, say, from 12 to 25 per cent.; a large percentage of illuminating-oil, from 50 to 60 per cent.; and a small percentage of lubricating-oil and coke; (3) oils containing about 15 to 30 per cent. of benzine; a small percentage of illuminating-oil; and a large percentage of pitch; and (4) those petroleums, which include practically no benzine nor illuminating-oil, but represent lubricating-oil only. The greater part of the oil produced in the Baku district of Russia belongs to the first class mentioned; the bulk of the American and Rumanian oils (Californian and Texan oils excepted) belong to the second group; the petroleum from Grosny and Alsace is placed in the third group; while in the fourth group oils produced in small quantities in Wyoming, and other parts of the world, are placed.

Refining Petroleum.—Owing to the different natures of the oils, the refining industry in various parts of the world has been developed on different lines and in various ways. In America, the industry has been chiefly directed towards producing the largest possible quantities of illuminating-oil, and in that country about 76 per cent. of the total crude oil has been turned into the illuminating-product, 11 per cent. into benzine, and 3 per cent. into lubricating-oil; while the remaining 10 per cent. is residuum. In Russia, however, only 25 or 30 per cent. of illuminating-oil is obtained, the remainder, with the exception of 3 or 4 per cent., which is lubricating-oil, being used for fuel-purposes.

The manufacture of the various products from crude oil can be divided into three stages:—(1) The separation of water, mud and sand, which is generally mixed with the crude oil; (2) the separation of the various component parts by means of heating; and (3) the treatment of the products, by chemicals, so as to remove all impure matter. Naturally, the system employed for separating the hydrocarbons depends upon the character of the oils. In America, practically the whole of the oil being turned into very valuable products, the intermittent system of distillation is employed, in which by the employment of a still, filled with crude oil, and heated, most of the products are taken off at various stages of temperature. The small percentage of residuum is then drawn off and separately treated, the still being again filled and the process of refining continued. In Russia, however, where the production of illuminating-oil averages only about 30 per cent., the intermittent process, which was in use 40 years ago, gave but unsatisfactory results. It required a long time to heat up 20 tons of crude oil, and as only about 25 or 30 per cent. of the product was illuminating-oil (which, when taken off, required the further tedious removal of the residuals) it will be apparent that the work was not only slow but very costly. But another difficulty presented itself: as the Russian oil contained water, there was overboiling in the still, and the refined product, eventually, contained pitchy matter.

About 25 years ago, however, Mr. Alfred Nobel, the celebrated inventor, found means to overcome the difficulty by turning the intermittent process into a continuous one. This was accomplished by means of a combination of several stills into one. Into the first still, the crude oil is admitted after it has been pre-heated and freed from water; here, a small percentage of the fractions is taken off, according to the prevailing temperature. The remainder of the oil travels to the next still, where, again, a percentage is removed, of the fractions with a higher boiling point. And so the process goes on, until the last still is reached and here the fractions with a very high boiling point are removed, and the residuals then pass to the still, in which the crude oil is pre-heated, and there they assist in the pre-heating. In this way, the Russian refiners have achieved remarkable success, for the Nobel system has enabled them to produce an oil which is uniform in its composition and very free from any tarry matter.

Illuminating-oil.—The total quantity of illuminating-oil produced in the world in 1903, the writer estimates as follows:—

		Tons.
America	... ...	6,000,000
Russia ...	... ...	2,400,000
Galicia ...	... ...	290,000
Rumania	... ...	76,400
Other countries	... ...	483,600
	Total ...	9,250,000

Thus if, on the average, one household petroleum-lamp burns 2 gallons per year, we arrive at the conclusion that over 200,000,000 such lamps (or their equivalents) must constantly be used throughout the world to consume the enormous output of illuminating-oil which is poured in ever-increasing quantities on the world's markets.

There can be no question that oil is the light of the poor in this country, and of the rich and poor alike in Russia and Eastern and Far-eastern countries, which altogether represent a population of about 1,000,000,000; and many years will elapse before it is financially possible for a more luxurious or more expensive illuminant to be used.

It will readily be understood that there is a wide opening for the extension of the petroleum-industry in the future; and it depends entirely upon mining and boring engineers, who have in hand the most important part of the work, namely the getting of the crude product from the bowels of the earth, how far that wide field, which is still left for expansion, shall be taken advantage of.

Lubricating-oil.—Until about 25 years ago, petroleum-oils were very little used for lubricating purposes. Attempts were made to introduce them as a substitute for vegetable oils, but the refining of the oil was a difficult subject, in which few persons were at all versed. At length, however, the great stumbling-block was successfully surmounted by the ingenuity of Mr. V. I. Ragosine, a Russian inventor, who, at that time, gave careful attention to the subject. He introduced a method of separation, by which the 65 to 70 per cent. of residuals, remaining after refining, were no longer a drug on the market or had to be burned in the open field; but from them he separated a product, which as time has proved, has been equal to any lubricant of vegetable origin, if not in many respects superior.

A lubricant to fulfil its duties properly must be viscous, in order to resist the pressure of the two metal-surfaces with which it comes into contact in various classes of machinery. The higher the pressure of these metal-surfaces, then the higher must be the quality of the lubricant used. The lubricant must necessarily be free from acids or any substance which will corrode or tend to corrode the metal that it is supposed to lubricate; and again, it must not be capable of oxidation; while, lastly, with the introduction of high-pressure steam-engines, the lubricant, to fulfil its duties satisfactorily, must not lose its lubricating-properties even at a high temperature. Mr. V. I. Ragosine was able by his invention to produce such oils as have just been mentioned, by means of the application of superheated steam in the course of the distillation of the residuals; and, by this means, he was able to prevent that decomposition which generally used to take place in petroleum-hydrocarbons when a high temperature was reached. However, the manufacture of lubricating-oils occupies a less important part in the petroleum-industry than the refining of petroleum for illuminating purposes; and, although there are no available data regarding the manufacture of lubricants throughout the world, the writer believes that the total quantity does not exceed more than 1,000,000 tons per annum.

Fuel-oil.—The utilization of petroleum as fuel rendered it a necessary duty, that the writer should refer to the important services of Mr. L. E. Nobel, who thirty years ago, among other subjects, paid much attention, indeed chiefly devoted himself, to the use of petroleum in furnaces as fuel. Thanks to his indefatigable efforts, and to his exceptional training, which admirably fitted him for the study of such a question and the solution of such a difficulty, his work was continued with much success by his collaborators and colleagues; and, as a result of their efforts in this direction, many appliances were invented with the aim of utilizing oil for the purposes of generating power. Mr. V. I. Ragosine, also, did a great work towards the utilization of residuals for fuel; for, although his invention for obtaining lubricating-oils from this product, did much to turn the residuals to profitable account, he was unable to utilize the whole of the residuals. He tried to use the remaining percentage as fuel, and eventually he succeeded in spraying it into furnaces by means of steam.

In the eighth century, an Arabian chronicler narrated that the people of Baku, being without wood, used earth saturated with petroleum, for cooking their food; and thus, although the idea of turning oil to account as a fuel dates back into almost pre-historic times, it was not until the last twenty years that its value has been utilized for commercial purposes. From the experimental stages it has slowly passed, until to-day its economic value, as well as its great power of generating steam, is admitted on all hands, and to how great an extent the inventive mind has been applied to render it the more widely used, the bulky record of the Patent Office is a witness.

The variety of liquid burners, designed for oil-fuel, can be easily classified into the following four groups:—(1) Apparatus

for injecting the fuel in a gasified form; (2) using the liquid-fuel
by passing it through fireproof porous material, such as pumice-
stone or asbestos, which act like wicks upon the oil, on the
same principle as petroleum standing in a lamp; (3) apparatus,
in which the liquid-fuel is turned directly into the furnace, by
means of burners or other corresponding appliances, in its liquid
state; and (4) liquid-fuel, divided and pulverized into the most
minute particles by means of spraying steam or compressed air
into the furnace under pressure.

With regard to the first-named apparatus, the writer need
only say that its introduction is based upon the same principle as
other gas-furnaces: the only difference being that petroleum
is gasified by a separate apparatus. The second apparatus has
found great difficulties in its application, for petroleum being
a mixture of various hydrocarbons, in its ascending course to the
top of the furnace, where it becomes ignited, the light oils are
consumed first, and the heavy oils remain chiefly unconsumed or
clog up the porous materials.

The direct introduction of petroleum into furnaces was, of
course, the natural method adopted when the first attempts were
made to use liquid-fuel. With this system, however, it has
been found impossible to obtain complete combustion, and a
thick volume of dense black smoke was vomited from the
chimneys, whenever the fuel was turned directly into the
furnaces. The smoke was so conspicuous when the refiners at
Baku tried the use of liquid-fuel, that it earned for the portion
of the district in which the refineries were situated, the name of
the " Black City."

The last type of liquid-fuel burners, has proved the most successful in application. This system consists in introducing into a
current of compressed air or steam, the liquid-fuel, which has been
pulverized and mixed with either air or steam. Once the supply
of fuel is properly regulated, the flame may be given any shape
or direction that may be required. The shape of the flame
depends entirely upon the shape given to the steam or air-outlet.
For instance, if the flame issues from a circular orifice, then the
flame is broom-shaped; if from a flat orifice the flame is flat;
if the slit be bent downward or upward, then the flame is convex
or concave, as the case may be. In short, according to requirements, the flame may be altered in every possible way, while it

can be endowed with oxidizing or reducing properties, which are most important for use in various metallurgical processes.

Investigations have been made to show the difference between the theoretical and actual steam-raising efficiencies of various fuels, and the results are recorded in Table I.

TABLE I.

Fuels.	Weight of Water Evaporated by 1 Pound of Fuel.	
	Theoretical Pounds.	Actual Pounds.
Petroleum	20·20	14·50
Anthracite	12·00	7.20
Average coal	10·84	6·50
Inferior and small coal	9·23	5·53
Lignite	9·23	5·53
Coke, with 4 per cent. of ash	10·46	6·27
,, 15 ,, ,, ...	9·23	5·53
Wood, perfectly dry	5·54	3.32
,, with 25 per cent. of water	4·31	2·58
Charcoal	10·00	6·00
Best peat, with 25 per cent. of water ...	4·62	2·77

Table I. clearly shows to how great an extent the heat-producing capacity of petroleum or liquid-fuel is superior to that of solid fuels, but in estimating the economic advantages of any system of heating, one has not merely to take into consideration the respective absolute thermal values, but one must also determine the amount of heat that is actually utilizable in that case. Thus, the figures show that less than 60 per cent. of the heat produced by a solid fuel is convertible into useful energy, and therefore that 40 per cent. is wasted. This is accounted for by the fact that with coal-fuel the loss of heat is four times higher than with petroleum. Even with the best furnace, dense volumes of black smoke issue from the chimney when coal is used: a fact indicating incomplete combustion. Then in addition to this, it must be also borne in mind that in burning solid fuel, it is always necessary to have a large excess of air, whereas, when petroleum is burnt with the aid of a spraying apparatus, such an excess is not required, in fact, the quantity of air admitted for complete combustion is practically the theoretical volume.

The rate at which heat is transmitted to water through the walls of the boiler is not the same with coal as with petroleum. When coal is used, a layer of soot is formed on the tubes as combustion proceeds, and greatly reduces the heat-conducting

capacity of the metal, but when liquid-fuel is used, the tubes are
perfectly clean, no soot is formed, and, consequently, less heat is
lost.

Further, in burning liquid-fuel only water and carbonic acid
are formed, and there is an absence of mechanical action on the
walls of the furnace. As is well known, the products of the
combustion of solid fuel carry with them small particles of the
fuel, which, rubbing against the walls, eventually remove the
oxidized layer with which the metal has been coated, and it is
consequently laid bare to the further action of oxidizing agents.
Again, there is the presence of sulphur in solid fuel, acting as a
corrosive on the fire-box and on the metal of the boiler.

In dealing with this extensive subject, the writer has only
been able to touch upon the most important points, omitting
details altogether; but he hopes that he has been able to convey
an adequate idea of it, and that he has expressed himself clearly
upon those points which are of interest to the members.

———

Mr. C. C. LEACH (Newcastle-upon-Tyne) said that he believed
that the trouble in drilling at Baku arose from the inability of the
riveted tube to keep back the water, and they had to use
screwed wrought-iron casing. Very little petroleum-fuel was
used in this country, and although there were locomotives at
present fired by petroleum, he believed that the Great Eastern
Railway Company was the only one using such fuel despite its
convenience. Although petroleum-fuel was nice and clean in
use, he was afraid that it was too dear for the coal-using people
of this country.

Mr. HERBERT A. JONES (Bradford) said that it was largely due
to inefficient methods of burning petroleum that it had not been
more largely used as a steam-raiser; there were, however, certain
industrial applications of it in this country which were coming
to the fore. He (Mr. Jones) was a motorist, and took a great
interest in the question of the supply of petrol. He believed
that petrol should have a specific gravity of 0·68. Could the
author tell him what percentage of petrol there was in a given
volume of crude petroleum, and whether the petrol sent from
Russia differed largely from that sent from America? Two

years ago, one could get good petrol, but now 60 per cent. more was charged for an inferior article; and they were at present very much troubled with residues, by which the valves of the engine were blocked up. Was this due to anything in the petrol or to adulteration?

Mr. R. D. NOBLE (London) said that they could not find any part of the world where the geological conditions would enable any one to get at the petroleum at so shallow a depth, and where the means employed to get it were of so simple a character, as in Canada. In Canada, the oil, found in the Carboniferous Limestone, contained large percentages of paraffin and lubricating oils, for which high prices were obtained. There, a well pumping 75 barrels a day was a valuable property, and the oil could be sold to the Standard Oil Company at a profit of 2 dollars a barrel, each barrel containing 35 gallons. The well and plant cost about £100, for a present return of 150 dollars or £30 per day. About 75 barrels a day would be pumped for six months, and the produce would then gradually diminish. He asked Britons to go to the Dominion, and give them the benefit of their capital.

Mr. T. E. FORSTER (Newcastle-upon-Tyne) asked the author for information as to the average depth to which wells were bored in Russia, their average cost, and the general character of the strata in which the oil was found.

Mr. C. C. LEACH (Newcastle-upon-Tyne) enquired as to the quantities of oil produced by Russian oil-wells, and the diameter of the holes.

The CHAIRMAN (Mr. George A. Mitchell, Glasgow) observed that Mr. Noble had stated that it only cost £100 to put down a Canadian well, but he had not stated how much deadwork had to be done before one of these wells was discovered. Did they know infallibly where petroleum existed, or had they to spend £3,000 or £4,000 in looking for it? It was surprising, if petroleum was so profitable in Canada, that the Dominion produced so little petroleum as to be included in the group of "other countries," producing the small quantity of 483,600 tons, mentioned in Dr. Dvorkovitz's paper. He would also like to ask the author whether the paraffin oil produced from oil-shales in Scotland was included in this small quantity. The oil-shale industry of Scotland had passed through times of depression, although fortunately at the present time it was enjoying a considerable measure of prosperity. It had been a wonderful industry and had been kept alive simply by continual improvements in the processes for the recovery of products and by reductions in the costs. He would like to ask Dr. Dvorkovitz why diamond drilling was not suitable for boring oil-wells in Russia. He was interested in the question of the price of petrol, and he hoped that the author would afford some information on the point. A small quantity of motor-spirit was produced in Scotland.

Mr. J. G. WEEKS (Bedlington) asked Mr. Noble whether any shafts had been sunk to ascertain the source of the oil, and, if so, whether the oil ran to the bottom of the shaft.

Mr. R. D. NOBLE said that drilling in Ontario was very simple. The surface-clay was penetrated by an auger, 10 or 12 inches in diameter; it was very easy to work, and its usual thickness of 100 feet was bored in a day. A wooden casing was put into the clay. The rock was struck at a depth of 100 feet, and then drilling commenced, with a drill 5 inches in diameter. After passing through 40 feet of rock, there was a soapstone, very easy to drill through, and then another layer of soft rock about 220 feet thick. After getting into the hard rock, which was struck at a depth of 360 feet, they drilled down to a further depth of about 110 feet, and struck the porous oil-bearing limestone beds at a total depth of 470 feet from the surface. He had never found any shale in Canada in oil-bearing strata; and, if it was found, they knew that they were not going to get oil. About 13,000 wells had been drilled in the oil-belt, and were in operation at the present time. No shafts had been sunk to ascertain the source of the oil. He questioned, however, whether it would be possible to do so, as there was so much gas present that sometimes the drillers were affected by it.

Mr. JOHN KIRSOPP, JUN. (Gateshead-upon-Tyne) wrote that research into the writings of American observers tended to show that asphalt was of vegetable origin, and derived (as the author seemingly endorsed) from the decomposition and distillation of plants. However, observations made by him in one of the West Indian islands led him to believe that such deposits therein were certainly derived from an outburst of organic matter of some sort, derived from animal rather than plant-remains. This theory was supported by Mr. A. Beeby Thompson, who, referring to the Baku field, disputed the theory that petroleum was derived from vegetable remains, and gave his opinion that it had originated from enormous quantities of fish-remains. He also pointed out that as the quantity of phosphates appeared to be commensurable with the oil, whatever action or organism had produced the oil, also yielded the phosphate* found in the surface-soil, and apparently derived from the underlying rocks. The older rocks, in which the asphalt-deposits were there found, were of Jurassic age, the beds lying more or less vertical, and the deposits were found in veins 1 or 2 inches thick, and never exceeding 12 inches thick. These rocks had been subjected to the action of metamorphism, but in parts of the island they were covered by Cretaceous limestone or sandstone from a few feet to several hundred feet thick. The rocks of Cretaceous age, although found lying in a more or less corresponding vertical position, did not seem to have been subjected to metamorphic action. Where found in the older and metamorphosed rocks, the veins contained a pure pitch-like asphalt, although hardly approaching the purity of manjak. But when found in the Cretaceous formation, the deposits occurred in patches, several feet in width and height, and of unproved length; and instead of being of a pure pitch-like composition, they could hardly be described otherwise than as mere deposits of bitumen, having three or four different appearances, the most common form being like coal, which had been mined close by a fault, or often like very light black coke, generally taking this form when found close to the surface. In some cases, the asphalt might be compared to the admixture of an oily liquid absorbed in sand, which had been afterwards pressed into a lump, and dried. With depth, and in proximity to the older rocks, it gained considerably in specific gravity, and, although far from being pure pitch-like in appearance, it was more solid and purer. No traces of fossil-remains had, so far as he was aware, been found associated with these rocks.

* *The Oil-fields of Russia*, 1904, page 83.

Several deposits were, at the present time, being worked in the island, but one illustration would suffice. An enormous deposit, discovered about 40 years ago, was covered by 5 to 20 feet of soil, and was worked by quarrying, but a shaft had been sunk to a depth of 80 feet without reaching the bottom. The axis of the deposit ran in an east-and-west direction, and its continuity had been proved for a distance of over a mile, with an average uniform width of about 70 feet. By blasting, two workmen could produce about 100 tons per day of 8 hours at the open face; but in tunnel-work (two tunnels being driven and kept a few feet in advance of the working-face) they could only produce about 2 tons. The material was filled into baskets, and carried on the shoulders of natives up steps cut in the side of the quarry, and emptied into railway-wagons on the surface.

He (Mr. Kirsopp) had not seen the following theory suggested before, but it occurred to him that these deposits were the result of the refuse of an oil-flow. The older rocks being broken by metamorphic action, the pent-up oil from unknown depths would rush into the fissures and ascend to the surface, where, after it had lost its hydrogen by atmospheric action, the remaining carbonaceous matter became solidified and formed pure asphalt. Where the Cretaceous formation (then presumably in the form of soft sand or silt) overlay the older rocks, as soon as the flow reached the surface-level of the older rocks, it would naturally expand on each side of the fissure through which it flowed, until such time as the rock on each side was thoroughly saturated with oil.

If these deposits were the result of outbursts of pent-up oil, it was difficult to conceive how they could have been formed by the distillation of vegetable matter, as these bitumen-deposits, unlike coal, did not extend over the entire area of the Cretaceous formation, but occurred in a few widely-separated areas. Had such deposits been derived from the distillation of a mass of decomposed vegetable matter, their area should have been almost as extensive as that of a coal-field.

It would be interesting if Dr. Dvorkovitz would state, as the result of his observation and experience, whether (1) a disturbed formation of the rocks and (2) considerable showings of bitumen and asphalt, found cropping out at the surface, were indications conducive to the expectation and reality of payable quantities of oil being found in such a locality by boring. It would almost seem, if bitumen and asphalt-deposits were the remains of oil-bursts and flows, that such showings would tend to prove the reverse, although at one time extensive oil-accumulations must have existed in such regions.

He (Mr. Kirsopp) understood that there were considerable surface-showings of ozokerite in the Galician oil-field, and he asked whether such showings did not detract from the profitable finding of oil lying below. It would be extremely interesting to have the results of Dr. Dvorkovitz's experience on this point; and also whether his experience of the oil-bearing districts of Europe led him to believe that the axis always ran along anticlines similar to the great Appalachian chain, which extended from the north to the south of the United States.

Mr. ERNEST VON GLEHN (London) wrote that the late Mr. V. I. Ragosine was in every respect a remarkable man, a born inventor, although possessing very little technical training. He was struck with the enormous waste that went on at the Baku oil-fields, where 60 to 70 per cent. of the product was allowed to flow away into the Caspian Sea, only the illuminating-oil being turned to account. Mr. Ragosine's first experiments with the residuum were made (as he told the writer) in a kettle over the kitchen-fire at home. Within a very few years, in 1876, he had succeeded in completing a large and well-equipped refinery at Balachna, on the banks of the Volga, and was turning out a well-assorted series of lubricating-oils, possessing a " body " or viscosity such as had never before been seen in mineral-oils. These oils were introduced into Great Britain in 1878, and since then the industry had rapidly developed.

Of the improvements introduced in the methods of distillation and refining, he (Mr. von Glehn) could not speak with technical knowledge, but the results were wellknown to him, as he had for 25 years been engaged in selling these oils. Perhaps the most remarkable point to be noticed was the enormous reduction in the cost of lubricating oils. In the seventh decade of the last century, a good engine-oil cost the consumer from 2s. 6d. to 3s. 6d. per gallon, and a better oil was now supplied for 1s. or 1s. 6d. per gallon.

Mr. ABR. GOUKASSOW (London) wrote that, within the limits of a few pages, it was hardly possible to treat the numerous data on petroleum in a more masterly manner, than had been done by the author.   It was to be regretted, however, that he was obliged to refer but slightly to many interesting questions deserving a more detailed discussion.   He (Mr. Goukassow) would not attempt to dispute Dr. Dvorkovitz's conclusions in general, but he would touch only on a question (which was of primary importance, though possibly on theoretical grounds only, to all interested in petroleum), namely, the origin of petroleum.

Strange to say, this question was still in the same position now as it was 25 years ago.   Whilst the origin of coal and the other more important minerals had been satisfactorily explained by men of science, no clear account had, up to the present time, been given of the origin of a product mined in such large quantities and playing so important a part in our daily life as petroleum.   It was very desirable, therefore, that authorities on the subject should be rather chary in expressing opinions, which might take us off the right path in attempting to solve the question.   The predilection of Dr. Dvorkovitz for the theory of the formation of petroleum by the decomposition of vegetation found no support in the case of the oil-deposits of the Apsheron peninsula.   Persistent boring on a very large scale for 35 years had not given a single positive indication of any signs of plant-life there, or any traces of the process of the distillation of coal.   The coal-deposits nearest to the Apsheron peninsula occurred in the Black Sea basin, and had nothing in common, either in the period of deposition or stratigraphically, with the oil-containing strata of the peninsula.   Moreover, there were many indications showing that the climatic conditions prevailing in the Tertiary period in the region of the Aralo-Caspian depression were not at all favourable to the formation of luxuriant vegetation, as Dr. Kroemer was inclined to believe in the case of the deposits in Pennsylvania.   It must therefore be assumed that in the Eocene and Oligocene periods there came into existence those climatic conditions, which were still found there to-day, namely, dry air, strong winds and great variations of temperature.   The receding sea had also left no indications of abundance of plant-life there.   At least, they were wanting, both on the shore-deposits of the Caspian and in the sea itself.

The investigations, made during the last few years by Dr. Lebedintzeff, indicated on the other hand an enormous wealth of animal-life in the Caspian; and those carried on in the Kara-bugas bay by Prof. Andrussow pointed to the exceptional conditions to which animal-life was exposed in the bays of the Caspian. These investigations might perhaps cast full light on the origin of the Apsheron petroleum. He (Mr. Goukassow) had no intention, however, of stating any positive data in favour of one hypothesis or the other, but he would only like to point out that:—(1) The Apsheron oil-deposits must be examined in connection with the whole series of the Tertiary strata of the Aralo-Caspian basin; (2) this series was characterized by an exceptional poverty of indications of plant-life; and (3) the climatic conditions prevailing during the Oligocene period did not support the theory of the vegetable origin of the oil in the Apsheron peninsula. However, he must limit himself to these somewhat vague assertions as the question had been touched upon only but slightly by Dr. Dvorkovitz, and might not be of much interest to the members.

Mr. H. T. BURLS (London) wrote that Dr. Dvorkovitz had hardly touched upon the geological occurrence of oil, and there was a point about this which he thought was of interest, as it had an important bearing on the economical value of an oil-deposit. In the United States, the two classes of oil, having an asphalt or a paraffin-base, had been found to be of different ages. The oil, having an asphalt-base, which was used for fuel, occurred as large deposits in Texas and California, and belonged to the Tertiary or even a later age; while the oil with a paraffin-base, containing vaseline and other valuable bye-products, and found in Pennsylvania, Alaska, Kansas, as well as in small quantities in California, was always older than the Cretaceous. This distinction was, so far as he knew, universal in the United States: whether it could be applied generally to other parts of the world he was not prepared to say, although the Russian and Borneo oils were also of Tertiary age.

Dr. P. DVORKOVITZ (London) said that he had only been able to touch upon the principal points in his paper: he could not go into geological questions, or costs, etc. Steel casing had lately been tried to shut off the water, screwed casing being used instead of riveted; but from his own experience during the five years he was in Baku at the beginning of the industry, and since then visiting there practically every year, he would say that if the old-fashioned riveted casing were used with the same care as formerly, instead of trying to hurry the work, they would not find so many wells producing nothing. Five years ago, he foretold that if they went on with the same speed and put down so many wells, the results would be disastrous, as they had found when they commenced to pump water instead of oil. He believed that the shutting off of the water by using riveted casing was the best method.

The sizes and depths of the wells varied to a large extent. In America, in Pennsylvania, he had seen a well 2,700 feet deep, starting with pipes 10 inches in diameter and finishing with pipes 4½ inches in diameter. In the South American fields, they generally started with pipes 12 inches in diameter and finished with a pipe 8 inches in diameter. In Russia, generally, the average depth was 500 or 600 feet, but some of them had gone down to 2,200 feet, starting with a diameter of 30 inches and finishing with 14 inches. In Galicia, the wells were 2,400 feet, and in Rumania, 1,500 to 2,000 feet deep. In certain places, the pressure of the oil was very high, and consequently there was no need to make the well sufficiently large to enable a pumping bucket to be lowered down, the oil being forced out of the well by the pressure below it. In Russia, the wells were all pumped, and therefore the well had to be made of large diameter.

A question had been asked with regard to petrol. Personally, he believed that if the members of the Institution of Mechanical Engineers invented a method by which motorists could use petroleum, they would confer a great benefit on that industry. He did not think that there were any theoretical objections, or that it was impossible to use that oil in motors. An oil with a specific gravity of 0·68 was received in small quantity from Pennsylvania; and the rest had a specific gravity of 0·72. The chief principle in all inventions relating to the carburreter consisted in passing compressed air through the carburreter in order to get rid of the volatile matters. The supply of light spirit was not sufficient for a hundredth part of the world's consumption for motor-car purposes, and therefore chemists had endeavoured to supply an addition to the light spirit, and lately they had brought out a large quantity of motor-car spirit, with a specific gravity of 0·710 and 0·720. They had tried to supply it from Rumanian oil with a specific gravity of 0·730, but none could come from Russia, as the oil did not contain any benzine. It was, however, possible with a pre-heater to heat up the heavy oils to 220° or 240° Fahr. and then pass in the air, and so obtain the same explosive effect. Personally, he looked to-day with the greatest pleasure on the manner in which the Scottish engineers had maintained their position, although they were compelled to mine the shale at a great cost.

The CHAIRMAN (Mr. George A. Mitchell, Glasgow), moved a vote of thanks to Dr. Dvorkovitz for his interesting paper.

Mr. C. C. LEACH seconded the resolution, which was cordially approved.

––––

Dr. G. P. LISHMAN's paper on " The Analytical Valuation of Gas-coals " was read as follows : —

THE ANALYTICAL VALUATION OF GAS-COALS.

By Dr. G. P. LISHMAN, D.Sc.

The manner in which various processes for valuing gas-coals were tried, different forms of apparatus used, and finally the standard-coal method adopted, was detailed in a previous paper[*] and need not be recapitulated here. A further two years' experience of the working of the process, therein described, has only served to confirm the writer's opinion of its value and of its necessity. It is now proposed to examine more closely into certain details.

On looking into the history of the subject, it is seen that there has been very much variety of procedure and also that, from time to time, considerable dissatisfaction has been expressed with regard to small-scale coal-testing methods. Mr. Andrew Scott,[†] at a meeting of the West of Scotland Gas-managers' Association in May, 1879, stated that published analyses were not realized in the works in the case of several shales and cannels. Later on, Mr. J. M'Crae[‡] stated that little heed should be paid to printed analyses. Some allowance can certainly be made for the difference between the clean sample sent to the analyst and the bulk forwarded to the gas-works, but this does not explain everything, and the tone of the discussions at these and other meetings is evidently against small-scale apparatus. The report of the 1893 Research Committee of the North British Association of Gas-managers stated that "all analysis should be conducted on a thoroughly practical scale;"[§] and previous to this

---

[*] *Trans. Inst. M.E.*, 1902, vol. xxiii., page 567.

[†] "Remarks on the Analysis of Gas-coals," *Journal of Gas Lighting*, 1879, vol. xxxiii., page 765.

[‡] "Coal-analysis," by Mr. J. M'Crae. *Journal of Gas Lighting*, 1892, vol. lx., page 246.

[§] *Journal of Gas Lighting*, 1893, vol. lxii., page 215.

date, the Association of German Gas-works Chemists had come
to a similar conclusion. Mr. J. Stellfox* quoted the following
analyses of the same coal from different sources:—

Analyst.	Yield per Ton. Cubic Feet.	Illuminating Power. Candles.	Sperm-value per Ton. Pounds.
A.	10,088	17·33	591·70
B.	10,691	18·36	627·00
C.	12,780	22·35	979·31

The continuous record of coal-testing, since the introduction
of gas-lighting, indicates that there is, and has always been,
a demand from practical men for information about gas-coals
at a less cost than a practical trial would entail. In 1815 and
1819, Mr. Accum† published lists of various coals and their
yield of gas, and described the arrangement of his apparatus;
although crude, it was very useful in distinguishing cannel
from bituminous coal. Mr. Samuel Clegg's improved form of
apparatus is figured in his treatise.‡ On the model of the plant
designed by Mr. W. T. Sugg,§ most subsequent forms have been
designed. Mr. L. T. Wright‖ put his condensers into a water-
tank; and Mr. J. T. Sheard¶ described another form, in which
the gas after passing through a washer-scrubber, spread itself over
a number of vertical tubes. Messrs. Clegg, Sheard, Wright,
and many others, have used a retort carbonizing 2 or 3 pounds;
but coal has been tested with all quantities from 50 grammes
to the "thoroughly practical scale." Dr. Wallace of Glasgow,
in whose laboratory an extensive series of analyses were made,**
used 22·4 pounds. Mr. Leicester Greville†† used $\frac{1}{160}$ ton or
14 pounds in two clay-retorts. Mr. T. Glover‡‡ in 1896 described
an arrangement in which two retorts were used, carbonizing
¼ cwt in each. Mr. J. G. A. Rhodin's apparatus§§ is designed

* *Journal of Gas Lighting*, 1892, vol. lx., page 120.

† *A Practical Treatise on Gas-light*, 1815, by Mr. Frederick Accum; and
Process of Manufacturing Coal-gas, 1819, By Mr. Frederick Accum.

‡ *Manufacture and Distribution of Coal-gas*, 1841, by Mr. Samuel Clegg,
page 38.

§ *Gas-manipulation and Analysis*, 1867, by the late Mr. Henry Bannister,
enlarged by Mr. W. T. Sugg.

‖ "The Analysis of Gas-coal," by Mr. L. T. Wright, *Journal of the Society
of Chemical Industry*, 1885, vol. iv., page 656.

¶ *Journal of Gas Lighting*, 1888, vol. li., page 369.

** *Ibid.*, 1877, vol. xxx., page 945.	†† *Ibid.*, vol. li., page 508.

‡‡ *Ibid.*, vol. lxviii., page 847.

§§ *Journal of the Society of Chemical Industry*, 1900, vol. xix., page 12.

to overcome the expense and inconvenience of other forms rather than to deal with the difficulties of the subject. Frequently, a full-sized retort in a set has been isolated by a bye-pass on the hydraulic main, and the gas tested.*

Abroad a similar state of affairs obtains. Prof. Stein, in 1858, used 20 pounds with two retorts in the same furnace, making duplicate estimations; he had great difficulty in estimating the illuminating-power, and was chiefly guided in this respect by the specific gravity. Mr. Schilling, in 1863, used 160 German pounds in a ⊂⊃ shaped retort, measuring 19 inches in width, 16 inches in height and 8 feet in length: his apparatus including a meter and an exhauster. Dr. H. Bunte has used an apparatus similar to that of Mr. Schilling, with an exhauster. Perhaps the chief form of laboratory-apparatus used abroad is that of Dr. Leybold, fully described by Mr. A. Schafer.† Another arrangement for carbonizing 1 kilogramme (2·204 pounds) of coal has been described by Mr. G. Jouanne,‡ but the purifier, in this case, seems much too small. Dr. E. Ste. Claire Deville has used the Leybold apparatus, and also a much larger one. Others§ have attempted to base a valuation of gas-coals on the estimation of the volatile matter, but the estimation is not accurate.

It will be recognized that there is no generally accepted manner of testing gas-coals. The fact is, as was stated in the writer's previous paper, that results from the various forms of laboratory coal-testing plant come out most irregularly; and Mr. E. Grahn‖ has emphasized these differences. A specimen is shown in Table I. The results, there given, were all obtained from one and the same coal, and for the most part on different days, during the first six months of 1903. The tests were all made from the coal used as a standard. The illuminating-value per ton of this coal is 580 pounds of sperm, that is, in any well-equipped gas-works this figure can be obtained from it. The list

---

* Mr. H. Veevers, *Journal of Gas Lighting*, 1880, vol. xxxv., page 985.

† *Einrichtung und Betrieb eines Gaswerkes*, page 24.

‡ *Le Gaz*, 1896, vol. xxxix., page 103.

§ *De Gasfabrikant in het Laboratorium*, by Mr. A. J. van Eyndhoven, page 14.

‖ "Versuchsanstalten für Gaskohlen," by Mr. E. Grahn, *Journal für Gasbeleuchtung*, 1869, page 59.

might be extended by several hundreds of tests made with the same coal, during the past six years, but the above are sufficient to give an ample idea of the variation to be expected. The yield of gas varied from 9,650 to 12,400 cubic feet per ton of coal; and the illuminating power, from 13·0 to 18·1 candles. The differences of sperm-value extend through the whole range of quality of ordinary gas-coals, that is, from 500 to 667 pounds of sperm per ton. The figures show, however, that, even when working with water-jacketed condensers (which may be assumed to be very suitable for such a purpose as this), the direct testing of gas-coals is useless for comparative purposes, and that any result may be obtained according to circumstances.

TABLE 1.

No. of Test.	Yield of Gas per Ton.	Illuminating Power.	Sperm-value per Ton	No. of Test.	Yield of Gas per Ton	Illuminating Power.	Sperm-value per Ton
	Cubic Feet.	Candles.	Pounds		Cubic Feet.	Candles.	Pounds.
1	9,700	16·6	552	30	12,100	14·4	598
2	10,600	15·0	545	31	11,400	13·7	536
3	10,800	14·8	547	32	10,700	15 6	572
4	10,300	16·5	583	33	10,300	16·3	575
5	9,650	17·8	589	34	12,100	14·4	593
6	10,050	17·2	593	35	12,400	15·1	643
7	11,000	15·4	581	36	12,300	15·6	658
8	10,600	16·1	585	37	12,300	13·0	548
9	10,800	16·9	626	38	11,400	15·6	610
10	11,300	16·0	620	39	10,650	16·2	592
11	11,000	15·7	593	40	10,450	18·0	645
12	10,500	16·3	586	41	11,100	15·2	573
13	10,400	14·8	528	42	10,900	16·6	621
14	10,500	16·8	605	43	10,000	16·2	556
15	12,100	15·0	622	44	10,200	15·6	546
16	11,100	15·5	590	45	10,200	14 3	500
17	10,700	16·8	616	46	11,100	15·0	571
18	10,500	17·4	627	47	11,500	15·4	607
19	11,200	16·7	642	48	11,400	15·8	618
20	12,100	13·8	573	49	11,500	15·2	599
21	11,600	15·2	604	50	11,000	16·5	622
22	10,500	15·8	569	51	11,500	16·9	667
23	11,600	15·0	597	52	11,500	16·0	631
24	11,800	13·6	550	53	11,300	14·3	554
25	11,200	13·8	530	54	11,600	15·7	625
26	10,200	16·6	581	55	11,050	14·8	561
27	9,950	18·1	617	56	10,700	16·4	601
28	10,800	17·0	630	57	10,700	16·6	609
29	11,200	16·2	622	58	11,100	15·7	598

The idea, therefore, so frequently found in the literature of the subject (Messrs. Clegg, Newbigging, Sheard, Wright and others) that the laboratory-apparatus gives results higher than those obtained in the gas-works, but comparable to each other, is not borne out by the present investigation. There are many apparent reasons why they should be higher, such as iron-retorts, no loss, perfect purification, etc., but they are all overborne by the differences of condensation which occur: the temperature of the apparatus being the main determining-factor in the quality of the gas. Mr. Greyson de Schodt, it may be noted, found that laboratory-tests were lower than those obtained in the works.[*] The differences in yield of gas, seen in Table I., are caused largely by differences of retort-temperature. When the retort is hottest it may happen that the condensers are also hottest, as was the case in No. 51 test; then a large yield is obtained, and it is ill-condensed and consequently of higher illuminating-power than it would otherwise have been: hence a high sperm-value is obtained. Then again, a new retort makes a difference: a cast-iron retort lasts about a month, towards the end of that time the yield of gas has begun to fall off; and, after a week, the difference is sometimes noticeable.

It has been observed that an especially high sperm-value is obtained on a warm morning, preceded by some days of cold weather: hydrocarbons deposited in the pipes, etc., during the cold days having been picked up again by the gas. Under conditions the reverse of the above, the results are much lower than the true value of the coal. It is probably for a similar reason that, on the retort becoming hotter during a series of tests and the yield of gas greater, the illuminating-power does not always immediately fall, for the gas, far from saturated with benzol, etc., picks up these illuminants on its way to the holder; and the larger the pipes the greater will be this effect. In testing a coal, whenever there has been, between two successive tests, a notable change in the quality of the gas, this effect can be looked for: it explains many an otherwise inexplicable result, and also the frequently observed fact of the comparative uniformity of illuminating-power during certain days of testing.

The state of the purifiers also affects the gas.

In a continuous series of tests on a gas-coal, no regularity is to be expected from day to day in the results.

These details indicate how essential it is that there should be some check on the indications of an instrument like a coal-testing apparatus.

[*] *Journal of Gas Lighting*, 1891, vol. lviii., page 444.

It will be desirable now to give some account of the manner in which the standard coal can be utilized in the valuation of others. It may be said that all extreme results, whether in yield of gas or in sperm-value, are rejected as standards of comparison with unknown coals; however many tests may have been made, the results in these cases are rejected and the coal tested again on another day. All the results recorded in Table I. have been obtained from the standard coal, but several have been rejected as standards of comparison. Although the sperm-values in the table are highly variable, they are in nearly every case explicable when one or more of the causes detailed above are taken into consideration, and very often each one was just what might have been expected at the time when it was obtained. Occasionally, unaccountable results will come out, but such are rare. If a series of tests be made from the same coal on any day, the results will, after the first two charges, generally be fairly constant, so long as the apparatus is not allowed to become warm and the retort-temperature remains the same: this being so, the testing of an unknown coal is an easy matter. Three tests will probably suffice, namely:—(1) Unknown coal, (2) standard coal, and (3) unknown coal. The two from the unknown coal being near each other, the mean is taken, and the figure corrected proportionately, according as the standard coal is giving a result higher or lower than its fixed normal value.

Sometimes, however, the case is less simple. It has often been noticed in a series of tests, made during one day and using the same coal, that the sperm-values continue to come out higher and higher for each successive test: this is commonly due to the apparatus becoming gradually warmer; and it is hopeless then to operate without a standard. In such a case, at least five tests are required to fix the unknown coal, the second and fourth being made with the standard coal. If a series of sperm-values, such as (1) 530, (2) 548, (3) 570, (4) 585 and (5) 605, be obtained, being a rise of about 18 each time, it may safely be inferred that the same figures would have been obtained had the same coal been used in all the tests, and that therefore the unknown coal is equal to the standard. The standard coal has a fixed value (in the writer's case) of 580 pounds per ton, therefore, the unknown coal is reported as having that value, although under the particular conditions of that day's testing the average

result was only 568 pounds. The same result (580 pounds per
ton) is obtained by taking the mean of tests (1) and (3) which
makes the sperm-value, 550 pounds, that is, on a par with the
intervening test from the standard coal, so that the unknown
coal would have to be corrected up to the normal value of the
standard coal, 580 pounds. But it is necessary to make the five
tests in order to be assured that the considerable difference
between (1) and (3) is due to a uniform rise in the sperm-value
and not to some other cause. When the results are rising in
this manner, and the unknown coal differs considerably from
the standard, the calculation of the unknown coal becomes more
complicated and frequently cannot be carried out, and then the
coal must be re-tested at a more suitable time.   Often, however, the quality can be gauged: as shewn by the results contained in Table II., taken from the writer's note-book.   In this

TABLE II.

No. of Test.	Sample of Coal.		Yield of Gas. Per Ton.	Illuminating Power.	Sperm-value.
			Cubic Feet.	Candles.	Pounds.
1	Unknown	...	10,700	13·5	494
2	Standard*	...	11,050	14·8	561
3	Unknown	...	10,400	15·3	546
4	Standard*	...	10,700	16·4	601
5	Unknown	...	10,700	16·4	601

* Sperm-value, 580 pounds.

case, the room was becoming warmer during the whole of the
time occupied by the tests. There is a difference of 107 pounds
in the sperm-value between the first test, taken from the
unknown coal, and the last test; and the third test occupies
the intermediate position at 546 pounds: there being a rise
of about 25 pounds in the sperm-value for each test.   In
order, therefore, to bring all the results to an equal footing,
50 pounds must be added to the first test, and subtracted from
the fifth test, and 25 pounds added to the second test and
subtracted from the fourth test. The following series is thus
obtained : —

Test.		1.	2.	3.	4.	5.
Sperm-value, unknown coal	...	544	—	546	—	551
Do. standard coal	...	—	586	—	576	—

The sperm-value of the standard coal is here seen to be practically at its true fixed value of 580 pounds, therefore, the other coal requires no further correction and may be taken as having a sperm-value of 547 pounds, or 10,500 cubic feet of gas of 15·2 candlepower.

It will be seen that considerable judgment is required in the interpretation of the actual results obtained from a coal-testing apparatus. This much has been admitted by some operators,* but the grounds upon which such judgment is to be based are often not gone into.

In making tests like the above, the last results can nearly always be anticipated before they come out. Over and over again, the same coal has been tested against the standard coal on different days, and while yielding totally different actual results the corrected figures are very much alike.

Sufficient has now been said to give an idea of the somewhat oblique paths by which it is necessary to proceed in order to arrive at the value of a gas-coal. No doubt, there are many objections to this method; its uncertainty, that is, the possibility of working all day and obtaining nothing; its laboriousness; and its cost, in time and gas, are all against it. Further, a standard coal having a sperm-value of 580 pounds is of no use for testing cannel: a separate cannel-standard being required for that purpose. Nevertheless, if ever it should be necessary to value gas-coals as between buyer and seller, it seems to the writer that, in the present state of knowledge, the only method of avoiding contention would be to adopt a common standard-coal obtained from a certain locality. The value of this standard once agreed upon, there would be no need for differences of more than about 15 pounds in the sperm-value obtained by different workers. Seams of coal may, of course, vary in quality, and some are well known to do so; but taking the top, the middle and the bottom coal together, many seams, with which the writer is acquainted, are very constant over areas, which may, in certain cases, be measured by miles. It is possible that this divergence of coal-seams has been exaggerated, and that differences of results really due to other causes may have been put down to variations in the seam.

* *Gas-manufacture*, by Mr. W. J. A. Butterfield, 1896, page 18.

In the case of a gas-company, which perhaps has no access to a seam of coal, the best way to obtain a standard-coal would probably be to select a colliery, preferably one which supplies gas-coal from only one or two seams, and to take the finest hand-picked coal obtainable from the wagons from that colliery. To this coal a fixed sperm-value of any reasonable figure could arbitrarily be given and utilized as above described. This method would certainly not be so good as always drawing coal from a single definite seam, but as an alternative it would be found of value.

With the increasing use of incandescent burners, illuminating-power as ascertained in the ordinary way, by burning raw gas at an Argand burner, is becoming of less moment, its place being gradually taken by calorific power. Whether, however, the quality of a coal be measured by its sperm-value (a function of the yield of gas and the illuminating-power) or by another figure (the product of the yield and the calorific power), it seems likely that a standard-coal will be required in testing. Experiments to determine this point are being made by the writer.

BIBLIOGRAPHY.

Anon., "List of Analyses of Gas-coals, with Authorities," *Journal of Gas Lighting*, 1874, vol. xxiii., page 795.

—, "A.B.C.D. du Gazier," *Le Gaz*, 1878, vol. xxii., page 51.

—, "Résumé des Travaux entrepris à l'Usine Expérimentale de la Villette sur la Composition des Charbons et les Propriétés du Gaz d'Éclairage obtenu par la Distillation en Vase clos de ces mêmes Charbons," *Journal de l'Éclairage au Gaz*, 1886, pages 201, 217 and 299.

—, "A Simple Method for the Analysis of Coal and Coke," *American Gas-light Journal*, 1891, page 887.

—, "Report of the 1893 Research Committee of the North British Association of Gas-managers on Coal Analysis," *Journal of Gas Lighting*, 1893, vol. lxii., page 214.

Bunte, Dr. H., "Versuche mit Gaskohlen," *Journal für Gasbeleuchtung*, 1886, page 589.

—, "Chemische Untersuchungen in Gasanstalten," *Journal für Gasbeleuchtung*, 1888, pages 570, 858 and 894.

Campredon, Louis, "Détermination expérimentale du Pouvoir agglutinant des Houilles," *Comptes-rendus hebdomadaires des Séances de l'Académie des Sciences*, 1895, vol. cxxi., page 820.

Davis, George E., "Testing the Illuminating-power of Coal-gas," *Journal of the Society of Chemical Industry*, 1892, vol xi., page 412.

Deville, E. Sainte Claire, "Étude sur le Gaz de Houille," *Journal des Usines à Gaz*, 1889, pages 71, 87, 102, 137, 161 and 178.

Euchene, M., "Thermic Reactions in the Distillation of Coal," a paper read before the International Gas-congress, Paris, *Journal of Gas Lighting*, 1900, vol. lxxvi., pages 1080 and 1141.

Fleck, Dr. —, Dr. — Geinitz and Dr. — Harteg, "Die Steinkohlen Deutschlands und andere Länder Europas," *Journal für Gasbeleuchtung*, 1866, pages 54, 81 and 118.

Foster, Prof. W., "The Effects of Specific Hydrocarbons on the Lighting-value of Combustible Gas," a paper read before the Incorporated Gas Institute, *Journal of Gas Lighting*, 1891, vol. lvii., page 1234.

Foullon, H. B. von and C. von John, "Technische Analysen und Proben aus dem chemischen Laboratorium der K. k. Geologischen Reichsanstalt " (Analyses of Gas-coals), *Jahrbuch der Kaiserlich-königlichen Geologischen Reichsanstalt*, Wien, 1892, vol. xlii., page 155.

Frankland, Dr. Percy F., "The Composition and Illuminating-power of Coal-gas," *Journal of the Society of Chemical Industry*, 1884, vol. iii., page 271.

—, "The Illuminating-power of Ethylene when burnt with Non-luminous Combustible Gases," *Journal of the Chemical Society*, 1884, vol. xlv., page 30.

Geinitz, Dr. --, Dr. — Fleck and Dr. — Harteg, "Die Steinkohlen Deutschlands und andere Länder Europas," *Journal für Gasbeleuchtung*, 1866, pages 54, 81 and 118.

Glover, T., "Commercial Testing of Gas-coal," a paper read before the Midland Association of Gas-managers, *Journal of Gas Lighting*, 1896, vol. lxviii., page 847.

Graham, D. A., "The True Principle of the Valuation of Gas-coal," *Journal of Gas Lighting*, 1874, vol. xxiii., page 359.

Grahn, E., "Versuchsanstalten für Gaskohlen," *Journal für Gasbeleuchtung*, 1869, page 59.

Harteg, Dr. —, Dr. — Fleck and Dr. — Geinitz, "Die Steinkohlen Deutschlands und andere Länder Europas," *Journal für Gasbeleuchtung*, 1866, pages 54, 81 and 118.

Hilt, Karl, "Die Beziehungen zwischen Zusammensetzung und technischen Eigenschaften der Steinkohle," *Zeitschrift des Vereines Deutscher Ingenieure*, 1873, vol. xvii., page 193.

—, "Ueber Zusammensetzung der Steinkohlen," *Zeitschrift des Vereines Deutscher Ingenieure*, 1876, vol. xxi., page 290.

Hilt, Karl Josef, "Ueber Eigenschaften und Zusammensetzung der Steinkohlen, *Glückauf*, 1873, vol. ix., No. 15.

Hoffmann, J., "Der Einfluss der Feuchtigkeit der Kohlen über die Qualität des Gases," *Journal für Gasbeleuchtung*, 1863, page 39.

John, C. von, und H. B. von Foullon, "Technische Analysen und Proben aus dem chemischen Laboratorium der K.k. Geologischen Reichsanstalt " (Analyses of Gas-coals), *Jahrbuch der Kaiserlich-königlichen Geologischen Reichsanstalt*, Wien, 1892, vol. xlii., page 155.

Jouanne, G., "Appareil de Laboratoire pour l'Essai de Charbons à gaz," *Le Gaz*, 1896, vol. xxxix., page 103.

Knublauch, Dr. O., "Ueber die Leuchtkraft des Benzols (Toluols) und Æthylens und deren Bestimmung im Leuchtgase," *Journal für Gasbeleuchtung*, 1879, page 652 ; and 1880, pages 253 and 274.

Lewes, Prof. Vivian B., "The Action of Heat upon Ethylene," *Proceedings of the Royal Society of London*, 1894, vol. lv., page 90 ; and 1895, vol. lvii., page 394.

—, "The Cause of Luminosity in the Flames of Hydrocarbon Gases," *Proceedings of the Royal Society of London*, 1895, vol. lvii., page 450.

Leybold, Dr. W., "Die Aufgaben des Chemikers im Gasanstaltsbetrieb," *Journal für Gasbeleuchtung*, 1895, page 625.

Lucian, —, "The Volatile Substances of Gas-coal," *Bulletin de l'Association Belge des Chimistes*, 1900, vol. xv., page 379.

M'Crae, J., "Coal-analysis," a paper read before the North British Association of Gas-managers, *Journal of Gas Lighting*, 1892, vol. lx., page 246.

McMillan, E., "Testing the Value of Coal for Gas-making, Seventh Annual Meeting of the Western Gas Association," *American Gas-light Journal*, 1884, page 4.

Malherbe, R., "Analysis of Gas-coals," *Moniteur Industriel Belge*, 1876, page 490.

Merkens, —, "Resultate über Vergasung," *Journal für Gasbeleuchtung*, 1881, page 385.

Newbigging, T., "Testing Coal for its Producing Qualities," *Transactions of the Gas Institute*, 1885, page 188.

Rhodin, J. G. A., "A Laboratory Method for the Analyses of Coals for Gas Manufacture," *Journal of the Society of Chemical Industry*, 1900, vol. xix., page 12.

Schilling, N. H., "Untersuchungen über Gaskohlen," *Journal für Gasbeleuchtung*, 1863, pages 120, 158, 219 and 317.

Schimming, G., "Die Ausnutzung der Brennstoffe," *Journal für Gasbeleuchtung*, 1891, pages 82 and 102.

Schodt, Greyson de, "The Testing of Coal," a paper read before the Belgian Association of Gas-managers, *Journal of Gas Lighting*, 1891, vol. lviii., page 444.

Scott, Andrew, "Remarks on the Analysis of Gas-coals," a paper read before the West of Scotland Gas-managers' Association, *Journal of Gas Lighting*, 1879, vol. xxxiii., pages 765.

Sheard, John T., "Coal Analysis and Analyses," *Journal of Gas Lighting*, 1888, vol. li., pages 233, 324 and 369.

Smithells, Prof. Arthur, "Flame-temperatures and the Acetylene-theory of Luminous Hydrocarbon-flames," *Journal of the Chemical Society*, 1895, vol. lxvii., page 1049.

Stein, Prof. W., "Untersuchungen über das Verhalten der Sächsischen Kohlen," *Journal für Gasbeleuchtung*, 1858, page 42.

Stevenson, G. Ernest, "The Relative Illuminating-value of the Hydrocarbon-vapours and Gaseous Hydrocarbons present in Coal-gas and their Quantitative Determination," *Journal of Gas Lighting*, 1880, vol. xxxvi., page 293.

Veevers, Harrison, "Testing Cannel and Coal," a paper read before the British Association of Gas-managers, *Journal of Gas Lighting*, 1880, vol. xxxv., page 985.

Verdier, M., "De l'Influence de l'État Physique de la Houille sur la Qualité des Produits Distillés," *Société Technique de l'Industrie du Gaz en France*, 1901, 28me Congrés, page 250.

Wright, Lewis T., "The Analysis of Gas-coal," *Journal of the Society of Chemical Industry*, 1885, vol. iv., page 656.

Young, W., "Condensation," a paper read before the West of Scotland Association of Gas-managers, *Journal of Gas Lighting*, 1876, vol. xxviii., page 665.

Some of the works, mentioned in the foregoing list, have only a remote connection with the subject.

DISCUSSION—THE ANALYTICAL VALUATION OF GAS-COALS. 527

Dr. H. SALVIN PATTINSON (Newcastle-upon-Tyne) wrote that Dr. Lishman attributed the great variations in the results which he had obtained with the same coal at different times to variations in the temperature of the retort and of the condensers. Although, in his experience, such differences in the sperm-value as those tabulated by Dr. Lishman very rarely occurred, yet he had no doubt that these causes might, at least in part, account for them when they did occur. If this were so, the obvious remedy was to keep the temperatures of retort and condensers constant; this was, perhaps, not possible in all cases; but the first step towards its attainment was to know the actual temperature of the retort. He had for some time past, in his practice, controlled this by means of a thermo-electric pyrometer. Failing this instrument, the time required for the distillation afforded a useful guide in keeping the temperature constant over a series of determinations. It was his opinion, based on a long and varied experience, that the direct laboratory-testing of gas-coals was of real value in ascertaining the quality of these coals for gas-making purposes; and that the use of a standard coal was unnecessary.

Dr. W. CARRICK ANDERSON (Glasgow) wrote that, in the discussion on Dr. Lishman's previous paper, he had already expressed approval of the lines on which he was working in the endeavour to fix a standard method for the valuation of gas coals.* In this further communication, Dr. Lishman had pointed out some of the difficulties that attended the practical working of a small-scale process of distillation. But there was nothing in these that invalidated the method for valuation-purposes. They only indicated the need for still closer and more rigorous adhesion to specified conditions, both in the apparatus itself and in the working of it. It was clear that small-scale working, which admitted of maintaining tolerably uniform and known conditions of distilling, was the only one that allowed the operator to ascertain what he was doing and was therefore the only one that could become the basis of a standard of comparison. The carefully controlled and uniformly conducted experiment must form the datum-line from which the varying results of large-scale practice were to be plotted as deviations; not the reverse. He (Dr. Anderson) would further take

* *Trans. Inst. M.E.*, 1902, vol. xxiv., page 171.

occasion to emphasize a previous criticism that he had made regarding the choice of a standard coal for reference. It was a matter of some importance that there should be a series (three or more) selected by wellknown responsible and competent bodies from seams of gas-coal in use in different parts of the country, which were known, or believed, to be of fairly constant quality and composition over large areas. By means of carefully conducted large and small-scale experiments on these coals, a set of relationships might be established once and for all in respect of gas-yield, sperm-value, calorific power, etc., which would enable any one interested in the buying and selling of gas-coals to have immediate access to a well authenticated standard or series of standards.

Dr. G. P. LISHMAN wrote that Dr. Anderson's suggestion as to the selection of certain seams, by competent and responsible bodies, for use as standards in gas-coal testing, is an excellent one and one which is perfectly practicable. In this direction lies the road to that uniformity in results so much desired not only in the coal-trade but by the gas-companies too.

The CHAIRMAN (Mr. G. A. Mitchell) moved a vote of thanks to Dr. Lishman for his interesting and valuable paper.

Mr. T. L. GALLOWAY seconded the resolution, which was cordially approved.

Mr. HORACE F. BROWN's paper on "A New Process of Chlorination for mixed Gold- and Silver-ores" was read as follows :—

A NEW PROCESS OF CHLORINATION FOR MIXED GOLD- AND SILVER-ORES.

By HORACE F. BROWN, San Francisco, California, U.S.A.

In ordinary practice for the recovery of silver from its native ore, the first step, after crushing, is to produce a chloridizing roast, that is, to roast the ore with salt for the purpose of reducing all the silver to a chloride. Where the ore contains a great excess of sulphur, the usual practice is to roast out a portion of the excess before charging in the salt, leaving a sufficient amount to decompose the salt and to liberate the chlorine gas. Where the sulphur is but little in excess for this purpose, the salt is often crushed with the ore.

During the operation of this chloridizing roast, there is usually a very great, if not practically a total loss of the gold-contents, from the fact that gold chlorides, under high temperature, are constantly broken up and much of the gold-values pass off with the excess chlorine vapours as volatile gold or as impalpable powder. For this reason, ores carrying commercial gold-values in connection with silver, are not suitable for the silver-chloridizing process. Where the gold-values predominate and gold chlorination is resorted to, the ores are roasted "sweet" without the addition of salt, and then subjected to the action of chlorine gas or chlorinated water, either of which renders the gold solvent, but has no appreciable effect upon the silver.

Some years ago, the writer had occasion to erect a roasting furnace for chloridizing an ore carrying both gold and silver. The ore was crushed to pass a mesh of 30 wires per inch (dry-stamp mill) and first roasted "sweet," then salt and iron sulphides, finely pulverized, were charged in, near the discharge-end of the furnace. The salt and sulphides were so mixed that there was ə per cent. of free sulphur in proportion to the salt. This mixture was charged into the furnace by means of a long spoon, made by splitting a gas-pipe along the seam so as to form a trough long enough to reach across the furnace-hearth. The reverberatory furnace was mechanically rabbled, the rabbles being carried by wheeled carriers at intervals of about 45 seconds; and 80 pounds of ore were charged in at the passing of each carriage. The amount of salt in the chloridizing mixture equalled 4 per cent. of the ore, this amount proving to be ample, while with the usual practice of charging the salt with the ore, from 8 to 12 per cent. would have been required. The mixture was charged in at a point about 9 feet from the furnace-discharge, the ore being at a bright, red heat and roasted down to $\frac{3}{10}$ per cent.: practically a "sweet" roast. The chloridizing mixture was distributed in the spoon so that it made an even line across the hearth just in front of the moving rubbling-carriage, and was ploughed under and mixed with the ore within 2 or 3 seconds after being charged in. The charging was maintained continually. Within 30 minutes after being mixed with the chloridizing mixture, the ore was discharged into a pit under the hearth, in which it was allowed to stand for 24 hours. The results were very good—90 to 95 per cent. of both the gold and silver being soluble in a strong solution of hyposulphite of soda. Apparently no abnormal gold-losses were sustained.

In the same year, the writer had occasion to test the siliceous telluride ores of South Dakota, and under the same treatment, despite the greatest care, the gold-loss was over 50 per cent. the silver-loss being very small. He also made tests of ores, with the same results, in other localities, and he accepted the statements made in various metallurgical works that roasting gold-ores with salt was a very uncertain operation, as to economic results, and that it was certain to result in unusual losses.

To overcome this difficulty, the writer conceived the idea of chloridizing the hot calcined ore in a closed receptacle, using chlorine gas under pressure. In the laboratory, this process gave very satisfactory results, not only with the South Dakotan ores above referred to, but with such other ores as were available. The apparatus used would hold 5 pounds of hot ore, or about the amount that could be handled in the muffle used for roasting. Chlorine gas was developed by the usual laboratory-methods and admitted to the chloridizing apparatus under very slight pressure, the whole being allowed to stand over night. No loss whatever was sustained, as far as could be determined by laboratory-tests. The extraction by strong hyposulphite of soda, by cyanide of potassium and by amalgamation with mercury (grinding in a mortar), was equally good as to both the gold and the silver, being about 95 to 96 per cent. of the values in both metals. The original ore used for making the test varied from 8s. to £3 2s. (2 to 15 dollars) of gold, and from 10 to 30 ounces of silver per ton.

The results were extremely satisfactory, and opened out so large a field for the economic treatment of low-grade ore of mixed values that the writer believes that it will prove of importance to the mining industries, and he will, therefore, briefly describe the process as applied to mill-work.

Fig. 1 (Plate XXIII.) shows the end of a common, hand-operated reverberatory roasting-furnace, A, with a hopper, B, formed in the hearth and close to the furnace-end. In the hopper, the hot calcined ores accumulate in any desired quantity, and are kept hot by their proximity to the fire-box, C.

A series of sheet-metal shells or upright cylinders, E, holding say 2 tons each, are situated so that they can be charged by a car, F, through the top. As soon as the charge is inserted, the top is closed, except a small pet-cock, thus forming an air-tight receptacle. Through pipes, G, at the bottom, chlorine gas is forced into the cylinders. This gas being heavier than atmospheric air, at the same temperature, fills all the interstices, completely displacing the air. As soon as the presence of chlorine gas is detected at the top, the pet-cock, a, is closed, and the pressure continued as long as may be found necessary. Under the combined action of chlorine and heat, the silver-contents are rapidly changed into a chloride. Such gold chlorides as are formed will be at once broken up by the heat to assume other forms, to be again broken up until the gradual reduction of the temperature reaches 180° Cent. or below, when all the volatile gold in the chlorine-atmosphere will come down as a stable chloride. After this takes place, the ore can be safely discharged and will carry with it all the values that it originally contained, as none can possibly escape.

Ample provision can be made to draw off the excess of chlorine

gas, and to confine and collect the dust raised by charging
and discharging the cylinders, so that practically nothing is
lost. The chloridized ore may be discharged on to an ordinary
brick cooling-floor and afterwards charged into tanks, pans or
barrels, as may be desired, by manual labour, or, as shown in
Fig. 1 (Plate XXIII.) on to a cooling and conveying floor, *H*,
where it is mechanically cooled, conveyed and discharged into
the leaching tank, *I*, or other receptacle.

There are several ways of developing chlorine gas that can be
used, and probably local conditions would govern the choice of
the apparatus to be employed for this purpose. By means of a
suitable exhauster and storage-tank, the necessary chlorine
gas can be generated from manganese, chloride of sodium and
sulphuric acid, in the same manner as now practised in the
process of gold-chlorination, where the gases are forced through
a slightly moistened bed of ore.   The same object might be
accomplished by burning a mixture of iron sulphide and
common salt, previously finely pulverized, in a closed chamber;
by forcing hot air through the mass in sufficient quantities to
burn practically all the sulphur into sulphurous acid gas, this
would effect the liberation of the chlorine in practically an
anhydrous state, and it can be stored under any desired pressure
until required for use.

Where it can be obtained, the simplest method would be to
use liquid chlorine. It is sold in iron-cylinders, under a pressure of from 6 to 12 atmospheres, and is anhydrous and pure.
Being under such a high pressure, it might appear that it would
be very difficult to handle the chlorine so as to prevent its
sudden escape when the valves are opened: in practice, however, there is no difficulty from this source. The evaporation
of the liquid chlorine and its expansion to gas produce an
intense degree of cold, which reduces the evaporation, keeping
the flow of gas to a comparatively small stream.   To produce a
large flow of gas, it is necessary to warm the neck or discharge-
orifice of the cylinder, by means of a jet of steam, by the
application of cloths saturated with boiling water, or by large
volumes of hot water. By regulating the heat in this manner,
any desired pressure can be obtained.   One pound of liquid
chlorine will yield 5·31 cubic feet of gas at 70° Fahr., or a

pressure of about 2 atmospheres at 800° Fahr., which is approximately the temperature required for the chloridization of the metallic oxides in the cylinders.

Supposing the chloridizing receptacle to contain 40 cubic feet or 2 tons of fine ore. Finely crushed ore represents about 66 per cent. of solid and 34 per cent. of space between the ore-particles, so about one-third of the contents must be filled with gas; and 2⅔ pounds of liquid chlorine will give 14·54 cubic feet of gas, which at 800° Fahr., will give a pressure of about 14 pounds to the square inch in the receptacle. This gas will force out all the free air, and will bring each particle of ore into contact with the hot chlorine gas more perfectly than in an open roasting-furnace, where so much of the gas passes on to the chimney. The amount of chlorine gas that will be absorbed depends almost entirely on the mineral contents of the ore. A little experimenting, in each case, will determine the length of time during which it will be necessary to maintain the pressure of the gas for the perfect chloridizing of the silver.

Anhydrous chlorine gas has no effect upon metallic iron, except at a very high temperature. The sheet-metal receptacles are lined with thin cast-iron staves or segments joined with graphite-cement. The heat radiated through this lining is not sufficient to cause any action of the chlorine on the outer shell, and the gradual decomposition of the cast-iron forms a semi-graphitic coating, which very largely prevents further action of the gas on the lining. The reason for using a cast-iron lining to protect the chloridizing cylinders from immediate contact with the ore, instead of using fire-brick or other refractory material, is, that owing to its being a much better conductor of heat, the cooling of the cylinders will be greatly facilitated.

The cooling can be accomplished by mechanical means, if so desired; but the preferable practice is to provide enough receptacles, so that by the time they are all filled, the first will be ready for discharging and use again. The provision of these cylinders is about the only expense to be incurred in addition to the cost of the present mills, and is but trifling compared with the extra saving, and the certainty of working without volatile losses.

The gist of the writer's improved method of treating ores of mixed gold-and-silver values consists in chloridizing highly heated ore in a closed receptacle under pressure, and preventing gold-losses by confining the volatile products until cooled, so that stable chlorides are formed. The after-treatment of the chloridized ores will in all cases be the same as that at present practised.

For fine gold and for a prepared chloride of silver, cyanide of potassium is a perfect dissolvent, but a rather poor one for a natural chloride of silver. A strong solution of hyposulphite of soda dissolves the gold, and is naturally a perfect dissolvent for silver. Pan-amalgamation is often the best possible means for recovering values from chloridized ore. Local conditions and the nature of the metals will, however, determine the best manner of extraction in all cases.

There are many districts and many mines that produce complex ores, so low in grade that no one of the metals will yield a profit, but by combining (providing the expense is not increased) the value would be sufficient to make a commercial success. Copper and lead, as well as other base chlorides, can be washed out and recovered as bye-products, while the percentage of precious-metal values that will be recovered will be greater than by the usual practice, as no ore can be roasted in an open furnace with salt, without 10 to 12 per cent. of loss of silver, and often many times that of gold.

Fig. 1 (Plate XXIII.) shows enough detail to give an idea of the application of this process in a small mill. The method of application to cylinders and other mechanical roasters, on a larger scale, will be readily apparent without an illustration.

––––––

Mr. JOHN S. MACARTHUR (Glasgow) wrote that he did not quite understand the expression that "the salt and sulphides were so mixed that there was 5 per cent. of free sulphur in proportion to the salt." He understood that this meant that there was 5 per cent. of sulphur, in the form of iron-pyrites, above the equivalent of salt used, for example, if 58·5 parts of common salt were used, there would be (32 plus 5 per cent. or) 33·6 parts of sulphur present in the form of iron sulphides. As a gold-

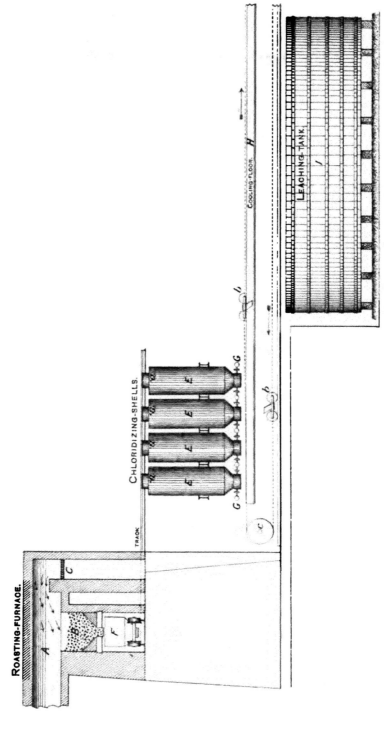

ROASTING-FURNACE.

CHLORIDIZING-SHELLS.

TRACK

COOLING-FLOOR. H

LEACHING-TANK.

extracting process, he did not see that this method presented any advantage over ordinary chlorination. In fact, it seemed unlikely that gold would combine directly with chlorine at any temperature, so long as there was no moisture present, and from the fact that no soluble gold was recorded as resulting from the laboratory-tests of this process, he inferred that no auric chloride had been formed. It was quite possible, however, that aurous (insoluble) chloride was present, although he thought that this was unlikely. As a method of chloridizing silver-ores, or gold-ores containing a notable proportion of silver, the method presented good points and laboratory-results would encourage the application of the process on a commercial scale.

The CHAIRMAN (Mr. G. A. Mitchell) moved a vote of thanks. to Mr. H. F. Brown for his valuable paper.

Mr. F. R. SIMPSON seconded the resolution, which was. cordially approved.

———

Mr. GEORGE A. STONIER's paper on "Graphite-mining in. Ceylon and India" was read as follows:—

GRAPHITE-MINING IN CEYLON AND INDIA.

By GEORGE A. STONIER, A.R.S.M.
LATE CHIEF INSPECTOR OF MINES IN INDIA.

According to figures given by Sir Clement Le Neve Foster in Part IV. of the *Mines and Quarries: General Report and Statistics,* the world's output of graphite for 1901 was nearly 77,100 tons, valued at about £785,000.   Ceylon furnished 29

FIG. 9.—SINHALESE OPENING THE BARRELS OF GRAPHITE.

per cent. of this quantity and 80 per cent. of the value, and India contributed 3 per cent.   The figures of export and value for 1902 are as follows:—Ceylon, 25,189 tons* of dressed ore valued at 10,516,366 rupees,† and India, 4,882 tons.

I. CEYLON.

Export Trade.—According to Mr. A. M. Ferguson,‡ graphite is mentioned in Sinhalese letters of the fourteenth century and

* Long tons of 2,240 pounds.   † A rupee is 1s. 4d.
‡ *Journal of the Royal Asiatic Society,* 1885, vol. ix., part ii., page 186.

in Dutch Government Records of 1675. British records for 1831 give figures of export, which must have commenced between 1820 and 1830, but it was not of importance until 1834, when it amounted to 129 tons, valued at 12,054 rupees. In 1869, the output reached 11,306 tons, valued at 889,620 rupees; in 1899, 31,761 tons, valued at 22,255,400 rupees; and in 1902 it was 25,189 tons, valued at 10,516,366 rupees. Details of the output for each year from 1834 to 1902 are given in Table I.

TABLE I. – OUTPUT AND VALUE OF GRAPHITE IN CEYLON FROM 1834 TO 1902.

Years.	Quantity.			Value.		Years.	Quantity.			Value.	
	Cwts.	Qrs.	lbs.	Rupees.	Cts.		Cwts.	Qrs.	lbs.	Rupees.	Cts.
1834.	2,582	2	14	12,054	00	1869.	226,131	3	8	889,620	00
1835.	4,952	3	0	11,082	50	1870.	85,248	3	18	345,622	00
1836.	12,644	0	5	14,663	50	1871.	125,257	1	5	620,953	50
1837.	3,700	0	0	4,293	00	1872.	136,051	2	23	438,366	64
1838.	1,164	1	12	1,379	00	1873.	173,996	0	17	1,479,395	44
1839.	423	1	18	490	00	1874.	149,938	1	3	1,440,166	87
1840.	981	0	0	1,225	00	1875.	110,023	1	0	1,100,232	53
1841.	2,002	2	7	2,684	50	1876.	117,361	1	2	1,173,612	64
1842.	7,285	0	3	12,317	00	1877.	96,792	1	21	967,924	37
1843.	3,677	3	20	5,238	50	1878.	84,634	3	15	846,348	84
1844.	9,914	3	21	12,946	00	1879.	162,495	2	24	1,624,957	15
1845.	19,245	0	15	24,519	50	1880.	205,738	2	9	2,057,385	81
1846.	25,036	3	7	30,361	00	1881.	259,909	0	16	2,599,091	42
1847.	9,248	3	11	10,583	50	1882.	260,166	1	6	2,601,663	3
1848.	6,787	0	0	7,062	00	1883.	262,773	3	1	2,627,737	58
1849.	3,329	2	20	3,302	00	1884.	182,425	3	10	1,824,258	40
1850.	23,021	1	6	38,330	00	1885.	196,399	2	23	1,963,997	06
1851.	23,865	1	2	52,554	00	1886.	241,760	0	13	2,417,601	15
1852.	13,110	1	21	26,281	00	1887.	238,599	3	0	2,385,997	50
1853.	19,577	2	25	40,572	00	1888.	223,277	3	4	2,232,777	82
1854.	17,451	2	19	39,162	00	1889.	486,138	3	1	4,861,387	59
1855.	6,129	3	16	11,448	50	1890.	392,577	2	13	3,925,776	16
1856.	13,380	2	27	33,380	00	1891.	400,540	0	15	4,005,401	34
1857.	33,497	0	4	83,850	00	1892.	430,666	3	20	4,306,669	28
1858.	19,432	3	12	33,841	50	1893.	332,168	3	16½	2,491,266	72
1859.	17,510	3	11	41,138	00	1894.	335,168	0	24	2,513,761	60
1860.	75,660	0	23	239,535	50	1895.	326,754	1	16	2,450,657	93
1861.	38,345	1	23	110,643	50	1896.	361,061	1	13	3,069,021	62
1862.	40,895	3	13	130,789	50	1897.	379,415	2	21	3,670,846	78
1863.	65,128	0	3	281,246	00	1898.	478,318	0	2	7,174,770	27
1864.	84,028	2	4	404,314	50	1899.	635,224	1	5	22,255,400	77
1865.	40,143	3	5	151,206	00	1900.	391,699	3	6	9,792,495	09
1866.	56,278	3	14	218,605	50	1901.	446,960	-	—	9,609,642	—
1867.	45,836	0	14	193,601	00	1902.	503,778	-	—	10,516,366	—
1868.	141,095	0	14	720,410	50	Totals	10,326,820	2	23½	125,286,282	90

NOTES.—The figures for the years 1834 to 1894 are taken from the *Journal of the Royal Asiatic Society*, 1885, vol. ix., part ii., page 240; and for the remaining years from the blue-books of the Ceylon Government. From 1836 to 1845, and from 1858 to 1868 inclusive, the Customs duty was 2½ per cent. *ad valorem*; from 1846 to 1857, inclusive, there was no Customs duty, by No. 9 ordinance of 1847; and from 1873 to 1883, inclusive, licenses to dig for plumbago cost 10 rupees each, with a royalty of 10 per cent. of the value of the plumbago dug, in the case of the Western province.

Of the total amount, 96½ per cent. is exported from Colombo and the remaining 3½ per cent. from Galle.

The foreign countries, to which graphite was despatched during the years 1885 and 1902, are shown in Table II.: the large increase in the amounts sent to the United States, Germany and Belgium will be noted.

TABLE II —EXPORTS OF GRAPHITE IN THE YEARS 1885 AND 1902.

Destination.	1885. cwts.	1902. cwts.	Destination.	1885. cwts	1902. cwts.
Great Britain ...	136,964	135,471	India	306	739
United States ...	54,891	272,219	Sweden	—	473
Germany	1,199	68,445	Turkey in Asia .	- .	103
Belgium ...	400	19,566	Italy ...		82
France	1,237	1,827	New Zealand ...	-.	24
Australia	1,176	1,497	Straits Settlements	—	1
Holland	—	1,157	Austria	226	—
Japan	—	1,100			
Russia	—	1,074	Totals ...	196,399	503,778

FIG. 10. · SCREENS FOR GRAPHITE.

General Features.—The island of Ceylon, a British Crown Colony divided into nine provinces, is pear-shaped in form, with a length of 271 miles from north to south and a maximum breadth of 139 miles. It has a central core of mountains rising up to 8,296 feet in height, a plain on the north occupying nearly half the area of the island, and flat or low undulating country on the east, west and south.

Occurrence of Graphite.—Graphite occurs in the western and south-western portion of the island, chiefly in the Western and Southern provinces and Sabaragamuwa.   The mineral area is 95 miles long in a north-and-south direction, with a width of 35 miles at the northern and 43 miles at the southern end.   There is one well-defined north-and-south belt, 18 miles from the coast and 5 miles wide at the northern end, touching the coast-line at the southern extremity where it is 20 miles wide.   A second, fairly well-defined belt is 40 miles in length and 4 miles in maximum width.   The payable mineral occurs in veins travers-

FIG. 11.—POLISHING GRAPHITE ON SACKING.

ing a normal granulite* with red garnets.   These veins generally have a small hade.   The hanging-wall is frequently well-defined, roughly polished and occasionally striated more or less horizontally.   The strike varies: in the Southern province it is generally meridional in direction, and in the northern end of the field it is frequently east and west.   No evidence of a main lode or of a series of lodes has been discovered, and their extension horizontally is limited.   Well-defined veins suddenly pinch out, and although one vein has been proved to

* "Ceylon Rocks and Graphite," by Mr. A. K. Coomára-Swámy, *Quarterly Journal of the Geological Society of London*, 1900, vol. lvi., page 590.

a depth of 720 feet, it is very doubtful whether they are true-fissure in character. . Fig. 3 (Plate XXIV.) shows the sudden disappearance of the veins along the strike, at a depth of 216 feet. The veins vary from a thin flucan to a width (including a horse) of 8 feet. The largest mass of graphite, yet discovered, is said to have weighed nearly 6 tons. A vein 4 inches thick is considered to be payable.

It seems to be clear that fissures were formed and then the graphite, quartz, etc., were deposited in the cracks. The quartz may have been derived from a siliceous fluid, and the graphite

FIG. 12.--POLISHING GRAPHITE ON A SCREEN.

introduced by sublimation,* not of the carbon itself but of hydro-carbons. A deposition of graphite-like material is often found in the cracks of the upper layers of coke made in closed ovens : the red-hot coke apparently robs the hydrocarbons (distilled from the uncoked coal below) of its carbon, which is deposited as a silvery-white layer. A somewhat similar substance is found in the flues of the retorts of gas-works. In Bengal, the coal, coked by mica-peridotite dykes, sometimes presents a graphitic lustre.

* " Beitrag zur Kenntniss der Gesteine und Graphitvorkommnisse Ceylons," by Mr. Max Diersche, *Jahrbuch der kaiserlich-königlichen geologischen Reichsanstalt*, 1898, vol. xlviii., page 231.

Graphite-mining.—In all, about 300 mines and quarries are at work, and are estimated to give employment to 10,000 persons. With the exception of three mines, they are all worked by natives of Ceylon, and, more or less, in a native fashion. European methods have been tried, but have generally failed on account of the inexperience of the manager, or because the company was unfortunate in the site of operations. Graphite-mining, like mica- and gem-winning, is very uncertain, but as a native owner's costs are low, he can afford to allow a mine to stand idle and await a rise in price. His methods are of two

FIG. 13.—WASHING GRAPHITE IN A PIT.

kinds:—If the ground is hard, the shoot of mineral is followed as far as water will allow; and, in a few cases, a shaft is sunk to the water-level, and the vein worked up to the surface. The curious ramifications of a mine worked by natives are shown in Fig. 4 (Plate XXIV.). In soft ground, a vertical pit, rectangular in section, is sunk for about 60 feet, and the mineral is followed in a series of winzes, 50 or 60 feet deep (Figs. 5 and 6, Plate XXIV.).

The mineral is wound up the shafts and winzes in barrels attached to each end of a locally-made rope (or occasionally an iron chain) which has two or three turns passed round a

wooden jack-roll (*dabare*), 7 feet long and 1 foot in diameter,
with iron handles (32 inches long, with cranks 13 inches long)
at each end : it is worked by six or seven men.

Round timber is used for securing the sides of the shafts
and winzes. The two end-pieces (*mukas*) are kept in position
by two side-pieces (*digangs*) notched out to fit against the round
mukas, and the *digangs* are kept apart and strengthened by three
dividers (one midway and the others at the ends) similarly
hollowed out. Horizontal boards, 8 inches wide and 1¼ inches
thick, and packing, or occasionally sticks with ferns (*kekilla*), are

Fig. 14. Graphite, being Sun-dried, on the Barbacue or Dressing-floor.

used to jam the set in position. Stulls are not used to support
the sets. The kibbles or barrels run on casing-boards nailed to
the dividers. There is no proper ladder-way, and the men
climb from set to set by the aid of a rope or a thin pole lashed to
the dividers.

Small fans, worked by hand, are frequently used to ventilate
the mines, but no endeavour is made to course the air except in
mines under European management.

The timber used is chiefly *alubo* and *hora*.

The labour employed is almost entirely Sinhalese, from Galle.
Only men are employed underground, but occasionally women

work at the surface. Men earn 8d. to 1s. (50 to 75 cents) per day, and women receive 3d. to 6d. (20 to 40 cents). Tamils, from Southern India, are employed on the tea-estates in Ceylon, but they have not taken to mining as a regular occupation.

The mineral is conveyed in bags by coolies or bullocks to a dressing-shed, where it is roughly picked, and packed in barrels for transit by road and rail or canal to Colombo or Galle for dressing.   The barrels, 22 inches in diameter at the ends and 3 feet long, are made of *hora*-wood (*Dipterocarpus zeylanicus*) and bound with four hoop-iron bands.

FIG. 15.—FLYING OR VANNING THE GRAPHITE-DUST.

Dressing.—On arrival at Colombo, the barrels are opened (Fig. 9) by Sinhalese men on an unroofed brick or asphalt dressing-floor (*barbacue*) rectangular in shape, averaging 30 feet wide and 80 feet long (at Mr. John Kotawala's sheds) and sloping to two sides for drainage. The big lumps are put on one side, and the remainder is carried in conically-shaped baskets, 17 inches in diameter at the base and 11 inches deep, and is thrown on to a series of stationary screens (with holes $\frac{5}{8}$, $\frac{3}{8}$, $\frac{1}{4}$ and $\frac{3}{16}$ inch in diameter, 10 holes to the inch and No. 60 mesh) 3 feet long and 2 feet wide, set at an angle of about 35 degrees from the horizontal (Fig. 10). The screened pieces are taken to sheds,

near by, which are open at the sides and roofed with coconut
leaf (*Cadjan, Phœnix zeylanica*), and women chop them up with
small iron-hatchets (Figs. 7 and 8, Plate XXIV.) and remove
the coarser impurities, such as quartz.   The valuable mineral
is then placed on sacking on boards or on the *barbacue*, water
is added, and the pieces are rubbed by hand (Fig. 11).   The final
polish is done by hand, on a screen which is placed flat on the
ground (Fig. 12).   At one time coconut-husks were used for
polishing.

FIG. 16.—FLYING OR VANNING THE GRAPHITE-DUST.

The poor material is reduced to powder by wooden cylindrical
mallets (3 inches in diameter and 5 inches long, with wooden
handles 1 foot long) or cylindrical beaters (2 inches in diameter
and 14 inches long, with the end reduced in size to form a
handle), and is hand-picked on sacking, 27 inches long and 20
inches wide, tacked on two strips of wood for carrying purposes.

At some establishments, further concentration is effected
by washing in a pit, 5½ feet long, 3½ feet wide and 2½ feet
deep (Fig. 13).   A circular motion is given to the mineral in
a saucer-shaped basket, immersed in the water of the pit, and
the graphite passes into the latter, while the heavier particles
remain behind : the graphite is sun-dried on the *barbacue*

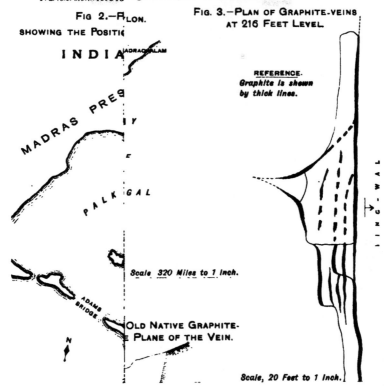

FIG 2.—Pₗₒₙ.

SHOWING THE POSITIC

INDIA
ᴬᴰᴿᴬᴰᴬᴸᴬᴹ

MADRAS PRES

ʏ

ᴇ

PALK GAL

Scale 320 Miles to 1 Inch.

ADAMS BRIDGE

N

OLD NATIVE GRAPHITE-
PLANE OF THE VEIN.

FIG. 3.—PLAN OF GRAPHITE-VEINS
AT 216 FEET LEVEL

REFERENCE.
*Graphite is shewn
by thick lines.*

IING. WAL

Scale, 20 Feet to 1 Inch.

(Fig. 14). To separate the very fine material, the powdered mineral is placed in a basket, 13 inches long, which is rectangular in section (12 inches by 4 inches), with the corners rounded at one end, and tapering to a line at the other. When the mineral is thrown into the air, the heavier particles fall back into the basket, but the fine graphite is blown forward and falls to the ground (Figs. 15 and 16).

The mineral, despatched to market, is classified according to size as lumps, ordinary, chips, dust and flying dust, and according to quality as x, xb, good b, b, bc and p.

The writer's thanks are due to Messrs. Hunter & Gill for the kind help afforded to him while visiting the mines, and in collecting the information given in these notes.

Figs. 1 and 2 (Plate XXIV.) show the positions of the mines and quarries in India and Ceylon. Fig. 3 (Plate XXIV.) is the plan of a graphite-vein at a depth of 216 feet. Fig. 4 (Plate XXIV.) is the section of an old graphite-mine. Figs. 5 and 6 (Plate XXIV.) are the section and plan of a graphite-mine now in operation. Figs. 7 and 8 (Plate XXIV.) are the side and end-elevations of an iron hatchet used to chop up pieces of screened graphite. Figs. 9 to 16 show the methods of dressing the graphite-ore at Colombo.

———

The CHAIRMAN (Mr. George A. Mitchell) moved a vote of thanks to Mr. G. A. Stonier for his interesting paper.

Mr. S. F. WALKER seconded the resolution, which was cordially approved.

———

Mr. A. R. SAWYER's paper on " The South Rand Gold-field, Transvaal," was read as follows :—

THE SOUTH RAND GOLD-FIELD, TRANSVAAL.

By A. R. SAWYER.

INTRODUCTION.

The writer read a paper before The Institution of Mining Engineers in 1897, on "The South Rand Coal-field, and its connexion with the Witwatersrand Banket Formation."[*]   He described the rocks of that formation which had then been located beneath the Coal-measures and underlying Breccia deposits.   In the discussion which followed, the writer stated that prospecting was proceeding:[†] this referred to the western portion of the gold-field, which, in part, underlies the coal-field.

The section (Fig. 2, Plate XXV.) shows the results obtained since the abovementioned paper was read, but the work was temporarily stopped in 1899, owing to the war.   The line of this section is shown on the plan (Fig. 4, Plate XXVI.).   Nos. 18, 19, 20 and 21 bore-holes were put down between March 14th 1898 and July 3rd 1899, when the work was unfortunately discontinued.

The writer has further described the granite-mass, which underlies this gold-field.[‡]

OVERLYING FORMATIONS.

The "Remarks on some Granite-masses of the Transvaal," referred to above, contains an account of some of the rocks of the overlying formations intersected in Nos. 19, 20 and 21 bore-holes.

It will be noticed in the section (Fig. 2, Plate XXV.) that the dolerite-sheet shares the same bend as the underlying coal-

---

[*] *Trans. Inst. M.E.*, 1897, vol. xiv., page 312.

[†] *Ibid.*, 1898, vol. xv., page 43.

[‡] "Remarks on some Granite-masses of the Transvaal," by Mr. A. R. Sawyer, *Transactions of the Geological Society of South Africa*, 1903, vol. vi., page 47.

seam; and this is strong evidence that (as the writer had previously pointed out) this sheet is subaërial or interbedded. The section also appears to indicate an unconformity between the Karroo beds and the Breccia or Dwyka beds. There was no unconformity visible in the core-sections between the Karroo beds and the underlying Breccia; they appear there to merge one into the other. The writer has included under the term "Breccia," all beds, whether containing pebbles or not, from the first pebble-bed struck down to the bottom. He also desires to draw special attention to the reddish, purple and chocolate coloration of some of the beds included in the Breccia deposit.

Igneous rock occurs in the Breccia in Nos. 18 and 20 bore-holes, but it is absent in No. 19 bore-hole. If correctly connected in the section (Fig. 2, Plate XXV.) this rock would appear to be intrusive. It is nearly 8 feet thick in No. 20 bore-hole, and 10 feet thick in No. 18 bore-hole. A microscope-slide of this rock at No. 20 bore-hole shows that it is an augite-andesite. The ground-mass is composed of small felspar crystals, granular augite, porphyritic crystals of labradorite, and a quantity of interstitial matter. It is a hard and heavy rock with calcite veins.

With regard to the suggested glacial origin of the Breccia beds, the writer may remark that scratchings have not been detected on any of the pebbles recovered in the bore-holes, but he had seen scratched pebbles from a shaft at Vereeniging. Owing to its importance in this connection, the section of the old shaft at Vereeniging, contained in the writer's paper on "The South African Coal-field"[*] is reproduced in Table I. It will be noted that the coal-seam lies immediately upon the Dwyka formation.

The coal-seam, which at the South Rand colliery and for a considerable distance around it is so wonderfully thick, thins out towards the region of the section (Fig. 2, Plate XXV.). There is also evidence that the coal-seam thins out towards the Orange River Colony: it therefore follows that the South Rand colliery possesses an unique tract of coal of considerable importance to the gold-mining and other industries.

[*] *Transactions of the North Staffordshire Institute of Mining and Mechanical Engineers*, 1889, vol. x., page 161.

TABLE I.—SECTION OF STRATA SUNK AND BORED THROUGH IN THE OLD PIT, VEREENIGING.*

No.	Description of Strata.	Thickness of Strata. Ft. In.	Depth from Surface. Ft. In.	No.	Description of Strata.	Thickness of Strata. Ft. In.	Depth from Surface. Ft. In.
	Karoo Beds: Shaft.				regularly formed, but rounded, polished, and striated, found on the floor, and partly in the coal. say	1 0	58 0
1	Water-worn gravel	2 0	2 0				
2	Sandstone, containing fossils, including Lepidodendron and Farularia† ...	3 0	5 0				
3	Loose sandstone and ferruginous clay ...	25 0	30 0	11	Black shale, slightly gritty, changing into sandstone ...	0 2	58 2
4	Grey sandstone ...	4 0	34 0				
5	Grey shale	8 0	42 0	12	Quartz-conglomerate	4 0	62 2
6	Black shale ...	3 0	45 0	13	Fire-clay	10 0	72 2
7	COAL	2 6	47 6	14	Conglomerate ...	10 0	82 2
8	Black shale... ...	4 0	51 6	15	Fire-clay	8 0	90 2
9	COAL	5 6	57 0	16	Sandstones, shales and carbonaceous matter	—	—
	Dwyka Conglomerate: Borehole.						
10	Basalt stones, ir-						

WITWATERSRAND AND HOSPITAL HILL SERIES.

With the exception of outcrops of rocks of the Hospital Hill Series, in which the Hospital Hill Slate is well developed, in the Orange River Colony (Fig. 4, Plate XXVI.) the rocks, other than those of the Karroo system and of diabase which outcrop within the area of the plan, belong to the Witwatersrand Series. Faulting occurs, as is shewn on the plan (Fig. 4, Plate XXVI.) and section (Fig. 2, Plate XXV.). The Witwatersrand formation, on the surface at Witkleifontein, and as revealed by boreholes at Boschkop and Braklaagte, is as regular as on the Rand or elsewhere. The dip of the strata in Nos. 18 and 20 boreholes, those nearest the granite-mass, is 45 degrees.

A dyke occurs in No. 19 bore-hole, accompanied by faulting, as shewn by the change of dip after passing below the dyke. West of this dyke, the beds dip at an angle of 15 degrees for some distance, they then flatten and resume the dip of 15 degrees beneath the amygdaloidal diabase, which occurs at the western boundary of Witkleifontein (Fig. 4, Plate XXVI.). The beds below the dyke in No. 19 bore-hole, dip at an angle averaging 25 degrees, and bend round, in the writer's opinion, as shown on the section (Fig. 2, Plate XXV.).

* *Transactions of the North Staffordshire Institute of Mining and Mechanical Engineers*, 1889, vol. x., page 167.

† "Mining at Kimberley," by Mr. A. R. Sawyer, *Transactions of the North Staffordshire Institute of Mining and Mechanical Engineers*, 1889, vol. x., page 85.

In addition to the numerous reefs, indicated on the section, a series of pebbly beds occurs in No. 19 bore-hole, between the depths of 1,471 feet and 1,827 feet.

The rocks intersected in No. 20 bore-hole are particularly interesting. The upper slate-bed is slightly siliceous in part, and it contains calcite veins and much pyrites, with numerous slickensides. The quartzite-bed contains highly indurated slate-bands, contorted and schistose in part. Quartz and calcite veins occur, together with joints containing green talc. The quartzite is dark in part. A piece of the core from this quartzite showed the following characteristics under the microscope:—A few mosaics of quartz and secondary felspar, numerous shreds and needles of a secondary hornblende, numerous granules of pyrites, and bands of polysynthetic quartz. It resembled mylonite.

The lower slate-bed consists of indurated slate, with numerous quartz and quartzite-laminæ; some of the quartzite and some of the slate is black. There are calcite veins, and much pyrites occurs throughout.

CORRELATION.

From Blue Sky eastwards, and through the Heidelberg district, several more or less thick beds of slate occur among the quartzites, the formation thus differing to some extent from that found in the Central Rand. The correlation of the various reefs in those districts with those in the Central Rand is therefore not always an easy matter.* The rocks and reef series on the East Rand vary approximately, as represented in some bore-hole sections, as shewn in Table II.

TABLE II.—SECTION OF STRATA ON THE EAST RAND.

No.	Description of Strata.	Thickness of Strata. Feet.	Total Thickness of Strata. Feet.
1	Kimberley Series, up to	330	330
2	Quartzites, from 114 feet to	200	530
3	Slate, where it is very thick, a few thin bands of quartzite are found, up to	600	1,130
4	Quartzite, with a few thin pebble-beds in places, from 50 feet to	350	1,480

* *Transactions of the South African Institute of Engineers*, July 29th, 1903; and "The Northern Extension of the Witwatersrand Gold-field," by Mr. A. R. Sawyer, *Transactions of the North Staffordshire Institute of Mining and Mechanical Engineers*, 1889, vol. x., page 124.

TABLE II.—*Continued.*

No.	Description of Strata.	Thickness of Strata. Feet.	Total Thickness of Strata. Feet.
5	Slate, interbedded with quartzite ; there are from two to three beds of slate, but in one instance the bed is all slate, from 108 feet to	372	1,852
6	Sandstone or quartzite, from *nil* up to	305	2,157
7	Amygdaloidal diabase, from *nil* up to	160	2,317
8	Bird Reef Series, from 90 feet to	416	2,733
9	Quartzite, with a few thin pebble-beds, from *nil* to	310	3,043
10	Slate, from *nil* to	27	3,070
11	Quartzite, including 68 feet of dolerite in one instance, from 156 feet to	597	3,667
12	Livingstone Series, from *nil* to	113	3,780
13	Quartzite, in one instance described as 113 feet thick including diabase, from 310 feet to	445	4,225
14	Reef or reefs in quartzite, generally assumed to represent the full thickness of the Main Reef Series, in one instance represented as 283 feet thick with dykes, from 2 feet to	49	4,274
15	Quartzite, with a zone of small pebbles in the centre, from *nil* to	200	4,474
16	Slate, quartzose rock and quartzite : the greatest thickness proved is	650	5,124

Table III. contains a description and gives the thickness of the rocks encountered below the reef, in the case of the greatest depth reached below the known reef on the East Rand.[*]

TABLE III.—SECTION OF STRATA BELOW THE LOWEST REEF ON THE EAST RAND.

No.	Description of Strata	Thickness of Strata. Ft.	Ins.	Total Thickness of Strata. Ft.	Ins.
1	Greenish talcose clay-slate, auriferous	2	7	2	7
2	Quartz-grit, auriferous	4	5	7	0
3	Quartz-grit	25	0	32	0
4	Quartzose rock	114	0	146	0
5	Slate	42	0	188	0
6	Quartz-grit, auriferous	1	0	189	0
7	Slate	1	4	190	4
8	Quartzose rock	2	10	193	2
9	Slate, alternating with grit	5	0	198	2
10	Quartzose rock	13	0	211	2
11	Quartzose rock, interbedded with slate	159	0	370	2
12	Quartzose rock, interbedded with quartzite ...	262	0	632	2
13	Slate	0	9	632	11
14	Quartzose rock, interbedded with quartzite ...	18	0	650	11

[*] "The Origin of the Slates occurring on the Rand and in other African Gold-fields," by Mr. A. R. Sawyer, *Transactions of the Geological Society of South Africa*, 1903, vol. vi., page 70.

The strata passed through below the Main Reef Series in the Bezuidenville bore-hole, Central Rand, were :—Quartzite, 400 feet; slates, shale, 78 feet; a total bore-hole thickness of 478 feet.* The dip varied from 20 to 25 degrees, and the actual thickness of the rocks exposed below the Main Reef Series would be only about 440 feet. The Rand colliery bore-hole, on the East Rand, after passing through a series of reefs, presumably belonging to the Main Reef Series, intersected further reefs, a very thick one being struck about 300 feet lower down.

It will be seen from Table II. that the Bird Reef Series is fairly thick. It consists of zones of pebbly beds with small pebbles. There is some resemblance between the thick pebble-zone encountered in No. 19 bore-hole and the Bird Reef Series, on the East Rand; and quartzite and slate lie below it, as on the East Rand. It would be premature to give further particulars, especially as to the gold-contents of the reefs so far encountered.

It is highly probable that the strata which occur in the region of Nos. 18 and 21 bore-holes include the same horizon as that of the Main, Van Ryn or Nigel Reef.

It will be admitted that, besides being an interesting section geologically, it is one of great promise industrially.

The Granite-mass or Bed-rock.

Dr. G. A. F. Molengraaff, who has made a special study of the Vredefort granite-mass in the Orange River Colony, states that this mass is intrusive.† The writer has traced this granite-mass further into the Orange River Colony.‡ This is the only granite-mass respecting which any authoritative statement as to its intrusive character has been made. The writer is not referring to the red granite—a much younger and intrusive granite—occurring north of Pretoria, and described by Dr. G. A. F. Molengraaff.

* "The Drilling of the Bezuidenville Bore-hole, near Johannesburg," by Mr. J. A. Chalmers, *Transactions of the Institution of Mining and Metallurgy,* 1896, vol. v., page 86.

† "Remarks on the Vredefort Mountain-land," by Dr. G. A. F. Molengraaff, *Transactions of the Geological Society of South Africa,* 1903, vol. vi., page 20.

‡ "Remarks on the South-eastern Extension of the Vredefort Granite-mass," by Mr. A. R. Sawyer, *Transactions of the Geological Society of South Africa.* 1903, vol. vi., page 75.

This granite-mass is one of several which occur in the Transvaal. The writer stated in his paper on the South Rand coal-field that it was "distinctly anterior to the Banket formation."[*] The writer has always been of opinion with regard to the granite masses lying east of Heidelberg, and the one lying north of Johannesburg,[†] that they formed part of the basement-granite, that it had been raised through earth-movements, affecting the overlying Hospital Hill and Witwatersrand beds, which had been thereby thrown into synclines and anticlines.[‡] This is the sense in which the writer used the word "protrusion" as distinct from "intrusion" in his "Remarks on some Granite-masses of the Transvaal";[§] and, as the context clearly shows, the term "bulging out," or "protuberance," would perhaps convey the writer's meaning better than "protrusion."

General Remarks.

The Coronation, a banket reef of great richness so far as opened out, has been discovered, to the north of the South Rand gold-field on the Vlakfontein syncline[||] (Fig. 1, Plate XXV.), north of Vlakfontein station. A valuable reef has also been discovered by boring on the Daaspoort farm, south of Vlakfontein station, in what may be termed the northern extremity of the South Rand gold-field. Other gold-bearing reefs also occur, and have been more or less explored.

There can be no doubt that this gold-field will, in time, when labour is more abundant than it now is, prove to be a gold-producer of considerable importance.

The plan (Fig. 4, Plate XXVI.) indicates the course, for a distance of about 13 miles, of a quartz-felsite dyke, the felspar of which has disintegrated in places into kaolin.[¶] The plan also shows the probable trend of an anticline, beneath the Karroo

* _Trans. Inst. M.E._, 1897, vol. xiv., page 323.

† "The Northern Extension of the Witwatersrand Gold-field," by Mr. A. R. Sawyer, _Transactions of the North Staffordshire Institute of Mining and Mechanical Engineers_, 1889, vol. x., plate X. _bis_, page 132.

‡ "Remarks on the Anticlinal Theory in Connection with Rand Deposits," by Mr. A. R. Sawyer, _Transactions of the Geological Society of South Africa_, 1899, vol. v., page 45.

§ _Transactions of the Geological Society of South Africa_, 1903, vol. vi., page 47.

Trans. Inst. M.E., 1897, vol. xiv., page 326, plate XIV.

¶ _Transactions of the Geological Society of South Africa_, 1903, vol. vi., page 48.

beds, accounting for the varying dip of the Hospital Hill and Witwatersrand beds on each side of the line representing its course.

A bed of fire-clay, 3 feet thick, was intersected in No. 15 bore-hole: good bricks might be made from it. It is probably the final decomposition-product of an igneous rock, occurring as a sheet. Excellent bricks are made from decomposed diabase at Eastleigh farm in the Bezuidenhout valley, east of Johannesburg.

The cost of boring the deeper holes was about 18s. per foot, including depreciation of drill, crowns, etc. Carbons were then cheaper than they are now, namely, £7 9s. per carat; the present price being about £14.

The Hospital Hill beds crop out in the Orange River Colony, near the junction of the Vaal and Wilge rivers, and dip to the south. Prospecting work through the Karroo beds, along the line CD, (Fig. 4, Plate XXVI.) has proved the occurrence of the Witwatersrand beds dipping regularly southward, at an angle varying from 25 to 30 degrees.

It will be seen by reference to the geological plan, which accompanied the writer's paper on "The South Rand Coal-field, and its connexion with the Witwatersrand Banket Forma-tion "[*] that at Hex river, the Hospital Hill and Witwaters-rand beds dip regularly to the northward. This points, in the writer's opinion, to the probable continuation of an assumed anticline taking the course shown in Fig. 4 (Plate XXVI.). The writer calls this the " Villiers anticline," from the name of a township in the Orange River Colony, on the Vaal river, about midway between the Wilge and Hex rivers. This anticline would, of course, only affect the Hospital Hill and Witwaters-rand beds, the Dwyka and Karroo beds having been deposited long afterwards, and assuming generally a more or less horizontal position. It is remarkable, however, that the Karroo beds in a freestone-quarry, shown on the writer's plan of the South Rand Coal-field,[†] and lying on the line of this assumed anticline, dip at an angle of 30 degrees, a most unusual occurrence.

From this new discovery it would appear that the Orange River Colony may have a " Rand " of its own, equally long, with

[*] *Trans. Inst. M.E.*, 1897, vol. xiv., page 326, plate XIV.
[†] *Ibid.*

the same strike and the same dip to the southward: the only difference being that there it would be covered by Karroo beds. These parallel synclines and anticlines point to a lateral thrust from the south, as indicated in the writer's "Remarks on the Anticlinal Theory in connection with Rand Deposits."[*]

In connection with the anticlinal theory, the writer had pointed out on previous occasions that, in his opinion, the Witwatersrand syncline does not form an uninterrupted curve from Johannesburg to Heidelberg, and that there is a possibility of the occurrence of folds and faults having the effect of bringing up the Main Reef Series, in part, to a workable depth.[†] This view had been amply corroborated by explorations on the Rand since that date. He had notified the occurrence of the Doornkop anticline,[‡] and other anticlines occur in the Witwatersrand beds. It was therefore reasonable to suppose that similar folds may occur beneath the large area of superincumbent strata lying between Johannesburg and Heidelberg, more especially to the west, where the distance between the Witwatersrand beds is so much greater than it is to the east (Fig. 4, Plate XXVI.).

The sketch-section (Fig. 3, Plate XXV.), taken between Langlaagte and Kaffirskraal and thence to Witkleifontein, shows the Klipriversberg amygdaloidal diabase sheet, which unconformably overlies the Witwatersrand beds. This sheet is exposed over a large stretch of country along the line of this section. It is exposed for a distance of $9\frac{1}{2}$ miles to the southeastward, and very nearly for the same distance to the northwestward. The occurrence of a gap, that is unconformity, between the amygdaloidal diabase and the Black Reef Series has been pointed out by the writer and others. Dr. F. H. Hatch pointed out that this gap was very great. A consideration of the facts indicated on the sketch-section (Fig. 3, Plate XXV.) tends to the conclusion that this gap was so great, that a con-

---

[*] *Transactions of the Geological Society of South Africa*, 1899, vol. v., page 45.

[†] "Remarks on the Banket Formation at Johannesburg, Transvaal," by Mr. A. R. Sawyer, *Trans. Inst. M.E.*, 1895, vol. ix., page 360; and "Remarks on the Anticlinal Theory in Connection with Rand Deposits," by Mr. A. R. Sawyer, *Transactions of the Geological Society of South Africa*, 1899, vol. v., page 45.

[‡] *Transactions of the Geological Society of South Africa*, 1896, vol. ii., page 15.

siderable extent of the sheet had been removed by denudation, as shown on the section, before the deposition of the Black Reef Series upon it.

The amygdaloidal diabase sheet occurs east of Meyerton station, 4¾ miles away, on the Vogelfontein farm, No. 317. There is also a sheet of diabase, about 3 feet thick, lying on the amygdaloidal diabase, and the plane of contact between these two portions of the Klipriversberg amygdaloidal diabase dips slightly eastward.

The writer is therefore of opinion that a bore-hole put down at the point indicated on the plan and sketch-section (Fig. 3, Plate XXV., and 4, Plate XXVI.) will not pass through the full thickness of the amygdaloidal diabase sheet which is about 3,000 feet.

A large area, lying to the west of the railway between Klipriver and Meyerton stations, is covered by Pretoria beds, forming the eastern end of the Gatsrand.   At their northern escarpment, these beds dip about 25 degrees southward; on their eastern escarpment, about 3 miles west of Klipriversberg station, they dip about 22 degrees westward; and farther to the southwest, they dip northward.   A thick sheet of diabase, forming high hills, overlies the quartzite-bed which covers the shale-beds within this area (Fig. 4, Plate XXVI.).

The basement-granite has been elevated, and is exposed north of Johannesburg and east of Heidelberg.   The writer has also located it south of Heidelberg, exposed so far as the Witwatersrand beds are concerned, but covered by Karroo and Dwyka beds. It is therefore also possible that the basement-granite may be exposed, so far as older beds are concerned, below, say, the Pretoria beds but directly covered by them.   The Witwatersrand beds would then undoubtedly lie on the slopes of the granite-mass and the Main Reef Series would then occur in places, within a workable depth.

———

The PRESIDENT (Mr. J. C. Cadman) moved a vote of thanks to Mr. A. R. Sawyer for his valuable paper.

Prof. H. LOUIS seconded the resolution, which was cordially approved.

Mr. G. A. MITCHELL moved a vote of thanks to the Royal Astronomical Society for the use of their rooms. The scientific societies of London had always been most willing to oblige The Institution of Mining Engineers by placing their rooms at the disposal of the members for the purposes of their meetings.

Mr. F. R. SIMPSON seconded the resolution, which was carried unanimously.

———

Mr. T. LINDSAY GALLOWAY (Glasgow) moved a vote of thanks to Mr. G. A. Mitchell for presiding at the meeting on that day.

Mr. M. WALTON BROWN seconded the resolution, which was cordially approved.

———

Mr. H. C. PEAKE moved a vote of thanks to the Geological Society of London for the use of their rooms, and to the owners of works, etc., to be visited by the members.

Mr. WILLIAM LOGAN seconded the resolution, which was cordially approved.

———

Mr. HENRY HALL moved a vote of thanks to the President (Mr. J. C. Cadman) for his services in the chair.

Mr. M. W. WATERHOUSE seconded the resolution, which was cordially approved.

———

The following notes record some of the features of interest seen by visitors to works, etc., which were, by kind permission of the owners, open for inspection during the course of the meeting on June 2nd, 3rd and 4th, 1904.

ALDWYCH AND KINGSWAY.

The general character of the subway which the London County Council is now constructing under its new streets—Aldwych and Kingsway—is represented in Figs. 1 and 2. Kingsway will be 100 feet wide, the roadway being 60 feet wide, and the two paths each 20 feet wide (Fig. 1). The roadway will be laid with wooden paving, 6 inches thick, and the pathways will be covered with York flagging, 3 inches thick. Underneath the sidewalks are arranged a series of vaults, each measuring 12 feet in length by 8 feet in height. These will serve, presumably, for coal-cellars; but they will, moreover, be connected by 12 inches earthenware pipes with the 12 feet by 7½ feet gas, water and electric-conduit subways constructed under the haunches of the roadway, in order that all house-connections can, after the

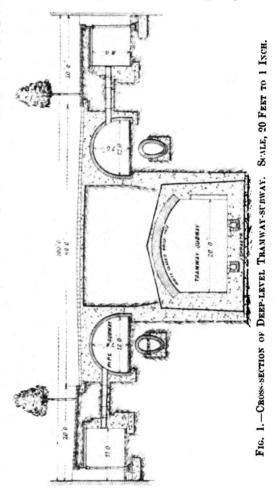

FIG. 1.—CROSS-SECTION OF DEEP-LEVEL TRAMWAY-SUBWAY. SCALE, 20 FEET TO 1 INCH.

erection of buildings, be made without disturbing the pathway. Underneath the pipe-subway, on each side, is a 4 feet 6 inches by 2 feet 8 inches egg-shaped sewer, with inverts of blue brick.

From this sewer branches are laid at every 30 feet, extending under the pathway-vaulting, so that connection to the sewer can be gained by sinking down through the floor of the vault to meet the aforesaid branch. The cost of the substructure of Kingsway, inclusive of the tramway-subway, is said to be close on £100 per lineal foot, or over £500,000 per mile.

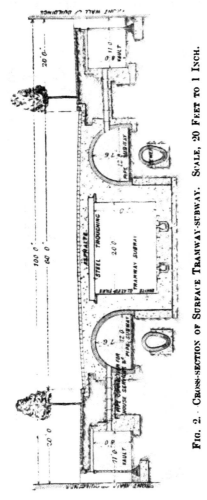

FIG. 2.—CROSS-SECTION OF SURFACE TRAMWAY-SUBWAY. SCALE, 20 FEET TO 1 INCH.

The tramway-subway, provided for two lines of rails, is 20 feet in width by 16 feet 3 inches in maximum height. The arching consists of five rings of brickwork, the inner ring of which is glazed. The side-walls are of concrete, and the whole tunnel is surrounded by a layer of asphalt ¾ inch thick, as indicated by the full line (Fig. 1). The tunnel will be fitted for traction on the conduit-system.

The tunnel is practically finished for a length of 400 feet north of the Strand, along the north-western side of the Gaiety Theatre. This portion has been constructed on the cut-and-cover system. In crossing the Strand and Holborn, however, it is intended to substitute a couple of tubes, driven by means of the Greathead shield, each 15 feet 10 inches in diameter, outside, for the single tunnel used elsewhere. The excavators will not have the benefit of working exclusively in the London Clay, as the upper portion of the shield will be driven through gravel, in which some water is likely to be encountered. On a portion of the line of route, the arch used in the deep-level section near the Strand, is replaced by steel troughing (Fig. 2).

The tunnel will come to the surface at Vernon-place, Theobald's-road, the terminating gradient being as much as 1 in 10; and the other end will similarly come to the surface at the Embankment. The depth of the subway below the street-surface varies, but it is generally shallow, though at the Strand and Holborn crossings it is increased to about 18 feet.

ADMIRALTY HARBOUR, DOVER.

This work is being carried out by Messrs. S. Pearson & Son, Limited, under the superintendence of Messrs. Coode, Son and Matthews, the engineers, to the plans described in their paper.*

A commencement was made in the spring of 1898 by reclaiming the foreshore alongside the South-Eastern Railway-station. On this ground a blockyard was laid out for making the blocks for the Admiralty pier-extension, and at the present time practically all the blocks have been made. The ballast-supplies have been brought by rail, and the concrete has been made by two fixed Messant mixers, and then run to the moulds in $\frac{1}{2}$ cubic yard side-tip Decauville wagons by hand.

The staging for the Admiralty pier-extension was commenced in January, 1899, and by the end of that year ten bays of 50 feet were erected with the plant, consisting of a 20 tons derrick of 50 feet radius, for handling the staging material, a 60 tons goliath for taking out the foundations with grab and bell, and two 40 tons goliaths for block-setting; the first block was set in December, 1899. The foundation of the head is now being completed, and it is anticipated that all block-setting will be finished in 2 or 3 months' time; seventeen bays of staging, or 850 feet, have been in use at one time, dismantling from the shore-end as it advanced seaward. There have been no losses of any consequence to the permanent work during its construction, nor any very serious damage to the temporary staging.

At the eastern section, there being no available site for a block-yard until the reclamation was made, the 3 tons blocks for the reclamation-work were made at Sandwich, and barged

---

* *Trans. Inst. M.E.*, 1900, vol. xix., page 448.

from there, a temporary timber-jetty being erected at the East
Cliff for the discharging and also for the cement, granite, etc., required on that section.   This wall was entirely built from staging,

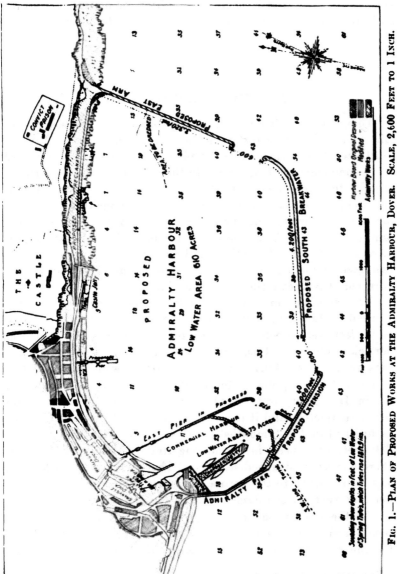

Fig. 1.—Plan of Proposed Works at the Admiralty Harbour, Dover.   Scale, 2,600 Feet to 1 Inch.

beds, accounting for the varying dip of the Hospital Hill and Witwatersrand beds on each side of the line representing its course.

A bed of fire-clay, 3 feet thick, was intersected in No. 15 bore-hole: good bricks might be made from it. It is probably the final decomposition-product of an igneous rock, occurring as a sheet. Excellent bricks are made from decomposed diabase at Eastleigh farm in the Bezuidenhout valley, east of Johannesburg.

The cost of boring the deeper holes was about 18s. per foot, including depreciation of drill, crowns, etc. Carbons were then cheaper than they are now, namely, £7 9s. per carat; the present price being about £14.

The Hospital Hill beds crop out in the Orange River Colony, near the junction of the Vaal and Wilge rivers, and dip to the south. Prospecting work through the Karroo beds, along the line CD, (Fig. 4, Plate XXVI.) has proved the occurrence of the Witwatersrand beds dipping regularly southward, at an angle varying from 25 to 30 degrees.

It will be seen by reference to the geological plan, which accompanied the writer's paper on "The South Rand Coal-field, and its connexion with the Witwatersrand Banket Formation"[*] that at Hex river, the Hospital Hill and Witwatersrand beds dip regularly to the northward. This points, in the writer's opinion, to the probable continuation of an assumed anticline taking the course shown in Fig. 4 (Plate XXVI.). The writer calls this the "Villiers anticline," from the name of a township in the Orange River Colony, on the Vaal river, about midway between the Wilge and Hex rivers. This anticline would, of course, only affect the Hospital Hill and Witwatersrand beds, the Dwyka and Karroo beds having been deposited long afterwards, and assuming generally a more or less horizontal position. It is remarkable, however, that the Karroo beds in a freestone-quarry, shown on the writer's plan of the South Rand coal-field,[†] and lying on the line of this assumed anticline, dip at an angle of 30 degrees, a most unusual occurrence.

From this new discovery it would appear that the Orange River Colony may have a "Rand" of its own, equally long, with

[*] *Trans. Inst. M.E.*, 1897, vol. xiv., page 326, plate XIV.
[†] *Ibid.*

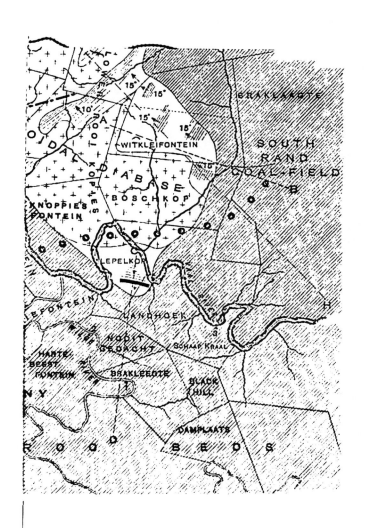

siderable extent of the sheet had been removed by denudation, as shown on the section, before the deposition of the Black Reef Series upon it.

The amygdaloidal-diabase sheet occurs east of Meyerton station, 4¾ miles away, on the Vogelfontein farm, No. 317. There is also a sheet of diabase, about 3 feet thick, lying on the amygdaloidal diabase, and the plane of contact between these two portions of the Klipriversberg amygdaloidal-diabase dips slightly eastward.

The writer is therefore of opinion that a bore-hole put down at the point indicated on the plan and sketch-section (Fig. 3, Plate XXV., and 4, Plate XXVI.) will not pass through the full thickness of the amygdaloidal-diabase sheet which is about 3,000 feet.

A large area, lying to the west of the railway between Klip-river and Meyerton stations, is covered by Pretoria beds, forming the eastern end of the Gatsrand. At their northern escarpment, these beds dip about 25 degrees southward; on their eastern escarpment, about 3 miles west of Klipriversberg station, they dip about 22 degrees westward; and farther to the southwest, they dip northward. A thick sheet of diabase, forming high hills, overlies the quartzite-bed which covers the shale-beds within this area (Fig. 4, Plate XXVI.).

The basement-granite has been elevated, and is exposed north of Johannesburg and east of Heidelberg. The writer has also located it south of Heidelberg, exposed so far as the Witwaters-rand beds are concerned, but covered by Karroo and Dwyka beds. It is therefore also possible that the basement-granite may be exposed, so far as older beds are concerned, below, say, the Pretoria beds but directly covered by them. The Witwatersrand beds would then undoubtedly lie on the slopes of the granite-mass and the Main Reef Series would then occur in places, within a workable depth.

———

The PRESIDENT (Mr. J. C. Cadman) moved a vote of thanks to Mr. A. R. Sawyer for his valuable paper.

Prof. H. LOUIS seconded the resolution, which was cordially approved.

Mr. G. A. MITCHELL moved a vote of thanks to the Royal Astronomical Society for the use of their rooms. The scientific societies of London had always been most willing to oblige The Institution of Mining Engineers by placing their rooms at the disposal of the members for the purposes of their meetings.

Mr. F. R. SIMPSON seconded the resolution, which was carried unanimously.

Mr. T. LINDSAY GALLOWAY (Glasgow) moved a vote of thanks to Mr. G. A. Mitchell for presiding at the meeting on that day.

Mr. M. WALTON BROWN seconded the resolution, which was cordially approved.

Mr. H. C. PEAKE moved a vote of thanks to the Geological Society of London for the use of their rooms, and to the owners of works, etc., to be visited by the members.

Mr. WILLIAM LOGAN seconded the resolution, which was cordially approved.

Mr. HENRY HALL moved a vote of thanks to the President (Mr. J. C. Cadman) for his services in the chair.

Mr. M. W. WATERHOUSE seconded the resolution, which was cordially approved.

The following notes record some of the features of interest seen by visitors to works, etc., which were, by kind permission of the owners, open for inspection during the course of the meeting on June 2nd, 3rd and 4th, 1904.

ALDWYCH AND KINGSWAY.

The general character of the subway which the London County Council is now constructing under its new streets—Aldwych and Kingsway—is represented in Figs. 1 and 2. Kingsway will be 100 feet wide, the carriageway being 60 feet wide, and the two footpaths each 20 feet wide (Fig. 1). The carriageway will be laid with wooden paving, 6 inches thick, in the city of Westminster, and with asphalt, 2 inches thick, in the borough of Holborn. The pathways will be covered with York flagging, 3 inches thick. Underneath the sidewalks are arranged a series of vaults, projecting 11 feet from the front of the buildings, and 8 feet in height. These will serve either for coal-cellars or for other storage purposes; and they will, moreover, be connected by 12 inches stoneware pipes with the 12 feet by 7½ feet subways constructed under the haunches of the roadway, and which will contain the gas, water and electric mains, in order that all house-connections can, after the

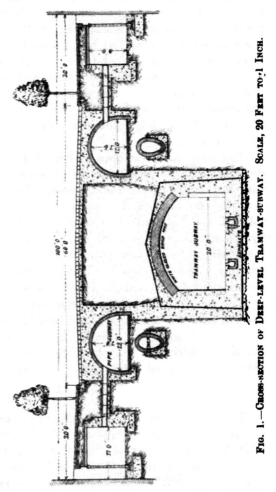

Fig. 1.—Cross-section of Deep-level Tramway-subway. Scale, 20 Feet to 1 Inch.

to the end of the chain fixed to the swivel below the stirrup-screw of the balance-beam (Figs. 1 and 2). The beam, m, is given a reciprocating motion from a cylinder, l, 40 inches in diameter, by a piston receiving steam on the upper surface only, and with a stroke of 40 inches. The frequency and length of the strokes is controlled by the master-borer, who manipulates the exhaust-steam and steam-admission valves in accordance with the character of the rock. Between each stroke, a slight turn is given to the trepan by eight men, in the boring-chamber, moving the levers, o, attached to the swivel. This rotation is continued until the length of the swivel-screw, n, about 1 foot, is exhausted. The rods are then detached from the balance-beam and withdrawn, or the stirrup-screw is reset, and the boring is continued with or without the insertion of a lengthening piece of rod. Ultimately, however, the sludging-barrel or spoon would be used.

The procedure has been to bore a hole, 9 feet 1½ inches in diameter, and 30 or 40 feet in advance of the full-sized shaft, with the small trepan : the centreing of the appliance being very carefully adjusted and maintained. The large trepan then follows, and the central guide-plate, fitting into the hole made by the small trepan, maintains the centre and the verticality of the boring. The sludging apparatus is worked by two of the winding-engines, i and k, which have been geared down to enable them to be used for this purpose.

The removal of the *débris* from the advance-hole is a simple operation, as the sludging-barrel readily removes about 6 tons of material. A grab can also be used for removing heavier and coarser rubbish. When the *débris* is removed, the sludging-barrel is detached, the trepan is replaced, and sinking is resumed.

The sinking by the Kind-Chaudron system started at the bottom of No. 2 pit, at a depth of 1,095 feet. At the depth of 1,126 feet, a steel cylinder, 20 feet high and 18½ feet in diameter, built of plates at the surface, was inserted so as to support the sides of the shaft. The boring was here reduced to a diameter of 18 feet. At the depth of 1,158 feet, a second steel cylinder, 20 feet high, was inserted. The boring was here reduced in diameter to 17½ feet; and the reduction was continued by degrees until, at the depth of 1,183 feet 8 inches, the boring was 16 feet 4¾ inches in diameter. At this point, the dark-grey shale will provide a suitable bed for the tubbing, the boring operations will cease, and the moss-box and the tubbing will be lowered into the shaft (Fig. 3). The successive reductions of the size of the shaft has entailed heavy work, in reducing the width of the massive steel-base of the large trepan so as to suit the altered dimensions. The shaft has been sunk 90 feet in somewhat soft strata (the clay proved especially difficult) in one year.

The rings of tubbing are each about 4 feet high, and about 15 feet in diameter; the total number is 278. They vary in thickness from 1½ inches in the top rings to 4½ inches in the lowest rings; and they range in weight from 7½ to 17½ tons each. They are provided at the ends with two inward flanges, and have, in addition, three circular strengthening ribs around the inside, the middle one being bigger than the other two. The total length of the column of tubbing will be 1,120 feet, and it weighs 3,675 tons. The driving-shoe and temporary crib-bed, in six segments, weigh 2½ tons. The moss-box, consisting of two con-

TABLE I.—SECTION OF STRATA SUNK THROUGH IN THE No. 2 SHAFT, DOVER COLLIERY.

No.	Description of Strata.	Thickness of Strata. Ft. In.	Depth from Surface. Ft. In.	No.	Description of Strata	Thickness of Strata. Ft. Ins.	Depth from Surface. Ft. Ins.
	Oolite:			11	Dark-grey marl ...	3 3	1,145 3
1	Calcareous grit ...	3 9	1,098 6	12	Grey clay	2 8	1,147 11
2	Dark-grey sandy marl	5 0	1,103 6	13	Dark-grey marl ..	6 0	1,153 11
3	Limestone and sandy grit ...	2 3	1,105 9	14	Fossiliferous clay	4 0	1,157 11
4	Calcareous grit and *lignite*	3 8	1,109 5		*Coal-measures:*		
5	Quartzose sand, with water ...	6 9	1,116 2	15	Grey shale ...	12 6	1,170 5
				16	Ironstone ...	0 2	1,170 7
6	Consolidated sand, with water ...	4 6	1,120 8	17	Grey shale ...	0 10	1,171 5
	Lias:			18	Dark-grey shale	17 2	1,188 7
7	Dark-grey clay ...	5 3	1,125 11	19	Bituminous shale and **COAL** ...	2 0	1,190 7
8	Calcareous sandstone	2 10	1,128 9	20	Dark-grey shaly clay	1 6	1,192 1
9	Hard grey limestone, with bands of clay	8 9	1,137 6	21	Bituminous shale and **COAL** ...	1 10	1,193 11
				22	Dark-grey shaly clay	3 7	1,197 6
10	Calcareous sandstone	4 6	1,142 0	23	Bituminous shale and **COAL** ...	0 6	1,198 0
				24	Argillaceous fossiliferous shale	8 7	1,206 7

centric cast-steel cylinders, each in six segments and arranged one to fit telescopically over the other, the inner one being about 14½ feet in diameter, weighs 9½ tons; and the outer one, about 15 feet in internal diameter, weighs 12½ tons; both measure

about 4 feet in height and $2\frac{3}{4}$ inches in thickness. The diaphragm or false bottom, an hemispherical dome, is about 4 feet
deep in the centre, where there is a round orifice, $3\frac{1}{2}$ feet deep,
and $31\frac{1}{2}$ inches in diameter at the base, diminishing to about 21
inches at the top. This diaphragm is strengthened inside and
outside with 12 ribs; on the outside they are $2\frac{3}{8}$ inches thick,
and as much as $15\frac{3}{4}$ inches deep towards the centre, the whole
casting weighing 27 tons. There are special attachments for
fixing the false-bottom inside the lowest ring of tubbing, and
permitting of its removal when the tubbing is installed in the
shaft, consisting of a hinged cast-steel ring in three segments
and sundry loops and eyes for supporting purposes. There are
120 flanged cast-iron pipes, 10 feet long by $6\frac{3}{4}$ inches in internal
diameter, $1\frac{1}{2}$ inches thick, with a reducing piece $3\frac{1}{4}$ feet high
and $2\frac{3}{4}$ inches thick. These pipes together will form the equi-
librium-column, which will rise from the central orifice of the
diaphragm during the sinking of the tubbing (Fig. 3).

The *grapin* or pinchers, d (Fig. 2), are used to pick up fragments of stone, which may lodge on the seat of the moss-box
at the shaft-bottom. The recovery of a broken rod is effected by
means of the crow-foot, e (Fig. 2), a specially shaped tool.

MESSRS. FRASER & CHALMERS, LIMITED, ERITH.

A Corliss winding-engine was seen in course of erection, intended for one of the Earl of Dudley's collieries. It is a four-
cylindered direct-acting winding-engine with two tandem compound cylinders, the high pressure being 26 inches in diameter,
and low pressure 45 inches in diameter by 66 inches stroke. The
winding-drum is double conical; it has a maximum diameter of
$17\frac{1}{2}$ feet, and is $10\frac{1}{4}$ feet wide. The maximum coal-load is 7
tons, and it will be lifted from a depth of 1,800 feet at a rope-
speed of 3,500 feet per minute.

Another winding-engine was being erected for the Tasmania
Gold-mining Company, Limited. The winding-engine in this
case will hoist from a depth of 2,000 feet at a maximum rope-
speed of 1,600 feet per minute, its normal capacity being 50
tons per hour. Other large winding-engines are being erected

for the Penrikyber Navigation Collieries, North's Navigation Collieries, and the Bridgewater Trustees.

A large Riedler pump was shown at work, pumping 175 gallons per minute, against a head of 600 feet, and working at 150 revolutions per minute with very little noise.

A number of cyanide-vats, 50 feet in diameter, were shown in course of erection, prior to their despatch to South Africa.

About 1,000 workmen are employed in the foundry, boiler-works and workshops. The workshops contain a large number of machine-tools in current use, and special mention may be made of a machine with six cones, placed obliquely, used to straighten steel shafts by tangential pressure.

Much interest was also taken in the system of keeping drawings and plans, a thoroughly efficient card-classification being employed.

CHISLEHURST CHALK-WORKINGS.

These old workings are situated close to Chislehurst station, and behind the Bickley Arms Hotel, from the grounds of which entrance is obtained. A small inlier of Chalk is exposed in the valley, owing to a slight anticlinal fold, so that the mines can be entered from the day.

There is an extensive range of workings in the upper portion of the Chalk and immediately under the Thanet Sand, a few feet of chalk having been left in the roof as a protection against the sand. Some of the workings are of considerable interest, owing to the care which has evidently been taken in the driving of the working-places.[*] They are, generally speaking, 7 to 10 feet in height and dry, in most places. Their age is not known, but it is probably not more than from 150 to 200 years.

---

[*] This has led to a local theory to the effect that they are of prehistoric origin. "The Chislehurst Caves and Dene-holes," by Mr. W. J. Nichols, *Journal of the British Archæological Association*, 1903, vol. x., page 64; and "The Chislehurst Caves," by Messrs. T. E. and R. H. Forster, *Ibid.*, 1904, vol. x., page 87.

THE NORTH OF ENGLAND INSTITUTE OF MINING AND MECHANICAL ENGINEERS.

EXCURSION MEETING,
Held in Cumberland, June 10th, 11th and 12th, 1903.

The following notes record some of the features of interest seen by the visitors to collieries, mines, works, etc., which were, by kind permission of the owners, open for inspection during the course of the meeting:—

SAINT HELENS COLLIERIES.

The St. Helens Colliery and Brickworks Company, Limited, have two collieries in the vicinity of Workington: the No. 3 pit being situated on the sea-shore at Siddick, about a mile to the north-east of Workington harbour; and the No. 2 pit, about 1½ miles farther east, near the village of Flimby. The former has an output of 250,000 tons per annum; and at the latter the shafts have only recently been re-opened and deepened to the lower seams.

The leasehold royalties have an area of about 9,000 acres. The coal-field is cut up by faults, into a series of irregular parallel lanes, with a northerly and southerly course and varying considerably in width. There are also occasional cross faults.

The seams worked include the Ten-quarter, 2 feet 8 inches thick; the Cannel-and-metal, 7 feet thick; the Little Main, 2 feet 2 inches thick; and the Lickbank, 2 feet 2 inches thick. The depths to the seams at the No. 3 pit are 504 feet, 714 feet, 918 feet and 1,020 feet respectively; and at No. 2 pit they are 316 feet, 508 feet, 697 feet and 802 feet respectively. Some of the workings extend beyond low-water mark. The general dip is about 1 in 10, westward towards the sea.

All the seams are worked on the longwall system. Until recently the Cannel-and-metal seam, in which there is a band of dirt about 3½ feet thick, separating the Metal from the Cannel seam, was worked by the pillar-and-stall method. It is now, however, worked successfully on the longwall system. The face of the lower or Cannel seam is kept in advance of the upper or Metal seam and the intervening stratum of dirt is left *in situ*.

No. 3 or Siddick Colliery.—The downcast shaft is 10 feet 9 inches in diameter and the upcast shaft 11 feet 8 inches in diameter. At the downcast shaft, the coupled horizontal winding-engine has two cylinders, each 30 inches in diameter by 5 feet stroke, the valves are of Cornish type, and the drum is 15½ feet in diameter. At the upcast shaft, the coupled horizontal winding-engine has two cylinders, each 26 inches in diameter by 5 feet stroke, the valves are of Cornish type, and the drum is 14 feet in diameter. The cages on the downcast shaft have four decks, and those on the upcast shaft three decks. Each deck carries one tub, having a capacity of 11¼ cwts.

The Cornish pumping-engine, at the upcast shaft, has a cylinder 60 inches in diameter with a stroke of 9 feet. The pumps are in two lifts, each 252 feet long. The ram of the top set is 18 inches in diameter, and of the bottom set 16 inches.

The Guibal fan, 36 feet in diameter and 12 feet wide, is driven by a non-condensing engine, with a duplicate cylinder.

The horizontal, cross-compound, condensing hauling-engine has two cylinders, 16 inches and 30 inches in diameter respectively by 30 inches stroke. It is placed on the pit-bank, and drives, by means of band-ropes, the haulage-drums at the underground stations; it also gives motion to the various endless-haulage ropes radiating from these stations.

Air, which is used underground for dip pumping, is compressed by a horizontal non-condensing engine having two steam and two air-cylinders, each 24 inches in diameter by 4 feet stroke.

The exhaust-seam from the non-condensing engines is passed through a Berryman heater, and raises the temperature of the feed-water to over 200° Fahr.

The coal-output is screened by four shaking-screens. Each screen discharges the screened coal upon its own travelling-belt, and the dirt is picked out by hand. The small coal is carried by a scraper-conveyor to the washery, where it is sized by a twin vibromotor-screen, and, afterwards, treated in a Coppée washery, the nut and coking coals being automatically discharged into the several bunkers, ready for loading into railway-wagons and coke-oven bogies.

The coking coal is treated in 24 Coppée ovens, and the gases therefrom are used to fire three Babcock-and-Wilcox water-tube boilers, which supplement eight double-flued Lancashire boilers, fired by hand.

The surface-plant and the various underground eyes and shaft-sidings are electrically lighted.

No. 2 Colliery.—The downcast shaft has a diameter of 17 feet 6 inches, and the upcast shaft is 10 feet 6 inches in diameter. The downcast winding-engine has two cylinders, each 36 inches in diameter by 5 feet stroke, with Cornish valves, and drums 16 feet in diameter. At the upcast shaft, the winding-engine has two' cylinders, each 24 inches in diameter by 4 feet stroke, and the winding-drum is 10 feet in diameter.

The Cornish pumping-engine, with overhead beam, has a cylinder 50 inches in diameter by 8 feet stroke. There are two lifts of pumps: the top set is placed in a small shaft inside the engine-house, and is coupled to the cylinder-end of the beam; while the bottom set is placed in the main shaft, and is attached to the pit-end of the beam.

The Guibal fan is driven by a horizontal non-condensing engine with a duplicate cylinder.

The surface-plant, underground eyes and shaft-sidings are electrically lighted.

The coal-tubs when they reach the surface are carried along an endless-rope tramway to the screens and washery, situated alongside the public railway at a distance of about 1,200 feet from the shafts. In the screen-house, there are two shaking-screens, which will ultimately be increased to six, with accompanying picking-belts. The coal-tubs run into and out of the tipplers by gravity-roads. The small coal from the screens is carried by a conveyor to the Coppée washery, where it is sized and washed.

At the brick-works, situated near the screens, common red wire-cut bricks are made from surface-clay.

WHITEHAVEN COLLIERIES.

The Croft, Wellington and William collieries stand as grimy outposts of industry to the north and south respectively, when seen from the northern pier of Whitehaven harbour. The position of the William pit recalls the position of the Kent collieries at Dover, but there all resemblance ends, for the William pit is drawing 1,000 tons a day.

The Croft, Wellington and William pits are each sunk 840 feet to the Six-quarter coal-seam, and each work the Main Band coal-seam at a depth of 600 feet. With the exception of a small district in the Croft pit, the whole of the coal is won under the sea, by the bord-and-pillar system, the average distance from the pit to the working-face being about 3 miles.

The Main Band seam, dipping westward at 1 in 12, with a roof of blue metal, has the following average section:—

	Ft.	In.				Ft.	In.	Ft. Iu
COAL, little	0	6	Metal			1	7	
COAL, bearing top ...	1	2	COAL, spar			1	2	
COAL, main top ...	3	1½	COAL, benk			2	6	
Black slate	0	0½	Metal			0	1	
COAL, laying-in ...	1	4	COAL, mother			0	3	
Metal	0	3						
COAL, 4 inches	0	6						12 6

CROFT PIT.

The Croft pit is situated at Sandwith, a short distance from Whitehaven. A stone-drift, 3,900 feet long, is being made to shorten the haulage-road. The shot-holes are made with Grant electric rock-drills. The drills are of the rotary type, and are carried on a steel column, which is set between the floor and the roof and securely fixed by tightening-screws. The drill is carried on a horizontal arm extending from the main column, and can be adjusted at any required angle. The motor rests on trunnion-bearings, and is placed on a turntable carried on an ordinary colliery-wagon. The power is transmitted from the motor to the drills by means of telescopic shafting, with universal joints at either end. Two drills are in use, and a single motor is placed between them; and while one is at work the other is being adjusted for the next hole. Usually, fourteen to eighteen holes, varying in depth from 5½ to 8 feet, are drilled by one setting of the machine, keeping four men employed from 3 to 4 hours. As soon as the machines have finished drilling the round of holes, the drilling-plant is removed outbye from the face of the drift. The actual size of the drift-cut is 16½ feet by 9 feet, and the average progress is 30 feet per week. The drift is being driven in the Mountain or Carboniferous Limestone, underlying the Coal-measures, which will be entered when the drift cuts a dip fault. Steel rails, weighing 80 pounds per yard, supported on wooden props or on steel plugs driven into the sides, are used to support the roof.

WELLINGTON PIT.

The vertical double-handled winding-engine has a cylinder, 34 inches in diameter by 6 feet stroke, fitted with double-lift steam-valves (8 inches in diameter), two discharge-valves (10 inches in diameter), a drum (13 feet in diameter), and a brake wheel (31½ feet in diameter). The flat winding-ropes are 4 inches wide by $\frac{11}{16}$ inch thick, and are fitted with King safety-hooks to prevent over-winding. The depth of the shaft to the Main Band coal-seam is 600 feet, it is 9 feet in diameter and fitted with wooden conductors. The weight of a cage and short chains is 1 ton 15 cwts. The cages have three decks, with a single tub on each deck. The wooden tubs, weighing 4⅝ cwts., carry 12 cwts. of coal. The output is about 800 tons per day of 10 hours.

The horizontal single-stage air-compressor has two steam-cylinders, each 26 inches in diameter by 4 feet stroke, and two air-cylinders, each 24 inches in diameter, cooled by water-jackets. The air is compressed to 60 pounds per square inch, and passes down a main pipe, 6 inches in diameter, placed in the downcast shaft; it is then conducted by branch pipes, 4 inches in diameter, a mile inbye to two pumps.

The horizontal hauling-engine has two cylinders, each 32 inches in diameter by 5 feet stroke, supplied with steam at a pressure of 60 to 80 pounds per square inch; it is geared 1½ to 1; the fly-wheel, 12 feet in diameter, weighs 13 tons; and the driving drum is 8 feet in diameter. The steel rope, 33,000 feet long and 1½ inches in diameter, weighs 34 tons, and has 4½ turns on the drum. This haulage-engine also works two branch ropes, 1¼ inches in diameter, each 3,000 feet long. The terminus wheel of the main rope and the driving wheels of the branch ropes are keyed to the same shaft.

The Guibal fan, 36 feet in diameter and 12 feet in width, is driven direct by a single cylinder, 30 inches in diameter by 30 inches stroke, with piston-valves. Steam is supplied at a pressure of 70 pounds per square inch, from two Galloway boilers, 26 feet long and 7 feet in diameter. The fan exhausts 54,000 cubic feet of air per minute at 3·2 inches of water-gauge, when running at 60 revolutions per minute.

The Cornish pump and engine has a vertical inverted cylinder, 90 inches in diameter by 10 feet stroke, and runs 5 strokes per minute. Steam is only admitted to the top side of the cylinder at a pressure of 40 pounds per square inch. There is a set lifting from a depth of 162 feet, with a bucket, 20 inches in diameter; and two forcing sets, each of 345 feet, with rams, 19 inches in diameter. The wet rods are 10 inches square, and the dry rods are 15 inches square, all of pitchpine. This engine raises 620 gallons per minute, to a height of 852 feet. The crab for lifting the rods, etc., has two cylinders, each 10 inches in diameter by 2½ feet stroke, and the drum, 6 feet in diameter, is geared 36 to 1.

The steam-generating plant comprizes six Lancashire boilers, 30 feet long and 7½ feet in diameter, hand-stoked, and fitted with Caddy movable fire-bars.

The screening and cleaning plant includes a shaking screen, driven by a vertical engine, with two cylinders, each 10 inches in diameter, and 18 inches stroke; a picking-table, 54 feet long, and 4½ feet wide, driven by a horizontal engine, with two cylinders, each 10 inches in diameter by 18 inches stroke; and four Shepherd bash-washers, driven by a horizontal engine, with a cylinder 16 inches in diameter by 32 inches stroke.

The fitting-shop contains two screw-cutting lathes, with 10½ inches and 11 inches centres, and as they are placed together end to end, any length can be turned up to 20 feet; a planing-machine, with a table, 10 feet by 3 feet, driven by a treble-thread screw and quick-return motion; a double-geared vertical drilling-machine; a screwing machine, to screw and tap from ⅜ inch up to 1½ inches, is fitted with an automatic release-motion for the dies; a slotting-machine, with 12 inches stroke, and quick-return motion; and a shaping-machine, with two travelling-tables, a stroke of 13 inches, and a quick-return motion. The various tools are driven by a horizontal engine, with a cylinder 12 inches in diameter.

The blacksmiths' shop contains a steam-hammer, with a

stroke of 18 inches: the force of the blow being equal to 4 cwts.; a Schiele fan, 2½ feet in diameter, to blow the fires; and a levered punching-and-shearing machine, with engine, will punch and shear plates up to ⅞ inch in thickness.

The saw-mill contains a circular saw and bench, 6 feet long by 3 feet wide. The saw, 3½ feet in diameter, is driven by a horizontal engine, with a cylinder 14 inches in diameter.

WILLIAM PIT.

The horizontal winding-engine has two cylinders, each 32 inches in diameter by 6 feet stroke, and a drum 18 feet in diameter. The shaft, 600 feet deep to the Main Band seam, is 12 feet in diameter, and fitted with steel-rail conductors. The round steel rope is 1¾ inches in diameter. The weight of the cages and short chains is 2 tons. The iron tub weighs 7 cwts., and the wooden tub, 5 cwts. The average load of the iron tub is 14½ cwts.; and that of the wooden tub, 11½ cwts. of coal. Each cage carries four tubs. The winding-engine is fitted with a steam-brake, and Walker detaching-hooks are fitted to prevent over-winding. About 800 tons are drawn per day of 10 hours.

The two-stage air-compressor, with two steam-cylinders, each 42 inches in diameter by 6 feet stroke, runs at 30 revolutions per minute. The low-pressure air-cylinder, 54 inches in diameter, compresses the air to 25 pounds per square inch; and the high-pressure air-cylinder, 36 inches in diameter, compresses the air to 60 pounds per square inch. The air is passed from the low-pressure cylinder into a cooler and from the cooler to the high-pressure air-cylinder, thence into the main, 20 inches in diameter, down the upcast shaft, whence it is passed for 3½ miles through a main, 15 inches in diameter, to the Countess district haulage; and thence a pipe, 4 inches in diameter, is used for a further distance of about 1 mile to the dip pumps.

The haulage-plant worked by compressed air comprizes a hauling-engine with two cylinders, each 20 inches in diameter by 30 inches stroke, geared 15 to 1, working an endless-haulage rope, 19,500 feet long and 1¼ inches in diameter; a hauling-engine, with two cylinders, each 9 inches in diameter by 12 inches stroke, geared 20 to 1, working an endless-haulage rope, 6,000 feet long and ¾ inch in diameter; two double-acting ram-pumps,

4 inches in diameter; and a hauling-engine, with a cylinder, 8 inches in diameter, used for driving the new engine-plane.

The main-road haulage is worked by a hauling-engine, with two cylinders, each 24 inches in diameter by 30 inches stroke, geared 4 to 1, with a drum 6 feet in diameter, working an endless rope, 25,000 feet long by 1¼ inches in diameter and weighing about 28 tons. This hauling-engine also drives an endless rope, about 10,500 feet long and 1¼ inches in diameter in the Delaval district: this rope is worked by a wheel and clutch on the terminal shaft of the main-road haulage.

A Walker fan, 22 feet in diameter and 7 feet wide, is driven by an engine with a cylinder 36 inches in diameter by 40 inches stroke, with cut-off valves. The driving wheel, 18 feet in diameter, is grooved for ten cotton driving-ropes, 1¾ inches in diameter; the pulley on the fan-shaft is 7½ feet in diameter, or speeded 2·4 to 1 of the engine. The engine is run at about 50 revolutions and the fan at about 120 revolutions per minute, produces 60,000 cubic feet per minute at 6 inches of water-gauge. The upcast shaft is 13 feet in diameter.

The horizontal duplex ram-pumps, 10 inches in diameter, are driven by two steam-cylinders, each 30 inches in diameter by 18 inches stroke. These pumps are placed at the shaft-bottom, working against a head of 600 feet.

An engine, placed at the shaft-bottom, with two cylinders, each 9 inches in diameter, works the tub-creeper for the transit of the empty tubs.

The coal is screened on two jigging-screens, driven by an engine, with two cylinders each 9 inches in diameter. The cleaning-table, about 70 feet long and 4 feet wide, is driven by an engine, with a single cylinder, 8 inches in diameter. The washery is of the Coppée type, with seven bashes.

The coke-ovens, 125 in number, are of the beehive type.

Seven Lancashire boilers, 30 feet long and 8 feet in diameter, are hand-fired: nine similar Lancashire boilers are gas-fired; and two Babcock-and-Wilcox water-tube boilers are fired by the waste-gases from the coke-ovens.

———

THE MONTREAL MINES.

The Montreal mines and colliery occupy lands in three adjoining parishes near Whitehaven. Previous to Mr. Stirling taking a lease of this royalty, it was in possession of Mr. Litt, who sank a shaft on the eastern boundary, the machinery consisting of a horizontal drum, or gin worked by a horse. This shaft, at a depth of about 72 feet, penetrated a gravel-bed, and tapped a large feeder of water which prevented further sinking. The shaft was timbered on the joggling system, notched at each corner. The present-day shafts are mostly close-timbered throughout their depth, thus preventing broken measures from falling from the sides.

In 1859, Mr. John Stirling leased the Montreal royalty, and immediately commenced several bore-holes, proving the presence of several ore-beds at various levels. In 1860, the first shaft was sunk, and passed through several ore-beds varying from 10 to 40 feet in thickness, and having a total thickness of 120 feet. Hæmatite was discovered at the Montreal mines on October 30th, 1861, and from that time onward it has been constantly found in large and small quantities.

The ore-deposits are clearly divided from the Coal-measures by an east-and-west fault, dipping northward at an angle of about 45 degrees, with a downthrow in the measures of between 1,400 and 1,650 feet. About a mile nearer the west coast. there is another fault; this reverses the situation by an upthrow of about 1,500 feet, and the Coal-measures crop out close at hand (Fig. 1, Plate XXVII.).

In the hæmatite-deposits, there are several faults, but only one is noteworthy: it runs almost at right-angles to the fault dividing the Coal-measures from the ore-measures. This fault hades at an angle of 65 degrees, and has an easterly downthrow of 293 feet.

On carefully examining the fault which divides the Coal-measures from the iron-ore, it is invariably found that the ore-measures approaching within 60 feet of the fault rapidly alter their gradients from 18 to about 60 degrees; and should a shale-bed, intervening between the limestone-beds, be tilted in the manner described, it is sometimes thought that it is a fault, unless the contents are carefully examined.

The surface-area of the Montreal mines embraces about 64 acres. Upward of 200 bore-holes, varying from 250 to 1,000 feet, have been put down from the surface by lever or diamond core-cutters at a considerable expense.

About twenty shafts have been sunk for the purpose of working the iron-ore and coal: five shafts have been sunk on the Coal-measures side of the divisional fault, averaging 480 feet in depth. Three shafts were sunk wholly in Coal-measures, two have passed through 120 feet of Coal-measures, and then passed through the fault into the Limestone-measures. Ten shafts have been sunk to an average depth of 300 feet in the Limestone-measures on the opposite side of the fault; and five shafts, averaging 120 feet, have been sunk for ventilating purposes and reducing the labour involved in carrying heavy timber into the rise-workings.

The mines now in operation comprise Nos. 4, 5, 6, 10 and 12. The No. 4 and principal plant is situated about 4 miles south-east of Whitehaven, and ¼ mile north of Moor Row Station on the Whitehaven, Cleator and Egremont Railway. The weekly output is about 1,000 tons of hæmatite and 500 tons of coal.

No. 4 Mine.—The winding and downcast shaft, at which coal and hæmatite are raised, was sunk in 1870-1871, through Coal-measures to a depth of 480 feet or thereabouts. It measures 14 feet by 7 feet, and is lined throughout with timber. The upcast shaft, about 150 feet north-east of No. 4 shaft, 12 feet by 6 feet and 480 feet deep, is lined with pitchpine, 4 inches thick. The winding shaft is divided into three compartments, two for cages and one for pumps.

The No. 4 winding-engine has two horizontal cylinders, each 22 inches in diameter by 5 feet stroke, and a cylindrical drum, 11 feet in diameter by 4 feet wide, with a double-breast brake worked by a hand-lever. Each cage carries two tubs on one deck. The wooden tubs, with iron axles and steel wheels, carry 5 cwts. of coal and 10 cwts. of iron-ore. The crucible-steel winding-ropes, 3¾ inches in circumference, are fitted with King hooks. There are two pitchpine conductors to each cage. The cages are fitted with safety-apparatus, consisting of a ratchet-plate, placed behind the whole length of each conductor, two spring-boxes each with 12 indiarubber-springs, and two pawls which rest on the ratchet-edges whenever the rope is slackened or a breakage occurs. The upcast winding-engine has two horizontal cylinders, 14 inches in diameter by 18 inches stroke, gearing 1 to 4, and two drums, 6 feet in diameter.

The pumping-engine has one horizontal cylinder, 28 inches in diameter by 5 feet stroke, steam at an initial pressure of 40 pounds is cut off at half stroke by an expansion-slide, and a fly-wheel, weighing 8 tons. This engine is arranged to work either high-pressed or condensing: a separate double-acting condenser is provided with a ram, 9 inches in diameter and a steam-cylinder 8 inches in diameter by 12 inches stroke. The pump-rods are actuated by a horizontal connecting-rod and two cast-iron quadrants. The engine raises water in two lifts from the Main Band seam: the top lift, 240 feet long, has a hollow ram, 12 inches in diameter by 4 feet stroke; and the bottom lift, also 240 feet long, has a hollow ram, 12 inches in diameter by 4 feet stroke. The hollow ram, fitted with a valve at the top with a vertical lift of 3 inches, and a packing-gland like an ordinary ram, raises the water at the upstroke. The engine is worked day and night at the rate of 10 revolutions per minute, giving a displacement of 190 gallons per minute, the maximum speed being 12 revolutions. In case of accident to this engine or the pumps, the water is raised in boxes by the winding-engine: a box of 350 gallons' capacity, connected to the bottom of the cage, being filled and discharged automatically. The crab-engine for the pumps, with a horizontal cylinder, 6 inches in diameter by 12 inches stroke, is double-geared, the drum is 30 inches in diameter, cleaded with timber, and has a plough-steel rope 4 inches in circumference.

The hauling-engine, on the surface at the upcast shaft, with two horizontal cylinders 14 inches in diameter by 18 inches stroke, geared 1 to 4, has two drums, 4 feet in diameter, for main- and tail-rope haulage. The crucible-steel hauling-ropes, 2¼ inches in circumference, are taken down the upcast shaft to work the haulage-road in the Main Band seam. This road is about 1,500 feet long, and sets of twenty tubs are run in and out alternately on the single road. Another hauling-engine, placed in the Main Band seam about 1,800 feet west of the shaft, has a horizontal cylinder, 8 inches in diameter by 18 inches stroke, geared 5 to 1, with two drums, 4 feet in diameter, and is actuated by compressed air. It hauls sets of five tubs from a double road dipping westward 1 in 2, the empty set descending while the full set ascends. A third hauling-engine, placed about 300 feet to the south of the main haulage-road, of the same dimensions as the second engine, works another road dipping heavily about 1 in 1¼ westward. It draws up two full tubs, and two empty tubs descend at the same time into the workings. The full tubs are taken to the branch-end siding to make up the train of twenty tubs, which is then conveyed to the shaft by the first-named engine. This engine is also worked by compressed air at a pressure of 58 pounds to the square inch.

A duplex pump, placed in the coal-pit at a distance of 720 feet down the first-mentioned dip-drift, is actuated by compressed air at a pressure of 58 pounds per square inch, with two cylinders, 7 inches in diameter, and rams, 4½ inches in diameter by 10 inches stroke. The water is raised a vertical height of 120 feet to the top of the incline, and flows by gravity to the shaft-sump. The pump is worked 6 hours daily.

The air-compressors have two steam-cylinders, 16 inches in diameter, and air-cylinders, 13 inches in diameter by 3½ feet stroke. The steam-pressure is 35 pounds, and the air-pressure 60 pounds per square inch. The front piston-rods are attached to cranks set at right-angles to each other on the flywheel-shaft, the flywheel weighing 5 tons. The compressed air is taken down the pit and inbye to the hauling-engines, a distance of 1,800 feet, through malleable-iron pipes, 4 inches in diameter, with Eadie joints.

The Guibal fan, 18 feet in diameter and 5 feet wide, is driven by a vertical engine having a cylinder 20 inches in diameter by 12 inches stroke. The fan produces 30,000 cubic feet per minute at 0·80 inch of water-gauge, when running at 60 revolutions per minute.

Seven cylindrical boilers raise steam at a pressure of 40 pounds per square inch. The upper side of the boilers is arched over with brickwork, so as to leave an annular space of 6 inches over the boiler, and part of the flame from the flash-flue passes and superheats the steam. The boilers are supplied with water by a self-acting high-pressure injector, with a donkey-pump, with a cylinder 8 inches in diameter, and a ram 4 inches in diameter by 10 inches stroke in reserve, to pump a cold-water feed when required.

No. 5 Mine.—The shaft measures 12 feet by 5 feet, and is 300 feet deep. The winding-engine has two horizontal cylinders 16 inches in diameter by 3 feet stroke, two drums 6 feet in diameter, and a breast-brake and hand-lever. A bogie, containing 10 cwts. of ore, is raised in each cage. The safety-cages in use are the same as at No. 4 pit. The winding-ropes of crucible steel, 3 inches in circumference, are fitted with King safety-hooks. Water boxes underneath the cages are used to assist the pumps in wet weather.

The horizontal compound condensing pumping-engine, with Hathorn-Davey differential gear, has a high-pressure cylinder, 15 inches in diameter by 4 feet stroke, supplied with steam at an initial pressure of 100 pounds per square inch, and a low-pressure cylinder 30 inches in diameter by 4 feet stroke. There is a separate condenser, similar to that at No. 4 mine. The engine is attached by a horizontal connecting-rod to two cast-iron quadrants and works two lifts of pumps, each 10 inches in diameter and 300 feet long, fitted with hollow rams, 12 inches in diameter by 4 feet stroke and similar to those at No. 4 mine. Each lift raises the water from the shaft-bottom to the surface. The engine is in operation for 14 out of 24 hours, and, at present, makes 10 strokes per minute. The crab-engine has one horizontal cylinder, 6 inches in diameter by 12 inches stroke, with double gear, and a rope-barrel 18 inches in diameter.

Two Lancashire boilers, each 30 feet long and 7 feet in diameter, with cross tubes, supply steam at a pressure of 50 pounds per square inch to the winding-engine. Two Cornish boilers 26 feet long and 6 feet in diameter, with cross tubes, supply steam at a pressure of 100 pounds per square inch to the pumping-engine. The upper side of each boiler is covered with brickwork, leaving an annular space of 3 inches over the boiler, and acts as a protection from the weather.

No. 6 Mine.—The shaft, 9¼ feet by 5 feet, is sunk to a depth of 264 feet. The horizontal winding-engine, with two cylinders, 14 inches in diameter by 3 feet stroke, and drums 6 feet in diameter, raises one bogie in each cage. The other particulars are the same as at No. 5 mine.

No. 10 Mine.—The shaft, 12 feet by 5 feet, is sunk to a depth of 498 feet. The horizontal winding-engine, with two cylinders,

14 inches in diameter by 3 feet stroke, has drums 6 feet in diameter. The ropes, safety-cages and conductors are similar to those in use at Nos. 5 and 6 mines.

Two cylindrical boilers, 30 feet long by 5 feet in diameter at No. 6 mine, and two of the same dimensions at No. 10 mine, are provided with flues and coverings similar to those at No. 4 mine.

No. 12 Mine.—The shaft, 14 feet by 7 feet, is sunk to a depth of 636 feet. A stone-drift driven between the coal workings of No. 4 mine and the hæmatite-workings of No. 12 mine, enables coal or hæmatite, or both, to be raised at No. 4 and No. 12 shafts at pleasure.

The horizontal winding-engine has two cylinders, each 20 inches in diameter by 5 feet stroke, and two drums, 10 feet in diameter by 4 feet wide, with a breast-brake in the centre. The cages, safety-apparatus, ropes, tubs and conductors are similar to those in use at No. 4 mine.

The horizontal compound condensing pumping-engine, with Hathorn-Davey differential gear, has a high-pressure cylinder 22 inches in diameter by 6 feet stroke, the initial steam-pressure being 50 pounds per square inch; the low-pressure cylinder, 40 inches in diameter by 6 feet stroke, is placed 4 feet behind the former. The engine works, by a horizontal connecting-rod, two quadrants made of wrought-iron plates, filled in with pitchpine, and these actuate three lifts of pumps. The top and middle lifts are each 228 feet long, respectively, and the bottom lift, 180 feet long, or 636 feet in all. Each lift is worked by two hollow rams, each 9 inches in diameter by 6 feet stroke, and the rising-main for each lift is 9 inches in diameter. The engine works, at present, for 13 out of 24 hours, at eight strokes per minute, equal to a displacement of 256 gallons. The water raised is chiefly derived from old workings, having communication with the surface. In winter, water-boxes are frequently attached below the cages, in order that the winding-engine may assist the pumps, which are then running at 10 strokes per minute during the whole of the 24 hours. The horizontal condenser is placed on one side and below the low-pressure cylinder; and the air-pump, 17 inches in diameter by 2½ feet stroke, is worked by a wrought-iron connecting-rod from an arm on the quadrant-spindle. The steam-crab, with two vertical cylinders, 7 inches in diameter by 12 inches stroke, is geared by means of a worm and a worm-wheel, and two teethed wheels. The drum, placed on the third-motion shaft, is 2 feet in diameter.

Figs. 4 and 5 (Plate XXVII.) represent the method of working a vertical deposit of hæmatite at the No. 4 pit. The deposit was worked, from below upwards, in horizontal slices, 6 feet high. The empty space, from which the hæmatite has been removed, was filled with stowing, poured in from the surface, and levelled so that the workmen standing upon it, could reach the working-face. The hæmatite was teamed down a chute, and filled into tubs at the lower level.*

ULLCOATS MINE.

The Ullcoats mine lies at a much higher elevation, nor is it so deep as the Egremont mine, and accordingly the water-feeder is not so heavy, although two hollow rams, 11½ inches in diameter, are kept constantly raising 9,000 gallons per hour from a depth of 50 fathoms. The direct-acting pumping-engine has a cylinder 24 inches in diameter by 4 feet stroke. The winding-engine has two cylinders, each 16 inches in diameter by 3 feet 8 inches stroke, and a drum 8 feet in diameter. The sinking of this pit occupied three years, and occasioned a good deal of difficulty, a bed of quicksand being encountered.

The No. 2 shaft, a few hundred feet to the south-west of the No. 1 shaft, will be sunk to a depth of about 480 feet. The winding-engine will have two cylinders, 20 inches in diameter by 4 feet stroke and a drum 10 feet in diameter. The engine will be supplied with steam by two Galloway boilers, working at a pressure of 120 pounds per square inch.

WYNDHAM MINES.

The Wyndham Mining Company, Limited, has been in existence since 1877, and the first hæmatite was raised in 1879.

---

* "Timbering in the Iron Ore-mines of Cumberland and Furness," by Messrs. J. L. Hedley and W. Leck, H.M. Inspectors of Mines, *Trans. Inst. M. E.*, 1898, vol. xvi., page 281.

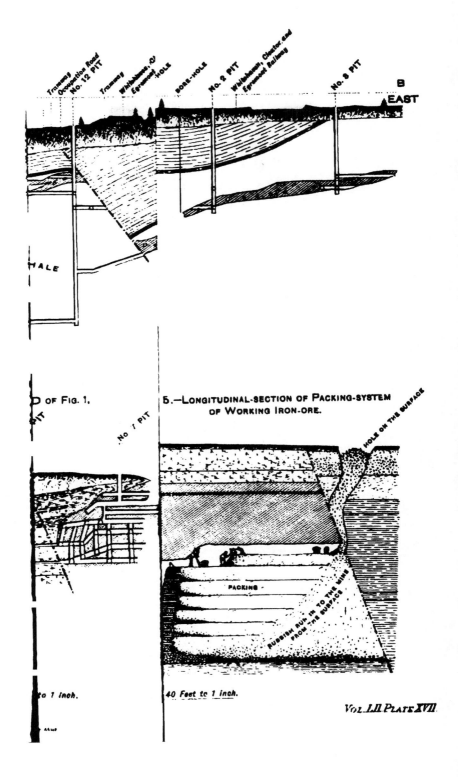

Trammay
Occupation Road
No. 12 PIT
Trammay
Whitehaven, &c
Egremont
HOLE
BORE-HOLE
No. 2 PIT
Whitehaven, Cleator and
Egremont Railway
No. 3 PIT
B
EAST

HALE

D OF FIG. 1,
PIT

No. 1 PIT

5.—LONGITUDINAL-SECTION OF PACKING-SYSTEM
OF WORKING IRON-ORE.

HOLE ON THE SURFACE

PACKING

RUBBISH RUN IN TO THE MINE
FROM THE SURFACE

to 1 Inch.

40 Feet to 1 Inch.

The early mining operations were confined to an area bounded on one side by the main street of Egremont and on the other side by the river; In course of time, the river was moved and a new course made for it; and then, as the town could not be moved, a considerable quantity of hæmatite was left beneath it. The ore, at the outcrop, is only covered by surface-measures of gravel and Boulder-clay, from 70 to 80 feet thick, and is almost vertical for a depth of about 180 feet. The ore-deposit flattens westward with a dip of about 1 in 10, and occasional drops of from 30 to 120 feet each, forming a series of steps downward from a depth of 250 feet on the east to 800 feet on the west, towards the Helder pit at Oregill. The ore-deposit, at a depth of 800 feet, is about ½ mile distant from the Falcon pit, where it is raised to the surface. About 1,500,000 tons of hæmatite have been worked, and the existence of ore having been proved in depth, the Helder shaft is being sunk to a depth of 1,260 feet.

At No. 3 pit, the winding-engine has two cylinders 16 inches in diameter by 4 feet stroke, with a drum 9 feet in diameter. The Cornish pumping-engine, with a cylinder 72 inches in diameter and 9 feet stroke, raises on an average 700 gallons of water a minute.

The Helder pit, at Oregill, will be sunk, 17 feet by 7½ feet, to a depth of 1,260 feet; and, when completed, it will be one of the deepest pits in West Cumberland. The water-feeders have given much trouble. The Evans pump has raised as much as 400 gallons per minute. A Cornish pumping-engine, with a cylinder, 80 inches in diameter, is in course of erection.

MESSRS. CHARLES CAMMELL, LAIRD & COMPANY, LIMITED: DERWENT IRON AND STEEL WORKS.

There are five blast-furnaces, with stoves, and ore, limestone and coke-bins on one side, and pig-beds on the other. A range of seventeen boilers is fired with blast-furnace gas, and a battery of self-stoking Babcock-and-Wilcox boilers is fired with coal. Six pairs of vertical blowing-engines are erected in a large building.

There are ten converters, eight for rail-steel and two for fish-plates; five semicircular casting-pits; two rail-mills and a

fish-plate mill: the ingots being transferred by locomotive
cranes and hand to re-heating furnaces. The rolls are arranged
in three groups, one in front of the other, the rail being completed from the ingot in one heat; rails weighing up to 120
pounds per yard can be rolled.

Beyond the mill are the great hot and cold banks for the
temporary storage and finishing of the rails.

The weekly output varies from 5,000 to 7,000 tons of rails,
including 450 tons of fish-plates; the blast-furnaces furnish
4,500 tons of pig-iron; and the remainder is obtained from the
Maryport furnaces, owned by the company, and in the open
market.

SANDWITH QUARRIES COMPANY, LIMITED.

Sandwith quarries are situated just over the ground rising
seaward from Croft pit; they are connected by a siding with the
railway, and occupy an imposing position at the brink of St.
Bees cliff, along which they extend for about ¼ mile. The
thickness of the stone is 300 feet. The quarry is equipped with
steam-cranes, steam-drills, steam dressing-machines and saws;
these are served by an overhead steam-gantry with a span of
50 feet. Some drilling is done by hand and some by steam. The
circular holes are nicked by a reamer on the line in which the
stone is to be cut, and just sufficient powder is used to cut the
stone without shattering it; consequently it breaks away in
fine blocks. About 13,000 tons of stone were loaded last year.
The stone is a red sandstone of agreeable and uniform colour
and a good grain, and is derived from the Permian formation.

THRELKELD QUARRIES.

These quarries and works are described in a paper written by
Mr. G. H. Bragg.*

---

* *Trans. Inst. M. E.*, 1903, vol. xxv., page 340.

THE THIRLMERE WATER-WORKS OF THE MANCHESTER CORPORATION.

Thirlmere, situated in Cumberland, at the foot of Helvellyn, alongside the road from Keswick to Grasmere, is about 3 miles long and ¼ mile across at the widest part. At the southern end, the land is flat, and appears to have been covered with water in olden times. At the north-eastern corner, there is a narrow gorge forming the outlet through which the St. John's beck finds its way to the river Greta, and ultimately through Derwent-water and by the river Derwent to the sea at Workington. The surrounding hill-sides, being steep, are covered with little verdure. They consist of Silurian rock, and the rain, falling from the clouds, finds its way at once into the lake. The gorge enabled an embankment or dam to be readily constructed, and the level of the water to be raised as required.

The natural level of the lake above sea-level is 533 feet 2¼ inches; the level, when raised 20 feet for the supply to Manchester of the first instalment of 10,000,000 gallons a day, is 554 feet; and the level, when raised to the full extent of 50 feet, is 584 feet. The length of the lake when raised 20 feet, is 3 miles 1,500 feet, and, when raised 50 feet, 3 miles 3,300 feet. The natural drainage-area is 7,400 acres, and the additional drainage-area to be hereafter diverted into the lake is 3,600 acres. The natural surface-area of the lake is 330 acres; when raised 20 feet, the area is 565 acres, and the capacity 2,534,000,000 gallons; and when raised to the full extent of 50 feet the area is 793 acres, and the capacity 8,135,000,000 gallons. The quantity of compensation-water discharged into the St. John's beck before the diversion of the additional drainage-area into the lake is 4,126,125 gallons per 24 hours; and when the lake is raised to the full extent of 50 feet, 5,520,487 gallons per 24 hours.

The level of the lake is raised by means of an embankment constructed at the outlet into the St. John's beck. The top of the embankment is 6¼ feet above the level of the lake when fully raised; the length of the embankment is 857 feet; the width at the top of the embankment is 18½ feet; and the greatest height of the embankment from the foundation is 104½ feet. The length of new roads constructed on the west side of the lake is 5 miles

2,895 feet; on the east side of the lake, 2 miles 726 feet, and across the embankment 5,049 feet.

The aqueduct, with a diameter of 7 feet, and a fall of 20 inches per mile, is constructed to convey 50,000,000 gallons a day. The length from Thirlmere to Manchester is about 96 miles, namely, tunnels, 14½ miles; cut-and-cover, 36¾ miles; and pipes, 45 miles. The diameter of the straining-well for the admission of water into the aqueduct is 37½ feet, and the depth 65 feet.

The rainfall over the watershed varies from 52 to 137 inches per annum. The storage, when all the works are completed, will provide 50,000,000 gallons of water per day for 160 days, even if no rain falls during that time, without drawing the water below the original margin of the lake. It is intended to provide the full supply by instalments of 10,000,000 gallons a day, at such intervals of time as may be found necessary.

The purchase of the watershed, including the lake; the way-leave for the tunnels, the cut-and-cover, and for the laying of five lines of siphon-pipes across the valleys from Thirlmere to Manchester, and the works at present carried out, including the tunnels, cut-and-cover or concrete tunnel, to convey 50,000,000 gallons per day; and one line of pipes to convey 10,000,000 gallons per day to Manchester, have cost over £2,700,000. When five lines of pipes have all been laid, and the level of the lake raised 50 feet, it is estimated that the cost will be about £5,000,000.

The laying of the second line of pipes of the aqueduct was completed and brought into use in November, 1904.

APPENDICES.

I.—NOTES OF PAPERS ON THE WORKING OF MINES, METALLURGY, ETC., FROM THE TRANSACTIONS OF COLONIAL AND FOREIGN SOCIETIES AND COLONIAL AND FOREIGN PUBLICATIONS.

EARTH-TEMPERATURES AND GEOLOGICAL PHENOMENA.

Der Wärmerégime der Erdoberfläche und seine Beziehungen zu geologischen Erscheinungen. By L. ZACZEWSKI. *Centralblatt für Mineralogie, Geologie und Palæontologie, 1904, pages 721-723.*

The author proposes to issue shortly a memoir on this subject, in the German language. He points out that the " so-called internal heat of the earth " played a quite insignificant part, in comparison with the heat-energy derived from the sun. The available geothermic data are confined to medium latitudes and to a comparatively small number of points on the earth's surface, and they really afford no justification for the assumption that a source of high temperature exists in the core of the globe. The author has drawn up a series of diagrams, based on the facts recorded by the meteorologists, Mr. Angot and Dr. Ekholm, and these lead him to the following conclusions:—

1.—Low latitudes form a region of continuous accumulation of energy derived from the sun; while high latitudes form a region of continuous loss of energy.

2.—The hypothesis of the existence of a stratum of constant temperature at a shallow depth all over the earth is based on a misapprehension: such a stratum can only exist within well-defined limits in medium latitudes.

3.—The existence of eternally frozen strata in high latitudes is precisely attributable to (loss by) radiation of heat-energy.

The author discusses such questions as the beginning of geological life, the primary causes of mountain-building, etc., and finally summarizes his views in the statement that " the tectonic life of the earth's surface is the result of the action of the sun's heat-energy on that surface." Seas and continents, rivers and valleys, glaciers and volcanoes, earthquakes and secular changes of level, all alike arise from the same source as that whence the entire organic life of the earth and all phenomena of movement derive their power, that source being (in the author's opinion) the sun. L. L. B.

THE INFLUENCE OF TEMPERATURE ON THE MAGNETIC PROPERTIES OF MAGNETITE.

Ueber das Verschwinden und Wiedererscheinen des Magnetismus beim Erhitzen und Abkühlen von Magneteisenerz. By F. RINNE. *Centralblatt für Mineralogie, Geologie und Palæontologie, 1902, pages 294-305, with 3 figures in the text.*

The author gives an account of the various experiments by which he has proved that if magnetite be raised to a red heat the ore loses its essential characteristic (essential at ordinary temperatures, at least) of being attracted by the magnet.   On cooling down again, it appears to regain possession of its magnetic properties.   The temperature at which the loss of magnetism takes place is estimated at 1,067° Fahr. (575° Cent.).   Previous to actual experimentation the author was unaware of this phenomenon, which, he says, is not mentioned in the mineralogical manuals.   But he finds that Michael Faraday had recorded it in the *Annalen der Physik und Chemie* as long ago as 1836.

Apart, however, from mere qualitative experiment, the present author resolved to follow up the investigation by quantitative methods, that is, to measure the actual progress of the phenomenon at different temperatures. His results show that the magnetic susceptibility of magnetite at first increases gradually with heightening temperature, then, however, decreases suddenly and very markedly.   With lowering temperature the phenomena follow a reverse order.   The magnetite does not, however, regain on cooling its magnetic properties in as full a measure as it once possessed them: this diminution is explicable by the chemical change which the powdered ore undergoes in the course of experiment, despite the precautionary use of asbestos-stoppers for closing the glass-tube in which it is heated.   The ingress of atmospheric air (which cannot be entirely prevented) alters in fact a portion of the magnetite from Fe_3O_4 into Fe_2O_3—a non-magnetizable oxide.   The attempt to determine exactly the (magnetically) critical temperature did not meet with success.   Certain theoretical considerations as to the distribution of magnetite in basalts are dealt with, and reference is made to the behaviour of metallic iron in regard to the magnet under varying conditions of temperature.   L. L. B.

UNDERGROUND TEMPERATURES IN THE PARUSCHOWITZ NO. 5 BORING, GERMANY.

(1) *Über die Temperaturen in dem Bohrloche Paruschowitz V.* By PROF. F. HENRICH. *Zeitschrift für das Berg-, Hütten- und Salinen-wesen im Preussischen Staate*, 1904, *vol. lii., Abhandlungen, pages* 1-11, *with 1 figure in the text and 1 Plate.*

(2) *Über die Temperaturverhältnisse in dem Bohrloch Paruschowitz V.* By PROF. F. HENRICH.   *Zeitschrift für praktische Geologie*, 1904, *vol. xii., pages* 316-320, *with 1 figure in the text.*

The Paruschowitz boring, now the deepest in the world, has been carried down to the enormous depth of 6,571 feet, and the temperature-observations were, as in the case of the celebrated Schladebach boring, taken with "outflow-thermometers."   These differ from the ordinary instruments in having their stems cut off at an angle of 60 degrees: several of them (six) are enclosed in a steel-cylinder which is sunk with the boring-rods down to the particular depth at which it is desired to record the temperature, and the thermometers are left there for 5 to 10 hours.   They are then brought up to the surface, and in due order are placed in a recipient containing water; into this warm water is stirred until a mercury-drop of medium size appears above the opening. Simultaneously with the appearance of the mercury-drop the temperature of the water is read off an ordinary thermometer, and in this way the underground temperature is arrived at.   The probable margin of error (either positive or negative) amounts to 0·2° Cent., but in practice errors three or four times as great have been made, and the author consequently recommends in future observations the use of maximum-thermometers.

At Paruschowitz, temperature-observations were taken at intervals of 102 feet (nearly) from the surface down to a depth of 6,425½ feet, at which point the unprecedented temperature of 69·3° Cent. or 156½° Fahr. was recorded. In order to avoid the errors of observation due to currents carrying warm water upward and cold water downward in the bore-hole, clayey mud or "slime" was pumped in, the frictional effect of which was sufficient to prevent the formation of such currents. At very great depths, where the rod occupies nearly the entire space of the bore-hole, the influence of these currents on temperature-observations is practically negligible. Other possible causes of error are enumerated; and, in order to avoid most of these, it is stated that temperature-observations should proceed *pari passu* with the progress of the boring, and should be carried out before pipes are inserted.

The boring was carried down through the Coal-measures, cutting on the way no less than eighty-two coal-seams ranging in thickness from 6 inches to 11½ feet, but one (at the depth of 3,698 feet) is about 33 feet thick. The total thickness of coal amounts to 290 feet, and the seams are interbedded with grey sandstones, sandy shales, quartz-grits, ironstone-shales, and conglomerates. The precautions taken to prevent inflows of water in the boring from vitiating the temperature-records are recited, and attention is then drawn both by word and by diagram to the important fact that the temperature was proved to increase constantly with the depth from the surface.

At first sight, the tabulated results appear to show that the ratio of increase is greater the farther one gets down from the surface, but the author hints that it would be rash to jump to this conclusion. The varying conductivity of the different strata accounts for much: thus, heat travels much more slowly through the compact 33 feet coal-seam than through the (comparatively) loose-textured sandstones overlying or underlying it.

The results are analysed and discussed by the author in great detail, and he sets forth his reasons for putting the increase of temperature as equivalent to 1° Cent. for every 31·82 metres or 104·37 feet of increase in depth from the surface. This ratio is greater than that calculated from the results obtained in the Schladebach and Sperenberg borings, but it would appear to be more reliable, the precautions taken at Paruschowitz and the experience previously gained having reduced the margin of probable error. It may be mentioned that the altitude of Paruschowitz is 833 feet above sea-level, and the mean annual temperature is 7·8° Cent. or 46° Fahr. L. L. B.

A FACTOR IN THE FORMATION OF SALT-DEPOSITS.

Über die Akkumulation der Sonnenwärme in verschiedenen Flüssigkeiten. By A. VON KALECSINSZKY. *Mathematischer und Naturwissenschaftlicher Anzeiger der ungarischen Akademie der Wissenschaften*, 1904, *vol. xxii.*, *pages 29-53.*

The author has been for some years engaged in the investigation of temperatures at various depths in the Hungarian salt-lakes (more especially in connection with the accumulation of solar heat therein), and he finds as the result of this investigation as well as of a series of elaborate experiments, that, if a thin stratum of fresh water overlies a saline solution, the deeper strata of that solution become warmer than the upper layers. This observation is confirmed by other temperature-records from salt-lakes in countries as far apart as Rumania, Norway, and Siberia.

Now, of the various salts which go to make up the Stassfurt deposits, for instance, some (such as polyhalite) are easily formed at a temperature of 77° Fahr., but others are formed only at very much higher temperatures. There, polyhalite is found interbedded with the upper layers of the deposits, while anhydrite occurs in the lower layers. The regular recurrence of this order of deposition is so marked that the miners compare it with the annual rings in tree-growth, and they connect it, as these are connected, with seasonal variation. Prof. Van't Hoff informed the author of his belief that anhydrite at Stassfurt represents a summer-deposit and common or rock-salt a winter-deposit. So, too, with the other salts, a critical point being reached in some cases where the temperature rises too high, in others where it falls too low, to admit of their remaining in solution. L. L. B.

FAULTS AND FOLDS.

Failles et Plis. By H. DOUVILLE. *Comptes-rendus hebdomadaires des Séances de l'Académie des Sciences*, 1904, *vol. cxxxviii., pages* 645-646.

The author defines faults as corresponding, in ultimate analysis, to a lengthening of a portion of the earth's crust which was in a state of tension: while folds correspond to a shortening of some part of the earth's crust, and are primarily due to compression. He then considers the distribution on the surface of the globe of fault-areas where the effects of tension are predominant, and fold-areas where those of compression dominate. The former category seems in some way to follow the parallels of longitude: take, for instance, the shores of the Pacific, marked out by the chain of the Andes on one side, and by the volcanic islands of Japan and the Javan-Sumatran group on the other. The belts marked by the Ægean, the Red Sea, and the Mozambique Channel are equidistant. The Atlantic, too, may represent an area of sagging due to faults. The fold-areas are transverse to those just considered, and generally follow the parallels of latitude: they are characterized by the Caledonian, Hercynian and Alpine chains.

To sum up: during geological time, the length of the Equator has not varied to any considerable extent, while the meridians have certainly diminished in length. Whence it may be inferred that the flattening of the terrestrial spheroid has become gradually more marked, and the rotational movement of the earth has been correspondingly accelerated. L. L. B.

THE CONCENTRATION-THEORY AND THE GENESIS OF ORE-DEPOSITS.

Die Bedeutung der Konzentrationsprozesse für die Lagerstättenlehre und die Lithogenesis. By DR. RUDOLF DELKESKAMP. *Zeitschrift für praktische Geologie*, 1904, *vol. xii., pages* 289-316.

The author begins with the remark that the concretionary processes going on within sedimentary material deserve far greater attention than has hitherto been paid to them, in connection with the study of ore-deposits and the formation of rocks in general. By the term "concentration" he implies the accumulation of any homogenous substance, for he assumes that in the primitive or original magma all substances were present, evenly distributed and intermingled.

This memoir is intended to supplement the work of Messrs. Vogt, Clarke and Hillebrand, Klockmann, and others, and deals at greatest length with those processes of concentration which originated ore-deposits other than veins, as well as various mineral formations of industrial importance although they do not fall within the definition of "ore-bodies."

The author classifies concentrations as (1) pre-existing, (2) primary, and (3) secondary: to the first category belong clay-galls in sandstone, the dolomite and limestone-nodules in the Rothliegende and the Bunter Sandstone, and placer-deposits, etc., formed by selective sedimentation.

The second category includes several subdivisions, such as (a) deposits formed by submarine mineral springs (under these Dr. Gümbel included the manganese-nodules dredged up in the *Challenger* expedition, but the author does not accept that explanation and ranges them under c); (b) concentration induced by local variation in the supply of detritus (such as the sphærosiderite-nodules in clay-slates, the Huelva pyrites, according to Dr. Klockmann, the Jurassic pisolitic iron-ores of Lorraine, etc.); (c) accretion and decomposition going on during the formation of a sediment; (d) mutual precipitation of locally-introduced metallic salts and of particles in suspension by adsorption; (e) formation of chemical sediments by the gradual dessication of gulfs shut off from the sea (lagoons), or in consequence of the increasing concentration of waters occupying depressions which had no outflow; and (f) successive precipitation from mineral-springs.

To the third category belong deposits originating (a) from the concentration of a substance which either was from the first distributed evenly throughout the sediment, or arose from the subsequent oxidation or reduction of such a substance; (b) from the concentration of a substance formed by the interaction of two or more bodies, which were primarily distributed evenly throughout the sediment; (c) from the concentration of substances, which were in part originally present in the sediment, or in the weathered residuum or decomposition-products of crystalline and sedimentary rocks, but were also in part introduced secondarily by thermal waters, etc.; (d) from the concentration of substances which were brought in by infiltration, and replaced metasomatically the primary material; (e) from concentrations initiated by secondary infiltration, but confined by adsorption to certain layers of a complex of bedded rocks; and (f) from concentrations similarly initiated, but in the form of aqueous solution, the infiltration arising from thermal springs or from the leach-waters of overlying beds.

Examples of deposits in all parts of the world are cited and described under these various headings; but it must suffice here to recite the author's main conclusions, in regard to epigenetic deposits which are precipitates from aqueous solution. They are as follows:—

(1) Ore-deposits may originate from solutions of the most diverse chemical composition, at various temperatures and pressures. Nevertheless, these solutions will have been preferably alkaline, and temperature and pressure will have been mostly high.

(2) The metal-bearing solutions may travel in any possible direction, yet in the main they will have travelled from below upward.

(3) Precipitation will have been induced by such variations in the conditions as decrease in pressure and temperature, increase of concentration by the evaporation of the water or by the escape of carbon dioxide and other gases; or perhaps by the mixture of various solutions travelling from different points. Moreover, the chemical nature of the country-rock must influence the character of the infilling of a vein.

(4) The solutions will have derived their metalliferous particles chiefly from "magmatic centres" (in the sense used by Dr. Stübel) or from the central region of the globe; those few, the components of which have been derived from the leaching-out of rocks, can in the majority of cases be traced to the lowermost rocks of the earth's crust (the so-called "thermosphere"). It is not denied, however, that the rocks of the upper regions of the crust and the country-rock of metalliferous veins have contributed their quota to the infilling of metalliferous fissures.   L. L. B.

THE ORIGIN OF PHOSPHORUS IN IRON-ORES.

Sur l'Association géologique du Fer et du Phosphore et la Déphosphoration des Minerais de Fer en Métallurgie naturelle. By L. DE LAUNAY. *Comptes-rendus hebdomadaires des Séances de l'Académie des Sciences,* 1904, *vol. cxxxviii., pages* 225-227.

The author has attempted to investigate the laws which determine either the association of iron and phosphorus in nature or their separation, with the view of making it possible to predict in what class of deposits we may *a priori* expect to find iron-ores free from phosphorus. His general conclusions are (1) that the association of the two elements is primarily of deep-seated origin, and he quotes in this connection the presence of phosphide of iron in meteorites: traces of phosphorus, too, are present in certain metalliferous veins; (2) that in the process of natural "scorification" which gave rise to the various silicates of alumina of which the earth's crust is chiefly made up, such phosphorus as existed in the state of a deep-seated phosphide was everywhere rapidly oxidized, and then absorbed by lime in the form of apatite; and (3) that the dephosphorization of sedimentary iron-ores by metamorphism is the more likely to occur in proportion as the iron-ores in question are of more ancient date.

In principle, very little phosphorus is found in ores of basic segregation (magnetites and titano-magnetites); very little also in those which, commencing by the metasomatic replacement of limestone, have passed through the stage of carbonates, before undergoing peroxidation and hydration nearer the surface. On the other hand, the author explains how sedimentary iron-ores must be expected to contain more or less phosphorus, unless subsequently purified by secondary reactions producing natural dephosphorization, as above-mentioned. In some districts, such as Norberg in Sweden, and the Vissókaya Gora in the Urals, siliceous ores highly charged with phosphorus are found alongside calcareous ores comparatively free from it. Generally speaking, it is found that those same phenomena which have had the effect of cleansing an ore of much of its phosphorus, have increased its original percentage of manganese; and less frequently, its percentage of barium or cobalt.   L. L. B.

THE PART PLAYED BY PHOSPHORUS IN ORE-DEPOSITS GENERALLY.

Sur le Rôle du Phosphore dans les Gîtes Minéraux. By L. DE LAUNAY. *Comptes-rendus hebdomadaires des Séances de l'Académie des Sciences,* 1904, *vol. cxxxviii., pages* 308-310.

In the processes of natural metallurgy which have resulted in the formation of igneous rocks and metalliferous ores, certain metalloids by combining with the metals seem to have enhanced the mobility of the latter and to have facilitated their ascent towards the surface. These metalloids, termed "mineralizers," are more especially chlorine and fluorine, sulphur and its homologues (selenium and tellurium), and finally arsenic and antimony. Attention has not been so largely directed to the part played by other metalloids, whose affinity for oxygen impels them to enter at once into combination with that element (as soon as approach to the surface affords an opportunity), and marks to a certain extent their relation to the metals. Chief among these is phosphorus, generally found in combination with lime in nature, because of the almost ubiquitous presence of that alkaline earth in the silicate-crust of the globe.

It appears probable that phosphorus, either in the elemental state, or in combination as phosphoric acid, has not infrequently acted as a "mineralizer" in the metallurgical processes of nature, simultaneously with chlorine and fluorine. This would seem to apply especially to the white-mica granites and to the stanniferous and uraniferous veins characteristic of them.

Vanadium, in the author's opinion, plays a part absolutely analogous to that of phosphorus, though more rarely. L. L. B.

THE FORMATION OF VANADIUM-ORES.

Sur la Formation dans la Nature des Minerais de Vanadium. By A. DITTE. *Comptes-rendus hebdomadaires des Séances de l'Académie des Sciences*, 1904, *vol. cxxxviii., pages* 1303-1308.

The only vanadium-ore that occurs in nature in any abundance is the vanadate of lead, occasionally combined into a chloro-vanadate. Vanadium itself is widely diffused in the earth's crust, occurring in minute, or rather infinitesimal proportions, in iron and other ores, and also in clays, grits, etc. Exactly in what form it so occurs is difficult to predicate; but it is at all events accessible to the dissolving action of pluvial or infiltrated waters, and so is converted into vanadic acid, and this again forms soluble vanadates. Waters charged with these and other salts possess the faculty of leaching out copper, zinc, lead, and other metals from the older rocks in which they are disseminated. If such waters come into contact with weak solutions of lead sulphate derived from the oxidation of galena, they form vanadates of lead, such as those found in the Wanlockhead and Beresov mines. In the presence of chlorinated solutions vanadinite or chloro-vanadate of lead may be formed, as Sir H. E. Roscoe and Prof. Hautefeuille have successfully proved by synthetic experiments. These ores do not even require for their genesis complete solution of the sulphate of lead, as the author's own experiments have shown. In the Santa Marta mine, in Spain, vanadate of lead occurs in association with cerussite on the top of galena. Of course, the action of pressure, in activating the solubility of certain intractable ores in the percolating waters, has to be taken into account. The author considers that his explanation of the origin of the ores of vanadium fits in exactly with their occurrence, as observed in various localities in Europe and America. L. L. B.

THE FIGURE OF THE EARTH IN RELATION TO EARTHQUAKES, ETC.

Relation des Volcans et des Tremblements de Terre avec la Figure du Globe. By CH. LALLEMAND. *Compte-rendu de la Trente-deuxième Session* (1903) *de l'Association française pour l'Avancement des Sciences, part ii.,* 1904, *pages* 157-168, *with 3 figures in the text.*

Under the term " earthquakes " the author includes not only the tremors perceptible to our ordinary senses (of which on an average about two occur daily in some part of the inhabited world, and possibly about fifty *per diem* if we take the surface of the entire globe), but also those continuous internal vibrations which instruments of extreme sensitiveness are alone capable of recording. These slight earth-movements, which seem to be amplified by barometric depression, are more frequent in winter than in summer, and usually increase in intensity at the approach of the equinoctial periods.

After pointing out that vulcanicity and seismicity have, during the last year or two, become vastly more active, after a period of comparative quiescence, the author discusses the tetrahedric theory of the figure of the earth in a favourable sense, and dismisses successively the principal objections which have been raised to it. He then urges that, although volcanic eruptions and earthquakes constitute two distinct orders of phenomena, they are both alike the natural and logical consequences of the movements of the lithosphere resulting primarily from its contraction on cooling. These phenomena are more likely to occur at those points where the crust has undergone greatest deformation, points which, in fact, form the zones of least resistance. We may expect to find these in the regions neighbouring the " crests " and apices of the tetrahedron, and especially in the great intercontinental depression where the torsion of the austral apex of the terrestrial " spinning-top " superimposes its influence on that of the plication of the tetrahedral ridges. Moreover, the possible existence of an internal " lunisolar tide," concordant with the great oceanic tides, might, in the vicinity of the Equator and in the entire tropical belt, cause on some occasions a rupture of equilibrium. These statements are confirmed by the examination of a map of the world, showing the distribution of seismic and volcanic phenomena, and their connection with the great inter-continental depressions.

L. L. B.

GEOSYNCLINALS AND SEISMICITY.

Sur la Coïncidence entre les Géosynclinaux et les Grands Cercles de Sismicité maxima. By F. DE MONTESSUS DE BALLORE. *Comptes-rendus hebdomadaires des Séances de l'Académie des Sciences,* 1904, *vol. cxxxix., pages* 686-687.

The author has already shown that the regions especially subject to earthquakes are comprized within two narrow belts coincident with two great circles of the globe. Now, these are found to coincide precisely with the geosynclinals of the Mesozoic era, which later, in the Kainozoic or Tertiary era, gave place to the great lines of corrugation of the earth's surface. It follows from this that 94 per cent. of the recorded earthquakes correspond to general movements of the crust, whereby the sediments of the ancient geosynclinals, folded and thrust up, have ultimately come to form the geanticlinals or great mountain-ranges, the upheaval of which is of comparatively recent date.

Seismic instability could hardly be uniform along these belts, because the movements do not synchronize and they differ in amplitude. In some cases. the seismicity of a given region is found to be proportionate to the mass of the upheaved sediments.

If we consider the older geosynclinals, as, for example, those in which the Coal-measure basins were laid down before the post-Carboniferous foldings. seismic areas may be plotted out, of moderate instability merely (because of the antiquity of the movements of the crust there), but strictly limited to those basins. This observation may be extended to all the Coal-measure regions of the globe, excepting those that lie within ancient folded massifs. The author concludes that there is a remarkable homology between ancient and recent geosynclinals, and that the general cause of earthquakes is intimately connected with the great movements of the crust, without thereby excluding secondary, that is, local and more immediate causes. L. L. B.

SEISMOLOGICAL PHENOMENA IN CENTRAL EUROPE.

Seismologische Untersuchungen. By S. GÜNTHER and J. REINDL. *Sitzungsberichte der mathematisch-physikalischen Klasse der königlich-bayerischen Akademie der Wissenschaften zu München,* 1903, pages 631-671, and 1 plate.

The first part of this paper is devoted to a careful collation and scientific investigation of the accounts which have been handed down to us of two of the most terrible earthquakes that have ever been recorded in Central Europe. The first of these took place on January 25th, 1348, and with Carinthia for its pleistoseismic area (including the towns of Villach and Klagenfurt) made its effects felt over a region extending from beyond Vienna in the east to Basel in the west, and from Frankfort-on-the-Main in the north to Naples in the south. The second took place in 1356, on St. Luke's day, and affected Bavaria among other countries.

The next part of the paper deals with the seismicity of the Ries basin, which in respect of earth-tremors is quite an exceptional district—so far as Bavarian territory on the right bank of the Rhine is concerned. A description is given of the principal earthquakes recorded in this district between the years 1471 and 1903; and it is pointed out that as none of these appear to have made themselves conspicuously felt far beyond the district, the epicentrum must lie within it. The latest recorded earth-tremors there took place between 5 and 6 a.m. on August 11th, 1903. Now, the Ries is a volcanic district, and the seismic phenomena noted there are classified by the authors as volcanic-tectonic or imperfectly tectonic, and are perhaps the long-continued results of an internal break-up of the earth's crust, due to vulcanicity in former ages.

The third part of the paper treats of the peculiar sounds variously known as "Barisal guns," "mist-pouffers," etc., but which the authors prefer to denominate "earth-reports" or "earth-bangs" (*Bodenknalle*). They do not concur in the view that any of these sounds are of atmospheric origin, if heard over wide areas: they admit indeed that such detonations may in some rare cases be traced back to atmospheric causes, but do not then possess sufficient inherent energy to travel far. On the whole, they regard most of these detonations are due to feeble, "embryonic" earth-shocks, not otherwise perceptible by the human organism. L. L. B.

EARTHQUAKES AND EARTH-TREMORS IN PORTUGAL IN 1903.

(1) *Les Tremblements de Terre de 1903 en Portugal.* By PAUL CHOFFAT. *Communicações da Commissão do Serviço geologico de Portugal*, 1904, *vol. r., pages* 279-306, *and* 1 *plate.*

(2) *Sur les Séismes ressentis en Portugal en 1903.* By PAUL CHOFFAT. *Comptesrendus hebdomadaires des Séances de l'Académie des Sciences*, 1904, *vol. cxxxviii., pages* 313-315.

Feeble shocks of earthquake are of frequent occurrence in Portugal, but the little kingdom does not yet boast a properly-organized seismological service, and the question of establishing recording-apparatus in the various meteorological observatories is still " under consideration."

On August 9th, 1903, a shock of unusual violence took place, which attracted universal attention, only three fairly comparable with it having occurred in Portugal within living memory. Feebler shocks followed in September and October. To return to the principal shock of August 9th: it was felt over nearly the whole area of Portugal, and across the Spanish frontier in the provinces of Huelva, Seville and Badajoz. It occurred about 10.15 p.m., and was accompanied by a barometric rise of 2 millimetres (observed at Lisbon). The temperature of the air at the time was rather over 64° Fahr. There was a preliminary vibration (horizontal) lasting 3 seconds, then an interval of 2 seconds, followed by a second vibration, stronger than the first, lasting 10 seconds, with a motion changing from vertical to horizontal. The usual earthquake-sound was heard, compared by some observers to a detonation, but by most others to the rumbling and rattling of a wagon laden with iron-bars. The shock appears to have travelled from east to west: the intensity was from 7 to 8 (in the De Rossi-Forel scale) over an area extending along the coast for 112 miles from north to south, that is from Lourinhã to the mouth of the Mira, but only for 20 miles or so from west to east. Within this area falls of chimneys, roofing-slates, etc. and the fissuring of walls, were recorded from many localities; and in Lisbon the effects (intensity 7) were such as to cause widespread terror among the audiences in the theatres, the patients in the hospitals, etc. The inhabitants of entire streets were panic-stricken, and some people rushed out of doors without stopping to dress themselves. The intensity was really greatest in the Arrabida range; as the same observation has been made in previous earthquakes, the cause may lie in the asymmetrical, all but monoclinal structure of that range, the southern flank of which plunges abruptly into a sea of considerable depth.

An after-shock made itself perceptible, over a much less extensive area, on September 14th, 1903, at 1·30 p.m. It is true that its intensity (7) was nearly as great as that of the principal shock (of August 9th), but as its duration was much shorter the effects were proportionately smaller. The direction of travel was again from east to west, and no vertical motion was observed. Earthquake-sounds were reported from Cezimbra and Cintra; material damage was noted only at the latter place and at Lisbon. Plotting out the isoseismals, the author obtains, as in the case of the earthquake of August 9th, two concentric zones open towards the Atlantic, the difference being that on September 14th the epicentral area coincided with the Serra de Cintra instead of coinciding with the Arrabida range.

Further after-shocks were recorded on September 28th (8 a.m.) and December 1st (6·40 a.m.), the second chiefly at Huelva; and on October 14th at 2·55 p.m. That of October 14th was noticed at localities lying north and south of the Serra do Monte Junto, and is believed to have been due to sagging along the fault-lines characteristic of the above-mentioned range.

A short description is given of previous earthquakes in Portugal, from 1755 onwards, and the author then lays down certain general conclusions. He distinguishes three categories of earthquakes in Portugal, the first of which includes those of considerable extent, one principal centre lying in the depths of the Atlantic opposite the western coast (generally in the latitude of the Arrabida range or else of the Cintra range). The zones of equal intensity cover areas consisting of rocks of the most diverse nature, and the records are neither sufficiently accurate nor sufficiently numerous to predictate the relationship between the seismic phenomena and the lithology of the country which they affect. Irregularities in the isoseismal curves may be traced to the effect of mountain-massifs; and there is, further, no doubt that a more intimate knowledge of the distribution of the seismic intensities will reveal a causal connection with great fault-lines, as well as with great ocean-depths.

On the whole, the earthquakes by which Portugal is shaken at intervals have two principal centres: that previously mentioned as lying out at sea, on the latitude of the mouths of the Tagus and the Sado, and another in Andalusia. Sometimes, an earthquake proceeds from one centre, sometimes from the other; and occasionally it proceeds alternately from both.

L. L. B.

THE EARTHQUAKE OF SHEMAKHA, SOUTHERN RUSSIA, 1902.

Shemakhinskoye Zemhtriasenie. By V. **Weber.** *Trudi geologicheskago Komiteta,* 1903, *series* 2, *No.* 9, 73 *pages, with* 56 *figures in the text and* 3 *plates.*

This most destructive earthquake took place on February 13th, 1902 (January 31st in the Russian calendar). The author's researches were confined to the area externally delimited by the isoseismal corresponding to 8 in the De Rossi-Forel scale; thorough investigation was rendered difficult by the low stage of civilization of the natives of the Caucasus, and by the complete absence of seismological stations provided with registering instruments such as would have furnished authentic records.

The isoseismal areas of the Shemakha earthquake, corresponding to 8, 9 and 10 in the De Rossi-Forel scale, are divisible into seven categories, as follows:—

X.—(1) Hardly a building left intact: walls demolished almost completely.

IX.—(2) Some buildings standing though badly damaged. Many entirely demolished.   (3) Solidly-built structures standing: less solid structures a heap of ruins. Many walls thrown down.   (4) Solidly-built structures damaged in part: some of the less solid ones left standing.

VIII.—(5) Less solidly-built structures: about 50 per cent. left standing. (6) Partial damage to buildings in places.   (7) Not much damage to buildings.

The district which suffered the most from the shocks lies on the southern slope of the main range, which gradually passes by a series of terraces down into the Kura plain. These terraces trend parallel with the strike of the rocks, and parallel with the main range, from west-north-west to east-south-east: they are cut through by three principal river-valleys and by a whole network of small streams and gullies. The isoseismal area over which buildings were destroyed was about 50 miles in length by 23 miles in width.

On the whole, the great river-valleys seem to have acted as barriers to the extension of the shocks, a part played in a lesser degree by some of the smaller valleys. The influence of rock-foundations in minimizing damage to buildings, as compared with those built on less compact strata, is also noticeable. On the other hand, villages built on thick deposits of alluvium suffered less than those built on comparatively thin alluvia overlying solid rock (an especially dangerous combination). Isolated hills and escarpments are also shown to be dangerous localities from the seismic point of view. The initial shock was preceded by a subterranean rumbling: the shock appeared to travel from west to north-west and back again, and was succeeded by undulations of " mixed character," during which people had difficulty in keeping their feet. General panic prevailed, and the confusion is described as terrible. Some springs ceased to flow and others flowed in increased volume. In some places fissures were opened up in the earth, and there were eruptions of small " mud volcanoes." This earthquake of 1902 appears to be identical in many respects with that recorded in the same area in 1859.   L. L. B.

EARTHQUAKES IN RUMANIA AND BESSARABIA.

Sur les Tremblements de Terre de la Roumanie et de la Bessarabie. By F. DE MONTESSUS DE BALLORE. *Comptes-rendus hebdomadaires des Séances de l'Académie des Sciences*, 1904, vol. cxxxviii., pages 830-832.

The systematic observations made by the Rumanian Meteorological Department during the past 12 years make it now possible to draw up a seismological chart of that kingdom and of Bessarabia. An inspection of this shows that the seismic epicentres are most thickly clustered along a line drawn from Kishinev to Bucharest.

Western Wallachia, the entire Danubian terrace between Turn-Severin and Kalarash, the Dobrudsha, and Northern Moldavia constitute a region of far greater stability than Bessarabia, or more especially than Central and Southern Moldavia, and Eastern Wallachia. It is curious to note that the Kishinev-Bucharest axis, ranging north-east and south-west, is parallel to the seismic axis of the Bohemian Erzgebirge, to the petroleum-belt of Ramnicu-Sarat, and to the strike of the volcanic rocks and principal metal-liferous veins of the western districts of Wallachia.

In former ages, the shore-line between Odessa and the mouths of the Danube was visited by violent earthquakes, the epicentres of which appear to have lain out somewhere in the middle of the Black Sea: perhaps, they represented a moribund stage of the Caucasian-Balkanian dislocation, which caused the comparatively-recent depression of the eastern portion of that sea.
      L. L. B.

EARTH-TREMORS IN SWEDEN IN 1902 AND 1903.

Meddelanden om Jordstötar i Sverige No. 13. By E. SVEDMARK. *Geologiska Föreningens i Stockholm Förhandlingar*, 1904, vol. xxvi., pages 201-209.

According to the newspapers, on April 29th, 1902, shortly after 2 p.m., violent earth-tremors were felt at several localities in Halland and the bordering districts. At certain points along the Halmstad and Nässjö rail-way there was a thunderous subterranean rumbling, houses quaked, and articles of furniture were set in motion: accounts agree in fixing the duration of the phenomenon at about 60 seconds. Curiously enough, the Central Meteorological Institute at Stockholm had, at the date of writing, received no reports in regard to these earth-tremors in Halland, but had, on the other hand, received much information as to earth-tremors having been observed on the same day and the following day in Western Gothland and Småland. These shocks were fairly severe, as they caused much rattling of windows and clattering of crockery, but they were not sufficiently so to cause damage to property or injury to persons. The times given range between 2 and 2·30 p.m., and the directions vary (on the whole) from east and west to north-east and south-west.

On some date between the 12th and 18th of October of the same year, two earth-tremors were observed at Ystad, accompanied by a short and sharp subterranean rumbling.

January of 1903 had been but two days under way when a similar phenomenon was reported from Umeå. Then on April 11th, at 8·30 p.m., in the Upland district, a shock of earthquake caused the houses in various localities to vibrate to such an extent that the terrified inhabitants ran into the street. Some averred to have seen the flash of a meteor.

On August 26th, at 6·55 p.m., a slight earthquake took place in the eastern portion of the districts of Filipstad and Örebro, and was attended with the customary subterranean rumblings. Houses shook, in certain cases, but no damage appears to have been done.

Earth-tremors were also recorded, of some severity, in Eastern Gothland (September 19th, 6·15 p.m. or thereabouts) and in Western Bothnia (October 4th, 5·25 p.m.). The author makes no suggestion as to the origin of all these earth-tremors—that is, whether they were due to sagging along fault-lines, or to some other cause. **L. L. B.**

THE SEISMICITY OF NORTHERN AFRICA.

Sur les Conditions générales de la Sismicité des Pays Barbaresques. By F. DE MONTESSUS DE BALLORE. *Comptes-rendus hebdomadaires des Séances de l'Académie des Sciences*, 1904, vol. cxxxviii., pages 1443-1445.

The districts comprized within ancient Barbary form a typically-unstable region as regards earth-tremors. The isobathic contour of 13,000 feet runs parallel to the north-western coast of Africa, not very far out to sea, and is probably the fault-line along which the basin of the Western Mediterranean has sunk, while on the other side of it, at a comparatively recent period also, has arisen the ridge of the Tellian Atlas. Parallel to these, too, run on the one hand a belt of eruptive rocks and on the other the dismantled remnants of an ancient Archæan and Palæozoic chain. Observation has shown that such tectonic conditions almost inevitably bring about seismic instability.

The frequent earthquakes of the Algerian coast-line may be traced, then, to the uprise of the Tellian Atlas. But several seismic areas, each independent of the other, may be plotted out on the map of Algeria. For instance, that of Oran does not overlap the volcanic district of the Lower Tafna, and the Shelif Valley constitutes an almost insurmountable obstacle to the propagation of earthquakes, either from the east or from the west of the colony. The most unstable portion of the Tellian Atlas is the southern rim of the Mitidja.

The ancient massif of Kabylia reveals the ordinary characteristics of highly-dislocated areas, that is, numerous epicentres, and earth-tremors more remarkable for their frequency than for their violence. East of the Shelif, the southern flank of the Tellian Atlas is just as unstable as the northern. On the other hand, the Alfa region between the two Atlas ranges is particularly stable, except on the east where the Hodna depression forms a seismic centre of some importance. The Saharan Atlas is very stable on the west up to Bu-Sanda, which marks the beginning of the remarkable seismic area of Aurès and Nemensha.

The Tripolitan area, in common with Cyrenaica and the Sahara, is extremely stable. In Tunisia, a seismic belt may be traced from Gabès to Tozeur, and there is probably some relationship between the seismic centre of Tunis itself and the Zaghuan faults.

Concerning Morocco next to nothing is known, but the region around Fez and Mequinez is much shaken by earth-tremors, along a belt corresponding to the former inflow of the Miocene Sea into the Oran district, and symmetrically situated in regard to the Guadalquivir Gulf (itself an area of high seismicity). Neither shore of the Straits of Gibraltar is remarkable for its seismic stability. The author observes, in conclusion, that most of the towns of Algeria have been built in such situations as to suffer the greatest possible damage from earthquakes.   L. L. B.

EARTHQUAKES IN MEXICO, 1902.

(1) *Informe sobre los Temblores de Zanatepec a fines de Setiembre de 1902, y sobre el Estado actual del Volcán de Tacaná. By* EMILIO BÖSE. *Parergones del Instituto geológico de Mexico,* 1903, *vol. i., pages 5-25 and 4 plates.*

Zanatepec is situated in the State of Oaxaca,. at the southern base of the range which forms the backbone of the Isthmus of Tehuantepec. Between September 24th and 30th, 1902, twenty shocks of earthquake were felt, accompanied by subterranean rumblings, which were reverberated from the circular lakes known as " the Sun and the Moon " (Sol y Luna). On the 23rd of the same month, the whole south-eastern region of Mexico had been shaken by an earth-tremor, the epicentrum of which lay somewhere in the State of Chiapas, and the shocks felt at Zanatepec are undoubtedly to be connected with this preliminary shock. The violence of the last-named is given as 6 in the De Rossi-Forel scale, and the pleistoseismic area' was a belt some 62 miles long and more than 15½ miles broad. Buildings suffered considerable damage, in some cases, far beyond this belt. A study of the tectonic structure of the region leads to the conclusion that the epicentral line of seismicity coincides with a succession of step-faults, and hence that the earth-tremors were due to a sagging movement along one or more of those faults.

The author visited the " Sun and Moon " lakes, and holds that they are not of volcanic origin (they were supposed to be crater-lakes, but their maximum depth does not exceed 82 feet). He describes the present condition of the volcano of Tacaná, on the frontier between Guatemala and Mexico, and concludes that, although not actually extinct, it resembles in many respects a moribund volcano. The last definitely-recorded eruption from the summit took place in 1855.

(2) *Informe sobre el Temblor del 16 de Enero de 1902 en el Estado de Guerrero. By* E. BÖSE *and* E. ANGERMANN. *Parergones del Instituto geológico de Mexico,* 1904, *vol. i., pages* 125-131.

On January 16th, 1902, at 4·15 p.m., an earthquake took place in the above-mentioned State, of which rather panic-stricken accounts were published in the newspapers. Taking into consideration the general flimsiness of the buildings in the towns where most damage was done, the authors infer that the intensity of the shock really lay somewhere between 5 and 6 in the De Rossi-Forel scale. In Chilapa, Chilpancingo and Tixtla, the direction of motion seems to have been approximately from north to south. In the more remote villages, the available evidence points to various directions of motion: this, coupled with the fact that the shock was most severely felt in the three towns above-mentioned, inclines the authors to the belief that the epicentrum lay somewhere in the neighbourhood of these towns, perhaps near Chilpancingo, and that the cause of the earthquake was the slipping of a "fault-block" over an east-and-west fracture (the general strike of the rock-fractures in that region ranging indeed east and west).

Investigation proved that the reports of yawning fissures having been opened up in the earth in two localities, and of a new volcano having sprung up near Atoyac, as a consequence of the earthquake, were, to all intents and purposes, unfounded.

In conclusion, it is pointed out that orogenic movements are still going on in that part of Mexico, and consequently that earthquakes may be looked for, from time to time, as a necessary accompaniment of the process of mountain-building. L. L. B.

VEGETABLE SOILS OF THE COAL-MEASURES.

Sur les Sols de Végétation fossiles des Sigillaires et des Lépidodendrons. By C. GRAND'EURY. *Comptes-rendus hebdomadaires des Séances de l'Académie des Sciences,* 1904, *vol.* cxxxviii., *pages* 460-463.

The shaly clays, which often form the roof and partings of coal-seams, when full of Stigmarian roots represent old vegetable soils, the more interesting that they are often overlain by prostrate Sigillarias and Lepidodendrons, which appear to have fallen in the place where they grew (as at Communay in Dauphiné, and at Hattingen in Westphalia). The author agrees with Prof. Potonié that *Stigmaria* represents an autochthonous vegetation, but adds that it was an aquatic plant and extended indefinitely by creeping along a submerged floor, where no other vegetation could take root. At Matallana, in the province of Leon (Spain), there is a shale-outcrop with *Stigmaria minor* radiating in quadripartite fashion from swollen rhizomes, from which stems would have doubtless sprung up if the water had not been too deep. Indeed, it would seem that *Stigmaria* did not prosper well in deep water, for at Rive de Gier, in the Loire coal-field, it finally disappears westward, in the direction towards which the old basin deepened. On the other hand, it flourished in the shallow waters of the northern rim, in company with Calamitean plants, and there began to put forth stems and grow trunks. In the old Coal-measure forests of St. Etienne and Alais, *Sigillaria* is succeeded downward by *Syringodendron*, and the trunks of the latter are based on *Stigmariopsis*. In point of fact, the author believes that many of the Stigmarian rhizomes helped to propagate *Sigillaria* in the Coal-measure marshes, and this probably holds good of *Lepidodendron* also. He was not always of this opinion, but he has recently accumulated evidence which has induced him to revise his former conclusions. In their ordinary state of preservation the Stigmarian rootlets traceable to *Sigillaria* cannot be distinguished from those traceable to *Lepidodendron*. It is true that different varieties of *Stigmaria ficoides* are known, and the author can differentiate the Westphalian form from that which occurs in the uppermost measures of St. Etienne. The extraordinary fixity in the character of the rhizomes of the Lepidophyta of the Coal-measures, when contrasted with the variability of the trunks and stems from species to species, seems to point to a common origin for them all.   **L. L. B.**

MARSH-VEGETATION AND THE ORIGIN OF COAL.

Sur le Caractère paludéen des Plantes qui ont formé les Combustibles fossiles de tout Âge. By C. GRAND'EURY. Comptes-rendus hebdomadaires des Séances de l'Académie des Sciences, 1904, vol. cxxxviii., pages 666-669.

The author begins with pointing out that recent research in the St. Etienne coal-field has shown that all the fossil plants from which its seams are derived were marsh-dwellers, or of distinctly aquatic habit. The later and rarer plants found in the upper beds, there and in the Autun coal-field such as *Pterophyllum, Nœggerathia, Zamites*, etc., are the ill-preserved fortuitous relics of a comparatively-monotonous dry-land flora, which played practically no part in the constitution of ancient fossil fuels.

During the last 4 or 5 years, the author has studied the brown coals and lignites of Keuper, Oolitic, Cretaceous and Tertiary age, and has come to the conclusion that just as Coal-measure coal, these newer deposits are derived from marsh-vegetation and aquatic plants. The lignites, it is true, can be generally traced to Coniferæ, but these conifers all belong to species which live best near water, or actually with their "feet in the water." It is true again that in very recent lignites, remains of palm, alder, laurel, etc., have been found, and so exceptions must be admitted in the case of this class of fossil fuel. The exceptions, however, do not appear to invalidate the general rule, laid down by the author, that marsh-vegetation was the main source of all varieties of coal.   **L. L. B.**

CONTACT-METAMORPHIC MAGNETITE-DEPOSITS.

Über kontaktmetamorphe Magnetitlagerstätten, ihre Bildung und systematische Stellung. By F. KLOCKMANN. Zeitschrift für praktische Geologie, 1904, vol. xii., pages 73-85.

The author considers that all magnetite-deposits of any consequence belong to one of the following categories:—(1) They are locally-concentrated constituents of eruptive rocks; or (2) they are a normal phase of the series of crystalline schists; or finally (3) they are so-called "contact-metamorphic" deposits, associated, as to position, with eruptive rocks. In regard to the origin of the first category all observers are now in practical agreement. With regard to the second, there is little doubt that they represent deposits originally of hæmatite or spathose iron-ore, which underwent the same processes of regional metamorphism as those to which the rocks that are now the crystalline schists were subjected. The question of the genesis of the third category, however, has not been settled beyond dispute. There is a widespread notion that such magnetite-deposits owe, not alone their chemical composition, but their very existence, to contact-metamorphism, and primarily, therefore, to the eruptive rocks which brought about that metamorphism. This view the author attacks as absolutely mistaken, and as being in contradiction with much of the available evidence. He goes step by step through the arguments upon which this view is usually based, and claims that they fail to fit in with the facts, as observed in the iron-ore deposits of Nassau, the Harz, the Sierra Morena, and Algeria, among other regions.

The author's opinion is that contact-metamorphic magnetite-deposits were originally either hæmatites, spathose iron-ores, or in some cases pyrites, or even the regionally-metamorphic magnetites of the Archæan Series. In the vast majority of instances there is no ground for assuming the intervention of any other agencies in the process of peroxidation than the high temperature and consequent heating effect of the eruptive rock, coupled with the presence of steam and the exertion of a certain amount of pressure. The typical "contact-metamorphic" deposits of magnetite are to be ranged alongside the magnetites of the crystalline-schist series, from which they differ merely in age and in the nature of the metamorphism which altered iron-ore deposits primarily of sedimentary (or, in some few cases, of metasomatic) origin.

Finally, the author enters a protest against the prevalent system of classifying ore-deposits according to their genesis, real or presumptive. He holds that, in the present state of knowledge, a comparative classification, laying chief stress upon geological and mineralogical similarities and relationships, is by far preferable.　　　　　　　　　　　　　　L. L. B.

THE INFLUENCE OF METAMORPHISM UPON THE MINERALOGICAL COMPOSITION OF PYRITES-DEPOSITS.

Über den Einfluss der Metamorphose auf die mineralische Zusammensetzung der Kieslagerstätten. By F. KLOCKMANN. *Zeitschrift für praktische Geologie,* 1904, *vol. xii., pages* 153.160.

The author considers that the part played by metamorphism in regard to mineral-deposits has been often misconceived, if not neglected, by investigators. He disagrees with the generally-accepted hypothesis of the epigenetic origin of typical pyrites-deposits, and favours the view that, so far from presenting any analogy with subsequently-formed metalliferous veins, they are (broadly speaking) contemporaneous with the surrounding "country-rock." They are divisible into two categories:—(1) Those which occur among regionally-metamorphosed rocks; and (2) those which are closely associated with or intercalated among normal clay-slates. To the second group belong the deposits of the Rammelsberg (near Goslar), of Meggen on the Lenne, and of the Huelva district. Under the first group are ranged the numerous pyrites-deposits of Scandinavia, more especially those of the western coast of Norway, and, in fine, the great majority of known pyrites-occurrences all the world over.

The author proceeds to show that Prof. Vogt's assimilation of the Rammelsberg and Huelva deposits to those of Norway, as having all alike originated from "pneumatolytic dynamometamorphism" is mistaken. So far as the metallic minerals on the whole are concerned (exceptions will be mentioned later) all the deposits above enumerated show indeed such similarities as inevitably lead to the inference that they belong to one great common family; but the differences are enormous when we come to consider the variety of the non-metallic minerals associated with them. It is partly by means of these non-metallic minerals, that the history and genesis and the proper grouping of the various pyrites-deposits may be traced out.

Metamorphism did not generate pyrites-deposits: it found them ready-made, and altered them in common with the immediately neighbouring "country-rock." On this point the author's conclusions are again diametrically opposed to those of Prof. Vogt. In metamorphosed pyrites-deposits, the presence of magnetic iron-ore and magnetic pyrites points to intense heat as having been the chief agent of alteration: these minerals do not occur in the non-metamorphic pyrites-deposits.

The author has discussed on a previous occasion the metamorphic nature of magnetite, and here he devotes some space to the question of the genesis of magnetic pyrites by alteration of ordinary pyrites, being careful, however, to point out that not all magnetic pyrites is metamorphic. The "natural roasting" or "thermal metamorphism" of pyrites results (1) in magnetic pyrites; (2) if carried a stage further, in magnetite; and (3) if carried to a rare extreme, in specular iron-ore.

The quantity of magnetic pyrites and magnetite associated with a pyrites-deposit should, therefore, serve as a measure of the amount or intensity of metamorphism which that deposit has undergone; although such exceptions as the unaltered deposit of Sain-Bel, immersed in metamorphosed schists, must be borne in mind. However, the numerous non-metallic silicates associated with metamorphic deposits will, as before hinted, help to standardize the intensity of the metamorphism.

If the author's views are correct, they should serve to throw light on much that has hitherto appeared to some minds inexplicable in connection with ore-deposits, as, for instance, the mode of origin of the lead, zinc, and silver-ores of Broken Hill, that of the ores of Bodenmais in Bavaria, and of Ducktown in Tennessee, etc.   L. L. B.

CONTACT-METAMORPHISM AND ORE-DEPOSITION IN SUMATRA.

Über Gesteinsumwandlung, hervorgerufen durch erzzuführende Processe (Beobachtungen an Gesteinen der Landschaft Ulu Rawas, Süd-Sumatra). By L. Milch. Neues Jahrbuch für Mineralogie, Geologie und Paläontologie, 1904, Beilage-Band xviii., pages 452-459.

In the course of a study of the rock-collection, which represents the fruits of Dr. W. Volz's journey to Southern Sumatra in the years 1899-1901, the author came upon some very peculiar rocks, palpably associated with ore-deposits, and altered by the same processes as those which had brought the metalliferous particles into their present position.

The matrix of the pyrites- and galena-vein of Sungei Tubo, in the district of Ulu Rawas, is a pale-grey, highly altered porphyritic eruptive rock. The mode of distribution of the ores, their tendency in association with quartz to replace the felspars of the original rock, and the peculiar coloration of the mica, suggested that the ores had originated from the activity of fumaroles. Further investigation, however, showed that this hypothesis could only hold good to a very small extent. North-west of the vein, a dark, dirty-green rock crops out, with pyrites distributed very evenly throughout its mass, and careful study shows that it originated from the same eruptive magma as the (apparently) quite different rock just described. In the immediate neighbourhood of the ore-body there is abundant evidence of the secondary formation of quartz and sundry silicates on a large scale. The presence of garnets, both fresh and decomposed, within the ore-body itself, is also noted; as, too, the speckled sandy mass next the vein, consisting of partly decomposed cerussite, iron-hydrates, finely-fibrous hornblende, and a little quartz.

A cherty phyllite (?) in the river-gravel of the Sungei Nitap, exhibits fissures infilled with metalliferous ore, fine-grained quartz, and hornblende— the two latter constituents varying considerably in relative proportions. Investigation shows that what was originally a phyllite was metamorphosed into a hornfels-like rock simultaneously with the arrival of the metalliferous particles upon the scene. What exactly was the nature of the metamorphic process the available evidence fails to establish. L. L. B.

ANOMALIES IN THE EARTH'S DENSITY IN THE NEIGHBOURHOOD OF ETNA.

Sur les Anomalies de la Gravité et les Bradysismes dans la Région orientale de l'Etna. By GAETANO PLATANIA. *Comptes-rendus hebdomadaires des Séances de l'Académie des Sciences*, 1904, vol. cxxxviii., pages 859-860.

Prof. de Lapparent, in regard to the anomalies in the law of gravity, has already pointed out that these do not increase as we pass from a terrestrial to a maritime area, but may be associated with a particular region of dislocation at the contact of an area which is sinking (and is being squeezed up as it sinks) with an area which is either stable or in process of elevation. Now, the eastern district of Etna, where the variations of gravity are especially noteworthy, is not only subject to frequent seismic phenomena (often purely local) but is conspicuous for deep fissures in the earth's crust; and while one part of the area is in process of elevation, the other part is gradually sinking. The evidence for this simultaneous depression and elevation is adduced in detail, and it is shown how it coincides with the anomalies there observed in the law of gravity. L. L. B.

BAUXITE.

La Bauxite. By G. AICHINO. *Rassegna Mineraria*, 1902, vol. xv., Nos. 15-18 (46 pages).

Premising that bauxite is now the most important source for the extraction of aluminium and its salts, the author points out that under one name it includes at least two distinct mineral species, one of which is a trihydrate (U.S.A.) and the other a dihydrate (France). He discusses at some length the analyses of bauxite which have been made on both sides of the Atlantic, and agrees with Mr. T. L. Watson, that the safest general chemical formula to adopt for the mineral is $Al_2O_3 nH_2O$. Its mode of occurrence is extremely variable, and it can hardly be said to possess any distinctive physical characteristic, unless that be its general pisolitic or concretionary form. In colour, it ranges from pure white to dark-red or even black; sometimes it seems to be a fine-grained, hard, compact rock, and sometimes it occurs as an earthy, clayey-looking substance.

The classic locality, which has given its name to the mineral, and where, until recent years, it was most extensively worked, is Baux in the department of the Rhône. There it is distinctly bedded, the strata varying in thickness from a few inches to 65 feet, and overlapping one geological formation after another, from the Infra-Lias to the Urgonian; the bauxite-deposits are overlain by rocks of Upper Cretaceous age. Mr. Francis Laur distinguished four types of bauxite in the French deposits; the other principal localities besides Baux being St. Chinian and Villeveyrac in Provence, Mende in the Lozère, and Pereilhes in the Ariège. The French analyses quoted by the author appear to show that the red or white varieties which are richest in alumina and poorest in sesquioxide of iron contain more silica than the red varieties in which the above-mentioned iron-oxide abounds, but this conclusion is not of universal application.

After a short description of the important deposits in county Antrim, and a reference to those of Austria and Vogelsberg in Germany, the author proceeds to deal with those of his own country, Italy. The recent discovery in the Central Apennines of considerable occurrences of bauxite has aroused the attention of industrialists. The most important locality from this point of view is Lecce ne' Marsi, where a deposit about 10 feet thick covers an area of 250 acres or more on the western slopes of the Monte Turchio: the bauxite is generally interbanded with Urgonian limestones, and is finely oolitic, while amid its mass are " porphyritically " scattered numerous more or less mis-shapen pisolites. It is asserted that analyses privately made show the mineral to contain 60 per cent. or more of alumina, but certain analyses made by the author himself of specimens submitted to him, yielded a somewhat smaller average percentage (say, rather more than 57): even so, the mineral would be of fairly high grade.

A lengthy description, based on American publications, is given of the bauxites of Georgia, Alabama and Arkansas, with tables of analyses made in the United States, and this is followed by a discussion of the origin of bauxite-deposits generally; but the author refrains from committing himself to any particular view, in the present incomplete state of knowledge with regard to the actual number and extent of such deposits. He points out, however, that perhaps the true line along which to work towards a solution may be found in the analogy between bauxite and certain iron-ore deposits.

Between 1895 and 1899 the annual output in France, Great Britain, and the United States, increased from 47,633 metric tons in the former year to 93,754 metric tons in the latter year. Although the United States rank second as producers of bauxite, they import a considerable quantity from France as well, and it may be reckoned that about a quarter of the world's output of the mineral is utilized for the extraction of metallic aluminium. The various processes of this extraction are briefly described, and attention is then drawn to another use for bauxite which may assume great importance in the future, namely, as a refractory or fire-resisting material, in those cases where the mineral does not contain too large a proportion of silica and iron.

On the whole, this somewhat lengthy paper constitutes rather a summary of previous knowledge, than a contribution of new facts or results.

L. L. B.

ANTIMONY-ORES OF KŘITZ, BOHEMIA.

Zur geologischen Kenntniss des Antimonitvorkommens von Křitz bei Rakonitz. By FRIEDRICH KATZER. *Verhandlungen der k. k. geologischen Reichsanstalt,* 1904, *pages* 263-268.

The antimonite-deposit near Křitz, a village 11 miles or so south of Rakonitz, was described as long ago as 1858. A fresh attempt was made to work it at the beginning of the 'nineties, but operations were soon suspended. However, the more recent excavations enabled the author to investigate the deposit more minutely than had been hitherto possible. The predominant rock of the neighbourhood is a greenish-grey phyllite, in places contact-metamorphosed by the eruptive " greenstones " (chiefly diabases) which traverse it. But regional metamorphism (due, that is, to phenomena of pressure, etc.), has had perhaps the greatest share in determining the present structure of the rocks of this district. A great belt of calc-schist is interbanded with the phyllite.

The antimony-ore occurs at or near the junction of the altered phyllite with the diabase, in the form of a brecciated vein striking about south-south-west, and dipping steeply to the east-north-east. The ore is of considerable purity, picked samples yielding an average of 85 per cent. of sulphide of antimony, and some as much as 92 per cent. The gangue is chiefly quartz, sometimes containing crystalline druses, wherein the quartz-prisms are coated with a limonite-film. The associated "torsion-fissures" and numerous complicated slickensides point to considerable thrusting and grinding movements having taken place among the rock-masses in bygone ages.

Nothing is said in the paper as to the industrial importance of the deposit.

L. L. B.

THE HOŘENSKO-KOSCHTIALOW COAL-BELT, BOHEMIA.

Der Hořensko-Koschtialower Steinkohlenzug bei Semil in Nordostböhmen. By FRIEDRICH KATZER. *Verhandlungen der k. k. geologischen Reichsanstalt, 1904, pages 150-159, with 4 figures in the text.*

In the Permian area on the south side of the Riesengebirge, besides bituminous shales, there occur coals which have been the object of rather spasmodic mining operations. The coal is of moderate quality, yields generally a great deal of ash, and the thickness of the seams rarely exceeds 20 inches. Some of the principal collieries were situated along a belt running east and west for several miles from Hořensko to Koschtialow-Œls, south of Semil, in North-eastern Bohemia, the coal-bearing beds occupying an anticline, which at a fault-line on the north is thrust over the Permian conglomerates and sandstones (Semil Group), while on the south the anticline is overlain by these conglomerates and sandstones. Intrusive sills of melaphyre are interbedded among the latter. The dip of the Permians is steep and almost uniformly to the south. The coal-seams have in some localities undergone an extraordinary amount of folding and squeezing, and the vertical section of the strata at one place is so different from that at another, that exact correlation of the seams is impracticable.

Another difficult problem is that of the age of the seams: to obtain sufficient evidence from the plant-remains it would be necessary to rake through all the old waste-heaps; but such evidence as there is points to a Carboniferous, rather than to a Permian, age. The author remarks, however, that it would be premature to consider the question as settled at present.

L. L. B.

THE MAGNETITE-DEPOSITS OF MALESCHAU AND HAMMERSTADT, BOHEMIA.

Die Magneteisenerzlagerstätten von Maleschau und Hammerstadt. By FRIEDRICH KATZER. *Verhandlungen der k. k. geologischen Reichsanstalt, 1904, pages 193-200, with 3 figures in the text.*

These magnetites, which occur among the gneisses south of Kuttenberg, were actively worked until the middle of the last century, and since then various unsuccessful attempts have been made to revive the mining industry in that neighbourhood. Of late, the near approach to the old mines of a new net-work of light railways has drawn attention to the possibility of a successful resumption of mining operations.

The gneiss-massif of Central Bohemia, which stretches southward from Kolin and Kuttenberg to the Sazawa river, is characterized by numerous intercalations of hornblendic rocks and granitic eruptives. The hornblendic rocks are of much more recent origin than the gneisses, and of still later date are the pegmatites, occurring in the form of bosses and innumerable dykes.

At Maleschau, the magnetite-deposit occurring on the Stimmberg plateau is intimately associated with hornblendic rocks, in which it is found in nests or with which it is interbanded. The purest ore always contains some hornblende and more especially garnet. The last-named mineral also occurs in considerable quantity, apart from the magnetite. That ore is mainly of magmatic origin, and was separated out from the original eruptive magma immediately after the apatite. Hornblende and (finally) felspar were separated out afterwards; but where garnet occurs felspar appears to be absent. The calcite, which generally impregnates the whole mass of the rocks, was probably derived from the overlying Cretaceous sedimentaries. Where garnet is the chief associate of the magnetite the latter occurs in finely-granular crystalline aggregates, but it is much coarser-grained when hornblende is its chief associate.

The geological conditions of the magnetite-deposit of Fiolnik or Fiovnik Hill, north-east of Hammerstadt on the Sazawa, are very similar, except perhaps that they are repeated on a larger scale. The magnetite is associated with the garnet-bearing hornblendite, but is rarely so finely granular as at Maleschau. Its chief impurity is hornblende, and the percentage of that is usually so small that the Fiolnik ore can be pronounced at the first glance as of high quality. For metallurgical purposes, however, it is best mixed with other iron-ores, with which view limonites were at one time worked in certain localities in the district. The old magnetite-workings of Fiolnik, as at Maleschau, were on the whole shallow, the deepest shaft having been sunk to a depth of 262 feet.   L. L. B.

THE ROOFING-SLATES OF EISENBROD, BOHEMIA.

Der Dachschiefer von Eisenbrod in Nordböhmen. By FRIEDRICH KATZER.
  Verhandlungen der k. k. geologischen Reichsanstalt, 1904, *pages* 177-182.

The slate-quarrying industry of Eisenbrod dates back to the first half of the nineteenth century, and although within the last decade the output has decreased considerably, it is still of economic importance, for the neighbouring district at least.

The phyllites north of the town, where the slate-quarries are situated, show greatly-disturbed bedding and are seamed by innumerable intrusive dykes (augite-rocks). The green roofing-slate occurs in lenticular masses among the phyllites, in such wise that 80 to 90 per cent. of the rock, that has to be quarried to get at the slate, must needs be tipped on to the waste-heaps. In some cases the quarry-owners have had to hew down rock for a twelvemonth before reaching a good bed of roofing-slate. These difficulties, naturally enhancing the cost of output, account for the decline that has set in since the end of the 'nineties.

The slate could be split up very thin, is of excellent quality, pre-dominantly pale-green in colour, occasionally violet or blue, and of ex-tremely-compact structure. As in many other occurrences of roofing-slate people distinguish between " dry " and " wet " slates, and therefore some of the Eisenbrod slate is best worked in the summer and some in the winter. The " winter-slates " are quarried night and day, as long as the frost holds. But when the slate has once been stacked and air-dried no difference can be detected, as between the two varieties. The best slate is exported to Saxony.

L. L. B.

GRAPHITE-DEPOSITS IN BOHEMIA.

Das Vorkommen von Graphit in Böhmen, insbesondere am Ostrande des südlichen Böhmerwaldes. By O. BILHARZ. *Zeitschrift für praktische Geologie*, 1904, *vol. xii.*, *pages 324-326.*

Bohemia possesses two widely-separated groups of graphite-deposits, the one lying at the foot of the Saaz Mountains close to the Moravian frontier, and the other in the southernmost portion of the great Bohemian Forest. The latter group is both geologically and genetically connected with the Passau graphites of Bavaria.

In the Saaz district, the phyllites and slaty schists, especially where they are rich in pyrites, pass into graphite-slates. There is, moreover, an unmis-takable genetic relationship between the occurrence of graphite and that of limestone-bands among the Archæan schists: five or six of these bands have been located in the Saaz area, and at their contact-surface with the schists, pure graphite is intercalated in lenticular masses, which seem to have bodily replaced schistose material. The graphite becomes harder as the depth from the surface increases, and finally passes into bedded graphite-slates. The most important deposits occur at Trpin, Wachteldorf, and at Predmesti, near Swojanow (where the graphitic deposit is seen to attain a thickness of about 40 feet).

Of far greater importance, however, both in regard to their extent and their regularity, are the graphite-deposits in the south of the Bohemian Forest. Starting at Schwarzbach, where they occur in association with limestone in lenticles among the gneisses, the deposits strike northward to Stuben, then eastward to Krumau, and finally due northward as far as the neighbourhood of Netolic. In the northern portion of this belt, the graphites occur rather in beds or seams than in lenticles, a whole succession of these seams following one on the other at short intervals. Exploration-work is at present concentrated on the township of Kollowitz, 12¼ miles north-west of Budweiss: the thickness of the " seams " is found to vary from 16 inches to 6¼ feet, roof and floor being in all cases made up of soft decom-posed gneiss. Borings have proved the deposits over an area a mile and a quarter long, and rather less than 3 furlongs broad. Pure massive graphite is not found in the uppermost seams, but occurs only in the deeper-lying beds. However, the top bed is being methodically worked: it yields a blackish-grey pulverulent mass, consisting of innumerable graphite-flakelets, and this is ground up and separated by winnowing into coarse and fine " flour," etc. The annual output averages at present from 400 to 500 wagon-loads of saleable stuff, and is shortly to be increased to 1,000 wagon-loads. According to the statistics published by the Austrian Ministry of Agriculture, the graphite-mines of the district around Schwarzbach-Krumau (in the southern portion of the belt) have a combined annual output of 1,200 wagon-loads. The selling price ranges between 27s. 5d. and 29s. 4d. per ton: and one gathers that, at such prices, the graphite-mining industry is very profitable.   L. L. B.

THE PITCHBLENDE OF JOACHIMSTHAL, BOHEMIA.

Les Filons de Pechblende de Joachimsthal (Bohême). By R. D'ANDRIMONT. *Annales de la Société géologique de Belgique, 1904, vol. xxxi., Bulletin, pages 91-93, with 1 figure in the text.*

The search for minerals containing radium has now become a matter of widespread commercial importance, and so the author feels justified in stating briefly the results of a recent visit to the Joachimsthal mines.   Premising that pitchblende is an extremely-complex mineral, containing, besides 40 per cent. of uranium oxide (UO_4), arsenic, molybdenum, sulphur, tungsten, vanadium, silver, lead, bismuth, magnesium, calcium, iron, aluminium, cobalt, nickel, silicium, and carbon dioxide, he says that the veins of it in the locality described are only 1 or 2 inches thick. They occur in a belt largely made up of biotite-schists, abutting on the west against a granitic massif: the country is traversed by two main systems of fissures, which cross one another at right angles. Those striking northward and southward are metalliferous, while the east-and-west fissures are barren.   The earliest eruption was that of the granite, followed by porphyry, then came the metal-liferous veins, and thereafter a flow of basalt closed that chapter of geo-logical history. The uranium-mineral occurs exclusively in the biotite-schists. and the gangue is invariably dolomite, with a large proportion of other mineral-substances.   These indications are perfectly well known to the Joachimsthal miners, who make use of them in looking for new occurrences of pitchblende.

It seems possible that the uranium first made its appearance in the form of a carbonate, which reacted with the magnesium of the biotite; indeed, it is noticeable that the biotite-schist is altered at the contact of the veins. A further inference is that, in a vein apparently consisting of dolomite alone, pitchblende may well be found to occur deeper down.   L. L. B.

GLAUBER'S SALT IN THE TRIAS OF BOSNIA.

Über ein Glaubersalzvorkommen in den Werfener Schichten Bosniens. By FRIEDRICH KATZER. *Centralblatt für Mineralogie, Geologie und Paläon-tologie, 1904, pages 399-402.*

In cutting the Jahorina tunnel (3,100 feet above sea-level) on the new railway from Sarajevo to the eastern frontier, a seam of impure brownish-black coal, only 2 inches thick, was found among the red Werfen Beds (equivalent to the Bunter Sandstone). Below the coal (which is the second known occurrence of the kind in the Werfen Beds of Bosnia) the first deposit of Glauber's salt was struck, in the form of a lenticle barely 4 inches thick; a few yards farther on, another deposit, about 3 inches thick, was discovered. The strata with which these deposits are interbedded are red, highly-mica-ceous clay-slates, but the Glauber's salt-bands are immediately overlain and underlain by a dark-brownish red micaceous clay.   The Glauber's salt is colourless, water-clear, non-crystalline, with conchoidal fracture. The deposits doubtless owed their origin to the same processes as those now going on in the shallow bays of the Caspian, where similar deposits are being laid down.

Nothing is said as to the possible industrial bearing of the discovery.

L. L. B.

PETROLEUM-DEPOSITS OF THE PUTILLA VALLEY, BUKOVINA

Über die Rohölführenden miocänen resp. oberoligocänen Schichten des Thales Putilla in der Bukowina. By DR. STANISLAW OLSZEWSKI. *Zeitschrift für praktische Geologie,* 1904, vol. xii., pages 321-324, with 1 figure in the text.

The Putilla is a tributary of the White Czeremosz River, which forms the boundary between Galicia and Bukovina. The views held in regard to the geology of the district were so conflicting, that the author, when he journeyed thither in May, 1904, in order to examine into its possibilities as a natural oil-reservoir, felt bound to exercise more than ordinary care in the investigation. In the Putilla valley, he found no trace of Cretaceous or Eocene rocks, nor any of the Upper Oligocene menilite-shales; but the high ridges of Rakova and Krasny-Dil, which wall in the valley, are mainly built up of Magura Sandstone, the massive beds of which dip north-eastward. The author has, on previous occasions, drawn attention to the importance of the Magura Sandstones or Dobrotow Beds, in connection with the occurrence of petroleum. It was because they had overlooked this point, that so recently as 1898, the Carpathian geologists declared what now proves to be the richest oilfield in Galicia, to be worthless from the petroleum-seeker's point of view. The western portion of the Putilla valley exhibits greater irregularity of configuration: the strike of the beds varies considerably, and they are much disturbed and faulted. The strata here are predominantly grey shales, with which are interbedded thin, micaceous fucoid-sandstones; and secondarily the saline clays (which form so characteristic a horizon in Galicia) make their appearance. All the beds mentioned belong to the Upper Oligocene-Miocene succession. The author discusses at some length the tectonics of the area, and then enumerates the localities where traces of petroleum have been recorded. At Dichtenitz, some 30 years ago, 5 barrels per day are said to have been got (from a well, 105 feet deep) of a pale-green oil, rich in paraffin, solidifying above the well-water into a green substance known as *kindibal* (intermediate between crude oil and ozokerite). Oil has also been got at three localities around Sergie, and at one point it would seem possible to correlate the oil-bearing bed with that of Dichtenitz. In view, however, of the extremely-broken character of this mountainous district, no such definitely-continuous oil-horizons as in Galicia can be looked for. The fact, however, that oil has been struck at such shallow depths in the Putilla valley admits of the expectation that richer oil-bearing beds will (as in Galicia) be struck at greater depths. The enormous development of the Magura Sandstones is also an encouraging indication. The nearest railway stations (Wisnitz and Brodina) are from 25 to 37 miles away; timber, both for structural purposes and for fuel, is cheap. The local labourers are skilful carpenters, but it is doubtful whether they would make good well-sinkers.

L. L. B.

THE BORYSLAV OIL-FIELDS, GALICIA.

Notiz über das geologische Profil durch die Œlfelder bei Boryslaw in Galizien. By
CARL SCHMIDT. *Verhandlungen der Naturforschenden Gesellschaft in Basel,*
1904, *vol. xv., pages* 415-424, *and* 1 *plate.*

This paper and the accompanying section are the outcome of a journey
which the author made to the petroleum-fields of Eastern Galicia in the
summer of 1903. He gives a bibliographical list of the works consulted by
him, beginning with 1877 and closing with 1904, and points out that the
Boryslav and Schodnica districts are now reckoned among the richest oil-
producing areas in Galicia. The annual output from the Boryslav district
alone averages 480,000 tons of petroleum and 2,000 tons of ozokerite. In
the Mraznica and Schodnica districts also industrial activity has increased
enormously of late: the wells in the former of these two have yielded a
steady, if not a very abundant, output of oil for the last 35 years.

The strata richest in oil are mainly associated with the secondary anticlinal folds, embraced in a huge highly-faulted Eocene-Oligocene syncline.
Two horizons in the Eocene, and a third in the *Inoceramus*-beds (Cretaceous)
are said to yield the greatest quantities of petroleum. The overthrusting and
overfolding of the rocks are briefly touched upon, and it is pointed out that
the most recent oil-bearing horizon in Galicia is the Saline Clay, of Miocene
age. The remark is here interjected that the sub-Carpathian Miocene
is known to be a great reservoir of oil in Rumania. Petroleum is got from
the Saline Clay along a belt of 12 miles or more in the neighbourhood of
Boryslav, and also at Bolochov and Solotvina to the south-east. The Boryslav oil, down to a depth exceeding 1,970 feet, contains from 6 to 10 per
cent. of paraffin—and there seems to be no doubt that there is a causal connection between the presence of bands and layers of ozokerite and this
amount of paraffin. Ozokerite usually occurs at depths not exceeding 850
feet or so below the surface, but has actually been struck at a depth of 2,280
feet. The question of the origin of mineral wax appears to belong to the
domain of chemistry; but it may be gathered that the author's own observations in Eastern Borneo (where occasionally there is so much ozokerite floating
about in the petroleum that the delivery-pipes are stopped) incline him to
favour the hypothesis that the wax is the residue of an evaporation, unaccompanied by oxidation, of oils rich in paraffin.

In the Carpathian region, as in other oil-fields all over the globe, the
proper application of the old anticlinal theory is of great importance from
the practical point of view; in confirmation of which the author cites his
experiences in Sumatra and Borneo, where all the borings for oil put down
in accordance with that theory were successful. He points out finally that
the Galician literature of the subject is unfortunately so taken up with polc-
mics on stratigraphical questions, and theories concerning the origin of petroleum, as to crowd out much wished-for information, such as exact sections of
the borings, etc.   **L. L. B.**

THE PETROLEUM-BELT OF SCHODNICA, EASTERN GALICIA.

Die geologischen Verhältnisse der Erdölzone Opaka-Schodnica-Uryez in Ostgalizien.
By DR. RUDOLF ZUBER. *Zeitschrift für praktische Geologie,* 1904, *vol. xii.,*
pages 86-94, *with* 9 *figures in the text.*

The author shows in the introduction that he has had quite exceptional
opportunities, during many years, of making himself thoroughly acquainted

with the geology of the district of which he treats. He describes briefly the rock-formations in ascending order, beginning with the Lower *Inoceramus*-beds or Ropianka beds of Neocomian age, passing on then to the upper *Inoceramus*-beds (or Gault), and the Jamna Sandstone (or Upper Cretaceous). These are followed by the Eocene clay-slates and conglomerates, and the menilite-shales of Lower Oligocene age, which latter are characterized in that particular district of the Carpathians by a prominent basement-bed of chert. Apart from drift and alluvial deposits, these menilite-shales are the most recent formation in the area, which is one of great tectonic disturbance, and is traversed by numerous folds and fissures striking generally north-west and south-east.

The geological structure of the district is described in some detail, much of the evidence being obtained from the numerous borings put down there for petroleum. Two chief oil-bearing horizons are recognized and worked in the Eocene formation, but the author feels no doubt that a third oil-bearing horizon will be struck ere long, this time, however, in the Lower Cretaceous. The most suitable locality to search for this probably productive horizon would be on the Pasieczki estate, the one drawback being that it will be necessary to go through the water-logged Jamna Sandstone before getting down to the indicated horizon.

The main oil-horizon in the north-western portion of the belt, towards Opaka, is almost untouched as yet, and industrial prospects here are very promising. The area to be worked covers at least 296 acres. Although the menilite-shales contain much bitumen, and fish-remains in abundance, nowhere in this district do they yield petroleum. Repeated search in them for oil has proved fruitless. The author concludes that the Eocene petroleum of this area occurs in the actual place where it was formed, and points out that no other conclusion is possible, in presence of the evidence which he has adduced. The entire length of the belt, from Opaka through Schodnica to Uryez, is about 7½ miles, and the average breadth is 1,300 feet or so,—the total workable area would therefore be about 1,186 acres, of which (as above-mentioned) 296 are still unworked. The year 1900 was the culminating-point in regard to output, which has since then tended to decrease. About 450 wells are yielding oil.   L. L. B.

THE OZOKERITE-DEPOSITS, ETC., OF BORYSLAV, EASTERN GALICIA.

Die geologischen Verhaltnisse von Boryslaw in Ostgalizien. By DR. RUDOLF ZUBER. Zeitschrift für praktische Geologie, 1904, vol. xii., pages 41.48, with 4 figures in the text.

In view of obviating the errors which necessarily attach to wide geological conclusions based on isolated observations, the author now puts forward his own results: these do not agree with those of certain writers whom he quotes, and he states that his conclusions are the fruit of a long-continued careful study of the Boryslav area.

The town of that name lies close to the north-eastern margin of the Carpathians, almost midway between the valleys of the Dniestr and the Stryj, at the point where the stream called the Tysmienica issues from the mountains. The outermost range of the Carpathians trends here nearly due north-west and south-east, the boundary between it and the pre-Carpathian foothills being well-marked: in point of fact, it delimits the overthrust of the older Carpathian formations over the younger strata peculiar to the foothills.

The strata are described in descending order.   First come alluvial and diluvial loams and gravels, in places very thick. Underlying these drifts are the ashen-grey Lower Miocene saline clays, with which are interbedded sandstones of varying texture, often containing ozokerite, natural gas, oil and carbonized plant-remains. Great rolled blocks of Eocene and Cretaceous limestones, etc. are irregularly distributed throughout the formation. Clays and sandstones alike are rich in gypsum and rock-salt, and brine-springs are very frequent.   Government salt-works are still in existence at Drohobyez (5 miles north-east) and at Stebnik (5 miles east) of Boryslav. In the same formation, blende, galena, and native sulphur have been recorded at Truska-wiec; and at Pomiarki an important deposit of pure Glauber's salt has been struck. The salt-formation is probably not of marine origin, but was laid down in one or more lake-basins under desertic conditions.

Ozokerite occurs, both in layers and as infillings of fissures: there is no question as to its occurrence in bedded layers, despite official theories to the contrary, but it is true that these layers are thin, and often pinch out. The greatest masses of ozokerite are concentrated in veins, nests and fissures. In the northern portion of the district the richest fissures, dipping steeply northward, have been opened up to a depth of about 1,000 feet below the surface; but in deep borings for petroleum more to the south, considerable masses of ozokerite have been found at depths as great as 2,300 feet.

The tremendous pressure of the strata in the Boryslav workings is notorious: the strongest timbering is snapped up like matchwood, whole shafts are twisted out of the perpendicular, boring-rods are flattened like straws, and yard-long ribbons of ozokerite are suddenly squeezed out of the fissures, in company with shapeless masses of rock-salt.

The plasticity of the entire formation and the recklessly-unmethodical fashion in which it was worked for many years have had the inevitable result, that the chief mining area is continually in motion: thus, too, the upper part of the salt-formation is in a state of chaos, in places completely turned upside down, to a depth of as much as 330 feet. The petroleum which occurs in the western portion of the district is very rich in paraffin, viscous, and blackish-brown in colour; in the eastern division (Tustanovice-Volanka), on the other hand, it is a considerably lighter and paler oil, and rich in benzine.

Next below the salt-formation comes a group chiefly of sandstones and conglomerates, which the author correlates with the Magóra Sandstones of Upper Oligocene age, and terms the Dobrotov group. The compact, argillaceous and micaceous sandstones are characteristically ripple-marked, and with them are interbedded narrow bands of dark shale: they are often full of carbonized plant-remains, and constitute the richest horizon for petroleum in this area. At Boryslav itself, they do not crop out at the surface, but an excellent exposure is seen 7¼ miles away to the north-west, at Nahujovice. On boring down to the underlying Lower Oligocene menilite-shales these are found to be barren of oil, as they are, indeed, almost everywhere in the Carpathian region.   With the progress of boring operations and the continual advance in technical matters, we may expect that the lower limit of the oil-bearing strata will be everywhere reached in the Boryslav district. It must be remembered that with the menilite-shales are associated water-bearing sandstones, and carelessness in conducting boring operations might conceivably result in the flooding-out of the overlying oil-horizons from these sandstones.

With regard to the output of oil, it amounts roughly to 40,000 tons a month, but it is very unevenly distributed, on account of the rapid alternation of rich and poor or even barren "ground." Towards the north-east (Volanka) the wells diminish continuously in yield, and finally the productive strata are abruptly cut off by what may be either a fold or a fault—a matter which still awaits investigation.

Below the menilite-shales are the Eocene red and green mottled slates and shales, with the lower beds of which especially are interstratified sandstones and conglomerates. These constitute in part the Carpathian Sandstone Series, and are known to be one of the most important oil-bearing formations in the Carpathian region. It seems, therefore, probable that deep below the strata now worked for oil at Boryslav yet another petroleum-reservoir lies untouched. The possibility of tapping it, however, is confronted by the necessity of boring to a minimum depth which might be anything between 5,000 and 6,500 feet, and of passing through the above-mentioned water-logged sandstones on the way down.

The author's diagrams illustrate the folding and faulting which have disturbed the strata in the district described, and he devotes the latter portion of his memoir to the vexed question of the origin of the ozokerite and petroleum. He regards both as primarily derived from vegetable organisms, the detritus of a luxuriant tropical flora.　　　　L. L. B.

COAL-FIELDS OF HUNGARY.

Die Mineralkohlen der Länder der Ungarischen Krone. By ALEXANDER VON KALECSINSZKY. [*Official publication, Budapest,* 1903], 324 *pages* 8vo, *with* 3 *figures in the text, and* 1 *map.*

We learn from the introduction to this work that the first memoir of any importance dealing with Hungarian coal was published in 1839. Since then the development of the mineral industry and the knowledge of the mineral resources of the Magyar kingdom have made great progress, as exemplified in the author's exhaustive survey. To his book has been awarded the prize of 1,000 florins founded by the Hungarian Natural History Society in 1892. He examined a vast number of specimens of coal, and tested their heating-power. No less than 22 pages are devoted to a description of his methods of analysis, and of the manner of using the Berthelot-Mahler calorimeter.

The only seams belonging to the Carboniferous system occur in the county of Krassó-Szörény. In the Permian a few unworkable seams are found, but the Liassic and Oligocene (older Tertiary) are the most important coal-bearing formations in Hungary. The Cretaceous coals are also of economic interest, and so too are the lignites and brown coals of the Neogene (newer Tertiary) groups. A full catalogue is given of the localities where coal is worked, with the number of workpeople employed in each case, and the average output. It may be noted that the total coal-output of the Hungarian kingdom steadily increased from 1,500,000 tons in 1875 to 5,500,000 in 1899: the proportionate increase of the Austrian output was not nearly so great. This catalogue is followed by a bibliography extending over 8 pages, and that again by a description of the various seams, with analyses of the coal, etc., arranged alphabetically under place-names. This, in fact, constitutes a sort of topographical dictionary or "coal-gazetteer," in which all the available information as to quantity and quality, methods of working, stratigraphy, etc., is concentrated. There are many useful cross-references, and it certainly appears as if no locality where coal has been actually found in Hungary is left undescribed. An appendix contains analyses of Bohemian, Moravian, Silesian, and other coals, for comparison with those of Hungary.   L. L. B.

MANGANESE-ORES OF DOLINA, ISTRIA.

Manganerzvorkommen von Kroglje bei Dolina in Istrien. By Dr. L. Karl Moser. *Verhandlungen der kaiserlich-königlichen Geologischen Reichsanstalt,* 1903, *pages* 380-381.

Chemical analyses of certain mineral-specimens sent to the author by the mayor of Dolina having revealed the presence in that neighbourhood of manganese-ores, the author proceeded to make an investigation on the spot in the latter part of 1903. The locality is situated at an altitude of about 650 feet above the small hamlet of Kroglje, some 3 miles distant from Borst station on the Istrian State Railway. Here a limestone-breccia, derived from the Cretaceous escarpment which towers above the Eocene sandstones of the valley, occurs in a highly-weathered condition, mostly altered into brown hæmatite, and is in places quite crumbly. Some experimental diggings which the mayor of Dolina had caused to be made brought to light thin bands of what is probably pyrolusite, passing by all sorts of gradations into limonite; among these is a pisolitic ore, the grains of which are uniformly of the size of poppy-seeds. The silica present in all the samples analysed is undoubtedly derived from the adjacent fossiliferous Cretaceous limestone. Nothing is said as to the industrial importance of the occurrence, except that, if the bands or veins of manganese-ore continue and thicken in depth, they might repay working.   L. L. B.

IRON-ORES OF THE ERZBERG, STYRIA.

(1) *Der steirische Erzberg.* By A. F. Reibenschuh. *Mittheilungen des Naturwissenschaftlichen Vereins für Steiermark,* 1903, *vol. xl., pages* 285-321.

The Erzberg, lying between the localities of Eisenerz and Vordernberg, belongs to the great belt of grauwackés which strikes east and west through the Austrian Alpine territories of Styria, the Salzkammergut and Tyrol. A series of iron-ore deposits, the most considerable of which is that of the Erzberg, characterize the basal portion of the grauwackés. The main ore-body is no less than 490 feet thick, with a height exceeding 2,132 feet: in the lower portion of the mountain the beds lie comparatively flat, while in the upper portion they are thrust up on end. Isolated on three sides by streams running in deep-cut gorges, the Erzberg has the form of a cone, standing clear of the steep ridges with which it is connected on the southeast by the *col* of the Platte. Ore has been mined there since time immemorial: it was with this Styrian iron that Celtic warriors, and then the Romans, fashioned their weapons, but the more continuous history of the mining industry on and around the Erzberg begins with the early Middle Ages, and is carried by the author down to the present day. Operations are now concentrated in the hands of a powerful syndicate, who have abandoned the system of underground workings (all the ore being got at present by gigantic opencast workings). Everything is done on a big scale.

by the most modern methods, and a narrow-gauge electric railway runs from the Erzberg to the single-track line (completed in 1891) built between Eisenerz and Vordernberg. Instead of the old charcoal-furnaces, most of which have been blown out, great new blast-furnaces, burning coke, have been erected at Donavitz and Eisenerz, and yield an output of pig-iron such as had never been dreamt of in the district. Although the average amount of ore got yearly has increased twelvefold within the last 30 years, the amount still in sight is reckoned to last for centuries to come.

The rock-succession consists, in ascending order, of (1) the Blasseneck Gneiss, the oldest formation in the district; (2) the unconformably overlying Lower Devonian limestones, etc.; (3) the Ironstone Series, made up of well-bedded, pale-grey or yellow, pure, small-grained, spathic iron-ores, with which are occasionally interstratified thin bands of impure pinkish limestone. Pale to mottled or dark micaceous clay-slates form the normal basement of this series, the actual age of which is unknown: all that can be said is that it is certainly newer than the Lower Devonian, but older than the lowermost Trias; (4) Werfen Slates [Triassic], red or greenish-grey sandy slates of enormous thickness, with much breccia and conglomerate at the base. The Werfen Slates are denuded away from the western portion of the district, in which, therefore, the great ore-mass is laid completely bare.

Among the minerals of the Erzberg, the chief part is played by the spathic iron-ore or siderite, termed *flinz* or *pflinz* by the local miners. The coloration varies through all shades from yellowish-grey to dark-brown, according to the degree of weathering. The portions of the ore-body nearest the surface are frequently decomposed to limonite, owing doubtless to the action of percolating waters from the surface. Ankerite is the most usual associate of the spathic iron-ore, and calcite and aragonite are of very common occurrence. A combination of the two last-named minerals in alternating tabulae, found only in one fissure in the deposit, received in 1891 the name of Erzbergite. Pyrites is found pretty frequently, but native sulphur, arsenical pyrites, cinnabar, galena, etc., are comparatively rare.

The output of ore in 1903 amounted to 482,543 tons, and that of pig-iron from the five furnaces in blast to 144,446 tons.

The paper is accompanied by a bibliography of the subject.

(2) *Le Gisement de Fer Spathique de l'Erzberg, près Eisenerz, en Styrie.* By J. TAFFANEL. *Annales des Mines,* 1903, *series* 10, *vol. iv., pages* 24-48, *and* 2 *plates.*

The author begins by observing that the spathic iron-ore of the Erzberg, near Eisenerz, is one of the most extensive and one of the most productive of the kind in the whole world. It was worked by the ancient Romans, and mining operations have been carried on in unbroken sequence from the Middle Ages down to the present day, the development having been especially great during the last half-century. The average annual output exceeds a million tons, and at that rate working may go on for centuries yet without exhausting the reserves of ore now in sight.

The deposit is certainly pre-Triassic in age, probably Upper Silurian or Lower Devonian: it occurs in the upper part of a very complex group of shales, grauwackés, and limestones—the stratigraphy and exact succession of which has given rise to considerable discussion. The ore is mainly a spathose carbonate of iron, sometimes associated with a decomposition-product in the shape of limonite. The miners give the name of *rohwand* or ankerite to a limestone too poor in iron to repay working, and there is every gradation from a rich ore to a perfectly-barren limestone. Thus far had the researches of Messrs. A. von Schouppe, D. Stur and E. Suess conducted geologists, when in 1900, Mr. Vacek published the results of a detailed investigation which he had made 14 years previously. He showed that there were great unconformities in the supposedly-homogeneous group of grauwackés, etc., and referred the components of that group to four different periods. The main iron-ore deposit he regarded as of Permian age, although he still assigned some of the ore-beds to the Lower Devonian, and he pointed out as one of the characteristic features of the Permian deposit, the absence of limestone. The exposures revealed in the cuttings of the Eisenerz-Vordernberg railway, constructed in 1889-1890, three years after Mr. Vacek's visit, yield, however, evidence which does not fit in with his conclusions. The ore-deposit must still be regarded as of more remote geological age, and of greater extent than he supposed. It is probably of metasomatic origin; whereas if Mr. Vacek's hypothesis were correct, the deposit would be of purely sedimentary origin. The author does not appear to claim that, even now, the last word in this discussion has been said.

(3) *Das Eisensteinvorkommen zu Kohlbach an der Stubalpe.* By DR. RICHARD CANAVAL. *Berg- und Hüttenmännisches Jahrbuch der k. k. Bergakademien zu Leoben und Přibram und der kön. ung. Bergakademie zu Schmeczbánya (Schemnitz), 1904, vol. lii., pages 145-158, with 1 figure in the text.*

The Kohlbach deposit consists chiefly of spathic iron-ore intermingled with, and passing into, crystalline limestone, and assuming the appearance of a bedded deposit. Garnet-mica-schists overlie conformably the group of ore-bearing rocks: an important member of this group is the yellowish-grey, banded micaceous quartzite, the chief constituent of the waste-heaps of the old mineworkings.

This ore-deposit, like that of the Hüttenberg, seems to have originated from metasomatic replacement of the limestone in the course of post-volcanic processes. It was worked towards the end of the eighteenth century, and again in the middle of the nineteenth, but mining operations were subsequently abandoned, as the quantity of available ore was not sufficient for a metallurgical industry on anything approaching a large scale.

The author devotes most of his paper to a detailed mineralogical description of the immediately-associated rocks.   L. L. B.

ORE- AND MAGNESITE-DEPOSITS OF THE STYRIAN ALPS.

Ueber das Alter und die Entstehung einiger Erz- und Magnesitlagerstätten der steirischen Alpen. By KARL A. REDLICH. Jahrbuch der kaiserlich-königlichen Geologischen Reichsanstalt, 1903, vol. liii., pages 285-294, and 4 figures in the text.

The conviction that a great number of the " bedded " ore-deposits did not originate at the same time as the adjacent " country-rock " is yearly gaining more adherents, and the author proceeds to adduce confirmatory evidence in favour of that opinion from the Styrian deposits. Thus the pyritic ores of Kalwang, Œblarn and the Veitsch, although apparently interbedded with hornblendic schists (metamorphosed diabases) are shown to be of later origin than these. Intimately associated, from the stratigraphical point of view, with the pyrites-deposits just mentioned are the siderites and ankerites of the Northern Alps, among which the author's most recent researches have been conducted. He quotes the Siluro-Devonian complex of limestones and slates of the Zeyritzkampel, among which occur copper-pyrites, fahlores, ankerite, less commonly siderite, and still more rarely cinnabar. He proposes later on to devote a monograph to the probable output and practical value of these deposits; meanwhile he points out that one characteristic feature is the intimate intermixture of the above-mentioned ores. The junction between ankerite and limestone is irregular, and the former seems to have invaded the latter after its deposition; but the pyritic invasion of the limestone was a still more rapid process than the ankerite-invasion; in fact, over the whole area of this portion of the Styrian Alps, from the Radmer to the Johnsbach valley, the various ores are shown to have invaded and impregnated the limestones, and in some cases the neighbouring slates as well. Just as siderites are seen to pass into ankerites, so the latter pass into magnesites by way of dolomite. All over the area described, carbonates of iron are associated with the sulphidic ores; and the general conclusion is that the deposits cannot be regarded as true beds, laid down simply by the ordinary processes of sedimentation, but that they are the outcome of the metamorphism of the pre-existing materials. On a period of submarine eruption appears to have followed a period of normal formation of clastic sediments, attended by the usual phenomena of moribund vulcanicity, gaseous exhalations and thermal springs; and these constituted the metamorphic agent that partly converted the pre-existing rocks into ore-deposits. In regard to age, the oldest are the pyritic ores of Kalwang and Œblarn; next come the Siluro-Devonian iron- and copper-ores of the Radmer, the Erzberg, and the Hintere Veitsch; and finally the Carboniferous deposits of the Dürsteinkogel. L. L. B.

THE CAMPINE COAL-FIELD OF NORTHERN BELGIUM.

Considérations géométriques sur le Bassin houiller du Nord de la Belgique. By ÉMILE HARZÉ. *Annales de la Société Géologique de Belgique*, 1903, vol. xxxi., *Mémoires, pages* 31-86, *with 2 figures in the text and 2 plates.*

The author, first of all, describes and figures the probable shore-line of the ancient Coal-measure gulf from Aerschot (south-east of Malines) to the Dutch boundary. This raises at once the question as to how near Antwerp the Coal-measures may be expected to occur, and borings have been put down to determine this point. At Santhoven, in December, 1902, two coal-seams were struck at a depth of about 2,300 feet, near the supposed edge of the basin; at Kessel on the other hand, south of Santhoven, the Carboniferous Limestone was reached at a depth of only 1,860 feet or so. But at Vlimmeren, north-east of Santhoven, the Coal-measures were struck, in August, 1903, at a depth of 2,939 feet: the boring was carried down to 3,372 feet, traversing four very thin seams of coal (3 to 8 inches each). The author thinks that at Vlimmeren there is a concealed ridge, separating two Coal-measure synclines.

He gives detailed tables of the various borings that have been put down in the newly-discovered coal-field, and discusses at some length such problems as the richness of the seams at different horizons, the limits of the barren belts, the probable course of the faults (a remarkable one has been recorded at Beeringen, where rock-salt and red marls were struck at a considerable depth); and then returns to the question of the possible prolongation of the coal-field north of Antwerp. As he believes that the general trend of the formation will ultimately prove to be towards the great

Yorkshire basin, he decides negatively in regard to the existence of workable Coal-measures in that particular portion of Eastern Flanders. The latest borings have done much towards determining the boundaries of the new coal-field, and have thereby reduced to more moderate proportions the extravagant anticipations which were formed regarding it a few years ago. Rich it is, nevertheless, and the Belgian Government are once again urged to reserve a portion of it for working by the State.   L. L. B.

IRON-ORES IN THE CAMPINE COAL-FIELD, BELGIUM.

Découverte d'un puissant Gisement de Minerais de Fer dans le Grand Bassin Houiller du Nord de la Belgique. By G. LAMBERT. Brussels, 1904, 24 pages and 11 plates.

The first half of this pamphlet is almost entirely taken up with a disquisition on the working and gradual exhaustion of the blackband ironstones of Staffordshire, and the similar exhaustion of the iron-ores formerly worked in the districts of Liége, Charleroi and Mons.

The author points out that the output of Cleveland iron-ores, which for many years compensated England for the decline of the Staffordshire iron-mining industry, is itself rapidly diminishing; further, that the reputedly-inexhaustible deposits of Bilbao are showing signs of decline, and that the same statement holds good of the Algerian magnetites.

The author then describes the six borings which he and his son Paul (with others) were concerned in putting down—the first three in 1900 near Sittard on the right bank of the Meuse, and the other three near Lanklaer, Leuth and Eysden respectively, on the left bank of the same river. (Six plates are devoted to the borings put down in Belgian territory.) All these struck the Coal-measures at depths varying from 920 to 1,810 feet. The author points out that the announcement of a mining concession, sought by him and his partners near Sittard, was published by the Dutch Government in May, 1901, six months before the first mining concession was announced on the Belgian side of the river. The specimens got from the cores show, according to the author, that over a thickness of several hundred feet, ironstones, practically identical with the Dudley blackband, are conformably interbedded with the Coal-measures. Various analyses of the ore made in Paris, Bonn and Berlin have yielded from 30 to 40 per cent. of metallic iron, and from 0·55 to 7 per cent. of manganese. After calcination the ore assumes magnetic properties.   The blackbands appear to increase in number and richness as the depth from the surface increases.

A new boring at Saerbeeck has recently struck thick seams of excellent coal at a depth of 3,936 feet. On the whole, the author believes that the new Campine coal-field is sufficiently rich and extensive to guarantee Central Europe against a shortage of coal for a long time to come.   L. L. B.

POSSIBLE EXTENSION OF THE LIÉGE COAL-FIELD.

Le Prolongement de la Faille eifélienne à l'Est de Liége. By P. FOURMARIER. Annales de la Société géologique de Belgique, 1903, vol. xxxi., Mémoires, pages 107-136, with 14 figures in the text and 1 plate.

The author has devoted much time and care to the study of the prolongation, east of Liége, of the great Eifelian fault, which brings the Liége Coal-measures against the Upper Coblentzian (Devonian) slates and quartzites. He describes in detail the results of his mapping, and shows that in the region studied by him, the accentuation of the folding movement gave rise to reversed faults of comparatively-low hade, dipping southward. He concludes that a great mass of Devonian and Lower Carboniferous strata has been thrust over the Coal-measures, concealing these over a considerable area. Whether they can be struck at anything short of very great depths, and whether they will prove workable even if struck at a reasonable depth, can only be determined by boring operations.

It will be remembered that, in the North of France, Coal-measures have been found far to the south of the southern boundary-fault; and in Hainault patches of Coal-measures have been discovered, faulted away miles to the north of that coal-field. In the Swiss Alps, to which the highly-folded region studied by the author bears a tectonic resemblance, overthrust masses have been traced more than 60 miles away from their original position. From these analogies, we may infer the probability of a more or less deep-lying extension of the Liége coal-field.   L. L. B.

GALENA IN THE BELGIAN COAL-MEASURES.

Découverte d'un Filon de Galène dans le Terrain houiller du Bassin de Charleroi. By J. SMEYSTERS. *Annales de la Société Géologique de Belgique,* 1903, *vol. xxx., Bulletin, pages* 120-122, *with* 1 *figure in the text.*

In the Amercoeur colliery, at Jumet, in the Charleroi basin, a grit (at the 2,300 feet level of the Bellevue pit) is traversed by a vertical fissure, about a couple of inches wide, which, in its lower portion, contains galena. The ore is perfectly crystallized in cubes, and appears to be very pure: with it is associated a small quantity of blende. It has also been found in a sort of pocket in the floor of the gallery: a certain amount of water welled out of this " pocket " as soon as it was struck, and from the above-mentioned fissure there is at present a flow of 1,300 to 1,700 gallons of water every 24 hours. As a heading is being driven in the rock below the 2,300 feet level and is likely to cut the fissure at a greater depth, further light will be thrown on the extent and possible importance of the metalliferous vein. At all events, this is believed to be the first-recorded occurrence of galena in the Coal-measures of Hainault.   L. L. B.

MINERALS FOUND IN THE BELGIAN COAL-MEASURES.

Sur des Minéraux du Terrain houiller de Belgique. By X. STAINIER. *Bulletin d la Société Belge de Géologie, de Paléontologie et d'Hydrologie,* 1904, *series* 2, *vol. viii., Procès-verbaux, pages* 173-177.

Three new localities for the occurrence of blende are described: namely, (1) the Burton colliery, at St. Georges, where in a hard grit forming the floor of the Flairante coal-seam, a band of whitish-grey quartzite, very like gannister, is traversed by veins of crystalline quartz in which are embedded big transparent crystals of blende. The zinc-ore deposits worked by the same company as that which owns the colliery (Société de la Nouvelle Montagne) are close at hand, at La Mallieue, on the southern rim of the coal-basin; (2) in No. 1 pit of the Ressaix colliery, a much dislocated mass of grit intervenes between the Louise and Sans-nom coal-seams: it is traversed

by fissures about 1 inch wide, forming miniature metalliferous veins, the
infilling consisting of successive bands of pyrites, blende, and yellow calc-
spar; (3) in the Biquet-Gorée colliery, blende is found in quartz-veins which
traverse a hard grit forming the floor of the Belle-et-Bonne coal-seam.   This
seam is the equivalent of the Flairante seam previously mentioned, and the
blende, like that of the Burton colliery, shows great resemblance to the ore
from the Asturias in Northern Spain.   It is a curious circumstance that all
the blende found in Belgian collieries occurs in grits at the floor of coal-
seams.

Galena, in the form of tiny crystals, has been found in the massive grit
which occurs below the Stenaye coal-seam, in the Fanny pit of the Marihaye
colliery.   This grit has been correlated beyond a doubt with a grit-bed containing plumbiferous veins in the Charleroi coal-basin.

Chalcopyrite has been recorded from the Abhooz colliery, where the fractured surfaces of a very hard, vitreous, grey grit, which forms the parting
between two coal-seams, are encrusted with that sulphidic ore.   In the Bois
d'Avroy colliery, minute crystals of iron- and copper-pyrites occur on calcspar
which (together with fine bipyramidal quartz-crystals) encrusts the fissures of
a calcareous grit lying some 5 feet above the Wicha coal-seam.

Heavy spar is fairly abundant in the Upper Coal-measures of the Mons
coal-field; it also occurs, at much lower horizons, in the Havré and Soye collieries.   The author thinks that there can be little doubt that all these
minerals, occurring as they do in the only Coal-measure rocks capable of
retaining open fissures (namely grits), are the results of precipitation in those
fissures (by evaporation or otherwise) of the metalliferous and other particles
carried in solution by percolating waters.   L. L. B.

ERECT TREE-TRUNKS IN THE BELGIAN COAL-MEASURES.

Découverte de Troncs d'Arbres-Debout au Charbonnage d'Oignies-Aiseau.   By
X. STAINIER.   Bulletin de la Société Belge de Géologie, de la Paléontologie et
d'Hydrologie, 1903, vol. xvii, Mémoires, pages 539-544 and 1 plate.

The author points out that the discovery of well-preserved tree-trunks in
the Belgian Coal-measures is not of very common occurrence, and yet he has
had the good fortune to record two discoveries of this nature within a very
short time.   He recently described fossil trees found in the Falizolle colliery
in the Lower Sambre district, and now he deals with that found in the
neighbouring colliery of Oignies-Aiseau, in the roof of a seam at the 1,050
feet level.   The seam is known as the Great St. Martin, and is 148 feet lower
down in the series than the Ahurie or Lambiotte seam, in the roof of which
tree-trunks were discovered at the Falizolle colliery. The trunk (20 feet high) is
cut by a fault, and is moreover bent and "nipped" at one point.   Fine
remains of *Calamites* occur in its immediate neighbourhood, but their anato-
mical connexion with the trunk could not be established.   Above this trunk
occurs a thinner one; and the discovery of a third tree was announced
while the author's paper was being passed through the press, though No. 3
is in all probability but the detached basal part of No. 1, displaced eastward
and upward by the fault already mentioned.

Occasion is taken to record the previously-unpublished discovery of a
Calamitean trunk in No. 14 pit of the Monceau-Fontaine colliery, but in this
case there is no doubt that the tree had been torn from its natal soil and
drifted into its present position.   L. L. B.

COAL-MEASURES IN FRENCH LORRAINE.

Le Terrain houiller en Lorraine française. By FRANCIS LAUR. *Comptes-rendus hebdomadaires des Séances de l'Académie des Sciences,* 1904, *vol. cxxxix., pages* 1048-1049.

A boring put down at Eply, north-east of Pont-à-Mousson, has struck carbonaceous shales of Coal-measure age, at a depth of 2,230 feet below the surface. Moreover, coal-seams have been passed through, and an average sample of the mineral yields the following analysis:—Fixed carbon, 48·77 per cent.; volatile matter, 36·12; ash, 13·23; and hygroscopic water, 1·28. Prof. Zeiller has determined the fossils submitted to him as belonging to the Westphalian stage, or middle sub-stage of the Saarbrücken coal-field. There is every probability of the lower stage of Saarbrücken, with its 117 seams of bituminous coking-coal, being met with below the measures already struck. The boring passed through the Keuper, the Muschelkalk, the Bunter Grits and the Vosges Grits, in regular succession before reaching the Coal-measures, but little or no Permian was found. Another boring, at Lesmesnils, has also struck the Coal-measures.

This new discovery confirms the conclusions published by the author in 1900, as to the extension of the Saarbrücken coal-field into French territory along the axis of Wenkirchen-Pont-à-Mousson; the Hercynian fold here is thus shown to be parallel with the Essen-Dover axis, the Villé-Autun axis, and the Ronchamp-Creusot axis. The author believes that the Coal-measure anticline extends from Pont-à-Mousson to Nancy, over a breadth of 15 to 20 miles; and that we have here, perhaps, the most considerable hidden coal-field yet discovered. He suggests its extension north of Commercy, its passage below the Parisian Cretaceous, and its re-emergence in the West of France, after a course of 375 miles or so. L. L. B.

LIGNITES OF THE AIN, FRANCE.

Sur les Lignites de l'Ain. By EMILE CHANEL. *Bulletin de la Société Géologique de France,* 1903, *série* 4, *vol. iii., pages* 67-73.

The department of the Ain is fairly rich in lignite-deposits, but at present only three of them are being worked. The main object of the author is to show that they were formed by phenomena analogous to those observed nowadays, namely:—(1) Formation of superposed layers by river-drifted plant-débris, which subside to the bottom when their saturation by water is sufficiently complete; and (2) accumulation and burial of plant-débris by torrents towards their "cone of dejection."

The orogenic movements which ridged up the Alps and the Jura had their natural effect on the relief of the Ain region, the configuration of which passed through several phases, conditioning a different "hydrographic régime" in each phase: thus, for instance, the rivers, among them the Saône, were at one time all shifted farther westward.

The lignites of Priay, Varambon, Mollon, Ambérieu, Ambronay, and St. Jean-le-Vieux are all regarded by the author as of Pliocene age. They belong to the first of the two categories above-mentioned, and are interbedded with marls. The Douvres deposit, south of Ambronay, belongs, on the other hand, to the second category. It is worked by adits: the lignite-seam, about 8 feet thick, dips into the hill, and the bottom of the deposit has not yet been reached. At Soblay, three thick seams of lignite (each from 6¼ to 10 feet) are worked opencast. They are interbedded with thick, highly fossiliferous marls. These are probably of Pliocene age. At La Rasa, a lignite-seam, barely 40 inches thick, is associated with pottery-clays and fire-clays. The fire-clay is the main object of mining operations at this locality. It is of a very pure white, after having been washed free of the accompanying sand, etc., and the annual output of 50 metric tons does not suffice to meet the demand. The selling price is about 15 francs per cubic metre (say 4½d. per cubic foot). The deposits here are possibly of Miocene age.

In all the lignite-deposits described in this paper, the greater portion of the material consists of branches and tree-trunks, the ligneous tissue of which is still clearly discernible. The remainder consists of débris of bark, leaves, etc.   L. L. B.

ANTIMONY-MINES OF LA LUCETTE, FRANCE.

Sur les Mines de la Lucette (Mayenne). By M. MICHEL. *Bulletin de la Société française de Minéralogie,* 1904, *vol. xxvii., pages* 79-80.

Since 1898 a stibnite-deposit has been worked at La Lucette, near Laval, in the Department of the Mayenne. It occurs in quartz-veins which cut through slates and micaceous grits of Upper Silurian age. The predominant ores are the stibnite and mispickel, with which are associated iron-pyrites, and (more rarely) blende and native gold. The gangue is mainly quartz, but calcite and fragments of slate and grit also occur in the infilling of the veins. These veins have a general north-easterly strike, and an all but vertical pitch; they vary in thickness from about 8 to 32 inches. The mispickel is highly auriferous, and so too is the pyrites. The antimony-ore is in the form of lamellar masses and acicular crystals, some of the latter equalling in size the big stibnite-crystals found in Japan.

From the point of view of their structure and mineralization these deposits of La Lucette may be regarded as analogous to the well-known auriferous stibnite-deposits of Bohemia and Hungary.   L. L. B.

KIESELGUHR-WORKINGS IN CENTRAL FRANCE.

Découverte et Exploitation de Gisements de Silice (Diatomées fossiles) dans l'Arron-dissement de Murat (Cantal). By J. PAGÈS-ALLARY. *Compte-rendu de la Trente-deuxième Session (1903) de l'Association française pour l'Avancement des Sciences,* 1904, *part i., pages* 217-218.

Since 1901, the author has discovered and worked four deposits of kieselguhr, all situated within easy reach of the railway-station of Neussargues; and in 1903, in conjunction with another capitalist, he opened up the immense deposit of Faufoulioux, barely 2 miles out of the town of Murat, in the department of the Cantal. The five workings above-mentioned occupy an area of some 30 acres, and the workable beds vary in thickness from 26 to 33 feet, the entire mass probably amounting to about 32 million cubic feet, of an estimated market-value which ranges between £1,000,000 and £1,200,000. To that extent France will in future, it is hoped, be independent of the importation from Germany and Italy.

The deposit consists of the frustules of fossil siliceous algae (Diatomaceæ); some of the stuff extracted is being ground up for polishing-powder, etc., but the author expects to find a great many other industrial uses for it, as it is a refractory material, a bad conductor of heat, and yet very light.
                                                    L. L. B.

STANNIFEROUS VEIN IN THE LOZÈRE, FRANCE.

Sur la Présence de l'Étain dans le Département de la Lozère. By MARCEL GUEDRAS. *Comptes-rendus hebdomadaires des Séances de l'Académie des Sciences,* 1904, *vol. cxxxviii., page* 1121.

Although the department of the Lozère is rich in metalliferous veins, tin-ore had not hitherto been recorded there. Recently, however, the author came upon a vein of it, striking south-east and north-west, close to the great Monastier fault, in the parish of Barjac. The country-rock is mica-schist, the vein dips 60 degrees, and is 7½ feet thick at the outcrop. The gangue consists chiefly of barytes and quartz: between these is a thin band of cassiterite, barely an inch thick, with which are closely associated pyrolusite and wolfram. Near by, another deposit of wolfram occurs, which, on analysis, yields 65 per cent. of tungsten trioxide with traces of gold. The stanniferous vein above described is now found to be intersected by another vein containing the same ore. All these occurrences are in the near neighbourhood of an important barytes-mine.   L. L. B.

ORE-DEPOSITS OF THE SCHAUINSLAND, BADEN.

Beitrag zur Kenntniss der Erzlagerstätte am Schauinsland. By IMMANUEL LANG. *Mittheilungen der grossherzoglich badischen Geologischen Landesanstalt,* 1903, *vol. iv., pages* 485-524, *with* 7 *figures in the text and* 1 *map.*

The chief rocks of the district are gneisses, traversed by granitic and pegmatitic dykes, and a dyke of mica-trap. These are described in some detail by the author. The workable ores are chiefly zinc-blende, and, secondarily, galena. The first-named ore is nearly always brown or black, but a paler mineral, so-called "honey-blende," has been observed (though rarely) in druses in the southern veins. A chemical analysis of the workable blende from Hofsgrund shows it to consist of 79·08 per cent. of zinc sulphide and 20·83 of iron sulphide: but the percentage, both of zinc and of silver, in the ore varies, as a matter of fact, considerably. Thus, six analyses show from 47·32 to 57·96 per cent. of metallic zinc, and from 0·066 to 0·105 per cent. of silver. On the whole, the northern veins, now being worked, are the poorest. Galena constitutes entire veins in the south and south-west of the area, but decreases northward, being found in that direction continuously more and more intergrown with blende, until it disappears entirely. The percentage of metallic lead in the galena is always about 70, but that of silver varies from 0·032 to 0·146. In the Willnau vein, a very small quantity of gold also occurs (1·9 parts in 10,000,000). The gangue-minerals are quartz (of two periods), baryta, calc-spar, dolomite or ankerite (rather abundant), and pyrites—the two last-named being invariable associates of the workable ores. Minerals of a later formation and of less frequent occurrence are pyromorphite, malachite, eusynchite (vanadinates of lead and zinc), anglesite, etc.

The metalliferous veins form two main reefs, striking north-north-eastward and south-south-westward through the area. The thickness in ore of the easternmost of the two main reefs averages 12 to 16 inches of blende, and about 2 inches of galena; that of the westernmost, 4 inches of galena, and only ⅜ inch of blende: but the veins thicken continuously in depth, while the gangue decreases. So, the deeper they are worked the more productive they are. Nearer the surface, much of the ore has been weathered away by atmospheric agencies, but there is no ferruginous gossan, properly so-called.

The curious brecciated structure of many of the veins is described and figured, and the author then proceeds to deal with the question of the genesis of the ore-deposits.   Their analogy in many respects to those of Saxony is pointed out, and their history is sketched as follows.   During the formation of the Rhine Valley in the Tertiary Era, fissures were opened up in the neighbouring gneiss-massif of the Black Forest; these did not, however, reach quite to the surface, while below they admitted of the free ingress of metalliferous solutions coming from great depths.   Through the sudden release of pressure, these solutions flowed up into the fissures, leached out the mica from the adjacent country-rock, and, on cooling, precipitated the particles of ore on the walls of clefts, and cracks and fissures, and on the rock-fragments which in places choked up the fissures (hence the " brecciated " veins).   This encrustation of the gneiss-fragments prevented in the latter case further chemical decomposition of the gneiss.   L. L. B.

IRON- AND COPPER-ORES OF THE KNOLLENGRUBE, NEAR LAUTERBERG-AM-HARZ, GERMANY.

Die Knollengrube bei Lauterberg am Harz.   By K. ERMISCH.   *Zeitschrift für praktische Geologie,* 1904, *vol. xii., pages* 160-172, *with 4 figures in the text.*

The Knollengrube (or Knollen mine) takes its name from the neighbouring boss of quartz-porphyry (2,253 feet above sea-level), the Grosse Knollen, which is intrusive among the slate-rocks that form the southern margin of the Harz.   The iron-ores, etc., occur as reefs or veins among the so-called Tanner Grauwackés, which, on the strength of the palæobotanical evidence, are assigned to the Culm-measures.   Among them occur parallel inliers of the siliceous, honestone-like Wieder Slates, which are probably Devonian. The Lauterberg district was for many generations, from the Middle Ages until about half-way through the nineteenth century, the scene of active mining operations directed to the cupriferous veins now practically worked out.   If the strike of one of these veins be prolonged north-westward, it will be found to coincide with the hæmatite-reef of the Knollen mine.   But it has been contended, by Prof. Kayser among others, that the barytes-*cum*-iron-ore veins belong to a different fissure-system from that to which we must assign the cupriferous quartz-veins.   However that may be, there are two very distinct types of veins formerly worked on the Knollengrube concession: (1) the main hæmatite-reef, averaging 40 inches in thickness, with barytes as the sole gangue-mineral; and (2) the copper-reef, the chief ores of which are copper-pyrites and some of its derivatives, with a little iron-pyrites and some hæmatite.   The gangue is mainly quartz, barytes playing quite a subsidiary part.   The second reef strikes at an acute angle with the first, and at one point attains the enormous thickness of 98 feet.   It has been suggested more than once that the hæmatite-reef is in reality merely the capel of a deeper-lying cupriferous deposit, but the author sets himself to point out that this hypothesis is untenable.   Crush-conglomerates consisting of striated grauwacké-pebbles, ranging in size from a pea to a man's fist (or even bigger), occur within the hæmatite-deposit itself, and bear witness to the tremendous frictional effects of earth-movements.   Indeed, a detailed study of the series of samples of ore obtained from the mine reveals the widespread cataclastic character of the infilling.   Evidence of faulting, folding and overthrust among the neighbouring rocks is abundant.   It may be observed that the extreme limits of the metalliferous district are marked by conspicuous bosses of quartz-porphyry, and sills, sheets, dykes, etc., of similar eruptives occur within the district. Whence one is led to infer some genetic association between the porphyry-eruptions and the formation of the fissures now in-filled with metalliferous deposits. Dr. Klockmann has suggested that these fissures were filled with material leached out from the overlying Permian rocks, but the author does not appear to agree with this view, which he regards as somewhat far-fetched.

He gives a description and sections of the workings of the Knollengrube, the latest deep level in which was driven in July, 1903, some 600 feet below the crest of the ridge which divides the Knollen valley from the Hübichen valley. The mining concession, worked up till 1870 by the Government, is now, after sundry vicissitudes, worked by the Hanover-Brunswickian Mining Company. The iron-ore is one of the best that is got in Germany, similar in quality to Cumberland hæmatite, and contains (crude ore) from 60 to 70 per cent. of iron oxides, while it is remarkably free from impurities such as sulphur, phosphorus, etc. A modern plant (put down in 1902) is used for sorting the ore, the picked hæmatite containing from 73 to 80 per cent. of iron oxides: a bye-product, in the shape of fine red ferruginous mud, is sold for dyeing purposes. After a long period of comparative stagnation, a slow upward tendency is perceptible in the industry of this mine: since August, 1903, it has been found necessary to increase month by month the number of workpeople, coincidently with an increasing output. Much un-worked ore is in sight, and operations, however actively conducted, are hardly likely to exhaust it for many years to come. One gathers that the copper-ores are no longer worked, or likely to be worked.   L. L. B.

WASHINGS FOR GOLD ALONG THE RHINE.

Die Gold-Wäscherei am Rhein. By DR. BERNHARD NEUMANN. *Zeitschrift für das Berg-, Hütten- und Salinen-wesen im Preussischen Staate,* 1903, vol. li., pages 377-420.

In compiling this paper, the author has made extensive use of the official records available in the Hessian, Baden and Bavarian Ministries of Finance. With the aid of the information thus obtained he sketches first of all the history of the gold-washing industry along the Rhine, from the times of the Romans down to our own day. The industry appears to have died of inanition somewhere about 1866, but the very last washing for gold recorded on the Rhine took place in the Speyer district in 1900, and yielded something like ½ ounce of the precious metal. The second chapter deals with the question of rights of ownership, royalties, etc.; and the third with the methods of obtaining the gold, as they varied from age to age. As recently as 1846, the illustrious Prof. Daubrée devoted a memoir in the *Annales des Mines* to the distribution of gold in the Rhenish plain " and the manner of extracting it."

The gold in the Rhine gravels and sands is derived from the detritus of quartzose rocks, most of which are of Alpine origin, though some, of course, have been brought down from the Black Forest on one side and the Vosges on the other. The richest gold-finds were made in quiet waters near low banks, or on the margin of islets, especially after a spate; and the best stretch of the river for the purpose was from Kehl to Daxlanden, no auri-

ferous beds being found below Mayence, and very little gold upstream between Waldshut and Basel. Annual statistics of the production are supplied,
extending, in the case of the Grand Duchy of Baden, from 1748 to 1874, and
in that of the other States concerned, over shorter periods.

The author sees no prospect of reviving the industry, which at one time
furnished a livelihood to several hundred persons.   L. L. B.

ORIGIN OF THE KUPFERSCHIEFER ORE-DEPOSITS, GERMANY.

*Formen, Alter und Ursprung des Kupferschiefererzes, zur Beurtheilung der
Mineralbildungen in Salzformationen.* By FERD. HORNUNG. *Zeitschrift
der Deutschen geologischen Gesellschaft,* 1904, vol. lvi., pages 207-217.

In controverting the theory that the metalliferous contents of the Kupferschiefer represent the results of an impregnation of the shales by mineral
springs subsequent to the deposition of the former, the author first of all
points out that the ores vary to such an extent in character as well as in
abundance, that the hypothesis would require a great number of thermal
springs of extremely-diverse composition, welling up from rocks of different
age and of every description, with a very restricted period of activity.

The author holds that the highly-concentrated mother-liquors resulting
from the evaporation of sea-water, at an epoch when most of the Rothliegende beds had been laid down (but when the Zechstein or Permian sedimentaries had not yet been deposited) induced deep-seated alteration in all the
rocks to which they could gain access. The most striking characteristics of
this metamorphic process are the widespread, nay, universal, evidence of
oxidation and the frequently-observed precipitation of red iron-oxide. The
effect of the oxidizing action was that the saline mother-liquors gradually
took up all sorts of heavy metals from the diabases, from the crystalline
schists, and from such metalliferous veins as may then have been in existence. Thereafter the ocean suddenly broke into the lagoons and swamps
below sea-level where these mother-liquors were stagnating, heaping up in
places the gravels and sands which constitute the Zechstein-conglomerate.
Meanwhile, owing in part to the decomposition of the marine organisms
swept in by the flood, and in part to the deposition of the sediments forming
the Marl-slates, the metalliferous particles were precipitated, and sorted
out or distributed by the currents. The consequence, of course, is that the
wealth of ore in the Kupferschiefer diminishes from below upward; since the
original "stock" of heavy metals present in the waters of the salt-lagoons
diminished *pari passu* with their precipitation, and the stock was not being
renewed by the supposititious agency of thermal springs. The relative permeability of the overlying or underlying sediments seems to have been of no
account in the process; nor is the metalliferous content of the Kupferschiefer
apparently determined by the amount of bitumen which it contains (from 8
to 30 per cent.). The author describes the occurrences of quartz and heavy-
spar in association with the ore-deposits of the Harz, and the intimate connection traced between manganiferous sediments and saline mother-liquors,
as so much additional evidence in favour of his contention. For the process
of alteration to which he traces the genesis of the Kupferschiefer ores, he
suggests the term "halurgometamorphism," as with the term "regional
metamorphism" geodynamic phenomena are generally associated.

L. L. B.

IRON AND MANGANESE-ORES IN THE GRAND DUCHY OF HESSE.

Eisen und Mangan im Grossherzogthum Hessen und deren wirthschaftliche Bedeutung.
By C. CHELIUS. *Zeitschrift für praktische Geologie,* 1904, *vol. xii., pages* 356-362.

The author first of all devotes a few paragraphs to the (industrially speaking) less important mineral-occurrences in the Grand Duchy, as, for instance, the cupriferous veins of the Odenwald; the plumbiferous slates which crop out in the spurs of the Taunus, at the margin of the Wetterau district; the brown-coal workings, the kieselguhr and bauxite-deposits of the Vogelsberg, etc. (the discovery of further brown-coal-fields is by no means unlikely); the saline deposits and brine-springs of Wimpfen, etc.; the barytes-mines of the Eastern Odenwald, and so forth.

The total output of manganiferous ores from Hesse-Darmstadt amounted for 1903 to some 150,000 tons, and that of iron-ores (pure and simple) to about 70,000 tons. The Odenwald manganese-ores are overlain by the red and mottled Lower Bunter Marls, while they are underlain by Zechstein (Permian) dolomite, Rothliegendes, or granite. These ores are very probably the residuum left from surface-decomposition of the Zechstein dolomite, and are most thickly bedded where they fill up depressions in the extremely-irregular surface of so much of the dolomite as still remains. The ores may be described as blackish, crumbly, manganiferous brown hæmatites, containing from 10 to 22 per cent. of metallic manganese and from 10 to 20 per cent. of iron: the relative percentages varying until the ores pass into a red or yellow material very much richer in iron than in manganese. Nodules, nests, reniform and stalactitic concretions of crystalline pyrolusite and barytiferous psilomelane (containing from 59 to 69 per cent. of manganese protoxide), occur in clefts or cavities, or irregularly strewn about in the mass of the deposits. In olden days the deposits were worked for the sake of the iron, but now they are more actively worked than ever, though for the sake of the manganese. Much hidden treasure, in the shape of unworked ore, lies in the Odenwald awaiting the touch of the miner's pick; but an indispensable preliminary to successful working will be a careful study of the stratigraphical conditions of the deposits, and more especially of the tectonics of that mountainous district. Neglect of such a study has led to waste of time and labour on practically-useless concessions lying in the Bunter-Sandstone area, where the ore, if it be there in workable quantity at all, lies buried beneath 1,000 or 1,500 feet of Trias. However, even if nine-tenths of the iron- and manganese-mining con-cessions in the Odenwald may be regarded as valueless from the industrial point of view, the remaining tenth constitute a valuable and easily-workable group of deposits. The building of certain new lines of railway is likely to contribute to the recrudescence of the mineral-industry in the district.

Turning then to Upper Hesse, the author refers to the active mining operations in the neighbourhood of Giessen, which have been carried on uninterruptedly for many generations; as also to the recent revival of such activity at Oberrosbach near Friedberg; and he further enumerates other localities where the prospects of successful mining are favourable. Here the manganese-ores are chiefly associated with the Devonian *Stringocephalus*- or *Pentamerus*-limestones, but the origin of part of them at least is, *mutatis mutandis*, lithologically and chemically the same as that of the post-Permian manganese-ores of the Odenwald, and their character and composition are also very similar. There are, however, rich crystalline pyrolusites not so closely associated with the limestones, which appear to have been deposited from solutions that percolated upward along the marginal fault-fissures of the limestones and the slates. At Giessen, the workings are opencast, while at Oberrosbach they are underground.

In the north-western part of the basaltic area of the Vogelsberg, patches of red soil, in vivid contrast with the usual grey and dark-brown of the loams, etc., resulting from the decomposition of the basalts, reveal the out-crops of iron-ore. Great opencast workings are in fair activity at Bernsfeld, Niederohmen, Flensungen, Ilsdorf and Stockhausen: the brown hæmatite which is mined at these localities occurs in nodular or pisolitic form, or in great banded lenticles, among highly-weathered masses of basalt. The ore-occurrences coincide with areas of great faulting, and it would seem that the overthrust and sheared basalts must have been partly brecciated and fissured, and that the hæmatite is the residuum from ferruginous thermal waters which welled up through the fissures. In some cases these original or primary hæmatites have been washed away and re-deposited in the form of gravels or alluvia, thus constituting secondary deposits. The quality of·the Upper Hesse hæmatite makes it a favourite at blast-furnaces where it has once been used, but the comparatively-heavy transport-charges to the Rhine and the Ruhr tell against it in competition with other ores. This is a diffi-culty which the proposed canalization of the Lahn may obviate, at some future time. Large areas of ore still await working, but the same blunders in regard to concessions have been made by too ardent prospectors as in the Odenwald. The bauxite-deposits of the Vogelsberg lie along the north-and-south and west-and-east iron-ore belts, and the bauxite is now being dug in large quantities at Grünberg and other localities.   L. L. B.

GENESIS OF THE DEVONIAN RED HÆMATITES OF NASSAU.

Sind die Rotheisensteinlager des nassauischen Devon primäre oder sekundäre Bildungen? By F. KRECKE. Zeitschrift für praktische Geologie, 1904, vol. xii., pages 348-355, with 1 figure in the text.

Nearly every author who has written on these deposits has explained their origin, in accordance with Prof. Bischoff's theory, as due to metasomatic replacement of limestone by iron-ore. Dr. Krecke holds, however, an entirely-different view, and concludes that the Nassau ores are neither more nor less than the outcome of primary deposition. One of the favourite conten-tions of the "metasomatists" is that the iron was derived from the neigh-bouring *schalsteins* and diabases: many facts, however, are adduced to show that this assertion is baseless. Much of the diabase is actually of later age than the iron-ores, and near the junction these have been altered by contact-metamorphism from hæmatite to magnetite. It is also difficult to under-stand why the beautifully-preserved calcitic shells of the numerous fossils in the so-called "Fluss-stein-beds," etc., should have survived a metasomatic process which is supposed to have dissolved away the impure limestones them-selves. Stress is laid on the constancy of the Nassau hæmatites to a particular horizon, at the precise boundary between the Middle and the Upper Devonian, characterized by a Goniatite known as *Prolecanites tridens*. It would be curious indeed if the supposed metasomatic change had been con-fined to that horizon throughout Nassau, while other limestone-beds had re-mained unaffected by the percolation of the (hypothetical) ferruginous solutions. It will be noted, too, that where the hæmatites are intercalated among the limestones, the ore-bands are sharply marked off from the underlying and overlying barren rock, there being no trace of a passage from one into the other. Finally, the chemical composition of the Nassau ores, their deficiency in magnesium carbonate and manganese, forbid the acceptance of the metasomatic theory, so far as they are concerned.

The author thinks that during and after the submarine eruption of the Middle Devonian diabases, iron-bearing solutions and vapours had access to the waters of the sea, and, as a result of chemical reactions which he describes, the iron-ores were precipitated from them on the sea-bottom.   L. L. B.

THE LORRAINE ANTICLINE AND THE SAARBRÜCKEN COAL-FIELD.

Der lothringische Hauptsattel und seine Bedeutung für die Aufsuchung der Fortsetzung des Saarbrücker Kohlensattels. By L. VAN WERVEKE. *Centralblatt für Mineralogie, Geologie und Paläontologie,* 1904, *pages* 390-395, *with a map in the text.*

The most striking feature in the geological structure of German Lorraine is the protrusion of the Bunter Sandstone south-westward into the Muschelkalk area, from the Saar up to Hargarten and Lubeln, and the corresponding protrusion of the Muschelkalk amid the Keuper rocks up to Vaucremont, near Rémilly.

In the sandstone-area or Kreuzwald plain, the extension of the coal-bearing anticline of Saarbrücken was already proved some 50 years ago by boreholes, and the above-mentioned protrusions are conditioned by the existence of what has been termed the "Buschborn anticline." This anticline has now been traced as far as Gézoncourt, and possibly extends yet farther to the south-west. In preparing a tectonic map of Alsace-Lorraine, the author has had recent opportunities of studying more closely the tectonic details of the country, and he finds that the chief anticline is divisible into two portions, a northern or Buschborn anticline proper, and a southern or Fletringen anticline: the two together constitute what he terms "the main anticline of Lorraine." The two portions are united in the Muschelkalk area by means of the Füllingen syncline, and in the Keuper area by the Rémilly syncline.

More than half a century ago it was pointed out that the westward extension of the Saarbrücken coal-field should be sought for, preferably along the summit of the anticline. Boring-operations to this intent, suspended since 1856, were renewed in 1900, both in French and in German territory. What are considered the most favourable indications in regard to coal have been reported from Baumbiedersdorf and Lubeln, localities which lie on the strike of the southern or Fletringen anticline. The author emphasizes the desirability of searching for coal along that line, and the importance of not confining the investigation to the Buschborn or northern anticline. He remarks that the results of the recent boring-operations are on the whole shrouded in an inpenetrable veil of secrecy, but the news has leaked out that borings put down in the most highly-disturbed area between the two anticlines, at Silbernachen (or Servigny) and Füllingen, have been unsuccessful.   L. L. B.

BORING FOR COAL AT THE POTZBERG, RHENISH PALATINATE.

Die Tiefbohrungen am Potzberg in der Rhein-Pfalz. By A. LEPPLA. Jahrbuch der königlich Preussischen geologischen Landesanstalt und Bergakademie zu Berlin für das Jahr 1902, vol. xxiii., pages 342-357.

About 1890, a boring was put down in the Spelgenbach valley, at the southern base of the Potzberg, in the hope of striking the Coal-measures of the Saarbrücken group at a workable depth.   Boring-operations were carried down to a depth of 1,312 feet from the surface, and coal was reported at several points in the cores, but it is not now definitely known whether the occurrence was of industrial importance.   Then a boring was put down in the Wildfrauenloch, in the Lochwiesgraben, and, owing to certain technical difficulties, was abandoned at a depth of 1,050 feet from the surface: traces of coal were found at the depth of 679 feet.

A third boring, some little way farther up the Lochwiesgraben valley, was carried down to a depth of 3,795 feet from the surface. Coal was first found in the cores between the depths of 2,397 and 2,552 feet, in strata of the Lower Ottweiler group, but no opinion can be expressed as to the thickness of the seams, if such they be. The lie and succession of the rocks, their character, and the nature of the coal and plant-remains contained in them, all go to prove that, from the depth of 3,275 feet downward, the boring was going through the measures of the Middle Saarbrücken or Flammkohlen group. Coal was found again and again after the depth just mentioned had been reached, and at 3,539 feet, a core of coal 3¼ feet thick was obtained. This is a sufficient indication that the coal-bearing beds of the Saarbrücken series continue in a north-easterly direction into the Palatinate, and they appear to dip less steeply the lower down from the surface that they occur: the dip shown for them in the boring is from 5 to 10 degrees only. It now rests with mining engineers to determine the best method of making industrial use of the information which this important boring has supplied.

L. L. B.

BORING FOR COAL IN SAXONY.

Wo könnte in Sachsen noch auf Steinkohlen gebohrt werden? By K. DALMER. Zeitschrift für praktische Geologie, 1904, vol. xii., pages 121-123.

This third paper apparently completes the survey of the subject which the author had begun in 1902.[*] With regard to the Döhlen basin, south of Dresden, the coal-bearing strata are shown by their fossil organisms to belong to the Lower Rothliegende, and not to the Coal-measures proper. The area between Nieder-Hermsdorf and Grumbach, forming the westernmost portion of that part of the coal-field which lies to the left of the Weisseritz, has not yet become the site of boring operations, but it appears very unlikely that workable seams will in any case be struck there. The exploration-work carried out at the Government Zaukerode colliery has proved that the main seam of the Döhlen coal-field becomes increasingly impure in the direction of Nieder-Hermsdorf, until it reaches the stage of complete unworkability.

It has been suggested that, in regard to that portion of the Döhlen coal-field which lies to the right of the Weisseritz, the neighbourhood of Quohren would form a good field for boring-operations. The author, however, adduces stratigraphical evidence to show that the prospect of finding workable coal there is very remote indeed.

[*] *Trans. Inst. M.E.*, 1903, vol. xxiv., pages 695-696.

He then considers the possibility of striking coal in the Elbe valley, close to Dresden itself. There is no doubt that Rothliegende strata do occur there, deep below the Cretaceous and later deposits. Deep borings alone can determine whether the coal-bearing shales of the Lower Rothliegende are also present.

Although the discovery of extensive coal-fields in the Erzgebirge is not to be hoped for, recent exploration-work in the Carboniferous area of Brandau has revealed the presence of a coal-seam, 6½ feet thick, at a locality where the rocks had been mapped as gneiss by the Saxon Geological Survey. The surface there is covered by an immense mass of gneissic débris, which makes it impossible to say exactly where the gneiss ends off and the Rothliegende begins. On the whole, however, expenditure on boring-operations in the region of the Erzebirge is unadvisable. The author concludes with the reiteration of his previously-expressed opinion that the area north of Leipzig and between Riesa and Elsterwerda would probably repay a search for coal, and he indicates the points where trial-borings should be put down.

L. L. B.

ORE-DEPOSITS OF THE SCHWARZENBERG DISTRICT, ERZGEBIRGE.

Über die Erzlager der Umgebung von Schwarzenberg im Erzgebirge. By Dr. R. Beck. *Jahrbuch für das Berg- und Hüttenwesen im Königreiche Sachsen,* 1904, *pages* A56-A96, *with 6 figures in the text and 2 plates.*

The first portion of this monograph was published in the *Jahrbuch* of the Saxon Department of Mines for 1902, and the second portion now published deals in the first place with the Breitenbrunn series of deposits, mineralogically and historically the most interesting of any in the district; but active working on these deposits is now confined to a single mine, that of St. Christopher. The consequence is that much of the author's descriptive matter in regard to the Breitenbrunn series is drawn from the memoirs of Messrs. Freiesleben, Schalch, and others. The majority of the metalliferous veins traversing the main deposits (magnetite) belong to the silver-cobalt group, while a few (and precisely the most considerable vein of all) belong to the pyrites-blende group; there is evidence also of the existence of veins of cassiterite. The main ore-deposits are intercalated conformably among mica-schists, dipping south-westward, near to the horizon where the schists pass into phyllites. The north-western portion of the Breitenbrunn series comes within the contact-area of the Eibenstock granite, while the south-eastern portion shows no trace of the influence of contact-metamorphism.

The author considers in detail the distribution of the ores in the St. Christopher mine: besides layers or bands of pure magnetite, there are others where the magnetite is much intermixed with sulphidic ores. The last-named are evidently impregnations from the veins which are mentioned above as traversing the main deposit, and it is noticeable that the impregnation is more pronounced and more extensive in the direction of the strike than in any other. The output from the mine, from the beginning of 1901 to June 1904, amounted to 356 tons of arsenical pyrites, 20 tons of zinc-blende, and 10½ tons of magnetite. But the sudden collapse in the output of magnetite imaged by these statistics is traceable to the shutting-down of the Zwickau blast-furnaces; and we are told that the stock of ore at the mine (apart from the above-mentioned output) includes 400 tons of magnetite, 300 tons of zinc-blende, and 10 tons of arsenical pyrites. It is proposed, in the case of the mixed iron-ore, to separate off the magnetite by the magnetic process, and to make up the "slime" thus obtained into briquettes.

The long-abandoned mine of St. Margaret was re-opened in the spring of 1904, as the demand for radium makes it now worth while to work the deposit for the uraniferous pitch-blende which it contains: its chief output in former times consisted, however, of magnetite, blende and iron-pyrites.

The series of deposits known as the Unverhofft Glück (unexpected good-luck), north-west of Antonsthal, also occur among mica-schists, just outside the zone of contact-metamorphism by which the granite-masses are fringed. The output of ore, at one time considerable, has been of no account since 1887. The predominence of galena and blende and the complete absence of magnetite are to be noted here, although the lithological conditions are other-wise similar to those of the St. Christopher deposit. Cross-veins belonging to the plumbic-spathic carbonate-group are frequent.

The last series of deposits to be described are those situated to the north-west and north of Schwarzenberg itself. They are not very numerous, and the abandoned workings are no longer accessible. Bed-like impregnations of tin-ore, and veins of the same, occur in the zone of intense contact-metamor-phism which borders the three granite-masses of Aue, Auerhammer, and Lauter, and the genetic relationship between the ore and the granite-intru-sion is undoubted. Pyrites and magnetite-deposits appear to have been worked at Bernsbach and Beierfeld respectively.

Some ore-deposits in other parts of the Erzgebirge are briefly described, in order to show their similarity to those of the Schwarzenberg district, and the author also claims to have observed analogous occurrences in Finland and Central Sweden. His general conclusion is that ore-deposits of the Schwarzen-berg type are, in every respect, a consequence of the intrusion of granite (in this instance, among crystalline schists).　　　　　　　　　　L. L. B.

RADIUM-BEARING MINERALS OF SCHLAGGENWALD, GERMANY.

Radium in Schlaggenwald. By J. HOFFMANN. *Zeitschrift für praktische Geologie,* 1904, *vol. xii., pages* 123-127.

A description is given of the following minerals from Schlaggenwald in the Saxon Erzgebirge, examined and analysed by the author in the laboratory of the Government Technical College at Elbogen:—Calc-uranite, uraninite (pitchblende), cupriferous uranite, uranium-ochre, gummite, uranotile (?) and "uranium-bloom" (a combination of copper and uranium oxides with phosphoric and silicic acids). Uranium-ochre, uranotile and gummite show much less radio-activity than the other four minerals; and, generally speaking, he finds that it is the crystallized compounds of uranium that show the greatest radio-activity.

Now that mining-operations have been restarted in the Schlaggenwald district, with the view of working the wolfram-, copper- and tin-ores which occur there, it seems likely that the radium-bearing minerals will constitute, as it were, a bye-product of some industrial importance. The tin-ores are associated with the granites, which are of probably later date than the surrounding gneiss of the district.　　　　　　　　　　L. L. B.

STANNIFEROUS DEPOSITS IN SAXONY.

Die Erzlagerstätten. By R. BECK. *Erläuterungen zur geologischen Specialkarte des Königreichs Sachsen,* 1903, *Section Fürstenwalde-Graupen, Blatt* 120, *pages* 40-57, *with 3 figures in the text.*

The district of Graupen is first dealt with: this has long been recognized as divisible into three mineral areas, that of Steinknochen, that of Mückenberg or Upper Graupen, and that of Knötl. The last-named area is left undescribed for the present, as it really comes altogether into a different sheet of the survey-map.

The best-known vein in the Steinknochen area is the Lucaszechner or Luxer vein, the only one still worked there, at the Martini mine. It strikes due north and south, and generally dips about 35 degrees westward. Most of the other veins of this area have an east-north-easterly strike. In thickness, the Luxer vein ranges between 8 and 12 inches, and often includes small fragments of highly-decomposed country-rock (biotite-gneiss). The vein-stuff is of exceedingly-variable composition, but its main constituent is a milk-white drusy quartz: coarse-grained orthoclase and violet fluorspar are of common occurrence. The principal ore is a brown or yellowish tinstone, developed in fairly-big compressed, columnar crystals. Copper-pyrites and galena are occasional associates, and the decomposition of the former has given rise to malachite and azurite. Later formed druses of barytes are mentioned as a somewhat unusual occurrence in conjunction with tin-ore.

The stockwork of the Preisselberg Hills occurs in immediate association with porphyritic microgranite and granitic porphyry: its most productive portions appear to have been worked out. An abortive attempt was made in 1863 to restart working the ore, but apparently the percentages obtained would only warrant working under exceptionally favourable conditions.

The underground workings of the Mückenberg area are no longer accessible, but mining operations were still in progress there in 1868. The description of these is now of little interest, from the industrial point of view.

Leaving the Graupen district, the author states that on the Zeidelweide, north-east of Fürstenau, a series consisting of at least four parallel veins, of moderate thickness, strike north-eastward through the biotite-gneiss. Two of these have proved eminently workable—indeed the mineral-industry was active in that neighbourhood, from the seventeenth century until the last quarter of the nineteenth. The principal ore is tinstone, but silver-bearing blende, copper-pyrites, and galena also occur. Shortly after 1876 the last mine was shut down. In the district of Tellnitz, veins of pyrites, galena and blende were worked for many generations, but the attempt to revive the mining industry there in 1853 was frustrated by the poor results yielded by the assays made in the Government laboratories at Freiberg. L. L. B.

RECENT DISCOVERIES IN THE COAL-FIELD OF UPPER SILESIA.

(1) *Der Flötzberg bei Zabrze: ein Beitrag zur Stratigraphie und Tektonik des Oberschlesischen Steinkohlenbeckens. By* FRIEDRICH TORNAU. *Jahrbuch der königlich Preussischen Geologischen Landesanstalt und Bergakademie zu Berlin für das Jahr* 1902, *vol. xxiii., pages* 368-524, *with 2 figures in the text and 4 plates.*

Mining operations on the Zabrze anticline have been conducted with such great activity and success, that a time-limit can now be foreseen, within which the thick coal-seams will be completely worked out. Therefore, the author considers that the moment is opportune for a general summary of all the available knowledge in regard to this Coal-measure anticline. He gives the details of no less than 177 borings and sections measured in shafts, a bibliography consisting of 33 entries, and a full list of the plant-remains and other fossils.

On the south and west, the Zabrze anticline is bounded by faults, while on the north and east the strata gradually flatten out into the basins of Beuthen and Ruda respectively. The Königin-Luise colliery, belonging to the Prussian Government, takes up the lion's share of the Zabrze portion of the coal-field, the remainder of it being worked by five private collieries. The five thick seams (varying from a minimum of 10 to a maximum of 46 feet in thickness) which form the main object of operations here, belong to the Saddle or Anticlinal Group, and form about one-sixth of the total thickness of that group; this group the author places in the stratigraphical succession between the Middle Coal-measures (equivalent to the Karwin Group, *sensu lato*) and the Lower Coal-measures (equivalent to the Ostrau Group, *sensu lato*). The Carboniferous system in the district here described, consisting, apart from the coal-seams, of sandstones, shales and clay-slates, occupies broad tracts at the surface, its outcrop being barely masked by a thin mantle of Drift-loams and sands: the Drift rarely approaches 7 feet in thickness. Immediately above the Anticlinal Group in the stratigraphical succession come the Ruda Beds, including seven or eight coal-seams, the thinnest of which is a 5 feet seam and the thickest a 20 feet seam. The Pochhammer, Reden and Heinitz seams, belonging to the lower division of the Anticlinal Group, yield gas-coal and coking-coal: all the other seams above them are of a different character, being described as non-coking, long-flame coals (*nicht backende Flammkohle*).

The Zabrze anticline is itself traversed by faults and overthrusts, which are, however, of comparative unimportance, and do not greatly affect its general structure. The age of the folding movements which originated the anticline is certainly post-Triassic, and possibly early Tertiary (Miocene).

(2) *Neues aus dem Oberschlesischen Steinkohlenbecken.* By D. GAEBLER. *Zeitschrift für das Berg-, Hütten- und Salinenwesen im Preussischen Staate,* 1903, vol. li., pages 497-519 and 1 plate.

The existence of a northern marginal basin had long been known, but that knowledge was not of a very detailed character, until in 1901 the third of three deep borings was put down near the railway-line north of the village of Mikultschütz, and determined to some extent the course of the main axis of the basin. Still more recent borings have shown a very steep dip of the strata along its southern boundary, and consequently a far greater depth for the basin than had been previously suspected, as well as a northward shifting of its axis. Fifteen workable seams, totalling a thickness of 131·3 feet of coal, were struck between the depths of 546 and 2,710 feet, in boring No. 1 of the Preussen colliery. About the same number of workable seams were passed through in the Vüllers shaft and boring, between the depths of 636 and 2,983 feet, with a total thickness of 155 feet of coal. In the last-named boring, brackish-water and marine shells were found at various horizons in the deeper-lying portion of the strata, and sphærosiderite was extraordinarily abundant.

As an example of the great caution needed in correlating, and theorizing on, the results of borings, the author points out the recent case of the

Anna shaft of the Frieden colliery. Here actual sinking operations have proved one seam fewer, but 13½ feet of coal more, than the boring which had been put down in 1900. Nine seams which the boring was thought to have passed through were not found in sinking the shaft, but eight seams missed in boring have been proved in sinking.

The author proceeds to show that his view of the Orlau disturbance, which has been the subject of controversy, has been upheld by the most recent results obtained from borings, and he then describes the favourable outcome of borings lately made on the Jastrzemb anticline. Across the frontier, the five borings put down at Gross-Kaniow, in Austrian Galicia,. on the banks of the Vistula, have shown the steepness of dip of the strata to increase continuously with the depth; and it is inferred that south, or south-east, of a line passing through Kurzwald, Bielany, Zator, and Tenczynek, only unproductive Coal-measures or Culm will be found, the latter even being, perchance, replaced by the Carboniferous Limestone, which in the east is known to underlie the Coal-measures.

The deep borings started some 50 feet above the floor of the Leopoldine seam in the Cons. Wanda colliery, at Brzezinka, is regarded as one of the most important borings lately put down in Upper Silesia, as it throws light on the variation in the Schatzlar beds from east to west of the Prussian part of the coal-field, and confirms the correlation of the Leopoldine seam with the Emmanuels-segen and Leopold seams. This boring was carried down to a depth of 3,572 feet, by which time 437 feet of Ostrau beds had been passed through, containing four coal-seams, of which the thickest did not exceed 2½ feet. In the overlying measures altogether twenty-six seams were proved, with a total thickness of 124½ feet of coal.

Recent sinkings at Woschczytz, Pallowitz, and Sohrau have shown the gradual rise of the strata to the south, thereby proving that the deepest portion of the great " inner basin " is north of, and near, the first-named locality. From this conclusion it now becomes possible to predicate the depth and lie of the successive groups of Coal-measures, from Chelm in the east to Binkowitz in the west.

Finally the author discusses the results of recent research, as applied to the correlation of the seams in the Jaworzno basin, or " eastern marginal basin," across the Russian frontier, with the seams proved in the German portion of the coal-field.

(3) *Über eine Erweiterung des Gebietes der produktiven Steinkohlenformation bei Landeshut in Schlesien. By J. HERBING. Centralblatt für Mineralogie, Geologie und Paläontologie, 1904, pages 403-405.*

Thanks to the rediscovery of long lost boring-cores from Reichhennersdorf, and thanks also to several new finds of Coal-measure plants, it has been ascertained that the Upper Carboniferous rocks extend at least 2 miles farther to the north-west of Landeshut than all hitherto-published maps allow. Coal may then be looked for in localities where there had been, up till now but little hope of finding it.

(4) *Das Oberschlesische Steinkohlenbecken und seine kartographische Darstellung. By R. MICHAEL. Zeitschrift für praktische Geologie, 1904, vol. xii., pages 11-20, with 2 figures in the text.*

The map of the seams of the Upper Silesian coal-basin on the scale of 1:10,000 is being issued by the Royal Bureau of Mines at Breslau, and the author takes occasion to give a detailed history of what has been done in regard to the geological mapping and description of that area since

1860. The extension of the coal-field into Russian territory has already been mapped on the new scale by the Russian Department of Mines, and identical symbols have been used, so as to coincide completely, and join up, with the German map now being issued. Similarly, the Austrians have published a map of the Ostrau seams, on the scale of 1:10,000.

In 43 sheets the map will cover an area of about 760 square miles, leaving Tarnowitz outside it on the north, Nicolai on the south, and Gleiwitz on the west. This is indeed, only a small portion of the entire Upper Silesian coal-field, which covers (so far as it is known) 5,600 square miles, whereof 4,000 belong to Prussia, 1,000 to Austria, and 600 to Russia. The coal-field is divisible into a central basin, separated by a zone of disturbance from a western basin, and by an anticline from a smaller northern basin. On the east, the coal-field is cut off by the Carboniferous Limestone and the Devonian strata, on the west and north-west by the Culm, but its southern boundary is more uncertain. A list is given of the seventeen localities where the coal crops out at the surface. The productive Coal-measures of this coal-field are divisible into three groups, whereof the middle one is characterized by the number, thickness, and excellence of its seams. These are all carefully distinguished on the new map by different tints and symbols, and six north-to-south sections accompany the map, which the officials arranged to complete in time for the St. Louis Exhibition.

Great efforts have been made of late, in the direction of economical working of the best seams; and many mines now use sand for packing. Meanwhile, more attention is being directed than formerly to the upper group of seams (which constitute a vast amount of available coal), and the centre of gravity of the coal-mining industry seems likely to shift gradually southward. Concessions have been registered and marked out over an extensive area that is still unworked. This will in time involve the preparation of another map, covering the region to the south of the present one.   L. L. B.

MAGNETITE-DEPOSITS OF SCHMIEDEBERG, GERMANY.

Die Magneteisenerzlager von Schmiedeberg im Riesengebirge. By GEORG BERG. Jahrbuch der königlich Preussischen geologischen Landesanstalt und Berg-akademie zu Berlin für das Jahr 1902, vol. xxiii., pages 201-267, with 10 figures in the text and 1 plate.

The little town of Schmiedeberg lies at the foot of the mountain-range of the Riesengebirge, in the province of Prussian Silesia. Mines were worked here in the Middle Ages, and Schmiedeberg iron enjoyed a widespread reputation in Germany. Various attempts were made in the first half of the nineteenth century to revive the industry, which had come to grief in the Thirty Years' War (1618-1648); but it was not until 1854 that it was found possible to resume really systematic, profitable working of the deposits.

Geologically, the Riesengebirge consists of a central granite-massif generally mantled over by gneisses and mica-schists. Nearly vertical where they abut on the granite, the bedding-planes of these become less and less highly inclined as one recedes from the granite. The crystalline rocks are conformably overlain by the older stratified formations, but are unconformably overlapped by the newer stratified formations, which are all but horizontal. The geographical centre of the range does not precisely coincide with its geological structure: for instance, its highest summit, the Schneekoppe, is at the junction of the granite with the schists.

The author describes in some detail the geological structure of the country round Schmiedeberg, having carefully mapped in 1901 an area of some 8 square miles. This is largely covered with forest: the northern portion is occupied by the porphyritic (newer) granite, and the central and southern portions by the gneisses and schists. Several chapters are devoted to the petrography of these rocks, and in the ninth chapter we learn that silicate-bearing crystalline limestone and hornblende-rock (or amphibolite) are characteristic associates of the ore-deposits, as also certain mica-schists. Quartzites are occasionally intercalated among these.

The magnetites vary in structure from fine-grained to coarse-grained, from compact to crystalline, but really well-formed crystals are not found. The best grade of ore contains about 79½ per cent. of iron peroxide (Fe_3O_4), 3·2 per cent. of silica, 6 of alumina, 7 of iron bisulphide, and 4 of calcium carbonate. Chlorite and biotite are intimately intermixed with the granules of magnetite, and sometimes the flakes of the former two minerals are so numerous as to crowd out the last-named mineral. In some few cases, the microscope has shown limestone passing into magnetite.

Just as there are magnetites in these deposits with which flakes of biotite and chlorite are closely intermingled, so too the author found biotite-schists and chlorite-schists, and also serpentinous decomposition-products, impregnated with magnetite.

The ore-beds alternately thicken and thin out in the most irregular fashion—the mean workable thickness averaging 6½ to 10 feet, though a maximum of 33 feet is sometimes attained. Both at the hanging-wall and at the foot-wall the ore-body is generally sharply marked off from the country-rock—this consists (as a rule) of calc-schists and biotite-schists or hornblende-schists. In the Bergfreiheit mine, ten ore-beds are distinguished, the farewell rock or basement being a limestone some 35 feet thick. Complications in the shape of folding, faulting, etc., are by no means wanting. At present, it is only the mine which has just been named that is being worked on a large scale: the sump is 1,295 feet below the surface.

Sulphidic ores not only occur as an impurity among the magnetites, but are widely spread among the limestones, hornblende-rocks, and other near neighbours of the ore-formation. They are predominantly iron-pyrites and magnetic pyrites, mispickel and copper-pyrites playing only a very subordinate part. These sulphidic ores occur in veins and "nests," and are in all probability the latest-formed minerals of the whole group. With them calcite is intimately associated. One does not gather that they would repay working, although an attempt has been made to work one exposure of them.

Another characteristic feature of the Schmiedeberg deposits is the occurrence of parallel pegmatitic dykes (termed *riegel* or "bolts" by the miners), some of which are more than 6 feet thick. These are probably connected with an after-phase of the granite-eruption, and are of undoubtedly later date than the magnetites.

With regard to the genesis of the ores, the author considers five different hypotheses. A very probable one connected them with the regional metamorphism of the Archæan strata into crystalline schists; and thus it is that they have been regarded by many investigators as analogous to the Scandinavian iron-ores. The author, however, adduces the evidence derived both from macroscopic and microscopic study of the rocks, to show that the Schmiedeberg ores are the result of contact-metamorphism set going by the eruption of the granite. L. L. B.

GENESIS OF THE ORE-DEPOSITS OF UPPER SILESIA.

Die Bildung der Oberschlesischen Erzlagerstätten. By A. SACHS. Centralblatt für Mineralogie, Geologie und Palæontologie, 1904, pages 40-49.

The author exposes in some detail the evidence upon which he bases the following conclusions:—(1) The ore-deposits of Upper Silesia in their present form are epigenetic. "Epigenesis" being understood to mean a later introduction of ores into the already existing country-rock. (2) The ores were introduced from above by concentration of the metalliferous material which was primarily in a very finely-divided state. (3) Dolomitization of the country-rock took place simultaneously with the introduction or percolation of the iron-, zinc- and lead-bearing solutions. (4) With regard to the enrichment of the ores in fissures, the Bernhardi reduction-theory may afford an explanation. This postulates the reduction of sulphates to sulphides by the gases given off from coal-deposits, an hypothesis which illustrates the close association existing between the coal-seams and the ore-deposits of Upper Silesia. It may be remarked that these ores are mainly sulphides (pyrites, blende, galena), but calamine, cerrusite, and a magnificent deposit of brown hæmatite also occur.   L. L. B.

JORDANITE IN UPPER SILESIAN ORE-DEPOSITS.

Über ein Vorkommen von Jordanit in den oberschlesischen Erzlagerstätten. By A. SACHS. Centralblatt für Mineralogie, Geologie und Palæontologie, 1904, pages 723-725.

This sulpharsenide of lead, in this instance containing a small percentage of iron-impurity, was found in a metalliferous mine near Beuthen: it shows the characteristic cleavage of jordanite, besides agreeing with it in chemical composition. The mineral is rare, only two other localities for it being known in the whole world; but its discovery at Beuthen is of some importance in regard to the much-discussed problem of the formation of the ore-deposits of Upper Silesia. It was found in a fissure, about 3 inches broad, lined with galena and zinc-blende and infilled with crumbly galena as well as with the jordanite. These conditions appear to bear out the author's contention that the above-mentioned ore-deposits are of epigenetic origin: that is, that the metalliferous particles, distributed in a state of fine division throughout a complex of rocks, were concentrated and deposited by downward-flowing waters percolating into fissures and cavities. Now, the arsenic which is a constituent of jordanite cannot be traced to the neighbouring masses of ore, and therefore the infilling of the fissures must be regarded as a primary, not a secondary phenomenon; that is, the accumulation of the great ore-bodies of Upper Silesia is due to infiltration into fissures occurring conjointly with metasomatic processes.

L. L. B.

ORE-DEPOSITS OF THE SORMITZ VALLEY, GERMANY.

Kontakterzlagerstätten im Sormitzthale im Thüringerwalde. By HANS HESS VON WICHDORFF. Jahrbuch der königlich Preussischen geologischen Landesanstalt und Bergakademie zu Berlin, 1903, vol. xxiv., pages 165-183, with 7 figures in the text.

Of late years, the long-abandoned mineral-industry of the Sormitz valley, in the Thuringian Forest, has been revived, and the re-opening of the ancient workings has facilitated close investigation of the contact ore-deposits. The three most important of these are:—(1) The magnetic pyrites of the Unter-hütte and of the western flank of the Goldkuppe at Leutenberg; (2) the galena and zinc-blende deposit of Weitisberga; and (3) the arsenical pyrites of the Grosse Silberberg at Gahma. The last-named ore-body has not been worked for many years. Active prospecting has revealed the presence of untouched ore-deposits in the immediate neighbourhood, at the eastern base of the Hainberg. All, without exception, lie within the contact-area of the eruptive mass of granite which has invaded the Palæozoic slates, and none are found outside that area.

The old records show that the galena-and-blende deposit of Weitisberga is of very even character, the ores being as rich in depth as at the surface. The new winnings had not reached the actual ore-body at the time of writing, and the author's description is chiefly based on specimens collected from the old waste-heaps. The galena appears to occur in thin bands inter-bedded among green *hornfelsen* (or cherty rocks), which themselves alternate with metamorphosed limestones. The occurrence of epidote, garnet, and other metamorphic minerals is characteristic. The ores are said to contain a fair percentage of silver, and they impregnate the whole of the Upper Devonian limestones in the contact-area; but the impregnation, as com-pared with the bedded deposit, is of small industrial importance.

The magnetic pyrites of the Goldkuppe too, occurs among contact-meta-morphosed Upper Devonian limestones; where the limestones are unaltered, no trace of the ore is to be seen. According to the original form of the limestone-deposits, so does the magnetic pyrites occur variously in bands or layers, in nodular concretions, in more or less spheroidal grains, or, finally, as a mere infilling of fissures. Beds rich in the ore show at the outcrop a rusty-brown to bluish-black coloration, with a greasy lustre. In conclusion, the author claims that the petrographical evidence brought forward by him, shows beyond doubt that the ores of the Sormitz valley owe their existence to the phenomena of contact-metamorphism. L. L. B.

———

DEEP BORINGS IN THE MÜNSTER BASIN, WESTPHALIA.

Das Ergebnis einiger Tiefbohrungen im Becken von Münster. By G. MÜLLER. *Zeitschrift für praktische Geologie,* 1904, vol. xii., pages 7-9.

The author describes a series of borings for coal which have been put down within the last decade in the district named. It is true that these borings have failed to reach the coal, but the results yielded by them are of importance, if only for the information which they afford as to the strati-graphy of the Münster basin.

In the Kreuzkamp boring, 2½ miles north of Lippstadt (begun on February 3rd, 1900, and stopped on February 26th, 1901, at a depth of 2,975 feet below the surface, when the Middle Devonian had been reached), the great mass of strata passed through belonged to the various horizons of the Cretaceous, and the Greensand was found directly overlying the Devonian.

As to the Metelen boring, put down in 1899, west of Burgsteinfurt, the cores down to a depth of 2,950 feet are still Upper Cretaceous; and therefore, if Coal-measures do occur in that locality, they would only be reached at a depth of 4,000 feet or so.

The Vreden boring, commenced in 1900, was stopped in 1901 at a depth of 4,038 feet in the Lower Zechstein (Permian), when probably it was just on the point of reaching the Coal-measures. At this locality, below the Drift and Tertiary Beds, come the Wealden and Lias (that is, all the Cretaceous and most of the Jurassic strata are absent) followed in descending order by Muschelkalk, Bunter Sandstones and Permian. The report that Wealden coal had been struck in this boring appears to be unfounded, but a pre-Glacial peat-bed was proved at a depth of 82 feet or thereabouts.

L. L. B.

THE OCCURRENCE OF PETROLEUM IN WESTPHALIA.

Das Vorkommen von Petroleum in Westfalen. By G. MÜLLER. *Zeitschrift für praktische Geologie,* 1904, *vol. xii., pages* 9-11.

An explosion of natural gas on June 28th, 1902, in the course of boring-operations carried out by the Rheinpreussen Company at Haus Sandfort, near Olfen, accompanied by a rush of water with a petroleum-like film on it, attracted great attention in mining circles, as it suggested the possibility of striking oil in the Westphalian coal-field.

At the locality above-mentioned, the highly-fissured *Brongniarti*-beds of the Cretaceous (Pläner) are reached at a depth of 2,100 feet from the surface, and below these come the Coal-measures, with a dip of about 22 degrees. At Walstedde, an explosion of natural gas occurred in the Mansfeld VIII. bore-hole, in the Cretaceous (Pläner), when a depth of 623 feet was attained. The Anneliese V. boring, put down at the cross-roads from Ahlen to Drensteinfurt and Sendenhorst, passed through 16½ feet of perfectly black Greensand, overlying 20 feet of Coal-measure sandstone. The latter dips 75 degrees, is fissured, and permeated with a yellowish-green, greasy substance, possessing a pungent odour of petroleum. It is to the occurrence of petroleum that the author attributes the blackening of the Greensand in this case, as he attributes its reddening in the Mansfeld VIII. and Ascheberg I. bore-holes.

Lately, a boring put down for salt, in Northern Westphalia, struck heavy oil in the porous anhydrite overlying the rock-salt, as well as in the cavernous dolomite below the salt.

The occurrence of bitumen in the upper beds of the Westphalian Cretaceous has long been recorded, but in order to get at the petroleum (of the presence of which the bitumen may or may not be an indicator) it would be necessary to carry borings down into the fissured Turonian strata, that is, down to considerable depths in the area under consideration.

The author does not commit himself to recommending boring-operations of so speculative a character as these would appear to be. L. L. B.

BRINE-SPRINGS FLOWING INTO A WESTPHALIAN COLLIERY.

Kohlensäureführende Solquellen im Schachte Robert der Zeche De Wendel bei Hamm i. W. By -- POMMER. *Zeitschrift für das Berg-, Hütten- und Salinen-wesen im Preussischen Staate,* 1903, *vol. li., Abhandlungen, pages* 375.377.

In the south-western district of the De Wendel colliery, near Hamm, are two shafts, one of which has been carried down for more than 650 feet into the Coal-measures, while the Robert shaft has not yet reached those measures. In the latter shaft, in February, 1903, at a depth of 1,495 feet from the surface, a brine-spring was struck which flowed into the sump at the rate of 88 gallons a minute. The salt-water is charged with carbon dioxide, and issues from a calcitic fissure in the white marl. Arrangements were made to draw off the water, in such a manner that 6 hours out of the 24 remained free for the haulage-work involved in continuing downward the masonry-lining of the shaft. But in July, 1903, 223 feet deeper down, a second fissure was struck, and salt-water flowed in from this at the rate of about 18 gallons a minute. Finally, on September 15th, 1903, at a depth of 1,780 feet (65½ feet short of the Coal-measures), a third spring was tapped, from which water welled forth at the rate of 44 gallons a minute, bringing up the total inflow of saline waters into the Robert shaft to 150 gallons a minute, all from fissures in the white marl. These waters are highly charged with salt and carbon dioxide, and appear to flow in under considerable pressure. Further sinking has been suspended, while the two winding-engines available are fully occupied in draining the shaft. The accumulation of carbon dioxide evolved by the water proved fatal to one of the workmen in the shaft, and nearly so to several others. The evolution of dangerous gas diminished greatly after August 14th, 1903.   L. L. B.

THE COPPER-ORES OF BOCCHEGGIANO, ITALY.

Note mineralogiche sul Giacimento cuprifero di Boccheggiano. By E. TACCONI. *Atti della Reale Accademia dei Lincei*, 1904, series 5, *Rendiconti*, vol. xiii., pages 337-341.

A full description of the mode of occurrence and formation of these cupriferous veins was published by Prof. Lotti in the memoirs of the Italian Geological Survey in 1893, the only important correction which now has to be made being in regard to the strike of the principal vein. This is due north-and-south, from the Farmulla valley to that of the Merse, and thence south-east and north-west. The dip is 45 to 50 degrees eastward, the mean thickness varies from 13 to 16 feet, and the most richly-mineralized portions are near the hanging-wall.

The most important ore in the deposit is chalcopyrite, which generally occurs in the form of finely-granular compact masses, more rarely as crystals within geodes. Its most frequent associate is iron-pyrites : this is also found alone in great masses, or in isolated cubic crystals, in the neighbouring Eocene limestone. Other associates are galena, blende and arsenical marcasite ; in conjunction with the second-named mineral and chalcopyrite, blackish-grey lustrous venules of tetrahedrite course through the limestone. Of the last-named ore a chemical analysis was made, with the result that it was shown to contain about 31 per cent. of copper, 28 per cent. of antimony, 7 per cent. of zinc, and 6·62 per cent. of silver. Its correct chemical formula, according to the author, would be

$$3 \ (CuAg)_2S. \ Sb_2S_3 + \tfrac{1}{4} \ (6 \ ZnFeS. \ Sb_2S_3).$$

The high percentage of silver differentiates this variety of tetrahedrite from any other known to occur in Tuscany. In the percentage of zinc it resembles the mercury-bearing tetrahedrite of Val di Castello, but it contains no mercury; while the tetrahedrites of Valle del Frigido contain from 12 to 13 per cent. of iron, but practically no zinc or silver.

The gangue of the principal vein consists of brecciated limestone (occasionally crystalline), quartz, baryta and fluorspar. Tiny prisms of the sulphidic ore, bismuthine, believed to be a so far unrecorded occurrence in Italy, have been found disseminated in a hæmatitic mass, in association with the chalcopyrite and pyrites.   L. L. B.

LIMONITE-DEPOSITS IN CAMPIGLIA MARITTIMA, ITALY.

I Giacimenti limonitici di Monte Valerio, di Monte Spinosa e di Monte Rombolo (Campiglia Marittima). **By G. RISTORI.** *Atti della Società Toscana di Scienze Naturali,* 1904, *Memorie, vol. xx., pages* 60-75.

With regard to the ore-deposits of Monte Valerio more especially, the author begins by remarking that the disappointing results, which have so frequently attended the exploration-work carried out there, are more justly attributable to the irrational fashion in which such work has been for the most part conducted, than to the comparatively-limited extent of the deposits, both of tin- and iron-ores.

The excavations made on the southern and south-western slopes of Monte Valerio, whether in ancient or in recent times, have been chiefly directed to the search for cassiterite; but veins and venules of limonite have been found to occur in almost constant association with that mineral, a circumstance which suggests a common origin for the two ores. However, ochre is there more abundant than limonite proper; and, moreover, the metalliferous veins thin away to nothing on reaching the deeper-lying strata of the Liassic limestone in which they occur. On the south-eastern slope of the same hill, limonite is seen to form with calcspar the infilling of fissures in an anticlinal fold of the limestone; cassiterite has also been found in the fissures, but it would seem as if the abundance of limonite were detrimental to that of the tin-ore. Analysis shows this limonite to be fairly pure, but it varies greatly in composition and in structure, and in the amount of calcite which in places forms a network of venules within it. The author does not appear to think that deep-level workings would ever pay with these deposits, but outcrops and shallow workings should furnish a reasonable amount of ore, the less to be despised that the industrial revival in Italy is daily becoming more marked. On the eastern slope of Monte Valerio the excavations for limonite have been attended with still less success than elsewhere: the veins are smaller and scarcer, and such ore as has been got is too drusy, and too full of calcitic and even siliceous impurities, to be of much value.

Eastward of these, occupying the slope towards the Pozzatello valley, are outcrops of a limonite which appears to be of excellent quality: it is covered with a sort of black varnish, due, presumably, to the reduction of a dioxide of manganese that forms a constituent of the ore. Exploration-work has revealed payable quantities of ore, and there are other outcrops which have not yet been investigated. The author has greater hopes of the Pozzatello deposits than of any others seen by him in the district. In the valley of Santa Caterina, north of the Pozzatello valley, there occurs among the highly-fractured and contorted limestones a vein which is infilled with a mixture of ilvaite, limonite, and magnetite. The specific gravity and the compact structure of the samples of ore give favourable indications; and, if these should be confirmed by chemical analysis, the mineral would probably rival the best Elban hæmatite. The course of the vein is so irregular, that although the excavations so far carried out have not shown it to be very extensive, it may well be struck again in the neighbouring fissured masses of limestone.

In Monte Spinosa the waxy Liassic limestone is no longer seen: pinkish-white limestones, of gradually-increasing crystallinity until they pass into absolutely-sacchoroidal limestones, take its place, the entire hill forming an anticlinal dome. The rocks are fissured in every direction, and with the fissures are associated numerous outcrops of limonite. Given modern methods of working, and a carefully-planned scheme of operations, the deposits would certainly prove to be of industrial value. It has been shown that the fissuring of the rocks can be traced to the volcanic eruptions which took place in the Eocene period, and with some phase of this vulcanicity the mineralization of the fissures is undoubtedly connected. This occurred over a vast area, including the neighbouring province of Massa Marittima, where so many mines are, or have been, worked.

The author then describes a vein of manganiferous iron-ore near the junction between white crystalline limestones and the lowermost Liassic marls (which overlie them unconformably) at Acqua Viva, near the Botro dei Marmi: it may be worth working for manganese. So too, the old mines of Campo alle Buche might prove to be worth re-opening, as their situation is favourable from the point of view of facility of transport, etc., and the limonite-deposits have by no means been exhausted by the ancient workings.

The most celebrated deposits, however, of the district studied by the author are those of Monte Rombolo. Here, as elsewhere, the metalliferous fissures are broadest near the outcrop and thin away in depth: the fact that occasionally they bulge out into pockets, or branch out into several veins, does not detract much from the general truth of the former statement. Nevertheless, the Monte Rombolo deposits are industrially important, and they are, on the whole, characterized by greater constancy or continuity than the others. The principal fissure strikes first from south-west to north-east, then sweeps round south-eastward in a curve, the convexity of which is mainly directed southward. Secondary veins, of small importance, branch off from it in various directions. The want of success which has frequently attended the spasmodic efforts made to work these limonites appears to be due here, as at Monte Valerio, in great part to the want of forethought with which such endeavours have been conducted, coupled with a disregard of the actual geological conditions of the deposits.   L. L. B.

ORIGIN OF THE MAGNESITE-DEPOSITS OF THE ISLAND OF ELBA.

La Formazione della Magnesite all'Isola d'Elba. By GIOVANNI D'ACHIARDI. *Atti della Società Toscana di Scienze Naturali,* 1904, *Memorie, vol. xx., pages* 86-134 *and* 3 *plates.*

Dispersed on the flanks of the Monte Capanna are great whitish cones which attract the traveller's attention, as he proceeds along the road from Marina di Campo to the hamlet of San Piero: these cones are, in fact, the tip-heaps of the magnesite-workings, the mineral being excavated from a friable, earthy, yellowish-green rock in which it forms an intricate network of interlacing white veins. The author's investigations were limited to a single mine, that of the Grotta d'Oggi; but he purposes to extend successively his researches to the others, more especially if they show any essential differences (from the mineralogical and geological point of view) from the mine just named.

A detailed petrographical description is given of the granitic rocks, the cornubianites, the peridotites, the serpentines, the hornblendites, and the pyroxenites, with numerous chemical analyses, as also of the magnesite itself. The typical mineral contains from 40 to 42 per cent. of magnesia, and the author considers that there is no room for doubt that it has originated from the alteration of the serpentinous rocks. These, in their turn, were derived from an anthophyllitic peridotite, but the alteration of serpentine into magnesite is, in all probability, of a much more recent date than the decomposition of peridotite into serpentine. It will be noted that the magnesite is apparently associated with the granitic contact-zone; but we are not dealing here with a case of direct contact-metamorphism: it seems probable that the contact-zone, forming necessarily a line of weakness, simply afforded an easier passage than other portions of the igneous complex to the decomposing agencies which transformed the serpentine into magnesite. These agencies were primarily thermal waters of high alcalinity, and in a very secondary degree pluvial waters: the part played by the latter consisting mainly in the partial dissolution and redeposition of magnesite-deposits already formed.　　　　　　　　　　　L. L. B.

ANTHRACITE-DEPOSITS OF THE WESTERN ALPS, ITALY.

Studio geologico-minerario sui Giacimenti di Antracite delle Alpi Occidentali Italiane. OFFICIAL. *Memorie descrittive della Carta geologica d'Italia*, 1903, vol. xii., pages xvi. and 232, with 31 figures in the text and 14 plates.

This exhaustive work comprises, in addition to an introductory synopsis, no less than seven descriptive memoirs by as many different authors, dealing with the general geology and stratigraphy of the deposits, with the methods of working and chemical analyses of the anthracites, and with the fossil flora found in association with them. It is preceded by a bibliography consisting of 60 separate entries and covering the literature of the subject from the year 1784 to 1903.

The attention of the Italian Government had been repeatedly drawn to the deposits, before the detailed investigations of which the volume under review is the outcome had been undertaken. But the reports furnished to them by the engineers of the Royal Corps of Mines concurred in representing the anthracites as of small industrial importance, both from the point of view of quantity and quality, and the most recent research now confirms this disappointing conclusion.

The fact that the carbonaceous material is widely diffused through the rocks, the considerable number of outcrops of anthracitic lenticles in different areas (especially in the Valle d'Aosta), the optimistic views of some geologists, and the more or less interested asseverations of certain prospectors, spread abroad the impression that vast unexplored treasures of coal lay somewhere hidden in these Alpine regions: an impression deepened by the rise in price of coal on the Italian market a year or two ago. There were also rumours of a possibility of the revival of the metallurgical industry in the Valle d'Aosta, and so the Government felt impelled to order an investigation that should be sufficiently exhaustive to rank as final.

Generally speaking, these North Italian anthracites all belong to the Permo-Carboniferous formation, which can be traced from the Mediterranean coast at Savona westward into the Ligurian Alps, and thence past Cuneo towards Monte Viso. Across the frontier, the Carboniferous belt is shown to stretch from Briançon north-eastward as far as Sion in the Rhône valley. The anthracite-deposits (in Italian territory) are generally in the form of thin, small lenticles, frequently nipping-out and then thickening again, extremely irregular in every way, owing doubtless to the intense plication and highly-disturbed condition of the rocks among which they lie. The mineral is generally somewhat graphitoid in character, rarely contains less than 15 per cent. of ash (the average percentage for a good seam being 18 to 20) and sometimes as much as 60 per cent. In the course of the great orogenic movements to which the rocks were subjected, the anthracite seems to have undergone a sort of recementation, and so can be got out in fairly-large blocks. At some localities in the valley of the Bormida di Calizzano, the mineral passes into an impure graphite, which was at one time actively worked.

It is estimated that the quantity of possibly-workable anthracite in the basin of La Thuile, where the best deposits on the Italian side of the frontier are concentrated, can hardly exceed one million tons; a trifle in comparison with the fabulous estimates previously put forward by some sanguine persons. The annual output in that basin at present is insignificant—some 250 tons or so. It may be here observed that, although on the French side of the frontier, in the Briançon, Maurienne and Tarentaise districts, the conditions (geological and economic) would appear at first sight to be more hopeful, the results so far achieved are not promising. The output there also is unimportant, in comparison with the enormous extent of country covered by the Carboniferous belt.

Considering, too, that the Italian mineral is useless for steam-raising or metallurgical purposes, the putting-down of plant on a large scale, or an attempt to send the coal away to any considerable distance, would be foredoomed to failure. It is held that these Italian anthracites can never be of more than local importance. It is quite possible, however, that workings on a small scale (especially in the basin of La Thuile) will prove remunerative, if judiciously managed.

The first two memoirs, of the seven above-mentioned, deal with the deposits in the Valle d'Aosta, the third with those of the Province of Cuneo, and the fourth and fifth with those of Western Liguria. In the sixth, the results of a chemical study of the anthracites of the upper valley of Aosta and of Western Liguria are set forth, 16 samples having been examined. The best yielded from 12 to 13½ per cent. of ash, and their heating-power was estimated at 6,400 to 6,500 calories. All contained a small proportion of sulphur: the proportions of hygroscopic water and fixed carbon were as extraordinarily variable as the proportions of ash. L. L. B.

MOLYBDENITE IN THE WESTERN ALPS, PIEDMONT.

Beitrag zur Kenntniss alpiner Molybdänitvorkommnisse. By G. LINCIO. *Central- blatt für Mineralogie, Geologie und Paläontologie,* 1905, *pages* 12-15.

The author describes molybdenite occurring in association with iron- and copper-pyrites and magnetic pyrites, in much-disturbed, fractured, schistose gneisses, preferentially near the contact with limestone-inclusions, in the ravine of the Cherasca river, near Varzo, province of Novara. He has also examined hand-specimens of molybdenite-gneiss obtained from rocks which crop out on the road from Crodo to Baceno in the Antigorio valley: here again the molybdenite is associated with pyrites, magnetic and non-magnetic. Stress is laid on the fact that the occurrence of the mineral is limited in these localities to gneiss, and to that rock alone. L. L. B.

UNCOMMON CONSTITUENTS OF SARDINIAN BLENDE.

Su alcune Blende di Sardegna. By C. RIMATORI. *Atti della Reale Accademia de Lincei, 1904, series 5, Rendiconti, vol. xiii., pages* 277-285.

The author made chemical analyses of several specimens of Sardinian blende, from various localities. That from the cupriferous mine of Bena de Padru, the only Sardinian mine which at present offers any hopeful prospects in regard to the extraction of copper, was found to contain 0·1 per cent. of cadmium. Two specimens from the Montevecchio mines were found to contain respectively 0·79 and 0·76 per cent. of cadmium. On the other hand, the blende of Masua contains mere traces of cadmium, manganese and copper, and with its percentages of 65·25 of metallic zinc, 32·21 of sulphur and 0·28 of iron, may be regarded as an approximately-typical blende.

Not content with chemical analyses, the author proceeded to make careful spectroscopic analyses of blendes from the above-mentioned localities, and from three or four others in Sardinia, in order to determine whether they contain such comparatively-rare elements as indium, gallium, etc. Only in one case, at first, did he succeed in finding indium. This was in the blende of Riu Planu Castangias, which contains as much as 0·1231 per cent. of that element. In a second series of experiments, he was able to ascertain that very minute quantities of indium and of gallium are present in the blendes of Bena de Padru and Montevecchio.   L. L. B.

FAHLORE OF PALMAVEXI, SARDINIA.

Il Fahlerz nella Miniera di Palmavexi (Sardegna). By C. RIMATORI. *Atti della Reale Accademia dei Lincei, 1903, series 5, Rendiconti, vol. xii., pages* 471-475.

The Palmavexi mine is about 3 miles distant from Iglesias, among limestones and calc-schists varying in colour from blue to white and in character from quartzose to dolomitic, but all lumped together locally as "the metalliferous limestone." The mine was formerly worked for lead- and zinc-ores, and recent investigations, made on behalf of the syndicate who now own the concession, have shown that deep-level workings would strike rich deposits of calamine and argentiferous galena. From east to west, a vein strikes through the property across these deposits: it appears to be parallel to the stratification of the limestone, is several feet wide, and may be termed a "bedded vein." In it was found the fahlore (which forms the chief subject of the author's remarks) associated with the predominating lead- and zinc-ores, and some copper-minerals. This is the first authenticated occurrence of fahlore in Sardinia: the mineral possesses a dark steel-grey colour and metallic lustre. Its percentage-composition is as follows: sulphur, 23·56; copper, 43·06; antimony, 23·66; zinc, 6·29; iron, 1·14; and silver, 1·64. It may be regarded as a sulph-antimonate of copper or tetrahedrite, most nearly approaching that variety of which the chemical formula is $4Cu_2S$. Sb_2S_3. One does not gather that it is present at Palmavexi in sufficient quantity to be of industrial importance.   L. L. B.

CINNABAR-DEPOSIT OF CORTEVECCHIA, TUSCANY.

Geologische Verhältnisse und Genesis der Zinnoberlagerstätte von Cortevecchia am Monte Amiata. By B. LOTTI. Zeitschrift für praktische Geologie, 1903, vol. xi., pages 423-427, with 4 figures in the text.

The Cortevecchia mine lies about 11 miles due south of the trachytic peak of Monte Amiata, half-way up the heights which form the right bank of the Fiora river. The strata of the neighbourhood consist largely of Eocene clays and marls, with great lenticles of Nummulitic limestone, overlying a series of Senonian limestones (white, red, and grey) with red fucoid-shales and ashen-grey marl-slates. The Cortevecchia cinnabar-deposit is especially associated with the faulted Ripacci lenticle of Nummulitic limestone. This is a granular rock, all but free from argillaceous constituents, much interbanded with chert and layers of nummulites: above and below, it passes into clayey and marly beds, and the chief concentration of metalliferous ore occurs near the junction in these passage-beds. Five different winnings have been opened up, and that on which the least progress has been made (on the Ripacci farmstead itself) promises to be the richest of all.

The cinnabar is found to have in part replaced the calcium carbonate of certain limestone-bands, and impregnates generally the marly layers interbedded with the limestones. The constant association with it of pyrites and gypsum suggests that it is genetically connected with sulphuric solutions. These probably welled up through a deep fissure, percolated through the crevices in the rock-mass, and in part corroded it. The upward and downward progress of the solutions would be arrested in places by the more or less impermeable argillaceous beds, and they would then spread out more or less parallel with the bedding, the cinnabar precipitated from them being chiefly found, as above stated, in the passage-beds at the hanging-wall and foot-wall of the limestone. The neighbouring ore-deposits of Cornacchino and Montebuono are analogous in this respect to the Cortevecchia deposit.   L. L. B.

―――――

KIESELGUHR AND COLOURING-EARTHS OF MONTE AMIATA, TUSCANY.

Kieselguhr und Farberden in dem trachytischen Gebiet vom Monte Amiata. By B. LOTTI. Zeitschrift für praktische Geologie, 1904, vol. xii., pages 209-211, with 2 figures in the text.

The kieselguhr or fossil-flour of Monte Amiata is made up of the débris of microscopic siliceous algae and diatoms. It contains, as a rule, about 85 per cent. of pure silica, 3 per cent. of iron oxides, lime, alumina and magnesia, and 12 per cent. of water. The specific gravity varies from 0·08 to 0·30. The average thickness of the kieselguhr-deposit of Castel del Piano is about 15 feet. That, and the deposit at Bagnolo are actively worked at the present time. Intimately associated with these deposits of fossil-flour, are iron-ochres, varying in colour from a very pale to a very deep yellow, and containing also remains of diatoms. Two varieties of ochreous earth chiefly are worked at Monte Amiata—the yellow-ochre, which on burning turns deep red, and the bole. The former is used in the manufacture of sienna-yellow and burnt sienna, and the latter in that of burnt umber. These deposits were, all of them, evidently laid down in freshwater basins from waters carrying silica and iron in solution. Siliceous and ferruginous springs are to this day active in the neighbourhood, and are probably the latest moribund phenomena of the vulcanicity which was at one time active there.

L. L. B.

COPPER-ORES IN TRANSCAUCASIA.

Le Cuivre en Transcaucasie: Notes de voyage. By P. NICOU. *Annales des Mines,* 1904, *series* 10, *vol.* ri., *pages* 5-54, *with* 4 *figures in the text and* 1 *plate.*

This paper is the outcome of a journey made by the author in the autumn of 1903, in that portion of the Russian Empire which stretches southward from the Caucasus, between the Black Sea and the Caspian, to the frontiers of Turkey and Persia. In that area, metalliferous deposits are numerous, copper-, lead-, zinc-, manganese-, nickel- and cobalt-ores being of frequent occurrence; but various circumstances have combined to prevent hitherto the establishment of workings on any considerable scale, except in the case of the cupriferous and manganiferous ores. Despite the recent opening of the railways from Tiflis to Kars (1898) and from Alexandropol to Erivan (1902) transport-facilities are still lacking in several districts; moreover, the only coal-deposit worked in Transcaucasia, that of Tkvibuli (output in 1900: 63,400 tons), yields a mineral unsuitable for coking purposes, and either wood or wood-charcoal has, in many cases, still to be used as being the cheapest fuel, or at least as being that which is most readily available. The introduction of foreign capital is, however, putting a new complexion on things; and the prospective raising by the Russian Government of the already high duties on imported copper, is likely to resuscitate many little smelting-works which have been long abandoned.

The three most productive districts at the present time are those of Kedabeg, 26½ miles from the Dalliar station on the Tiflis and Baku railway; Allah-Verdi, on the railway from Tiflis to Alexandropol; and thirdly, the comparatively remote district of Zanghezur, 130 miles or so distant from Evlak station on the first-named railway. About 44 per cent. of the total output of copper in the Russian Empire is credited to Transcaucasia, the remaining 56 per cent. being supplied by the Urals, Siberia, Finland and the Kirghiz steppes; for 1902, the copper-ore got in Transcaucasia amounted to 106,718 tons, and the quantity of metallic copper produced there was 3,438 tons.

Passing then to the detailed description of the ore-deposits, the author commences with those of Zanghezur, which appear to have been worked, more or less, since time immemorial. Remote though the district be from the Russian centres of industry, it is but a journey of 65 miles (along the track followed by the camel-caravans) to Tabriz, still one of the richest towns in Persia; and it is curious to note that, in fact, the Zanghezur copper is still mostly absorbed by the Persian market. Some of the ancient workings went down to depths of 130 feet below the surface, but as the old miners only worked ores containing more than 15 per cent. of copper, their mines may well be re-opened with profitable results. The Zanghezur deposits consists mostly of quartz-veins mineralized in depth with various sulphides of copper and native copper, and with iron-pyrites (an invariable associate of the cupriferous sulphides); blende and galena are rare, but gold and silver occur in notable proportions. Towards the outcrops, that is, nearer the surface, the sulphides give place to oxides and carbonates. The quartz-veins traverse, in some cases, andesites, in others porphyries; and there are also brecciiform cupriferous deposits intercalated between syenites and diorites. The methods of working and treating the ores are in some parts of the district old-fashioned and costly; in other parts, everything is being brought up to date, and there is some prospect of the abundant water-power available being utilized for electric installations. The author depicts in the most roseate hues the future prospects of the mineral-industry in this region, mentioning (among other factors) the existence there of a whole series of ore-deposits as yet untouched.

At Kedabeg, one may see the most considerable smelting-works at present in activity in Transcaucasia: but it is not a very accessible locality, and the road thither from Dalliar railway-station (26½ miles) rises in that distance from an altitude of barely 100 feet to one of 4,600 or more. The necessary fuel (a mixture of refined petroleum and *massiut*, or petroleum-residues) has, since 1897, been conveyed from Dalliar to the Kedabeg works by a pipe-line. The ore-deposits occur in the Copper mountain or Mis Dagh, near the contact of the quartz-porphyries with andesites or diorites, all of which rocks are believed to date back to the Jurassic period: the ore-bodies are thought to have been precipitated at a subsequent period, from waters percolating through cavities and fissures in the volcanic rocks. The chief ores are copper- and iron-pyrites, with blende and galena, occasionally magnetite, and always considerable proportions of gold and silver. The methods of extracting and refining the copper are described at some length, and we are told that in 1899 the amount of metal turned out by the Kedabeg smelting-works was 83 per cent. of the total Transcaucasian output, while in 1902 the competition of Allah-Verdi and Zanghezur brought it down to 44 per cent.

The district of Allah-Verdi lies in a very mountainous region, some 50 miles south of Tiflis. The mineral-industry, flourishing there in ancient times, was interrupted by the internecine wars which raged throughout the Caucasus for many generations; it has now taken a new lease of life, and coke is brought by railway from the far-off Donetz basin for the purposes of the smelting-works. In order to diminish the consumption of this fuel, the cost of which is extravagantly heightened by the freight-charges, a modified form of pyritic smelting has been adopted, the character of the ores not allowing of the application of that process in all its integrity. The crude ore is got from the three deposits of Akthala, Allah-Verdi and Chamluk: it occurs in pockets, among quartziferous andesites or among dacites, a notable enrichment being observed in the neighbourhood of the masses of gypsum which are frequently associated with the ore-deposits. At Akthala, near the uppermost portion of the deposits, gold- and silver-bearing galena is abundant. The method of treating the ores is described in detail; and the paper concludes with a brief description of other cupriferous deposits in Transcaucasia, some of which are perhaps of less importance, while the remainder must await the provision of transport-facilities before their undoubtedly vast wealth can be utilized.   L. L. B.

THE ORE-DEPOSITS OF MOUNT DSYSHRA, WESTERN CAUCASIA.

Die Erzlagerstätten des Berges Dzyschra in Abchasien. By A. G. ZEITLIN. *Zeitschrift für praktische Geologie, 1904, vol. xii., pages 238-242.*

The Dsyshra ridge, one of the spurs of a lesser mountain-range which runs along the eastern coast of the Black Sea, parallel with the mighty main chain of the Caucasus, is about 22 miles distant from the harbour of Gudout: from which, owing to the present condition of the tracks, it cannot be reached in less than 12 hours. The average altitude of the ridge above sea-level exceeds 8,000 feet, and the rocks of which it is made up are Jurassic limestones and dolomites. In these rocks occur veins and nests of argentiferous galena, zinc-blende, calamine, pyrites, and reefs of red and brown hæmatite. One of the calamine-deposits is practically 16 feet thick. In the fissures of the limestones, moreover, the occurrence of bitumen of very pure quality has been observed: it is exceedingly tough, extraordinarily free from sulphur, has a high melting-point, and has been experimentally applied with great success in the manufacture of patent-fuel (briquettes made from coal-dust). The various metalliferous ores have been analysed in several laboratories, with the following results: the brown hæmatite contains nearly 55 per cent. of metallic iron, and the red 68·5 per cent. The galena contains nearly 50 per cent. of lead and 8¼ of zinc: whereas the blende contains 8¼ per cent. of lead and 22·9 of zinc. With regard to the percentage of silver in the lead-ores, contradictory results have been obtained; but there is no question as to the presence of silver, the only dispute being as to its average proportion.

There seems to be no doubt that the ore-deposits would repay careful prospecting-work; many of the lead-ores, for instance, although not occurring in very thick veins on the whole, have a considerable horizontal and vertical extent. The calamine-blende ore-bodies are of such mass as to give every hope of successful working, and the lead-, zinc- and iron-ores of the Dsyshra river-valley are of very high purity. The lead- and zinc-ores are not intimately intermixed (an intermixture which often gives rise to metallurgical difficulties), but occur in separate bands or veins; and the treatment of the ores (more especially as scarcely any quartz or other refractory mineral is associated with them) in the smelter should be comparatively simple.

Much of the ore could be worked opencast, and even where underground workings are necessary these could be arranged by means of horizontal adits. The country-rock is soft, and therefore easily disposed of; the adjacent forests would furnish an abundant timber-supply; the rapid current and high fall of the mountain-streams could be made available for power-distribution. The one thing needed is the conversion of the present track along the banks of the Chüpsty into a good cart-road. The population of the district is scanty, but this is a drawback to which time will, bring a remedy.   L. L. B.

PETROLEUM-RESOURCES OF EUROPEAN AND ASIATIC RUSSIA.

Die Erdölvorkommen im europäischen und asiatischen Russland. By F. Thiess. Zeitschrift für das Berg-, Hütten- und Salinen-wesen im Preussischen Staate, 1904, vol. lii., pages 12-16, with a map in the text.

The richest natural reservoirs of oil within the mighty Russian Empire lie in the country surrounding the Caucasus. Official estimates, based on geological investigations, of the probable productive area of which Baku is the centre, put it at 16,205 acres. Of this, barely one-sixth has been tapped so far. In common with other writers, however, the author points out that the average output of the individual wells has, on the whole, decreased: and, as, in order to make up for this diminishing productiveness, the bore-holes are put down year by year to greater depths, the costs of installation, maintenance and working have risen enormously. The *Official Gazette of Industry and Commerce* (*Torgovo Promyshlenaya Gazeta*) states the total output for 1900, 1901 and 1902, as 9,839,466; 11,049,948; and 10,434,060 tons respectively. The diminution recorded for 1902 has, however, but little bearing on the future of the Baku oil-field, as the workable area of it will undoubtedly be increased by draining certain bays of the Caspian, not to speak of the lakes of Romany and Sabunchy. On that portion of the first-named lake, which is already reclaimed, a Moscow syndicate has recently erected five derricks for oil-boring purposes.

The area of the rich oil-field of Grozny, 100 miles north-west of the harbour of Petrovsk, on the railway from Rostov to Vladikavkas, extends over 5,400 acres, and the output of petroleum in 1902 amounted to 556,920 tons. The chief difficulty met with in this field is the waterlogged condition of the strata that have to be passed through before striking oil.

In Daghestan, about 3 miles from the western shore of the Caspian, and 62 miles south of Petrovsk, an abundant oil-spring with a continuous flow was tapped in 1902, on the Bereki estate of the Imperial Treasury. This circumstance, and other evidence as well, points to the existence of a good reservoir of oil here: the conditions of working are very favourable. Other occurrences of petroleum are reported in Daghestan, as for instance, in the Talgin ravine, where out of a known oil-field of 6,752 acres, barely a sixth has been worked over as yet.

In the Kakhetinsky hills, which range parallel with the Caucasus from Tiflis south-eastward into the Apsheron Peninsula, petroleum has been tapped at several localities (the Chatma and Signakh deposits will soon be in full working by British companies), as also in the districts of Kutais and Elisavetpol, and at Anapa on the Black Sea. As a matter of fact all these occurrences of oil in the Caucasian region lie along a belt which stretches from the Sea of Azov to Batúm on the Black Sea.

It is a far cry from this region to the extreme north-east of European Russia, where, at Ukhta in the Pechora district that abuts on the Arctic Ocean, petroleum has been found, and the quality of the crude oil examined. The yield of illuminating-oils from it will not prove very large, but it will furnish a big proportion of residues suitable for heating purposes.

Turning now to Asia proper, the island of Cheleken, which lies opposite the long narrow tongue of land that stretches southward from the harbour of Krasnovodsk, the western terminus of the Central Asian Railway, proves to contain a quite considerable natural reservoir of oil. Petroleum, too, is reported from the provinces of Samarkand and Ferghana, and is actually worked for local use in the Nafta Dagh and the Buja Dagh. The province of Ferghana is especially rich in oil and ozokerite: petroleum is seen trickling out at the base of the hills, forming small pools, and covering with an iridescent film the waters of the streams. At Chemion, 12½ miles from the Vanovskaya railway-station, oil-springs were tapped by the Chinese in former days; and, thanks to the completion of the Samarkand railway, working has been restarted there, but by a European syndicate this time. At Mailissai, in the Namagan district, a newly-proved oil-field, covering an area of about 1¾ square miles, is being put up at auction by the Ministry of Agriculture.

The occurrence of petroleum on the Island of Sakhalin, north of the island-empire of Japan, is of importance for the Russian dependencies in the Far East. On the eastern shore, the Nutowo deposits were discovered in

1898: they include seven great oil-lakes, yielding an oil very similar to that of Baku, and quite free from sulphur and benzine. Petroleum-deposits in the north of the island were discovered and described in 1893.

Altogether, Russia possesses so many natural oil-fields that she can afford to regard with equanimity the eventual exhaustion of the Baku oil-wells.   L. L. B.

TERTIARY BROWN-COALS AND ORE-DEPOSITS OF KRIVOI ROG, RUSSIA.

Materialien zur Geologie der Tertiär-Ablagerungen im Rayon von Krivoi Rog. By A. FAAS. *Mémoires du Comité géologique, St. Pétersbourg,* 1904, *new series, No.* 10, 140 *pages,* 24 *figures in the text,* 1 *map and* 2 *plates.*

This memoir is primarily devoted to the geology of the Tertiary deposits of Krivoi Rog, and consideration of the useful minerals is exclusively confined to those which occur in these deposits. The oldest unaltered sedimentaries that crop out in the area, overlying the Archæan (?) crystalline schists and the gneisses, granites, etc., which form the original framework of the country, are almost certainly of Lower Tertiary age. These Palæogene deposits are chiefly grey and white sands and marls, and greenish or bluish clays. Next above come the Neogene shelly sands and oolitic shelly limestones, the uppermost of which date from Pliocene times. Overlying these is a great succession of reddish-brown clays (with gypsum- and marl-concretions) and loess-like loams of post-Tertiary age, as also fluviatile gravels and sands.

The cavernous brown hæmatites got in the southern part of the district, of excellent quality and easily smelted, and containing 56 per cent. of metallic iron, with only 0·12 of phosphorus and 0·11 of sulphur, are probably derived from the mechanical erosion and leaching-out of the ferruginous Archæan rocks that form the Ingulez ridge, at a time when that ridge was washed by the surf of the Lower Tertiary sea. Blocks of iron-ore are found scattered along the flanks of the ridge, but more abundantly on the steeper western flank than on the eastern. Actual beds of the ore vary enormously in thickness (from 20 inches to 30 feet) and in depth from the surface (3 to 140 feet), wherefore many of those that are shown on the map are not of immediate industrial importance.

The manganese-ores occur chiefly at the contact between the Lower Tertiary sands and clays and the Sarmatic Beds, that is, at a higher horizon than the brown hæmatites. Like the latter, the manganese-ores occur only in the neighbourhood of the ancient ridge of Ingulez, but perhaps extend over a larger area than the hæmatites: these manganese-ores, too, must have been primarily derived from certain Archæan crystalline rocks. The deposits so far known are not particularly conspicuous for abundance or quality, but as they have not been studied very minutely and have merely been discovered in the course of prospecting for iron-ores, it would be premature to assert that workable manganiferous deposits do not exist in the area. A sample, got a little distance away from the Kamienkowicz mine, assayed to 33·42 per cent. of metallic manganese, and the insoluble residues amounted to 22·72 per cent.

The brown-coal or lignite-bed (6 to 22 feet thick, with a general south-easterly dip), associated with some of the Lower Tertiary deposits, covers an area of 57 acres or so on the Novo-Pavlovskoye estate, and is there estimated to contain 400,000 tons of mineral. But it is of poor quality, feeders of water are numerous, and the depth from the surface varies from 60 to 200 feet. A sample of lignite got from another locality yielded 42·32 per cent. of carbon, 12·8 per cent. of ash, and 3·66 per cent. of sulphur, with 5·36 per cent. of hydrogen (?). Volatile substances amounted to 52 per cent., and hygroscopic water to 11·78. The heating-power is estimated at 4,325 calories, and the yield of very pulverulent coke at 36·25 per cent.

The gypsum-deposits do not occur in such conditions as to make it likely that it would ever pay to work them. L. L. B.

IRON-ORE DEPOSITS OF THE BAKAL REGION, RUSSIA.
Bakalskiaya Miyestorozhdeniaya Zheliyeshikh Rud. By L. KONIUSHEVSKY and P. KOVALEV. *Mémoires du Comité géologique, St. Pétersbourg,* 1903, *series* 2, *No.* 6, 126 *pages, with* 83 *figures in the text and* 2 *plates.*

The first-named author describes the deposits of the Bulandikha and Shuida hills in the district of Zlatóust, known to fame for more than a century.

The strata are mainly shales and limestones seamed by diabase-dykes, and folded into anticlines. Faults are numerous. The ore-deposits present the appearance of bedded masses intercalated among the shales with a roof of quartzite. The minerals are mainly brown hæmatite and spathose iron-ore, but turgite in crumbly masses also occurs. These ores are all very pure, and are indeed remarkable for their high percentage of metallic iron. The thickness of the ore-bodies is considerable, averaging 165 feet in the State Bakalsky mine and 100 feet in the Bulandinsky mine: in the latter, a diabase-dyke, decomposed to clay, occurs in the very midst of the ore. In the Uspensky mine, spathose iron-ore is seen passing into brown hæmatite and turgite. A short distance east, dolomitic limestones are seen immediately overlying the same shales as those which form the floor of the ore-deposit, an observation the importance of which is evident to the geological reader; especially when we learn that in the south-western part of the mine crystalline dolomitic limestones also occur, on the same strike as the ore-deposit. Other evidence is brought forward to establish the conclusion that the ore-deposits are the result of the chemical replacement of limestones, more or less dolomitic, by iron carbonates, the oxidation of which has yielded hæmatite, etc. In the absence of fossils the age of the rocks cannot be determined with certainty, but it is presumably Lower Devonian.

The ore-deposits of Mount Irkuskan, a ridge running east of and parallel to the Bulandikha and Shuida hills, are described by the second author as inexhaustible. They occur in a group of shales of various colours and limestones, believed to be also of Lower Devonian age. The brown hæmatite contains up to 65 per cent. of metallic iron, and is almost completely free from silica. The spathose iron-ore does not contain more than 60 per cent. of metallic iron, and impurities in the shape of magnesia and sulphur are present. The turgites are of a dark brownish-red, and contain still more magnesia than the spathose ore, but on the whole they approximate in composition to the typical formula $2Fe_2O_3H_2O$. The spathose iron-ore occurs generally immediately below the quartzite and passes into dolomitic limestones. Where the other two ores are associated with it, the succession in ascending order is (1) spathose ore, (2) turgite, and (3) hæmatite. There appears to be no doubt that turgite represents the intermediate stage of the transformation of spathose ore into brown hæmatite. At the Elnichny mine, a deposit of spathose ore, 40 feet thick, is worked, while in the Ivanovsky mines nests of all three varieties of ore enclosed in the limestones are the chief object of mining operations. The richest deposits of the neighbourhood, bedded hæmatites and turgites (the two main beds are respectively 33 and 92 feet thick, and there are 9 other beds ranging in thickness from 10 inches to 6½ feet), are worked at the Tiazholy mines, while at the Gayevsky mine, a spathose ore-deposit (analogous to that of the Elnichny) is worked.

All the ore-deposits of the Bakal region may be regarded as belonging to Dr. Beck's category of *epigenetische Erzstöcke*: they are the outcome of the chemical reactions, which took place at some remote period between iron-bearing solutions and the carbonates of lime and magnesia of the limestones. **L. L. B.**

PLATINUM-DEPOSITS IN THE URALS, RUSSIA.

Notice préliminaire sur les Gisements de Platine dans les Bassins des Rivières Iss, Wyia, Toura et Niasma (Oural). By N. VISSOTZKI. *Bulletins du Comité géologique, St. Pétersbourg,* 1903, *vol. xxii., pages* 533-559 *and 2 plates.*

The region studied by the author is situated on the eastern slope of the Ural range; it includes the most considerable platinum-deposit hitherto discovered in the world, seeing that the placer extends uninterruptedly over a length of 80 miles, and yields annually from 40 to 50 tons of the precious metal.

The platiniferous belt of the Urals, geologically speaking, consists of four parallel bands striking roughly north and south: the westernmost of these, made up of crystalline schists, forms the watershed between Europe and Asia. The next band to the east comprizes olivine- and mica-gabbros, diallage-peridotites, diorites and altered syenites—all of them rocks which have been erupted from a great depth. The third band is made up of Lower Devonian sedimentaries, partly torn into, shattered and buried by diabasic eruptives: farther eastward eruptives of deep-seated origin (gneissose granites?) reappear; and finally we reach a belt of ancient rocks eroded by the action of the advancing sea of the Lower Tertiary age.

The area emerged from the waves as early as the Carboniferous period; consequently the accumulations of platinum, and in some localities of gold, in the surface-deposits were not swept away. They were concentrated later on in the alluvia—perhaps at the time of the most intense glaciation (Pleistocene?).

As is the case throughout the Urals, the primary source of the platinum is associated with the eruptive basic rocks; among these the platiniferous and auriferous dunite forms three great masses, shown on one of the author's maps. The other map illustrates the distribution, over the greater part of the Urals, of the platinum-bearing alluvia and the outcrops of platiniferous dunite—these being connected with the second of the four parallel bands mentioned above. Towards the south, the band becomes discontinuous, and finally dies out altogether: simultaneously with it disappear all the placers that can be reckoned as having any industrial importance. There is, it is true, in the Southern Urals, a small number of outcrops of platiniferous olivine-rock, but the percentage of platinum therein is small, and other metals of the group are generally associated with it, notably osmiridium. The annual output of the last-named is stated to average 385 troy ounces. **L. L. B.**

THE IRON-ORE DEPOSITS OF SCANDINAVIA.

(1) *L'Origine et les Caractères des Gisements de Fer Scandinares. By* L. DE LAUNAY. *Annales des Mines*, 1903, series 10, vol. iv., pages 49-106 and 109-211, with 23 figures in the text and 6 plates.

It is impossible in a brief abstract to do full justice to this exhaustive memoir, which is divided into five chapters, preceded by an introduction and followed by an appendix and a bibliography. The author visited Scandinavia in 1890 and again in 1899, with the view of studying on the spot the principal ore-deposits, and he acknowledges his indebtedness to the Swedish and Norwegian geologists and engineers, both for written and for verbal information.

The most important iron-ore deposits of Scandinavia form two well-defined groups—one situated north-west of Stockholm, including the Dannemora, Norberg, Persberg, Grängesberg mines, etc.; and the other in Northern Bothnia or Swedish Lapland, comprising the much more recently-opened mines of Gellivaara, Kiirunavaara, Svappavaara, Rutivaara, etc. The deposits of the first group are mutually similar in many respects, while those of the second group are, geologically speaking, very dissimilar. The industrial primacy long enjoyed by the first group (the Dannemora mines dating from the thirteenth century) is manifestly shifting to the second group. In 1895, the annual iron-ore output for the whole of Sweden averaged 1,517,000 tons; but in 1903 the Kiirunavaara mines alone are reckoned to produce 1,200,000 tons, and the output from Gellivaara will probably have amounted to little less.

Geologically, the Scandinavian iron-ore deposits may be classified as follows:—

1.—Directly-segregated masses in immediate association with the basic rocks (gabbro, olivine-hyperite, nepheline-syenite) from which they are derived. The ores are frequently magnetites, sometimes titaniferous magnetite or titanic iron. Typical localities are Taberg, Kragerö, Ekersund, Lofoten, Rutivaara, Alnö, etc.

2.—Masses of debatable origin, possibly related in some way to the porphyries amid which they occur, as at Kiirunavaara and Luossavaara.

3.—Lenticular bodies occurring in " strings " amid metamorphic, Archæan, or Silurian rocks, probably originating from the recrystallization of sedimentary deposits, as at Svappavaara, Gellivaara, Grängesberg, Norberg, Persberg, Dannemora, Dunderlandsdal, etc.

The author gives a detailed description of these three classes of deposits in the order named. He compares the conditions, under which the segregation-deposits originated, to metallurgical processes, taking place in the depths of the earth. He imagines a molten mass of metal being churned up with steam in the presence of hydrocarbons, sulphuretted hydrogen, or hydrochloric acid, with the result that iron-ore free from phosphorus was generated, the phosphorus having passed off in the slag. He suggests that native iron probably exists within the globe at inaccessible depths.

After showing how very exceptional are the characteristics of the Kiirunavaara-Luossavaara deposits, differing indeed entirely from any others in Scandinavia, the author remarks that a chronological succession can be traced in them. It looks as if the ore had been deposited by sedimentation between two periods of submarine eruption of porphyries, these porphyries themselves being intercalated in a pre-Cambrian (?) series of conglomerates, shales and quartzites. The minimum quantity of iron-ore in sight at

Kiirunavaara and Luossavaara is said to amount to 233,000,000 tons. The ore is an almost absolutely pure mixture of magnetite and apatite, with a small proportion of titanium, and very compact in structure. The percentage of phosphorus generally ranges from 1 to 4; but it is said that a great quantity of ore can be got, containing only from 0·1 to 0·8 per cent. of phosphorus. The percentage of sulphur varies between 0·05 and 0·08, and that of titanic acid between 0·32 and 1·5.

A very detailed description is given in the third chapter of the well-known deposits of Gellivaara, Grängesberg, Norberg, etc., and the author then devotes a few pages to the newly opened-up deposits of the Dunderlandsdal, etc., in Northern Norway. It is expected that the export of ore from these mines will, in 1905, be at the rate of 750,000 tons per annum. The ore-body, a mixture of quartz, hæmatite and magnetite, with which are associated garnet, epidote, some hornblende, pyroxene, etc., is defined as a ferruginous quartzite, passing gradually into barren quartzites comparable with the itabirites of Brazil. It occurs amid a complex of mica-schists and marbles, which are believed to represent a syncline of metamorphosed Cambro-Silurian strata. The percentage of metallic iron varies between 30 and 40, rarely attaining 55: the hæmatite is 10 times more abundant than the magnetite. Manganese occurs in very variable proportions, ranging from zero in the southern area to 3 or 5 per cent. at Ofoten and Salangen. Phosphorus, in the form of apatite, averages 0·2 per cent.; sulphur occurs in very small proportions, and titanium is absent.

The fourth chapter is taken up with the consideration of the part played by phosphorus in the Scandinavian iron-ores generally. The author points out that in most sedimentary formations the circumstances which have brought about the precipitation of iron have also favoured the concentration of phosphorus, a concentration which appears to attain its maximum in the ores of Lapland and Grängesberg. Perhaps both iron and phosphorus are derived from the erosion of very ancient folded rocks, which were really the "slag" remaining over from those deep-seated processes of natural metallurgy to which reference has already been made. In the appendix, a brief description is given of the methods by which the iron-ores and the apatite are separated and sorted out at Luleå and Grängesberg.

(2) *Bericht über eine im Sommer 1903 nach den Eisenerzvorkommen an der Ofotenbahn ausgeführte Studienreise.* By Dr. Hecker. *Zeitschrift für das Berg-, Hütten- und Salinen-wesen im Preussischen Staate,* 1904, vol. lii., *pages* 61-85 *and* 5 *plates.*

Of the total output of iron-ore in Sweden for the year 1901, that is, 2,793,566 tons, the Gellivaara mines produced 38·6 per cent. and the Grängesberg mines 23·6 per cent. (Nothing had yet been forthcoming from Kiirunavaara in 1901 and 1902, regular systematic working not having been started there until 1903). Of the above-mentioned total output 89·7 per cent. consisted of magnetite and 10·3 per cent. of hæmatite. The export of iron-ore from Sweden during the same year amounted to 1,761,007 tons, no less than 1,477,124 tons of which were taken by Germany, who seems to be in this respect consistently Sweden's largest customer. But the near future may see a shifting of the balance.

The deposits of Kiirunavaara, Luossavaara and Gellivaara constitute by far the most important occurrences of the kind in Scandinavia; the working of the first two on a large scale could not well get into full swing until completion of the railway to Narvik on the Ofoten Fiord, at the end of 1902. The first cargo of ore was shipped from there on January 2nd, 1903, and within the first six months of the same year the Grängesberg Company bought up most of the shares of the Gellivaara Mining Company and of the Luossavaara-Kiirunavaara Company, thus forming a huge trust, managed from the central offices in Stockholm.

The great Kiirunavaara deposit strikes from north to south along a range of hills, in such wise that all the hilltops consist of ore, which has better withstood erosive agencies than the neighbouring country-rock: it really consists of three belts separated by narrow partings of rock, and the total length is about 3 miles, while the average breadth exceeds 300 feet, and the average thickness or depth is about 230 feet. The dip, invariably eastward, ranges from 50 to 70 degrees, and borings have proved that the thickness of the ore-body diminishes with the depth. Taking this into account, the minimum estimated mass of the deposit above the level of the neighbouring Luossajärvi Lake (the surface of which lies at an altitude of 1,640 feet) is 215,000,000 tons, but the more probable figure is now believed to approach 300,000,000 tons. The country-rock is porphyry, which itself forms a sort of island amid bedded sedimentary rocks, mainly slates and conglomerates. The ores (magnetites) are very compact, much harder than the generality of Swedish iron-ores, and practically free from all impurities, except apatite. The proportion of this varies greatly, but so far four well-recognized grades of ore are worked, and are distinguishable without any special sorting: the best of these grades contains from nil to 0·04 per cent. of phosphorus and 69 to 70·5 of metallic iron, the corresponding percentages in the worst grade being from 3·5 to 6 per cent. of phosphorus and 54 to 57 of metallic iron. Dr. Lundbohm regards the deposit as the product of magmatic differentiation, that is, of eruptive origin; while Prof. Vogt holds that it was deposited first of all by chemical precipitation as hæmatite, and was subsequently altered into magnetite by contact-metamorphism. The author adduces reasons in favour of the former view, and shows how the latter is in conflict with the available evidence.

The Luossavaara deposit lies on the opposite (northern) bank of the Luossajärvi Lake: on account of the thick covering of drift by which it is largely mantled, it has not been so minutely studied as that of Kiirunavaara, and investigations conducted with the magnetic needle have failed to reveal a connection between the two deposits. That of Luossavaara is believed to constitute a huge lenticle, nearly a mile long and from 100 to 165 feet thick; the easterly dip is very steep, and the minimum mass of ore above lake-level is estimated at 18,000,000 tons. Most of the magnetite corresponds in quality to type No. 2 of the Kiirunavaara ore; containing from 0·05 to 0·1 per cent. of phosphorus and from 68 to 69 per cent. of metallic iron, and it also contains rather more titanium. The country-rock is again porphyry, and the ore-deposit is undoubtedly of the same eruptive origin as that of Kiirunavaara.

The climatic conditions are not quite so unfavourable as might have been expected at so high a latitude (68 degrees north), but the plague of the millions upon millions of mosquitoes which breed in the neighbouring marsh-lands is almost unbearable in the summer.

The ores are worked opencast, and are got mainly by shot-firing with dynamite, the Ingersoll compressed-air drill being the appliance finally selected for use. The haulage and loading arrangements are of the most modern description. For distribution of power and lighting purposes there is a power-station provided with three Lancashire boilers, and with room for two more. The exciters of the two dynamos are driven by De Laval steam-turbines, which make 3,000 revolutions in a minute. The two air-compressors furnish air at 5 atmospheres, and this is distributed through the workings by wrought-iron pipes 8 inches in diameter. The electric current is at a tension of 2,000 volts, and besides driving machinery of various kinds, it serves to light the mines and the railway-station and a portion of the town of Kiiruna. The number of people employed is from 500 to 600, working in 8 hours shifts. The prime cost of a ton of ore works out at about 2s. 3d., but with a larger output it is hoped to reduce this soon to about 1s. 10½d.

Short descriptions are given of the Ofoten railway, and of the increasing facilities for shipping at the harbour of Narvik, and the author then proceeds to give an account of his visit to the well-known mines of Gellivaara. Only one of these (Koskullskulle) lies outside the scope of the great Swedish trust previously mentioned and belongs to a company, most of whose shares are held in Austria. The ore-deposits (magnetites and subsidiary hæmatites) lie in a hill about 3 miles north of Gellivaara, and 46½ miles north of the Arctic Circle; they conform to the normal type of Swedish iron-ore deposits, and the amount of ore in sight is estimated at 100,000,000 tons. The country-rock is a distictly-bedded, pinkish hälleflinta-gneiss, but neither it nor the ore is often seen to crop out at the surface, being generally mantled over by the all-pervading Glacial Drift; the whole mass of rock is seamed by numerous intrusive dykes, mostly pegmatitic and granitic, but also syenitic at the Koskullskulle mine. The lenticles of ore form two distinct belts, the northernmost being the main one: the greatest thickness of ore (maximum, 492 feet) is observed in the Tingvallskulle lenticle, and the dip varies between 50 and 70 degrees south-ward. The structure of the ore is generally crystalline-granular, and crumbles easily into its constituent granules, but at Koskullskulle the mineral is finer-grained and tougher. Apatite is almost the sole impurity, and as its proportion in the ore diminishes, so does the mineral gain in toughness. There are six recognized grades of ore, the highest of which contains 69·09 per cent. of metallic iron and 0·021 per cent. of phosphorus, while the lowest contains 59·10 of iron and 1·9 per cent. of phosphorus. Most of the output is fifth- and fourth-grade ore, but the Koskullskulle mine produces 70 per cent. of first-grade and 30 per cent. of third-grade ore. As at Kiirunavaara, the ores are sorted by mere inspection on being loaded into the haulage-wagons, and here again this rough sorting, when tested by chemical analysis, proves to be remarkably accurate.

The climatic conditions are somewhat more favourable than at Kiiruna-vaara, and the country is covered with a thick growth of conifers; but the depressions are occupied by vast swamps, and the mosquito-plague becomes really terrible on hot summer-days. The system of working the ores is very simple, " almost primitive "; hand-drilling has been found more suitable and cheaper than electric drills, and the mass of ore is got down by shot-firing. Although the workings are mainly opencast, deep-level workings have been started at various points (especially for the cold season) as in them the miners are better shielded from the rigours of the Arctic winter, and the levels are lighted by electric arc-lamps. The total output of ore in 1902 exceeded 1,000,000 tons. The mines have direct railway com-

munication with Gellivaara, and thence with the seaport of Luleå on the
Gulf of Bothnia: this harbour is, however, free from ice for only about
5 months in the year, wherefore the winter-output is kept stacked at the
mines until navigation is open, and the condition of the ore is not improved thereby.   L. L. B.

SABERO COAL-FIELD, SPAIN.

Descripción de la Cuenca Carbonífera de Sabero (Provincia de León). By L.
MALLADA. *Boletín de la Comisión del Mapa geológico de España,* 1900,
series 2, *vol. vii., pages* 1-65, *with 8 figures in the text and 1 plate.*

The abundance of the coal-outcrops in this district of the province of
Leon, together with the occurrence in the same neighbourhood · of considerable iron-ore deposits, gave rise to the hope, some 60 years ago, that
Sabero would ultimately become the centre of a great metallurgical industry. The mines were most actively worked between 1850 and 1860; yet
in 1868 mining operations were abandoned all but completely, and it was
only from 1890 onwards that new projects of railway-construction encouraged
the capitalists of Bilbao to form syndicates for the purpose of re-starting
active work in the Sabero coal-field. From the railway-station of La Ercina
to the central point of this coal-basin, which is one of the longest and
narrowest in Castile, the distance is barely 2 miles. From the neighbourhood of Las Bodas the coal-belt extends for some 12 miles to beyond
Fuentes, while its breadth varies from a few feet to rather more than
¼ mile, the total area being estimated as equivalent to 5,263 acres. To
the north and south, the Coal-measures are bounded by Devonian strata, but
in one or two localities they abut direct upon the underlying Cambrian. It
is chiefly in the Devonian quartzose grits that occur the iron-ores referred
to above: they contain between 20 and 40 per cent. of metallic iron. The
occurrence of these ores is connected by some geologists with certain intrusive igneous rocks, scattered bosses of which are seen in the very centre
of the coal-basin, as well as among the Devonian sedimentaries. In the last-
named, copper, lead, and zinc-ores have also been proved. All the .
Palæozoic strata in this region have been subjected to much disturbance
and dislocation. On the north-east and west, the coal-field is bounded by
newer strata, belonging to the Cretaceous series. The lowest of these newer
beds, described as " kaolin," is of great industrial importance, as it has not
only been proved in the area dealt with here, but is a characteristic associate of the Spanish coal-basins for a distance of 93 miles or more, and
occupies a belt varying in breadth from 70 to 330 feet. The Cretaceous beds
show signs of great dislocation also, and it is pretty certain that the coal-
field will be found to extend below them over a greater area than is
generally suspected at present. The coal-measures themselves consist of
alternating sandstones and shales, with conglomerates towards the base,
made up of more or less angular fragments of Devonian limestone. A
detailed description is given of the coal-basin, with many records of the
succession and lie of the strata and the thickness of the seams, as revealed
in the mine-workings, etc. The richest portion of the field lies between the
meridians of Sotillos and Saelices, but the seams vary greatly as to quality
and texture of coal. More than 50 per cent. of them yield a mineral which
has to be carefully sorted into different grades, and gives much trouble
also in the matter of washing. The northern group of seams are of regular

occurrence and vary in thickness from 2¼ to 8¼ feet; the other group comprizes seams which are very irregular, but attain extraordinary thicknesses
(from 13 to 80 feet or more).   There is not a mine which does not show
repeated dislocation of the strata, with changes of strike and dip.   This.
latter is generally steep (between 50 and 70 degrees, and in some cases.
steeper than that); the general strike is east and west.

After tabulating analyses published in 1856 and in 1885 respectively, the
author gives further analyses of four specimens of coal which he himself
obtained.   In these the volatile matter varies from 18 to 23·05 per cent.,
and 'the ash from 2·3 to 8·10 per cent.   The coal is considered to be the
best that is found in Castile.   As to quantity of available mineral, the
author discusses at great length the various calculations entered into
by geologists and engineers, and finally decides on 48,000,000 tons as
the probable amount contained in the Sabero basin.   He regards this,
however, as a very moderate estimate.   Including sorting and washing, the
prime cost of the coal is 11 pesetas, or 8s. 3d. per ton; of coke, 16 to 18
pesetas, or 12s. to 13s. 6d.; and of patent fuel or briquettes, 18 to 19
pesetas, or 13s. 6d. to 14s. 3d.   The statistics of output appear to be incomplete, but in the case of one of the most important companies in the
field (owning three-quarters of the total area) the average annual output
is stated at 110,000 tons.   The author considers that a fusion of the various.
companies would not only bring about great economies in the cost of
management, etc., but would conduce to greater efficiency of working and.
hence to an increase of output.   Certain northerly extensions of the coal-
field (Santa Olaja, Ocejo and Argovejo) are described in detail.   The last-
named of the three alone is reckoned to contain 2,000,000 tons of coal.

L. L. B.

MESOZOIC BROWN-COAL- AND ORE-DEPOSITS IN SPAIN.

Explicación del Mapa geológico de España : Sistemas Infracretáceo y Cretáceo. By
L. MALLADA. *Memorias de la Comisión del Mapa geológico de España,*
1904, *vol. v., pages* 458-502, *with 8 figures in the text.*

The author of this magnificent memoir remarks that, of the whole Mesozoic series in Spain, the Cretaceous and Infra-Cretaceous systems are the
richest in useful minerals.

He describes first of all the coal-fields of the province of Teruel: here
coal occurs at three distinct horizons, the lowermost being associated with
Orbitolina-beds, the middle with *Trigonia*-beds and the uppermost with
equivalents of the Cenomanian sandstones, which are overlain by limestones
containing *Ostrea flabellata*.   The coal-basin of Utrillas is 10 miles in length
and 3 miles in utmost breadth, and of elliptical shape.   The seams mostly crop.
out in the southern portion, and have a dip varying between 20 and 25
degrees.   They are 10 in number, with a total thickness of 60 feet, including
two seams of jet.   Two or three seams of lignite of good quality have also
been proved at some localities.   North-north-east of the basin just described
lies that of Gargallo, occupying a triangular area of about 10 square miles,.
and containing fewer coal-seams, and those of later date (basal Cenomanian).   There are at most four seams, very near together, each about 40.
inches thick, and several bands of pyritous lignite varying in thickness.
from 3¼ to 5 feet.   Cenomanian coal also characterizes the basin of Ariño:
there are three seams, of a thickness varying between 10 and 23 feet, yield-

ing a mineral which burns with a long flame, but crumbles away to dust on exposure to atmospheric agencies. The coal-seams of the basin of Aliaga, about seven in number, are mostly of earlier date than those of Utrillas, and are of good quality. On the whole, although the coal-fields of Teruel give hopes of a brilliant future, the amount of mineral in sight has been much exaggerated. Seven specimens from Utrillas and five from Gargallo have been analysed, and are found to contain from 41 to 45 per cent. of carbon, and 48½ to 49½ of volatile substances; they yield from 6 to 9 per cent. of ash; their heating-power varies between 4,700 and 5,000 calories, and their specific gravity between 1·3 and 1·4.

Amid the uppermost Cretaceous (Danian) of the province of Barcelona occur four coal-basins, the most important of which is that of Vallcebre: it contains as many as 15 seams of brown coal, and one of these seams alone is calculated to yield 40,000,000 tons or more of the mineral. The coal from this seam contains from 52 to 53 per cent. of carbon, 41 to 43 of volatile substances, and yields from 4 to 7 per cent. of ash. Its heating-power is estimated at 5,200 calories, and it burns easily with a long flame. The mineral from some of the seams appears capable of being turned into coke; but, as a rule, it is found necessary to mix it with bituminous coal for coke-making purposes.

Lignite-seams are worked in the province of Santander; they were also worked at one time in Álava, and the seams there are so thick that it will probably be found profitable one day to restart the workings. Brief mention is also made of the lignites of Guipúzcoa, Soria, and other provinces.

The principal deposits of jet occur in the Turonian basins of Utrillas and Gargallo, in the form of lumps of varying size scattered through clay-beds some 2 or 3 feet thick. The output has of late years been reduced to quite insignificant proportions.

Although, in the province of Álava, the main occurrences of asphalt are in the Eocene limestone, there are also deposits of it in the Cretaceous sands and sandstones. These are apparently worked with some success.

With regard to iron-ores, the first rank is of course taken by the province of Biscay. The *vena*, *campanil* and *rubio* grades of hæmatite of Somorrostro have formed the subject of a considerable literature; but the formerly less famous spathic iron-ore, which occurs along with them, is now being worked on a gigantic scale, and after calcination, is exported to foreign countries. The author gives many quotations from Mr. Adán de Yarza's well-known *Geological and Physiographical Description of the Province of Biscay*, and horizontal sections taken in sundry mines. Estimates of the amount of iron-ore in sight at Somorrostro vary considerably, but it certainly far exceeds the 40,000,000 tons estimated by Mr. Goenaga. Certain other iron-ore deposits of Biscay are described, and the author then refers to their westward prolongation into the province of Santander, the most important localities in this connection being Setares and Dicido. The hæmatites of the Asturias and of Soria were worked in times gone by, and deposits that might well be worked occur among sandstones in the Aliaga district of Teruel.

The best zinc-ores (blende and calamine) in the province of Santander occur at Reocín and Mercadal. At both localities, calamine is mined, containing 40 per cent. of the metal, and is enriched by calcination on the spot to 56 per cent. The zinc-ores of Biscay are of less importance: three vertical veins carrying blende, calamine and galena, and ranging in thickness from 2 to 10 feet, are worked at Lanestosa, and a similar one is worked near Matienzo. The blende from these contains 50 per cent. of zinc, and the more abundant calamine assays (after calcination) to 45 per cent.

The principal lead-ore deposits occur at Villareal, in the Province of Álava, where several veins of galena associated with blende are or have been worked. Those of Biscay, Guipúzcoa, Soria, etc., are apparently too poor to repay systematic working.

Very pure pyrolusite occurs in pockets in the Crivillín district of Teruel.

Petroleum occurs, in comparatively small quantities, about 1¼ miles distant from Huidobro, in the province of Burgos, and is utilized locally. The other mineral-occurrences enumerated by the author are, on the whole, either of very little industrial importance, or else have been worked out.

L. L. B.

ORE-DEPOSITS IN THE SIERRA MORENA, SPAIN.

Die Erzlagerstätten von Cala, Castillo de las Guardas, und Aznalcollar in der Sierra Morena (Prov. Huelva und Sevilla). By C. SCHMIDT *and* H. PREISWERK. *Zeitschrift für praktische Geologie,* 1904, *vol. xii., pages* 225-238, *with* 7 *figures in the text.*

The western portion of the Sierra Morena, in Southern Spain, consists largely of steeply-dipping Palæozoic slates and limestones, striking on the whole east and west, and traversed in many places by masses of eruptive rock. The far-famed ore-deposits belong to two distinct belts: (1) the southern, 143 miles long and 12½ miles broad, in which lenticles of pyrites and manganese-ores predominate; and (2) the northern, in which oxidic iron-ores, such as magnetite and hæmatite, predominate, but are in places associated with pyrites.

The authors describe first of all the deposits in the Sierra del Venero, near Cala (province of Huelva) belonging to the northern belt, which they visited in September, 1903. The summit of the Sierra (2,380 feet above sea-level) actually consists of hæmatite-crags, and the outcrops of iron-ore extend for some two-thirds of a mile through Cambrian slates, close to the southern margin of a granite-massif. The slates here have been much altered by contact-metamorphism, whereas on the northern margin of the intrusion the slates are far less altered, and are barren of iron-ore. The ore is chiefly magnetite associated with pyrites, and is a typical example of a contact-metamorphic deposit, disposed in lenticles conformably interbedded with the altered slates. The outcrops at the actual surface are almost universally made up of hæmatite, considerable surface-outcrops of magnetite and pyrites being observed only at two points. The average percentage in metallic iron of the surface-hæmatite is 50. Four metalliferous zones in this area are plotted out and described, on the strength of the evidence obtained from the mine-workings, and the amount of ore in sight is estimated at 4,000,000 tons of hæmatite (50 per cent. of metallic iron, 4,000,000 tons of magnetite (about 60 per cent. of metallic iron) and 1,000,000 tons of cupriferous pyrites (about 6 per cent. of metallic copper). But the available evidence leads the authors to believe that, over and beyond the mass of ore already proved, fully as much again will be found to exist in the Cala district, and of quality in no wise inferior to the ore which is now worked. The conditions of working are extremely favourable, and at least 4,000,000 tons of iron-ore can be got by opencast. No difficulties, even in the underground workings, are likely to be experienced in the matter of ventilation or from water-inflows. Much space is devoted to a discussion of the contending views of various writers in regard to the genesis of contact-metamorphic ore-deposits, the present authors finally concluding that the iron of the Sierra del Venero was originally derived from the intrusive granite.

They then describe the pyrites-deposits of Castillo de las Guardas and Aznalcollar (province of Seville) belonging to the southern belt: these are the easternmost mines of that portion of the Sierra Morena, and, although mentioned by Mr. J. H. Collins, they have on the whole received scant attention from the majority of writers.

Three great pyrites-lenticles intercalated among Silurian slates dipping steeply northward, with a great granite-massif close by on the south-east, are found at Castillo de las Guardas. The westernmost lenticle is accompanied along its northern margin by a sill of diabase. The outcrops are indicated very clearly by the ferruginous gossans, and from the north-western lenticle alone (the only one worked so far), more than 2,000,000 tons of ore could be got by opencast working merely, if the adjacent country-rock were blasted away (as is done at Rio Tinto). The average composition of the pyrites is 50 per cent. of iron, 48 per cent. of sulphur, and 2 per cent. of copper, and the mineral is remarkably free from arsenic.

About 2 miles east of the village of Aznalcollar, which itself lies about 19 miles north-west of Seville, two great belts of pyrites are intercalated among the Palæozoic slates; in their immediate vicinity are numerous sills of quartz-porphyry, sometimes conformably interbedded with the slates, sometimes cutting down through them or overlying them. The whole formation is overlapped by fossiliferous Miocene limestones and conglomerates, which have been eroded away over large areas. The ore-bodies are discoidal or lenticular, and comparatively flat-bedded: each consists of two distinct portions, the southern and smaller portion being rich in copper, and the northern and bigger portion being poor in that metal. The highly-cupriferous masses soon pinch out in depth, and have been all but worked away; while the less cupriferous masses increase with the depth in considerable proportions, say from a thickness of 50 feet near the surface to one of 85 feet at a depth of 360 feet. The mass of ore still unworked is estimated at about 7,000,000 tons, containing only about ½ per cent. of copper, with 0·6 per cent. of arsenic. Evidence is adduced to prove the association of the ores with the quartz-porphyry intrusions, and then a discussion is opened on the genesis of the pyrites-deposits of Southern Spain generally. This question assumes the greater importance that the Spanish deposits are typical of whole groups of deposits occurring in various parts of the world. It is to be noted, by the way, that the contact-zone of metamorphism due to the Cala granite-intrusion is far more extensive and "intense" than that determined by the porphyries of the Rio Tinto belt. The Spanish deposits are compared with those of the Island of Elba, Campiglia Marittima, Massa Marittima and Gavorrano, in Italy, and final stress is laid on two points: (1) the pyrites-magnetite reefs in the north of the Sierra Morena must be taken into account when one is studying the origin of the pyrites-lenticles in the south; and (2) it has been ascertained beyond doubt, in one case at least, that the quartz-porphyry [of Aznalcollar] with the accompanying bands of pyrites cuts through the slates.   L. L. B.

ROCK-SALT DEPOSIT OF CARDONA, SPAIN.

Criadero de Sal de Cardona. **By** D. L. M. VIDAL. *Boletin de la Comisión del Mapa geológico de España,* 1900, *series* 2, *vol. vii., pages* 149-155, *and* 2 *plates.*

This deposit, associated, as is so frequently the case, with gypsum, forms a great mass in the neighbourhood of Cardona, in the district of Manresa, Catalonia. The beds are highly folded and contorted, but the salt is very pure, especially in the lower portion of the mass. The question of age is still in dispute. While many geologists believe that the deposit is Triassic, others, and the author among them, hold that it probably dates from the Oligocene division of the Tertiary Era. It is pointed out that the salt-deposit of Vilanova de la Aguda, in the province of Lérida, some 22 miles away, is of Oligocene age, and that the splendid rock-salt of Remolinos, in the province of Zaragoza, is Miocene. In fact, all the known rock-salt deposits in Spain are of early Tertiary age.

The paper contains no particulars as to output, methods of working, etc.

L. L. B.

BITUMEN-MINES IN ALBANIA, TURKEY. .

Note sur les Mines de Bitume exploitées en Albanie. **By** A. GOUNOT. *Annales des Mines,* 1903, *series* 10, *vol. iv., pages* 5-23.

These mines are situated in the sandjak of Bérat, in the vilayet of Janina, and the village of Selenitza (the local centre of the industry) is 9½ miles or so distant from the harbour of Vallona on the Adriatic. The district is rugged and mountainous in character; the strata are of Tertiary age, comprizing thick beds of clay (abundantly fossiliferous), shelly sands, and bands of gypsum, resting upon a basement of very tough, blackish limestone.

The author distinguishes four varieties of bitumen in these deposits. First of all, the "dull solid bitumen" occurring in scattered, irregular pockets, generally near the surface of the ground, and ranging in thickness from a few inches to several feet. A chemical analysis shows its percentage composition to be as follows:—Bitumen, soluble in carbon bisulphide, 72·69; water and loss at 212° Fahr., 9·12; ash, 17·19; and organic substances, 1. This may be compared with the Trinidad mineral, which contains on an average 34 per cent. of bitumen.

Next he describes the "lustrous solid bitumen." Unlike the variety just dealt with, it occurs over a restricted area, and only in the dark limestone. It is always in perfectly-distinct, mutually-parallel, nearly-vertical veins, striking north 50 degrees east. That it was originally deposited from above, and not injected from below, is proved by the fact that nowhere does it continue in depth. At 50, or at most 70, feet below the surface, all trace of it disappears. The lustrous bitumen has a still finer conchoidal fracture, and is of a richer black, than the dull variety. No mineral exactly like it is known to occur elsewhere; its composition is as follows:—Bitumen, soluble in carbon bisulphide, 98 per cent.; water given off at 212° Fahr., 1·4; carbonaceous substances, 0·2; loss and undetermined substances, 0·4. The mineral is often called after the village of Romsi, near which most of it is got.

The third variety is the liquid bitumen, which occurs in a low-lying area, at the base of the mountain-massif, near the Viussa river: the workings were flooded out after heavy rains in the winter of 1895-1896, and the river has practically obliterated all traces of them. Fortunately the author had thoroughly investigated the area in 1894; and he is able to state that the liquid bitumen came up through cracks in the ground, where it partly solidified in the form of miniature volcanic cones, from a viscous mass of bitumen flowing along a sort of subterranean pebble-bed down from higher altitudes. The composition of the liquid bitumen was found to be as follows:—Bitumen, soluble in carbon bisulphide, 96·4 per cent.; insoluble matter, 1·9; carbonaceous substances, 1; loss and undetermined substances, 0·7.

The fourth variety, asphalt, is abundant at Selenitza, but does not repay working, on account of the difficulties and cost of transport. However, it comes in handy as fuel in a locality where coal costs 32s. a ton: this asphaltic fuel, known locally as *javor*, has the following composition:— Bitumen, soluble in carbon bisulphide, 39 per cent.; water given off at 212° Fahr., 2·85; carbonaceous substances, 16·8; insoluble residues, 40·7; loss and undetermined substances, 0·65.

The wasteful and careless methods of former generations of miners have made it practically impossible to work over the same ground again in a systematic manner, and many thousand tons of bitumen must needs remain untouched. At other points, methodical working on a small scale is going on, under the author's direction apparently: in some cases by means of adits, in others by means of small shafts. The mineral is hand-picked at the mines, and again re-sorted at Selenitza. The best grades are sent off as they are, while the lower grades are freed of their impurities by fusion. Labour is cheap, the daily wage of a workman averaging 5 piastres, or say 10d. The author is of opinion that the Albanian miner is exactly worth his pay, and no more. The transport of the mineral by pack-mules or pack-horses to the harbour of Vallona, along rough paths, costs about 6s. 10d. per ton: there, the bitumen is mostly loaded on Italian sailing-vessels. It may be mentioned that roads suitable for wheeled traffic are non-existent in the Selenitza district.   L. L. B.

———

CINNABAR-DEPOSITS OF KARA-BARUN, ASIA MINOR.

Notizie sul Giacimento Cinabrifero di Kara-Barun nell' Asia Minore. By G. D'ACHIARDI. *Atti della Società Toscana di Scienze Naturali, Processi verbali,* 1903, *vol. xiii., pages* 173.176.

In the north-eastern portion of the peninsula of Kara-Barun, in the province of Smyrna, is a highly-metamorphic dark schist, traversed by a quartzose brecciated rock, which contains from $2\frac{1}{4}$ to $2\frac{1}{2}$ per cent. of mercury. A shaft was sunk some 65 feet down into this rock, which strikes north and south; and about halfway down a heading was driven east and west. It cuts through about 26 feet in cinnabar-bearing rock, and then enters the metamorphic schist (on the west), which itself is found to contain about 0·3 per cent. of mercury. On the east, the cinnabar-bearing breccia is flanked by some 10 feet of yellow and red ochreous deposits, poor in cinnabar but very rich in nodules of pyrites. The schist has been proved for more than 300 feet to the eastward, and then abuts against a Cretaceous Hippurite-limestone, while on the west it is suddenly cut out by basaltic eruptives. Small cinnabar-bearing quartz-veinlets (cross-veins or *filons-croiseurs*) occur in the schist in the neighbourhood of the limestone. Specimens of the various rocks were examined by the author under the microscope, and he feels no doubt that the cinnabar was deposited simultaneously with the siliceous cement of the breccia. This breccia was undoubtedly mineralized by siliceous waters containing in solution salts of mercury and iron, which found their way through rock-fissures originated by the basaltic eruptions. The yellow and red ochres are alteration-products of the pyrites. In appearance, the breccia recalls the mercury-bearing *frailesca* of Almaden, but its mineralogical composition differentiates it completely from that rock. As in the case of other cinnabar-deposits, that of Kara-Barun will possibly be found to grow richer in depth; a hope which appears to be borne out by the discovery, in a stream-bed close to the exploration-workings, of quartzose fragments containing 45 per cent. of mercury. L. L. B.

COAL AND OTHER MINERALS IN THE TAPASHAN MOUNTAINS, CHINA.

Reise durch das Gebirgsland der Ta-pa-shan; Provinzen Hupeh, Shensi und Szechuan. By KARL VOGELSANG. *Petermann's Mittheilungen,* 1904, *vol. l., pages* 11-19 *and* 1 *plate.*

Starting from Ichang, on the Yangtsze river, on May 9th, 1900, the author reached, 9 days later, Panchui-ho, 5 miles north-west of which place is a copper-ore deposit, some 5,000 feet above sea-level. It consists merely of a few thin veins seaming pinkish-yellow quartzites and dark slates, and is of no industrial importance. The same statement holds good of the copper-glance, etc., occurring near Paofeng and Chihluiwan. Yet Chinese prospectors had endeavoured to induce European business-men in Shanghai to purchase concessions for working these useless deposits.

The author then turned southward from Chuhsi, finally reaching the Yangtsze valley again, at Wushan, on June 11th, 1900. At the Luliya pass in Shensi, he observed a thin coal-seam associated with flaggy limestones, probably of Carboniferous age. At Taumushanpen, in Szechuan, the Chinese work thin seams of a highly-sulphureous coal, of inferior quality. It is taken on boats down the river (a left-bank tributary of the Yangtsze), and is mostly used for the evaporation of brine.

It is in the Yenchang valley at Lungchsinmiao, that the brine-spring occurs, from which the salt is obtained for consumption over a wide area. The Government levies a tax proportionate to the amount evaporated from the brine-pans.

The author's general conclusion is that none of the mineral-deposits seen by him on his journey are such as to warrant any attempt at mining operations on a large scale. The coal, especially, can only be of local importance.

It may be observed that the Tapashan mountains include portions of the provinces of Hupeh, Shensi and Szechuan. L. L. B.

THE MINERAL WEALTH OF MANCHURIA AND KOREA.

Constitution géologique et Ressources minérales de la Mandchourie et de la Corée. By L. PERVINQUIÈRE. *Revue Scientifique,* 1904, *series* 5, *vol. i., pages* 545-552, *with* 2 *maps in the text.*

Dealing first with Manchuria, the author calls attention to the Coal-measures in the neighbourhood of Mukden, where the seams are in the usual fashion intercalated among grits and shales. Small coal-basins are fairly numerous in the Liaotung peninsula; as at Wuhoshin, on Society Bay, where several seams, varying from 3 to 13 feet in thickness, of friable coal have been worked; at Talienwan; at Saimaki, on the Korean frontier, where a seam (3 to 5 feet thick) of crumbly bituminous coal occurs. At Pönnhsihu, south-east of Mukden, five or six seams of coal, each 1 foot or 2 feet thick, have been proved: the conditions appear favourable for working, but the strata are much faulted. The mineral is friable, though not specially bituminous, and yields coke of middling quality. None of these coals are suitable for naval purposes or for railway-locomotives, nor would it pay to export them. Good anthracite, however, has been found in Liaohsi.

In Northern Manchuria, the true Coal-measures are not so well developed, but coal-seams of later age (Triassic, Jurassic and Tertiary), crop out at several localities in the Upper Sungari valley, in the neighbourhood of Kirin, in the Argun valley (where a seam of lignite [?], 23 feet thick, occurs among post-Pliocene sands), etc.

Magnetic iron-ore is of common occurrence in the Liaotung peninsula, and copper- and lead-ores have been discovered in Eastern Manchuria. Gold-placers are of widespread occurrence; and rich auriferous quartz-reefs, some of which are worked, occur in the Santaoku district and at Tsitskuho, near Kirin.

In Korea, the true Carboniferous coals are of less importance than those of Tertiary age. The latter occur over large areas in the north of the peninsula. Unfortunately, the high percentage of ash (31) quoted by the author seems to preclude any hope of remunerative working on a large scale. In the Cambrian series of various localities, good hæmatites, copper-ores, and argentiferous galena are found. Auriferous reefs have been noted in several places, as also rich gold-placers (these latter, apparently, are actively worked). The export of gold from Korea was, in 1900, five times as considerable as it was in 1894.   L. L. B.

COAL-MEASURE PLANTS IN NORTH AFRICA.

Le Terrain Houiller dans le Nord de l'Afrique. By ED. BUREAU. *Comptes-rendus hebdomadaires des Séances de l'Académie des Sciences*, 1904, vol. cxxxviii., pages 1629-1631.

Already, south of the Algerian province of Constantine, red grits had been reported with carbonaceous deposits containing *Lepidodendron Veltheimianum*, a species characteristic of the lowermost Coal-measures; but recently Lieut. Poirmeur has found in the far south of the province of Oran a specimen of the same *Lepidodendron* and one of *Stigmaria ficoides*: the latter had evidently not been drifted very far from its position of growth before fossilization. These specimens both occur in ferruginous grit. It may be observed that the Carboniferous Limestone, with its characteristic fossils, is of widespread occurrence in the region traversed by Lieut. Poirmeur; and it is shown to be practically contemporaneous with the Carboniferous Limestone of England and of Tournay in Belgium.

The plant-bearing beds seen by Lieut. Poirmeur undoubtedly underlie the marine beds, and if they are in their original position (that is, not overfolded or jammed in) they probably belong to the Dinantian or Culm stage. If this be so, the prospect of striking workable coal-seams is, of course, less hopeful. Yet it is not uncommon to find later and richer

Coal-measures in the near neighbourhood of the Culm, and the industrial importance which would attach to a coal-basin, situated close to the projected great trunk-railway from Oran to Timbuctoo, should encourage further investigation in that part of Africa.   L. L. B.

––––––

THE COPPER-ORES OF SOUTH-WEST AFRICA.

Kupfererzvorkommen in Südwestafrika. By J. Kuntz. *Zeitschrift für praktische Geologie,* 1904, *vol. xii., pages* 199-202, *with 6 figures in the text.*

In Little Namaqualand, as elsewhere in South Africa, granite or granitic gneiss forms the basement-rock and is overlain by mica- and hornblende-schists, quartzites, etc. Great masses of iron-ore, especially pyrites and magnetite, occur among these, and may be traced by their gossans over long distances. Although exposed along the coastal belt, this Archæan series is inland mantled over by the sandstones, shales and limestones of the Cape formation, which form tabular hills and have a generally-horizontal lie. However, the greater part of Little Namaqualand, and especially the Ookiep-Concordia mining district, lies within the Archæan belt. This is traversed by two fault-systems, trending respectively east-north-eastward and south-eastward, while the crystalline schists strike east and west. The first-named fault-system defines the fissures along which a plagioclase-hornblende-rock was erupted, while, so far as the author can ascertain, the second fault-system shows no trace of infilling with eruptive material.

The copper-ore is an accessory constituent of the plagioclase-hornblende-rock (which may, perhaps, be termed a diorite). It is noticeable that the proportion of hornblende diminishes as that of copper-ore increases, which rather looks as if the latter had metasomatically replaced the former. The ore is mainly bornite and copper-pyrites, and the form and continuance of the ore-body is irregular: it can be followed, for instance, uninterruptedly as a string of nests in a dark dyke amid the pale-grey gneissose country-rock for miles and miles between Springbok, Kupferberg and Carolusberg, while no outcrop is seen between Nababiep and Ookiep, or between the latter locality and Narap. However, it is only at certain spots that the ore occurs in really workable quantity. The rich nests are, in fact, accumulated at points where the two fault-systems cross one another; points, too, which often mark the occurrence of the older iron-bearing schists, as a comparison of dips and strikes shows that the eruptive rocks cut these schists at a sharp angle.

The Ookiep deposit was 1,082 feet in extreme length, its greatest breadth was 230 feet and its greatest depth 330 feet, and it has now been all but worked out. At Narap, the ore-occurrence is of small extent, and is not being worked at present; the Nababiep deposit, however, is apparently enormous. There the diorite forms a great hill, but forewinning had not proceeded far enough, at the time of the author's investigation, for him to say definitely what proportion of the mass consists of copper-ore. At Tweefontein, near Concordia, three " nests " of ore occur, one on the top of the other, at the second line of eruptive dykes north of Ookiep. The mines formerly worked by the Namaqualand Copper Company lie on the first line of eruptives north of that place: they yielded too little copper-ore, but plenty of magnetite.   L. L. B.

GOLD-DEPOSITS OF THE KHAKHADIAN MASSIF, FRENCH SUDAN.

Sur les Gîtes aurifères du Massif du Khakhadian (Soudan occidental). By H. ABSANDAUX. *Bulletin de la Société française de Minéralogie,* 1904, *vol. xxvii., pages* 81-86.

This massif rises in the centre of the Bambuk country, which is bounded on the east by Upper Senegal and on the west by the Falemé country: the core consists of steep mountains rising to heights of 2,500 and 2,600 feet, girdled round by lesser hills of the tabular form so common in Western Africa. The region is built up of sedimentary rocks of indeterminate age, but probably of ancient date, amid which are intercalated andesitic tuffs, the whole being traversed by igneous intrusions, and more especially by granite. The last-named has given rise to phenomena of contact-metamorphism, while all the other rocks exhibit the effects of powerful dynamic action which has altered the most basic among them into hornblendic schists. On the site of the old gold-mines of Kenieba, this dynamo-metamorphism appears to have been peculiarly intense, and the hornblende-schists are heavily charged with pyrites, a mineral which occurs also in the microgranites in the immediate neighbourhood of the schists, but there only. Both kinds of rock are auriferous, the gold being derived from the pyrites, and their ultimate decomposition-product is an unctuous clay. Some exploration-work was done among these, half a century ago, by the Senegalese Government. An assay of the microganite recently made by the author showed 85 parts of gold per 1,000,000.

At Yatella, lateritic earths derived from an uralitized ophitic gabbro were at one time washed for gold, but the industry has been long defunct. The most important deposit seems to be that of Sadiola, where the natives work the auriferous laterites: here the proportion of gold in the decomposed rock increases with the depth, until a maximum is reached somewhere about the hydrostatic level of the immediate neighbourhood (a level which is subject to considerable variations, in all tropical countries such as that here described). The gold got from the laterites is but partly capable of amalgamation.

The surface of the Khakhadian massif is covered by a thick crust of reddish-brown conglomerate, consisting of quartzose rock-fragments cemented by a ferruginous clay: this diminishes in thickness as one comes down from the hills to the low country, giving gradually place to a clayey loam. The laterites washed for gold underlie this conglomerate.　　　　L. L. B.

GOLD-DEPOSITS OF KATANGA, CONGO FREE STATE.

Les Dépôts aurifères du Katanga. By H. BUTTGENBACH. *Bulletin de la Société Belge de Géologie, de Paléontologie et d'Hydrologie,* 1904, *vol. xviii., Mémoires, pages* 173-186, *with 5 figures in the text.*

Gold occurs in this province in three forms: firstly, in association with silver in the numerous and important copper-ore deposits which the author proposes to describe fully at a future date. Shales, grits and quartzites are in these occurrences so intensely impregnated with carbonates of copper that the ore may be frequently said to constitute the cementing-material of the rocks. The greatest proportion of gold in these hardly exceeds 3 parts per 1,000,000, but the average proportion of silver is 42 parts per 1,000,000: the precious metals occur, however, in nearly every rock-sample, no matter what the level from which it may have been taken.

Secondly, gold occurs in the alluvia of innumerable streams in the south of Katanga, the most important deposits of the kind being those of Kambove, at which locality is also found one of the richest copper-ore deposits in Katanga. Here are four precipitous gorges, where the torrent-beds are bare in the dry season, exposing a gravel largely made up of pebbles of specular iron-ore and malachite; below this gravel comes a sand almost invariably auriferous, richest in its lower portion: a good deal of native gold, too, is accumulated in the hollows and fissures of the bed-rock. The biggest "nuggets" weigh only from 3 to 6 grammes, the gold being usually in the form of flakes, grains and irregular lamellæ. There is also a gold-placer on the top of the plateau in which the above-mentioned gorges have been cut. The author believes that all this gold is derived from the leaching-out, by atmospheric agencies, of the uppermost portions of the adjacent copper-ore deposits. These uppermost portions are probably but the gossan of deeper-lying veins of magnetite, chalcopyrite and auriferous pyrites. At Funguruma, north-west of Kambove, there is another great cupriferous deposit in grit- and quartzite-hills above the river Dipeta: the detritus brought down from these in the rainy season, and in the rainy season only, is found on washing to contain grains and flakes of native gold; and yet, if the precious metal be sought for in the rocks themselves, traces of it, invisible to the naked eye, can be discovered in the cupriferous bands alone. A similar observation holds good of the fine copper-ore deposit at Likasi, south-south-east of Kambove: this has an average content of 22 per cent. of metallic copper, with 41 parts of silver per 1,000,000 and less than 1 part of gold.

The third form of occurrence, which the author believes to be unique among the auriferous deposits so far known, is exemplified at the Ruwe hill, 9 miles or so west of the Lualaba river. Here the gold is found in undoubted sedimentary strata, which nevertheless bear no resemblance to the Witwatersrand conglomerates: one stratum is made up of a limonite with microscopic quartz-spherules. Five exploration-shafts are being sunk in the plateau-like bed of the hill, which slopes more or less abruptly down to a streamlet known as the Kurumashiwa. The surface is more or less thickly mantled by a sort of drift, formed by the decomposition of the underlying rocks, in which gold is always found in the free state. Below comes a succession of quartzites and grits, with a ferruginous conglomerate at the base. Except the two uppermost quartzites, all these contain gold, silver and platinum, the average proportions (as tested by 24 analyses) being 12·3 parts per 1,000,000 of the first-named metal, 8·3 of the second, and 3·4 of the third. In a banded grit, small, absolutely spherical globules of gold, about $\frac{1}{25}$ inch in diameter, occur. In the basement limonitic conglomerate, some samples have assayed as much as 50 parts of gold per 1,000,000.

The strata of Ruwe are apparently conformable with the slightly cupriferous beds of the Konkolo and Kitambala hills; the former are possibly of Carboniferous age, while the copper-ores generally impregnate rocks provisionally regarded as Silurian.   L. L. B.

SUGGESTED ORIGIN OF THE GOLD-BEARING CONGLOMERATES OF THE TRANSVAAL.

Expériences sur la Formation de certains Conglomérats: Origine des Poudingues aurifères du Transvaal. By P. FOURMARIER. *Annales de la Société Géologique de Belgique,* 1904, *vol. xxx., Bulletin, pages* 124-128.

With the view of determining the origin of the cementing-material of certain conglomerates, such as the Palæozoic ferruginous puddingstone of the neighbourhood of Spa, the author has been for some time engaged in a series of experiments. These have led him to enunciate a new theory in regard to the genesis of the gold-bearing conglomerates of the Witwatersrand. According to this hypothesis, the rolled pebbles of quartz and quartzite are derived from the erosion of rocks traversed by veins of auriferous and ferruginous quartz. The association of the gold with pyrites is invariable; whence it is concluded that the rolled pebbles themselves contain gold and pyrites. Now, just as in the author's experiments, when lumps of ferric sulphate, contained within spheres of plastic clay, laid amid a bed of sand, gradually dissolved away, and iron-compounds were deposited in the sand around the spheres of clay—so the gold and the iron in the Witwatersrand pebbles were dissolved away particle by particle, and were reprecipitated outside the pebbles, forming the cementing-material, from which, by metamorphism, was derived the crystallized pyrites.

The author points out that the foregoing hypothesis fits in exactly with all the characteristic features of the Transvaal deposits, as described by Prof. L. de Launay; whereas no other theory so far suggested does fit in exactly with them.   L. L. B.

NEW DIAMOND-DEPOSITS IN THE TRANSVAAL.

Über einige neue Diamantlagerstätten Transvaals. By A. L. HALL. *Zeitschrift für praktische Geologie,* 1904, *vol. xii., pages* 193-199, *with 3 figures in the text.*

The six diamond-mines opened up in the colony at the time of writing, are situated about 22 miles east-north-east of Pretoria in a bare, hilly, quartzite-country, cut off to the south by a valley eroded in softer shaly rocks. The diamantiferous deposits occur among the Magaliesberg quartzites and associated eruptives of the Pretoria series, a subdivision of the Cape system, which is unconformably overlain by the dark-red grits and conglomerates of the Waterberg Sandstone series. The eruptive rocks consist chiefly of sills of so-called " diabase " striking parallel with the Pretoria beds; with this diabase an acidic rock (provisionally termed " felsite ") is associated, especially where " blue ground " occurs. The general dip of the beds is 19 degrees north-eastward, and the strike is south-east and north-west.

The hard " blue ground " is found in the form of pipes, and in a decomposed condition as " yellow ground "; and diamantiferous deposits also occur as alluvial placers. At the Premier mine, " yellow ground," at a depth of 45 feet from the surface, passes into hard " blue ground." A boring put down in the neighourhood of the south-western boundary of the concession passed through " blue ground " down to a depth of 1,000 feet, but other two bore-holes struck barren ground after depths of only 190 and 260 feet respectively. The area of this concession alone exceeds that of all the De Beers claims at Kimberley put together, and the possible output of the " yellow ground," not to speak of the "blue ground," is estimated at more than £10,000,000 sterling. The breccia-like "blue ground" resembles the "hard bank" of Kimberley; it undoubtedly will yield diamonds, but whether so many in proportion as the "yellow ground" remains to be seen.

On the whole, up to March, 1904, in an area of 38 square miles of the Pretoria district, four "pipes" of diamantiferous "blue ground" had been discovered, the biggest being 2,950 feet and the smallest 240 feet in diameter. At least three diamond-bearing placers have been proved in the same district. The hard "blue ground" is, in all probability, a highly-serpentinized peridotite, at one time rich in olivine: where the slates, quartzites, etc., have been burst through, the "blue ground" takes on all the characters of a breccia.

The output of the mines has increased enormously during the past year or so, and there seems every reason to believe that the diamond-mining industry of the Transvaal has a brilliant future before it. The author supplies detailed statistics from July, 1902, up to March, 1904.   L. L. B.

METALLIFEROUS AND OTHER DEPOSITS IN THE ARGENTINE REPUBLIC.

Comunicaciones mineras y mineralógicas. By GUILLERMO BODENBENDER. *Boletin de la Academia Nacional de Ciencias de Córdoba,* 1903, *vol. xvii., pages* 359-381, *with 1 figure in the text.*

The author first of all describes the onyx-marble of the provinces of San Luis and Mendoza, the thickest bed of which measures about 5 feet. The prevailing colour is pale-green, often handsomely streaked with yellow, pink and grey. The onyx contains a good deal of carbonate of iron, and its specific gravity ranges between 2·7 and 2·9. It no doubt originated in the thermal waters, the upflow of which followed on the andesitic and basaltic eruptions of late Tertiary or early Quaternary time. The onyx of the Sierra de San Luis has been worked for many years, and finds a ready sale: but the author believes that unworked deposits still await the explorer in that region. The onyx of San Rafael, on the Arroyo Salado, is of equally-fine quality, and the transportation-facilities which the railway now being built to Mendoza will afford, will doubtless permit of the mineral being worked with profit.

The best-known auriferous veins in the province of Jujuy are those of La Rinconada, a small village situated at a height of about 13,000 feet above sea-level, in a ravine which cuts the eastern face of the Sierra de Cabalonga. Facilities of transport are promising, the timber-supply is good, and water can be got in abundance from the lake of Pozuelos. The gold-quartz reefs occur among black slates and greywackés of Silurian or Devonian age, generally striking north and south and pitching almost perpendicularly. The thickness varies from 1 to 80 inches, though in the Blanca mine a group of reefs attains a combined thickness of 50 feet. A heading has been driven below ground, to a depth of some 52 feet and for a length of 525 feet, cutting at least seven of the reefs: those in the black slates are, generally speaking, richer than those in the greywackés. The inflow of water constitutes a great difficulty, even at so shallow a depth as 33 feet, and the conditions do not permit of draining it off through shafts. Placers also are worked in this locality, yielding in the *batea* 3 parts of gold per 1,000,000.

The author claims to have found "native lead" in the alluvium of these placers, in the form of small nodules, grains, flakes, etc., covered by an earthy-grey pellicule of carbonate of lead. He also describes and gives analyses of the nodules of ulexite (boro-natro-calcite), about the size of a man's fist, which occur to a depth of about 5 feet in the Salinas Grandes of Jujuy. The formation of these nodules, sufficiently important from the industrial point of view, is going on at the present day, owing to the continuous evaporation of the saline waters containing borates in solution. From a locality to the west of Tinogasta, in the province of Catamarca, selenides, sulphides and carbonates of copper have obtained, and it seems probable that the ores there will prove to be of industrial importance.

Cassiterite has lately been recorded from the Cerro de las Minas in the province of Rioja, in ferruginous quartz-veins in a country-rock of gneiss, greisen and granulite, and associated with arsenical pyrites. Water-supply for mining purposes is readily available in the district. The nearest railway-station is Chumbicha, 14 leagues away, on the Recreo-Catamarca line.

New occurrences of metalliferous ores in various other localities in the Argentine Republic are described, but we are not told whether they are in quantity sufficient to repay working. L. L. B.

BRAZILIAN TITANIUM-MAGNETITES.

Über die Mikrostruktur einiger brasilianischer Titanmagneteisensteine. By E. HUSSAK. *Neues Jahrbuch für Mineralogie, Geologie und Paläontologie,* 1904, *vol. i., pages* 94-113, *with* 2 *figures in the text and* 1 *plate.*

The author made an elaborate chemical and microscopic investigation of magnetites from 15 different localities in Brazil. As regards their origin, these ores are divisible into two distinct groups: (1) magnetites derived from acidic eruptives (such as granites and gneissose granites), and generally occurring only as thin reefs or narrow veins; and (2) magnetites derived from basic eruptives (pyroxenites, etc.), especially noticeable in the nepheline-syenite areas, and occurring in great masses, the outcome probably of magmatic differentiation.

The author summarizes the results of the investigation as follows:—A titanium-mineral analogous to ilmenite, corresponding to the formula $(FeMgMn)O.TiO_2$, occurs intergrown with the magnetite, generally following the octahedral faces of the latter, sometimes accumulated irregularly and sometimes in the form of denticulated granules. The last-named prove to be built up of lamellar portions, which are unequally attackable by hydrochloric acid. In addition to these titanium-lamellæ, the residue insoluble in acids contains predominantly a green iron-spinel and a whole series of other accessory minerals, which respectively characterize the derivation of the iron-ore either from an acid or from a basic eruptive. Thus, in titaniferous magnetites of granitic origin, zircon, corundum and monazite appear; while in those of basic origin, perowskite (frequent), baddeleyite, secondary anatase, and a pyrochlore-mineral are found. The titanium-content of titaniferous magnetites, and probably of many ilmenites, is assignable to the usually very regular intergrowth of the aforesaid titanium-lamellæ with the magnetic iron-ore. L. L. B.

ROCK-SALT DEPOSITS, ETC., IN THE REPUBLIC OF COLOMBIA.

Minerales alcalinos y terrosos de Colombia. By R. LLERAS CODAZZI. *Trabajos de la Oficina de Historia Natural : Seccion de Mineralogia y de Geologia*, 1904, *pages* 1-27.

The richest salt-deposit in Colombia, and one of the most important in the whole world, is that of Zipaquirá. Here the mass of rock-salt is stated to be such as to require centuries of working to exhaust it. The mineral is of high quality, being on the whole exceptionally pure, and is worked by means of horizontal galleries. Of equally good quality is the salt of Nemocón, a short distance away, but the methods of working are very defective. Rock-salt deposits are also worked at Sesquilé and Upín. In all the localities mentioned, the associated strata are sandstones, limestones and coal, striking south-westward and north-eastward, and exhibiting a fairly-steep dip (45 to 50 degrees).

Brine-springs are tapped at a great number of localities, and there are vast efflorescent salt-deposits on the Atlantic coast of the Republic.

Deposits of potassium nitrate have been worked from time immemorial in the south of the department of Santander.

Gypsum is one of the minerals most commonly met with in Colombia, and anhydrite occurs in considerable masses at Zipaquirá.

Marble of commercial value is recorded at several localities, especially in the department of Santander, where there is a fine-grained shell-marble, which admits of a high polish, and is comparable with the famous *lumachello* of Italy.

The occurrence of wavellite (hydrated phosphate of alumina) may be mentioned here, as its fine green coloration misled ignorant prospectors, in some localities, into pegging out claims for emerald-mining.

Monazite is found in the sands of the Rio Chico, in the department of Antioquía.   L. L. B.

PETROLEUM-DEPOSITS OF PICHUCALCO, CHIAPAS, MEXICO.

Criaderos de Petroleo de Pichucalco. By MAXIMINO ALCALÁ. *Memorias de la Sociedad Científica " Antonio Alzate,"* 1903, *vol. xiii., pages* 311-326 *and* 1 *plate.*

The Guadalupe estate, where the petroleum is found, lies 7¼ miles south-west of Pichucalco, the chief town of the department of that name, in the state of Chiapas, some 660 feet above sea-level. There is river-communication with San Juan Bautista (the capital of the state of Tabasco) some 50 miles away to the north-east, but the road leading to the same place is only practicable in the dry season, and even then simply for packhorses and mules, not for wheeled traffic.

The petroleum wells out at the surface, on the northern flank of the Cerros del Diablo, one of the most easterly spurs of the great Sierra Madre : the oil-bearing beds strike generally north-east and south-west, dipping 15 degrees north-westward. At the San José oil-well, on the left bank of the Chapopote river, the petroleum is seen to come from a loosely-compacted conglomeration of clayey gravel and sand, intercalated among blue clays. The oil impregnates this bed over a thickness of 7 feet, the impregnation being more pronounced in the lower half. There are indications of another but less important petroliferous band at a higher geological horizon. The actual extent of the principal deposit has not yet been determined, but it has been traced at various points in the area situated between the Chapopote and Guineo rivers.   The crude oil does not contain paraffin or the light benzines, but it is used locally for illuminating purposes. It could be more profitably utilized as fuel, just as it comes from the well.   The author believes that when deep wells are sunk in the district, deeper-lying reservoirs, containing oil which is not bereft of the above-mentioned constituents, will be tapped.   He points out in some detail the striking analogy between the geological (tectonic and stratigraphical) conditions of the district and those of the oil-fields of North America and Baku, the one difference being the clayey nature of the oil-bearing bed already struck near Pichucalco. Naturally, sandy beds are better absorbers of petroleum, and such beds may be found deeper down, possibly much nearer the surface than is the case, say, in Pennsylvania.

A light railway is being built to Cosoayapa; timber and water-supplies are abundant; provisions are cheap, but skilled mining labour is scarce in the immediate neighbourhood. Wages average from 2s. 6d. to 3s. (1¼ to 1½ Mexican dollars) *per diem.*   L. L. B.

ORIGIN OF THE PETROLEUM OF ARAGÓN, MEXICO.

Estudio de la Teoría química propuesta por el Sr. D. Andrés Almaraz para explicar la Formación del Petróleo de Aragón. By JUAN D. VILLARELLO. *Parergones del Instituto geológico de Mexico,* 1904, *vol. i., pages* 95-111.

This paper is an elaborate refutation, on chemical grounds, of an official report published in 1903 by Mr. Andrés Almaraz, on the oil and natural gas found in a well at Aragón in the Federal District of Mexico. That author stated that the petroleum consisted of hydrocarbons, ranging from decane $(C_{10}H_{22})$ to octodecane $(C_{18}H_{38})$, and contained no unsaturated hydrocarbons; further, that the well-water yielded abundance of carbon dioxide and variable quantities of ferrous orthocarbonate. He reviewed several of the current theories as to the origin of petroleum, concluded that none of them were applicable to the Aragón well, and propounded the following chemical theory: The existence of an unstable salt of iron being proved in this well, could not that salt act as a reducing agent on the water and the carbon dioxide, the result of the reaction being the combination of the hydrogen of the water with the carbon of the carbon dioxide? He terms this an " exothermic " reaction, which the present author shows it is not, but distinctly " endothermic."   Now, in order to bring about an endothermic combination, the intervention of some extraneous source of energy is indispensable.   The idea, therefore, that the above-mentioned reaction is going on and will continue indefinitely, of itself, in the Aragón well, is shown to be baseless, the proof that such a reaction takes place at all in nature being still to seek. Thus far, then, there is no evidence of the existence of a petroleum-deposit of any great importance, still less of an " inexhaustible " one, in the neighbourhood of Aragón. It is impossible to give in a short abstract an adequate idea of the closely-reasoned arguments and long series of chemical equations, by means of which the author establishes his main contention.   L. L. B.

THE COAL-FIELDS OF LAS ESPERANZAS, COAHUILA, MEXICO.

The Coal-fields of Las Esperanzas, Coahuila, Mexico. By EDWIN LUDLOW. *Transactions of the American Institute of Mining Engineers,* 1901, *vol. xxxii., pages* 140.156, *with* 6 *figures in the text.*

The Las Esperanzas coal-field was discovered in the spring of 1899. It is situated in the state of Coahuila, about 85 miles south-west of Eagle Pass, Texas. The coal is found in the Upper Cretaceous, and corresponds with the Laramie measures of the United States.

The area of the basin at the south-eastern end, where the seam is of good thickness and not too steep for economical operations, is about 6,000 acres. Work was commenced in November, 1899, with the view of developing the field for an output of 5,000 tons per day. In the course of 2 years, the necessary development-work has been done and machinery erected for obtaining this output, but great difficulty has been experienced in obtaining suitable labour.

The coal is a soft bituminous coking coal, containing 9·8 per cent. of ash. The seam is from 8 to 9 feet thick, and contains various bands of shale, aggregating 14 to 17 inches in thickness.

A coking-plant, consisting of 224 ovens of the bee-hive type, has been erected, so as to supply coke to the smelters of Northern Mexico. The present output from the mines is 1,200 tons per day.   R. W. D.

THE IRON-ORES OF IRON MOUNTAIN, DURANGO, MEXICO.

The Iron Mountain, and the Plant of the Mexican National Iron and Steel Company, Durango, Mexico. By T. F. WITHERBEE. *Transactions of the American Institute of Mining Engineers,* 1901, *vol. xxxii., pages* 156.163, *with* 1 *figure in the text.*

The Iron mountain, situated ¾ mile north-east of the limits of the city of Durango, rises abruptly from a level plain. The mass of solid ore, about 1¼ miles long, ¼ mile wide and from 200 to 400 feet high, is estimated to amount to 360,000,000 tons of iron-ore containing from 60 to 65 per cent. of iron. A blast-furnace has been erected in close proximity to the mountain. The fuel used is principally charcoal, but coal is occasionally mixed with the charcoal. The rolling-mill has 5 double puddling-furnaces. The iron produced is of good quality. All the labour, skilled and unskilled, is Mexican throughout, excepting the heads of departments. The iron-ore is also used as a flux by Mexican lead-smelters.   R. W. D.

THE ORE-DEPOSITS OF THE SIERRA MOJADA, COAHUILA, MEXICO.

The Sierra Mojada, Coahuila, Mexico, and its Ore-deposits. By JAMES W. MALCOLMSON. *Transactions of the American Institute of Mining Engineers,* 1901, *vol. xxxii., pages* 100-139, *with* 14 *figures in the text and* 1 *plate.*

The Mojada mines, situated in the midst of an extensive desert, were discovered in 1878. At first, they were worked in a very primitive way, and the ore was carted 75 miles to the railroad. The average grade of ore now being worked is:—10 ounces of silver per ton and 15 per cent. of lead. The production in 1900 was 191,000 tons.

The mountains forming the Sierra Mojada are composed entirely of limestone. The ore-deposits may be divided into three main groups:—(1) The contact-deposits; (2) the lime-impregnations; and (3) the lead-carbonate ores in the limestone. Detailed descriptions are given of the deposits.

In the early workings very little timber was used, and the miners contented themselves with following up the rich ore-streaks. The excavations are now timbered with square sets, and by the introduction of modern methods it is found profitable to re-work the old and but partly exhausted stopes throughout their entire extent. Labour is fairly abundant, and costs 3s. (1½ Mexican dollars) per day. The climate is healthy and pleasant.

R. W. D.

GENESIS OF THE MERCURY-DEPOSITS OF PALOMAS AND HUITZUCO, DURANGO, MEXICO.

Genesis de los Yacimientos mercuriales de Palomas y Huitzuco, en los Estados de Durango y Guerrero de la Republica Mexicana. By JUAN D. VILLARELLO. *Memorias de la Sociedad científica "Antonio Alzate,"* 1903, vol. xix., pages 95-136.

The author shows, in the first place, how thermal waters containing in solution carbonic acid, sulphide, sulphate, thiosulphate and carbonate of soda, are perfectly capable of dissolving all the minerals usually found associated together in mercury-deposits; and, after devoting some pages to the purely-chemical side of the question, he proceeds to give a brief description of the Palomas deposit in the state of Durango. It occurs at the zone of contact between rhyolites and basalts, some 50 miles west of the capital city of the state. Contact-metamorphism has partly altered the rhyolites into a white clay, raddled in places by iron peroxide. Within the clay occur irregular veins of amorphous silica, with which in places cinnabar is closely intermixed: occasionally the whole vein is coloured red by the ore, and the neighbouring clay, sometimes for a breadth of about 3 feet on each side of the vein, is impregnated with cinnabar. Bituminous substances are found in various parts of the deposit, and also a small quantity of native mercury. The horizontal extent of the ore-body is small, say barely 1,000 feet, and in depth it does not usually go down beyond 20 to 25 feet. The richness of the crude ore is variable, from 0·05 to 5 per cent. of metallic mercury being the usual assay, and very few specimens assay to as much as 40 per cent. of the metal. The ore-body is bounded both on the east and west by very slightly-altered rhyolites; it is manifestly later in date than the eruption of these, and is consequently of Tertiary age. It was evidently formed by thermal waters which deposited, in fissures opened up in the rocks, some of the metalliferous material that they held in solution. It cannot possibly have been formed by sublimation.

The Huitzuco deposit lies about two-thirds of a mile south of the locality of that name, and 17 or 18 miles east of Iguala. It occurs in the Middle Cretaceous limestones, which are bounded on the west by the andesitic eruptives of Noxtepec and Taxco. The irregular cavities (fissures enlarged by the action of thermal waters) in the limestone are in filled with cinnabar, livingstonite or sulphantimonite of mercury, pyrites, sulphur, gypsum and occasionally metacinnabarite or black sulphide of mercury, also ferruginous clay impregnated with cinnabar. Both the ore-body and the contiguous country-rock are impregnated with organic substances. At various levels, but more especially at the depth of 590 feet, emanations of sulphuretted hydrogen are observed. Nearer the outcrop, the sulphantimonite of mercury and the pyrites appear to have undergone oxidation. The whole deposit forms a series of pockets occupying a horizontal breadth of 260 feet or so, and extending down to a depth of at least 820 feet (the lowest level of the La Cruz mine in 1896). There is a gossan of red clay impregnated with cinnabar, seamed with veinlets of calcite. The distribution of the minerals previously enumerated is very irregular throughout the deposit. The amount of metallic mercury and antimony obtained from the ores appears to diminish as the workings are pushed deeper downward, varying from 5 to 10 per cent. down to ¾ and 1 per cent. of mercury, and from 40 per cent. down to 10 per cent. of antimony. It is evident that the Huitzuco deposit, like that of Palomas, is the outcome of the percolation of thermal waters, and in this case too, the sublimation-theory must be dismissed as impossible. The deposition is probably connected with the volcanic phenomena of Tertiary times, whereof the andesites of Noxtepec and Taxco are the silent witnesses. The *dictum* of Prof. Fuchs and Prof. De Launay may be re-called, in regard to mercury-deposits all over the world, that they are, geologically speaking, of recent formation as a rule.

The author pictures, in considerable detail, the connection between the phenomena of moribund vulcanicity and the various chemical processes and reactions whereby the mercury and other associated ores were finally precipitated in their present *locus*.	L. L. B.

THE ORE-DEPOSITS OF ANGANGUEO, MICHOACÁN, MEXICO.

El Mineral de Angangueo, Michoacán. By EZEQUIEL ORDÓÑEZ. *Parergones del Instituto geológico de Mexico*, 1904, *vol. i., pages 59-74, with 4 figures in the text and 1 plate.*

The mining district described in this memoir is situated on the western spurs of an extensive mountain-range on the borders of the states of Mexico and Michoacán. The massif is built up of an enormous thickness of more or less metamorphosed slates and limestones, presumably of Cretaceous age: with these are associated two groups of eruptive rocks:—(1) Miocene andesitic " greenstones," and (2) basalts, the outpourings of the volcanoes which have been active in the region from Pliocene down to recent times. Both the Cretaceous slates and the Miocene andesites are traversed by metalliferous veins of varying composition and structure. Two localities are the scene of great industrial activity: El Oro, where gold-bearing quartz-reefs occur in the slates, buried up at some points by the later basalt-flows; and Tlalpujahua, where thick silver-bearing veins traverse similar slates and highly-metamorphosed slaty " greenstones." Third in order of importance, and probably of antiquity, comes Angangueo, with its comparatively-thin metalliferous veins seaming the andesitic eruptive rocks. These veins are generally poor in silver, and the chief ores are sulphides of iron, lead and zinc. Four main veins are recognized, striking between 10 and 40 degrees north-east, and dipping variously east and west: they are mutually connected by numerous branch-veins, which do not strike off from the main leaders at angles much exceeding 30 degrees. There is every reason to infer that they all belong to one system of fissures, originated by a single tectonic " spasm." The gangue is essentially a milky-white quartz, the transparent form of the mineral occurring but seldom. Manganese occurs in the shape of pink carbonates and silicates, and the commonest ore is a very pale pyrites, sometimes mixed with marcasite. Blende is abundant, usually black, and its increasing abundance is generally an indicator of the impoverishment of the vein: on the other hand, the darkest blende always yields a larger percentage of silver. Galena occurs in big masses, but has diminished to such an extent in the veins most recently worked, that not enough is got to keep the smelters going. A mass of pure galena is being worked on the San Cristobal vein, and such masses usually yield an average of 0·2 per cent. of silver. The three sulphides which have been described were evidently deposited contemporaneously in the fissures, as they are intermixed in the most irregular manner. The topographical situation of Agangueo and the abundance of water in the mines have prevented the workings from being carried to any great depth so far, but the modern miner is now equipped with more powerful weapons for fighting these difficulties. The average silver-content of the ores varies from 0·09 to 0·1 per cent., and the output of ore amounts to about 500 tons weekly. Some 1,800 persons are employed in and about the mines.

L. L. B.

ASBESTIFORM MINERAL FROM MICHOACÁN, MEXICO.

Estudio de una Muestra de Mineral asbestiforme procedente del Rancho del Ahuacatillo, Distrito de Zinapécuaro, Michoacán. By JUAN D. VILLARELLO. *Parergones del Instituto geológico de Mexico,* 1904, *vol.* i., *pages* 133-149.

This yellowish-white mineral, described as being of foliated texture, the folia being flexible but not elastic, resembles in its general aspect " fossil cork " or " mountain-leather." Its hardness is 2·5 and its specific gravity 2·18; and analysis shows it to be a hydrosilicate of alumina with carbonate of lime, the chemical composition being expressed by the formula $H_{10}Al_2Si_3O_{14} + CaCO_3$. It belongs to the kaolin-group and approximates to montmorillonite. Despite its outward resemblance to many asbestiform minerals, it differs widely from them in its composition and in its properties. The industrial uses for which it could be made available do not cover a very wide field: it might be used for boiler-covering, or it might be manufactured into inferior firebricks, but this would be an expensive process. The sample analysed was obtained on the Ahuacatillo Ranch, in the district of Zinapécuaro, state of Michoacán.

L. L. B.

THE MINING DISTRICT OF PACHUCA, MEXICO.

The Mining District of Pachuca, Mexico. By EZEQUIEL ORDÓÑEZ. *Transactions of the American Institute of Mining Engineers,* 1901, *vol.* xxxii., *pages* 224-241.

The mining district of Pachuca was discovered in 1522, and mining operations have been carried on there since that date with varying energy and success. It is estimated that Pachuca has produced more than 8,000,000 pounds of silver. The Pachuca range is formed of volcanic Tertiary rocks, and the crest of the mountains runs approximately north-west and south-east. One system of fissures, running more or less east and west, comprises the principal veins of the Pachuca district. In these, lode-quartz forms the principal part of the mass. The width seldom exceeds 22 feet, and some of the lodes extend for a distance of 10 miles. The rich ore occurs in *bonanzas* of varying shape and size. One of the largest of these, at San Rafael, is elliptical in form, the greatest axis being more than 3,000 feet long and the smaller 1,200 feet, with an average thickness of 8 feet; this *bonanza* has produced nearly £2,800,000 during a period of 10 years. In Pachuca, the impoverishment of the veins at great depths is admitted to be a fact, but the author believes that by carrying on investigations at a greater depth, new *bonanzas* might be discovered.   R. W. D.

THE VANADIUM-ORE OF CHARCAS, MEXICO.

El Vanadio de Charcas. By GUSTAVO DE J. CABALLERO.   *Memorias de la Sociedad cientifica " Antonio Alzate,"* 1903, *vol. xx., pages* 87-98.

The ore occurs in the form of waxen-yellow crystalline needles of vanadate of lead, set closely parallel one to the other in fairly-continuous streaks, in a mine hitherto worked for silver-ores. As a matter of fact, the vanadate may be regarded as an incrustation on the walls of the argentiferous reef. The mineral has a hardness of 3·5 in the accepted scale, and a specific gravity varying between 6·20 and 6·25. Analysis shows it to contain 54 per cent. of lead, 13·3 per cent. of vanadium, 4·2 per cent. of zinc, 4·1 per cent. of copper, 2 per cent. of phosphorus, and 4·8 per cent. of arsenic. Sesquioxide of manganese occurs in small quantities as an impurity, evidently mixed, but not combined with, the other constituents. The chemical composition is expressed by the formula $PAs(VO_4)4Pb_4CuZn + H_2O$, and the author defines the mineral as true ramirite, a variety of descloizite. The mineral is regularly worked and exported to France; thus, between January 1st and August 28th, 1903, about 3¼ tons of ramirite were exported from Charcas (in the state of San Luís Potosí), with an average content of 10¼ per cent. of vanadium.

                                                         L. L. B.

MINERAL RESOURCES OF CAJATAMBO, PERU.

La Minería en Cajatambo. By MARIANO LEZANIBAR.   *Boletin del Ministerio de Fomento,* 1903, *No.* 9, *pages* 14-18.

The argentiferous mines of Chanca and Anamaray were worked with successful results until quite lately; indeed the reason of their recent abandonment has not been made known, to the Peruvian Government at any rate. Smelters remain actively employed in the province, and new lixiviation-works are in course of erection on the Ututo Estate. The fall in the price of silver and difficulties of transportation appear to be the chief obstacles to the development of the mineral industry in the province of Cajatambo. On the other hand, the cheapness of labour, the abundance of fuel, the mildness of the climate, and the wealth of its minerals in lead, copper and silver give promise of a brilliant future, so soon as the above-mentioned obstacles are surmounted.

In addition to the ores just enumerated, wolframite is largely worked in the district of Chiquian; graphite occurs abundantly at Auquimarca and Andajes (localities noted also for iron-ores, more especially hæmatite); gypsum is found in abundance at Churin, and good fire-clays at Pumahuaín and Ututo. Silver-bearing antimony-ores are reported from the village of Roca, district of Ticllos, and gold-bearing veins occur at Ambar, 14 leagues from the Pacific seaport of Huacho.

Finally, it is asserted that the coal-bearing belt of Oyón (a locality distant 30 leagues from the same port) is one of the most considerable of the kind in the world. Over an area measuring 48 miles from south-east to north-west, and 34 miles from north-east to south-west, range three continuous mighty seams of coal, each varying in thickness from 65 to 165 feet. They are intercalated among Jurassic sandstones, and are well exposed in the five great ravines which traverse the coal-field and converge on Oyón. At some localities, the seams are divided by so thin a parting of sandstone as to form one huge mass of coal, 2,000 feet thick and more. (In reality, we are dealing here with seams repeated over and over again by tightly-packed folding.) The coal is of good quality, and varies in character from hard anthracite to a soft bituminous mineral; it never yields more than 5 to 7 per cent. of ash. The author recommends his Government to undertake a scientific investigation of a coal-field which is evidently destined to prove a source of great wealth to Peru.   L. L. B.

MINING IN THE PROVINCE OF DOS DE MAYO, PERU.

(1) *La Minería en la Provincia Dos de Mayo.* By Estenio J. Pinzás. *Boletín del Ministerio de Fomento*, 1903, *No.* 11, *pages* 32-44.

This province occupies a high plateau flanked by the Western and Central Cordilleras, and watered by the Río Marañon and its tributaries. In the Cordilleras and in the spurs which run down from them, walling in many a deep valley, are countless metalliferous reefs and veins, belonging in part to a north-and-south group and in part to an east-and-west one, and therefore crossing each other at right angles. Between the Western Cordillera and the Marañon basin occur outcrops of anthracite, bituminous coal, and petroleum-bearing limestone in exceptional abundance and of excellent quality. The seams strike north-west and south-east, and dip south-westward, coinciding as to their strike with the silver-bearing metalliferous veins.

The author devotes the major part of his memoir to the western portion of the region, because it lies nearer to Huallanca, and has been explored in greater detail. Mines have been worked here for pyrites, blende, galena, cinnabar and richly-argentiferous grey copper-ore. The great Lluiyag reef, of an average thickness of 6½ feet, has at the outcrop an infilling of iron-pyrites, blende and galena: it is, however but little worked, and one gathers that the conditions do not permit of shaft-sinking. West of this, in a central belt bounded by great fault-fissures, range:—(1) The Mercedes reef, 1 to 6½ feet thick, yielding iron- and copper-pyrites, and richly argentiferous tetrahedrite, with an appreciable proportion of gold; (2) the Pozo Rico reef, very similar to the former in thickness and in mineralization; and (3) the gigantic Pucacyacu or Huanzalá reef, attaining at some points a thickness of 131 feet, with a visible extension of 75 miles, and having parallel to it a sort of accessory reef about 6½ feet thick: the infilling consists mainly of sulphidic ores of copper, zinc and lead. Five-hundred mining concessions have been taken out in this central belt, but a large area remains still untouched. West of the central belt again, and coincident in fact with the chain of the Western Cordillera, is a strip some 80 miles long and 1½ miles broad, where the ores (although rich enough in copper) are too poor in silver to repay the cost of export. It may be observed that, for many years, the mines in that part of Peru have been worked, simply with the view of getting ores rich enough in silver to justify their export in the crude state to

Great Britain and Germany. East of the central belt before-mentioned is a strip about 4 miles wide, characterized by numerous seams of anthracite, varying in thickness from 4 to 10 feet. In the very midst of the strip is situated the town of Huallanca, and the anthracite crops out in the surrounding hills. The mineral is of good quality, and is used both for industrial and for household purposes. Farther eastward, the metamorphic quartzites are folded into synclines, and anthracite is replaced by a noncoking coal, as far as Huacoto. Here the quartzites disappear, and are replaced by limestones, folded into anticlines and synclines, and extending over a distance of nearly 100 miles. Among them occur petroleum-bearing beds, and seams of a bituminous, highly-tumescent coal, which yields a very light porous coke. Thick seams of good bituminous coal have been recorded in other districts, as, for instance, in the Magapata range, where no less than seven seams occur, varying in thickness from 2 to 6½ feet. Gold-placers are worked in various localities, and the author points out that the following deposits have never yet been worked at all: the titaniferous iron-ore of Yanas; the stibine of Llama-ragra and Ventanilla; the magnetite of Chuspi; the native copper of Huamash; and the native sulphur and elaterite (elastic bitumen) of Chonta; etc.

On the whole, the mineral industry of the province bears at present no sort of proportion to its enormous mineral resources. The needful capital, opines the author, can hardly be attracted thither until a Government commission of engineers has instituted a thorough and impartial enquiry into the facts, noting carefully whatever appears to offer a favourable opening to the capitalist.

(2) *Informe preliminar sobre la Veta Pozo Rico, en la Provincia Dos de Mayo.* By M. A. DENEGRI. *Boletin del Ministerio de Fomento,* 1903, *No.* 7, *pages* 80-91, *with 1 figure in the text and 1 plate.*

The mines which work this reef are situated in the mountains west of the village of Huallanca (11,625 feet above sea-level), in the province of Dos de Mayo, department of Huánuco. The district of Huallanca has been farfamed for its mineral wealth for well-nigh 130 years, and abandoned mines and broken-down smelting-works are seen on every hand. That they are not now in active operation is largely due to the absence of roads, whereby the ores could be transported to the Pacific seaboard.

On the Pozo Rico reef, however, seven concessions are being worked: it really consists of two parallel bands, with a barren parting of 330 feet, interbedded, to all appearance conformably, with quartzites and greywackés. Thin bands of black shale are in direct contact with the reef, and their presence should rather tend to facilitate the working thereof. The reef thins out and thickens again with such extraordinary irregularity, that the author does not venture to estimate its average capacity. Both parts of the reef consist of a quartzose gangue impregnated with silver-bearing sulphantimonite of copper (tetrahedrite) and iron-pyrites. That portion of the ore which is sent away for export, represents the richest tenth of the whole output, and contains from 15 to 20 per cent. of silver and 14 to 17 per cent. of copper. The ore worked upon the spot, on the other hand, contains only 0·15 to 0·25 per cent. of silver and 1½ to 2 of copper.

So far the workings have only touched the surface of the deposit, and the author, in contradistinction to the popular opinion, holds that it continues in depth. He is incredulous, however, of its real conformity with the country-rock, and believes that lower down the reef cuts across the bedding.

The smelting-works of La Florida treat locally 70 tons of ore per month. Outcrops of coal (anthracite, bituminous gas-coals and coking coals) are abundant in the neighbourhood, but no one troubles to work them at present. The moment the cart-road, now under construction, is completed, a good trade can and will be done with Supe harbour on the Pacific. Labour is dear and unsatisfactory, and the author speaks severely of the indiscipline and immorality of the native *peones*. Timber and building materials, provisions and good pasturage for cattle are easily available. Salt, sulphur and chemicals are brought over the mountains by pack-mules from the Pacific seaboard, and the price of such articles is consequently much enhanced. The author advocates the syndicating of the mines at present at work, so as to enable operations to be conducted more economically and on a larger scale. L. L. B.

MINERAL RESOURCES OF THE PROVINCE OF HUÁNUCO, PERU.

Recursos minerales de la Provincia de Huánuco. By NICANOR G. OCHOA. *Boletín del Cuerpo de Ingenieros de Minas del Perú*, 1904, No. 9, pages 1-43 and 6 plates.

In the district of Huánuco itself are reefs of ferruginous gold-quartz, in a country-rock of mica-schist, partly worked in former days, but apparently of no great industrial interest now. The district of Ambo seems to be rich in metalliferous ores, and in thermal springs yielding medicinal waters. The ores are silver (with varying proportions of gold), iron- and copper-pyrites, galena, etc. Near Chauchac are outcrops of anthracite, which is, however, so full of impurities that it cannot give rise to any hope of working it at a profit. Reasons are also given for the author's opinion that it would not be practicable to work the gold-placers of the El Valle district on a large scale (by hydraulicking): the occurrence of gold is very widespread in the alluvial deposits of this district. The Rondoní mountain in the Cayna district is seamed with narrow veins of brown hæmatite, galena and blende, and there is just a chance that skilful prospecting may hereafter reveal some metalliferous deposit of importance. Outcrops of coal of excellent quality are known to exist about 2 leagues away from the village of Margash, but the author was unable to visit the locality on account of the difficulty of securing guides. Magnificent specimens of the coal were shown to him.

On the whole, however, it is rather a melancholy tale that he has to tell, of the province and its mineral resources: wasteful working in the ancient days, so that mine after mine has caved in, and could not now be reopened; scores of mining concessions, which are practically valueless from the industrial point of view; and so forth. The staple industry at present is agriculture, which is flourishing. L. L. B.

THE MINERAL FIELD OF HUAROCHIRÍ, PERU.

La Minería en Huarochiri. By J. L. ICAZA. *Boletín del Ministerio de Fomento,* 1903, No. 7, pages 91-92.

Out of 200 metalliferous mines opened up in this area, 18 only are now in active operation. Most of them work silver, copper and lead-ores, and in a very few gold is got. Coal of excellent quality has been proved in the districts of San Lorenzo de Quinte and San Mateo y Carampoma;

numerous other outcrops are known to exist, but remain for the present unexplored. A cinnabar-mine, long worked in the Carampoma district, is now abandoned, and the workings have fallen in. Tin, zinc and iron-ore deposits are abundant, but, under present conditions, fail to attract the serious attention of prospectors. The want of capital is the sole obstacle to the revival and further development of a prosperous mineral-industry in the province of Huarochirí. L. L. B.

NICKEL AND TUNGSTEN-ORE DEPOSITS IN PERU.

Los Yacimientos de Niquel de Rapi y los de Tungsteno de Lircay. By EDUARDO A. V. DE HABICH. *Boletin del Cuerpo de Ingenieros de Minas del Perú, No. 11, 1904, 39 pages, with 23 figures in the text and 2 plates.*

The nickeliferous deposits are situated in the rugged, mountainous province of La Mar, in one of the most remote districts in Peru, the journey thither from Lima occupying from 12 to 14 days. The outcrops are not always easy to locate, on account of their being masked by the luxuriant vegetation, but they have been traced over a distance of 12 miles or more on the Rapi estate, near the Rio Pampas. The general strike is at first east and west, and thereafter invariably north-north-east and south-south-west; the dips of the reefs or veins are variable both in amount and in direction. In thickness, the metalliferous deposits range from 2 to 50 inches, and the most abundant are the sulphantimonides, sulpharsenides, arseniates and hydrosilicates of nickel, with which are variously associated smaltine, native silver, pyrites, galena and spathose iron-ore. The gangue is almost invariably quartz, and in some cases this is gold-bearing. At one locality, the "country" is spoken of as slaty rock, but, as a rule, it would appear to be a "contact-rock," the outcome presumably of metamorphism induced by some igneous intrusion. Only one analysis is cited, of a sample which yielded 20¼ per cent. of metallic nickel with traces of silver.

Water-supply is abundant, and the fall of the mountain-torrents and rivers is so considerable as to be capable of furnishing adequate motive power for mining operations and ore-treatment on a large scale: it may be mentioned that the Rapi estate lies at a level of 11,480 feet above the sea. Skilled labour for the mines would be doubtless available from the neighbouring province of Huancavelica. Such roads as there are may be said to have made themselves: the opening-up of proper means of communication is imperative, not only on account of the nickeliferous deposits, but on account of the other minerals (gold, silver, copper, coal, etc.), which are reported as existing in abundance in the mountainous region here described.

The tungsten-deposits of Julcani are situated some 3 leagues distant from Lircay, the chief town of the province of Angaraes, which is reached (partly by railway, partly by road) in some 5, 8 or 10 days, according to circumstances, from Lima. The deposits consist essentially of two great reefs, the upper called the Rosary and the lower the Souls (Las Animas), the intervening mass of rock being full of metalliferous veins cutting each other at all angles. The strike is west-north-westerly and east-south-easterly, sensibly parallel with the crests of the mountain-ridges, and the pitch is practically vertical. The name "Rosary" is applied to the upper deposit, because it consists of a series of "pockets," about 5 feet broad, connected by metalliferous "stringers," like huge beads loosely strung together: in it the wolfram-ore is more abundant than in the lower. In both the gangue is quartz, sometimes auriferous, and pyrites is locally abundant. The ore is a tungstate of iron and manganese, steely grey, with a metallic lustre; the hardness is 5·5, and the specific gravity varies from 7 to 7·5. On analysis the ore is shown to contain 55·9 per cent. of tungstic acid and 5·9 ounces of gold per ton. The country-rock is diorite, abundantly pyritiferous, and containing in many places masses of very pure kaolin. East of the tungsten-deposits are considerable outcrops of somewhat impure pyrolusite.

The climate is extremely cold, the water-supply is not precisely abundant, but the roads (although of steep gradient) are fairly good. As mining is the principal industry of the province, skilled labour at moderate wages is easily available. So far, the Julcani deposits have only been worked for the sake of the gold, which is best extracted from the ore by the chlorination-process.

L. L. B.

THE MINERAL RESOURCES OF THE PROVINCES OF MOQUEGUA AND TACNA, PERU.

Informe sobre la Provincia Litoral de Moquega y el Departamento de Tacna. By FRANCISCO ALAYZA Y PAZ-SOLDÁN. *Boletín del Cuerpo de Ingenieros de Minas del Perú, No. 3, 1903, 123 pages and 6 plates.*

This memoir embodies the results of an official mission, with which the author was entrusted by the Peruvian Government, to enquire into the mineral resources of the regions mentioned in the title.

Taking first the district of Moquegua itself, beds of rock-salt, frequently associated with gypsum, appear to be of fairly-widespread occurrence. Some 13½ leagues south-east of the town, in the ravine of Toquepala, are mines working ferruginous quartz-reefs in the granite, which are mineralized with copper-ores (chrysocolla, malachite, azurite, cuprite, etc.). The percentage of metallic copper varies from 10 to 11, with mere traces of silver. The investment of a small amount of capital in judicious prospecting is recommended, as likely to secure a profitable return. The 70 miles of railway from Ilo to Moquegua, constructed in 1873, was destroyed by the Chilian troops in the war of 1880, and its reconstruction is now imperative.

In the district of Ilo, cupriferous veins are, and have been, worked in the granites, 4 leagues north of the town that gives its name to the district. The ore from the San Juan vein assays on an average to 20 per cent. of metallic copper; that from the San Juanito mine yields 28 per cent. of metallic copper and 20 parts of gold per 1,000,000. The facilities in regard to transport and fuel-supply are excellent: the two main requisites are modern plant and methods of working, and a reasonable amount of fresh capital. The last statement applies with additional emphasis to the mines formerly worked by a Chilian company, in the Cilatilla range, 2½ leagues north-east of Ilo.

The most important rock-salt deposit in the whole of Peru is opened up at a distance of 7 leagues from Pacocha, and 1½ leagues from the Pacific seaboard. The salt is of prismatic structure and of great purity. Other rock-salt deposits occur at Loreto, Osmore and in the Pampa Colorada.

In the district of Torata, prospecting work has been done in the Talabaya gulch, on what appeared to be very promising veins of argentiferous galena, but the quantity of ore in sight is too small to repay working. The same remark applies to the metalliferous deposits of Huairurí.

In the district of Carumas the author draws particular attention to the coal-bearing beds, which extend, indeed, into the districts of Ichuña, Ubinas,

Puquina and Omate.   The general strike in Carumas is north-east and south-west, with a south-easterly dip varying from 0 to 90 degrees.   The strata have been much dislocated by volcanic phenomena.   Nine analyses of Carumas coal are given, including in that term lignite, bituminous coal and anthracite.   The quality is good, and the amount of ash (in nearly all cases) surprisingly small.   But before there is any chance of successful mining operations in the district, a railway must needs be built and tolerable cart-roads provided.

The people of the neighbourhood work the native sulphur-deposits associated with the extinct volcano of Ticsane, but the sulphur is not of particularly excellent quality, nor does it occur in such quantity as to repay working on a large scale.   Hot springs of medicinal value occur at Putina and Cadena.

The district of Ichuña appears to be characterized by roads nearly as bad as those of Carumas, and by the absence of bridges.   The most distinctive outcrops of the coal-bearing beds are in the Pubaya gulch, where three seams are seen, of a total thickness of 7 feet.   Five analyses of coal are given, the fixed carbon averaging 70 per cent.; volatile matter, 17 per cent.; and ash, 14 per cent.   The coal is of a semi-bituminous character, and less pyritous than that of Carumas: the strata are, moreover, less disturbed than in the last-named district.   Silver, lead and copper-ores were worked in the decade from 1840 to 1850, but were then abandoned, and the workings have mostly fallen in.

The same tale of bad roads is recorded from the district of Ubinas as from the others just described.   The author devotes a good deal of space to considerations on the great volcano of Ubinas, which rises to a height of more than 16,000 feet above sea-level.   Deposits of native sulphur of excellent quality are worked in a primitive fashion by the Indians of the neighbourhood, and the great borate (ulexite)-deposits of the Pampa de Salinas are well-known.   Three analyses of the boro-natro-calcite are given, and the author states that the important workings of the British company, suspended for some time past, will now be restarted.

No serious work has yet been done on the anthracite and other coals of Querala: nor, of late, on the blende and galena of the Chimbuyo range.   The question of transport is, in this connexion, a difficulty which appears likely to prove insurmountable for a long time to come.

The coal-bearing beds of the districts of Puquina and Omate are regarded by the author as unworkable.   At Omate and Ullucán are thermal springs of medicinal value.

The usual evidences of moribund vulcanicity, in the shape of geysers, thermal springs, and immense deposits of native sulphur, are not wanting. In the department of Tacna, copper-ores of good quality occur in the Tojenes gulch, and would repay working on a large scale, and concentrating on the spot.   Similar ores, some very rich in silver, occur in other districts, but the workings of Mecalaco (after a spasmodic attempt at revival in 1898) have been finally abandoned.   L. L. B.

PETROLEUM IN THE PROVINCE OF PAITA, PERU.

El Petróleo en Paita.   *By* P. I. SEMINARIO *and* F. O. LOPEZ.   *Boletín del Ministerio de Fomento,* 1903, *No.* 2, *pages* 16-19.

The Tertiary strata of this province are rich in sulphur, salt, gypsum and petroleum.   The last-named is the only mineral product worked there at present, and the industry has already attained great dimensions. There is evidence to show that the oil-belt extends right across the province northward into the neighbouring province, where the oil-wells of Zorritos are in full yield. The oil occurs in coarse sands at very various depths, and is associated with a large amount of bitumen, veins of which are traced into the hills, from 2 to 4 leagues distant from the Pacific seaboard. In some deep ravines between these and the ocean are outcrops yielding a petroleum, which, owing to the extreme heat of the sun in that climate, has lost its volatile constituents and consequently its inflammable properties.

It is only in the northern portion of the province that the petroleum-deposits are being worked at present; at Negritos, $6\frac{1}{4}$ miles south of Talora, and at Lobitos, north of that town. At Negritos, the oil is got from 79 wells (many of which are on the sea-beach) at depths varying from 300 to 1,400 feet below the surface. The oil got from the greatest depths is rich in kerosene and benzine. Natural gas occurs in great abundance, and is used on the spot as fuel. At present, the output of oil exceeds the local demand for it, but the rapid development of manufacturing industries on the Pacific coast is expected to change this condition of affairs before long.

L. L. B.

IRON-ORES OF TAMBO GRANDE, PIURA, PERU.

Los Yacimientos de Fierro de Tambo Grande. By PEDRO C. VENTURO. *Boletín del Cuerpo de Ingenieros de Minas del Perú,* 1904, *No.* 8, 37 *pages and* 6 *plates.*

Tambo Grande, the chief town of the district of that name, in the northernmost department of Peru, namely Piura, is 22 miles east of Sullana, where the railway to Payta (the nearest harbour on the Pacific seaboard, 38 miles distant) may be reached. Water is, on the whole, scarce in the district; really abundant rains occur only every 5 or 7 years, and then much of the land usually given up to goat-pasturage, is temporarily turned into arable. Tambo Grande itself, however, lying on the banks of the Piura river, is surrounded by fields of cereals and market-gardens (wherefore provisions are cheap), and labour for mining purposes should not be difficult to obtain.

The iron-ore forms an entire hill astride of the road from Piura, causing the road to bifurcate; one of the branches leads straight into the town, and the other leads to the Tambo Grande farmstead. The ore crops out in the very streets of the township, and is exposed in the central square, the town being really built over part of the ore-deposit. The ore is a red hæmatite varying in structure, for it is in places compact, in others earthy, and in yet others cavernous. It often contains quartzitic and jaspery inclusions, or concretionary masses of calcite. Two samples of ore enclosing such extraneous material, nevertheless, yielded 36·4 and 26·2 per cent. of metallic iron respectively. These results are far from representing the average quality of the hæmatite, which is generally pure, containing (as shown by assay) 51·5 per cent. of metallic iron. It is free from phosphorus, and its specific gravity averages 3·21. The scarcity of good stratigraphical sections makes it difficult to dogmatize as to the origin of the ore-deposit; but, on the whole, the author inclines to conclude that it is the outcome of sedimentation in a lake-basin, or at the bottom of an ancient sea. The ore was probably deposited in the form of brown hæmatite, which has been altered and dehydrated in the course of ages to red hæmatite. The amount of ore in sight is estimated at about 1,180,000 tons; but it seems probable that the deposit extends over a much larger area than can be proved at present.

The coal-mines of Jibito, yielding an excellent coal, appropriate for metallurgical purposes, are about 84 miles distant. Unworked coal-deposits are said to exist at Tangarara and Mallares, but those previously reported to occur around Tambo Grande itself are mythical. Use might be made of the lignite-deposits of Tumbes, which formed the object of official investigation as long ago as 1866 and 1867.   L. L. B.

MINERALOGY OF THE MUSO EMERALD-DEPOSITS, NEW GRANADA.

Sur les Minéraux associés à l'Émeraude dans le Gisement de Muso (Nouvelle-Grenade). By H. HUBERT. *Bulletin du Muséum d'Histoire Naturelle,* 1904, *pages* 202-208.

In this paper the author describes minutely such minerals associated with the Muso emeralds as had not hitherto formed the subject of any detailed investigation.

In the midst of an abundantly-fossiliferous bituminous limestone and of black shales of Neocomian age, occur veins the constituents of which are essentially crystalline. In some cases these are fine-grained and positively form a rock (" Emerald-limestone "); in others, they are built up of big crystals which merely coat the walls of the fissures.

The mass of the Emerald-limestone is made up of calcite, amid which the unaided eye may detect innumerable tiny crystals of pyrites, and, less frequently, fair-sized crystals of dolomite and parisite. Microscopic examination reveals the presence of quartz, albite and emerald, and, in a secondary degree, that of rutile and limonite. From the crystallographic study of the albite-felspar the author is enabled to arrive at the important conclusion that the action of mineralization at Muso was strictly confined to the fissures in the strata, and that the crystalline minerals were not developed by " imbibition in the limestones."

The emeralds for which the deposit is worked are found in the geodes: they are well-known for their splendid coloration and their uniform transparency. The biggest crystals attain a length of several inches. Calcite is the most abundant associate of the emerald in the geodes; and quartz bulks largely too, whereas it is not very abundant in the Emerald-limestone. The author describes also the different types of pyrites, and enumerates, besides the minerals already mentioned, anthracite, gypsum, fluorspar, pyrophyllite and allophane as (less common or rare) associates of the Muso emeralds.

At present the deposits are worked opencast, the deep workings started by the Spaniards having been abandoned since the time when the country declared itself independent of Spain. The known quarterly output is extremely variable: it is more than suspected that the losses by theft are enormous, and many veins are not worked at all.   L. L. B.

NEW QUICKSILVER-MINERALS FROM TEXAS, U.S.A.

1. *Das Vorkommen der texanischen Quecksilbermineralien.* By B. F. HILL; and
2. *Eglestonit, Terlinguait und Montroydit, neue Quecksilbermineralien von Terlingua in Texas.* By A. J. MOSES. *Zeitschrift für Krystallographie und Mineralogie,* 1904, vol. xxxix., pages 1-13, with 6 figures in the text.

The first paper gives a brief description of the quicksilver ore-deposits of Terlingua, Brewster County, Texas. These lie partly in the massive Lower Cretaceous limestones, and partly in the thinly-bedded marls, slates and impure limestones of the Upper Cretaceous. All these strata are cut by dykes, or broken into by bosses of ancient volcanic rock (phonolites, andesites and basalts).

Those ore-deposits which are at present of the greatest industrial importance occur chiefly as brecciated belts (primarily connected with fissures) in the Lower Cretaceous Edwards and Ouachita limestones. The fissures are frequently metalliferous, and always infilled with calcite: at some points aragonite is abundant, and gypsum, iron oxides and wad are among the associated minerals. Quartz is never found in association with the ores. The chief ore is cinnabar, which occurs sometimes in magnificent ruby-red crystals up to ¾ inch long, intimately intergrown with calcite and native mercury; sometimes it occurs in great semicrystalline granular masses, or again in compact masses, varying in colour from vermilion to dark reddish-brown. Native mercury is of fairly common occurrence, and some cavities in the limestone have yielded more than 20 pounds in weight of it.

The Upper Cretaceous ore-deposits form reefs in the Eagleford Slates, and are not so conspicuously associated with calcite as those just described. Pyrites, on the other hand, occurs in considerable quantity.

The second paper deals with the new quicksilver-minerals, which appear to be confined to the Lower Cretaceous deposits, and were in fact all obtained from one drusy cavity in a calcite-vein.

The mineral called eglestonite, after the late Prof. Th. Egleston, is an oxychloride of mercury, corresponding in chemical composition to the formula $Hg_6Cl_3O_2$. The tiny brownish-yellow crystals, of a resinous lustre, are hexakisoctahedra: some exhibit cavities containing native mercury; they rarely measure as much as $\frac{1}{25}$ inch in diameter, and blacken rapidly on exposure to the sun. The hardness is between 2 and 3, and the specific gravity is 8·327.

Terlinguaite is the name restricted by Dr. Moses to the monoclinic oxychloride of mercury (Hg_2ClO) here described. It occurs in very small sulphur-yellow crystals, of hardness 2 to 3, and specific gravity 8·725. The colour slowly darkens to olive-green on exposure to the air. It is found associated with the eglestonite, from which, however, it is easily distinguishable. A third mineral, montroydite, so named after Mr. Montroyd Sharpe, one of the Terlingua mine-owners, is found with the two foregoing, in the form of velvety incrustations of orange-red acicular crystals, with which are associated some dark-red larger crystals. These belong to the orthorhombic system, possess a vitreous lustre, and a hardness less than 2. Their specific gravity has not been determined, but analysis shows the mineral to be a simple oxide of mercury (HgO). In addition, crystallized calomel, and an undetermined yellow quicksilver-mineral, were obtained from the same drusy cavity.

L. L. B.

THE MINERAL RESOURCES OF THE MENADO DISTRICT, CELEBES.

Geologische en mijnbouwkundige Onderzoekingen in de Residentie Menado gedurende het Jaar 190?. By M. KOPERBERG. *Jaarboek van het Mijnwezen in Nederlandsch Oost-Indië*, 1903, *vol. xxxii., pages* 170-178, *and* 1 *plate.*

The search for ore-deposits in this district continues unabated, but appears to have been confined in 1902 to the north-western portion of it. In the Kaidipang division, prospecting revealed veins of iron and copper-pyrites, more or less auriferous, associated with quartz. Along the Pagoejama river, in Gorontalo, occur gold-placers; and this statement applies to several other rivers in the same division. In the hilly country south of the Pagoejama, prospectors for ores report unsatisfactory results, the only occurrences worth mentioning being some veins of pyrites.

In the andesite- or porphyrite-breccias along the Bodi valley, quartz veins mineralized with pyrites, galena and blende, and yielding a variable proportion of gold, occur. Some manganese-ore has been found at Cape Torawitan, but one does not gather that it is proved to occur in sufficient quantity to be of industrial importance.   L. L. B.

THE MINERAL WEALTH OF NEW CALEDONIA.

Rapport à M. le Ministre des Colonies sur les Richesses minérales de la Nouvelle-Calédonie. By E. GLASSER. *Annales des Mines*, 1903, *series* 10, *vol. iv., pages* 299-392 *and* 397-536, *and* 3 *plates ; and* 1904, *series* 10, *vol. v., pages* 29-154, 503-620 *and* 623-701, *and* 3 *plates.*

In this lengthy memoir the author details the results of the investigations which he made, in the course of a journey undertaken in 1902 on behalf of the French Government, into the mineral resources of their greatest island-colony in the Pacific. The absence of railways and of regular coasting-steamers, the small extent of really-good roads and the bad condition of the bridlepaths, were the main difficulties which beset him.

The first few chapters are devoted to a description of the physiography and general geology of the island, which was probably separated off from Eastern Australia before the end of the Mesozoic era, and from New Zealand in late Cretaceous or early Tertiary times.

Archæan rocks, mainly represented by mica-schists, form the core of the northern portion of New Caledonia, running up there indeed into high peaks. Granite crops out among the serpentines at two localities, at any rate, in the southern portion of the island. The Archæan rocks are flanked on either side by a series of sedimentary deposits, ranging from pre-Cambrian to Tertiary. Sheets and dykes of eruptive rocks are intruded among the sedimentaries; of these eruptives, the comparatively-recent serpentines are the most widespread and the most interesting from the industrial point of view, since with them are associated all the nickel, cobalt and chrome-ores, and the more important accumulations of iron-ore known in the colony. Coal occurs in the less ancient sedimentaries [Cretaceous] and gold, copper, argentiferous lead-ore, ores of zinc, antimony, mercury, tungsten, etc., in the Archæan massif. Coral-reefs, living and dead, form an almost continuous belt round the island.

We are reminded that it was in New Caledonia that hydrosilicates of nickel, since then discovered in various other regions of the globe, were for the first time pronounced to be workable metalliferous ores; and after various crises, which the author describes in detail, the annual output rose in 1901 to 133,676 tons. Exports have increased almost continuously from 37,467 tons in 1896 to 129,653 tons in 1902. The green silicates (garnierite, noumeite) were alone worked in the earlier years, until it was discovered that the brown mineral (" nickel-chocolate ") frequently associated with them, was just as rich in nickel. All the varieties of nickel-ore in the island are exclusively confined to the peridotite-massifs, but not to one particular type of peridotite: they are found equally in fresh crystalline rock and in highly-serpentinized material, decomposed by weathering into clays. In all cases, the ore occurs in such portions of the massif as have been near enough the surface to be subject to the influence of atmospheric agencies. The difficulty of describing the different kinds of nickel-ore is pointed out, as there are innumerable instances of the passage of one variety into the other.

Nowadays, no ores are worked nor are indeed saleable unless they contain a minimum of 7 per cent. of metallic nickel, after dessication at 212° Fahr. It is true that there are exceptions to this rule, the Nepoui mine having recently exported to America ores containing only 6 and 6½ per cent. of the metal. The minimum has been fixed, not so much in view of metallurgical exigencies, as in consequence of the heavy cost of transport. The general practice is to mix low-grade and high-grade ores in such proportions as to reach the desired tenour. This, of course, necessitates very careful sampling and assaying. The dried 7 per cent. ore, on the spot, is worth from £1 15s. to £2 per ton (see, however, below): prices were slightly lower in 1902 than the previous year and showed a tendency to go down lower still. New mines, of late years, have been opened up chiefly on the western coast of the island, while operations on many rich deposits of the eastern coast have been conducted with renewed activity. Nevertheless, it was expected that the total exports of nickel-ore during 1903 would show a diminution, as compared with those of the two preceding years. Canadian competition and the shrinkage of the general demand for nickel count for much.

About 40 pages are devoted to a detailed description of the nickel-mines on the eastern and western coasts, and 8 pages to a description of the workings that are now abandoned. The question of the reserves of untouched nickel-ore in the interior of the island is then discussed at some length, and the author marshals the evidence in favour of the view that the nickeliferous deposits are, speaking generally, superficial. However this may be, the area over which they may be still expected to occur is vast enough to furnish materials for an extensive mineral-industry for many years to come. The methods of working, transport-facilities, cost of labour, etc., are then dealt with in detail, and taking every possible item into consideration, the author reckons the minimum prime cost of a ton of dry ore at 25 to 29 shillings. At the time of his visit the sale-prices left an ample margin of profit.

The author then proceeds to consider the statistics of export and consumption, the available markets, and the possibilities of development of the nickel-industry in New Caledonia. He thinks that it would be advantageous to establish smelting-works in the island, and enters into minute calculations on this subject. He holds that, as against Canadian competition, all the natural advantages are in favour of New Caledonia, whose nickel-ores are richer, purer, more cheaply worked, and more easily smelted. He admits that Canada has in her favour abundant and excellent labour, appropriate methods of working on a large scale, good transport-facilities, and well-equipped smelting works. Improvement in all these respects is by no means impracticable, however, even in the remotely-situated French island-colony.

The third part of the memoir deals with the other ores that are associated with the nickeliferous serpentines.

Cobalt, the all but inseparable companion of nickel, has in New Caledonia been redeposited, in conjunction with manganese, in nodules or concretions in the red clays; while the nickel, obedient to its marked affinity for magnesium, has combined with it in the form of hydrosilicates precipitated among the more or less serpentinized peridotites. The red clays represent a further stage of decomposition of these rocks. It is rare to find ores which, even after careful sorting and washing, yield more than 10 per cent. of oxide of cobalt. A considerable amount of iron peroxide and monoxide and dioxide of manganese is always intimately associated with it. On the whole, these ores may be defined as a cobaltiferous asbolite or wad. Their occurrence in the colony was hardly noticed until 1876, and it was only in 1883 that they began to figure regularly among the exports, at the rate of 2,000 to 3,000 tons *per annum*. The total output in 1901 from 35 workings amounted only to 2,552 tons, but in 1902, on account of the continuous rise in prices, no less than 74 mines were at work, producing a total of 7,512 tons. Some of the principal cobalt-mines are described in detail, as also the methods of working, sorting and washing the ores. Hand-labour, in respect of this preliminary preparation of the ore, appears to be both somewhat costly and wasteful; but the introduction of mechanical screeners and washers, on account of the great variation in grade of the ore, might not in the long run prove much more advantageous. The cost-price of a ton of washed ore put down at Nouméa ranges from £5 16s. 10d. to £6 5s. per ton, while the selling price varied (at the time of the author's visit) from £13 5s. per ton of 4 per cent. ore to £30 per ton of 8 per cent. ore. The supply on European markets has since then exceeded the demand, and prices have gone down considerably. New Caledonia remains by far the greatest cobalt-producer in the world, the output from all other regions being by comparison insignificant, reaching indeed only a tenth of the tonnage got from the French colony. A word of warning is given as to the improvident methods of working which obtain there.

Chrome-iron-ore occurs with remarkable constancy in the great serpentine-formation of New Caledonia. It is found either as a rock-mass among the peridotites, weathered and unweathered alike; or in the form of grains among the red clays which are their ultimate decomposition-product. Consequently, the workings may be severally described as reef-workings or placer-workings. Exports began with 500 tons in 1880, and increased enormously in the 'nineties, the total export for 1901 amounting to 17,600 tons, got from 10 placer-workings. The ore is of such excellent quality, that, without picking or washing. it averages 50 to 55 per cent. of chromium sesquioxide, or even more. One gathers, however, that at present prices, the smaller placer-workings cannot be making very great profits. Mining operations, if pursued more methodically and on a larger scale, would spread the prime cost over a greater output and hence ensure a steadier and higher profit. The most serious rivals of New Caledonia in the production of chrome-iron-ore are Russia, Turkey in Asia, New South Wales and Canada.

Red hæmatite, in the form of " scoriaceous blocks," occurs on the flanks and crests of the peridotite-massifs, and is the material with which the Kanakas build their cairns. Analysis shows it to be a good and rich ore. The author does not think that it can be regarded as the gossan of a deep-lying vein-deposit; but its relation in certain localities to the chrome-iron-ore suggests that both ores were originally derived from a massive or bedded deposit of chromite and magnetite, which was the outcome of igneous segregation. Hæmatite also forms the nucleus of the rough grains which lie in thick and extensive beds on the gentler hillslopes: the pulverulent shell or crust of these grains consists of a hydrated oxide of iron, and they yield, after calcination, something like 50 per cent. of metallic iron. Further, the serpentinous massifs are in many places mantled by thick masses of a red, impermeable, unctuous, more or less plastic material which may be regarded as consisting of iron-ores, highly siliceous at some points, highly ochreous at others. The author discusses the reasons which now and for some time to come are likely to forbid any serious attempt at working the hæmatites, etc., of New Caledonia.

Copper-ores occur both on the eastern and western coasts, and in the interior of the island, but working was not seriously begun until 1873. From that date until 1902, more than 50,000 tons of ore and about 1,000 tons of rich matte have been exported. The La Balade group of deposits are mainly copper-pyrites interbedded with and impregnating hornblende and glaucophane-schists. These mines were abandoned, perhaps rather too hastily, in 1884. In the La Pilou group, cupriferous quartz-reefs are seen to cut vertically through black slates: they carry, besides various ores of copper, lead, zinc, silver and iron-ores. Abandoned for a time, working on these deposits was re-started in 1897 by a British company. But capital was frittered away in useless expenditure, and in a disastrous attempt to treat the ore on the spot, instead of exporting it to Australia; bankruptcy ensued in 1902. However, a third endeavour is to be made to work these mines.

Gold is not being mined in New Caledonia at the present time, but between 1871 and 1873, and again from 1876 to 1878, the gold-mining industry was fairly active, the value of the total output for the years mentioned being about £26,000. At Fern Hill mine, a quartz-reef at the contact of black schists and mica-schists was worked in 1872, mining was twice abandoned and twice restarted; but the conclusion was arrived at in 1888, as the result of further exploration-work, that the distribution of the precious metal was too capricious and the deposit on the whole too poor, to offer favourable prospects. Deep down, the native gold at Fern Hill gives place to auriferous pyrites, and it is suggested that the facilities offered by modern methods of treatment should largely modify the unfavourable judgment arrived at in 1888. Gold occurs among the glaucophane-schists of the Tiari massif, and in the sands brought down by the rivers in the northern district of the island. Auriferous deposits have also been explored in the centre and south; some have been worked more or less spasmodically, and then abandoned.

Argentiferous lead-ores occur at various localities in the north of New Caledonia, but have only been mined at one place—the Mérétrica mine, discovered towards the end of 1884. Work was started here in 1886, and finally abandoned in 1898: the proportion of silver averaged more than 1 part per 1,000 of pig-lead. The deposit, yielding cerussite at the outcrop, and galena (with pyrites and blende) deeper down, occurs among black slates which are traversed by a greenstone-dyke. Hand-specimens of the cerussite are spangled with native silver, and the amount of the precious metal therein is said to have averaged 12 to 15 ounces per ton of ore: the assays of the sulphidic ores were far less favourable, the amount of silver averaging from 3 to 6 ounces per ton. The author tabulates the results of 21 analyses made in 1898, but he has not been able to check them himself.

Zinc occurs as an unwelcome impurity in the copper- and lead-ores of the island, but actual zinc-ore deposits have not been discovered there so far.

Antimony-ores were worked at Nakety for a year or two in the early 'eighties. Under present economic conditions, however, it will be long ere the workings can be restarted with any hopeful prospect.

Cinnabar and *native mercury* are known to exist at several localities, but no information is available at present as to the actual extent or value of the deposits.

Platinum has only been reported so far from one locality in the island: it occurs, in very small quantity, in association with gold, in the sands of the Andam streamlet. Its original matrix is unknown. The sands of the rivers which flow down from the great serpentine-massifs have not as yet been systematically searched for platinum, and so it is possible that future exploration will reveal unsuspected deposits of the precious metal.

Manganese-ores occur in New Caledonia, although they are not of immediate industrial importance.

Tungsten in the form of scheelite (tungstate of lime) has been proved at Kouaoua, in the Faja valley, but it is questionable whether the deposit is sufficiently rich to repay working.

Coal is considered in the fifth part of the memoir: it was the first useful mineral to be discovered in the colony, as the coal-bearing rocks extend over coastal districts of comparatively-easy access. The seams, however, have not been worked so far and some observers are doubtful as to the possibility of utilizing them industrially. The coal-bearing strata (alternating shales and grits) are the uppermost of the sedimentary formations of the island, and are overlain by the peridotites: the palæontological evidence points to an Upper Cretaceous age, and so the coal-seams of New Caledonia date from about the same period as those of Westport in New Zealand. There is otherwise no sort of analogy or resemblance between them. Underlying the coal-formation unconformably are felspathic schists of Triassic age: between the two there sometimes intervene contact-metamorphosed rocks of Liassic and Upper Jurassic age. The coal-belt can be traced afar off by its characteristic bare escarpments of white to pink felspathic grit (a rock which is very friable), and extends along the western coast from Mont Dore to Mount Kaala, a distance of 180 miles. Despite this great extent, however, the author is inclined to agree with the opinion expressed by Mr. Heurteau in 1876, that many years must elapse before coal-mining can be carried on with real success in New Caledonia. The coal may prove a valuable asset later on.

The three chief coal-fields are those of Nouméa, Moindou, and Poya: there are, besides, six smaller fields known, one only of which (that of Voh) is of any considerable extent. A detailed description is given of all these, followed by analytical tables, showing that the mineral varies in character from anthracite, through bituminous coal, to lignite. The bituminous coal of the Portes de Fer (Nouméa) burns with a short flame, has in some instances a heating-power of 7,200 calories, is quite equal to Australian coal for boiler-firing, and yields fairly-good coke. Other coals in the same field have also been shown experimentally to be suitable for industrial purposes, but it is thought that the coals of the Moindou basin will not prove easy to use. The little evidence at present available in regard to the regularity of the seams is in many cases unfavourable, and steep and rapidly-changing dips are to be expected. On the whole, the natural conditions and the difficulties of transport are such as to prevent New Caledonian coal from ever competing for shipping purposes with the coal of New South Wales or New Zealand. The annual consumption of coal in the French island at present ranges from 10,000 to 12,000 tons, and a gradual increase to 18,000 tons may be looked for when the railway from Nouméa to Bourail is completed (a somewhat remote contingency at present). If, however, the colony were to proceed to smelt its own nickel-ores, the coal-consumption would be sextupled.

The sixth and last part of this exhaustive memoir deals with the general economic conditions of the mineral-industry in New Caledonia: these are declared to be the reverse of favourable. Of the 3,000 men working on the mines, one-half are whites (nearly all ticket-of-leave men), and most of the remainder belong to one or other of the yellow races. The mining camps are of the roughest description, and drunkenness is the sole form of recreation. Pay is largely swallowed up in drink and in purchases (at exorbitant prices) at the camp-stores. The average daily wage ranges from 4s. to 5s. 6d. It is curious to note that the author is at one with the Government of the colony in favouring increased importation of Chinese labour. Commodities in the island are dear, and the taxation of industry and the dues levied from shipping are heavy. New roads and railways are among the most imperative requirements of the colony.　　　L. L. B.

THE MINERAL RESOURCES OF SOUTHERN SUMATRA.

Einige Notizen zur Geologie von Südsumatra. By AUGUST TOBLER. *Verhandlungen der naturforschenden Gesellschaft in Basel*, 1904, *vol. xv., pages 272-292, and* 1 *plate.*

Acting as geologist for two Dutch petroleum-syndicates, the author was enabled, in 1900-1903, to make very detailed observations on the rocks of Southern Sumatra and their fossil contents, and brought back with him to his native Switzerland considerable collections.

Describing first the older schists, and following Mr. R. D. M. Verbeek's classification, he points out that in the crater of the Ringgit volcano, near Bajur, petroleum wells up, and is undoubtedly derived from these schists. Hitherto, the occurrence of petroleum in Sumatra had never been recorded in strata older than the Tertiaries. Some of the green schists contain metalliferous ores, a fact already known, but the author cites new occurrences of them.

The richest oil-wells have been struck at very various horizons in the Lower Pliocene (4,900 feet thick), consisting in this region of blue clays passing locally into sandy shales and argillaceous fine-grained sandstones. Calcareous septaria, absent in the Upper Pliocene, are very characteristic of this lower division, which is also extremely rich in fossil mollusca.

Three groups of brown coal-seams are markedly characteristic of the Middle Pliocene division (2,000 feet thick), which consists of blue and brown clays, shales and (at the top) fine-grained, soft, shaly, pale-blue and white sandstones. Some of the lignitic seams are 40 to 50 feet thick, and some contain slabs of silicified coal 4 to 12 inches thick. The author found, of course, plant-remains in the Middle Pliocene, but practically no marine fossils. Petroleum occurs in this division too, in workable quantities, and appears to be more especially concentrated at particular horizons, at the base of each of the brown-coal groups. The topmost brown-coal group is immediately overlain by the vast mass of the tufaceous sediments of the Upper Pliocene (3,300 to 5,000 feet thick), which bear witness to long-continued vulcanicity. No oil has been struck in this division of the Pliocene, nor is it likely to be. The other strata described by the author do not appear to be of economic interest, but his paper constitutes an excellent summary of the geology of Southern Sumatra.   L. L. B.

SEARCH FOR COAL IN THE ATJEH VALLEY, SUMATRA.

Verslag eener geologisch-mijnbouwkundige Verkenning der Atjeh-Vallei gedurende het Jaar 1902. By P. J. JANSEN. *Jaarboek van het Mijnwezen in Nederlandsch Oost-Indië,* 1903, *vol. xxxii., pages* 179-184, *and 2 plates.*

Reports of the occurrence of coal having reached the Governor of the district, an official investigation was ordered, with the view of reserving the working of any such deposits to the state, and supplying the mineral so worked to the Atjeh light railway.

The oldest (sedimentary) geological formation in the region appears to be the Carboniferous Limestone, beneath which appears an ancient crystalline rock (gabbro). There is a series of younger sedimentaries (sandstones, conglomerates and limestones), probably of Tertiary age; and of more recent volcanic rocks (trachytes, andesites, tuffs, etc.). One of the reported coal-outcrops turned out to be the dark gabbro. In another case, a patrol is said to have brought back from a spot near Data Teureubéh several lumps of coal, but no seam has been found in place; and the " coal " is thought to have been semi-carbonized wood. Semi-lignitized plant-remains have been recorded on a former occasion from this river-valley.

On the whole, the evidence is all against the probable occurrence of workable deposits of coal, whether of Carboniferous or Tertiary age.   L. L. B.

THE AURIFEROUS PLACERS OF THE BATAK PLATEAU, SUMATRA.

Über die Goldvorkommen auf der Bataker Hochfläche. By H. BÜCKING. *Sammlungen des geologischen Reichsmuseums in Leiden ; Beiträge zur Geologie Ost-Asiens und Australiens, vol. viii.,* 1904, *pages* 97-99.

From time immemorial these placers have been worked in a very primitive manner by the natives, the gold being fashioned by them into bracelets, finger-rings and ear-rings. Masses of auriferous sand and silt fill up the valley-floors, and the hollows and depressions in and between the limestone-rocks, which crop out on the hillsides. The " gold-sand," generally black, contains a quantity of magnetite and small crystals of zircon; on washing and re-washing this, the precious metal is obtained in the form of thin flakes and little grains.

With regard to the original *locus* of the gold, it is observed that a part, at least, must be derived from the decomposition of the auriferous pyrites contained in the eruptive rocks, the weathered residuum of which latter occurs in the form of laterite in many of the placers. But pyrites is also found in the quartz-veins which traverse several of the limestones of the district; and there is not much doubt that the quartz is itself gold-bearing, since the natives wash for gold the detritus (often forming a deposit 10 feet thick) derived from the atmospheric degradation of the veins, while they cast aside the semi-lateritized, easily-crumbling blocks of eruptive rock.

On the whole, the deposits, so far as their primary origin is concerned, may be compared with the auriferous occurrences of Hungary and Transylvania; in both cases they are associated with quartz and iron-pyrites, precipitated from fumaroles and thermal springs, which were connected with the later phases of andesitic and trachytic eruptions. Auriferous deposits of considerable importance are said to exist on the south-western flank of the Longsuwattan mountain, but are at present inaccessible, as they lie within the territory of a warlike hostile tribe.   L. L. B.

THE CANNEL-COAL FLORA OF NÝŘAN, BOHEMIA.

Beitrag zur Kenntniss der Cannelkohlenflötzen bei Nýřan. By F. RYBA. *Jahrbuch der kaiserlich-königlichen Geologischen Reichsanstalt*, 1903, *vol. liii., pages* 351-372, *and 3 plates.*

This paper is devoted to a description of plants not before found in the Nýřan cannel-coal, or otherwise but little known. The flora on the whole, if interesting, is poor. It consists of 79 species in all, 34 of which are ferns, 16 are Calamarians, and 12 Lycopods. Those species which are most abundantly represented in the cannel-coal belong to such forms as are known to range from the Middle or Upper (very rarely from the Lower) Carboniferous into the Lower Rothliegende. True Carboniferous types are numerous among the Nýřan species: and among the Lepidodendra are some forms which are exclusively Carboniferous. The evidence for the existence of purely Permian plants in the Nýřan flora is practically negligible. Indeed, so far as the palæobotanical evidence goes, the Nýřan cannel-coal belongs to the Upper Productive Coal-measures, and may be correlated with the uppermost Ottweiler beds of the Saar and Rhenish coal-fields.   L. L. B.

PALÆONTOLOGY OF THE HERVE COAL-MEASURES, BELGIUM.

Note préliminaire sur les Caractères paléontologiques du Terrain Houiller des Plateaux de Herve. By A. RENIER. *Annales de la Société Géologique de Belgique*, 1904, *vol. xxxi., Bulletin, pages* 71-73.

The results, to which the author's investigations have so far led him, are, briefly, as follows:—(1) The Coal-measures of the Herve plateau contain a fairly-rich fauna and flora, of distinctly Westphalian character, since they include the various species which are regarded by most authors as typical of the several horizons of that division of the Coal-measures. (2) The order of succession of the floras is practically the same as that established by Prof. Zeiller in the Valenciennes coal-basin, and repeated in the north of Belgium. (3) The presence of *Dictyopteris* in the Florent and Théodore seams, above which still come several workable seams, permits of the inference that the Flénus horizon (of which that genus is characteristic) extends, contrary to the generally-received opinion, into the east of Belgium. It may be noted, by the way, that the seams with *Dictyopteris* yield a coal containing, on an average, only from 15 to 16 per cent. of volatile substances and 3 per cent. of ash. (4) The thickness of the strata between the Sidonie or Florent seam and the (basement?) Beaujardin seam, measured in the only-available continuous section (at the Hasard colliery), is about 1,640 feet. This comparatively-low figure points to a thinning of the Coal-measures in that district, and, if the same horizons can be proved at Liége, it will serve as a measure of the amplitude of the tectonic disturbances which cut off the Herve coal-field from that of Liége.   L. L. B.

FOSSIL FLORA OF THE TONGKING COAL-FIELDS.

Flore Fossile des Gîtes de Charbon du Tonkin. By R. ZEILLER. *Etudes des Gîtes Minéraux de la France: Colonies*, 1903, 328 pages, and 62 plates.

The detailed investigations recorded in this magnificent memoir have enabled the author to confirm triumphantly the inference, hotly disputed by others, which he had drawn as long ago as 1882, that the coal-formations of Lower Tongking are of Rhætic age. This conclusion is strengthened by the independent observations of Mr. Leclère in Southern China, who, it will be remembered by readers of the *Annales des Mines,* has shown that the analogous and neighbouring coal-basins of Yunnan and Kweichau are also of Rhætic age. It is true that there are also in Southern China seams of Coal-measure or Permian age. The Yen-bai coal-field of Tongking is, however, of much more recent date, its flora and fauna showing it to be of Upper Miocene or Lower Pliocene age at the earliest, if not Middle Pliocene.

The memoir opens with a brief description of the coal-fields from which most of the specimens studied by the author were obtained, and is accompanied by maps and sections, as well as by photographic illustrations of the fossils.

In Lower Tongking, the Hongay mines, with enormously-thick seams, producing anthracitic coals of a heating power varying from 6,900 to 7,800 calories, yielded in 1901 an output of 249,000 tons. The Kebao mines (said to include about 30 seams), with a much smaller output thus far, produce a similar coal. In both basins, a small portion of the output is mixed with 20 to 25 per cent. of Japanese coal, and bound together with pitch into briquette-fuel. The Dong-trieu anthracite-mines have hardly yet got into working order. In Annam, the Nong-sön coal-mine shows a single seam of workable coal varying in thickness from 20 to 40 feet: the output in 1902 amounted to about 24,000 tons. The mineral is still more anthracitic than that of Lower Tongking, containing only from 6 to 10 per cent. of volatile substances.   L. L. B.

THE COAL-MEASURE FLORA OF THE DONETZ, RUSSIA.

Végétaux fossiles du Terrain Carbonifère du Bassin du Donetz: I. Lycopodiales. By M. D. ZALIESSKI. *Mémoires du Comité géologique, St. Pétersbourg, 1904, new series, No. 13, 126 pages, with 11 figures in the text and 14 plates.*

This memoir is the first instalment of a full description of such fossil plants from the Donetz coal-field as are to be found in the collections of the Imperial Russian Geological Survey. In it are described eleven species of *Lepidodendron* (three of which are new to science), one of *Lepidophloios,* two of *Ulodendron,* one of *Lycopodites,* three of *Lepidostrobus* (one new), two of *Lepidophyllum*, two of *Bothrodendron*, one of *Bothrostrobus*, twenty-four of *Sigillaria* (six new), two of *Syringodendron* (one new), and two of *Stigmaria*. So much for the Lycopodiales: the other groups are to be dealt with in succeeding memoirs, and a final fasciculus is to embody the general conclusions which arise from the palæobotanical evidence. **L. L. B.**

CRYOLITHIONITE, A NEW LITHIUM-FLUOR MINERAL, FROM GREENLAND.

Sur la Cryolithionite, Espèce minérale nouvelle. By N. V. USSING. *Oversigt over det kongelige Danske Videnskabernes Selskabs Forhandlinger*, 1904, *pages* 3-12..

This mineral was found in 1903, in the cryolite-mines of Ivigtut (Greenland), in the form of big transparent rhomboidal dodecahedra, some of which are as much as 7 inches in diameter. They are completely enveloped by cryolite, and contain many vacuoles with bubbles of gas and liquid. The mineral is colourless, its specific gravity is little short of 2·8, its hardness lies between 2½ and 3, and its chemical composition is represented by the formula $Li_3Na_3Al_2F_{12}$ (fluoride of lithium, sodium and aluminium). Its discovery adds another factor, in the shape of the presence of lithium, to the remarkable resemblance between the mineral associates of cryolite and those of tin-ore deposits. Recent research has, moreover, strengthened the evidence in favour of the eruptive origin of cryolite. The new mineral recrystallizes on cooling after fusion, and also recrystallizes without difficulty by evaporation from an aqueous solution of it. It contains a larger proportion of lithium than any mineral so far known (11·4 per cent.). In cleavage and crystalline habit, in molecular build, in a word, it shows unmistakable analogies with the garnets. **L. L. B.**

DETERMINATION OF MINERALS BY THEIR SPECIFIC ELECTRIC RESISTANCE.

Sur un Caractère spécifique des Minéraux opaques. By G. CESÀRO. *Bulletin de la Classe des Sciences de l'Académie Royale de Belgique*, 1904, *pages* 115-122.

The author points out that it does not appear to have occurred to anyone, so far, to make use, as a distinctive character of minerals, of the greater or less resistance which they offer to the passage of an electric current. Yet this is a property which it is easier to test than specific gravity or hardness, and can be utilized in the microscopic study of cast-iron, steel, and other alloys containing a metalloid which reduces the conductivity of the metal pure and simple.

The author's experiments were made with a dry battery, which ordinarily supplied an incandescent lamp: he attached two platinum-wires to the poles, and placed the free ends of the wires on the minerals which he examined. It seems probable that, in the case of cubic crystals of, say, mispickel, the specific electric resistance will be found to vary from one face to the next, or from one direction to the other along the same face. But this does not invalidate the practical utility of the method, whereby the author was able, for instance, to identify as native copper what might, to all appearances, have been chalcopyrite. Generally speaking, all transparent minerals, all silicates, " sulpho-salts " (including the silver-antimony ores,. stephanite and polybasite) and oxides (excepting magnetite) are non-conductors. With few exceptions, the conductors are confined to the opaque simple bodies, amalgams, sulphides, antimonides, arsenides, tellurides, etc. The author gives a detailed list of these, according to their degree of conductivity.

Cinnabar is a non-conductor, while metacinnabarite (of the same chemical composition) is a very good conductor. Pyrites varies in conductivity, according to the direction of the striations on its crystal-faces, while marcasite is a non-conductor. Moreover, it is found that the measurement of electric resistance in galena constitutes a quick method of determining the percentage of silver in the ore. Another advantage of the method is that the specimens thus tested are in no way spoilt or injured.   L. L. B.

METHOD OF CALCULATING ANALYSES OF COAL.

(1) *Sur une Cause fréquente d'Erreurs dans l'Analyse centésimale des Houilles.*   By JUST ALIX and ISIDORE BRAY. *Comptes-rendus hebdomadaires des Séances de l'Académie des Sciences,* 1904, *vol. cxxxix., pages* 215-216.

Premising that nearly all coals contain, in addition to iron-pyrites, a certain amount of calcium carbonate, the authors point out that the presence of this impurity must constitute a source of error in determining the percentage of carbon in the mineral. In order to form an idea of the proportions which this error may assume, they conducted a series of experiments on the gas-coal from the Montrambert mines, in the St. Étienne coal-field. They found that it contained on an average 4·99 per cent. of carbonate of lime, which would correspond to 0·56 per cent. of carbon. Now, this is a normal coal; but, in the case of such coals as the Tongking anthracite, practically seamed with calcitic veins, nearly ¼ inch thick, the possible error would be much greater.

As this error leads one to suppose that a coal is richer in combustible carbon than it really is, the presence of carbonate of lime is a factor which should also be taken into account in determining the calorific value or heating-power of a coal.

(2) *Über die Berechnung der Elementaranalysen von Kohlen mit Bezug auf den Schwefelgehalt derselben, und den Einfluss der verschiedenen Berechnungsweisen auf die Menge des berechneten Sauerstoffes und die Wärmeeinheiten.*   By C. VON JOHN. *Verhandlungen der kaiserlich. königlichen geologischen Reichsanstalt,* 1904, *pages* 104-111.

The author points out that a possible cause of error which vitiates the elementary analysis of coal, is that, as a rule, the hygroscopic water alone is determined, and that no account is taken of the chemically-combined water originally present in the constituents of the ash. The freer a coal is from ash, the more insignificant is this error; the general effect of which is, to minimize the proportion of ash and to magnify the proportions of hydrogen and oxygen.

Turning then to the calculation of sulphur, the author says that Prof. F. Schwackhöfer* reckons the so-called "detrimental" or combustible sulphur in his analyses, apart from the totals making up 100. This is of little consequence in cases where a coal does not contain much sulphur, but may lead

---

* *Die Kohlen Œsterreich-Ungarns und Preussisch-Schlesiens* (coals of Austro-Hungary and Prussian Silesia), second edition, 1901.

to considerable error in the case of a highly-sulphureous coal. An example is given, wherein the result is that the heating capacity of a coal has been tabulated too low by 300 to 400 calories or more.

An important question, which the author hopes will be determined by the Analysis Committee of the Fifth International Congress for Applied Chemistry, is the value to be set on the heat-equivalent of hydrogen, that is, whether the lower formula, 29,000 (wherein water is reckoned as steam) or the higher formula, 34,500 (wherein water is reckoned as a fluid) is in future to be the standard when making the necessary calculations in coal-analysis. In Germany, the first figure usually finds especial favour; but in France, as in the Laboratory of the Austrian Geological Survey, the second figure is adopted. These differences involve considerable discrepancies in the tabulation of the heat-units of any given coal. The same observation applies, especially in the case of brown coals and lignites, to the omission or inclusion of the percentage of water corresponding to the amount of oxygen determined as present in these minerals. International uniformity on the subject is a consummation devoutly to be wished. L. L. B.

ANALYSES OF HUNGARIAN COAL, ETC.

Mittheilungen aus dem Chemischen Laboratorium der königlich-ungarischen geologischen Anstalt. By ALEXANDER VON KALECSINZKY. *Jahresbericht der königlich-ungarischen geologischen Anstalt für* 1901, *pages* 174-183.

The Upper Mediterranean (Tertiary) coal of Kistapolcsány, County Bars, formerly known as Fenyökosztolány coal, yielded 58·52 per cent. of carbon, 0·77 per cent. of sulphur and 7·38 per cent. of ash. Its heating-power is estimated at 5,107 calories.

The coal from seam No. 1, at Jakfálva, County Borsód, yielded 6·48 per cent. of ash, and contained as much as 4 per cent. of sulphur. Its heating-power was estimated at 5,409 calories. The coal from seam No. 2 contains less sulphur, but yields double the amount of ash, and its calorific power is inferior.

Lignite from the neighbourhood of Derna, County Bihar, yielded 9·76 per cent. of ash, 43·49 per cent. of carbon, 1·84 per cent. of sulphur and 23·15 per cent. of water. Its calorific power was estimated at 4,281 calories. The lignite from Kartal, County of Pest, is superior in heating-power and in percentage of carbon (45 to 46), but contains, on the other hand, much more sulphur (3 to 4 per cent.).

An analysis is also given of the kaolin worked at Beryszász, and the results of an investigation of the salt-lakes of Transylvania are summarized. L. L. B.

ANOMALIES OF TERRESTRIAL MAGNETISM IN THE NEIGH-BOURHOOD OF LIÉGE, BELGIUM.

Anomalies dans la Déclinaison Magnétique aux Environs de Liége. By — DEHALU. *Mémoires de la Société royale des Sciences de Liége,* 1904, *series* 3, *vol. v., No.* 3, *43 pages and a map.*

The author has recently made a series of careful determinations of the magnetic declination at 17 localities in and around Liége, and his results lead him to the conclusion that there are in the Liége coal-field causes which undoubtedly influence the direction of the compass-needle; and further that one source of this deviation must be sought in the colliery-heapsteads. Curiously enough, the deviation appears to be greatest in the case of pit-heaps which have been burnt.

With regard to other causes of deviation, a table drawn up by the author shows magnetic declination to increase generally eastward, a direction of increase which does not fit in with the normal distribution of the magnetic elements. The difference of declination between the extreme points plotted out on his map exceeds 20 minutes, more than 4 times what it should be. Consequently, there is no escape from the conclusion that the Liége coal-field is a region of magnetic anomalies, which are, perhaps, not unconnected with the great faults that are known to traverse it.

The author proposes to pursue his investigations along this line of research, if circumstances permit of his so doing.   L. L. B.

MAGNETIC ANOMALIES OF THE PARIS BASIN, FRANCE.

L'Anomalie magnétique du Bassin de Paris. By TH. MOUREAUX. *Comptes-rendus hebdomadaires des Séances de l'Académie des Sciences,* 1903, *vol. cxxxvii., pages* 918-920.

The observations carried out at 617 localities in French territory have revealed numerous irregularities in the normal distribution of the magnetic elements. Apart from the well-known anomaly of the Central Massif, traced directly to the influence of volcanic rocks, others have been discovered amid strata usually considered as quite innocent of any action upon the compass-needle. The most considerable and the most unexpected of these anomalies proves to be in the region defined geologically as the Paris basin.   The records collected from 130 observing-stations, ranging from Upper Normandy to the Nivernais, indicate very marked deformation of the isomagnetic curves. Maps have therefore been drawn up, showing the differences between calcu-lated and ascertained values. For instance, in regard to declination, the divergences are all positive to the east, and negative to the west of a line drawn from Fécamp south-eastward to Moulins, and passing through or near Rouen, Rambouillet, and Gien, at an angle of about 30 degrees with the geographical meridian. In regard to the horizontal component, the positive divergences form three zones alternating with zones of negative divergence.

A study of the maps based on these and other observations leads to the conclusion that, if the anomaly of the Paris basin be assignable to the action of magnetic rocks, the upper surface of the " perturbing mass " takes probably the form of a mountain-massif, overlain by more recent deposits, with peaks and summit-ridges at those points or in those zones which a con-sideration of the magnetic anomalies has shown to be centres of attraction.
L. L. B.

THE MAGNETIC STORM OF OCTOBER 31st, 1903.

Das magnetische Ungewitter vom 31. Oktober 1903.   By J. B. MESSERSCHMITT. *Sitzungsberichte der mathematisch-physikalischen Classe der königlich bayerischen Akademie der Wissenschaften zu München,* 1904, *pages* 29-39, *and* 1 *plate.*

The period of minimum disturbances of the compass-needle came to an end a year or two later than had been calculated, and it was not till the latter half of 1902 that a marked accentuation of these disturbances showed that the turning-point had at last been reached. Observation had previously shown that the curve of magnetic variations rises much more quickly to a maximum than it descends to a minimum, and so it has proved in the present instance. The great magnetic storm of October 31st, 1903, was heralded by considerable disturbance of the compass-needle on the previous night, though by 6 a.m. this seemed to have died down. Only an hour later, however, a magnetic storm commenced, such as had not been observed for more than 40 years or more. In the first and second phases, declination altered to and fro in waves of great amplitude, and the end of the oscillations in value of the horizontal component appears to have coincided both with the last great swing of the declination and with the end of the second phase of the storm (about 8 p.m.). The third phase was characterized by a gradual decrease in the anomalous magnetic conditions, and the storm came finally to an end between 5 and 6 a.m. on November 1st. Slight disturbances continued, however, for several days thereafter, and it was only on November 14th that the recording instruments showed absolute freedom from disturbance.

Complete tables and diagrams of the records are given by the author, and he then proceeds to compare the intensity of the magnetic storm with those which took place in 1847, 1859, 1872 and 1882. It would appear that the storm of 1859 approached most nearly that of 1903: the former was accompanied by nightly apparitions of *aurora borealis* from August 28th to September 3rd; and similarly in 1903 the " northern lights " were observed in many European localities about the time of the storm, also in North America, besides " southern lights " in Australia. The connection between maximum magnetic disturbance and a maximum sun-spot period is again made evident: the maximum sun-spot period was late in its arrival, coincidently with the lateness of the magnetic-storm period. On October 31st, a particularly large group of sun-spots was seen to travel across the central meridian of the sun. With regard to telegraph-wires, etc., those running underground were more affected by the storm than those carried on poles aboveground, and the disturbance was greater from north to south than from east to west, although there appear to have been exceptions to this rule in America.   L. L. B.

VARIATION OF TERRESTRIAL MAGNETISM WITH ALTITUDE.

Sulla Variazione del Campo magnetico orizzontale terrestre coll' Altezza sul Livello del Mare. By A. POCHETTINO. Atti della Reale Accademia dei Lincei, 1904, Rendiconti, series 5, vol. xiii., pages 96-101.

Dr. Schmidt has already shown that, although terrestrial magnetism is mainly due to influences seated within the globe itself, about one-fortieth part of the total magnetic force is probably derived from the electrical phenomena taking place in the atmosphere. But, according to Prof. Gauss' theory, the variation of the horizontal component of the earth's magnetism should be about 0·00001 for every 1,000 metres (3,280 feet) above sea-level.

In the summer of 1899, the author made a first series of comparative measurements in the mountain-group of the Gran Sasso d'Italia, and found a diminution in the horizontal component of 0·00005 for every 1,000 metres (3,280 feet) rise in altitude. Not regarding these measurements, however, as final, on account of certain disturbing conditions, he chose another district, in the Rocciamelone group of the Graian Alps, where he should be able to record measurements at localities of very diverse altitude indeed, but not too far apart. These measurements were carried out in October, 1902, one observing-station being 900 metres (2,952 feet), and the other 3,400 metres (11,152 feet) above sea-level. The results obtained prove once again that the value of the horizontal component of the earth's magnetism, does decrease as the altitude above sea-level increases, and the author finds that the decrease approximates very closely to 0·00004. Dr. Sella had stated it to be 0-00002 in publishing the results of his investigations in 1896, while the figures arrived at by two Austrian investigators differ both from those of Dr. Sella and Dr. Pochettino: thus Dr. Kreil puts the decrease at 0·000147, and Dr. Lisnar puts it at 0·00003.   L. L. B.

THE PETROLEUM-INDUSTRY OF BAKU IN 1902.

Zur Lage der Naphta-Industrie in Baku im Jahre 1902.   By P. J. Scharow
   Zeitschrift für praktische Geologie, 1904, *vol. xii., pages* 263-267.

Although the crisis which had depressed this industry was perceptibly passing away in the second half of 1902, the output for the whole year was less by 5·2 per cent. than even that of the unlucky year 1901; and the average prices for crude and refined oils were also lower.   Some firms were compelled to close their wells down, rather than to continue working at a loss.

An analysis of the numerous statistical tables supplied by the author leads to the inference that the oil-bearing strata of the Sabuntshi district are showing unmistakable signs of gradual exhaustion, and the average depth to which borings had to be carried increased from 1,369 feet in 1901 to 1,426 feet in the following year.   The contrary statement holds good of the comparatively-recently opened-up Bibi-Eibat district, where the average depth of the borings was less by 137 feet: owing, no doubt, to the immense productivity of certain oil-bearing strata, which continuously yield, despite extremely-active working, great fountains of naphtha.   The average output of an artesian naphtha-well in 1902 was 5,217,000 puds (84,108 tons) as compared with 6,373,000 puds (102,744 tons) in the preceding year.   The total quantity of oil got by pumping amounted to 542,094,852 puds (8,739,512 tons), or 31,062,776 puds (500,786 tons) less than in 1901. The number of bore-holes in use in 1902 was 1,840, being 84 fewer than in the previous year.   Nevertheless, the export of lubricating-oil and oil-residues was more considerable in 1902 than ever before, though that of illuminating-oil and crude-oil was less by 9·6 per cent. than in 1901.

                                                                        L. L. B.

THE GOLD-MINING INDUSTRY OF THE RUSSIAN EMPIRE.

Die staatliche Förderung der Goldindustrie in Russland.   By B. Simmersbach.
   Zeitschrift für das Berg-, Hütten- und Salinenwesen im preussischen Staate,
   1904, *vol. lii., B. Abhandlungen, pages* 491-493.

The continuous diminution in the annual output of placer-gold from 1,319,725 ounces in 1892 to 1,127,983 ounces in 1899, is largely attributed by the author to the vexatious legal enactments with which the Russian Government hedged round the industry until quite lately.   In opening up a new gold-mine endless formalities had to be gone through, and even then the miner was not allowed to have the free control of the output, or to so arrange matters as to diminish the cost of working. In addition, the taxes on the industry were very unequal in their incidence; wages were high, and stores were expensive, on account of the generally-considerable distance of the mines from the great centres of population; and the climatic conditions in such regions as Siberia and the Amur country are, of course, terribly severe.

A change of policy on the part of the Imperial Government found drastic expression in the law of March 12th, 1901, whereby the taxes on gold-mines were abolished and the free sale of placer-gold was authorized. Under the old *régime* a large portion of the gold-output had been secretly smuggled across the frontier into China, thus cheating the Russian Treasury of its dues, in defiance of penal enactments. In view of the primitive methods of mining, so primitive, indeed, even yet, that the rich placers of Siberia and the Urals may be spoken of as having been plundered rather than mined, the Government has decreed the free importation until 1909 of all machinery intended for use in the gold-industry. This concession is of great value for such mining-centres as lie anywhere within reasonable distance of a railway, but it will not go far towards helping the more remotely-situated mines in the heart of Siberia or the Urals. Potassium cyanide has also been placed on the list of free imports. As it is further recognized that the difficulties in the way of private prospecting on a large scale are enormous, the Government itself has undertaken a detailed scientific and economic survey of the gold-fields of Siberia, which will include the preparation of geological memoirs and maps. Moreover, a special committee has been appointed to draw up schemes directed to the further advancement of the gold-mining industry; it has already issued maps of gold-mines now being worked or capable of being worked, together with detailed statistics as to their wealth, and full information as to the most advantageous means of transport. All this activity is doubtless reflected in the increased output of placer-gold, which amounted for 1901 to 1,272,818 ounces. It may be noted, in conclusion, that this Russian gold averages 900 parts per 1,000 of fine gold: the cost of production varies enormously according to the district, but the further spread of regular methods of working will put an end to many anomalies.   L. L. B.

PYRITES AND IRON-ORES OF SCANDINAVIA.

Über den Export von Schwefelkies und Eisenerz aus Norwegischen Häfen. By J. H. L. Voor. *Zeitschrift für praktische Geologie*, 1904, *vol. xii.*, *pages* 1-7. *with 2 diagrams in the text.*

The author points out that since pyrites began to be exported from Norway in 1860, the average annual export of that mineral has risen from 9,300 to 105,000 tons. Before long, the annual amount is expected to reach 150,000 tons, as measures are being taken to extend greatly the Sulitjelma mines in Northern Norway, and to open up new ones. More than half of the exported Norwegian pyrites is free from anything but a mere fraction per cent. of zinc, and so when the sulphur is roasted off it yields very good "purple ore." The mines that formerly worked pyrites containing a comparatively-high percentage of zinc-impurities are now abandoned. Moreover, in contradistinction to the Spanish and Portuguese pyrites, that of Norway is practically free from arsenic.

Still more important than the exports of pyrites from that northern region are, however, the exports of iron-ore, which began to make giant strides at the opening of the last decade of the nineteenth century. Turning first to Sweden, we are reminded that the ancient metallurgical industry of that country was based on the possession of iron-ores free from phosphorus. But the new export-industry derives its impetus from the formerly-untouched deposits of iron-ore containing an appreciable percentage of phosphorus. Some statistics and other particulars are given as to the Grängesberg, Gellivaara, Kiirunavaara and Luossavaara deposits. The amount of ore available in Northern Sweden is reckoned at 1,000,000,000 tons containing from 62 to 63 per cent. of iron. In Northern Norway, the most important deposits yet opened up are those of the Dunderland valley, within the Arctic circle, and a whole series of similar but smaller deposits are scattered along the coast between 65 degrees 50 minutes and 69 degrees 10 minutes of north latitude. The author considers that the mistrust expressed in Continental trade-periodicals as to the probable success of the Edison method of magnetic separation is unfounded: he has seen similar Swedish methods in operation, and they work smoothly and cheaply. Although iron-ore has been exported from Southern Norway to the extent of 50,000 tons or so *per annum*, no considerable development of the export from that part of the country need be looked for.

The Swedish exports of iron-ore have risen from 120,000 tons *per annum* in 1889 to 1,750,000 tons in 1901 and 1902. From Ofoten, in Norway, where exports began only in January, 1903, the export is soon expected to average 2,000,000 tons *per annum*, and the railway can furnish transport-facilities for 3,000,000 tons. Within a very few years the united annual export from Sweden and Norway will probably average 4,000,000 tons. In comparison with the Spanish ores, which contain a smaller percentage of iron, the Scandinavian will, in time, assume at least equal importance from the point of view of European industry. The Norwegian harbours, being free from ice, will send out more than the Swedish harbours on the Baltic: England and Scotland are likely to remain the principal customers for the Bessemer-ore; while the Thomas-ore is chiefly taken up by German buyers, and the more Germany takes of this ore the higher will wax the prosperity of the Norwegian harbour of Ofoten.   L. L. B.

KLONDYKE GOLD-MINES, CANADA.

Gîtes aurifères du Klondike, Yukon, Canada. By J. M. Bell. Comptes-rendus Mensuels des Réunions de la Société de l'Industrie Minérale, 1904, pages 76-79.

The mean temperature of the region is 23° Fahr. The alluvial deposits are frozen down to the bed-rock, and mining is only carried on during the very short summer, when water can be obtained for treating the "dirt." The auriferous veins are not workable to profit. Copper-ores are being worked in the White Horse district, near to recently-discovered deposits of bituminous coal, probably of Carboniferous age. The alluvial gold is chiefly found in frozen gravels of Pliocene or Quaternary age, generally overlying the crystalline schists of the Klondyke series.

The Quaternary alluvia attain a thickness in the recent valleys varying from a few feet to 60 feet; and on the slopes and the hills they form terraces at a higher level than the present valleys, with thicknesses up to

150 feet. In the valley and terrace-deposits, the rich zone, which is alone remunerative, follows the ancient channels with a width that does not generally exceed 350 feet. The workable thickness varies from 3 to 7 feet, including about 1 foot of bed-rock, more or less altered and enriched by natural concentration and the action of running water. The gold, always found in grains, sometimes flattened, is deposited near the bed of the subjacent rock; and it appears to have been rolled for only a slight distance. The gold is accompanied by the usual minerals, including titanic iron and sometimes cassiterite.

The royalty to the state has been diminished from 10 to 2½ per cent. of the gold exported. The granular nature of the native gold permits of very simple treatment, by washing in very short " sluices," without mercury, although the absence of fall in the district and the obligation of leaving the water at the disposal of one's neighbours often necessitates the use of pumps.

In the Klondyke during 1902, the cost of extracting a cubic yard of dirt varied from 10s. to 15s. The gold-production of the region, which attained a value of £500,000 in 1897, increased gradually until it realized £4,600,000 in 1900, but fell to £2,540,000 in 1903, the decrease being attributed to the rapid exhaustion of the richer deposits. J. W. P.

THE MINING DISTRICT OF POTRERILLO, CHILE.

El Mineral de Potrerillo. By Enrique Kaempffer. *Boletin de la Sociedad Nacional de Mineria, 1903, series 3, vol. xv., pages 408-415 and 1 plate.*

This district lies in the south-eastern portion of the department of Chañaral (province of Atacama), some 54 miles distant from Pueblo Hundido, the present eastern terminus of the Chañaral State Railway. It includes the Hueso mountain-range, and the neighbouring massifs, at an average altitude of 10,440 feet above sea-level. The climate is that common to the Andine region, involving brusque changes of temperature and occasional snow-storms. Timber and water, if not exactly available on the spot, are to be obtained easily enough in the near neighbourhood of the mines.

The author visited about two dozen mines, and he mentions about a score of others which he had not the time to see. The ores are mainly carbonates and oxides of copper, slightly auriferous and argentiferous. Native copper is of comparatively rare occurrence. The gangue varies greatly in character and composition; and in some cases the ore can be separated from it with the greatest ease, while in other cases exactly the reverse statement applies. The ore-deposits are generally " bedded," occurring less commonly in the form of veins.

A table of assays is given, and one company worked between December 31st, 1899, and May 31st, 1903, a total of 14,564 tons, assaying on an average 14·94 per cent. [of metal]. The author considers that the conditions are favourable for erecting the necessary plant for treating the ores on the spot in the wet way. L. L. B.

METALLIFEROUS MINES OF NAICA, MEXICO.

Apuntes sobre el Mineral de Naica. By Leopoldo Salazar. *Memorias de la Sociedad científica " Antonio Alzate," 1903, vol. xix., pages 71-83.*

Naica lies some 33 miles north-west of La Cruz station on the Central Mexican Railway, in an immense arid plain, out of which rise a few small hills.   The construction of a good road for wheel-traffic, bringing the mines into direct communication with the railway, should offer no difficulty. As it is, a road communicates with the Conchos station, farther north, and a longer road with Santa Rosalía, the chief town of the district of Camargo, to which Naica belongs. The climate is, on the whole, cool, but the region is a desert which does not grow a stick of timber. Mines were worked there in the olden time, although not much trace of them remained at the period of the author's visit in 1899. Difficulties of water-supply account for the former abandonment of mining operations; but the face of things has changed completely within the last 4 years, and the district has now become one of the most important mining centres of the state of Chihuahua. The predominant rock of the Sierra de Naica is a compact grey limestone of Cretaceous age, containing lenticular cavities of comparatively-low dip (10 to 30 degrees) and of very irregular breadth, infilled with metalliferous ores, from which nearly-vertical metalliferous seams or veins run up to the surface; it is quite possible that similar ramifications extend from the main ore-bodies downward. Several assays made by the author show the ores to be chiefly silver-lead ores, in some cases auriferous and cupriferous, the percentage of lead varying from 40 to 60, and of silver from ¼ to nearly 1¾.   L. L. B.

THE MINES OF THE PROVINCE OF ANGARAES, PERU.

La Minería en Angaraes.   *By* W. A. ZUMAITA.   *Boletín del Ministerio de Fomento,* 1903, *No.* 8, *pages* 62-67.

The writer, making an official report to the Peruvian Government, points out that to enumerate the number of abandoned mines in this province would be a wearisome task: at one time or another in the course of ages, several thousands have been worked there, but few were on any considerable scale. A list is given of half-a-dozen fairly-important copper, blende and lead-mines abandoned within recent years, and it is stated that the chief reasons why the mineral industry does not prosper in that remote province are:—(1) Difficulties of transport, obstructing both the export of ores and the introduction of machinery; and (2) the want of smelters for obtaining copper-silver mattes on the spot.   If a smelting-works with reverberatory furnaces could be established, it would give the mineral industry of the whole province a fresh lease of life.

A catalogue is given of abandoned smelters and amalgamation-works, but there is at any rate one of the former still in active operation, at the San Pedro mine. The Sullac establishment, where the lixiviation-process was formerly carried on, is now being converted into an amalgamation-works.

Besides innumerable auriferous and argentiferous veins, there are in the Caruapata district and in the neighbourhood of Ingalmasi, abundant deposits of native sulphur. Coal, of middling quality, occurs on the Yana-hututo estate and in the highlands of Pirca. In the Acobamba district, coal-seams yielding remarkably-little ash have been found, but the attempt to work them has been given up, as the mineral crumbles so easily that transport is difficult. In the neighbourhood of the coal are great " reefs " of perfectly pure iron-ore. Moreover, oxides and carbonates of copper and antimony-ores are abundant, and of widespread occurrence. So, too, are the deposits of silver-bearing zinc-blende and of gypsum. A thick vein of orpiment and realgar occurs among the highlands of Yanahututo, and "native arsenic" is reported from Caruapata. In the neighbourhood of Julcani, there is a tungsten-deposit associated with gold, and there are abundant deposits of pyrolusite. Mercury and bismuth-ores have been found, but apparently not in payable quantity. L. L. B.

MINING IN THE PROVINCE OF HUALLANCA, PERU.

La Minería en Huallanca. By ESTENIO J. PINZÁS *and* FELICIANO PICÓN. *Boletín del Ministerio de Fomento,* 1903, *No.* 12, *pages* 37-53.

The city, which gives it name to the province, is situated at an altitude of 11,320 feet above sea-level, and two smelting-works are now in full activity there. Very brief descriptions are given of the various mines in the ravines of Capacog, Tucapag and Contaicocha; these apparently work metalliferous ores containing handsome proportions of silver and gold. In the Cerro de Huanzalá, argentiferous and auriferous copper-ores, in places associated with iron-pyrites, are worked.

Anthracite, often of unexceptionable quality, occurs in thick seams (3½ to 6½ feet) among the sandstones of the Queropalca district, and extends over a considerable area. It is stated that "hundreds of millions of tons" of coal are in sight. At present four mines are working the mineral. Abundant seams of bituminous coal have been struck in the districts of El Cercado and Pachas.

Gold-placers are known to exist in many localities, and gold is washed in the dry season by the natives ("Indians") on a small scale.

Bituminous limestones "sweating petroleum" (?) have been recorded in two localities.

The roads appear to be in a lamentable condition, and the management of the mines generally is defective in technical knowledge and skill. L. L. B.

THE CINNABAR OF HUANCAVELICA, PERU.

El Cinabrio de Huancavelica. By A. F. UMLAUFF. *Boletín del Cuerpo de Ingenieros de Minas del Perú,* 1904, *No.* 7, *pages* 1-62, *with* 16 *figures in the text and* 2 *plates.*

The first chapter of this memoir is devoted to a survey of the economic conditions affecting the output and market-price of mercury generally, which leads to the conclusion that there is no present prospect of any notable increase in the consumption of that metal, but on the contrary some probability that there will be a distinct diminution in the demand for it. The present annual import into Peru averages 30 tons, free of duty, but it is thought possible by some that the deposits about to be described may well render that country in the future independent of foreign supplies, nay, even convert Peru from an importer into an exporter of mercury.

The cinnabar-mines of Huancavelica have a lengthy history behind them, the Santa Barbara mine having been purchased by the Spanish Government as long ago as the year 1570. Mining operations, attended with more or less administrative mismanagement, incompetence and peculation, continued on the Santa Barbara property until 1785, when "great reforms" were introduced; and by the year 1800, the number of "managers" and superintendents connected with that mine alone amounted to 70. The Spanish viceroys claimed that the Huancavelica ore-deposit was "the greatest wonder of the world," and they seem to have regarded it as the most considerable asset of the dependency over which they ruled. Its great importance lay, of course, in the utilization of the mercury for the amalgamation of the silver got from the innumerable mines worked during the period of the Spanish colonization. After the liberation of Peru from the Spanish yoke, various spasmodic attempts were made, at intervals of many years, to re-open the workings at Huancavelica, but up to the present time nothing has been done on a scale worth mentioning.

The city of Huancavelica lies in 12 degrees 53 minutes south latitude and 75 degrees 4 minutes west longitude of Greenwich, at a height of 12,438 feet above sea-level; the climate is fairly rigorous, abundant rains falling from December to April, and heavy frosts beginning in May. Mineral springs of medicinal value abound. On the south extend the ancient workings of the Santa Barbara and other mines: in none of these, opines the author, will it ever be practicable to resume mining operations, for these would prove dangerous as well as unduly expensive. Many parts of the old workings have collapsed, or are obstructed by falls of rock; and, apart from this, one is never certain in the area described, where one is going to come upon old workings or landslips. Consequently, a certain quantity of cinnabar is irrecoverable; but outcrops of the ore are recorded from eleven localities south and east of Huancavelica, many of which are untouched. The cinnabar appears in these to occur indifferently in limestone, sandstone, and conglomerate, occasionally in the form of veins. In the San Jerónimo or Quishura ridge, however, the cinnabar is found as the partial infilling of a fissure (10 to 13 feet broad) in a mighty mass of basalt. The occurrences of mercury-ore north of Huancavelica do not appear to be of much commercial importance.

The sandstones and limestones above-mentioned are shown, by the fossils (which the latter rocks especially contain in abundance), to be of Cretaceous age. The coarse conglomerates, largely made up of angular fragments, mantle over a considerable portion of the surface, and are evidently of post-Cretaceous age. Some shales are in places interbedded with the sandstones and limestones, and andesite-dykes and bosses (with which iron-pyrites is almost invariably associated) traverse the sedimentary Cretaceous rocks. In the basalt-ridge already mentioned, south-west of Huancavelica, the fissure-veins are mineralized with silver, lead and copper, as well as mercury. In the sandstones, the cinnabar is, on the whole, an impregnation-deposit; in the limestones, it appears to be the outcome of sublimation; and in the conglomerates, that of sedimentation. So it occurs in every form that can well be imagined: isolated granules, veinlets perpendicular to the bedding, lenticular masses, gash-veins, stockworks, pockets, etc. The cinnabar in the conglomerates really consists of particles surviving from the erosion and disintegration of the original matrix-rocks, and washed into their present position by the acid waters which continuously percolate through the conglomerates. The ore is manifestly of later formation than the sandstones and limestones which are its matrix, and is genetically associated with the volcanic phenomena of which the andesites and basalts are the silent witnesses.

Native mercury has been found on the river-banks and in the tributary glens in this area, and is evidently derived from the decomposition of portions of the cinnabar-deposits. Choice specimens of the cinnabar yield from 7 to 10 per cent. of the metal, while the average assay (taking all grades together) is 2 per cent. Enormous waste-heaps remain from the old mines, but it is uncertain whether they would repay working.

With regard to fuel, the department of Huancavelica possesses abundant resources, in the shape of coal-seams, bituminous shales, peat-deposits, and growing timber (the last-named being also available for structural purposes). Very little exploration-work has been done on the coal-seams so far. Labour is easily obtainable locally, and is cheap; and many watercourses could be utilized for hydraulic power. From Huancavelica, the distance by road to Oroya railway-station, the present terminus of the line from Callao harbour, is 146½ miles; the distance to Ica, whence a railway runs to Pisco, another harbour on the Pacific seaboard, is about 186 miles, and the road is in many respects not so good. In an appendix, the author describes the 103 hand-specimens of rocks, fossils and minerals which he collected in the district and brought back with him.   L. L. B.

USE OF CALCIUM CARBIDE AS AN EXPLOSIVE.

Etude sur le Carbure de Calcium employé comme Explosif dans les Travaux miniers.
By M. P. S. Guédras. *Comptes-rendus hebdomadaires des Séances de l'Académie des Sciences,* 1904, *vol. cxxxix., pages* 1225.1226.

The author has made experiments, with the view of showing that this substance may be regarded as a possible substitute in mining operations for dynamite or gunpowder, and that it may be similarly used for military purposes. He first of all granulates the carbide, then loads it in a water-cartridge, and in a specially-shaped cavity of the cartridge introduces an electric fuse. A drill-hole being máde in the rock in the usual fashion, the cartridge is thrust in, the remaining space is stemmed with wood, and the percussion-rod (which breaks through the thin partition that separates the water from the carbide) having been struck, an interval of 5 minutes is allowed for the evolution of the acetylene gas; then the cartridge is fired by electricity. It is found that the effect of the explosion is not to project masses of rock in every direction, but the rock in which the shot has been fired is easily brought down by the miner's pick. The quantity of calcium carbide used weighs about 1½ ounces, from which about ½ cubic foot of acetylene gas is evolved.   L. L. B.

STORAGE OF NITRO-GLYCERINE EXPLOSIVES.

Essais de Congélation des Explosifs à Base de Nitro-glycérine. By Léon Saclier. *Comptes-rendus Mensuels des Réunions de la Société de l'Industrie Minérale,* 1904, *pages* 37.42, *and* 1 *plate.*

Mine-owners who have used nitro-glycerine explosives have experienced great difficulty in preventing their congelation, and also in providing for their storage. When first employed in French mines, these explosives were stored in surface-depôts, unwarmed, where they froze in winter; and it became necessary before using them to thaw them in a water-bath, which caused so many accidents that attempts were made to substitute underground for surface-depôts. In underground magazines, with constant temperature, the frozen explosive returned to its plastic state, and that which was stored at the temperature suitable for use retained all its good qualities, so that underground magazines constituted a step in advance, while for 15 to 20 years accidents were avoided; but the explosion of the Aniche magazine caused the quantity of explosives stored in underground magazines to be limited to 2 cwts. (100 kilogrammes).

This limit of quantity in mines where the daily consumption varies from ⅓ to 1¼ cwts., obliges the mine-owner to renew his supply daily, involving the daily carriage, handling, letting down, and storage of the explosives, in addition to which dynamite, which is stored underground in the frozen state, has not time to thaw before being used, so that the advantages of underground magazines are lost. This carriage and handling, with the absolute impossibility of thawing naturally, constitute a far greater danger than that which it is intended to avoid by the above-named limit of deposit; and the author hopes that a serious accident will not occur, to demonstrate the correctness of his opinion on the subject. It is therefore desirable that the question of storing dynamite in underground magazines should be thoroughly thrashed out, with a view to permitting the storage of 4 cwts. (200 kilogrammes) of explosives in such magazines, so that their thawing may take place naturally, thus avoiding dangerous carriage and handling.

J. W. P.

DAMMING OFF WATER BY CEMENT-GROUTING OF THE NATURAL WATER-CHANNELS.

Die Wasserabdämmung beim Abteufen des Pöhlauer Schachtes der Gewerkschaft Morgenstern in Reinsdorf (in Sachsen) durch Versteinung der natürlichen Wasseradern. By ALFRED WIEDE. *Berg- und Hüttenmännisches Jahrbuch der k.k. Bergakademien zu Leoben und Přibram und der königlich ungarischen Bergakademie zu Schemnitz,* 1902, *vol. l., pages* 173-182 *; and Jahrbuch für das Berg- und Hüttenwesen im Königreiche Sachsen,* 1901, *pages* 66-73, *and 2 plates.*

The Morgenstern Colliery Company started, on September 3rd, 1900, sinking a new shaft at Pöhlau, near Reinsdorf, in Saxony. This shaft, intended chiefly for ventilating purposes and for taking the miners up and down the pit, was put down about the centre of a newly opened-up coal-district, immediately north of the high road from Zwickau to Liechtenstein, on the summit of the Pöhlau ridge, which strikes east and west. The shaft was designed to be circular, 13½ feet in internal diameter. The uppermost strata were found to consist of a greyish-yellow impermeable loam, passing deeper down to a dark-grey, containing some rolled pebbles of brown coal and many chalk-covered flint-nodules. At the depth of 72 feet, weathered Rothliegende conglomerate was struck, but soon the unweathered tough rock was got into, and walling of the shaft with hard-burnt bricks was begun. Meanwhile, as they went down through the conglomerates, water began to come in, the flow varying from 3 to 50 gallons per minute. Iron tubing was inserted in the walling at the points where the water-feeders were observed, and the water was led down to a sump whence it was pumped out of the pit by an electrically-worked differential plunger-pump, the maximum efficiency of which was reckoned at 132 gallons per minute. The possibility of a larger flow was not foreseen, as nothing approaching that had been experienced in the other shafts of the Morgenstern colliery. But, at a depth of 206 feet, another feeder was met with, the water from this alone averaging 132 gallons per minute, and as the pumping-engine was already dealing with 44 gallons a minute from points nearer the surface, it could no longer suffice to meet all the calls made upon it.

Under these circumstances, the engineers resolved to see whether the inflow of water could not be stopped by forcing cement-grouting under pressure into the natural water-channels as far in as possible, the "grouting," as it hardened, forming an impenetrable stone-filling of these same chanels. The subterranean watercourses were traced parallel to the stratification with a northerly dip of 8 degrees, in a soft sandy layer intercalated among the conglomerates; at the three main points of inflow, grouting was pumped in (to the amount of 292 bushels of dry cement). The shaft was then walled round, and through the walling more grouting was pumped into the strata. The operation was completely successful, and when, on a still worse feeder at a depth of 270 feet being encountered, the process was repeated, equally-satisfactory results were obtained. For fairly-broad fissures a mixture of a volume of dry cement with an equal volume of water was used; with narrow fissures it was found advisable to dilute this rather more, on account of the increased resistance opposed by friction to the flow of the mixture.

The process does not involve any considerable expenditure of time or money. It is not yet possible to say whether, under such circumstances, pumping operations may with safety be completely abandoned, but at all events they may be reduced to a minimum. Many springs in the village of Pöhlau, which had dried up during the inflow of water into the above-described shaft, began immediately to flow again, after the petrification of the subterranean water-channels had been successfully accomplished.

L. L. B.

_____

COLLAPSE OF A SHAFT AT GLADBECK, WESTPHALIA.

Der Einsturz des im Abteufen begriffenen Schachtes IV der Zeche Graf Moltke am 16. März 1903. By — FREUND. Zeitschrift für das Berg-, Hütten- und Salinenwesen im Preussischen Staate, 1904, vol. lii., pages 195-199.

The Nordstern Colliery Company had been instructed by the Royal Mining Bureau at Dortmund, to sink a new shaft at their Graf Moltke pit, near Gladbeck, in order to improve the ventilation of the mine, and so sinking was begun in November, 1902. A depth of 259 feet had been reached when, on March 16th, 1903, the lower portion of the shaft fell in, and an overman and 6 workmen were killed.

The condition of the strata having been ascertained when sinking the closely-adjacent shaft, No. 3, in 1900-1901, no great difficulties were expected; and in fact the 26 feet of superficial deposits were sunk through to the comparatively-dry Emscher Marl without any hitch. In this the total inflow of water was found to amount to 132 gallons per minute; it was dealt with by means of a pulsometer-pump, and there was every hope of reaching the Coal-measures at a depth of 890 feet or so.

The disaster was preceded by the appearance of two almost vertical cracks in the strata, then a wedge of rock, 6½ feet broad, fell forward and exerted a very noticeable pressure on the provisional timbering. Thereafter, small masses of rock began to break down continually as the tubbing was being carried on, the timbering cracked ominously, and though none of these signs escaped the attention of the overmen who were in charge of the work, no precautionary measures were taken, for these overmen were anxious to get the work over as quickly as possible, apparently at any risk. Walling had been built to a height of about 50 feet, when at 6 a.m. on March 16th,

1903, the cracking of the upper timbering became so loud and continuous
as to cause genuine alarm, and the men engaged in the sinking began
to clamber up towards the cage.   But before they could reach it, the
timbering gave way, the strata between the wall and the wedging-ring
fell in and smashed down the staging, and several men were hurled, with
the flying débris, to the bottom. All lamps went out, and a continuous
shower of fragments of marl began, under which the victims of the disaster
were subsequently found buried. One of the two overmen, to whose neglect
the accident was due, perished in it, and the other fled from the district,
with warrants out against him. Stringent regulations have been drafted
by the colliery-officials to prevent the recurrence of a similar disaster in
future sinking-operations.

At the Mönopol colliery in Westphalia, an accident, resembling in many
respects that just described, occurred in October, 1893, and the conclusion is
drawn that in sinking through the soft, crackly Emscher Marl, special
attention must be given to the strengthening of the walls of the shaft.

L. L. B.

ROCK-THRUSTS IN WESTPHALIAN COLLIERIES.

*Die in den letzten Jahren auf Steinkohlengruben des Oberbergamtsbezirkes Dortmund
vorgekommenen Gebirgs-stösse und die hierdurch herbeigeführten Unfälle.
By — DILL.   Zeitschrift für das Berg-, Hütten- und Salinenwesen im
Preussischen Staate, 1903, vol. li., pages 439-466, with 7 figures in the text
and 3 plates.*

For some time past, the Dortmund mining-district has had to record
almost every year a number of accidents due to so-called thrust-movements
in the rocks. Fortunately, the victims claimed by these accidents are far
fewer in number than in the case of ordinary falls of stone and coal. Such
accidents deserve, however, investigation, as in many instances it is the
actual progress of mining operations which brings them about, by gradual
destruction of the natural support of a given mass of rock.

In the Ruhr coal-field, the name of rock-thrust (*gebirgs-stoss* or *gebirgs-
schlag*) has been applied to a hitherto unexplained form of earthquake,
apparently started by sudden release of tension in a rock-mass, and occurring more especially when safety-pillars are being removed. With a report
like the boom of artillery and a violent gust of air, the coal-face is torn
asunder, and the coal is shot out in fragments along the levels: the floor
buckles up, the timbering is knocked over (though not shattered), and every
object near the site of the accident is whirled along or struck down by
the falling masses of rock.   So sudden and violent a phenomenon as this
generally implies death for any workman near the spot. Subsequent investigation sometimes reveals the presence of a considerable quantity of fire-
damp and shows the roof to be generally uninjured.   Above-bank, these
rock-thrusts, so far as they make themselves felt, resemble an ordinary earthquake. Curiously enough, they usually occur in those parts of the workings
which are in the immediate neighbourhood of the goaf, or form a sort of
island in the midst of it, or where considerable activity in mining operations
has been attended by partial or complete neglect of packing, and where
the seams have an unusually-sound roof, by no means liable to crack.

The first three accidents of this kind described by the author occurred in
the Frederick-the-Great pit, at Herne, in 1896 and 1897. In two out of the

three, a man was killed in each case, and in the third a hewer was disabled for life. Packing, previously considered unnecessary, was afterwards. resorted to. An accumulation of fire-damp was reported in one case.

Then, two such accidents took place at the Victor pit, at Rauxel, in 1898. and 1899. Here the same Sonnenschein seam was being worked, and under the same conditions, as at the first-mentioned colliery, but in the eastern district of the workings packing was customary. Not so in the western district, where the two accidents occurred. In both cases, some men were badly injured. The considerable evolution of fire-damp which followed upon the second accident may not have been a mere invasion of gas from the goaf, but may have arisen from the sudden crushing and spreading out of great masses of coal (characteristic of a rock-thrust) which would liberate the gases. occluded in the pores of the mineral. Precautions were afterwards taken,. in the way of safety-pillars, timber-supports, and packing where necessary.

The rock-thrusts, which took place at the Shamrock I. and II. collieries, at Herne, in 1897 and 1899, differed from those so far mentioned, in that their effects were especially marked above-bank and not so much below. Here again, they seem to have been due to the working-away of the Sonnenschein seam. The shock felt in the town of Herne, in 1897, was so violent that the affrighted inhabitants ran out of their houses into the streets, and the thunderous rumbling which accompanied the earthquake heightened the general feeling of insecurity and terror. Walls were fissured, chimneys. and roof-slates came crashing down, and the effects were felt over an area which would be represented by a circle with a diameter of 2¼ miles. Packing, in that district of the colliery which was mainly affected, had only been begun a short time before the accident here described.

The most destructive in its effects of all the rock-thrusts, so far recorded. in Westphalian collieries, was that which took place on July 14th, 1899, at the Recklinghausen I. colliery, at Bruch. Many of the workings in the Sonnenschein seam were totally or partly destroyed, 4 workmen were killed, and a number of others more or less seriously injured. Aboveground, the shock was felt throughout the parishes of Bruch, Herne, Recklinghausen, Gelsenkirchen, Bochum, Castrop, etc., in fact over an area which would be covered by a circle not less than 6¼ miles in diameter. In the pleistoseismic portion of this area, from 100 to 150 separate buildings suffered damage.. Full details are given of the destruction wrought within the colliery itself. Special regulations were afterwards drawn up, as to the precautions to be observed in future in working the Sonnenschein seam there and elsewhere in the Herne district. No further rock-thrusts have been recorded in that particular district since then.

On March 16th, 1900, at the Steingott colliery, Laura shaft, near Altendorf, a sudden rock-thrust, again in the Sonnenschein seam, cost two workmen their lives.

The author then describes similar accidents which occurred in 1902 and 1903, at collieries near Eppendorf and Hugo respectively, in levels where the Finefrau and Bismarck seams are worked. These were also attended with loss of life. Stress is again laid on the partial or total absence of packing.

The possible causes of such rock-thrusts are discussed at considerable length, but although the author evidently believes them to be the indirect result of very active mining operations, he does not commit himself definitely. to an hypothesis which would exclude other possible causes.　　L. L. B.

WATER-PACKING OF SEAMS.

Bericht über das Schlammversatzverfahren auf den oberschlesischen Bergwerken sowie auf der Zeche Sälzer-Neuack bei Essen.—Erörterungen über die mögliche Anwendbarkeit dieses Verfahrens im Ruhrkohlenbezirke unter besonderer Berücksichtigung der Verhältnisse auf Schachtanlage Alma. By KARL MÜLLER AND — HUSSMANN.   *Glückauf,* 1903, *vol. xxxix., pages 927-941, with* 18 *figures in the text and* 1 *plate.*

The authors describe a method of filling goaves in coal-mines by fine sediment carried into place by water. The working of thick seams, without stowing the goaf, is attended by great damage to the surface, the loss of up to 35 per cent. of the coal, and damage to adjacent seams. The cost is also high, due chiefly to the large size and quantity of timber needed. Moreover, the height of the working-places is a danger to the workmen. Stowing the goaf has not been much adopted, as it costs too much to bring material from bank, and the material having been got into the mine, it will not gravitate into place when the inclination of the seam is only from 10 to 15 degrees. The sluicing system was introduced 2 years ago at the Kattowitz and Myslowitz collieries, and has been found cheaper than stowing by hand. Whether the sand, etc., will give permanent support, time alone will show. But, up to the present, although the surface-level has been slightly lowered, the adjacent seams have not been damaged by settlement.

Briefly, the system is as follows:—The goaf is packed by some fine material, such as sand, breeze, cinders, etc., carried into place through pipes by a flow of water. The proportion of water to material is from 2 to 1 part to equal parts, according to the nature of the material used. The mixture is made either at bank or near the goaf to be filled. When the material has been deposited, the water is drained off, filtered and collected. It is then pumped to bank, and in some cases used over again.

Myslowitz Pit.—Three seams of coal are being worked, from 10 to 60 feet thick. Near one of the shafts there is a large sand-bed upon the surface, from which the sluicing material is taken. Tip-wagons are filled by a sand-dredger, and taken by an electric locomotive to the shaft. The sand is poured down into a recess, which has been formed in the shaft-side, on to a grid. Above this grid, a pipe-nozzle is fixed, which sprays water on to the sand. Below the grid is a funnel to catch the mixture of sand and water, and convey it into the range of pipes. The mixture used here is 1½ parts of water to 1 part of sand, but the sand is very clean and free from clay. This leads them to hope to manage with an equal parts mixture. The pipes used are 20 feet (6 metres) long, 6·61 inches (168 millimetres) in diameter and 0·24 inch (6 millimetres) thick. The pipes convey the sluicings 300 feet (90 metres) vertically and 2,000 feet (600 metres) along the level underground. The pipes soon wear out, especially on the under side, but they are turned four times, through a quarter-arc each time, and last from 2 to 2½ years. They are best placed on the ground to facilitate turning, and all T pieces and bends should be made thickest on the outer side of the bend, as they wear most there, lasting only 3 or 4 months. To overcome stoppages, short pipes are inserted. These can be readily taken out, and give convenient access. The range is always flushed before stopping, except on the steep lengths. There is telephone-communication between the chief valves underground and the mixing-station at bank.

The lower seam, 36 feet (11 metres) thick, is won by pairs of levels; the space between is cut up into pillars, which are taken out one by one, beginning with the lowest. As soon as a pillar is taken out, the empty space left is dammed off by timber, caulked with dried horse-dung, and the sluicings turned on. When the space is filled, the sluicings are turned off, and the water drained away, leaving a solid pack. The water, while flowing along to the shaft through a deep level 3,300 feet (1,000 metres) long, deposits any fine dirt held by it, and arrives at the shaft clean enough to be easily pumped. These levels are cleaned every 3 months.

The timber used, while taking out the pillars, is not drawn before sluicing the goaf, but less is put in than when filling the goaf with stone by hand. The timber for the dams is all drawn and used again. By this system, barrier-pillars to protect the temporary incline-banks are not needed to such an extent, and more coal can be got from each district.

Ferdinand Pit.—The sand used here is not so clean, and requires a slightly-different form of nozzle and mixing-tray. The proportion is 2 parts of water to 1 part of sand. The surface here is built over, and has to be supported. The seam is 8·2 feet (2½ metres) thick, and lies at an angle of 15 degrees to the horizontal. The pillars are formed 40 feet by 50 feet (12 by 15 metres), and the coal is taken from them by lifts of 26 to 33 feet (8 to 10 metres). All the coal is extracted. The timber in the brokens is lost, but the timber forming the dams is regained. The drain-water is cleaned, by allowing it to settle in a standage. The clay in the water seems to have a cleaning action.

Alfred Shaft, Hohenlohe.—At this colliery, a bore-hole 19·7 inches (500 millimetres) in diameter and 100 feet (30 metres) deep is used. The dry sand is poured down this hole into the mine, where it is mixed with the water. The hole is only walled at the top and partly piped, but it has never been stopped. There is nothing else differing from the previously mentioned arrangements.

Arnold Shaft, Borsig Works.—The seams here are flat, and the water is forced through the pipes under a pressure of 5 atmospheres.

Concordia Pit.—The mixture used here is 1½ parts of water to 1 part of breeze. This kind of sluicing is expected to squeeze more in the goaf, and takes four weeks to become solid, but the method has not been long enough on trial to give definite results. The breeze gives off fumes when first put in place, and is apt to give the men headaches, if the ventilation is at all slack.

Salzer-Neuack-Krupp Colliery.—Here again the surface has to be supported, it being built over by Messrs. Krupp's works. The mixture used is 2 parts of water to 1 part of material. The pillars are made 60 to 66 feet (18 to 20 metres) long, and removed by 10 feet (3 metres) lifts. The sand is taken down the pit dry, after being screened through a mesh of 3·15 inches (80 millimetres) and over a mesh of 1·18 inches (30 millimetres). The sluicings shrink about 10 per cent., so that the goaf is only filled 90 per cent. The cost of hand-stowing was 1s. 8d. (1·65 marks) per ton. The cost of water-packing, less the costs of the pipes and pumping is 9·6d. (0·80 mark) per ton. The estimate for these last items is 2·4d. (0·20 mark), which makes the total cost of water-packing the goaf 1s. (1·00 mark) per ton. But this cost could be reduced, if the mixing was done at bank instead of in the pit.

The writer discusses the feasibility of adopting this system in the Ruhr coal-field. He is of opinion that it cannot be generally adopted, but might be used in cases where special support is required. No sand is available, but some collieries have washeries, which might provide suitable material. The seams are inclined up to 50 degrees, which will make the pressure on the dams very great. A suggestion is made that a trial could be made in the Matthew seam of the Alma pit; and details of the requirements are given, including an estimate of the probable cost per ton.   C. H. M.

SUBSIDENCES DUE TO SALT-WORKINGS IN FRENCH LORRAINE.

Note sur les Affaissements produits en Meurthe-et-Moselle par l'Exploitation du Sel.
By L. BAILLY. *Annales des Mines*, 1904, series 10, vol. v., pages 403-494, *with 14 figures in the text and 3 plates.*

In the department of Meurthe-et-Moselle, during the last 30, and more especially during the last 10, years, phenomena, remotely comparable with those that are so well known in the Cheshire salt-field, but on a much smaller scale, have made their appearance, and they seem likely to become still more frequent as time goes on. In what remains to France of Lorraine, the Middle Keuper salt-belt extends uninterruptedly for about 19 miles from Tonnoy to the frontier, with a breadth of over 9 miles from Rosières-aux-Salines to Nancy: the mining concessions taken out, so far, cover little more than a third of the area. The thickness of the salt-deposit varies between 33 and 230 feet; it is said to constitute one of the finest deposits of the kind in the whole world, and to be " ten times richer than those of Cheshire." For various reasons, the western limits of the field are unknown, but the Madon valley on the south-west might repay exploration-work. The rock-salt lies deeper down (300 feet or more below ground) than in Cheshire, is consequently better protected from the action of surface-waters, and is not overlain (except in a few instances) by a natural brine-lake, as is so often the case in the above-mentioned English county.

A short history is given of the salt-industry in Lorraine; and the author then remarks that his investigations do not apply to the actual rock-salt mines of Rosières-Varangéville, St. Nicolas and St. Laurent, which a recent Government enquiry has shown to be in no way concerned with any important subsidence in the area since 1873. The output of rock-salt, mined as such, in French Lorraine averages 100,000 tons yearly, while that extracted in the form of brine averages 450,000 tons *per annum*. The author then deals with the method of working by " dissolution-chambers "; where no natural brine-lake covers the salt, fresh water is introduced from the surface by means of pipes, and after a time a cavity of sufficient dimensions is formed to allow of regular brine-pumping for several hours daily. On account of the different densities of the water according to its more or less complete saturation with salt, these cavities tend to enlarge asymmetrically, extending more to the rise than to the dip of the strata; and in plan they form roughly an ellipse, of which the major axis trends parallel with the dip. The intervention of marly bands, even when saliferous, constitutes a great obstacle to the enlargement downward of the " dissolution-chambers." Nevertheless, the soft Keuper Marls, which form the roof of the salt-deposits, crumble away rapidly on contact with water, with frequently-resulting damage to the bore-holes and the pumping-apparatus. With the slow subsidences, actual or possible, that are traceable to the natural brine-lakes of the Sanon area, the author does not concern himself; but to the dip of that area, where the Middle Keuper sinks below the very compact dolomites of the Upper Keuper, and these in turn are succeeded by the Infra-Liassic grits and the Sinemurian limestones, lies the valley where the only rapid subsidences so far recorded in the region have occurred. One took place at Ars-sur-Meurthe in 1876, another at St. Nicolas (only a third of a mile away) in 1879: elsewhere in the Meurthe valley there have been subterranean landslips, but none the effect of which has been evident at the surface.

The third chapter of the memoir is devoted to the workings connected with the natural underground brine-lakes of Dombasle and Einville, and to the subsidences which energetic brine-pumping has brought about in that district. The actual damage done to buildings, roads, canals and railways does not seem, however, to have been very great.

The fourth chapter deals with preventive and curative measures, some decreed by the French Government on expert advice, others suggested for future use. On the whole, the subsidences observed in Meurthe-et-Moselle bear no sort of comparison, in regard to their destructive effects, with those of Cheshire; in England it has been found necessary to invoke the aid of the "heavy Parliamentary traction-engine." The French engineers hope to avoid that necessity in their own case, by means of friendly negotiation between the interested parties. Meanwhile, they will avail themselves to the full of the experience so dearly bought in Cheshire. The strict methodizing of the pumping operations, involving an unsleeping control of that "blind and fickle mining-tool—water," appears to furnish the key to most of the difficulties connected with the subject.

L. L. B.

KARLIK REGISTERING SPEED-INDICATOR FOR WINDING-ENGINES.

Tachymètre Enregistreur Karlik. By H. Kuss. *Comptes-rendus Mensuels des Réunions de la Société de l'Industrie Minérale,* 1904, *pages* 35-36, *and* 1 *plate.*

Thanks to its simplicity, the Karlik tachymeter has rapidly come into general use. This instrument keeps a rapid and stringent check upon the working of winding-engines and the attention of the enginemen. The graphic diagrams show clearly the difference in speed for winding coal and that for men. Each caging operation is truly registered; and even the raising of the cage the height of one deck to another is exactly indicated. Every stoppage in winding, any slackening of speed and any excess of speed, is immediately put on record. In a well-regulated mine, manuscript notes will explain, as far as possible, any apparent anomaly; and the manager is able to check, day by day, by merely examining the cards, which are very clear and easily handled, all particulars of the winding.

J. W. P.

GUIBAL FAN WITH COLLECTING VOLUTE.

Résultats d'Expériences sur un Ventilateur Guibal à Volute collectrice, établi au Siège No. 2 des Charbonnages de Fontaine-l'Evêque. By E. Lagage. *Publications de la Société des Ingénieurs sortis de l'Ecole Provinciale d'Industrie et des Mines du Hainaut,* 1904, *series* 3, *vol. xiii., pages* 106-113, *and* 1 *plate.*

Owing to the low mechanical efficiency of the existing fan and the air-volume having become insufficient, it was determined to erect a Guibal fan,

21 feet (6·4 metres) in diameter, 5·9 feet (1·8 metres) wide, with an inlet 8·85 feet (2·7 metres) in diameter and a volute collecting case, making 107 revolutions per minute, driven by a belt from an engine of 110 horsepower. As the shutter of the ordinary Guibal fan had been retained, notwithstanding the volute form of the case, experimental trials were made for determining its most favourable position, with the result that, for equal speeds, the largest volume was obtained when the shutter was raised 15·74 inches (40 centimetres).

The tests for efficiency, often repeated, with the same results, justify the conclusion that the Guibal fan thus modified can bear comparison with the best ventilators, as regards both efficiency and water-gauge. It was found that, when the equivalent orifice diminished, the efficiency diminished appreciably, while the water-gauge varied but slightly. These results prove that the Guibal fan is specially suitable for mines with large orifices; but it does not follow that this type is not also suitable for mines with small orifices. All that is required is to fix the dimensions of the fan in accordance with the orifice under consideration.   J. W. P.

IMPROVED VENTILATION THROUGH SINKING A THIRD SHAFT.

Note sur l'Amélioration d'Aérage obtenue au Siège No. 1 de la Société des Mines de Liévin par la Création d'un Puits spécial de Retour d'air. By — Morin. Comptes-rendus Mensuels des Réunions de la Société de l'Industrie Minérale, 1904, pages 45·48, and 2 plates.

The workings of No. 1 pit, Liévin colliery, were ventilated by No. 1 or the upcast pit and No. 1 *bis* or the downcast pit. The former was provided with a Guibal fan, 29·53 feet (9 metres) in diameter, and a second in reserve having a single inlet, 9·84 feet (3 metres) in diameter, running normally at 60 revolutions per minute and delivering 1,942 cubic feet (55 cubic metres) per second with a water-gauge of 2·36 inches (60 millimetres), the equivalent orifice being about 30 square feet (2·8 square metres).

With a daily output of 1,250 tons, the fire-damp content in the main return-airway was 0·4 per cent. The content in the most highly charged return-airway did not as a rule exceed 1 per cent., but that of several return-airways in the upper levels, which ventilated old workings, attained 1·2 per cent. generally and 1·5 per cent., or even 1·6 per cent. with a low barometer. The enlargement of the numerous return-airways, however, did not appear capable of affording any appreciable improvement; the increase of sectional area, long and costly to carry out, no longer producing sufficient increase in the ventilation; and, in some cases, owing to the large sectional area, the speed became insufficient to carry away the fire-damp issuing from old workings.

Accordingly, in 1897, further experiments were undertaken to determine the resistance due to the shafts and the workings, with the following approximate results:—Out of 2·05 inches (52 millimetres) of water-gauge produced by the fan at No. 2 shaft, extracting about 1,836 cubic feet (52 cubic metres) per minute, only 0·87 inch (22 millimetres) was due to the resistance of the workings, so that the remaining 1·18 inches (30 millimetres) represented the resistance of the shafts. The tubbing of No. 1 *bis* pit, 12 feet (3·65 metres) in diameter, having a travelling compartment, had a resistance of 0·59 inch (15 millimetres) over a height of 312 feet (95 metres); the timber tubbing of No. 1 pit, 13·12 feet (4 metres) in diameter, also having a travelling compartment, and in addition compressed-air and water pipes, only absorbed 0·22 inch (from 5 to 6 millimetres) and the remainder, 0·39 inch (10 millimetres), represented the resistance of the walled portion, about 1,640 feet (500 metres) of a pit 14·76 feet (4·5 metres) in diameter, so that more than 60 per cent. of the total water-gauge was absorbed by the shafts. The resistance caused by the tubbing of No. 1 *bis* pit showed clearly the disadvantage of a small sectional area, and demonstrated the impossibility of improving the ventilation of No. 1 pit under the existing conditions; also that the improvement could only be obtained by widening the shafts, or sinking a third shaft of large sectional area; and this last solution was evidently the most rational, while being the only practical one.

No. 1 *ter* pit, used as an upcast-shaft, should afford, in addition to the improvement in ventilation due to its sectional area, the advantage of suppressing variations in the water-gauge and the leakage of air at the valves on the passage of the cages, while making No. 1 a downcast-pit, a very favourable state of things in the event of a fire-damp explosion or underground fire. The sectional area of the new shaft was made equal to that of the other two together, namely, 19·68 feet (6 metres) in diameter, both of these latter becoming downcast-shafts. The same fans were employed at the new shaft, with a single modification, namely, that the inlet was increased to 18 feet (5·5 metres) in diameter; and the same engine was used: the lower water-gauge being set off by the increased volume. The new shaft, brought into operation on January 1st, 1903, delivered 3,287 cubic feet (93·879 cubic metres) per second at a fan-speed of 55 revolutions per minute with 1·58 inches (40 millimetres) of water-gauge. The fire-damp content of the main return-airway is 0·2 per cent., and that of the most highly charged 0·8 per cent., while the equivalent orifice is 59 square feet (5·5 square metres).

Curves representing the resistances, measured in accordance with the different volumes in the three shafts, show at once the necessity of a third shaft for obtaining large volumes at great depths. Calculating these resistances for a depth of 3,280 feet (1,000 metres), leads approximately to a total water-gauge of 4·72 inches (120 millimetres) for ventilation with two shafts, but only 0·79 inch (20 millimetres) for one with three shafts. The advantage of the third shaft is also seen for smaller volumes; and especially as it would only have required, for taking off 1,942 cubic feet (55 cubic metres), a water-gauge of 0·55 inch (14 millimetres), while the work of the fan, which was formerly 44 horsepower, would have been reduced to one-fourth of that power. A saving would therefore have been effected of 33 horsepower or of 70 steam horsepower, with an efficiency of 45 to 50 per cent., which roughly represents the interest on the capital expended, about £32,000 (800,000 francs). In addition to the daily saving effected by the diminished resistance of the air, if it be considered that the new shaft permits of increasing the output to 475,000 tons without diminishing the safety, it will be seen that the proportion per ton of the expense incurred is less than that which would have been involved by a new plant.   J.W.P.

EXPLOSIONS IN PRUSSIAN COLLIERIES DURING 1902 AND 1903.

(1) *Mittheilungen über einige der bemerkenswerthesten Explosionen beim Preussischen Steinkohlenbergbau im Jahre 1902.* OFFICIAL. *Zeitschrift für das Berg-, Hütten- und Salinenwesen im Preussischen Staate* 1903, *vol. li., Abhandlungen, pages* 421-431, *with 2 figures in the text and 2 plates.*

On February 13th, 1902, at 12·15 p.m., a coal-dust explosion took place in the third level, working No. 2 seam, in the Königsborn pit, near Heeren,

Dortmund. Of the 4 victims of the explosion, all more or less badly burnt, 2 died after they had been brought to bank. No. 2 seam is about 4 feet thick, dips to the south 70 degrees, has a roof and floor of fairly-compact shale, and gives off extremely-little fire-damp. On the other hand, it yields an enormous amount of dust; and, in view of this, the precautions prescribed by the Government ordinance of December 12th, 1900, were observed throughout the pit: these precautions, of course, included systematic watering. No dynamic effects of the explosion, such as damage to the timbering or the spray-pipe system were observed; but a thick layer of fine coal-dust was traced for some distance upon every exposed surface. A shot, for the purpose of bringing down obstructive masses of stone and coal, had been fired with two cartridges of rock-dahmenite, and the disaster followed immediately thereupon, contrary to all previous experience with that safety-explosive. It would seem, however, that there is more than a suspicion that some dynamite had been surreptitiously mixed with the charge by the shot-firers, in contravention of the regulations. Fire-damp appears to have played no part in the matter at all.

On Tuesday, July 29th, 1902, at 12·30 p.m. a fire-damp explosion took place at the Government colliery of Camphausen, near Saarbrücken, whereby 14 persons were injured, 6 of them mortally. It occurred in the second level, working No. 3 seam, the lower portion of which is there about 5 feet thick, and separated by a parting, of variable thickness, from an upper layer of coal 2½ feet thick. From fissures in the strata evolution of gas had been noticed at rare intervals; but an examination of that portion of the pit where the explosion took place, carried out only 2½ hours before the disaster, that is, at 10 a.m., failed to reveal a trace of gas. The dynamic effects of the explosion appear to have been inconsiderable: crusts and beads of coke were traceable in some places, and a brattice or two were torn down. Coal-dust played a subordinate part in the disaster, the immediate cause of which has not been ascertained with certainty. It would seem, however, that while one man was engaged in watering a particular working-place, his mate, who was meanwhile resting, had hung up his lamp near a fissure in the roof, and a sudden unnoticed evolution of gas may have brought the inner gauze to incandescence (of which there is evidence). This being noticed, the man may, in his haste to remove the lamp, have dropped it, and the explosion will have arisen from the flame, which was then burning between the inner and the outer gauze, striking through.

On Saturday, August 30th, 1902, at 10 p.m., an explosion took place above bank in the boiler-house of Anselm II. shaft of the Hutschin collieries, at Petrzkowitz, Ratibor, Silesia, whereby a fireman was killed. This shaft serves as an upcast for Anselm I. shaft, which latter is a downcast and is used for haulage-purposes; Anselm II. shaft is provided with a Witkowitz fan, which can deal with 122,500 cubic feet of air per minute, and this had been stopped at 6 a.m., and the boiler-fires drawn for the purpose of repairs. No one had been left down the pit. Several masons and the above-mentioned fireman had been busied all day in repairing the foundations of the fan-engine, and when darkness set in they used naked lights. The above-mentioned fireman went with one of these into the boiler-house, and a few minutes later a tremendous concussion took place, and the fireman rushed out with his clothes all aflame. About a ¼ hour afterwards a dull thud was heard coming from the pit, but this second explosion produced no flames discernible at the surface. In the pumping-engine room, windows were shattered and doors burst in; the latter caught fire, as did also the erections at the pit-head itself. An investigation, made the following day, showed that the dynamic effects of the explosion were traceable especially in the second, and down to the third, level of the pit. It is presumed that the stoppage of the fan must have caused a complete cessation of the circulation of air-currents in the pit; the weather was excessively sultry at the time, and the barometer was low. Fire-damp, evolved in great quantity from the Bruno seam, in all probability finally found its way into the engine-house above-bank, as well as into the second level, etc., below. The second explosion (within the pit) may have been started by the accidental dropping down the shaft of some burning substance [*sic auctor*].

On December 13th, 1902, at the Minister-Achenbach pit, near Dortmund, several shots, fired with gelatine-dynamite in a cross-cut, started a fire-damp explosion whereby 5 persons were slightly injured. The dynamic effects were inconsiderable, and coal-dust played no part in the matter. Electric time-igniters were used, and these do not appear to furnish quite the same elements of safety as ordinary electric ignition. New regulations, more stringent than before, have been drawn up for shot-firing in the Minister-Achenbach colliery.

(2) *Mittheilungen über einige der bemerkenswerthesten Explosionen beim Preussischen Steinkohlenbergbau im Jahre 1903.* OFFICIAL. *Zeitschrift für das Berg-, Hütten- und Salinenwesen im Preussischen Staate*, 1904, *vol. lii.*, *B. Abhandlungen*, *pages* 483-490, *with 6 figures in the text.*

In the Hillebrand pit of the Gottessegen colliery, at Antonienhütte in Upper Silesia, an explosion took place during the night-shift of Saturday to Sunday, April 4th to 5th, 1903, with fatal results. Three shots had been fired about midnight in one of the workings where the temperature was exceedingly high (on account, chiefly, of leakage of the cool air brought into the mine through defective brattices), and two shots more were to be fired. Eye-witnesses do not quite agree as to what then happened. At all events, there appears to have been carelessness on the part of the persons concerned. Either a fuse was ignited accidentally by too close contact with a lamp, or a spark falling from a lamp was blown by the current from a compressed-air drill on to gunpowder and dynamite cartridges which had been laid on the floor preparatory to shot-firing. Eight persons suffered severe burns, which proved fatal in some cases. There appears to be little doubt that the coal-dust, produced in abundance by the drilling-machine, played a considerable part in the explosion. The most careful investigation has failed to reveal the slightest trace of fire-damp in the workings. Orders have since been issued to spray the area neighbouring a bore-hole with water before shot-firing; and an accessory ventilator has been set up at the foot of the inclined plane, so as to improve the ventilation of that portion of the workings.

At 2 a.m. on April 26th, 1903, an explosion occurred in the sump of No. 4 level of the Friedlicher Nachbahr colliery, in the mining-district of Hattingen, whereby 5 workmen were more or less severely burnt, 1 dying of his injuries on the following day. The safety-lamps carried by the victims were found properly locked and undamaged; the site of the explosion was never dry, owing to the constant drip of water, and it need hardly be added that coal-dust was conspicuous by its absence; the ventilation of the mine was excellent. The cause of the disaster appears to have been the sudden collapse of part of the sump into a cavity formerly occupied by a mass of coal, but at that moment filled with an accumulation of fire-damp:

this, of course, was immediately ignited by the lamps carried by the pitmen.

Between 5 and 6 p.m. on October 30th, 1903, a fire-damp explosion took place at the 2,132 feet level of the Werne colliery, in the mining-district of Hamm, whereby 3 persons lost their lives. The part of the workings where the disaster occurred had been abandoned for about 1 month, but 2 men and 1 lad had been told off to restart working there that afternoon, and the overman who went with them (to show them what had to be done, and to inspect the place) found it cool, well-ventilated and somewhat damp. The cause of the explosion cannot be determined with certainty, but the available evidence negatives the supposition that the victims had been tampering with their lamps: these were of the single-gauze type, and made by Mr. Seippel of Bochum. There had been a fire in another seam in the same colliery, but this again appears to have had no causal connection with the disaster.   L. L. B.

GAS-EXPLOSIONS IN CLAY-PITS, AND THEIR PREVENTION.

Gasexplosionen in Thongruben und deren Verhütung. By H. E. MÜLLER. Jahrbuch für das Berg- und Hüttenwesen im Königreiche Sachsen, 1904, pages A3-A18.

The presence of inflammable mixtures of gases in the underground clay-workings of France and Belgium has given rise, within the last decade or so, to a series of accidents, the attendant circumstances of which were exhaustively dealt with in the *Annales des Mines* for 1895. History is now repeating itself in Saxony, outbursts of inflammable gas being frequently recorded in certain clay-pits. The clay of the neighbourhood of Löthain appears to be especially characterized by the evolution of such gases, and no less than 8 pits in that locality have been scheduled as dangerous by the Saxon Department of Mines. (It is mentioned, by the way, that the use of safety-lamps is made obligatory at depths exceeding 50 feet.) Various accidents which occurred in these pits, with inflammable gas, during the years 1886 to 1903, are briefly touched upon, and it is stated that analysis of a sample of the gas made at Freiberg showed it to contain 64·5 per cent. of methane, 19 per cent. of carbon dioxide, and 15·9 per cent. of nitrogen. The general opinion now is that the gas originates from the decomposition and fermentation of the timbering in the wet clay, in old portions of the workings from which the outer air is excluded. The curious fact that the evolution of methane is in all three countries (France, Belgium and Saxony) more especially characteristic of a certain number of clay-pits, is explicable, on the assumption that the particular bacterium, through the agency of which the decomposition of the timber is effected, is present in those pits only, and not in the others. Compare, in this regard, the exhaustive researches of the late Prof. Renault on the agency of bacteriaceæ in the formation of coal.

In long-drowned metalliferous mines, on the other hand, the decomposition of the timbering appears to have been attended in many cases with the evolution of hydrogen rather than with that of methane; and a more careful study of the question leads the author to the conclusion that the presence of finely-divided organic particles within the substance of the clay itself is quite as important a factor as, if not a more important factor than, the decomposition of the timbering in the evolution of pit-gases. Impregnation of the timbering or its complete removal from old workings are not, there-

- .IV.

fore, the remedies called for by the situation: the only reliable preventives are thorough ventilation and the use of safety-lamps. These are now prescribed by the Department of Mines in the Saxon clay-pits, as also the construction in every case of a second winding-shaft (to permit of the men escaping in the event of an explosion). L. L. B.

EXPLOSION IN A GOLD-MINE IN NEW SOUTH WALES.

Explosion at Lucknow. By A. JAQUET. *Australian Mining Standard*, 1903, *vol. xxx., page* 828, *and* 1904, *vol. xxxi., page* 8.

An explosion of fire-damp occurred in the Reform shaft-workings of the D'Arcy Wentworth mines, Lucknow, whereby 2 miners received superficial burns on the face and arms. Inflammable gas, consisting essentially of carburetted hydrogen, produced during the rotting of old mine-timbers under water, was carried by flowing water through a bore-hole into the gallery, where it was mixed with air, and ignited at a naked light.
M. W. B.

FIRE AT THE LAURAHÜTTE COLLIERY, SIEMIANOWITZ, GERMANY.

Der Grubenbrand in der Ficinus-Schachtanlage des Steinkohlenbergwerks Laurahütte bei Siemianowitz am 26. September 1903. By — JAEKEL. *Zeitschrift für das Berg-, Hütten- und Salinenwesen im Preussischen Staate*, 1904, *vol. lii., pages* 264-269, *and* 1 *plate.*

At the Laurahütte colliery, in Upper Silesia, a fire took place on September 26th, 1903, with the result that the manager and 3 workmen perished by suffocation, and 43 other persons were more or less seriously injured. Pit-fires being very frequent in that coal-field, the system of precautionary and preventive measures has attained a sufficient degree of perfection to be, as a rule, successful in quickly damming off areas of conflagration without injury to life or limb. In most cases, fires arise from spontaneous combustion (to which the Upper Silesian coal is peculiarly liable) in the goaf. Open conflagrations of considerable extent are, fortunately, rare, but when they do occur the inflammability of the coal constitutes an additional source of danger and a vehicle for the spread of the mischief: the disaster here described is attributed to an open conflagration, which escaped notice until too late.

Three seams are worked at the Laurahütte colliery, to which access is gained by 3 independent winding-shafts (the Richter, Ficinus and Knoff shafts): the sole communication between these consisting of certain air-courses and waterways. It was in the district served by the second-named shaft that the fire occurred; this shaft is both the main winding and downcast-shaft, and it has as auxiliaries the Aschenborn shaft for drainage and 4 ventilating-shafts (Ernst, Saara, Holz and Therese). The Caroline seam is 16 to 23 feet thick; it is worked by the pillar-and-stall system customary in Upper Silesia; and fires are extremely rare in the goaf, as this is immediately filled up (after being worked out) with the soft sands and shales which overlie the coal, and is thus made all but air-tight. The number of men at work on the seam in that district of the colliery averaged 250, and the ventilation was so arranged that about 124 cubic feet of fresh air flowed past every man per minute. Shortly after 6 a.m. on the fatal day, an overman was warned that a fire had broken out in the particular area that was under his charge; and, pending the arrival on the spot of the manager, he took measures to roll back the smoke and gases of combustion by opening up doors in masonry-dams to admit opposing air-currents. He also attempted to limit the area of conflagration by means of a timber-dam. Meanwhile the fire raged so furiously that these measures, and others yet, proved inadequate. Rescue-parties provided with pneumatophores and smoke-masks were beaten back, and even the attempt to leave a still more extensive area to the conflagration and dam it off securely, at about 11 a.m., had to be given up. The Ficinus shaft had to be completely closed off, operations to this end being carried out from the main cross-cut at the 460 feet level and from the Saara shaft. The rescue-parties were finally successful in recovering the bodies of 3 out of the 4 victims who died from suffocation, and in bringing to bank the 43 injured persons who were suffering from burns, contusions, and in some cases carbon-monoxide poisoning.

The cause of the fire has not been ascertained, but one lesson taught by the disaster is that the precautionary measures, which are sufficient to stop such small fires as are frequent in the goaf of Upper Silesian collieries, are inadequate to cope with an extensive conflagration. The coal of the Caroline seam is so tough, that most of the levels driven in it are unprovided with pit-props; nevertheless, there happened to be at the site of the fire a certain amount of woodwork in the shape of wooden tubs, scaffolding, etc., to give plausibility to the supposition that the outbreak was in some unexplained way connected with these.   L. L. B.

———

A NEW SYSTEM OF COMBATTING FIRES IN MINES.

Nouveau Système pour Combattre les Incendies dans les Mines. By J. KRZYZANOWSKI and ST. WYSOCKI. *Paris, 1904, 41 pages, with 6 figures in the text and 3 plates.*

The authors first consider the causes of fires in mines:—(1) Fires due to inflammation of timber and other easily combustible materials, such as hay and lubricating-oils. They cite several cases in which such fires have caused great loss of life and damage to property. (2) Fires due to inflammation of coal: (a) by contact with a flame, and (b) by spontaneous combustion. (3) Fires caused by inflammation of coal-dust. And (4) fires caused by explosions of fire-damp.

The authors then contrast the precautions against fires which are taken in large establishments on the surface and in towns, with the general lack of such precautions in mines, notwithstanding that in mines the risk of fire is greater than on the surface, owing to the profusion of inflammable materials and the constant use of artificial lights. The danger from the gaseous products of combustion is also much greater underground than on the surface.

The only means of defence actually available in most mines are: (1) Partial changes in the direction of the air-currents, where doors or ventilating appliances allow of such changes. (2) The modification of the whole system of ventilation, which is a complicated operation requiring much time. (3) The construction of wooden dams after the fire has broken out, in order to arrest the propagation of the products of combustion and to isolate the fire. These measures have been in use for nearly four centuries, and they are inadequate to meet the dangers of fires, owing to the incessant development of mines and the increased number of persons employed therein, nor are they in accord with the progress made in mining science.

It is difficult to estimate, even approximately, the speed with which a fire is propagated in timbered mines. The speed varies with the state of the ventilating current and the physical condition of the mine and the timber; the amount and direction of the inclination of the roadways are also factors in the speed of propagation of fires. It is possible, however, to estimate with some precision the route and velocity of the gases of combustion, which is incomparably greater than that of the fire. In a given mine, the velocity of the ventilating current may vary greatly, according to the sectional area of the passages through which it moves. The average speed of the main air-current in different mines may be taken at 400 feet per minute (2 metres per second). It follows that the rate of advance of the smoke and gases from a fire may also be estimated at 400 feet per minute, between the fire and the point of exit of the current. The fire and smoke also advance in the opposite direction where their rate of advance is equal, but the speed in this direction is insignificant, except where the passage is vertical or highly inclined.

It follows from the foregoing observations that in a mine-fire the persons in greatest danger are those between the fire and the point of exit of the air-current. If the length of the main air-current is taken at from 6,000 to 12,000 feet and the speed of the air-current at 400 feet per minute, it follows that the current traverses 3,000 feet of the airway in 7½ minutes, and traverses the whole route in from 15 to 30 minutes. This is about the speed and time in which the gases and fumes traverse the main airway when the fire occurs near the entrance to the mine. A more complete idea of the greatness of the risk in such circumstances is obtained by considering that the mass of noxious gases produced by the burning of the timber in a few feet of roadway, or by the inflammation of a limited amount of fire-damp and coal-dust, or by the combustion of a few hundred pounds of coal, would be sufficient to fill all the mine; and that the fatal effects of these gases on men are produced, not at the end of 1 hour or ½ hour, but, according to the greater or lesser density of the poisonous mixture, at the end of some minutes. Considering the rapidity of the progress of the smoke and gases, it is evident that miners, surprised by the poisonous current, have no chance of escaping. The system of warning by word of mouth is not sufficient to reach the persons interested in all parts of the mine, or the warning only reaches them when the intake airway near their working-places is saturated with poisonous gases.

The system proposed by the authors for dealing with fires in mines consists in the pre-establishment of dams capable of being instantly closed in case of fire, and situated in such positions as to divide the mine into sections, which could be isolated by the closure of the dams. The dams would be fitted with doors of such size, that when open, as in ordinary times, they would not impede the ventilation. To illustrate the system a diagram is given showing the airways of a mine producing 500,000 tons per annum, with the positions of the preparatory dams, 32 in number. These dams are placed as far as possible near points where currents of air divide or come together, so that access to one side of the dams may be easily maintained by ventilation. Where this is not practicable, special means for ventilation should be provided. The dams may be made of wood or bricks, the openings being kept as large as practicable, so as not to impede the ventilation. The doors may be of wood or sheet-iron, made as air-tight as possible, and there should always be a supply of mortar or moist clay to make them quite air-tight. Places which should be specially guarded, as being by their nature extra dangerous in case of fire, are vertical shafts or staples, timbered engine-houses, stables, hay-depôts and timbered roads containing steam-pipes.

The organization of the service for closing the doors in case of fire is stated to have been elaborated in another work entitled *Incendies Miniers.* Certain officials are entrusted with the work, and special signals arranged. A method for operating the doors by means of electricity from a central station and a system of electric signals are described. The estimated cost of a dam is £5, and of the electrical apparatus £7 per dam.   W. N. A.

CENTRIFUGAL PUMPS.

Pompe Centrifuge à Haute Pression: Système de Laval.   By K. Sosnowski. *Mémoires et Compte-rendus des Travaux de la Société des Ingénieurs Civils de France,* 1904, *vol. lvii., pages 233.241, with 8 figures in the text.*

The volume delivered by centrifugal pumps can be considerably varied without greatly impairing the efficiency; and the consumption of a steam-turbine, coupled with a centrifugal pump, may be reduced to 20 pounds (8·5 to 9 kilogrammes) of steam per horsepower per hour in water raised. Tests made during the last few years have shown that centrifugal pumps are capable of affording high pressures with a good mechanical yield.

Trials of the Laval high-pressure pumps date from the beginning of 1899, although the actual applications are more recent. Lifts of 328 feet (100 metres) and upwards are obtained by means of high speed; and it may be considered generally that the height of lift is in direct ratio with the speed. As regards the motive power, steam-turbines afford the highest speeds, so that it was quite usual, for high lifts, to couple the pump directly to the turbine; but, on account of the high number of revolutions, the revolving portion of the pump becomes very much reduced in diameter, and this necessarily restricts the sectional area of the suction-pipe, only permitting of the delivery of small volumes. Accordingly, in order to raise large volumes of water, the main pump received the addition of an auxiliary one, mounted on the intermediate shaft of the turbine, and revolving at a lower speed. This auxiliary pump may have suitable dimensions for raising by suction the quantity of water required, and delivering it under slight pressure to the high-speed pump. A Laval steam-turbine of 50 horsepower, making 20,000 revolutions per minute, is mounted on the same shaft as a high-pressure centrifugal pump, capable at this speed of raising 220 gallons (1,000 litres) per minute to a height of 492 feet (150 metres) in a single lift. The auxiliary pump, of low pressure, is only intended to supply the high-pressure pump with the quantity of water which it could not, on account of the small size of the orifices, raise by its own suction.

The first application in France of these high-speed, high-pressure pumps to raising mine-water was made at the Lens colliery. This turbine-pump, capable of raising 22,000 gallons (100 cubic metres) per hour to the height of 853 feet (260 metres), subsequently increased to 1,148 feet (350 metres), is constituted by a turbine of 150 horsepower directly driving a high-pressure centrifugal pump, and also, by means of a speed-reducer, a low-pressure centrifugal pump, the former making 13,000 revolutions and the second 650 revolutions per minute. The low-pressure pump raises the mine-water about 10 feet (3 metres) and forces it, with a pressure of about 33 feet (10 metres) to the high-pressure pump, which raises it to the surface in a single lift; there is a retaining-valve on the delivery-pipe and also, in case it should fail, a safety-valve on the low-pressure pump. As this plant is underground, the moisture from the steam and the hot water of condensation have to be dealt with; but steam-traps and good cleading of the pipes ensure sufficient dryness of the steam, and the water of condensation is removed without mixing with the mine-water. The condenser has its own feed-pump (mounted on the same shaft as the low-pressure pump) which supplies it with mine-water at nearly constant temperature, so that the vacuum is very uniform. If the efficiency of centrifugal pumps alone is a little lower than that of plunger-pumps, such is not the case when the centrifugal pump is driven directly by a steam-turbine, while they are smaller and less costly, and work without shock.

J. W. P.

METHOD OF PREPARING PEAT FOR BRIQUETTES.

Ein neues Verfahren zur Aufbereitung von Torf für Briquettirungs- und andere Zwecke. By GUSTAV KROUPA. *Œsterreichische Zeitschrift für Berg- und Hüttenwesen,* 1902, *vol l., pages* 57-60 *and* 79-81, *and 4 figures.*

The fresh peat, containing 80 to 85 per cent. of moisture, is passed through a Schlickeysen or Dollberg press, converted into a homogeneous mass, and delivered through a copper-mouthpiece as an endless band, about 24 inches wide and 2 inches thick. It is then carried on a belt-conveyor, and is cut into lengths of about 24 inches; these are delivered into a series of revolving chambers, which discharge the lengths of peat into an appliance where it is enclosed in strong sacking, without having been touched by hand at any stage of the process. From this appliance, the filled bags fall through a hopper into a double rotary chamber, where the mouth of the bag is automatically closed, and the bag is then delivered into a truck holding about 30, each in a separate compartment. When the truck reaches the hydraulic press, which serves to expel the moisture from the peat, a slotted plate in the truck-bottom is caused to slide by means of a lever, so that the bags fall through into the press-plates, which are connected with one another and with the press-head by means of chains, so that when the press-head is lifted the plates are returned to their original position. From the press, the peat (which now contains only 50 to 60 per cent. of moisture) is discharged on to a vibrating conveyor, which carries it to a floor where the bags are emptied by youths. It is then fed to a breaker, and transferred to a drying-oven, where the water is further reduced to 12 or 15 per cent.; finally the dried product is conveyed to the briquette-press. The apparatus will deal with 1,145 cubic feet of crude peat *per diem* of 20 hours. C.S.

MEASURING THE VOLUMES DELIVERED OF GASEOUS SUBSTANCES.

Mesurage des Débits des Corps gazeux. By N. FRANÇOIS. *Annuaire de l'Association des Ingénieurs sortis de l'École de Liége,* 1904, *vol. xvii., pages* 243-271, *with* 13 *figures in the text.*

Now that every effort is being made to turn blast-furnace gases to the best possible account, the importance of measuring the gas-volumes delivered has specially asserted itself. Of the various methods for this purpose, namely, direct measuring, the calorimetric method, the chemical method and indirect measurement by determining the speed of flow, the most correct and trustworthy is direct measurement by the gasometer; but, unfortunately, this requires so cumbersome and costly a plant, that it can only be applied in exceptional cases.

The indirect measurement, which consists in determining the speed of the current, in accordance with the *vis viva* of the gas, was first suggested by Mr. Pitot, towards the middle of the eighteenth century, for gauging rivers; and the method was subsequently improved by Mr. Darcy. It has been in turn tried, abandoned and lastly re-employed with success by the author, whose greatest difficulty was to render it applicable to comparatively slight speeds of off-flow. Its principle is the following:—If to a pipe, in which a gas passes, a pressure-gauge be applied, with the end dipping into the pipe and terminating in a plane perfectly parallel with the axis of the pipe, the gauge will indicate a pressure which is not influenced by the flow of the gas; but if the end of the tube be modified so that the plane be set at right angles to the axis of the pipe and facing the current, the pressure will increase, and the increase of pressure will correspond with the dynamic pressure to which the plane is subjected owing to the resistance that it opposes to the current, this resistance corresponding with the *vis viva* of the gas. If the second branch of the gauge, instead of being left free, be also put in connection with a tube dipping into the pipe, the mouth of which tube is perfectly parallel with the direction of the current, we shall obtain the Darcy tube.

In order to ascertain the volume delivered, it is sufficient to multiply the sectional area of the pipe by the speed of flow; but, to obtain absolutely correct results, such speed must be very uniform over the whole sectional area of the pipe. This, however, is never the case, if only on account of friction due to the sides; but, for large metal-pipes, this influence is slight. As the difference of level will generally be insignificant, recourse must be had to a special gauge, and that best known for slight pressures is the manometer with an inclined tube.

The first attempts made by the author with the Darcy tube and a manometer with an inclined tube were not encouraging; and he has introduced various modifications into the method with the object of (1) rendering possible the determination of low velocities; (2) only requiring portable instruments of easy management; and (3) avoiding certain causes of error, especially those due to variations in the speed of flow.

The method, thus improved, has been applied with equal success to gauging blast-furnace gas at the furnace-mouth, at the hot-air stoves, at the boilers, or at the gas-motors; and this success naturally induced an extension of its application to other purposes, especially the passage of air through mine-workings. In that case, what is chiefly to be sought is a method, only requiring apparatus easily moved about and manipulated, while giving approximately correct results for speeds varying from 7 to 40 feet per second; and, lastly, it is important that the readings be taken as instantaneously as possible, without being too much influenced by the presence of the observer in the air-way. The method is specially applicable to testing fans; and it has also been applied to measure the air delivered by blowing-engines and air-compressors, for which latter purpose the following process was adopted. In the absence of a compressor, the air of which could be gauged in a gasometer, the apparatus was compensated by means of receivers. On the delivery-column of an air-compressor, between two receivers, was mounted the Pitot tube; and this communication-pipe of the receivers was provided with a set of valves permitting either the escape of the compressed air and the evacuation of the second receiver, or of sending this air into the second receiver. Before beginning an observation, the latter was, therefore, emptied; and then, when the time was taken, the valve open to the air was closed, and the air sent into the gauging-receiver. Thanks to a pressure-gauge mounted on the pipe, the opening of the valve between the pipe and the receiver was so regulated that the pressure in the pipe remained almost constant. As the air-compressor worked with a constant pressure, its running was also constant; and in this manner, while the pressure varied from nil to x atmospheres in the receiver, the air in the pipe, maintained at a constant pressure, also circulated with an almost constant speed. The capacity of the gauging-receiver, as also the initial and final pressures and temperatures being known, it was easy to determine the quantity of air forced into the receiver during the observation, and to deduce from it the mean speed of flow in the pipe. The concordance between the results obtained by gauging with the receiver and those afforded by the Pitot method gives confidence in this method, as regards its application to compressed air.   J. W. P.

PATIO PROCESS OF SILVER-AMALGAMATION.

The Patio Process of Amalgamation of Silver-ores. By MANUEL VALERIO ORTEGA. *Transactions of the American Institute of Mining Engineers*, 1901, *vol. xxxii., pages* 276-285.

The ore from the mine, after being hand-picked, is reduced in *arrastras* or *tahones* to fine slime. After the slime has acquired a suitable consistency by the evaporation, through the sun's heat, of a part of the water which it contained, it is spread upon the *patio* or amalgamating-floor, where it is mixed with 5 or 6 per cent. of common salt. The next day, a certain amount of cupric sulphate is added, and immediately afterwards mercury, in the proportion of 8 units to each unit of silver contained in the mineral. These chemicals are thoroughly mixed with the slime by means of horses or mules trampling over it, and this is continued from 2 to 5 weeks. Amalgamation being finished, the slime is transferred to deep, circular stone vats, through which water is passing agitated by a revolving paddle. The amalgam and other heavy metalliferous materials collect at the bottom, while the light earthy impurities are held in suspension and carried away. The author discusses in detail the chemical theory on which the process is based.
                                                   R. W. D.

THE SALTPETRE-INDUSTRY OF CHILE.

Die Salpeterindustrie Chiles. By DR. — SEMPER *and* DR. — MICHELS. *Zeitschrift für das Berg-, Hütten- und Salinenwesen im Preussischen Staate*, 1904, *vol. lii., B. Abhandlungen, pages* 359-482, *with* 13 *figures in the text and* 12 *plates.*

The first portion of this memoir deals with the conditions of occurrence, form, and origin of the nitrate-deposits; the second, with the mode of winning and treating the nitrates, and with the cost of plant, labour, transport, etc.;

while the third portion deals with the history and the economic and legal aspects of the industry.

The Chilian nitrate-fields are practically confined to the two northernmost and rainless provinces of Tarapacá and Antofagasta. The region is orographically divisible into four distinct belts, parallel with the Pacific seaboard, these being, from west to east: (1) The coastal Cordilleras, which abut steeply on the ocean, but fall eastward in gentler slopes—on which latter the nitrate-deposits have been mainly formed and preserved; (2) the high tableland of the Pampa de Tamarugal; (3) the foothills of the Andes; and (4) the great Cordilleras of the Andes themselves. Dust-storms are of almost daily occurrence in the Pampas; and earth-tremors are observed in the Atacama district two or three times a week, on the average. The authors believe, indeed, that the network of fissures which have split up the saltpetre-conglomerates into a chaos of blocks (tablas) is the outcome of seismic phenomena. Many of the features of the country are characteristic of æolian agencies—dunes have accumulated in favourable localities, exposed faces of rock show unmistakable marks of wind-erosion, and so forth. The rainfall, when it does occur, assumes somewhat the character of a waterspout, washes great cavities in the soil, and piles up huge masses of detritus on the floor of the Pampas.

After pointing out the unsatisfactory want of precision about the term caliche, both from the lithological and the stratigraphical point of view, the authors proceed to show that the nitrate-deposits may be subdivided into four essentially-distinct categories, as follows: (a) Beds, nests and lenticles of the impure sodium nitrate, associated with chloridic and sulphatic salts and stony and earthy substances, overlying loose Quaternary rubble and gravel, and underlying a series of saline conglomerates, gravels, etc., of variable thickness; (b) impregnations of the weathered crust of Mesozoic eruptive rocks by sodium nitrate and the associated salts; (c) infillings of cavities in the Jurassic limestone; and (d) efflorescences at the surface of the saline steppes (salares). The first group is perhaps the only one of real importance, from the industrial point of view. A detailed description of the deposits is followed by a discussion of the various theories mooted as to their origin. Without committing themselves to a definite conclusion in this regard, the authors appear to think that, among the genetic agencies to be reckoned with, is electric tension occurring in a damp atmosphere, combined with the action of wind-borne guano-dust on the saline deposits of shrunken lagoons.

The method of winning the saltpetre has undergone but little alteration since the start of the industry. The workings are all but universally open-cast, and take the form of long, straight trenches, the direction of which is conditioned by the levels and greatest extension of the assigned area. As a rule, a trench is begun by shot-firing with blasting-powder in some of the bore-holes put down to test the value of the deposit. The gunpowder is usually manufactured on the spot; the stemming consists of earth and stone, very closely packed; and the fuse is of the ordinary type, with a gunpowder-core. Wages, and the very currency in which they are paid, are subject to continual variation: they vary, too, from district to district of the nitrate-fields. The range of the prime cost of winning a metric centner or quintal (50 kilogrammes or 110 pounds) of crude saltpetre has been variously estimated as lying between a minimum of 4·3d. and a maximum of 14·6d., or between a minimum of 7·7d. and a maximum of 11·7d.

The nitrate is extracted from the crude stuff or *caliche* by leaching at a high temperature; the insoluble residues are separated off from the mother-liquor, and the pure nitrate is crystallized out from this. The cost of treatment depends on so many factors, that it varies within wide limits. Among the bye-products obtained are iodine, perchlorate of potash, common salt and sulphate of soda. The question of water-supply, not only for the nitrate-factories, but for domestic use as well, is one that bristles with difficulties in these rainless regions. For instance, the water obtained from subterranean streams in the Pampa de Tamarugal is so full of compounds of lime, magnesia and alumina that it cannot be utilized in the nitrate-works until it has undergone treatment with quicklime and soda; then, in the case of very many springs, the water is so excessively saline that it is useless for domestic purposes. In the Toco district a good supply of slightly brackish water is yielded by the Loa river. The fuel, without which the works could not exist, is almost entirely British and Australian coal; in some cases, a small proportion of the inferior coal got in Southern Chile is mixed with the imported coal. The native sulphur, used in the process of extracting the iodine, was at one time imported from Sicily, but is now got from deposits in the Andes.

In 1902, the number of workpeople employed in the nitrate-fields amounted to 24,538 (nearly 6 times as many as in 1887)—of whom 71 per cent. were Chilians, mainly immigrants from the southern provinces. The output per head diminished continuously from 1900 to 1902. Wages have risen of late years, as the demand for suitable labour is greater than the supply. The owners of the works generally make a good profit out of the stores connected with them, as, in view of the absence of ordinary shops and the remoteness of the nitrate-fields from the chief centres of population, the workpeople are necessarily driven to make most of their purchases from such stores. Some of the statements advanced by the authors in this connection verge on the libellous.

Nearly half of the ordinary revenue of Chile (49 per cent.) is derived from the export-duties on nitrate and iodine; while the sales by auction of nitrate-fields belonging to the Government form a big item in the non-recurring revenue. In short, the finances of that State may be said to depend all but entirely on the saltpetre-industry nowadays, an industry carried on in territory that belonged mainly to Peru and Bolivia, until its annexation by Chile as the result of the war of 1879-1883. At the present rate of consumption, it would take about half a century to exhaust the nitrate-fields; but the demand shows signs of more rapid expansion than had been foreseen, and it is quite within the bounds of possibility that the fields will be exhausted long ere another 40 years have elapsed.

The memoir, of which a brief sketch has been given, is a store-house of facts bearing on every aspect of the Chilian nitrate-industry, and is likely to rank as a classic (at any rate among German readers) for a long time to come.　　　　　　　　　　　　　　　　　　　　　　　　　　L. L. B.

BOILER-ACCIDENTS IN FRANCE DURING 1902.

Bulletin des Accidents d'Appareils à Vapeur survenus pendant l'Année 1902. ANON. *Annales des Mines*, 1903, series 10, vol. iv., pages 564-576, and 3 plates.

The number of boiler-accidents recorded in France during 1902 amounted to 34, no less than 23 lives being lost thereby and 32 persons greviously injured. With regard to the causes, 6 accidents were traced to defective installation, 17 to defective upkeep, 11 to improper usage, and the actual cause of 5 remained undetermined. Smoke-tube boilers were involved in 16 cases, water-tube boilers in 3, non-tubular boilers in 10, and ordinary " recipients " in 5.

The only accident recorded at a mine took place on June 26th, at the iron-mine of Diélette, in Lower Normandy. Here, a horizontal cylindrical non-tubular boiler was suddenly rent over a length of 5 feet, and the fireman was killed on the spot. The dynamic effects were otherwise not as considerable as might have been expected, and the bursting of the boiler was undoubtedly due to insufficiency of water. The water-gauges appear to have been badly planned.

On November 3rd, at a slate-quarry of Pierrie in Britanny, a horizontal semi-tubular boiler, fed with decidedly acid water, suddenly burst, killing a fireman and badly injuring the engineman. The accident was evidently due to long-standing internal corrosion of the boiler-plates.

Most of the other accidents occurred in various factories (chemical and metallurgical, weaving and spinning industries, threshing-mills, etc.).

L. L. B.

ELECTRIC MINE-PLANT AND LOCOMOTIVES AT LA MURE COLLIERY, FRANCE.

Les Installations Électriques de la Compagnie des Mines d'Anthracite de La Mure, Isère. By LÉON DE CHARENTENAY. Bulletin de la Société de l'Industrie Minérale, 1904, series 4, vol. iii., pages 79-117, with 13 illustrations in the text and 2 plates.

In 1895, a scheme of electric power-transmission was prepared for (1) substituting electric motors for insufficient steam-boilers and engines at various points, (2) haulage, ventilation, rock-drilling, and all purposes for which electricty affords a simple power-agent, (3) driving the new workshops, and (4) lighting. As the slack and impure fuel is disposed of to advantage, and the region abounds in water-courses, the company had every reason to employ " white coal " instead of black for producing its electric current, and has acquired water-rights that can yield 50,000 horsepower. Two electricity companies, the first with a generating-station for 10,000 horse-power, have been installed; and a third company will supply additional power, as a reserve for the principal motors.

As the electricity supply-companies had adopted triphase current of 50 periods per second, at pressures of 10,000 and 15,000 volts, it was natural that the mining company should arrange for a distribution of triphase current of 50 periods, but with uniform and suitable secondary tensions. For the surface-works a 500 volts compound tension was chosen, as best satisfying the conditions of efficiency and insulation, always critical for conductors and motors in such dusty places as a screening-plant; and for the motors in the underground workings, quite free from gas, triphase current was also adopted, with a compound tension of 185 volts, as best fulfilling the safety-conditions imposed by various mine-regulations.

The 500 volts current supplied by the electric companies is received by two white marble switch-boards, provided with the appliances for measurement, distribution and safety, and connected by a common line which, in the event of break-down at either of the generating-stations, ensures the working of all the motors. There are at present twenty triphase asynchronous motors of 3, 10, 15, 20, 25, 30, 50, 120 and 230 horsepower, and all are well earthed, so that, if contact should be made between the winding and the frame, the latter, of the same potential as the earth, can be touched with impunity. The 230 horsepower motor gives out this power, at a speed of 420 revolutions per minute, with an efficiency of 92 per cent.

The air-lines are constituted by three naked tinned-copper wires or cables, hung by porcelain-insulators on brackets, or on posts about 35 feet (10 to 12 meters) high; but, inside buildings, the wires or cables are insulated and carried by porcelain sheaves, well out of reach. The current is led underground by cables, constituted by three cords or strands of copper-wire, insulated from one another and further very carefully insulated, and also armoured by steel ribands for the horizontal, and by steel wires, capable of sustaining a length of 328 feet (100 metres), for the vertical portions.

The 500 volts pressure permits of taking the current to the farthest extent of the workings without too great loss; and, as a tension of 185 volts has been adopted for the locomotives and hauling-engines, underground transformers became necessary. The latter, with closed magnetic circuits, are provided with cast-iron casings of barrel-form, for protection against dust and damp, while they can be filled with oil for completely immersing the transformers if the necessity should arise. The metallic masses, carefully earthed, so that they may be touched without danger, are enclosed in masonry-lined chambers, access to which can only be gained by those in charge. Each transformer has its switchboard, and each of the three motors has a separate conductor for leading up the 185 volts current. The conductors, similar to those in buildings on the surface, are carried by porcelain-sheaves along the sides of the workings.

The Ganz hauling-engines, each capable of lifting 1 ton vertically at a speed of 3¼ feet (1 metre) per second, are placed at the top of inclines, about 820 feet (250 metres) distant from the transformers. The spur-wheels on the motor-shaft are of compressed green hide; and the gear is cased in. The motors, which make 965 revolutions per minute, can exert 23 horsepower with a current of 185 volts, showing an efficiency of 84 per cent. with full load, or of 50 per cent. including losses by transformation.

The erection of a large screening-plant in 1898 brought about such an increase of output, that the old haulage-roads became insufficient; and mechanical haulage imposed itself, with the following conditions:—Saving of labour by concentrating all the haulage on a single road; rapidity and regularity in conveying the coal from the working-faces to the screening-plant; better utilization of the rubbish from preparatory workings; and safety for the workmen.

As the roads could only receive a single line of way, the electric locomotive alone appeared to afford a practical solution of the problem; and a single engine should suffice to take off 1,200 tubs of coal in 10 hours over a distance of 0·62 mile (1 kilometre) at a speed of 10 miles (15 kilometres) per hour. It is supplied with triphase current at 185 volts, and is provided with three current off-takes, all return by the rails being avoided, not only on the score of safety, but also because iron opposes great resistance to an alternating current, owing to induction. With 40 tubs in a set, the locomotive can make 30 double journeys in the day of 10 hours, giving 10 minutes for each run and leaving 6 minutes for shunting and contingencies.

The road, of 31 inches (785 millimetres) gauge, is laid with 28 pounds (14 kilogrammes) flange-rails to a maximum gradient of 1 in 64 (15·5 per thousand)

and with a minimum curve-radius of 33 feet (10 metres). The tub, weighing 5 cwts. (250 kilogrammes) holds 10 cwts. (500 kilogrammes) of coal or 14 cwts. (700 kilogrammes) of rubbish; and the gradient is in favour of the coal, which gives the coal-train a weight of 35 tons including engine, against 22 tons for the rubbish, the quantity of which is about one fourth of the volume of coal extracted. The maximum resistance to be overcome occurs, therefore, when the 22 tons train mounts a gradient of 1 in 64.

The locomotive (with another in reserve) is calculated for a pull of 4,918 footpounds (671 kilogrammetres) corresponding, at a speed of 10 miles (15 kilometres) per hour and with a loss of 10 per cent. in the gear, to 41 effective horsepower; and a Ganz vertical-shaft 25 horsepower electromotor, liberally designed, so as to be capable of giving out 47 horsepower and of starting a train under the most unfavourable circumstances, was adopted.

The three-wire distribution required an arrangement permitting the engine to pass easily, without stopping at the points (switches) in the track, and to run in both directions without turning and without the trolley being shifted. The hardened copper-wires, 0·39 inch (10 millimetres) in diameter, are placed one at the centre-line of the roof, and the other two laterally, in the same vertical plane, and at a perfectly regulated distance from the centre-line of the road.

The engine, which has been running regularly since January 12th, 1903, has given every satisfaction.   J. W. P.

MINING LAWS OF MEXICO.

A Synopsis of the Mining Laws of Mexico. By RICHARD E. CHISM. *Transactions of the American Institute of Mining Engineers*, 1901, *vol. xxxii.*, *pages* 4-55.

Prior to 1884, mining in Mexico was governed mainly by the " Ordinances of Mining " enacted by the King of Spain in 1783. In 1884, these ordinances were repealed, and the laws of all the states were made uniform. The mining code of 1884 was based on a working tenure of mining property, under conditions hard to fulfil and fruitful of litigation. A new law was introduced in 1892, giving the miner the property of his mines in an irrevocable, perpetual and secure form, through the payment of a yearly tax, with full liberty as to methods of work and the amount of work to be done. The author gives a synopsis of the 1892 regulations, arranged in an orderly manner under 17 different heads.   R. W. D.

II.—BAROMETER, THERMOMETER, Etc., READINGS FOR THE YEAR 1903.

By M. WALTON BROWN.

The barometer, thermometer, etc., readings have been supplied by permission of the authorities of Glasgow and Kew Observatories, and give some idea of the variations of atmospheric temperature and pressure in the intervening districts in which mining operations are chiefly carried on in this country.

The barometer at Kew is 34 feet, and at Glasgow is 180 feet, above sea-level. The barometer readings at Glasgow have been reduced to 32 feet above sea-level, by the addition of 0·150 inch to each reading, and the barometrical readings at both observatories are reduced to 32° Fahr.

The statistics of fatal explosions in collieries are obtained from the annual reports of H.M. Inspectors of Mines, and are also printed upon the diagrams (Plates XXVIII. and XXIX.) recording the meteorological observations.

The times recorded are Greenwich mean time, in which midnight equals 0 or 24 hours.

TABLE I.—SUMMARY OF EXPLOSIONS OF FIRE-DAMP OR COAL-DUST IN THE SEVERAL MINES-INSPECTION DISTRICTS DURING 1903.

Mines-inspection Districts.	Fatal Accidents.			Non-fatal Accidents.	
	No.	Deaths.	Injured.	No.	Injured.
Cardiff	2	2	4	2	2
Durham	0	0	0	8	9
Ireland	0	0	0	0	0
Liverpool	0	0	0	3	5
Manchester	0	0	0	0	0
Midland	0	0	0	4	4
Newcastle-upon-Tyne ...	1	1	0	7	16
Scotland, East	4	5	4	22	29
Do. West	2	3	1	47	65
Southern	0	0	0	2	2
Staffordshire	1	1	1	8	9
Swansea	1	2	0	22	29
Yorkshire	0	0	0	9	13
Totals	11	14	10	134	183

TABLE II.—LIST OF FATAL EXPLOSIONS OF FIRE-DAMP OR COAL-DUST IN COLLIERIES IN THE SEVERAL MINES-INSPECTION DISTRICTS DURING 1903.

1903.		Colliery.	Mines-inspection Districts.	Deaths	No. of Persons Injured.
Feb.	9, 1·30	Stanrigg	Scotland, East ...	1	1
,,	24, 22·30	Aldridge (No. 1)... ...	Staffordshire ..	1	1
,,	25, 22·30	Darran	Swansea	2	0
Mar.	23, 19·30	Ferndale (No. 5 Pit) ...	Cardiff	1	3
June	10, 12·30	Biggarford	Scotland, East ...	1	3
July	17, 6·45	Boglesbole (No. 4 Pit) ...	Scotland, West ...	2	1
,,	24, 11·15	Garriongill	Scotland, East ...	1	0
Oct.	5, 0·30 to 3·30	Westburn (No. 2 Pit) ...	Scotland, West ...	1	0
Nov.	9, 21·40	Llwynypia (No. 3 Pit)...	Cardiff	1	1
,,	12, 2·0	Stargate Pit	Newcastle-upon-Tyne	1	0
Dec.	11, 22·30	Braidhurst	Scotland, East ...	2	0
				14	10

TABLE III.—LIST OF NON-FATAL EXPLOSIONS OF FIRE-DAMP OR COAL-DUST IN COLLIERIES IN THE SEVERAL MINES-INSPECTION DISTRICTS DURING 1903.

1903.		Colliery.	Mines-inspection Districts.	No. of Persons Injured.
Jan.	7, 4·45...	West Tees (Railey Fell) ...	Durham	1
,,	7, 8·30...	Dewshill	Scotland, East ...	1
,,	8, 17·30...	Wednesbury Oak (No. 22 Pit)	Staffordshire	1
,,	9, 7·30...	Glangarnant	Swansea	1
,,	9, 15·0 ...	Noyadd	Do.	1
,,	13, 19·30...	Hook	Do.	1
,,	19, 3·0 ...	Walkmill	Newcastle-upon-Tyne	3
,,	21. 16·30...	Auchenharvie (No. 5 Pit)	Scotland, West ...	1
,,	21, 20·0 ...	Mountain	Liverpool	2
,,	22, 6·30...	Newbattle	Scotland, East ...	1
,,	23, 3·15...	Duffryn Rhondda	Swansea	1
,,	30, 19·30...	Dalziel and Broomside ...	Scotland, East ...	1
Feb.	1, 10·30...	Newbattle	Do. ...	1
,,	3, 8·40...	Clyde (Townlands No. 1 Pit)	Scotland. West ...	1
,,	6, 15·0 ...	Millburn	Do. ..	2
,,	9, 6·30...	Pumpherston (oil-shale) ...	Scotland, East ...	4
,,	10, 10·0 ...	Manners	Midland	0
,,	15, 23·0 ...	Pennant Hill	Staffordshire	2
,,	19, 2·0 ...	Onllwyn	Swansea	1
,,	23, 8·0 ...	Limefield	Scotland, East ...	1
,,	23, 10·0 ...	Auchincruive (Mossblown No. 2 Pit)	Scotland, West ...	1
,,	26, 7·0 ...	Struther (No. 5 Pit) ...	Do. ...	1
,,	26, 15·45...	Dillwyn	Swansea	1
,,	27, 5·30...	Inkermann (Walkinshaw No. 2 Pit)	Scotland, West ...	1
Mar.	6, 8·0 ...	Oakley	Scotland, East ...	1
,,	12, 10·0 ...	Lanemark (No. 2 Pit) ...	Scotland, West ...	3
,,	12, 19·30...	Holytown (No. 12 Pit) ...	Do. ...	1
,,	13, 3·0 ...	Mwrwg Vale	Swansea	1
,,	16, 7·30...	Glengarnock (No. 6 Pit) ...	Scotland, West ...	2
,,	16, 8·30...	Blackwell (A Pit) ...	Midland	1
,,	17, 15·15...	Boglesbole (No. 4 Pit) ...	Scotland, West ...	1
,,	19. 9·0 ...	Pumpherston (oil-shale) ...	Scotland, East ...	1

TABLE III.—*Continued.*

1903.	Colliery.	Mines-Inspection Districts.	No. of Persons Injured.
March 27, 9·0 ...	Whitehill (No. 1 Pit) ...	Scotland, West ...	1
„ 31, 15·30..	New Cwmgorse	Swansea	1
April 7, 10·0 ...	Garriongill	Scotland, East ...	3
„ 9, 12·0 ..	Crigglestone	Yorkshire	1
„ 11, 11·0 ..	New Cwmgorse	Swansea	2
„ 27, 8·0 ...	Cynon	Do. ...	1
May 2, 13·0 ...	Over Dalserf	Scotland, West ..	2
„ 11, 17·0 ...	Caerbryn	Swansea ...	1
„ 12, 1·0 ...	Gartness	Scotland, East ...	1
„ 16, 9·30..	Dunaskin (No. 2 Pit) ...	Scotland, West .	1
„ 21, 10·10...	Ravensworth	Newcastle-upon-Tyne	1
„ 28, 14·15...	Throckley	Do. ...	1
„ 29, 19·0 ...	Blantyre Ferme	Scotland, West ..	1
„ 30, 12·0 ...	New Cwm Mawr	Swansea	1
June 4, 8·30..	Auchincruive (Mossblow No. 1 Pit)	Scotland, West	1
„ 8, 6·50...	Tareni	Swansea	3
„ 8, 22·0 ...	St. John's	Midland	1
„ 13, 7·30...	Park Oval	Liverpool	1
„ 14, 20·0 ..	Dunnikier	Scotland, East ...	1
„ 20, 12·0 ..	Victoria	Swansea	2
„ 22, 11·0 ...	Butterknowle	Durham	1
„ 23. 9·0 ...	Bourtreehill (Capringstone No. 7 Pit)	Scotland, West ...	1
„ 25, 6·30...	Philpstoun (oil-shale) ..	Scotland, East ...	1
„ 26, 22·20 .	Dumbreck (No. 2 Pit) ...	Scotland, West ...	1
July 2, 10·0 ..	Wednesbury Oak (No. 22 Pit)	Staffordshire	1
„ 2, 17·15...	Gilwen	Swansea	1
„ 8, 21·30..	Haugh (No. 1 Pit)... ...	Scotland, West ..	1
„ 10. 10·0 ...	Westburn (No. 2 Pit) ...	Do. ...	1
„ 10, 14·30...	Auchenharvie (No. 1 Pit)	Do. ...	1
„ 16, 17·30...	Wester Queenslie (No. 2 Pit)	Do. ...	1
„ 17, 7·30...	Gordon Navigation ...	Swansea	1
„ 17, 23·0 ..	Wallyford	Scotland, East ...	1
„ 22, 15·30...	Blaencaegurwen	Swansea	1
„ 23, 9·30.	Rheola	Do. ...	1
„ 24, 15·45 ..	Old Roundwood	Yorkshire	2
Aug. 14. 19·15...	Black Brook	Liverpool	2
„ 15, 7·0 ...	Bonvilles Court	Swansea	2
„ 18, 2·0 ...	Silverwood	Yorkshire	1
„ 18, 10·30...	Calderbank (No 2 Pit) ...	Scotland, West ...	1
„ 20, 15·0 ...	Lingdale	Durham	1
„ 21. 2·0 ...	Stonelaw (No. 1 Pit) ...	Scotland, West ...	2
„ 27, 5·0 ...	Benwell	Newcastle-upon-Tyne	3
„ 28, 9·0 .	Llwyn	Swansea	1
„ 28. 14·0 ...	Woodhall	Scotland, West ..	1
Sept. 1, 9·15...	Crigglestone	Yorkshire	2
„ 2, 18·0 ...	Charles	Do. ...	1
„ 2, 22·30...	Throckley	Newcastle-upon-Tyne	1
„ 7, 6·0 ...	Crigglestone	Yorkshire	1
„ 7, 8·0 ...	Souterhouse	Scotland, West ...	1
„ 9, 16·0 ...	Drumpeller (No. 4 Pit) ...	Do. ...	1
„ 10. 9·30...	Wester Queenslie (No. 2 Pit)	Do. ..	
„ 10, 13·0 ...	Wednesbury Oak (No. 6 Pit)	Staffordshire	1
„ 10, 17·40...	Madeley Court (No. 3 Pit)	Do. ...	1
„ 10, 19·30...	Schoolfield (No. 2 Pit) ...	Do.	1
„ 13, 2·30...	Lodge Mill	Yorkshire	1

TABLE III.—*Continued*.

1903.		Colliery.	Mines-Inspection Districts.	No. of Persons Injured
Sept.	25, 13·0 ..	Meiros	Cardiff	1
„	25, 18·0 ...	North Skelton	Durham	1
„	25, 19·30...	Bartonholm...	Scotland, West ...	4
„	28, 14·30..	Rosehall (No. 5 Pit) ...	Do.	1
Oct.	1, 5·30..	Bridgeness	Scotland, East ...	1
„	6, 10·15..	Fishley (No. 4 Pit) ..	Staffordshire	1
„	8, 14·10..	Nitshill (No. 2 Pit) ..	Scotland, West ...	2
„	12, 13·15...	Preston	Newcastle-upon-Tyne	5
„	12, 13·45...	Kirkwood (No. 1 Pit) ..	Scotland, West ..	2
„	12, 15·0 ...	Dumbreck (No. 2 Pit) ...	Do.	1
„	19, 16·30...	Charles	Yorkshire ...	1
„	21, 16·30	East Plean (No. 4 Pit) ...	Scotland, West ..	2
„	21, 18·30...	Prestongrange	Scotland, East .	1
„	22, 7·0 ...	Kingswood	Southern	1
„	25, 18·0 ...	Carriden	Scotland, East	2
„	25, 20·30...	Swinhill (No. 5 Pit) ...	Scotland, West ...	1
„	26, 5·0 ...	Bowhill	Scotland, East ...	2
„	30, 0·30...	Whitehill (No. 2 Pit) ..	Scotland, West ...	2
Nov.	3, 7·30...	Kilton	Durham	2
„	9, 0·15..	Burnhope	Do.	1
„	9, 7·30...	Morningside	Scotland, East ...	1
„	9, 8·0 ...	Hilton	Staffordshire	1
„	9, 8·0 ..	Rheola	Swansea	2
„	9, 13·30...	Varteg Deep Black Vein	Southern	1
„	13, 7·5 ...	Canderigg (No. 1 Pit) ..	Scotland, West ..	2
„	14, 5·0 ...	Cleland	Scotland, East ..	1
„	14, 13·0 ...	Brayton	Newcastle-upon-Tyne	2
„	19, 12·0 ...	International	Swansea ...	2
„	25, 10·0 ...	Lochwood (No. 3 Pit) ..	Scotland, West ...	2
„	25, 23·30...	Rosehall (No. 14 Pit) ...	Do. ...	1
„	26, 7·30...	Walkinshaw (No. 2 Pit) ..	Do. ...	2
„	27, 9·30...	North Skelton	Durham ...	1
„	27, 15·0 ...	Belhaven	Scotland, East ..	1
Dec.	2, 7·0 ..	Boglesbole (No. 4 Pit) ...	Scotland, West ...	1
„	4, 7·30...	Meiros	Cardiff	1
„	9, 6·30...	Pollington	Midland	2
„	9, 11·0 ...	Moorfield (No. 1 Pit) ...	Scotland, West ..	2
„	10, 7·30...	Toftshaw Bottom	Yorkshire	3
„	10, 8·0 ...	Rigfoot	Scotland, West ...	1
„	11, 15·30...	Gartshore (No. 9 Pit) ...	Do. ...	1
„	14, 2 0 ...	Fairlie (No. 3 Pit) ...	Do.	1
„	15, 16·30...	Rosehall (No. 13 Pit) ...	Do.	2
„	16, 9·15...	Butterknowle	Durham	1
„	23, 22·0 ..	Motherwell	Scotland, East .	1
„	24, 19·0 ...	Sundrum (No. 3 Pit) ...	Scotland, West ...	1
„	29, 13·0 ...	Bourtreehill (Capringstone No. 6 Pit)	Do.	1
„	29, 19·30 ..	Niddrie	Scotland, East ..	1

TABLE IV.—BAROMETER, THERMOMETER, ETC., READINGS, 1903.

JANUARY, 1903.

	KEW.							GLASGOW.						
	BAROMETER.				TEMPERA-TURE.		Direction of wind at noon.		BAROMETER.				TEMPERA-TURE.		Direction of wind at noon.
Date.	4 A.M.	10 A.M.	4 P.M.	10 P.M.	Max	Min.		Date.	4 A.M.	10 A.M.	4 P.M.	10 P.M.	Max	Min.	
1	29·538	29·623	29·667	29·660	38·3	29·8	W	1	29·406	29·437	29·406	29·300	39·1	29·0	SSW
2	29·575	29·395	29·388	29·382	51·1	31·8	SW	2	29·107	29·059	29·079	29·048	43·5	37·3	WSW
3	29·494	29·462	29·601	29·732	49·7	40·0	SW	3	28·964	29·00d	29·083	29·194	42·1	36·3	W
4	29·732	29·808	29·772	29·633	47·2	39·9	SW	4	29·245	29·272	29·400	29·495	45·2	38·1	WSW
5	29·649	29·654	29·642	29·640	53·6	44·4	WSW	5	29·542	29·544	29·474	29·316	42·6	37·2	E
6	29·542	29·486	29·387	29·346	52·4	50·5	SW	6	29·096	28·952	29·013	28·964	44·9	37·0	W
7	29·277	29·302	29·393	29·573	50·9	41·7	SW	7	28·776	28·812	29·034	29·268	43·6	35·7	W
8	29·668	29·730	29·648	29·590	44·5	34·6	SW	8	29·416	29·555	29·570	29·522	36·5	31·2	W
9	29·478	29·412	29·303	29·296	51·4	44·0	S	9	29·399	29·271	29·077	28·979	35·2	31·4	E
10	29·318	29·385	29·377	29·391	50·7	34·1	WSW	10	28·994	29·299	29·531	29·683	37·2	30·2	NE
11	29·475	29·661	29·777	29·929	38·6	32·0	N	11	29·748	29·866	29·963	30·102	34·6	26·8	W
12	30·016	30·152	30·223	30·299	36·1	29·6	NNE	12	30·170	30·266	30·327	30·390	32·2	24·6	SSE
13	30·325	30·413	30·435	30·486	32·6	27·0	NE	13	30·494	30·474	30·502	30·548	28·7	17·9	NNE
14	30·507	30·550	30·505	30·490	32·7	28·9	E	14	30·557	30·552	30·506	30·514	32·3	19·6	E
15	30·453	30·439	30·399	30·389	33·1	26·9	E	15	30·509	30·495	30·421	30·343	31·6	22·1	ENE
16	30·340	30·330	30·321	30·176	32·4	25·0	ESE	16	30·258	30·179	30·085	30·050	33·4	29·1	ESE
17	30·124	30·129	30·115	30·134	34·2	26·3	E	17	30·026	30·068	30·074	30·064	32·9	28·8	ESE
18	30·113	30·135	30·152	30·189	39·0	34·4	E	18	30·050	30·068	30·073	30·099	35·8	31·3	E
19	30·173	30·184	30·175	30·192	42·3	37·8	E	19	30·094	30·132	30·124	30·162	39·1	34·0	E
20	30·206	30·270	30·273	30·235	43·7	35·3	ESE	20	30·202	30·267	30·245	30·221	37·4	33·4	E
21	30·274	30·316	30·290	30·285	41·7	35·2	SE	21	30·145	30·095	30·096	29·956	43·1	35·8	S
22	30·195	30·103	29·996	29·851	45·3	38·7	S	22	29·825	29·651	29·581	29·635	43·7	35·9	WSW
23	29·979	30·057	30·121	30·165	45·4	33·4	W	23	29·605	29·762	29·800	29·710	48·3	35·2	WSW
24	30·115	30·075	30·052	30·068	48·2	38·0	SW	24	29·483	29·639	29·723	29·892	46·1	40·9	SW
25	30·070	30·064	30·011	30·094	49·2	46·9	SW	25	29·496	29·361	29·404	29·697	50·7	41·0	SW
26	30·148	30·192	30·178	30·175	51·6	46·9	SW	26	29·589	29·597	29·620	29·608	51·8	44·1	SW
27	30·121	30·113	30·046	30·009	52·3	44·8	SW	27	29·606	29·642	29·665	29·672	47·7	40·4	WSW
28	30·021	30·115	30·118	30·202	48·9	50·4	WSW	28	29·509	29·594	29·586	29·590	44·3	39·8	WSW
29	30·293	30·385	30·326	30·302	48·9	39·5	WSW	29	29·930	29·862	29·703	29·758	46·4	40·3	WSW
30	30·268	30·276	30·237	30·190	49·7	45·5	WSW	30	29·739	29·760	29·736	29·697	48·3	45·7	WSW
31	30·092	29·967	29·763	29·511	46·5	41·0	S	31	29·621	29·505	29·261	29·031	47·3	44·3	SW

FEBRUARY, 1903.

Date.	4 A.M.	10 A.M.	4 P.M.	10 P.M.	Max	Min.		Date.	4 A.M.	10 A.M.	4 P.M.	10 P.M.	Max	Min.	
1	29·218	29·347	29·315	29·388	46·7	32·1	W	1	28·823	28·921	29·081	29·368	45·3	33·3	WSW
2	29·620	29·903	30·082	30·237	42·0	31·6	N	2	29·676	29·906	29·909	29·967	40·4	33·8	W
3	30·266	30·302	30·317	30·358	47·1	32·5	W	3	29·973	30·00d	30·076	30·050	48·3	40·0	WSW
4	30·364	30·393	30·363	30·372	47·7	45·0	WSW	4	30·015	29·974	29·942	29·963	49·1	45·6	WSW
5	30·349	30·369	30·321	30·272	47·4	41·0	SW	5	29·976	29·996	29·941	29·881	49·4	46·0	SW
6	30·133	30·054	29·957	30·029	49·5	40·9	SW	6	29·769	29·623	29·391	29·623	49·0	40·4	S
7	30·016	30·053	29·980	30·016	52·5	44·6	SW	7	29·651	29·487	29·209	29·385	52·9	39·1	SW
8	30·051	30·126	30·144	30·224	56·3	50·4	SW	8	29·663	29·561	29·561	29·807	52·6	40·6	W
9	30·263	30·336	30·440	30·595	55·8	49·2	W	9	29·874	30·073	30·286	30·313	50·8	38·5	SW
10	30·595	30·620	30·538	30·539	54·6	34·6	W	10	30·208	30·144	30·045	30·117	51·5	46·2	SW
11	30·494	30·506	30·430	30·416	54·2	44·1	WSW	11	30·151	30·230	30·228	30·190	48·4	42·9	WSW
12	30·347	30·291	30·256	30·371	53·9	39·3	WNW	12	30·139	30·171	30·255	30·349	47·3	39·1	W
13	30·434	30·472	30·395	30·358	46·6	35·5	WNW	13	30·381	30·344	30·187	30·097	46·6	35·1	SW
14	30·296	30·130	30·041	30·030	49·3	36·5	W	14	29·957	29·917	29·900	29·899	49·2	44·5	WSW
15	30·007	30·020	30·051	30·137	47·7	43·9	NW	15	29·969	29·953	30·043	30·143	46·9	39·9	ESE
16	30·213	30·350	30·436	30·573	44·8	34·2	E	16	30·196	30·252	30·326	30·380	44·5	39·3	E
17	30·597	30·638	30·555	30·564	45·1	29·7	SSW	17	30·348	30·339	30·305	30·300	48·4	41·8	SW
18	30·530	30·525	30·440	30·435	47·7	25·4	S	18	30·261	30·231	30·098	30·006	47·1	42·6	SW
19	30·355	30·317	30·247	30·213	54·5	34·5	W	19	29·808	29·695	29·578	29·589	53·3	44·4	SW
20	30·192	30·393	30·367	30·201	57·3	48·4	W·W	20	29·672	30·030	29·799	29·530	50·6	41·2	·W
21	30·104	30·079	29·905	30·171	55·9	45·1	SW	21	29·546	29·514	29·406	29·672	52·1	38·2	W
22	30·181	30·017	29·827	29·673	53·9	43·0	SW	22	29·708	29·619	29·248	29·150	50·7	36·8	SSW
23	29·670	29·706	29·759	29·951	50·9	36·0	W	23	29·139	29·210	29·315	29·537	44·1	33·3	W
24	30·005	29·920	29·718	29·577	49·0	35·3	SSW	24	2·407	29·259	28·964	28·925	47·3	34·6	SW
25	29·618	29·799	29·784	29·692	52·2	44·3	SW	25	29·091	29·241	29·249	29·092	43·3	37·6	SW
26	29·526	29·707	29·867	29·788	49·2	40·6	SW	26	28·997	29·276	29·431	29·291	43·8	35·2	WSW
27	29·379	29·490	29·473	29·460	51·1	43·2	SW	27	28·431	28·641	29·096	29·260	44·2	36·2	W
28	29·382	29·441	29·657	29·872	46·1	37·5	NW	28	29·314	29·377	29·381	29·479	41·5	34·1	SW

MARCH, 1903.

	KEW.							GLASGOW.						
	Barometer.				Tempera-ture.		Direction of wind at noon	Barometer.				Tempera-ture.		Direction of wind at noon
Date	4 A.M.	10 A.M.	4 P.M.	10 P.M.	Max	Min.		4 A.M.	10 A.M.	4 P.M.	10 P.M.	Max	Min.	
1	29·853	29·639	29·388	29·537	48·6	35·6	S	29·316	29·004	29·014	29·133	42·8	35·9	SW
2	29·521	29·187	28·809	28·756	51·0	35·0	SSW	29·084	28·805	28·561	28·646	37·2	32·1	E
3	38·824	29·179	29·514	29·761	47·4	37·8	NW	29·952	29·232	29·460	29·536	45·7	33·9	W
4	29·844	29·804	29·774	29·840	53·7	34·3	SW	29·384	29·319	29·393	29·410	46·2	37·6	W
5	29·705	29·777	29·834	29·823	50·3	39·2	WNW	29·434	29·536	29·576	29·640	40·1	32·9	WSW
6	30·001	30·127	30·132	30·144	48·1	34·3	W	29·682	29·781	29·780	29·805	40·4	34·1	WSW
7	30·002	29·820	29·885	30·053	45·2	36·5	WSW	29·705	29·681	29·766	29·879	39·6	33·0	N
8	30·179	30·285	30·320	30·353	47·6	33·7	NW	30·050	30·139	30·078	29·963	44·8	33·7	SW
9	30·270	30·170	30·047	29·974	49·1	31·8	SSW	29·748	29·637	29·645	29·739	39·0	30·0	WSW
10	29·908	29·933	29·914	29·963	45·1	33·6	SSW	29·735	29·774	29·742	29·700	44·9	34·9	SW
11	29·958	29·984	29·937	29·961	49·3	28·2	WSW	29·634	29·650	29·695	29·606	50·6	40·1	SW
12	29·924	29·933	29·901	29·937	52·6	30·7	S	29·562	29·642	29·663	29·706	40·5	38·5	SW
13	29·905	29·918	29·882	29·006	56·0	35·6	S	29·674	29·727	29·781	29·901	40·1	35·6	SW
14	29·857	29·875	29·855	29·873	53·8	37·3	SE	29·963	30·049	30·000	29·947	45·1	33·1	NE
15	29·837	29·794	29·718	29·700	50·8	30·1	SW	29·804	29·671	29·483	29·444	44·1	33·1	E
16	29·603	29·663	29·755	29·869	50·2	37·5	SW	29·340	29·275	29·306	29·419	43·1	37·9	ESE
17	29·855	29·778	29·642	29·551	51·0	37·5	S	29·206	29·171	29·272	29·354	45·8	38·9	W
18	29·416	29·736	29·967	30·097	51·0	42·9	W	29·320	29·434	29·489	29·445	44·3	34·6	SW
19	30·061	30·064	30·107	30·183	53·1	45·5	SW	29·395	29·500	29·512	29·588	49·1	43·3	SW
20	30·150	30·186	30·192	30·255	55·9	48·9	SW	29·566	29·722	29·778	29·892	46·9	40·1	WSW
21	30·212	30·181	30·106	30·089	55·4	47·6	S	29·687	29·629	29·519	29·424	51·5	42·3	SSW
22	30·059	30·068	30·016	29·974	60·6	49·1	SW	29·473	29·461	29·470	29·482	53·6	43·0	SW
23	29·880	29·789	29·691	29·556	58·8	41·1	SW	29·355	29·281	29·315	29·412	42·1	36·1	W
24	29·750	29·811	29·790	29·741	53·4	40·1	S	29·462	29·531	29·469	29·356	48·9	36·8	SW
25	29·552	29·453	29·302	29·359	65·3	46·2	S	29·173	29·011	28·973	28·953	53·7	44·6	S
26	29·281	29·381	29·423	29·421	56·5	43·0	S	28·990	28·909	28·913	29·127	48·1	41·0	SSE
27	29·356	29·381	29·500	29·622	53·4	45·6	NW	29·179	29·202	29·219	29·285	46·6	40·1	WSW
28	29·513	29·548	29·507	29·754	57·0	43·9	SW	29·154	28·859	28·997	29·287	50·3	40·4	SW
29	29·944	30·060	30·062	30·009	54·5	41·1	WSW	29·461	29·616	29·617	29·373	47·5	36·1	SW
30	29·814	29·831	29·803	29·851	51·7	43·4	W	29·390	29·172	29·372	29·684	46·1	36·6	SW
31	29·989	30·110	30·157	30·157	53·3	41·6	NNW	29·914	30·043	30·046	30·019	52·9	40·7	W

APRIL, 1903.

Date	4 A.M.	10 A.M.	4 P.M.	10 P.M.	Max	Min.	Dir.	4 A.M.	10 A.M.	4 P.M.	10 P.M.	Max	Min.	Dir.
1	30·068	30·013	29·863	29·776	51·1	40·6	WSW	29·910	29·821	29·784	29·816	49·6	40·1	NE
2	29·778	29·859	29·998	29·997	49·0	42·0	NNW	29·850	29·918	29·918	29·928	48·1	35·7	W
3	30·048	30·090	30·022	29·977	53·6	35·3	W	29·852	29·750	29·642	29·569	49·1	38·1	SSW
4	29·884	29·837	29·741	29·871	54·8	43·2	WSW	29·419	29·541	29·603	29·756	49·1	37·6	W
5	29·951	30·060	30·086	30·131	52·6	40·0	N	29·831	29·923	29·963	29·923	49·1	37·1	WNW
6	30·096	30·061	29·953	29·912	54·6	40·0	WSW	29·747	29·641	29·565	29·534	52·8	43·1	W
7	29·807	29·785	29·766	29·821	53·6	46·5	WNW	29·470	29·537	29·639	29·819	48·1	39·4	NW
8	29·940	30·077	30·074	30·148	55·0	39·6	N	29·965	30·077	30·114	30·157	54·1	39·9	NW
9	30·147	30·203	30·224	30·264	49·9	42·8	N	30·163	30·185	30·177	30·204	52·9	42·8	W
10	30·282	30·309	30·258	30·268	53·7	35·7	N	30·200	30·198	30·092	29·963	51·3	44·3	WSW
11	30·187	30·105	29·986	29·943	52·0	45·9	W	29·806	29·740	29·712	29·776	49·4	37·4	W
12	29·989	30·009	29·955	29·957	49·1	36·9	WNW	29·857	29·856	29·831	29·829	43·1	33·6	W
13	29·908	29·871	29·847	29·918	16·4	33·7	NW	29·785	29·789	29·801	29·843	44·3	30·4	NW
14	29·940	30·000	30·023	30·021	48·3	31·9	NW	29·859	29·902	29·869	29·785	42·1	31·6	W
15	29·855	29·927	29·942	30·069	47·1	33·8	NNW	29·859	29·963	30·018	30·133	43·1	27·6	NNW
16	30·113	30·166	30·184	30·278	44·4	30·0	N	30·187	30·247	30·268	30·335	42·1	31·4	NNW
17	30·304	30·321	30·302	30·364	45·1	30·2	N	30·370	30·387	30·376	30·438	42·3	30·3	NNW
18	30·372	30·384	30·283	30·317	48·9	30·1	NW	30·383	30·353	30·276	30·296	47·3	29·6	WNW
19	30·304	30·259	30·116	30·070	53·0	30·0	N	30·237	30·191	30·091	30·027	53·5	34·2	W
20	29·994	29·914	29·762	29·694	54·0	30·0	W	29·929	29·835	29·712	29·727	49·1	34·4	W
21	29·567	29·531	29·507	29·520	48·1	37·3	NNE	29·701	29·688	29·619	29·663	44·9	34·9	NE
22	29·473	29·467	29·441	29·486	48·9	34·9	NE	29·635	29·604	29·534	29·510	44·1	30·6	E
23	29·454	29·476	29·491	29·583	46·9	32·5	NNE	29·482	29·538	29·608	29·738	46·7	34·1	E
24	29·601	29·611	29·646	29·682	47·8	35·1	N	29·741	29·746	29·687	29·676	53·8	30·1	NNW
25	29·660	29·656	29·578	29·590	57·0	32·5	W	29·569	29·541	29·498	29·522	56·0	34·8	S
26	29·558	29·485	29·357	29·315	48·0	42·4	S	29·408	29·491	29·416	29·447	49·5	37·8	ESE
27	29·249	29·286	29·360	29·445	58·0	45·9	S	29·384	29·361	29·313	29·341	48·8	39·3	E
28	29·497	29·563	29·550	29·396	65·6	45·1	SW	29·347	29·387	29·406	29·424	52·1	40·9	S
29	29·245	29·338	29·378	29·396	59·8	48·6	SW	29·384	29·336	29·310	29·293	51·1	41·9	E
30	29·394	29·430	29·415	29·385	59·0	47·6	SSW	29·283	29·330	29·345	29·417	55·3	43·1	N

MAY, 1903.

| | KEW | | | | | | | | GLASGOW | | | | | |
| | Barometer | | | | Tempera-ture | | Direction of wind at noon. | | Barometer | | | | Tempera-ture | | Direction of wind at noon. |
Date	4 A.M.	10 A.M.	4 P.M.	10 P.M.	Max	Min		Date	4 A.M.	10 A.M.	4 P.M.	10 P.M.	Max	Min	
1	29·362	29·395	29·406	29·562	56·4	46·6	SE	1	29·440	29·490	29·470	29·505	52·2	44·6	E
2	29·618	29·642	29·638	29·587	60·1	45·4	S	2	29·507	29·495	29·472	29·488	52·3	43·1	E
3	29·453	29·452	29·412	29·357	53·7	48·6	S	3	29·400	29·427	29·384	29·388	48·6	43·3	E
4	29·246	29·224	29·183	29·218	63·3	48·7	SW	4	29·356	29·373	29·311	29·242	48·6	43·1	E
5	29·196	29·216	29·234	29·345	59·6	42·8	N	5	29·183	29·231	29·275	29·371	48·5	43·4	E
6	29·389	29·461	29·456	29·518	60·5	49·6	SSW	6	29·397	29·454	29·472	29·507	49·1	42·7	N
7	29·547	29·613	29·643	29·703	58·3	42·8	SW	7	29·505	29·535	29·547	29·605	48·1	41·3	SSE
8	29·674	29·685	29·653	29·686	55·5	38·3	E	8	29·630	29·657	29·700	29·761	51·6	41·8	E
9	29·678	29·695	29·674	29·704	57·8	46·0	E	9	29·779	29·809	29·798	29·850	53·3	39·1	E
10	29·692	29·730	29·710	29·745	57·1	45·0	NW	10	29·850	29·885	29·895	29·947	48·9	41·2	ESE
11	29·742	29·790	29·767	29·816	50·6	42·8	N	11	29·933	29·908	29·841	29·837	50·3	33·9	W
12	29·816	29·842	29·832	29·886	53·0	39·9	NNE	12	29·819	29·837	29·791	29·758	55·1	37·7	WNW
13	29·879	29·867	29·847	29·898	58·0	34·9	W	13	29·636	29·540	29·468	29·454	58·5	44·1	SW
14	29·931	30·011	30·064	30·128	59·0	48·4	WSW	14	29·494	29·614	29·733	29·726	49·8	44·3	W
15	30·134	30·141	30·125	30·176	60·0	47·9	SW	15	29·691	29·798	29·894	29·944	52·1	43·6	W
16	30·208	30·181	30·083	29·943	55·4	43·6	W	16	29·885	29·827	29·685	29·604	52·0	43·8	WSW
17	29·715	29·682	29·739	29·933	53·9	44·9	W	17	29·611	29·774	29·845	29·942	55·0	44·1	E
18	30·046	30·109	30·045	30·045	58·3	39·9	N	18	29·962	30·019	29·989	29·946	52·1	39·1	NNE
19	30·015	29·996	29·934	29·948	62·3	38·2	SW	19	29·961	29·929	29·859	29·850	56·1	39·3	SW
20	29·941	29·949	29·951	30·043	64·1	42·3	SW	20	29·794	29·736	29·763	29·802	56·1	41·0	WSW
21	30·088	30·161	30·164	30·193	69·9	47·8	WSW	21	29·769	29·880	29·975	29·988	57·4	48·8	WNW
22	30·163	30·119	30·055	30·120	75·0	47·8	S	22	29·914	29·908	29·797	30·000	59·2	47·6	SW
23	30·247	30·354	30·351	30·405	66·6	50·3	NNE	23	29·817	30·322	30·364	30·379	57·9	43·3	WSW
24	30·390	30·382	30·327	30·320	62·1	44·5	E	24	30·373	30·355	30·300	30·276	67·0	46·9	SW
25	30·308	30·298	30·233	30·293	68·2	40·5	NE	25	30·278	30·283	30·282	30·316	69·2	45·6	E
26	30·276	30·257	30·209	30·213	67·6	46·7	NE	26	30·357	30·380	30·340	30·348	73·1	47·3	ENE
27	30·177	30·134	30·041	30·034	68·5	47·6	NE	27	30·357	30·331	30·257	30·255	69·1	46·1	E
28	29·930	29·869	29·788	29·801	65·6	55·2	SE	28	30·215	30·123	30·005	29·979	69·5	44·3	ENE
29	29·777	29·748	29·702	29·745	69·2	53·7	E	29	29·936	29·880	29·845	29·861	65·9	47·9	E
30	29·734	29·734	29·734	29·743	75·6	56·0	SE	30	29·842	29·828	29·779	29·815	55·9	46·3	ENE
31	29·715	29·770	29·738	29·780	74·3	57·2	N	31	29·812	29·837	29·828	29·845	67·8	48·4	W

JUNE, 1903.

Date	4 A.M.	10 A.M.	4 P.M.	10 P.M.	Max	Min	Wind	Date	4 A.M.	10 A.M.	4 P.M.	10 P.M.	Max	Min	Wind
1	29·772	29·778	29·741	29·833	78·7	51·6	N	1	29·847	29·848	29·863	29·943	59·0	48·6	W
2	29·878	29·964	30·046	30·153	60·5	45·7	NNE	2	30·021	30·074	30·124	30·227	58·5	42·8	WNW
3	30·203	30·257	30·281	30·346	55·1	43·5	NE	3	30·290	30·337	30·325	30·356	65·6	43·2	NW
4	30·354	30·373	30·342	30·374	65·6	42·0	NE	4	30·374	30·385	30·354	30·347	68·4	47·4	NW
5	30·343	30·306	30·293	30·346	66·7	48·5	NE	5	30·345	30·348	30·333	30·382	69·3	53·1	NE
6	30·341	30·365	30·302	30·323	61·3	46·8	NE	6	30·397	30·386	30·326	30·349	75·1	51·1	SSE
7	30·296	30·284	30·213	30·212	57·3	45·8	NE	7	30·374	30·375	30·316	30·344	68·4	52·3	ESE
8	30·171	30·134	30·014	29·898	63·1	48·7	NE	8	30·338	30·332	30·263	30·262	65·3	49·5	E
9	29·785	29·768	29·740	29·793	65·6	50·5	ENE	9	30·201	30·136	30·016	30·009	63·4	45·6	F
10	29·759	29·771	29·731	29·755	59·3	55·2	E	10	29·946	29·942	29·914	29·946	64·7	49·3	ESE
11	29·756	29·829	29·894	29·961	55·7	48·5	N	11	29·961	29·989	29·982	30·047	59·2	48·3	E
12	29·985	30·031	30·048	30·069	54·0	42·0	NNE	12	30·049	30·041	29·981	29·968	56·1	43·1	NW
13	30·019	29·942	29·890	29·853	58·0	39·6	SW	13	29·970	30·002	29·964	30·090	57·3	46·5	SE
14	29·807	29·806	29·763	29·740	51·1	45·4	NE	14	30·050	30·065	30·058	30·055	58·7	41·5	E
15	29·678	29·673	29·625	29·579	51·1	47·3	NE	15	30·004	29·955	29·889	29·849	61·4	45·1	NE
16	29·552	29·588	29·610	29·665	56·9	46·8	ENE	16	29·775	29·781	29·786	29·797	53·1	48·7	E
17	29·677	29·713	29·647	29·677	66·2	46·5	W	17	29·758	29·725	29·709	29·731	53·1	47·1	E
18	29·690	29·677	29·623	29·623	58·2	43·5	NNE	18	29·712	29·710	29·685	29·683	53·4	44·1	N
19	29·581	29·544	29·539	29·624	50·2	46·7	NE	19	29·668	29·695	29·748	29·850	55·4	45·2	ESE
20	29·669	29·756	29·887	30·011	52·8	45·4	NNE	20	29·905	29·968	29·974	30·024	59·2	40·1	W
21	30·075	30·148	30·163	30·122	60·4	45·4	NNE	21	30·058	30·079	30·087	30·119	64·1	41·3	WSW
22	30·222	30·245	30·201	30·228	66·5	41·4	S	22	30·118	30·128	30·105	30·118	65·2	51·0	SW
23	30·197	30·109	30·093	30·047	67·2	44·3	SE	23	30·074	30·071	30·014	29·992	61·2	50·3	SSE
24	29·963	29·970	29·967	30·011	68·1	54·4	ESE	24	29·952	29·940	29·906	29·938	62·4	49·5	S
25	30·089	30·098	30·080	30·138	71·2	51·0	WNW	25	29·938	29·972	29·993	30·027	66·0	53·6	SSW
26	30·159	30·180	30·166	30·165	73·9	55·0	SSW	26	29·997	29·962	29·935	29·897	59·2	51·2	SE
27	30·119	30·093	30·052	30·031	80·8	55·4	SSW	27	29·893	29·943	29·927	29·870	60·9	56·4	WNW
28	29·975	30·007	30·037	30·108	80·0	61·0	W	28	29·775	29·813	29·935	29·975	65·7	57·4	WNW
29	30·158	30·205	30·196	30·255	71·8	55·2	WNW	29	29·934	29·936	30·020	30·112	61·9	54·4	W
30	30·268	30·315	30·299	30·309	71·9	53·8	W	30	30·127	30·128	30·113	30·078	62·8	51·3	W

JULY, 1903.

| | KEW | | | | | | | | GLASGOW | | | | | | |
| | Barometer | | | | Temperature | | Direction of wind at noon | | Barometer | | | | Temperature | | Direction of wind at noon |
Date	4 A.M.	10 A.M.	4 P.M.	10 P.M.	Max	Min		Date	4 A.M.	10 A.M.	4 P.M.	10 P.M.	Max	Min	
1	30·299	30·295	30·204	30·204	76·9	53·2	W	1	30·042	30·060	30·041	30·051	66·8	54·8	W
2	30·196	30·060	29·913	29·843	79·9	52·8	S	2	30·032	29·966	29·794	29·608	71·3	51·8	SW
3	29·921	30·016	30·031	30·086	69·7	54·6	NW	3	29·663	29·690	29·683	29·714	63·7	52·6	W
4	30·089	30·126	30·087	30·055	70·4	51·3	W	4	29·766	29·870	29·931	29·930	60·2	52·1	W
5	30·008	29·963	29·823	29·755	70·2	56·7	S	5	29·821	29·687	29·585	29·405	57·2	52·8	SW
6	29·684	29·767	29·832	29·939	63·7	52·1	WNW	6	29·534	29·708	29·849	29·960	50·0	48·3	N
7	30·006	30·036	30·062	30·150	64·2	47·1	NW	7	30·015	30·055	30·076	30·130	65·3	45·6	NW
8	30·196	30·207	30·181	30·214	69·3	46·3	N	8	30·121	30·104	30·094	30·136	62·1	46·6	W
9	30·216	30·266	30·263	30·295	76·1	61·7	N	9	30·162	30·208	30·190	30·168	70·9	55·8	WNW
10	30·298	30·292	30·211	30·207	83·2	60·0	WSW	10	30·160	30·160	30·132	30·124	63·2	57·6	W
11	30·171	30·140	30·045	30·001	83·3	60·3	SW	11	30·145	30·131	30·070	30·050	64·2	49·1	WNW
12	29·927	29·921	29·887	29·880	67·5	54·0	N	12	29·990	29·942	29·890	29·867	50·5	46·4	NW
13	29·873	29·906	29·906	29·948	64·4	47·8	W	13	29·822	29·838	29·834	29·855	61·7	48·1	NW
14	29·956	29·961	29·921	29·931	68·8	48·4	W	14	29·847	29·840	29·793	29·748	61·5	43·9	WSW
15	29·881	29·897	29·821	29·821	75·0	54·6	WSW	15	29·646	29·670	29·703	29·727	57·1	50·2	ESE
16	29·709	29·705	29·627	29·588	66·4	55·8	S	16	29·663	29·690	29·584	29·484	56·9	52·7	ESE
17	29·553	29·561	29·544	29·585	71·5	56·9	SSW	17	29·454	29·495	29·529	29·608	57·5	53·3	E
18	29·568	29·583	29·632	29·725	68·0	56·7	W	18	29·612	29·682	29·694	29·780	61·1	51·0	E
19	29·768	29·818	29·847	29·890	67·2	54·7	NW	19	29·814	29·880	29·876	29·902	61·8	49·5	ENE
20	29·917	29·981	30·030	30·099	61·6	55·4	NNE	20	29·942	29·974	29·978	29·992	64·5	49·6	W
21	30·099	30·099	30·038	30·007	69·7	49·0	SW	21	29·950	29·893	29·806	29·761	61·0	48·3	SSW
22	29·930	29·898	29·823	29·810	72·0	61·1	W	22	29·768	29·659	29·652	29·657	60·0	56·5	WSW
23	29·748	29·738	29·657	29·638	65·8	55·1	W	23	29·646	29·637	29·624	29·671	67·5	54·4	W
24	29·659	29·787	29·857	29·969	69·2	52·4	W	24	29·676	29·708	29·721	29·811	67·3	48·4	W
25	29·990	29·982	29·909	29·840	69·8	48·2	S	25	29·842	29·839	29·731	29·639	69·1	51·1	S
26	29·736	29·742	29·778	29·891	69·7	55·1	S	26	29·614	29·666	29·683	29·731	66·1	55·9	WNW
27	29·904	29·846	29·888	29·649	63·7	48·1	S	27	29·717	29·714	29·650	29·632	60·0	50·4	N
28	29·679	29·665	29·532	29·530	64·5	56·3	S	28	29·604	29·588	29·570	29·548	56·1	52·8	E
29	29·873	29·621	29·633	29·615	65·1	54·1	W	29	29·492	29·490	29·481	29·538	63·4	50·3	W
30	29·576	29·668	29·774	29·838	63·4	52·5	NW	30	29·575	29·609	29·675	29·738	64·0	50·9	W
31	29·836	29·902	29·953	30·017	62·0	54·5	W	31	29·753	29·806	29·838	29·863	61·2	53·2	WNW

AUGUST, 1903.

| | KEW | | | | | | | | GLASGOW | | | | | | |
Date	4 A.M.	10 A.M.	4 P.M.	10 P.M.	Max	Min	Wind	Date	4 A.M.	10 A.M.	4 P.M.	10 P.M.	Max	Min	Wind
1	30·018	30·018	29·965	29·977	70·9	54·5	SW	1	29·838	29·809	29·804	29·803	61·4	50·7	WNW
2	29·962	29·963	29·902	29·847	67·3	56·0	W	2	29·789	29·796	—	29·495	62·6	51·3	W
3	29·635	29·657	29·787	29·907	69·1	56·0	NW	3	—	29·542	—	29·704	60·7	55·0	WNW
4	29·926	29·890	29·890	29·862	70·0	54·5	SW	4	29·599	29·573	29·626		61·3	52·2	W
5	29·963	30·013	30·035	30·127	67·5	54·1	W	5	29·687	29·731	29·825	29·905	60·2	52·5	WNW
6	30·161	30·205	30·179	30·179	67·1	50·4	WNW	6	29·905	29·990	29·935	30·001	58·7	50·8	WNW
7	30·179	30·162	30·095	30·068	70·5	46·9	SW	7	30·008	30·018	29·970	29·892	62·2	49·3	WSW
8	29·972	29·863	29·731	29·732	76·3	49·1	SSE	8	29·733	29·664	29·492	29·342	63·3	53·0	SW
9	29·758	29·785	29·732	29·718	69·4	55·8	SW	9	29·283	29·310	29·316	29·342	50·3	55·2	SW
10	29·623	29·701	29·769	29·988	66·0	52·8	WSW	10	29·336	29·464	29·606	29·705	50·8	50·4	WNW
11	29·912	29·949	29·529	29·685	64·0	46·1	S	11	29·744	29·755	29·735	29·748	62·1	46·0	WSW
12	29·711	29·860	29·884	29·919	69·0	55·0	W	12	29·739	29·774	29·774	29·803	63·1	48·4	WNW
13	29·917	29·941	29·884	29·831	71·2	50·1	W	13	29·768	29·788	29·779	29·716	62·7	51·3	SW
14	29·616	29·437	29·292	29·246	69·7	53·4	S	14	29·519	29·284	29·019	28·971	62·8	51·3	ESE
15	29·198	29·246	29·405	29·559	64·3	56·0	WSW	15	28·853	28·791	28·911	29·069	56·9	49·9	NE
16	29·675	29·783	29·832	29·834	65·0	52·2	W	16	29·137	29·389	29·564	29·690	59·5	51·1	WNW
17	29·652	29·613	29·683	29·727	69·3	55·5	W	17	29·617	29·671	29·666	29·637	61·5	46·4	WNW
18	29·504	29·490	29·601	29·612	63·3	54·2	NW	18	29·552	29·554	29·587	29·546	61·4	46·3	E
19	29·569	29·682	29·774	29·856	66·6	53·3	NW	19	29·549	29·577	29·591	29·569	60·3	49·7	NW
20	29·811	29·769	29·588	29·427	65·3	53·6	SW	20	29·513	29·451	29·356	29·303	58·7	47·6	SW
21	29·441	29·506	29·684	29·794	65·3	52·0	W	21	29·310	29·434	29·589	29·604	57·9	45·8	WNW
22	29·820	29·862	29·835	29·882	67·9	48·9	WSW	22	29·612	29·678	29·737	29·781	60·3	47·4	W
23	29·864	29·834	29·742	29·789	66·0	45·0	W	23	29·793	29·805	29·741	29·753	62·1	47·5	WNW
24	29·775	29·679	29·620	29·605	65·2	49·6	S	24	29·743	29·775	29·771	29·856	61·5	40·5	E
25	29·604	29·771	29·981	30·143	60·2	50·7	NW	25	29·887	29·959	29·977	30·036	62·1	43·3	NW
26	30·205	30·223	30·188	30·159	66·3	46·3	WSW	26	29·950	29·872	29·761	29·735	59·8	50·2	SW
27	30·107	30·082	30·021	29·985	67·7	55·0	SW	27	29·597	29·612	29·572	29·560	63·1	55·7	WSW
28	29·976	30·015	29·967	29·848	67·3	55·2	SW	28	29·571	29·612	29·637	29·593	60·9	51·9	W
29	29·730	29·866	29·964	30·085	66·9	55·8	NW	29	29·577	29·669	29·796	29·938	61·0	49·9	NW
30	30·100	30·132	30·045	29·950	64·6	52·5	SSW	30	29·961	29·913	29·584	29·334	58·9	46·0	SE
31	29·895	30·019	30·057	30·084	69·8	56·3	WNW	31	29·458	29·801	29·934	29·968	59·3	50·6	WNW

SEPTEMBER, 1903.

	KEW.								GLASGOW.						
	BAROMETER.				TEMPERA-TURE.		Direction of wind at noon		BAROMETER.				TEMPERA-TURE.		Direction of wind at noon
Date.	4 A.M.	10 A.M.	4 P.M.	10 P.M.	Max	Min		Date.	4 A.M.	10 A.M.	4 P.M.	10 P.M.	Max	Min	
1	30·099	29·960	29·926	30·023	79·0	53·4	SSW	1	29·828	29·659	29·770	29·895	62·9	48·8	SW
2	29·907	29·750	29·679	29·821	78·1	57·6	S	2	29·911	29·873	29·741	29·647	55·9	47·6	E
3	29·984	30·078	30·067	30·123	68·3	52·0	SW	3	29·653	29·713	29·814	29·908	59·8	48·7	WSW
4	30·111	30·059	29·922	29·847	70·9	50·0	SE	4	29·893	29·900	29·796	29·657	62·0	50·1	S
5	29·754	29·798	29·876	30·015	69·0	51·4	SW	5	29·469	29·477	29·684	29·795	60·8	52·4	W
6	30·042	30·096	30·094	30·143	60·8	47·6	W	6	29·845	29·908	29·958	30·007	59·7	49·3	WSW
7	30·179	30·236	30·178	30·179	65·8	7·1	WSW	7	29·978	29·885	29·764	29·774	56·5	47·1	SW
8	30·133	30·113	30·031	29·873	65·0	53·9	SW	8	29·739	29·763	29·657	29·326	56·7	49·6	W
9	29·635	29·726	29·794	29·923	63·4	47·9	NW	9	29·396	29·535	29·612	29·720	54·3	46·6	WNW
10	29·985	29·960	29·632	28·957	57·7	44·7	WSW	10	29·727	29·626	29·384	29·306	51·6	45·0	W
11	29·149	29·445	29·501	29·543	58·6	43·7	WNW	11	29·322	29·368	29·376	29·420	55·4	41·1	WNW
12	29·547	29·596	29·621	29·672	56·6	40·2	NW	12	29·488	29·540	29·512	29·567	55·2	43·9	WNW
13	29·666	29·755	29·838	30·023	57·4	40·8	N	13		29·831	29·992	30·178	56·1	38·9	N
14	30·107	30·233	30·323	30·435	56·6	45·0	N	14	30·291	30·401	30·442	30·499	54·0	36·7	ESE
15	30·431	30·451	30·387	30·377	55·5	42·3	NNE	15	30·501	30·500	30·434	30·420	56·9	38·0	WSW
16	30·274	30·235	30·136	30·161	58·0	43·4	N	16	30·350	30·296	30·158	30·113	52·1	37·1	ESE
17	30·182	30·251	30·258	30·319	62·7	37·0	SSW	17	30·061	30·085	30·117	30·175	57·3	37·4	S
18	30·302	30·304	30·200	30·140	63·3	44·0	E	18	30·198	30·283	30·183	30·208	63·1	48·1	SSE
19	30·063	30·074	30·004	30·010	64·0	52·7	E	19	30·181	30·173	30·175	30·231	60·7	48·0	E
20	29·974	29·989	29·971	29·974	66·1	57·8	E	20	30·205	30·214	30·157	30·152	58·1	52·5	E
21	29·922	29·911	29·865	29·890	66·2	55·9	E	21	30·105	30·062	30·011	30·010	61·7	52·9	E
22	29·873	29·960	30·020	30·137	63·0	53·0	E	22	29·972	30·000	30·000	30·072	59·8	52·1	SE
23	30·187	30·265	30·263	30·279	67·6	55·6	E	23	30·092	30·154	30·142	30·143	64·1	52·6	E
24	30·256	30·253	30·172	30·115	66·9	56·3	ESE	24	30·127	30·159	30·123	30·105	64·1	54·3	SW
25	30·046	30·096	30·107	30·203	66·7	53·6	SW	25	30·057	30·063	30·025	30·075	58·7	48·8	E
26	30·209	30·225	30·129	30·065	66·1	53·0	W	26	30·093	30·117	30·039	29·988	60·0	50·	SW
27	29·971	29·925	29·873	29·889	66·0	53·9	SW	27	29·829	29·768	29·691	29·717	62·6	51·	SE
28	29·868	29·884	29·817	29·760	67·5	55·7	S	28	29·713	29·718	29·651	29·601	62·1	50·	SE
29	29·697	29·735	29·731	29·796	67·7	58·3	S	29	29·441	29·553	29·576	29·602	63·3	56·	SSW
30	29·835	29·913	29·937	29·958	66·1	52·8	SSW	30	29·581	29·616	29·641	29·651	60·9	54·	SW

OCTOBER, 1903.

Date.	4 A.M.	10 A.M.	4 P.M.	10 P.M.	Max	Min	Dir	Date.	4 A.M.	10 A.M.	4 P.M.	10 P.M.	Max	Min	Dir
1	29·880	29·834	29·729	29·716	65·1	50·9	W	1	29·601	29·610	29·540	29·505	61·1	54·1	SSW
2	29·732	29·755	29·715	29·702	63·0	51·5	WSW	2	29·436	29·442	29·477	29·517	58·1	51·6	W
3	29·695	29·673	29·708	29·737	64·3	54·5	W	3	29·443	29·437	29·478	29·452	57·4	50·5	NW
4	29·644	29·639	29·749	29·792	63·5	54·5	W	4	29·368	29·433	29·465	29·41	55·7	48·5	ENE
5	29·593	29·544	29·563	29·653	61·9	54·1	W	5	29·109	29·067	29·032	29·150	55·6	50·6	SW
6	29·702	29·658	29·489	29·593	62·5	53·3	SW	6	29·346	29·371	29·210	29·34	53·1	40·3	E
7	29·763	29·882	29·840	29·793	63·2	53·6	SW	7	29·579	29·639	29·685	29·709	44·8	40·1	E
8	29·644	29·569	29·473	29·351	62·0	54·7	SSW	8	29·651	29·615	29·588	29·579	49·3	43·5	ENE
9	29·463	29·535	29·600	29·746	57·6	44·6	W	9	29·559	29·617	29·692	29·760	50·0	41·9	NE
10	29·825	29·890	29·859	29·820	54·7	39·8	W	10	29·776	29·812	29·759	29·695	49·3	35·5	SE
11	29·706	29·569	29·401	29·296	57·1	45·3	S	11	29·538	29·399	29·230	29·123	51·7	41·3	ESE
12	29·151	29·011	28·945	29·016	61·3	52·2	S	12	28·959	28·824	28·788	28·906	55·9	48·5	SSE
13	29·230	29·476	29·589	29·687	57·8	50·0	WSW	13	28·858	29·040	29·184	29·310	53·2	42·5	W
14	29·716	29·745	29·671	29·604	58·9	49·4	SW	14	29·360	29·378	29·342	29·116	51·9	43·2	WSW
15	29·417	29·429	29·613	29·687	57·9	46·1	W	15	29·017	29·029	29·144	29·246	50·8	43·4	WNW
16	29·678	29·711	29·628	29·659	55·7	45·1	SW	16	29·237	29·202	29·222	29·388	49·2	43·6	WNW
17	29·640	29·784	29·936	30·012	54·9	44·9	NNW	17	29·652	29·828	29·906	29·951	53·5	45·0	NW
18	30·019	30·054	30·073	30·066	52·2	45·1	NW	18	29·967	29·972	29·931	29·86	48·1	39·3	NE
19	30·047	30·023	29·982	29·962	56·0	47·0	S	19	29·737	29·721	29·665	29·660	56·1	45·4	SW
20	29·920	29·900	29·801	29·747	56·0	51·3	S	20	29·623	29·590	29·516	29·48	53·5	49·3	SW
21	29·693	29·711	29·658	29·682	55·0	46·2	WSW	21	29·433	29·432	29·380	29·383	51·3	46·3	SW
22	29·603	29·430	29·314	29·401	55·1	45·3	SSW	22	29·341	29·302	29·302	29·25	49·1	44·8	NE
23	29·447	29·470	29·566	29·736	52·0	42·1	WNW	23	29·278	29·397	29·514	29·59	49·9	39·9	WNW
24	29·797	29·775	29·634	29·503	55·1	48·3	S	24	29·392	29·400	29·222	29·064	52·1	39·0	SW
25	29·470	29·434	29·349	29·248	58·3	47·3	S	25	29·122	29·146	29·026	28·958	52·7	46·9	SW
26	29·285	29·214	29·243	29·387	57·6	50·0	SW	26	28·925	28·975	28·980	29·004	51·5	46·6	SW
27	29·430	29·375	29·195	29·112	57·0	44·1	SSE	27	29·109	29·226	29·230	29·034	48·2	40·4	SW
28	29·033	29·136	29·375	29·464	55·6	45·9	W	28	28·970	28·954	29·068	29·249	51·2	40·2	WSW
29	29·515	29·576	29·571	29·570	57·0	46·7	SSW	29	29·275	29·300	29·178	29·114	49·2	39·3	SSE
30	29·613	29·747	29·825	29·925	54·8	41·4	W	30	29·262	29·323	29·480	29·596	51·4	46·2	WNW
31	29·968	30·008	29·947	29·841	53·2	38·7	SW	31	29·653	29·735	29·744	29·746	51·2	44·1	SW

NOVEMBER, 1903.

	KEW								GLASGOW					
	Barometer				Tempera-ture		Direction of wind at 6h.		Barometer				Tempera-ture	Direction of wind at 6h.
Date	4 A.M.	10 A.M.	4 P.M.	10 P.M.	Max	Min		Do.	4 A.M.	10 A.M.	4 P.M.	10 P.M.	Max Min	
1	29·841	29·993	30·062	30·134	55·4	48·9	NW	1	29·823	29·949	29·957	29·971	50·1 44·8	SSW
2	30·139	30·157	30·120	30·124	54·4	40·3	SSW	2	29·944	29·921	29·899	29·995	49·1 44·5	SSW
3	30·128	30·254	30·292	30·358	53·9	42·9	N	3	30·051	30·125	30·140	30·187	50· 42·4	W
4	30·372	30·429	30·425	30·495	52·3	35·7	NNE	4	30·239	30·342	30·417	30·495	52· 45·0	WNW
5	30·527	30·582	30·542	30·562	51·3	34·1	NE	5	30·514	30·525	30·512	30·511	50·0 40·4	WSW
6	30·525	30·544	30·494	30·502	52·4	37·9	E	6	30·500	30·513	30·469	30·476	45·1 39·0	W
7	30·485	30·499	30·442	30·417	48·2	32·9	E	7	30·443	30·428	30·347	30·284	45·8 35·3	NE
8	30·334	30·278	30·182	30·127	49·6	33·4	S	8	30·147	30·046	29·858	29·747	50·1 44·0	SSW
9	30·036	29·984	30·018	30·096	55·1	34·9	WSW	9	29·711	29·717	29·704	29·748	49·5 40·8	W
10	30·111	30·127	30·137	30·187	54·6	42·0	WNW	10	29·739	29·857	29·944	30·066	48·3 43·8	W
11	30·237	30·313	30·309	30·317	53·0	49·3	WNW	11	30·115	30·170	30·153	30·094	51·0 47·2	W
12	30·295	30·269	30·228	30·215	52·2	45·4	WSW	12	29·927	29·979	30·001	30·035	53·3 43·1	WNW
13	30·194	30·190	30·082	29·961	54·3	48·5	SW	13	29·950	29·877	29·733	29·585	53·0 40·1	SW
14	29·821	29·726	29·674	29·725	54·2	42·0	WSW	14	29·399	29·461	29·353	29·377	51·5 40·1	S
15	29·691	29·713	29·714	29·761	48·3	36·6	W	15	29·427	29·503	29·607	29·722	47·3 38·3	NW
16	29·771	29·817	29·806	29·855	44·4	31·0	W	16	29·741	29·790	29·857	29·904	43·3 35·3	NNW
17	29·902	29·860	29·786	29·922	42·1	37·5	W	17	29·750	29·901	29·993	30·118	44·0 36·1	NNE
18	30·030	30·131	30·128	30·155	43·9	36·3	N	18	30·157	30·180	30·162	30·150	41·3 30·7	SW
19	30·167	30·235	30·225	30·254	45·3	32·2	NNE	19	30·147	30·153	30·100	30·035	47·1 41·6	SW
20	30·187	30·052	29·871	29·863	49·0	29·0	W	20	29·802	29·691	29·677	29·651	48·6 43·8	NNW
21	29·821	29·753	29·775	29·985	55·1	44·2	W	21	29·488	29·453	29·685	29·829	50·8 39·7	NW
22	30·095	30·143	30·110	30·146	53·1	44·4	W	22	29·863	29·803	29·798	29·864	50·9 40·1	W
23	30·186	30·208	30·159	30·073	53·2	46·3	SW	23	29·840	29·789	29·788	29·714	51·3 46·9	WSW
24	30·008	30·159	30·223	31·282	54·0	36·7	N	24	29·853	30·026	30·039	30·055	51·2 39·3	W
25	30·254	30·234	30·145	30·103	47·0	34·0	W	25	29·993	29·941	29·850	29·691	41·7 34·2	W
26	30·095	30·113	30·115	30·122	46·5	39·7	NW	26	29·917	29·975	29·978	29·941	38·2 34·1	W
27	29·992	29·739	29·178	29·028	50·0	40·6	SW	27	29·700	29·179	29·234	29·004	36·2 32·4	NNE
28	29·053	29·116	29·062	29·103	46·4	42·6	WNW	28	29·047	29·129	29·225	29·382	40·7 35·1	E
29	29·186	29·268	29·289	29·300	44·1	34·1	ENE	29	29·458	29·468	29·438	29·419	37·9 30·4	N
30	30·304	29·337	29·348	29·412	35·5	31·9	NNE	30	29·439	29·479	29·527	29·611	36·1 26·6	NNW

DECEMBER, 1903.

Date	4 A.M.	10 A.M.	4 P.M.	10 P.M.	Max	Min		Do.	4 A.M.	10 A.M.	4 P.M.	10 P.M.	Max Min	
1	29·424	29·492	29·572	29·714	37·9	31·4	N	1	29·599	29·650	29·729	29·864	37·2 29·0	WNW
2	29·818	29·968	30·022	30·087	38·9	28·1	N	2	29·895	29·903	29·833	29·714	37· 26·0	ENE
3	30·055	29·978	29·811	29·633	43·2	26·9	SW	3	29·487	29·239	29·197	29·141	47· 36·9	WSW
4	29·447	29·421	29·378	29·360	45·7	34·2	WSW	4	29·073	29·101	29·067	29·104	41 34·1	W
5	29·254	29·182	29·134	29·222	35·0	25·9	W	5	29·079	29·089	29·103	29·207	37· 33·7	S
6	29·322	29·495	29·587	29·643	35·7	28·4	NW	6	29·256	29·343	29·368	29·347	38· 31·4	NNW
7	29·547	29·331	29·162	29·211	44·9	30·6	S	7	29·130	28·816	28·811	28·787	43· 35·1	SW
8	29·255	29·282	29·290	29·385	47·5	37·4	SW	8	28·744	28·855	28·973	29·069	44· 40·4	SSW
9	29·362	29·136	29·206	29·231	52·3	43·0	WSW	9	29·010	28·904	28·874	28·888	44· 37·4	S
10	29·179	29·108	29·050	29·153	48·2	41·4	S	10	28·888	28·936	28·914	28·931	42· 38·2	E
11	29·246	29·361	29·405	29·475	46·6	37·7	SW	11	28·948	29·020	29·094	29·182	42·9 39·0	SW
12	29·490	29·517	29·390	29·338	44·2	35·9	SE	12	29·214	29·355	29·462	29·536	41·5 35·6	S
13	29·281	29·278	29·250	29·404	48·9	43·3	SSE	13	29·482	29·415	29·362	29·331	43·1 35·2	ESE
14	29·536	29·646	29·675	29·718	47·8	39·3	S	14	29·274	29·408	29·525	29·602	46·1 42·1	SW
15	29·728	29·736	29·691	29·670	44·5	39·5	SSE	15	29·616	29·641	29·652	29·676	43·4 35·1	ESE
16	29·614	29·600	29·590	29·654	43·2	37·5	E	16	29·663	29·685	29·678	29·674	39·7 34·8	E
17	29·682	29·783	29·797	29·841	45·2	37·8	SE	17	29·641	29·636	29·676	29·760	39·3 30·4	SE
18	29·793	29·798	29·756	29·805	44·1	41·2	E	18	29·818	29·890	29·895	29·933	37·5 30·6	E
19	29·824	29·916	29·896	29·948	43·9	39·2	E	19	29·903	29·923	29·898	29·879	41·6 37·1	E
20	29·966	30·058	30·081	30·188	40·5	37·7	E	20	29·883	29·961	29·985	29·992	42·5 38·1	ENE
21	30·213	30·272	30·260	30·260	46·5	37·8	S	21	29·985	29·923	29·812	29·796	48·2 40·2	SW
22	30·236	30·228	30·165	30·126	40·6	45·2	S	22	29·736	29·691	29·728	29·774	52·3 45·2	WSW
23	30·058	30·042	29·944	29·886	46·1	37·1	S	23	29·791	29·917	29·967	29·973	46·4 39·6	WNW
24	29·853	29·886	29·919	29·970	40·5	36·8	N	24	29·962	29·984	29·979	30·032	40·2 33·1	WNW
25	29·993	30·028	30·010	30·013	41·0	37·5	N	25	30·049	30·068	30·098	30·135	38·1 29·3	SE
26	29·968	29·959	30·014	30·058	40·1	34·9	E	26	30·125	30·149	30·143	30·153	36·5 27·5	ESE
27	30·061	30·103	30·069	30·096	38·4	35·3	NE	27	30·149	30·180	30·170	30·189	38·1 33·2	ESE
28	30·105	30·156	30·120	30·107	37·5	34·4	NE	28	30·196	30·205	30·149	30·131	38·1 35·2	SE
29	30·036	30·069	30·046	30·099	34·2	30·9	ENE	29	30·082	30·129	30·145	30·205	36·8 29·9	S
30	30·069	30·075	29·964	29·932	34·2	28·0	E	30	30·189	30·163	30·049	29·994	33·4 25·4	ESE
31	29·848	29·829	29·796	29·847	31·4	27·5	E	31	29·928	29·911	29·859	29·900	37·6 31·8	ESE

IIMA TEMPERATURES AND THE DIRECTION
3ETHER WITH THE EXPLOSIONS
AND.

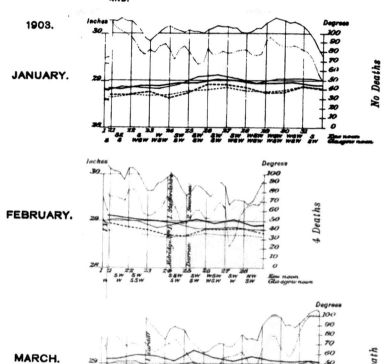

1903.

JANUARY.

FEBRUARY.

MARCH.

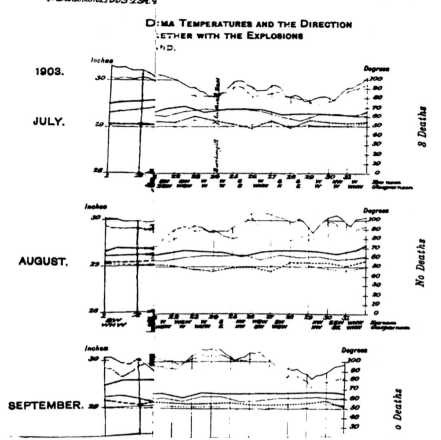

D:MA TEMPERATURES AND THE DIRECTION
.ETHER WITH THE EXPLOSIONS
ND.

INDEX TO VOL. XXVII.

EXPLANATIONS.

The — at the beginning of a line denotes the repetition of a word ; and in the case of Names, it includes both the Christian Name and the Surname ; or, in the case of the name of any Firm, Association or Institution, the full name of such Firm, etc.

Discussions are printed in *italics*.

The following contractions are used :—

M.—Midland Institute of Mining, Civil and Mechanical Engineers.

M.C.—The Midland Counties Institution of Engineers.

S.—The Mining Institute of Scotland.

N.E.—The North of England Institute of Mining and Mechanical Engineers.

N.S.—North Staffordshire Institute of Mining and Mechanical Engineers.

S.S.—The South Staffordshire and East Worcestershire Institute of Mining Engineers.

A.

B.

Broomhill colliery, over-winding prevention gear, 484.
— —, underground sanitary conveniences, 13, 14.
BROUGH, B. H., *underground temperatures, especially in coal-mines*, 367.
BROWN, ADAM, councillor, election S., 68.
BROWN, HORACE F., new process of chlorination for mixed gold- and silver-ores, 529.—Discussion, 534.
BROWN, M. WALTON, barometer, thermometer, etc., readings for 1903, 743.
- , *coal-fields of Faröe Islands*, 342
—, *miners' anæmia*, 14.
BROWN, THOMAS, election, S., 68.
Brown coal, Austria, Bilin, changed to anthracite, 353.
— —, —, Falkenau, analyses, 357.
— —, Russia, Krivoi Rog, 658.
- —, Spain, 666.
BROWNE, J. T., election, S.S., 269.
Brüx basin, Austria, Bohemia, underground temperatures, 352.
BÜCKING, H., auriferous placers of Batak plateau, Sumatra, 702.

BUCKLEY, PHILIP, election, M., 38.
Bukovina, Austria, petroleum-deposits, 615.
BULL, HENRY MATTHEWS, election, N.E., 125.
BUNDE, DR. H., quoted, 357.
Bundy, Straits Settlements, 343.
BUNNING, THEOPHILUS WOOD, quoted, 323.
BUNTE, DR. H., quoted, 518.
BUREAU, ED., coal-measure plants in north Africa, 673.
BURKART, - -, quoted, 178.
BURLS, H. T., *petroleum and its uses*, 513.
Burners, oil-fuel, 503.
Burnt coal, 92.
— —, Faröe Islands, 342.
— —, Lanarkshire, Douglas colliery, 251.
BURT, THOMAS, election, N.E., 126.
Buscones, Mexico, 170.
BUSSE, —, quoted, 111.
BUTTGENBACH, H., gold-deposits of Katanga, Congo Free State, 675.
Butts, longwall workings, 223.
Bye-laws, xvii.

C.

Cab or chalcedonic quartz, Cornwall, 184.
CABALLERO, GUSTAVO DE J., vanadium-ore of Charcas, Mexico, 686.
Caballetes or horses, Mexico, 174.
Cables, electric, armoured, liability to firing, 314.
—, —, mines, 399.
—, —, proposed rules, 299.
—, —, Shirebrook colliery, 284.
-, wire, 277.
Cableways, Derbyshire, Derwent Valley water-scheme, 276.
Cadeby colliery, wet slurry for boiler firing, 61.
CADELL, H. M., *gold in Great Britain and Ireland*, 127.
CADMAN, JOHN, *fires in mines*, 123.
—, *miners' anæmia*, 262, 264, 472.
CADMAN, J. C., presidential address, 323.
Cadmium, Sardinia, blende, 652.
Cal y Canto vein, Mexico, Pachuca, structure, 183.
Calcareous coal, occurrence in a Lanarkshire coal-field, 92.—Discussion, 94, 241.
Calcium carbide, use of, as an explosive, 717.
Caliche, Chile, 738.
California, U.S.A., petroleum, 500, 513.
—, —, quicksilver-deposits, 468.
CALLEY, JOHN, quoted, 103.
Calyx drill, Russian oil-fields, 498.
CAMDEN, W., quoted, 99.
CAMMELL, LAIRD & COMPANY, CHARLES, Derwent iron and steel works, 587.

Campbeltown coal-field, Scotland, calcareous coal, 94.
— colliery, three-phase electrical plant, 418.
Campine coal-field, Belgium, 623.
— — —, —, iron-ores, 624.
Canada, Klondyke gold-mines, 712.
—, petroleum, 507.
Canadian system, boring for petroleum, 498.
CANAVAL, DR. RICHARD, iron-ores of Kohlbach, Austria, 622.
Cannel-coal, flora, Bohemia, Nýřan, 703.
— —, unaltered, in contact with altered coal, 93.
CAPELL, REV. G. M., *comparison of electric and compressed-air locomotives in American mines*, 436.
Capping, winding-ropes, 256, 260, 261, 262.
Caps, locked-coil winding-ropes, 260.
Capulin vein, Mexico, Sultepec, 175.
Carbide, iron, formation of petroleum by action of water on, 495.
Carbon, occurrence in nature, 495.
— dioxide, evolved during formation of coal, 355.
— —, Witwatersrand mines, 24.
- monoxide, Witwatersrand mines, 24.
Carbonates, calcareous coal, 242.
Carboniferous limestone series, Scotland, Dumbarton, coal, 70, 71.
CARLOW, C. A., *Duddingston shale-mines*, 18.
CASHMORE, S. H., election, S.S., 16.

D.

E.

F.

G.

H.

L.

M.

N.

O.

P.

T.

U.

V.

W.

Y.

Z.

Printed by BoD™in Norderstedt, Germany